Handbook of Exact Solutions to Mathematical Equations

This reference book describes the exact solutions of the following types of mathematical equations:

- Algebraic and Transcendental Equations
- Ordinary Differential Equations
- Systems of Ordinary Differential Equations
- First-Order Partial Differential Equations
- Linear Equations and Problems of Mathematical Physics
- Nonlinear Equations of Mathematical Physics
- Systems of Partial Differential Equations
- Integral Equations
- Difference and Functional Equations
- Ordinary Functional Differential Equations
- Partial Functional Differential Equations

The book pays special attention to equations found in various fields of natural and engineering sciences (in the theory of heat and mass transfer, wave theory, hydrodynamics, gas dynamics, combustion theory, elasticity theory, general mechanics, theoretical physics, nonlinear optics, biology, chemical engineering sciences, ecology, and others) and equations of a reasonably general form, which depend on free parameters or arbitrary functions.

The *Handbook of Exact Solutions to Mathematical Equations* is unique in world literature and contains extensive new material. The exact solutions given in the book, being rigorous mathematical standards, can be used as test problems to assess the accuracy and verify the adequacy of various numerical and approximate analytical methods for solving mathematical equations, as well as for checking and comparing the effectiveness of exact analytical methods.

Advances in Applied Mathematics
Series Editor: Daniel Zwillinger

Boundary Value Problems on Time Scales, Volume I
Svetlin Georgiev, Khaled Zennir

Boundary Value Problems on Time Scales, Volume II
Svetlin Georgiev, Khaled Zennir

Observability and Mathematics
Fluid Mechanics, Solutions of Navier-Stokes Equations, and Modeling
Boris Khots

Handbook of Differential Equations, Fourth Edition
Daniel Zwillinger, Vladimir Dobrushkin

Experimental Statistics and Data Analysis for Mechanical and Aerospace Engineers
James Middleton

Advanced Engineering Mathematics with MATLAB®, Fifth Edition
Dean G. Duffy

Handbook of Fractional Calculus for Engineering and Science
Harendra Singh, H. M. Srivastava, Juan J. Nieto

Separation of Variables and Exact Solutions to Nonlinear PDEs
Andrei D. Polyanin, Alexei I. Zhurov

Advanced Engineering Mathematics
A Second Course with MATLAB®
Dean G. Duffy

Quantum Computation
Helmut Bez and Tony Croft

Computational Mathematics
An Introduction to Numerical Analysis and Scientific Computing with Python
Dimitrios Mitsotakis

Delay Ordinary and Partial Differential Equations
Andrei D. Polyanin, Vsevolod G. Sorkin, Alexi I. Zhurov

Clean Numerical Simulation
Shijun Liao

Multiplicative Partial Differential Equations
Svetlin Georgiev and Khaled Zennir

Engineering Statistics
A Matrix-Vector Approach with MATLAB®
Lester W. Schmerr Jr.

General Quantum Numerical Analysis
Svetlin Georgiev and Khaled Zennir

An Introduction to Partial Differential Equations with MATLAB®
Matthew P. Coleman and Vladislav Bukshtynov

Handbook of Exact Solutions to Mathematical Equations
Andrei D. Polyanin

Introducing Game Theory and its Applications, Second Edition
Elliott Mendleson and Dan Zwillinger

Modeling Operations Research and Business Analytics
William P. Fox and Robert E. Burks

https://www.routledge.com/Advances-in-Applied-Mathematics/book-series/CRCADVAPPMTH?pd=published,forthcoming&pg=1&pp=12&so=pub&view=list

Handbook of Exact Solutions to Mathematical Equations

Andrei D. Polyanin

CRC Press
Taylor & Francis Group
Boca Raton London New York

CRC Press is an imprint of the
Taylor & Francis Group, an **informa** business

A CHAPMAN & HALL BOOK

First edition published 2025
by CRC Press
2385 Executive Center Drive, Suite 320, Boca Raton, FL 33431

and by CRC Press
4 Park Square, Milton Park, Abingdon, Oxon, OX14 4RN

CRC Press is an imprint of Taylor & Francis Group, LLC

© 2025 Andrei D. Polyanin

Reasonable efforts have been made to publish reliable data and information, but the author and publisher cannot assume responsibility for the validity of all materials or the consequences of their use. The authors and publishers have attempted to trace the copyright holders of all material reproduced in this publication and apologize to copyright holders if permission to publish in this form has not been obtained. If any copyright material has not been acknowledged please write and let us know so we may rectify in any future reprint.

Except as permitted under U.S. Copyright Law, no part of this book may be reprinted, reproduced, transmitted, or utilized in any form by any electronic, mechanical, or other means, now known or hereafter invented, including photocopying, microfilming, and recording, or in any information storage or retrieval system, without written permission from the publishers.

For permission to photocopy or use material electronically from this work, access www.copyright.com or contact the Copyright Clearance Center, Inc. (CCC), 222 Rosewood Drive, Danvers, MA 01923, 978-750-8400. For works that are not available on CCC please contact mpkbookspermissions@tandf.co.uk

Trademark notice: Product or corporate names may be trademarks or registered trademarks and are used only for identification and explanation without intent to infringe.

Library of Congress Cataloging-in-Publication Data
Names: Políànin, A. D. (Andreĭ Dmitrievich), author.
Title: Handbook of exact solutions to mathematical equations / A.D. Polyanin.
Description: First edition. | Boca Raton, FL : CRC Press, 2025. | Series: Advances in applied mathematics | Includes bibliographical references and index.
Identifiers: LCCN 2024005775 | ISBN 9780367507992 (hardback) | ISBN 9781032807232 (paperback) | ISBN 9781003051329 (ebook)
Subjects: LCSH: Differential equations, Partial--Numerical solutions--Handbooks, manuals, etc. | Differential equations, Nonlinear--Numerical solutions--Handbooks, manuals, etc.
Classification: LCC QA377 .P5678 2025 | DDC 515/.35--dc23/eng/20240412
LC record available at https://lccn.loc.gov/2024005775

ISBN: 978-0-367-50799-2 (hbk)
ISBN: 978-1-032-80723-2 (pbk)
ISBN: 978-1-003-05132-9 (ebk)

DOI: 10.1201/9781003051329

Typeset in CMR10 font
by KnowledgeWorks Global Ltd.

Contents

Preface	xi
Author	xv
Notations and Remarks	xvii

1 Algebraic and Transcendental Equations — 1
- 1.1. Algebraic Equations — 1
  - 1.1.1. Linear and Quadratic Equations — 1
  - 1.1.2. Cubic Equations — 2
  - 1.1.3. Equations of the Fourth Degree — 5
  - 1.1.4. Equations of the Fifth Degree — 9
  - 1.1.5. Algebraic Equations of Arbitrary Degree — 12
  - 1.1.6. Systems of Linear Algebraic Equations — 19
- 1.2. Trigonometric Equations — 20
  - 1.2.1. Binomial Trigonometric Equations — 20
  - 1.2.2. Trigonometric Equations Containing Several Terms — 21
  - 1.2.3. Trigonometric Equations of the General Form — 27
- 1.3. Other Transcendental Equations — 29
  - 1.3.1. Equations Containing Exponential Functions — 29
  - 1.3.2. Equations Containing Hyperbolic Functions — 30
  - 1.3.3. Equations Containing Logarithmic Functions — 34
- References — 35

2 Ordinary Differential Equations — 37
- 2.1. First-Order Ordinary Differential Equations — 37
  - 2.1.1. Simplest First-Order ODEs — 37
  - 2.1.2. Riccati Equations — 38
  - 2.1.3. Abel Equations — 41
  - 2.1.4. Other First-Order ODEs Solved for the Derivative — 46
  - 2.1.5. ODEs Not Solved for the Derivative and ODEs Defined Parametrically — 49
- 2.2. Second-Order Linear Ordinary Differential Equations — 50
  - 2.2.1. Preliminary Remarks and Some Formulas — 50
  - 2.2.2. Equations Involving Power Functions — 51
  - 2.2.3. Equations Involving Exponential and Other Elementary Functions — 69

	2.2.4. Equations Involving Arbitrary Functions	72
2.3.	Second-Order Nonlinear Ordinary Differential Equations	74
	2.3.1. Equations of the Form $y''_{xx} = f(x,y)$	74
	2.3.2. Equations of the Form $f(x,y)y''_{xx} = g(x,y,y'_x)$	77
	2.3.3. ODEs of General Form Containing Arbitrary Functions of Two Arguments	82
2.4.	Higher-Order Ordinary Differential Equations	86
	2.4.1. Higher-Order Linear Ordinary Differential Equations	86
	2.4.2. Third- and Fourth-Order Nonlinear Ordinary Differential Equations	100
	2.4.3. Higher-Order Nonlinear Ordinary Differential Equations	102
References		109

3 Systems of Ordinary Differential Equations — 111

- 3.1. Linear Systems of ODEs 111
  - 3.1.1. Systems of Two First-Order ODEs 111
  - 3.1.2. Systems of Two Second-Order ODEs 114
  - 3.1.3. Other Systems of Two ODEs 120
  - 3.1.4. Systems of Three and More ODEs 121
- 3.2. Nonlinear Systems of Two ODEs 123
  - 3.2.1. Systems of First-Order ODEs 123
  - 3.2.2. Systems of Second- and Third-Order ODEs 133
- 3.3. Nonlinear Systems of Three or More ODEs 141
  - 3.3.1. Systems of Three ODEs 141
  - 3.3.2. Equations of Dynamics of a Rigid Body with a Fixed Point 144
- References 147

4 First-Order Partial Differential Equations — 149

- 4.1. Linear Partial Differential Equations in Two Independent Variables 149
  - 4.1.1. Preliminary Remarks. Solution Methods 149
  - 4.1.2. Equations of the Form $f(x,y)u_x + g(x,y)u_y = 0$ 150
  - 4.1.3. Equations of the Form $f(x,y)u_x + g(x,y)u_y = h(x,y)$ 153
  - 4.1.4. Equations of the Form $f(x,y)u_x + g(x,y)u_y = h(x,y)u + r(x,y)$ 155
- 4.2. Quasilinear Partial Differential Equations in Two Independent Variables 158
  - 4.2.1. Preliminary Remarks. Solution Methods 158
  - 4.2.2. Equations of the Form $f(x,y)u_x + g(x,y)u_y = h(x,y,u)$ 159
  - 4.2.3. Equations of the Form $u_x + f(x,y,u)u_y = 0$ 161
  - 4.2.4. Equations of the Form $u_x + f(x,y,u)u_y = g(x,y,u)$ 164
- 4.3. Nonlinear Partial Differential Equations in Two Independent Variables 167
  - 4.3.1. Preliminary Remarks. A Complete Integral 167
  - 4.3.2. Equations Quadratic in One Derivative 168
  - 4.3.3. Equations Quadratic in Two Derivatives 171
  - 4.3.4. Equations with Arbitrary Nonlinearities in Derivatives 173
- References 178

5 Linear Equations and Problems of Mathematical Physics — 179

5.1. Parabolic Equations — 179
- 5.1.1. Heat (Diffusion) Equation $u_t = au_{xx}$ — 179
- 5.1.2. Nonhomogeneous Heat Equation $u_t = au_{xx} + \Phi(x,t)$ — 181
- 5.1.3. Heat Type Equation of the Form $u_t = au_{xx} + bu_x + cu + \Phi(x,t)$ — 183
- 5.1.4. Heat Equation with Axial Symmetry $u_t = a(u_{rr} + r^{-1}u_r)$ — 183
- 5.1.5. Nonhomogeneous Heat Equation with Axial Symmetry
  $u_t = a(u_{rr} + r^{-1}u_r) + \Phi(r,t)$ — 185
- 5.1.6. Heat Equation with Central Symmetry $u_t = a(u_{rr} + 2r^{-1}u_r)$ — 186
- 5.1.7. Nonhomogeneous Heat Equation with Central Symmetry
  $u_t = a(u_{rr} + 2r^{-1}u_r) + \Phi(r,t)$ — 188
- 5.1.8. Heat Type Equation of the Form $u_t = u_{xx} + (1-2\beta)x^{-1}u_x$ — 189
- 5.1.9. Heat Type Equation of the Form $u_t = [f(x)u_x]_x$ — 190
- 5.1.10. Equations of the Form $s(x)u_t = [p(x)u_x]_x - q(x)u + \Phi(x,t)$ — 191
- 5.1.11. Liquid-Film Mass Transfer Equation $(1-y^2)u_x = au_{yy}$ — 193
- 5.1.12. Equations of the Diffusion (Thermal) Boundary Layer — 195
- 5.1.13. Schrödinger Equation $i\hbar u_t = -\frac{\hbar^2}{2m}u_{xx} + U(x)u$ — 196

5.2. Hyperbolic Equations — 198
- 5.2.1. Wave Equation $u_{tt} = a^2 u_{xx}$ — 198
- 5.2.2. Nonhomogeneous Wave Equation $u_{tt} = a^2 u_{xx} + \Phi(x,t)$ — 199
- 5.2.3. Klein–Gordon Equation $u_{tt} = a^2 u_{xx} - bu$ — 200
- 5.2.4. Nonhomogeneous Klein–Gordon Equation
  $u_{tt} = a^2 u_{xx} - bu + \Phi(x,t)$ — 201
- 5.2.5. Wave Equation with Axial Symmetry
  $u_{tt} = a^2(u_{rr} + r^{-1}u_r) + \Phi(r,t)$ — 202
- 5.2.6. Wave Equation with Central Symmetry
  $u_{tt} = a^2(u_{rr} + 2r^{-1}u_r) + \Phi(r,t)$ — 204
- 5.2.7. Equations of the Form $s(x)u_{tt} = [p(x)u_x]_x - q(x)u + \Phi(x,t)$ — 205
- 5.2.8. Telegraph Type Equations $u_{tt} + ku_t = a^2 u_{xx} + bu_x + cu + \Phi(x,t)$ — 206

5.3. Elliptic Equations — 207
- 5.3.1. Laplace Equation $\Delta u = 0$ — 207
- 5.3.2. Poisson Equation $\Delta u + \Phi(x,y) = 0$ — 210
- 5.3.3. Helmholtz Equation $\Delta u + \lambda u = -\Phi(x,y)$ — 212
- 5.3.4. Convective Heat and Mass Transfer Equations — 216
- 5.3.5. Equations of Heat and Mass Transfer in Anisotropic Media — 221
- 5.3.6. Tricomi and Related Equations — 228

5.4. Simplifications of Second-Order Linear Partial Differential Equations — 231
- 5.4.1. Reduction of PDEs in Two Independent Variables to Canonical Forms — 231
- 5.4.2. Simplifications of Linear Constant-Coefficient Partial Differential Equations — 233

5.5. Third-Order Linear Partial Differential Equations — 235
- 5.5.1. Equations Containing the First Derivative in t and the Third Derivative in x — 235
- 5.5.2. Equations Containing the First Derivative in t and a Mixed Third Derivative — 236

                5.5.3. Equations Containing the Second Derivative in t and a Mixed Third Derivative 242
        5.6. Fourth-Order Linear Partial Differential Equations 244
                5.6.1. Equation of Transverse Vibration of an Elastic Rod $u_{tt} + a^2 u_{xxxx} = 0$ 244
                5.6.2. Nonhomogeneous Equation of the Form $u_{tt} + a^2 u_{xxxx} = \Phi(x,t)$ 245
                5.6.3. Biharmonic Equation $\Delta\Delta u = 0$ 247
                5.6.4. Nonhomogeneous Biharmonic Equation $\Delta\Delta u = \Phi(x,y)$ 248
        References 249

6 Nonlinear Equations of Mathematical Physics 251
        6.1. Parabolic Equations 252
                6.1.1. Quasilinear Heat Equations with a Source of the Form $u_t = a u_{xx} + f(u)$ 252
                6.1.2. Burgers Type Equations and Related PDEs 256
                6.1.3. Reaction-Diffusion Equations of the Form $u_t = [f(u)u_x]_x + g(u)$ 260
                6.1.4. Other Reaction-Diffusion and Heat PDEs with Variable Transfer Coefficient 268
                6.1.5. Convection-Diffusion Type PDEs 272
                6.1.6. Nonlinear Schrödinger Equations and Related PDEs 277
        6.2. Hyperbolic Equations 282
                6.2.1. Nonlinear Klein–Gordon Equations of the Form $u_{tt} = a u_{xx} + f(u)$ 282
                6.2.2. Other Nonlinear Wave Type Equations 287
        6.3. Elliptic Equations 295
                6.3.1. Heat Equations with Nonlinear Source of the Form $u_{xx} + u_{yy} = f(u)$ 295
                6.3.2. Stationary Anisotropic Heat/Diffusion Equations of the Form $[f(x)u_x]_x + [g(y)u_y]_y = h(u)$ 298
                6.3.3. Stationary Anisotropic Heat/Diffusion Equations of the Form $[f(u)u_x]_x + [g(u)u_y]_y = h(u)$ 299
        6.4. Other Second-Order Equations 302
                6.4.1. Equations of Transonic Gas Flow 302
                6.4.2. Monge–Ampère Type Equations 303
        6.5. Higher-Order Equations 306
                6.5.1. Third-Order Equations 306
                6.5.2. Fourth-Order Equations 324
        References 329

7 Systems of Partial Differential Equations 335
        7.1. Systems of Two First-Order PDEs 335
                7.1.1. Linear Systems of Two First-Order PDEs 335
                7.1.2. Nonlinear Systems of the Form $u_x = F(u,w)$, $w_t = G(u,w)$ .. 336
                7.1.3. Gas Dynamic Type Systems Linearizable with the Hodograph Transformation 341
        7.2. Systems of Two Second-Order PDEs 347
                7.2.1. Linear Systems of Two Second-Order PDEs 347

- 7.2.2. Nonlinear Parabolic Systems of the Form
  $u_t = au_{xx} + F(u,w)$, $w_t = bw_{xx} + G(u,w)$ 349
- 7.2.3. Nonlinear Parabolic Systems of the Form
  $u_t = ax^{-n}(x^n u_x)_x + F(u,w)$, $w_t = bx^{-n}(x^n w_x)_x + G(u,w)$. 363
- 7.2.4. Nonlinear Hyperbolic Systems of the Form
  $u_{tt} = au_{xx} + F(u,w)$, $w_{tt} = bw_{xx} + G(u,w)$ 369
- 7.2.5. Nonlinear Hyperbolic Systems of the Form
  $u_{tt} = ax^{-n}(x^n u_x)_x + F(u,w)$, $w_{tt} = bx^{-n}(x^n w_x)_x + G(u,w)$ 376
- 7.2.6. Nonlinear Elliptic Systems of the Form
  $\Delta u = F(u,w)$, $\Delta w = G(u,w)$ 380
- 7.3. PDE Systems of General Form 384
  - 7.3.1. Linear Systems 384
  - 7.3.2. Nonlinear Systems of Two Equations Involving the First Derivatives with Respect to t 385
  - 7.3.3. Nonlinear Systems of Two Equations Involving the Second Derivatives with Respect to t 389
- References 393

8 Integral Equations 395
- 8.1. Integral Equations of the First Kind with Variable Limit of Integration .. 395
  - 8.1.1. Linear Volterra Integral Equations of the First Kind 395
  - 8.1.2. Nonlinear Volterra Integral Equations of the First Kind 403
- 8.2. Integral Equations of the Second Kind with Variable Limit of Integration 406
  - 8.2.1. Linear Volterra Integral Equations of the Second Kind 406
  - 8.2.2. Nonlinear Volterra Integral Equations of the Second Kind 424
- 8.3. Equations of the First Kind with Constant Limits of Integration 428
  - 8.3.1. Linear Fredholm Integral Equations of the First Kind 428
  - 8.3.2. Nonlinear Fredholm Integral Equations of the First Kind 437
- 8.4. Equations of the Second Kind with Constant Limits of Integration 439
  - 8.4.1. Linear Fredholm Integral Equations of the Second Kind 439
  - 8.4.2. Nonlinear Fredholm Integral Equations of the Second Kind 450
- References 455

9 Difference and Functional Equations 457
- 9.1. Difference Equations 457
  - 9.1.1. Difference Equations with Discrete Argument 457
  - 9.1.2. Difference Equations with Continuous Argument 465
- 9.2. Linear Functional Equations in One Independent Variable 480
  - 9.2.1. Linear Functional Equations Involving Unknown Function with Two Different Arguments 480
  - 9.2.2. Other Linear Functional Equations 488
- 9.3. Nonlinear Functional Equations in One Independent Variable 492
  - 9.3.1. Functional Equations with Quadratic Nonlinearity 492
  - 9.3.2. Functional Equations with Power Nonlinearity 496
  - 9.3.3. Nonlinear Functional Equation of General Form 497
- 9.4. Functional Equations in Several Independent Variables 501
  - 9.4.1. Linear Functional Equations 501

　　　　9.4.2. Nonlinear Functional Equations 507
　　References 516

10 Ordinary Functional Differential Equations   519
　10.1. First-Order Linear Ordinary Functional Differential Equations 519
　　　10.1.1. ODEs with Constant Delays 519
　　　10.1.2. Pantograph-Type ODEs with Proportional Arguments 522
　　　10.1.3. Other Ordinary Functional Differential Equations 524
　10.2. First-Order Nonlinear Ordinary Functional Differential Equations 525
　　　10.2.1. ODEs with Constant Delays 525
　　　10.2.2. Pantograph-Type ODEs with Proportional Arguments 527
　　　10.2.3. Other Ordinary Functional Differential Equations 530
　10.3. Second-Order Linear Ordinary Functional Differential Equations 531
　　　10.3.1. ODEs with Constant Delays 531
　　　10.3.2. Pantograph-Type ODEs with Proportional Arguments 533
　　　10.3.3. Other Ordinary Functional Differential Equations 535
　10.4. Second-Order Nonlinear Ordinary Functional Differential Equations . . . 536
　　　10.4.1. ODEs with Constant Delays 536
　　　10.4.2. Pantograph-Type ODEs with Proportional Arguments 537
　　　10.4.3. Other Ordinary Functional Differential Equations 538
　10.5. Higher-Order Ordinary Functional Differential Equations 539
　　　10.5.1. Linear Ordinary Functional Differential Equations 539
　　　10.5.2. Nonlinear Ordinary Functional Differential Equations 542
　References 543

11 Partial Functional Differential Equations   545
　11.1. Linear Partial Functional Differential Equations 546
　　　11.1.1. PDEs with Constant Delay 546
　　　11.1.2. PDEs with Proportional Delay 554
　　　11.1.3. PDEs with Anisotropic Time Delay 560
　11.2. Nonlinear PDEs with Constant Delays 562
　　　11.2.1. Parabolic Equations 562
　　　11.2.2. Hyperbolic Equations 581
　11.3. Nonlinear PDEs with Proportional Arguments 590
　　　11.3.1. Parabolic Equations 590
　　　11.3.2. Hyperbolic Equations 600
　11.4. Partial Functional Differential Equations with Arguments of General Type 605
　　　11.4.1. Parabolic Equations 605
　　　11.4.2. Hyperbolic Equations 613
　11.5. PDEs with Anisotropic Time Delay 617
　　　11.5.1. Parabolic Equations 617
　　　11.5.2. Hyperbolic Equations 619
　References 621

Index   **623**

Preface

Linear and nonlinear mathematical equations arise in almost all areas of natural and technical sciences, as well as in some areas of economics and humanities. This book is devoted to a brief description of exact solutions to a large number of mathematical equations of various types; it also contains numerous reductions and transformations leading to simpler equations.

A solution is called *exact* if, when substituted into the mathematical equation under consideration, it turns it into an identity. In this case, no approximations or simplifications of the equation are allowed, and no a priori assumptions are used. For different types of mathematical equations, the concept of an exact solution allows for various clarifications and modifications.

Exact solutions of mathematical equations have always played and continue to play a massive role in the formation of a correct understanding of the qualitative features of many phenomena and processes in various fields of natural science. In particular, exact solutions to nonlinear equations of mathematical physics clearly demonstrate and allow us to understand better the mechanisms of such complex nonlinear effects as the spatial localization of transport processes, the multiplicity or absence of stationary states under certain conditions, the existence of blow-up regimes, the possible non-smoothness or discontinuity of the unknown quantities, and others. Simple solutions of linear and nonlinear differential equations are widely used to illustrate theoretical material and some applications in university and technical college courses (in applied and computational mathematics, asymptotic methods, theoretical physics, theory of heat and mass transfer, hydrodynamics, gas dynamics, wave theory, nonlinear optics, etc.).

Exact solutions of equations play an essential role as rigorous "mathematical standards," which are appropriate to assess the accuracy of various numerical, asymptotic and approximate analytical methods. Exact solutions are also crucial for the development and improvement of the relevant sections of computer programs intended for symbolic computations (in computer algebra systems such as Mathematica, Maple, Maxima, and others).

This text deals with the following types of equations:

- algebraic equations;
- trigonometric, hyperbolic, and other transcendental equations;
- ordinary differential equations;
- systems of ordinary differential equations;
- first-order partial differential equations;
- linear equations of mathematical physics;

- nonlinear equations of mathematical physics;
- systems of partial differential equations;
- integral equations;
- difference and other functional equations;
- ordinary differential equations with delay and other ordinary functional differential equations;
- equations of mathematical physics with delay and other functional differential equations with partial derivatives.

Importantly, solutions to different types of mathematical equations are often related, either explicitly or implicitly. For example, solving linear ordinary differential equations with constant coefficients reduces to solving algebraic equations, and solutions to higher-order algebraic equations are constructed using the relevant ordinary differential equations and expressed in terms of special functions. Solving first-order partial differential equations reduces to integrating systems of ordinary differential equations. Solutions to many linear and nonlinear equations of mathematical physics are sought by the methods of separation of variables and generalized separation of variables and are expressed via solutions of ordinary differential equations (so-called one-dimensional reductions) or systems of such equations. The solution of some nonlinear equations of mathematical physics reduces to solving functional or functional-differential equations, while solutions of various functional equations are expressed via solutions of ordinary differential or partial differential equations. Solutions to some classes of integral equations are constructed using the appropriate classes of differential equations and vice versa. Solutions to equations of mathematical physics with constant or proportional delay are often expressed in terms of solutions to ordinary differential equations or ordinary differential equations with delay. The close connections between solutions to mathematical equations of different types have served as one of the main motivations for writing this book, which deals with the most common and many other exact solutions to various equations.

When selecting material, the author gave the most significant preference to the following three essential classes of mathematical equations:

- equations that arise in various fields of natural and engineering sciences (the theory of heat and mass transfer, wave theory, hydrodynamics, gas dynamics, combustion theory, elasticity theory, general mechanics, theoretical physics, nonlinear optics, chemical technology, biology, ecology, and others);
- equations of a reasonably general form, which depend on arbitrary functions or several free parameters (exact solutions to such equations are of great interest for testing numerical and approximate analytical methods);
- equations that are studied in universities and technical colleges.

In general, the book has no analogues in world literature and contains a lot of new material not published previously.

To maximize the range of potential readers with different mathematical backgrounds, the author tried to avoid using special terminology whenever possible. The presentation of the material pursues the principle "from simple to complex." Many sections can be read independently, making it easier to work with the material. A detailed Table of Contents lets the reader find the required information quickly.

The author thanks A. V. Aksenov and A. I. Zhurov for discussions and useful comments.

The author hopes that the book will be helpful to a wide range of researchers, university teachers, and graduate and postgraduate students specializing in the fields of applied and computational mathematics, mathematical and theoretical physics, mechanics, control theory, biology, biophysics, biochemistry, medicine, chemical technology, and ecology. Individual equations and their solutions can be used as illustrative material in lectures and seminars on applied and computational mathematics, differential equations, equations of mathematical physics, and integral equations.

Andrei D. Polyanin
April 2024

Author

Andrei D. Polyanin, D.Sc., Ph.D., is a well-known scientist of broad interests and is active in various areas of mathematics, theory of heat and mass transfer, hydrodynamics, and chemical engineering sciences. He is one of the most prominent authors in the field of reference literature on mathematics. Professor Polyanin graduated with honors from the Department of Mechanics and Mathematics at the Lomonosov Moscow State University in 1974. Since 1975, Professor Polyanin has been working at the Ishlinsky Institute for Problems in Mechanics of the Russian (former USSR) Academy of Sciences, where he defended his Ph.D. in 1981 and D.Sc. degree in 1986.

Professor Polyanin has made important contributions to the theory of differential and integral equations, mathematical physics, applied and engineering mathematics, the theory of heat and mass transfer, and hydrodynamics. He has developed analytical methods for constructing solutions to mathematical equations of various types and has obtained a huge number of exact solutions of ordinary differential, partial differential, delay partial differential, integral, and functional equations.

Professor Polyanin is author of more than 30 books and over 270 articles and holds three patents. His books include: V. F. Zaitsev and A. D. Polyanin, *Discrete-Group Methods for Integrating Equations of Nonlinear Mechanics*, CRC Press, 1994; A. D. Polyanin and V. V. Dilman, *Methods of Modeling Equations and Analogies in Chemical Engineering*, CRC Press/Begell House, Boca Raton, 1994; A. D. Polyanin and V. F. Zaitsev, *Handbook of Exact Solutions for Ordinary Differential Equations*, CRC Press, 1995 (2nd edition in 2003); A. D. Polyanin and A. V. Manzhirov, *Handbook of Integral Equations*, CRC Press, 1998 (2nd edition in 2008); A. D. Polyanin, *Handbook of Linear Partial Differential Equations for Engineers and Scientists*, Chapman & Hall/CRC Press, 2002 (2nd edition, co-authored with V. E. Nazaikinskii, in 2016); A. D. Polyanin, V. F. Zaitsev, and A. Moussiaux, *Handbook of First Order Partial Differential Equations*, Taylor & Francis, 2002; A. D. Polyanin, A. M. Kutepov, et al., *Hydrodynamics, Mass and Heat Transfer in Chemical Engineering*, Taylor & Francis, 2002; A. D. Polyanin and V. F. Zaitsev, *Handbook of Nonlinear Partial Differential Equations*, Chapman & Hall/CRC Press, 2004 (2nd edition in 2012); A. D. Polyanin and A. V. Manzhirov, *Handbook of Mathematics for Engineers and Scientists*, Chapman & Hall/CRC Press, 2007; A. D. Polyanin and

V. F. Zaitsev, *Handbook of Ordinary Differential Equations: Exact Solutions, Methods, and Problems*, CRC Press, 2018; A. D. Polyanin and A. I. Zhurov, *Separation of Variables and Exact Solutions to Nonlinear PDEs*, CRC Press, 2022; and A. D. Polyanin, V. G. Sorokin, and A. I. Zhurov, *Delay Ordinary and Partial Differential Equations*, CRC Press, 2023.

Professor Polyanin is editor-in-chief of the international scientific educational website *EqWorld—The World of Mathematical Equations* and a member of the editorial boards of several journals.

Notations and Remarks

Latin Characters

$C, C_1, C_2, \ldots$ arbitrary constants;
- t time (independent variable);
- u unknown function (dependent variable); for PDEs in two independent variables x and t, $u = u(x,t)$;
- x unknown variable in algebraic equations, independent variable in ODEs, and space variable in PDEs;
- y unknown function (dependent variable) in ODEs, $y = y(x)$, and space variable in PDEs.

Greek Characters

- Δ Laplace operator, $\Delta = \frac{\partial^2}{\partial x^2} + \frac{\partial^2}{\partial y^2}$;
- τ delay time ($\tau > 0$); it can be constant or time dependent, $\tau = \tau(t)$.

Short Notations for Derivatives and Operators

Ordinary derivatives of a function $y = y(x)$:

$$y'_x = \frac{dy}{dx}, \quad y''_{xx} = \frac{d^2 y}{dx^2}, \quad y'''_{xxx} = \frac{d^3 y}{dx^3}, \quad y''''_{xxxx} = \frac{d^4 y}{dx^4}, \quad y_x^{(n)} = \frac{d^n y}{dx^n} \quad \text{for} \quad n > 4.$$

Partial derivatives of a function $u = u(x, t)$:

$$u_x = \frac{\partial u}{\partial x}, \quad u_t = \frac{\partial u}{\partial t}, \quad u_{xx} = \frac{\partial^2 u}{\partial x^2}, \quad u_{xt} = \frac{\partial^2 u}{\partial x \partial t}, \quad u_{tt} = \frac{\partial^2 u}{\partial t^2}, \quad u_{xxx} = \frac{\partial^3 u}{\partial x^3}, \quad \ldots$$

Remarks

1. The book often uses the abbreviations ODE (or ODEs) and PDE (or PDEs), which stand for "ordinary differential equation(s)" and "partial differential equation(s)".

2. When referring to a particular equation, we use notation like 2.4.1.8, which denotes equation 8 in Section 2.4.1.

3. The coefficients, functions, independent variables, and unknown quantities appearing in the equations are considered real (unless otherwise specified).

4. If a formula or a solution involves derivatives of some functions, it is assumed that these derivatives exist.

5. If a formula or a solution involves indefinite or definite integrals, these integrals are assumed to exist.

6. In formulas and solutions that involve expressions like $\frac{f(t)}{a-2}$, it is often omitted but implied that $a \neq 2$.

Chapter 1

Algebraic and Transcendental Equations

▶ **Preliminary remarks.** An algebraic equation is a mathematical equation of the form $P(x) = 0$, where $P(x)$ is a univariate polynomial in the unknown quantity x. The degree of an algebraic equation is the highest power of x in $P(x)$.

This chapter deals with algebraic equations whose solutions (roots) are representable in radicals. It gives higher-order algebraic equations whose solutions are expressible in terms of special functions. It also specifies transformations that make it possible to obtain simpler equations.

We also consider transcendental equations of the form $f(x) = 0$, where $f(x)$ is an elementary function other than a polynomial and x is the unknown (desired) quantity. Furthermore, we describe trigonometric, hyperbolic, logarithmic, and other transcendental equations whose solutions can be expressed in terms of solutions to algebraic equations or obtained by substituting the defining parameters of the equation into elementary functions. In addition, we discuss some transcendental equations whose solutions are special functions.

This chapter assumes that the coefficients of all the equations under consideration are real numbers.

1.1. Algebraic Equations

▶ *The coefficients of all algebraic equations in Section 1.1 are assumed to be real numbers (unless otherwise specified).*

1.1.1. Linear and Quadratic Equations

1. $ax + b = 0 \quad (a \neq 0)$.

Linear algebraic equation (or *algebraic equation of the first degree*).
   Solution:
$$x = -\frac{b}{a}.$$

2. $ax^2 + bx + c = 0 \quad (a \neq 0)$.

Quadratic equation (or *algebraic equation of the second degree*).
   Solutions (roots):
$$x_1 = \frac{-b - \sqrt{b^2 - 4ac}}{2a}, \quad x_2 = \frac{-b + \sqrt{b^2 - 4ac}}{2a}.$$

DOI: 10.1201/9781003051329-1

The existence of real or complex roots is determined by the sign of the discriminant $D = b^2 - 4ac$:

Case $D > 0$. There are two distinct real roots.
Case $D < 0$. There are two distinct complex conjugate roots.
Case $D = 0$. There are two equal real roots.

Viète's theorem (Vieta's formulas). The roots of a quadratic equation satisfy the relations
$$x_1 + x_2 = -\frac{b}{a}, \quad x_1 x_2 = \frac{c}{a}.$$

1.1.2. Cubic Equations

1. $x^3 - a = 0.$

Simplest two-term cubic equation.
Solutions (roots):

$$x_1 = a^{1/3}, \quad x_2 = -a^{1/3}\left(\frac{1}{2} + i\frac{\sqrt{3}}{2}\right), \quad x_3 = -a^{1/3}\left(\frac{1}{2} - i\frac{\sqrt{3}}{2}\right) \quad \text{if } a > 0;$$

$$x_1 = -|a|^{1/3}, \quad x_2 = |a|^{1/3}\left(\frac{1}{2} + i\frac{\sqrt{3}}{2}\right), \quad x_3 = |a|^{1/3}\left(\frac{1}{2} - i\frac{\sqrt{3}}{2}\right) \quad \text{if } a < 0,$$

where $i^2 = -1$.

2. $x^3 + ax + ab + b^3 = 0.$

For any a and b, this equation has the root $x = -b$. The other two roots are determined from the quadratic equation $x^2 - bx + a + b = 0$.

3. $x^3 - 3a^2 x + 2a^3 = 0.$

This equation has the roots:
$$x_1 = x_2 = a, \quad x_3 = -2a.$$

4. $x^3 - (a^2 + ab + b^2)x + ab(a + b) = 0.$

This equation has the roots:
$$x_1 = a, \quad x_2 = b, \quad x_3 = -a - b.$$

5. $y^3 + py + q = 0.$

Incomplete cubic equation.

$1°$. *Cardano's solution.* The roots of the incomplete cubic equation are expressed as
$$y_1 = A + B, \quad y_{2,3} = -\frac{1}{2}(A + B) \pm i\frac{\sqrt{3}}{2}(A - B),$$

where
$$A = \left(-\frac{q}{2} + \sqrt{D}\right)^{1/3}, \quad B = \left(-\frac{q}{2} - \sqrt{D}\right)^{1/3}, \quad D = \left(\frac{p}{3}\right)^3 + \left(\frac{q}{2}\right)^2, \quad i^2 = -1,$$

with A and B being arbitrary values of the cubic roots such that $AB = -\frac{1}{3}p$.

The number of real roots of the incomplete cubic equation depends on the sign of the discriminant D:

Case $D > 0$. There is one real and two complex conjugate roots.

Case $D < 0$. There are three real roots.

Case $D = 0$. There are one real root and one real double root (a real root of multiplicity 2).

$2°$. *Trigonometric solution.* If the incomplete cubic equation has real coefficients p and q, then its solutions can be found using the trigonometric formulas given below.

(a) If $p < 0$ and $D < 0$, then

$$y_1 = 2\sqrt{-\frac{p}{3}} \cos\frac{\alpha}{3}, \quad y_{2,3} = -2\sqrt{-\frac{p}{3}} \cos\left(\frac{\alpha}{3} \pm \frac{2\pi}{3}\right),$$

where the values of the trigonometric functions are calculated from the relation

$$\cos\alpha = -\frac{q}{2\sqrt{-(p/3)^3}}, \quad \sin\alpha = \frac{\sqrt{-D}}{\sqrt{-(p/3)^3}} > 0.$$

(b) If $p > 0$ and $D \geq 0$, then

$$y_1 = -2\sqrt{\frac{p}{3}} \cot(2\alpha), \quad y_{2,3} = \sqrt{\frac{p}{3}}\left[\cot(2\alpha) \pm i\frac{\sqrt{3}}{\sin(2\alpha)}\right],$$

where the values of the trigonometric functions are calculated as

$$\tan\alpha = \left(\tan\frac{\beta}{2}\right)^{1/3}, \quad \tan\beta = \frac{2}{q}\left(\frac{p}{3}\right)^{3/2}, \quad |\alpha| \leq \frac{\pi}{4}, \quad |\beta| \leq \frac{\pi}{2}.$$

(c) If $p < 0$ and $D \geq 0$, then

$$y_1 = -2\sqrt{-\frac{p}{3}}\frac{1}{\sin(2\alpha)}, \quad y_{2,3} = \sqrt{-\frac{p}{3}}\left[\frac{1}{\sin(2\alpha)} \pm i\sqrt{3}\cot(2\alpha)\right],$$

with the values of the trigonometric functions obtained as

$$\tan\alpha = \left(\tan\frac{\beta}{2}\right)^{1/3}, \quad \sin\beta = \frac{2}{q}\left(-\frac{p}{3}\right)^{3/2}, \quad |\alpha| \leq \frac{\pi}{4}, \quad |\beta| \leq \frac{\pi}{2}.$$

In the above three cases, the real value of the cubic root should be taken.

$3°$. *Viète's method.* The substitution

$$y = t - \frac{p}{3t}$$

reduces the incomplete cubic equation to the equation

$$t^3 + q - \frac{p^3}{27t^3} = 0,$$

which after multiplication by t^3 and replacement $z = t^3$ is transformed to the quadratic equation $z^2 - qz - \frac{1}{27}p^3 = 0$.

4°. For further information on the incomplete cubic equation, see the books King (1996), Korn & Korn (2000), Weisstein (2002), and Bronshtein & Semendyayev (2015) and the article *Cubic equation* (from Wikipedia).

6. $ax^3 + bx^2 + bx + a = 0 \quad (a \neq 0).$

Reciprocal algebraic (polynomial) equation of the third degree. This equation has the root $x_1 = -1$, and the other two roots are determined from the quadratic equation $ax^2 + (b - a)x + a = 0.$

7. $ax^3 + bx^2 + bx + a - b + c = 0 \quad (a \neq 0).$

This equation has the root $x_1 = -1$, and the other two roots are determined from the quadratic equation
$$ax^2 + (b - a)x + a - b + c = 0.$$

8. $ax^3 + bx^2 + bx - a - b - c = 0 \quad (a \neq 0).$

This equation has the root $x_1 = 1$, and the other two roots are determined from the quadratic equation
$$ax^2 + (a + b)x + a + b + c = 0.$$

9. $ax^3 + (ab + c)x^2 + (a + bc)x + c = 0 \quad (a \neq 0).$

1°. Suppose $x = -a/c$ is not a root of this equation. Multiplying both sides of the equation by $(cx + a)$, we get a fourth-degree reciprocal equation of the form 1.1.3.7:
$$acx^4 + (a^2 + abc + c^2)x^3 + (a^2b + 2ac + bc^2)x^2 + (a^2 + abc + c^2)x + ac = 0. \quad (*)$$

The substitution
$$y = x + \frac{1}{x}$$
takes $(*)$ to a quadratic equation. Having thus obtained the four roots of equation $(*)$, one should keep only three roots, discarding the extra root $x = -a/c$.

2°. Let $x = -a/c$ be a root of this equation. The other two roots are determined from the quadratic equation
$$ax^2 + \left(ab + c - \frac{a^2}{c}\right)x + bc - \frac{a^2b}{c} + \frac{a^3}{c^2} = 0.$$

10. $x^3 + a(b + 1)x^2 + (a^2b + c)x + ac = 0.$

One root of this equation is $x = -a$. The other two roots are determined from the quadratic equation $x^2 + abx + c = 0$.

11. $ax^3 + bx^2 + cx + d = 0 \quad (a \neq 0).$

Complete cubic equation (algebraic equation of the third degree).
The roots of the complete cubic equation are calculated as
$$x_k = y_k - \frac{b}{3a}, \quad k = 1, 2, 3,$$

where y_k are the roots of the incomplete cubic equation 1.1.2.5 with the coefficients

$$p = -\frac{1}{3}\left(\frac{b}{a}\right)^2 + \frac{c}{a}, \quad q = \frac{2}{27}\left(\frac{b}{a}\right)^3 - \frac{bc}{3a^2} + \frac{d}{a}.$$

Viète's theorem (Vieta's formulas). The roots of the complete cubic equation satisfy the following relations:

$$x_1 + x_2 + x_3 = -\frac{b}{a}, \quad x_1x_2 + x_1x_3 + x_2x_3 = \frac{c}{a}, \quad x_1x_2x_3 = -\frac{d}{a}.$$

1.1.3. Equations of the Fourth Degree

1. $x^4 - a = 0.$

Simplest two-term quartic equation.

Solutions (roots):

$$x_{1,2} = \pm a^{1/4}, \qquad x_{3,4} = \pm a^{1/4}i \qquad \text{if } a > 0;$$

$$x_{1,2} = |a|^{1/4}\left(\frac{\sqrt{2}}{2} \pm i\frac{\sqrt{2}}{2}\right), \quad x_{3,4} = |a|^{1/4}\left(-\frac{\sqrt{2}}{2} \pm i\frac{\sqrt{2}}{2}\right) \quad \text{if } a < 0,$$

where $i^2 = -1$.

2. $x^4 + ax + ab - b^4 = 0.$

For any a and b, this equation has the root $x = -b$.

3. $x^4 - 4a^3x + 3a^4 = 0.$

This equation has the double root $x_1 = x_2 = a$. The other two (complex) roots are determined from the quadratic equation $x^2 + 2ax + 3a^2 = 0$.

4. $x^4 - (a^3 + a^2b + ab^2 + b^3)x + ab(a^2 + ab + b^2) = 0.$

This equation has the roots $x_1 = a$ and $x_2 = b$. The other two (complex) roots are determined from the quadratic equation

$$x^2 + (a+b)x + a^2 + ab + b^2 = 0.$$

5. $ax^4 + bx^2 + c = 0 \quad (a \neq 0).$

Biquadratic equation.

This equation can be reduced to the quadratic equation 1.1.1.2 with the substitution $z = x^2$. Therefore, the roots of the biquadratic equation are given by

$$x_{1,2} = \pm\sqrt{\frac{-b - \sqrt{b^2 - 4ac}}{2a}}, \quad x_{3,4} = \pm\sqrt{\frac{-b + \sqrt{b^2 - 4ac}}{2a}}.$$

6. $x^4 + 2ax^3 + a^2x^2 - b = 0.$

All the roots of this equation are determined by solving the two quadratic equations:

$$x^2 + ax - \sqrt{b} = 0,$$
$$x^2 + ax + \sqrt{b} = 0.$$

7. $ax^4 + bx^3 + cx^2 + bx + a = 0$ $(a \neq 0)$.

Reciprocal algebraic (polynomial) equation of the fourth degree.

The substitution
$$y = x + \frac{1}{x}$$
leads to a quadratic equation of the form 1.1.1.2:
$$ay^2 + by + c - 2a = 0.$$

8. $ax^4 + bx^3 + cx^2 - bx + a = 0$ $(a \neq 0)$.

Modified reciprocal algebraic equation of the fourth degree.

The substitution
$$y = x - \frac{1}{x}$$
leads to a quadratic equation of the form 1.1.1.2:
$$ay^2 + by + c + 2a = 0.$$

9. $ax^4 + bx^3 + cx^2 + b\lambda x + a\lambda^2 = 0$ $(a \neq 0)$.

Generalized reciprocal algebraic equation of the fourth degree (generalizes the previous two equations).

The substitution
$$y = x + \frac{\lambda}{x}$$
leads to a quadratic equation of the form 1.1.1.2:
$$ay^2 + by + c - 2a\lambda = 0.$$

10. $ax^4 + bx^2(x+k) + c(x+k)^2 = 0$.

Let us divide the equation by $(x+k)^2$ and then substitute $z = x^2/(x+k)$ to arrive at the quadratic equation
$$az^2 + bz + c = 0.$$

11. $(x+a)^4 + (x+b)^4 = c$.

This is a special case of equation 1.1.5.21 with $n = 2$.

1°. Let us make the substitution $y = x + \frac{1}{2}(a+b)$ to obtain
$$(y+k)^4 + (y-k)^4 = c, \quad k = \tfrac{1}{2}(a-b).$$

Using the identities $(y \pm k)^4 = y^4 \pm 4ky^3 + 6k^2y^2 \pm 4k^3y + k^4$, we get a biquadratic equation,
$$y^4 + 6k^2y^2 + k^4 - \tfrac{1}{2}c = 0, \quad k = \tfrac{1}{2}(a-b),$$
which becomes a quadratic equation after substituting $z = y^2$.

2°. The identical transformations of the left-hand side

$$(x+a)^4 + (x+b)^4 = [(x+a)^2 + (x+b)^2]^2 - 2[(x+a)(x+b)]^2$$
$$= \{2[x^2 + (a+b)x] + a^2 + b^2\}^2 - 2[x^2 + (a+b)x + ab]^2$$

and the substitution $\xi = x^2 + (a+b)x$ lead to a quadratic equation:

$$(2\xi + a^2 + b^2)^2 - 2(\xi + ab)^2 = c.$$

12. $(x+a)^4 + (x+b)^4 + c(x+a)(x+b) = d.$

This is a special case of equation 1.1.5.24.

Let us make the substitution $y = x + \frac{1}{2}(a+b)$ to obtain

$$(y+k)^4 + (y-k)^4 + c(y^2 - k^2) = d, \quad k = \frac{1}{2}(a-b).$$

Using the identities $(y\pm k)^4 = y^4 \pm 4ky^3 + 6k^2y^2 \pm 4k^3y + k^4$, we get a biquadratic equation,

$$y^4 + (6k^2 + \tfrac{1}{2}c)y^2 + k^4 - \tfrac{1}{2}ck^2 - d = 0, \quad k = \tfrac{1}{2}(a-b),$$

which becomes a quadratic equation after substituting $z = y^2$.

13. $(ax^2 + bx + c_1)(ax^2 + bx + c_2) = d.$

The substitution $z = ax^2 + bx + c_1$ leads to a quadratic equation: $z^2 + (c_2 - c_1)z - d = 0$.

14. $(ax^2 + b_1x + c)(ax^2 + b_2x + c) = dx^2.$

Let us divide the equation by x^2 and substitute $z = ax + b_1 + c/x$ to arrive at the quadratic equation

$$z^2 + (b_2 - b_1)z - d = 0.$$

15. $(x+a)(x+b)(x+a+c)(x+b+c) = d.$

Multiplying the first and fourth factors on the left-hand side together as well as the second and third factors, we get

$$[x^2 + (a+b+c)x + a(b+c)][x^2 + (a+b+c)x + b(a+c)] = d.$$

Substituting

$$z = x^2 + (a+b+c)x + a(b+c)$$

leads to the quadratic equation

$$z[z + (b-a)c] = d.$$

16. $(x+a)(x+b)(x+ac)(x+bc) = dx^2.$

Multiplying the first and fourth factors on the left-hand side together as well as the second and third factors, we get

$$[x^2 + (a+bc)x + abc][x^2 + (b+ac)x + abc] = dx^2.$$

Dividing both sides of the resulting equation by x^2 and substituting $z = x + abc/x$, we arrive at the quadratic equation

$$(z + a + bc)(z + b + ac) = d.$$

17. $a(x^2 + p_1 x + q)^2 + b(x^2 + p_2 x + q) = dx^2.$

Dividing both sides by x^2 and substituting $z = x + q/x$, we arrive at the quadratic equation

$$a(z + p_1)^2 + b(z + p_2)^2 = d.$$

18. $x^4 + ax^3 + bx^2 + cx + d = 0.$

General algebraic (polynomial) equation of the fourth degree (general quartic equation).

1°. *Reduction to a simpler equation.* The substitution

$$x = y - \tfrac{1}{4}a$$

leads to a simpler equation,

$$y^4 + py^2 + qy + r = 0, \tag{1}$$

where

$$p = \frac{8b - 3a^2}{8}, \quad q = \frac{8c - 4ab + a^3}{8}, \quad r = \frac{256d - 64ac + 16a^2 b - 3a^4}{256}.$$

Equation (1) is known as an *incomplete equation of the fourth degree*.

2°. *Descartes–Euler solution.* The roots $y_1, \ldots, y_4$ of the fourth-degree incomplete equation (1) are equal to one of the following expressions:

$$\pm\sqrt{z_1} \pm \sqrt{z_2} \pm \sqrt{z_3}, \tag{2}$$

where z_1, z_2, and z_3 are the roots of the auxiliary cubic equation

$$z^3 + \tfrac{1}{2}pz^2 + \tfrac{1}{16}(p^2 - 4r)z - \tfrac{1}{64}q^2 = 0. \tag{3}$$

The combinations of the signs in (2) are chosen in such a way that the relation

$$(\pm\sqrt{z_1})(\pm\sqrt{z_2})(\pm\sqrt{z_3}) = -\tfrac{1}{8}q$$

holds. It is noteworthy that according to Viète's theorem, the product of the roots z_1, z_2, and z_3 is equal to $\tfrac{1}{64}q^2 \geq 0$.

The roots of the fourth-degree equation (1) are determined by those of the auxiliary cubic equation (3); see Table 1.1.

TABLE 1.1. Relations between the roots of the fourth-degree incomplete equation and the roots of the auxiliary cubic equation (3).

Auxiliary cubic equation (3)	Fourth-degree equation (1)
All roots are real and positive	Four real roots
All roots are real, one positive and two negative	Two pairs of complex conjugate roots
One real root and two complex conjugate roots	Two real roots and two complex conjugate roots

3°. *Ferrari's solution.* Let u_0 be any of the roots of the auxiliary cubic equation

$$u^3 - bu^2 + (ac - 4d)u - a^2d + 4bd - c^2 = 0.$$

Then the four roots of the original equation are found by solving the following two quadratic equations:

$$x^2 + \tfrac{1}{2}ax + \tfrac{1}{2}u_0 = \pm\sqrt{(\tfrac{1}{4}a^2 - b + u_0)x^2 + (\tfrac{1}{2}au_0 - c)x + \tfrac{1}{4}u_0^2 - d},$$

where the radicand on the right-hand side is a perfect square.

4°. *Viète's theorem.* The roots of the general equation of the fourth degree satisfy the following relations:

$$x_1 + x_2 + x_3 + x_4 = -a,$$
$$x_1x_2 + x_1x_3 + x_1x_4 + x_2x_3 + x_2x_4 + x_3x_4 = b,$$
$$x_1x_2x_3 + x_1x_2x_4 + x_1x_3x_4 + x_2x_3x_4 = -c,$$
$$x_1x_2x_3x_4 = d.$$

4°. For further information on the general quartic equation, see the books King (1996), Korn & Korn (2000), Weisstein (2002), and Bronshtein & Semendyayev (2015) and the article *Quartic equation* (from Wikipedia).

Remark 1.1. *It is noteworthy that the quartic is the highest-order polynomial equation that can be solved by radicals in the general case (i.e., one in which the coefficients can take any value).*

1.1.4. Equations of the Fifth Degree

1. $x^5 + ax - 1 = 0.$

This is a special case of equation 1.1.5.14 with $n = 5$ and $m = 1$.

1°. For $a = b - b^{-4}$, the original equation has the root $x = 1/b$.

2°. The substitution $x = a^{1/4}y$ leads to an equation of the form 1.1.4.7:

$$y^5 + y - a^{-5/4} = 0.$$

3°. The three leading terms in the expansions of the real root of this equation for small and large a are given by

$$x = 1 - \tfrac{1}{5}a - \tfrac{1}{25}a^2 + O(a^3), \quad a \to 0;$$
$$x = a^{-1} - a^{-6} + 5a^{-11} + O(a^{-16}), \quad a \to \infty.$$

2. $x^5 + ax + ab + b^5 = 0.$

For any a and b, this equation has the root $x = -b$.

3. $x^5 - 5a^4x + 4a^5 = 0.$

This equation has a double root $x_1 = x_2 = a$. The other three roots are determined from the cubic equation

$$x^3 + 2ax^2 + 3a^2x + 4a^3 = 0.$$

4. $x^5 - \dfrac{a^5 - b^5}{a - b}x + \dfrac{ab(a^4 - b^4)}{a - b} = 0, \quad a \neq b.$

This equation has the roots $x_1 = a$ and $x_2 = b$.

5. $x^5 - (a^4 - 3a^2b + b^2)x + ab(a^2 - 2b) = 0.$

$1°$. Two roots are determined by the quadratic equation $x^2 - ax + b = 0$. The other three roots are determined from the cubic equation

$$x^3 + ax^2 + (a^2 - b)x + a(a^2 - 2b) = 0.$$

$2°$. For the special case $a = b = 1$, the left-hand side of the original equation admits factorization:

$$x^5 + x - 1 = (x^2 - x + 1)(x^3 + x^2 - 1).$$

In the special case $a = 1$ and $b = -\frac{1}{2}$, the left-hand side factorizes as

$$x^5 - \tfrac{11}{4}x - 1 = (x^2 - x - \tfrac{1}{2})(x^3 + x^2 + \tfrac{3}{2}x + 2).$$

$3°$. The following statement is true (Lee & Spearman, 2011). Let $p(x) = x^5 + cx - 1$, where c is a rational number. Then $p(x)$ factors into the product of an irreducible quadratic and an irreducible cubic if and only if $c = -1$ or $c = -\frac{11}{4}$.

The respective factorizations are given in Item $2°$.

6. $x^5 + \dfrac{5e^4(3 - 4\varepsilon c)}{c^2 + 1}x - \dfrac{4e^5(11\varepsilon + 2c)}{c^2 + 1} = 0.$

If $c \geq 0$ and $e \neq 0$ are rational numbers, and $\varepsilon = \pm 1$, then the roots of this equation are determined by the following formulas (see Spearman & Williams, 1994, 1996):

$$x_k = e(\omega^k u_1 + \omega^{2k} u_2 + \omega^{3k} u_3 + \omega^{4k} u_4), \quad k = 0, 1, 2, 3, 4,$$

where ω is a fifth root of unity,

$$\omega = \cos(\tfrac{2}{5}\pi) + i\sin(\tfrac{2}{5}\pi), \quad i^2 = -1,$$

and

$$u_1 = (v_1^2 v_3 / D^4)^{1/5}, \quad u_2 = (v_3^2 v_4 / D^4)^{1/5}, \quad u_3 = (v_2^2 v_1 / D^4)^{1/5}, \quad u_4 = (v_4^2 v_2 / D^4)^{1/5},$$
$$v_1 = D + (D^2 - \varepsilon D)^{1/2}, \quad v_2 = -D - (D^2 + \varepsilon D)^{1/2}, \quad v_3 = -D + (D^2 + \varepsilon D)^{1/2},$$
$$v_4 = D - (D^2 - \varepsilon D)^{1/2}, \quad D = (c^2 + 1)^{1/2}.$$

7. $x^5 + x - a = 0.$

This equation has the roots (Ritelli & Spaletta, 2021):

$$x_1 = Y_1(a),$$
$$x_2 = -e^{3\pi i/4}Y_0(a) - \tfrac{1}{4}Y_1(a) - \tfrac{5}{32}e^{\pi i/4}Y_2(a) - \tfrac{5i}{32}Y_3(a),$$
$$x_3 = e^{3\pi i/4}Y_0(a) - \tfrac{1}{4}Y_1(a) + \tfrac{5}{32}e^{\pi i/4}Y_2(a) - \tfrac{5i}{32}Y_3(a),$$
$$x_4 = -e^{\pi i/4}Y_0(a) - \tfrac{1}{4}Y_1(a) - \tfrac{5}{32}e^{3\pi i/4}Y_2(a) + \tfrac{5i}{32}Y_3(a),$$
$$x_5 = e^{\pi i/4}Y_0(a) - \tfrac{1}{4}Y_1(a) + \tfrac{5}{32}e^{3\pi i/4}Y_2(a) + \tfrac{5i}{32}Y_3(a),$$

where $i^2 = -1$ and

$$Y_0(a) = {}_4F_3\left(-\tfrac{1}{20}, \tfrac{3}{20}, \tfrac{7}{20}, \tfrac{11}{20}; \tfrac{1}{4}, \tfrac{2}{4}, \tfrac{3}{4}; -\tfrac{3125}{256}a^4\right),$$
$$Y_1(a) = a\, {}_4F_3\left(\tfrac{4}{20}, \tfrac{8}{20}, \tfrac{12}{20}, \tfrac{16}{20}; \tfrac{2}{4}, \tfrac{3}{4}, \tfrac{5}{4}; -\tfrac{3125}{256}a^4\right),$$
$$Y_2(a) = a^2\, {}_4F_3\left(\tfrac{9}{20}, \tfrac{13}{20}, \tfrac{17}{20}, \tfrac{21}{20}; \tfrac{3}{4}, \tfrac{5}{4}, \tfrac{6}{4}; -\tfrac{3125}{256}a^4\right),$$
$$Y_3(a) = a^3\, {}_4F_3\left(\tfrac{14}{20}, \tfrac{18}{20}, \tfrac{22}{20}, \tfrac{26}{20}; \tfrac{5}{4}, \tfrac{6}{4}, \tfrac{7}{4}; -\tfrac{3125}{256}a^4\right).$$

Here ${}_pF_q(\ldots)$ is the *generalized hypergeometric function*, which is defined as

$$_pF_q(\alpha_1,\ldots,\alpha_p; \beta_1,\ldots,\beta_q; z) = \sum_{k=0}^{\infty} \frac{(\alpha_1)_k \ldots (\alpha_p)_k}{(\beta_1)_k \ldots (\beta_q)_k} \frac{z^k}{k!},$$

$$(\alpha)_k = \alpha(\alpha+1)\ldots(\alpha+k-1).$$

For $p = q + 1$, the generalized hypergeometric series converges for $|z| < 1$ and diverges for $|z| > 1$.

The generalized hypergeometric function $F(z) = {}_pF_q(\alpha_1,\ldots,\alpha_p; \beta_1,\ldots,\beta_q; z)$ satisfies the ODE (Bailey, 1935):

$$\delta(\delta + \beta_1 - 1)\ldots(\delta + \beta_q - 1)F(z) = z(\delta + \alpha_1)\ldots(\delta + \alpha_p)F(z),$$

where $\delta = z(d/dz)$ is a differential operator.

8. $x^5 - \tfrac{5}{2}a^3x^2 + \tfrac{3}{2}a^5 = 0.$

This equation has a double root $x_1 = x_2 = a$. The other three roots are determined from the cubic equation

$$x^3 + 2ax^2 + 3a^2x + \tfrac{3}{2}a^3 = 0.$$

9. $x^5 - \dfrac{a^5 - b^5}{a^2 - b^2}x^2 + \dfrac{a^5b^2 - a^2b^5}{a^2 - b^2} = 0, \quad a \neq b.$

This equation has the roots $x_1 = a$ and $x_2 = b$.

10. $x^5 - a(a^2 - 3ab + b^2)x^2 + a^2b^2(a - b) = 0.$

Two roots of this equation are determined from the quadratic equation $x^2 - ax + ab = 0$. The other three roots are determined from the cubic equation

$$x^3 + ax^2 + a(a-b)x + ab(a-b) = 0.$$

11. $x^5 + ax^3 - 1 = 0.$

This is a special case of equation 1.1.5.14 with $n = 5$ and $m = 3$.

12. $x^5 + 5ax^3 + 5a^2x + b = 0.$

De Moivre quintic.

Solutions:
$$x_k = \omega^k(y_j)^{1/5} - a\omega^{-k}(y_j)^{-1/5}, \quad k = 1,\ldots,5,$$

where y_j is any root of the auxiliary quadratic equation

$$y^2 + by - a^5 = 0,$$

and ω is a fifth root of unity,
$$\omega = \cos\left(\tfrac{2}{5}\pi\right) + i\sin\left(\tfrac{2}{5}\pi\right), \quad i^2 = -1.$$

13. $x^5 + ax^4 - 1 = 0.$

This is a special case of equation 1.1.5.14 with $n = 5$ and $m = 4$.

14. $ax^5 + bx^4 + cx^3 + cx^2 + bx + a = 0 \quad (a \neq 0).$

Reciprocal algebraic (polynomial) equation of the fifth degree.

This equation has a root $x = -1$, and the remaining roots are found from the reciprocal equation of the fourth degree
$$ax^4 + (b-a)x^3 + (a-b+c)x^2 + (b-a)x + a = 0,$$
which reduces to a quadratic equation by substituting $z = x + \tfrac{1}{x}$.

15. $ax^5 + adx^4 + bx^3 + bdx^2 + cx + cd = 0 \quad (a \neq 0).$

This equation has a root $x = -d$, and the remaining roots are found from the biquadratic equation of the fourth degree
$$ax^4 + bx^2 + c = 0,$$
which reduces to a quadratic equation by substituting $z = x^2$.

16. $x^5 + ax^4 + bx^3 + cx^2 + dx + e = 0.$

General algebraic equation of the fifth degree (or the *general algebraic quintic equation*).

$1°$. *Ruffini–Abel theorem.* The general algebraic equation of the fifth degree is undecidable in radicals, meaning that its solutions cannot be expressed in terms of its coefficients using a finite number of arithmetic operations and the root extraction operation.

$2°$. A Tschirnhaus transformation (King, 1996), which may be computed by solving a quartic equation, reduces the general algebraic equation of the fifth degree to the Bring–Jerrard normal form $y^5 + py + q = 0$ (or $z^5 \pm z + k = 0$). Earlier, Euler showed that the general quintic equation can be reduced to another simpler form $Y^5 + p_1 Y^2 + q_1 = 0$.

$3°$. The solution to the original equation can be expressed in terms of the Jacobi theta functions and their associated elliptic modular functions (Hermite, 1858; King, 1996; Weisstein, 2002) as well as in terms of the generalized hypergeometric functions (see Cockle, 1860; Harley, 1862; Weisstein, 2002).

1.1.5. Algebraic Equations of Arbitrary Degree

In this section, n and m are positive integers.

1. $x^n - a = 0.$

Simplest binomial algebraic equation of the nth degree.

Solutions (roots):

$$x_{k+1} = \begin{cases} a^{1/n}\left(\cos\dfrac{2k\pi}{n} + i\sin\dfrac{2k\pi}{n}\right) & \text{for } a > 0, \\ |a|^{1/n}\left(\cos\dfrac{(2k+1)\pi}{n} + i\sin\dfrac{(2k+1)\pi}{n}\right) & \text{for } a < 0, \end{cases}$$

where $k = 0, 1, \ldots, n-1$ and $i^2 = -1$.

2. $ax^{2n} + bx^n + c = 0 \quad (a \neq 0)$.

Trinomial algebraic equation of a special form of arbitrary degree.

The substitution $y = x^n$ leads to a quadratic equation of the form 1.1.1.2.

Remark 1.2. *In equations 1.1.5.2, 1.1.5.5, and 1.1.5.8, n can be noninteger.*

3. $x^{2n} + 2ax^{n+1} + a^2 x^2 - b = 0$.

All the roots of this equation are determined by solving two simpler nth-degree equations:

$$x^n + ax - \sqrt{b} = 0,$$
$$x^n + ax + \sqrt{b} = 0.$$

4. $x^{2n} + 2ax^{n+m} + a^2 x^{2m} - b = 0 \quad (n > m)$.

All the roots of this equation are determined by solving two simpler nth-degree equations:

$$x^n + ax^m - \sqrt{b} = 0,$$
$$x^n + ax^m + \sqrt{b} = 0.$$

5. $ax^{3n} + bx^{2n} + cx^n + d = 0 \quad (a \neq 0)$.

The substitution $y = x^n$ leads to a complete cubic equation of the form 1.1.2.11.

6. $ax^{4n} + bx^{3n} + cx^{2n} + dx^n + e = 0 \quad (a \neq 0)$.

The substitution $y = x^n$ leads to a general quartic equation of the form 1.1.3.18.

7. $a_n x^{2n} + a_{n-1} x^{2n-2} + a_{n-2} x^{2n-4} + \cdots + a_2 x^4 + a_1 x^2 + a_0 = 0 \quad (a_n \neq 0)$.

Algebraic equation of the $2n$th degree containing only even powers. The substitution $y = x^2$ leads to an algebraic equation of the nth degree.

8. $a_0 x^{2n} + a_1 x^{2n-1} + a_2 x^{2n-2} + \cdots + a_2 x^2 + a_1 x + a_0 = 0 \quad (a_0 \neq 0)$.

Reciprocal algebraic equation of even degree. The substitution

$$y = x + \frac{1}{x}$$

reduces the original equation to a simpler algebraic equation of the nth degree.

Note that the left-hand side of the original equation is called a *reciprocal polynomial*.

9. $a_0 x^{2n} + a_1 x^{2n-1} + \cdots + a_{n-1} x^{n+1} + a_n x^n$
$\qquad + \lambda a_{n-1} x^{n-1} + \lambda^2 a_{n-2} x^{n-2} + \cdots + \lambda^{n-1} a_1 x + \lambda^n a_0 = 0 \quad (a_0 \neq 0).$

Generalized reciprocal algebraic equation of even degree.

The first $n+1$ terms of this equation (shown in the top line) coincide with the respective terms of the reciprocal equation 1.1.5.8. The remaining terms (bottom line) only differ by multipliers of the form λ^m from the respective terms of the reciprocal equation.

The substitution
$$y = x + \frac{\lambda}{x}$$
reduces the original equation to a simpler algebraic equation of the nth degree.

10. $a_0 x^{2n+1} + a_1 x^{2n} + a_2 x^{2n-1} + \cdots + a_2 x^2 + a_1 x + a_0 = 0 \quad (a_0 \neq 0).$

Reciprocal algebraic equation of odd degree.

This equation has a root $x = -1$, and its left-hand side admits the representation

$$[\text{left-hand side of the equation}] = (x+1) P_{2n}(x),$$

where $P_{2n}(x)$ is a reciprocal polynomial of degree $2n$. Therefore, the solution of the original equation reduces to the solution of the reciprocal equation of even degree, $P_{2n}(x) = 0$.

11. $x^n - n a^{n-1} x + (n-1) a^n = 0.$

This equation has a double root $x_1 = x_2 = a$.

12. $x^n - \dfrac{a^n - b^n}{a - b} x + \dfrac{a^n b - a b^n}{a - b} = 0, \quad a \neq b.$

This equation has the roots $x_1 = a$ and $x_2 = b$.

13. $x^n + x - a = 0.$

Trinomial algebraic equation of arbitrary degree.

1°. This equation has the following real root (Ritelli & Spaletta, 2021):

$$x(a) = a\, {}_{n-1}F_{n-2}\left(\frac{1}{n}, \frac{2}{n}, \ldots, \frac{n-1}{n};\, \frac{2}{n-1}, \frac{3}{n-1}, \ldots, \frac{n-2}{n-1}, \frac{n}{n-1};\, -\frac{n^n a^{n-1}}{(n-1)^{n-1}} \right),$$

where ${}_pF_q(\ldots)$ is the generalized hypergeometric function, which is defined as

$$ {}_pF_q(\alpha_1, \ldots, \alpha_p; \beta_1, \ldots, \beta_q; z) = \sum_{k=0}^{\infty} \frac{(\alpha_1)_k \cdots (\alpha_p)_k}{(\beta_1)_k \cdots (\beta_q)_k} \frac{z^k}{k!},$$

$$(\alpha)_k = \alpha(\alpha+1) \ldots (\alpha+k-1).$$

For $p = q + 1$, the generalized hypergeometric series converges for $|z| < 1$ and diverges for $|z| > 1$.

See also Glasser, 2000; Kato & Noumi, 2003; Passare & Tsikh, 2004; Mikhalkin, 2006, 2012; and Botta & Silva, 2019.

2°. The original equation will not change if we simultaneously transform the unknown x and parameter a according to the formulas

$$x = \bar{x}\varepsilon^j, \quad a = \bar{a}\varepsilon^j, \quad \varepsilon = e^{2\pi i/(n-1)}, \quad i^2 = -1,$$

where j is an integer. Therefore, the original equation has roots that can be expressed in terms of the solution specified in Item 1° according to the formulas

$$x_j(a) = \varepsilon^j x(\varepsilon^{-j} a), \quad \varepsilon = e^{2\pi i/(n-1)}, \quad i^2 = -1, \quad j = 1, \ldots, n-2.$$

14. $x^n + ax^m - 1 = 0, \quad n > m.$

Trinomial algebraic equation with two arbitrary degrees.

1°. For $a = b^m - b^{m-n}$, the original equation has the root $x = 1/b$.

2°. This equation has the following real root (Mellin, 1921):

$$x(a) = \frac{1}{n} \sum_{k=0}^{\infty} \frac{(-1)^k \Gamma\left(\frac{1}{n} + \frac{mk}{n}\right)}{\Gamma\left(\frac{1}{n} - \frac{(n-m)k}{n} + 1\right)} \frac{a^k}{k!},$$

where $\Gamma(z)$ is the gamma function. This solution satisfies the normalization condition $x(0) = 1$, and the above series converges for sufficiently small coefficient a (see Item 4° below for details).

3°. The original equation will not change if we simultaneously transform the unknown x and parameter a by the formulas

$$x = \bar{x}\varepsilon^j, \quad a = \bar{a}\varepsilon^{-mj}, \quad \varepsilon = e^{2\pi i/n}, \quad i^2 = -1,$$

where j is an integer. Therefore, all the roots of the original equation can be expressed in terms of the solution given in Item 2°, by the formulas (Mellin, 1921):

$$x_j(a) = \varepsilon^j x(\varepsilon^{jm} a), \quad \varepsilon = e^{2\pi i/n}, \quad i^2 = -1, \quad j = 0, 1, \ldots, n-1.$$

Here the value of $j = 0$ corresponds to the solution from Item 2°.

4°. The solution given in Item 2° can be expressed in terms of generalized hypergeometric functions by the formulas (Mikhalkin, 2012):

$$x(a) = -\frac{1}{n} \sum_{s=0}^{n-1} \frac{\Gamma\left(s - \frac{1+ms}{n}\right)}{\Gamma\left(1 - \frac{1+ms}{n}\right)} \frac{a^s}{s!}$$

$$\times {}_nF_{n-1}\left(\alpha_s, \ldots, \alpha_s + \frac{m-1}{m}, \beta_s, \ldots, \beta_s + \frac{n-m+1}{n-m}; \gamma_s, \ldots, \hat{1}, \ldots, \gamma_s + \frac{n-1}{n}; z\right),$$

$$\alpha_s = \frac{s}{n} + \frac{1}{mn}, \quad \beta_s = \frac{s}{n} - \frac{1}{n(n-m)}, \quad \gamma_s = \frac{s}{n} + \frac{1}{n}, \quad z = \frac{m^m (n-m)^{n-m}}{n^n} a^n.$$

Here we use the following short notations: the entry $\alpha_s, \ldots, \alpha_s + \frac{m-1}{m}$ denotes the sequence of numbers $\alpha_s, \alpha_s + \frac{1}{m}, \ldots, \alpha_s + \frac{m-1}{m}$; the entry $\gamma_s, \ldots, \hat{1}, \ldots, \gamma_s + \frac{n-1}{n}$ denotes the sequence of numbers $\gamma_s, \gamma_s + \frac{1}{n}, \ldots, \gamma_s + \frac{n-1}{n}$, in which the number 1 is omitted. The series representing generalized hypergeometric functions (the definition of these functions is given in the solution of equation 1.1.5.13) converge in the region $|a| < nm^{-m/n}(n-m)^{-(n-m)/n}$.

5°. The three leading terms in the expansions of the real root of this equation for small and large a are given by

$$x = 1 - \frac{1}{n}a + \frac{2m-n+1}{2n^2}a^2 + O(a^3), \quad a \to 0;$$

$$x = a^{-\frac{1}{m}}\left(1 - \frac{1}{m}a^{-\frac{n}{m}} + \frac{n}{m^2}a^{-\frac{2n}{m}}\right) + O\!\left(a^{-\frac{3n+1}{m}}\right), \quad a \to \infty.$$

6°. For integral representations of solutions and more detailed information about this and related equations, see the articles by Glasser, 2000; Kato & Noumi, 2003; Passare & Tsikh, 2004; Mikhalkin, 2006, 2012; and Botta & Silva, 2019.

15. $(x^n + a)^n + a = x.$

The left-hand side of the equation represents the second iteration of the binomial $P(x) = x^n + a$. The roots of the simpler equation $x^n + a = x$ are also roots of the original equation. Note that for $n = 3$, all the roots of the original equation can be expressed in radicals.

16. $[(x+a)^n + a]^n = x.$

The left-hand side of the equation represents the second iteration of the binomial $P(x) = (x+a)^n$. The roots of the simpler equation $(x+a)^n = x$ are also roots of the original equation. Note that for $n = 3$, all the roots of the original equation can be expressed in radicals.

17. $x^n - \frac{n}{m}a^{n-m}x^m + \frac{n-m}{m}a^n = 0, \quad n > m.$

This equation has the double root $x_1 = x_2 = a$.

18. $x^n - \frac{a^n - b^n}{a^m - b^m}x^m + \frac{a^n b^m - a^m b^n}{a^m - b^m} = 0, \quad a \ne b, \ n > m.$

This equation has the roots $x_1 = a$ and $x_2 = b$.

19. $ax^{2n} + bx^n(x+k) + c(x+k)^2 = 0.$

Let us divide the equation by $(x+k)^2$ and then make the substitution $z = x^n/(x+k)$ to arrive at the quadratic equation

$$az^2 + bz + c = 0.$$

Having solved this equation, we get two equations of degree n for x.

20. $ax^{2n} + bx^n(x+k)^m + c(x+k)^{2m} = 0.$

Let us divide the equation by $(x+k)^{2m}$ and then make the substitution $z = x^n/(x+k)^m$ to arrive at the quadratic equation $az^2 + bz + c = 0$.

21. $(x+a)^{2n} + (x+b)^{2n} = c.$

This is a special case of equation 1.1.5.24.

1°. The equation remains unchanged if x is replaced by $-a - b - x$. Hence, with the change of variable $z = [x + \frac{1}{2}(a+b)]^2$ followed by removing the brackets on the left-hand side, the original equation reduces to an equation of degree n.

2°. For $a \ne b$ and $c = (a-b)^{2n}$, the original equation has also two roots: $x_1 = -a$ and $x_2 = -b$.

22. $(x+a)^{2n} + (x+b)^{2n} + c(x+a)(x+b) = d.$

This is a special case of equation 1.1.5.24. The equation remains unchanged if x is replaced by $\lambda - x$ with $\lambda = -a - b$. Hence, with the change of variable $z = [x + \frac{1}{2}(a+b)]^2$ followed by removing the brackets on the left-hand side, the original equation reduces to an equation of degree n.

23. $(x+a)(x+b)[(x+a)^{2n} + (x+b)^{2n}] = c.$

This is a special case of equation 1.1.5.24. The equation remains unchanged if x is replaced by $\lambda - x$ with $\lambda = -a - b$. Hence, with the change of variable $z = [x + \frac{1}{2}(a+b)]^2$ followed by removing the brackets on the left-hand side, the original equation reduces to an equation of degree $n+1$.

24. $P_{2n}(x) = 0.$

Let $P_{2n}(x)$ be a polynomial of even degree $2n$ that, for some λ and any x, satisfies identically the relation $P_{2n}(x) = P_{2n}(\lambda - x)$. Then, with the substitution $z = (x - \frac{1}{2}\lambda)^2$, the original equation reduces to a simpler equation of degree n.

25. $x^n + a_1 x^{n_1} + \cdots + a_p x^{n_p} - 1 = 0, \quad n > n_1 > \cdots > n_p > 0.$

1°. This equation has the following real root (see Mellin, 1921 and Passare & Tsikh, 2004):

$$x(a_1, \ldots, a_p) = \frac{1}{n} \sum_{|k| \geq 0} \frac{(-1)^{|k|} \Gamma\left(\frac{1}{n} + \frac{n_1}{n} k_1 + \cdots + \frac{n_p}{n} k_p\right)}{k_1! \ldots k_p! \Gamma\left(\frac{1}{n} - \frac{n_1'}{n} k_1 - \cdots - \frac{n_p'}{n} k_p + 1\right)} a_1^{k_1} \ldots a_p^{k_p},$$

$$|k| = k_1 + \cdots + k_p, \quad k_m \geq 0, \quad n_m' = n - n_m, \quad m = 1, \ldots, p,$$

where $\Gamma(z)$ is the gamma function. This solution satisfies the normalization condition $x(0, \ldots, 0) = 1$, and the above series converges for sufficiently small coefficients $a_1, \ldots, a_p$.

2°. The original equation will not change if one simultaneously transforms the unknown x and parameters a_s according to the formulas

$$x = \bar{x}\varepsilon^j, \quad a_s = \bar{a}_s \varepsilon^{-n_s j}, \quad \varepsilon = e^{2\pi i/n}, \quad i^2 = -1, \quad s = 1, \ldots, p,$$

where j is an integer. Therefore, all the roots of the original equation are expressed in terms of the solution specified in Item 1° as

$$x_j(a_1, \ldots, a_p) = \varepsilon^j x(\varepsilon^{jn_1} a_1, \ldots, \varepsilon^{jn_p} a_p), \quad \varepsilon = e^{2\pi i/n}, \quad i^2 = -1, \quad j = 0, \ldots, n-1.$$

Here the value of $j = 0$ corresponds to the solution from Item 1°.

26. $a_n x^n + a_{n-1} x^{n-1} + a_{n-2} x^{n-2} + \cdots + a_2 x^2 + a_1 x + a_0 = 0 \quad (a_n \neq 0).$

General algebraic equation of the nth degree.

The following statements about the roots of the general algebraic equation with real coefficients hold:

1. Any algebraic equation of degree n has exactly n roots (real or complex), considering their multiplicity. This also holds true for any algebraic equations with complex coefficients.

2. Any algebraic equation of odd degree has at least one real root.

3. The number of complex roots of an algebraic equation of any degree is always even. If $x_1 = \alpha + i\beta$ ($i^2 = -1$) is a complex root of an algebraic equation, then $x_2 = \alpha - i\beta$ is also a (complex conjugate) root of this equation.

4. The left-hand side of any algebraic equation can be represented as the product of linear and quadratic polynomials with real coefficients.

5. A general algebraic equation of degree $n > 4$ with arbitrary coefficients is unsolvable in radicals; i.e., no solution can be expressed in terms of the equation coefficients using finitely many arithmetic operations (addition, subtraction, multiplication, and division) and root extraction operations (*Abel's theorem*).

6. If α is a root of an algebraic equation $P_n(x) = 0$ of degree n, then the left-hand side of this equation can be represented as the product $P_n(x) \equiv (x - \alpha) Q_{n-1}(x)$, where $Q_{n-1}(x)$ is a polynomial of degree $n - 1$.

7. Any integer root of an algebraic equation with integer coefficients is a divisor of the free term a_0.

8. If the coefficient of the leading term of an algebraic equation with integer coefficients is equal to one ($a_n = 1$), then all rational roots of this equation (if exist) are integer.

9. The solutions of an algebraic equation of any degree with arbitrary coefficients can be expressed in terms of higher-genus theta functions (Umemura, 1984). The solutions of algebraic equations of any degree can also be obtained in terms of generalized hypergeometric functions (see Mellin, 1921 and Sturmfels, 2000).

10. *Viète's theorem.* The roots of the general algebraic equation of the nth degree satisfy the relations

$$x_1 + x_2 + \cdots + x_n = -\frac{a_{n-1}}{a_n},$$

$$x_1 x_2 + x_1 x_3 + \cdots + x_1 x_n + x_2 x_3 + \cdots + x_{n-1} x_n = \frac{a_{n-2}}{a_n},$$

$$x_1 x_2 x_3 + x_1 x_2 x_4 + \cdots + x_{n-2} x_{n-1} x_n = -\frac{a_{n-3}}{a_n},$$

$$\cdots\cdots\cdots\cdots\cdots\cdots\cdots\cdots\cdots\cdots\cdots\cdots\cdots$$

$$x_1 x_2 x_3, \ldots, x_n = (-1)^n \frac{a_0}{a_n}.$$

27. $(x-a)p(x) + (x-b)q(x)$
$$= \frac{a-b}{r(a) - r(b)} \{[p(b) + q(a)]r(x) - p(b)r(a) - q(a)r(b)]\}.$$

Here $p(x)$, $q(x)$, and $r(x)$ are some polynomials, $a \neq b$ and $r(a) \neq r(b)$.

This equation has the roots $x_1 = a$ and $x_2 = b$.

28. $a_n p^n(x) + a_{n-1} p^{n-1}(x) q(x) + \cdots + a_1 p(x) q^{n-1}(x) + a_0 q^n(x) = 0.$

Algebraic equation homogeneous with respect to polynomials. Here $p(x)$ and $q(x)$ are some polynomials.

Let the roots of the polynomial $q(x)$ not be roots of the polynomial $p(x)$. Then after the term-by-term division of the equation by $q^n(x)$ and substitution $z = p(x)/q(x)$, one arrives at the algebraic equation

$$a_n z^n + a_{n-1} z^{n-1} + \cdots + a_1 z + a_0 = 0.$$

Remark 1.3. *In equations 1.1.5.27 and 1.1.5.28, $p(x)$, $q(x)$, and $r(x)$ can also be any (non-algebraic) functions.*

1.1.6. Systems of Linear Algebraic Equations

1. $a_1 x + b_1 y = c_1$, $\quad a_2 x + b_2 y = c_2$.

System of two linear algebraic equations in two unknowns.

Depending on the coefficients a_k, b_k, and c_k, the following three cases are possible:

1°. If $\Delta = a_1 b_2 - a_2 b_1 \neq 0$, then the original system has a unique solution
$$x = \frac{c_1 b_2 - c_2 b_1}{a_1 b_2 - a_2 b_1}, \quad y = \frac{a_1 c_2 - a_2 c_1}{a_1 b_2 - a_2 b_1}.$$

2°. If $\Delta = a_1 b_2 - a_2 b_1 = 0$ and $b_1 c_2 - b_2 c_1 = 0$ (the case of proportional coefficients), then the original system has infinitely many solutions described by the formulas
$$x = t, \quad y = \frac{c_1 - a_1 t}{b_1} \quad (b_1 \neq 0),$$

where t is arbitrary.

3°. If $\Delta = a_1 b_2 - a_2 b_1 = 0$ and $b_1 c_2 - b_2 c_1 \neq 0$, then the original system has no solutions.

2. $a_{11} x_1 + a_{12} x_2 + \cdots + a_{1k} x_k + \cdots + a_{1n} x_n = b_1$,
$\quad a_{21} x_1 + a_{22} x_2 + \cdots + a_{2k} x_k + \cdots + a_{2n} x_n = b_2$,
$\quad \cdots\cdots\cdots\cdots\cdots\cdots\cdots\cdots\cdots\cdots\cdots\cdots\cdots$,
$\quad a_{n1} x_1 + a_{n2} x_2 + \cdots + a_{nk} x_k + \cdots + a_{nn} x_n = b_n$.

System of n linear algebraic equations in n unknowns.

Let $\mathbb{A} = [a_{ij}]$ denote the matrix of coefficients of the left-hand side of the system of linear equations under consideration. Let the determinant of the matrix,

$$\Delta \equiv \det \mathbb{A} \equiv \begin{vmatrix} a_{11} & a_{12} & \cdots & a_{1k} & \cdots & a_{1n} \\ a_{21} & a_{22} & \cdots & a_{2k} & \cdots & a_{2n} \\ \vdots & \vdots & \vdots & \vdots & \ddots & \vdots \\ a_{n1} & a_{n2} & \cdots & a_{nk} & \cdots & a_{nn} \end{vmatrix},$$

be nonzero. Then the system has a unique solution expressed as (*Cramer's rule*):

$$x_1 = \frac{\Delta_1}{\Delta}, \quad x_2 = \frac{\Delta_2}{\Delta}, \quad \ldots, \quad x_n = \frac{\Delta_n}{\Delta},$$

where Δ_k ($k = 1, \ldots, n$) is the determinant of the matrix obtained from $\mathbb{A}$ by replacing its kth column with the column of free terms:

$$\Delta_k = \begin{vmatrix} a_{11} & a_{12} & \cdots & b_1 & \cdots & a_{1n} \\ a_{21} & a_{22} & \cdots & b_2 & \cdots & a_{2n} \\ \vdots & \vdots & \vdots & \vdots & \ddots & \vdots \\ a_{n1} & a_{n2} & \cdots & b_n & \cdots & a_{nn} \end{vmatrix}.$$

See also Korn & Korn, 2000; Weisstein, 2002; Polyanin & Manzhirov, 2007; Bronshtein & Semendyayev, 2015.

1.2. Trigonometric Equations

▶ Throughout Section 1.2, it is assumed that the coefficients of all trigonometric equations are real numbers and only real solutions are considered.

1.2.1. Binomial Trigonometric Equations

1. $\sin x = a.$

Solutions for $|a| \leq 1$:
$$x = (-1)^n \arcsin a + \pi n, \quad n = 0, \pm 1, \pm 2, \ldots$$
There are no solutions for $|a| > 1$.

2. $\cos x = a.$

Solutions for $|a| \leq 1$:
$$x = \pm \arccos a + 2\pi n, \quad n = 0, \pm 1, \pm 2, \ldots$$
There are no solutions for $|a| > 1$.

3. $\tan x = a.$

Solutions:
$$x = \arctan a + \pi n, \quad n = 0, \pm 1, \pm 2, \ldots$$

4. $\cot x = a.$

Solutions:
$$x = \operatorname{arccot} a + \pi n, \quad n = 0, \pm 1, \pm 2, \ldots$$

5. $\cos(a_1 x + b_1) = \cos(a_2 x + b_2).$

Using the trigonometric identity $\cos X - \cos Y = -2\sin[\frac{1}{2}(X+Y)]\sin[\frac{1}{2}(X-Y)]$, one can obtain the following solutions.

1°. Two groups of solutions for $a_1 \neq \pm a_2$:
$$x_n = -\frac{b_1 + b_2}{a_1 + a_2} + \frac{2\pi n}{a_1 + a_2}, \quad n = 0, \pm 1, \pm 2, \ldots;$$
$$x_m = -\frac{b_1 - b_2}{a_1 - a_2} + \frac{2\pi m}{a_1 - a_2}, \quad m = 0, \pm 1, \pm 2, \ldots$$

2°. Solutions for $a_1 = a_2$:
$$x_n = -\frac{b_1 + b_2}{2a_1} + \frac{\pi n}{a_1}, \quad n = 0, \pm 1, \pm 2, \ldots$$

3°. Solutions for $a_1 = -a_2$:
$$x_m = -\frac{b_1 - b_2}{2a_1} + \frac{\pi m}{a_1}, \quad m = 0, \pm 1, \pm 2, \ldots$$

6. $\cos(a_1 x + b_1) = -\cos(a_2 x + b_2).$

This equation is equivalent to equation 1.2.1.5, in which b_2 should be replaced by $b_2 + \pi$,
$$\cos(a_1 x + b_1) = \cos(a_2 x + b_2 + \pi).$$

7. $\sin(a_1 x + b_1) = \sin(a_2 x + b_2).$

Using the trigonometric identity $\sin X - \sin Y = 2\cos[\frac{1}{2}(X+Y)]\sin[\frac{1}{2}(X-Y)]$, one can obtain the following solutions.

$1°$. Two groups of solutions for $a_1 \neq \pm a_2$:

$$x_n = -\frac{b_1+b_2}{a_1+a_2} + \frac{2}{a_1+a_2}\left(\pm\frac{\pi}{2}+2\pi n\right), \quad n=0,\pm 1,\pm 2,\ldots;$$

$$x_m = -\frac{b_1-b_2}{a_1-a_2} + \frac{2\pi m}{a_1-a_2}, \quad m=0,\pm 1,\pm 2,\ldots$$

$2°$. Solutions for $a_1 = a_2$:

$$x_n = -\frac{b_1+b_2}{2a_1} + \frac{1}{a_1}\left(\pm\frac{\pi}{2}+2\pi n\right), \quad n=0,\pm 1,\pm 2,\ldots$$

$3°$. Solutions for $a_1 = -a_2$:

$$x_m = -\frac{b_1-b_2}{2a_1} + \frac{\pi m}{a_1}, \quad m=0,\pm 1,\pm 2,\ldots$$

8. $\sin(a_1x+b_1) = -\sin(a_2x+b_2)$.

This equation is equivalent to equation 1.2.1.7, in which b_2 should be replaced by $b_2+\pi$:

$$\sin(a_1x+b_1) = \sin(a_2x+b_2+\pi).$$

9. $\cos(a_1x+b_1) = \sin(a_2x+b_2)$.

This equation is equivalent to equation 1.2.1.7 with b_1 replaced by $b_1+\tfrac{1}{2}\pi$:

$$\sin(a_1x+b_1+\tfrac{1}{2}\pi) = \sin(a_2x+b_2).$$

10. $\cos(a_1x+b_1) = -\sin(a_2x+b_2)$.

This equation is equivalent to equation 1.2.1.5, in which b_2 should be replaced by $b_2+\tfrac{1}{2}\pi$:

$$\cos(a_1x+b_1) = \cos(a_2x+b_2+\tfrac{1}{2}\pi).$$

11. $\tan(a_1x+b_1) = \tan(a_2x+b_2)$.

Solution for $a_1 \neq a_2$:

$$x_n = -\frac{b_1-b_2}{a_1-a_2} + \frac{\pi n}{a_1-a_2}, \quad n=0,\pm 1,\pm 2,\ldots$$

12. $\tan(a_1x+b_1) = -\tan(a_2x+b_2)$.

This equation can be written in the form of equation 1.2.1.11:

$$\tan(a_1x+b_1) = \tan(-a_2x-b_2).$$

1.2.2. Trigonometric Equations Containing Several Terms

▶ **Equations linear with respect to trigonometric functions.**

1. $a\sin x + b\cos x = c.$

This equation is equivalent to the equation

$$\cos(x-\varphi) = \frac{c}{\sqrt{a^2+b^2}}, \tag{*}$$

where the auxiliary angle φ is determined by the relations

$$\cos\varphi = \frac{b}{\sqrt{a^2+b^2}}, \quad \sin\varphi = \frac{a}{\sqrt{a^2+b^2}}.$$

Equation $(*)$ is reduced by substituting $y = x - \varphi$ to an equation of the form 1.2.1.2.

2. $a\sin x + b\sin(2x) + c\sin(3x) = 0.$

Using the trigonometric identities

$$\sin(2x) = 2\sin x \cos x, \quad \sin(3x) = 3\sin x - 4\sin^3 x, \quad \sin^2 x = 1 - \cos^2 x,$$

it can be shown that the solution of the original equation reduces to the solution of two simpler equations:

$$\sin x = 0, \quad 4c\cos^2 x + 2b\cos x + a - c = 0.$$

With the substitution $z = \cos x$, the second equation reduces to a quadratic equation (one should look for its real solutions that satisfy the inequality $|z| \leq 1$).

3. $a\cos x + b\cos(2x) + c\cos(3x) = 0.$

Using the trigonometric identities

$$\cos(2x) = 2\cos^2 x - 1, \quad \cos(3x) = -3\cos x + 4\cos^3 x,$$

it can be shown that the solution of the original equation reduces to the solution of a cubic equation with respect to $z = \cos x$ (one should look for real solutions that satisfy the inequality $|z| \leq 1$).

4. $a\sin(px) + b\cos(px) + c\sin(qx) + d\cos(qx) = 0, \quad a^2 + b^2 = c^2 + d^2.$

Let us move the last two terms to the right-hand side of the equation and then divide all the terms by $\rho = \sqrt{a^2+b^2} = \sqrt{c^2+d^2}$. Using the trigonometric identity $a\sin X + b\cos X = \rho\sin(X+\varphi)$, we obtain an equivalent equation of the form 1.2.1.7:

$$\sin(px + \varphi_1) = \sin(qr + \varphi_2),$$

where the angles φ_1 and φ_2 are found from the relations

$$\cos\varphi_1 = a/\rho, \quad \sin\varphi_1 = b/\rho, \quad \cos\varphi_2 = -c/\rho, \quad \sin\varphi_2 = -d/\rho.$$

5. $\cos(ax) + \cos(bx) + k\{\cos(cx) + \cos[(a+b-c)x]\} = 0.$

Using the sum-to-product identity $\cos X + \cos Y = 2\cos[\tfrac{1}{2}(X+Y)]\cos[\tfrac{1}{2}(X-Y)]$, one arrives at the equivalent equation

$$2\cos[\tfrac{1}{2}(a+b)x]\{\cos[\tfrac{1}{2}(a-b)x] + k\cos[\tfrac{1}{2}(2c-a-b)x]\} = 0.$$

As a result, we get two groups of solutions that are determined from simpler equations

$$\cos[\tfrac{1}{2}(a+b)x] = 0,$$
$$\cos[\tfrac{1}{2}(a-b)x] + k\cos[\tfrac{1}{2}(2c-a-b)x] = 0.$$

For solutions of the second equation with $k = \mp 1$, see equations 1.2.1.5 and 1.2.1.6.

6. $\cos(ax) - \cos(bx) + k\{\cos(cx) - \cos[(a+b-c)x]\} = 0.$

With the difference-to-product trigonometric identity $\cos X - \cos Y = -2\sin[\frac{1}{2}(X+Y)] \times \sin[\frac{1}{2}(X-Y)]$, one arrives at the equivalent equation

$$-2\sin[\tfrac{1}{2}(a+b)x]\{\sin[\tfrac{1}{2}(a-b)x] + k\sin[\tfrac{1}{2}(2c-a-b)x]\} = 0.$$

As a result, we get two groups of solutions that are determined from simpler equations

$$\sin[\tfrac{1}{2}(a+b)x] = 0,$$
$$\sin[\tfrac{1}{2}(a-b)x] + k\sin[\tfrac{1}{2}(2c-a-b)x] = 0.$$

For solutions of the second equation with $k = \mp 1$, see equations 1.2.1.7 and 1.2.1.8.

7. $\sin(ax) + \sin(bx) + k\{\sin(cx) + \sin[(a+b-c)x]\} = 0.$

Using the sum-to-product identity $\sin X + \sin Y = 2\sin[\frac{1}{2}(X+Y)]\cos[\frac{1}{2}(X-Y)]$, we arrive at the equivalent equation

$$2\sin[\tfrac{1}{2}(a+b)x]\{\cos[\tfrac{1}{2}(a-b)x] + k\cos[\tfrac{1}{2}(2c-a-b)x]\} = 0.$$

As a result, we get two groups of solutions that are determined from simpler equations

$$\sin[\tfrac{1}{2}(a+b)x] = 0,$$
$$\cos[\tfrac{1}{2}(a-b)x] + k\cos[\tfrac{1}{2}(2c-a-b)x] = 0.$$

For solutions of the second equation with $k = \mp 1$, see equations 1.2.1.5 and 1.2.1.6.

8. $\sin(ax) - \sin(bx) + k\{\sin(cx) - \sin[(a+b-c)x]\} = 0.$

Using the difference-to-product trigonometric identity $\sin X - \sin Y = 2\sin[\frac{1}{2}(X-Y)] \times \cos[\frac{1}{2}(X+Y)]$, we arrive at the equivalent equation

$$2\cos[\tfrac{1}{2}(a+b)x]\{\sin[\tfrac{1}{2}(a-b)x] + k\sin[\tfrac{1}{2}(2c-a-b)x]\} = 0.$$

As a result, we obtain two groups of solutions determined from simpler equations

$$\cos[\tfrac{1}{2}(a+b)x] = 0,$$
$$\sin[\tfrac{1}{2}(a-b)x] + k\sin[\tfrac{1}{2}(2c-a-b)x] = 0.$$

For solutions of the second equation with $k = \mp 1$, see equations 1.2.1.7 and 1.2.1.8.

9. $\sum_{k=1}^{n} \sin(2kx) = 0.$

Obviously, the equation has the trivial solution $x = 0$. To find other roots, we transform the sum on the left-hand side into a product using the formula (Gradshteyn & Ryzhik, 2000):

$$\sum_{k=1}^{n} \sin(2kx) = \sin[(n+1)x]\sin(nx)\csc x.$$

Therefore, other roots are determined by solving two simple equations of the form 1.2.1.1: $\sin[(n+1)x] = 0$ and $\sin(nx) = 0$.

10. $\sum_{k=0}^{n} \cos(2kx) = 0.$

We transform the sum on the left-hand side into a product by the formula (Gradshteyn & Ryzhik, 2000):

$$\sum_{k=0}^{n} \cos(2kx) = \sin[(n+1)x]\cos(nx)\csc x.$$

Therefore, roots of the equation are determined by solving two simple equations of the form 1.2.1.1 and 1.2.1.2: $\sin[(n+1)x] = 0$ $(x \neq 0)$ and $\cos(nx) = 0$.

11. $\sum_{k=1}^{n} \sin[(2k-1)x] = 0.$

Obviously, the equation has the trivial solution $x = 0$. To find other roots, we transform the sum on the left-hand side into a product by the formula (Gradshteyn & Ryzhik, 2000):

$$\sum_{k=1}^{n} \sin[(2k-1)x] = \sin^2(nx)\csc x.$$

Therefore, other roots are determined by solving a simple equation of the form 1.2.1.1: $\sin(nx) = 0$ $(x \neq 0)$.

12. $\sum_{k=1}^{n} \cos[(2k-1)x] = 0.$

We transform the sum on the left-hand side into a product according to the formula (Gradshteyn & Ryzhik, 2000):

$$\sum_{k=1}^{n} \cos[(2k-1)x] = \frac{1}{2}\sin(2nx)\csc x.$$

Therefore, roots of the equation are determined by solving a simple equation of the form 1.2.1.1: $\sin(2nx) = 0$ $(x \neq 0)$.

13. $\sum_{k=0}^{n-1} \sin(kx + a) = 0.$

We transform the sum on the left-hand side into a product by the formula (Gradshteyn & Ryzhik, 2000):

$$\sum_{k=0}^{n-1} \sin(kx + a) = \sin\left(\frac{n-1}{2}x + a\right)\sin\frac{nx}{2}\csc\frac{x}{2}.$$

Therefore, roots of the equation are determined by solving two simple equations of the form 1.2.1.1: $\sin(\frac{n-1}{2}x + a) = 0$ and $\sin\frac{nx}{2} = 0$ $(x \neq 0)$.

14. $\sum_{k=0}^{n-1} \cos(kx + a) = 0.$

We transform the sum on the left-hand side into a product by the formula (Gradshteyn & Ryzhik, 2000):

$$\sum_{k=0}^{n-1} \cos(kx + a) = \cos\left(\frac{n-1}{2}x + a\right) \sin\frac{nx}{2} \csc\frac{x}{2}.$$

Therefore, roots of the equation are determined from two simple equations of the form 1.2.1.1 and 1.2.1.2: $\sin\frac{nx}{2} = 0$ ($x \neq 0$) and $\cos(\frac{n-1}{2}x + a) = 0$.

15. $\sum_{k=0}^{2n-1} (-1)^k \cos(kx + a) = 0.$

We transform the sum on the left-hand side into a product by the formula (Gradshteyn & Ryzhik, 2000):

$$\sum_{k=0}^{2n-1} (-1)^k \cos(kx + a) = \sin\left(\frac{2n-1}{2}x + a\right) \sin(nx) \sec\frac{x}{2}.$$

Therefore, roots of the equation are determined by solving two simple equations of the form 1.2.1.1: $\sin(nx) = 0$ and $\sin(\frac{2n-1}{2}x + a) = 0$.

16. $\sum_{k=1}^{n} (-1)^{k+1} \sin[(2k-1)x] = 0.$

We transform the sum on the left-hand side into a product by the formula (Gradshteyn & Ryzhik, 2000):

$$\sum_{k=1}^{n} (-1)^{k+1} \sin[(2k-1)x] = (-1)^{n+1} \frac{\sin(2nx)}{2\cos x}.$$

Therefore, roots of the equation are determined by solving a simple equation of the form 1.2.1.1: $\sin(2nx) = 0$.

17. $a \tan x + b \cot x + c = 0 \quad (ab \neq 0).$

The substitution $z = \tan x$ leads to a quadratic equation of the form 1.1.1.2:

$$az^2 + cz + b = 0.$$

▶ **Other equations.**

18. $a \sin^2 x + b \sin x \cos x + c \cos^2 x = 0.$

$1°$. For $a = 0$, the solution of this equation reduces to solving two simpler equations, $\cos x = 0$ and $b \sin x + c \cos x = 0$, which were considered earlier (see equations 1.2.1.2 and 1.2.2.1).

2°. For $a \neq 0$, the equation reduces, after dividing by $\cos^2 x \neq 0$ and substituting $z = \tan x$, to a quadratic equation of the form 1.1.1.2:

$$az^2 + bz + c = 0.$$

19. $a_1 \sin^2 x + a_2 \sin x \cos x + a_3 \cos^2 x + b_1 \sin(2x) + b_2 \cos(2x) + c = 0.$

Using the identities $\sin(2x) = 2\sin x \cos x$, $\cos(2x) = \cos^2 x - \sin^2 x$, and $c = c(\sin^2 x + \cos^2 x)$, we reduce this equation to an equation of the form 1.2.2.18:

$$(a_1 - b_2 + c)\sin^2 x + (a_2 + 2b_1)\sin x \cos x + (a_3 + b_2 + c)\cos^2 x = 0.$$

20. $\sin(ax)\sin(bx) = \sin(cx)\sin[(a + b - c)x].$

Using the product-to-difference trigonometric identity $\sin X \sin Y = \frac{1}{2}[\cos(X - Y) - \cos(X + Y)]$, we get the equivalent equation

$$\cos[(a - b)x] = \cos[(2c - a - b)x],$$

which coincides with equation 1.2.1.5 at $a_1 = a - b$, $a_2 = 2c - a - b$, and $b_1 = b_2 = 0$.

21. $\sin(ax)\sin(bx) = -\sin(cx)\sin[(a + b + c)x].$

Renaming $a = -a_1$ and $b = -b_1$, we get an equation of the form 1.2.2.20:

$$\sin(a_1 x)\sin(b_1 x) = \sin(cx)\sin[(a_1 + b_1 - c)x].$$

22. $\cos(ax)\cos(bx) = \cos(cx)\cos[(a + b - c)x].$

Using the product-to-sum trigonometric identity $\cos X \cos Y = \frac{1}{2}[\cos(X - Y) + \cos(X + Y)]$, we get the equivalent equation

$$\cos[(a - b)x] = \cos[(2c - a - b)x],$$

which coincides with equation 1.2.1.5 at $a_1 = a - b$, $a_2 = 2c - a - b$, and $b_1 = b_2 = 0$.

23. $\sin(ax)\cos(bx) = \sin(cx)\cos[(a + b - c)x].$

Using the product-to-sum trigonometric identity $\sin X \cos Y = \frac{1}{2}[\sin(X - Y) + \sin(X + Y)]$, we get the equivalent equation

$$\sin[(a - b)x] = \sin[(2c - a - b)x],$$

which coincides with equation 1.2.1.7 at $a_1 = a - b$, $a_2 = 2c - a - b$, and $b_1 = b_2 = 0$.

24. $\sin(ax)\cos(bx) = -\sin(cx)\cos[(a - b + c)x].$

Using the product-to-sum trigonometric identity $\sin X \cos Y = \frac{1}{2}[\sin(X - Y) + \sin(X + Y)]$, we get the equivalent equation

$$\sin[(a + b)x] = \sin[(b - a - 2c)x],$$

which coincides with equation 1.2.1.7 at $a_1 = a + b$, $a_2 = b - a - 2c$, and $b_1 = b_2 = 0$.

25. $a \sin x + b \sin^2 x + c \cos^2 x + d \cos(2x) + k = 0.$

Expressing $\cos^2 x$ and $\cos(2x)$ in terms of sine, we arrive at a quadratic equation for $z = \sin x$:

$$(b - c - 2d)z^2 + az + c + d + k = 0.$$

One should only take the roots that satisfy the condition $|z| \leq 1$.

26. $a\cos x + b\cos^2 x + c\sin^2 x + d\cos(2x) + k = 0.$

Expressing $\sin^2 x$ and $\cos(2x)$ in terms of cosine, we arrive at a quadratic equation for $z = \cos x$:
$$(b - c + 2d)z^2 + az + c - d + k = 0.$$
One should only take the roots that satisfy the condition $|z| \leq 1$.

27. $a\cos(2x) + b\cos^2(2x) + c\cos(4x) + d\cos^4 x + k = 0.$

Using the trigonometric identities
$$\cos(4x) = 2\cos^2(2x) - 1,$$
$$\cos^4 x = (\cos^2 x)^2 = \left[\tfrac{1}{2} + \tfrac{1}{2}\cos(2x)\right]^2 = \tfrac{1}{4} + \tfrac{1}{2}\cos(2x) + \tfrac{1}{4}\cos^2(2x),$$
we transform the original equation into a quadratic equation with respect to $z = \cos(2x)$:
$$(b + 2c + \tfrac{1}{4}d)z^2 + (a + \tfrac{1}{2}d)z + k - c + \tfrac{1}{4}d = 0.$$
One should only take the roots that satisfy the condition $|z| \leq 1$.

28. $a\tan^2 x + b\cot^2 x + c = 0 \quad (ab \neq 0).$

The substitution $z = \tan^2 x$ leads to a quadratic equation of the form 1.1.1.2:
$$az^2 + cz + b = 0.$$

1.2.3. Trigonometric Equations of the General Form

1. $\sum_{k=1}^{n} a_k \sin(kx) = 0.$

With the multiple-angle trigonometric formula (Weisstein, 2002):
$$\sin(kx) = \sin x \left[\sum_{j=0}^{[(k-1)/2]} \frac{(-1)^j 2^{k-2j-1}(k-j-1)!}{j!(k-2j-1)!} \cos^{k-2j-1} x \right],$$
where $[A]$ stands for the integer part of the number A, the equation in question can be reduced to the simple trigonometric equation $\sin x = 0$ and an algebraic equation of degree $n - 1$ with respect to $z = \cos x$ (one should only look for real solutions that satisfy the inequality $|z| \leq 1$).

2. $\sum_{k=0}^{2n+1} a_k \sin[(2k+1)x] = b.$

With the multiple-angle trigonometric formula (Gradshteyn & Ryzhik, 2000):
$$\sin[(2k+1)x] = (2k+1)\left\{ \sin x + \sum_{j=1}^{k} (-1)^j \right.$$
$$\left. \times \frac{[(2k+1)^2 - 1][(2k+1)^2 - 3^2]\ldots[(2k+1)^2 - (2j-1)^2]}{(2j+1)!} \sin^{2j+1} x \right\},$$

the equation in question can be reduced to an algebraic equation of degree $2n + 1$ with respect to $z = \sin x$ (one should only look for real solutions that satisfy the inequality $|z| \leq 1$).

3. $\sum_{k=1}^{n} a_k \cos(kx) = b.$

Using the multiple-angle trigonometric formula (Weisstein, 2002):

$$\cos(kx) = k \sum_{j=0}^{[k/2]} \frac{(-1)^j 2^{k-2j-1}(k-j-1)!}{j!(k-2j)!} \cos^{k-2j} x,$$

where $[A]$ stands for the integer part of the number A, the equation in question can be reduced to an algebraic equation of degree n with respect to $z = \cos x$ (one should only look for real solutions that satisfy the inequality $|z| \leq 1$).

4. $\sum_{k=0}^{n} a_k \sin^k x \cos^{n-k} x = 0.$

$1°$. For $a_0 \neq 0$, after dividing by $\cos^n x \neq 0$ and substituting $z = \tan x$, the equation reduces to an algebraic equation of the form

$$\sum_{k=0}^{n} a_k z^k = 0 \qquad (z = \tan x). \qquad (*)$$

$2°$. For $a_0 = 0$, one must add the solutions of the equation $\cos x = 0$ to the solutions obtained from equation $(*)$.

5. $P(\sin x + \cos x, \sin x \cos x) = 0.$

Here $P(y, z)$ is a polynomial in two variables. By substituting $y = \sin x + \cos x$ and taking into account the identities

$$y^2 = (\sin x + \cos x)^2 = \sin^2 x + \cos^2 x + 2 \sin x \cos x = 1 + 2z,$$

we arrive at an algebraic equation for the unknown y:

$$P\left(y, \frac{y^2 - 1}{2}\right) = 0.$$

6. $P(\sin x - \cos x, \sin x \cos x) = 0.$

Here $P(y, z)$ is a polynomial in two variables. By substituting $y = \sin x - \cos x$ and taking into account the identities

$$y^2 = (\sin x - \cos x)^2 = \sin^2 x + \cos^2 x - 2 \sin x \cos x = 1 - 2z,$$

we arrive at an algebraic equation for the unknown y:

$$P\left(y, \frac{1 - y^2}{2}\right) = 0.$$

7. $P(\sin x, \cos x) = 0$.

Here $P(y, z)$ is a polynomial in two variables. Using the trigonometric identities

$$\sin x = \frac{2\xi}{1+\xi^2}, \quad \cos x = \frac{1-\xi^2}{1+\xi^2}, \quad \xi = \tan\frac{x}{2},$$

we obtain the equation

$$P\left(\frac{2\xi}{1+\xi^2}, \frac{1-\xi^2}{1+\xi^2}\right) = 0,$$

which is reduced by multiplying by $(1+z^2)^m$, where m is a suitable positive integer, to an algebraic equation.

Note that the points $x = \pi + 2\pi n$, where $n = 0, \pm 1, \pm 2, \ldots$, must be checked separately.

8. $P(\tan x, \cot x) = 0$.

Here $P(y, z)$ is a polynomial in two variables of maximum degree m in z. The substitution $y = \tan x$ leads to an algebraic equation for the unknown y:

$$y^m P(y, 1/y) = 0.$$

1.3. Other Transcendental Equations

▶ *Throughout Section 1.3, it is assumed that the coefficients of all equations are real numbers, and only real solutions are considered.*

1.3.1. Equations Containing Exponential Functions

1. $a^{kx+m} = b^{px+q}$ $(a, b > 0)$.

Solution:
$$x = \frac{q \ln b - m \ln a}{k \ln a - p \ln b} = \frac{q \log_c b - m \log_c a}{k \log_c a - p \log_c b},$$

where c is any positive number other than one. In particular, one can set $c = a$ or $c = b$.

2. $ap^{2\beta x} + bp^{\beta x} + c = 0$ $(p > 0, p \neq 1)$.

The substitution $z = p^{\beta x}$ leads to the quadratic equation

$$az^2 + bz + c = 0.$$

3. $ap^{2\beta x} + bq^{2\beta x} + c(pq)^{\beta x} = 0$ $(p, q > 0)$.

Let us divide this equation term by term by $q^{2\beta x}$. As a result, we arrive at a quadratic equation for $z = (p/q)^{\beta x}$:

$$az^2 + cz + b = 0.$$

4. $\sum_{k=0}^{n} a_k e^{k\beta x} = 0$.

The substitution $z = e^{\beta x}$ leads to the algebraic equation $\sum_{k=0}^{n} a_k z^k = 0$.

5. $ax + b = e^{-cx}$.

Solution:
$$x = -\frac{b}{a} + \frac{1}{c}W\left(\frac{ce^{bc/a}}{a}\right),$$

where $W(z)$ is the *Lambert W function*, which is implicitly defined by the relation $We^W = z$. For real $z = t$, the function $W(t)$ is single-valued for $t \geq -1/e$ and $W \geq -1$ (see equation 10.1.1.2 in Section 10.1.1 for details on the Lambert W function).

6. $x^k e^x = a$.

Here $a > 0$ and $x > 0$. Solution:
$$x = kW\left(a^{1/k}/k\right),$$

where $W(z)$ is the Lambert function.

7. $(ax)^{bx} = c$.

Here $a > 0$, $c > 0$, and $x > 0$. Solution:
$$x = \frac{1}{a}\exp\left[W\left(\frac{a}{b}\ln c\right)\right],$$

where $W(z)$ is the Lambert W function.

1.3.2. Equations Containing Hyperbolic Functions

▶ **Binomial hyperbolic equations.**

1. $\sinh x = a$.

Solution for any a:
$$x = \ln\left(a + \sqrt{a^2 + 1}\right) \equiv \operatorname{arsinh} a.$$

2. $\cosh x = a$.

Solution for $a \geq 1$:
$$x = \ln\left(a + \sqrt{a^2 - 1}\right) \equiv \operatorname{arcosh} a.$$

There are no real solutions for $a < 1$.

3. $\tanh x = a$.

Solution for $|a| < 1$:
$$x = \frac{1}{2}\ln\frac{1+a}{1-a} \equiv \operatorname{artanh} a.$$

4. $\coth x = a$.

Solution for $|a| > 1$:
$$x = \frac{1}{2}\ln\frac{a+1}{a-1} \equiv \operatorname{arcoth} a.$$

5. $\cosh(a_1 x + b_1) = \cosh(a_2 x + b_2)$.

Using the hyperbolic identity $\cosh X - \cosh Y = 2\sinh[\frac{1}{2}(X+Y)]\sinh[\frac{1}{2}(X-Y)]$, one can obtain the solutions given below.

$1°$. Two solutions for $a_1 \neq \pm a_2$:
$$x_1 = -\frac{b_1+b_2}{a_1+a_2}, \quad x_2 = -\frac{b_1-b_2}{a_1-a_2}.$$

$2°$. Solution for $a_1 = a_2$:
$$x = -\frac{b_1+b_2}{2a_1}.$$

$3°$. Solution for $a_1 = -a_2$:
$$x = -\frac{b_1-b_2}{2a_1}.$$

6. $\sinh(a_1 x + b_1) = \sinh(a_2 x + b_2)$.

Solution for $a_1 \neq a_2$:
$$x = -\frac{b_1-b_2}{a_1-a_2}.$$

For $a_1 = a_2$ and $b_1 \neq b_2$, there are no solutions.

▶ **Hyperbolic equations involving three or more terms.**

7. $a \sinh x + b \cosh x = c$.

$1°$. For $c = 0$ and $|b/a| < 1$, the solution is
$$x = \frac{1}{2} \ln \frac{1-(b/a)}{1+(b/a)}.$$

For $c = 0$ and $|b/a| \geq 1$, there is no solution.

$2°$. For $c \neq 0$, this equation reduces to the quadratic equation
$$(a+b)z^2 - 2cz + b - a = 0, \quad z = e^x.$$

One should look for real roots satisfying the condition $z > 0$.

8. $a \sinh x + b \sinh(2x) + c \sinh(3x) = 0$.

Using the hyperbolic identities

$\sinh(2x) = 2\sinh x \cosh x, \quad \sinh(3x) = 3\sinh x + 4\sinh^3 x, \quad \sinh^2 x = \cosh^2 x - 1,$

one can find that the solution of the original equation reduces to the solution of two simpler equations:
$$\sinh x = 0, \quad 4c \cosh^2 x + 2b \cosh x + a - c = 0.$$

The first equation has the trivial solution $x = 0$. With the substitution $z = \cosh x$, the second equation reduces to a quadratic equation (one should only consider the roots satisfying the condition $z \geq 1$).

9. $a \cosh x + b \cosh(2x) + c \cosh(3x) = 0$.

Using the hyperbolic identities
$$\cosh(2x) = 2\cosh^2 x - 1, \quad \cosh(3x) = -3\cosh x + 4\cosh^3 x,$$

one can find that the solution of the original equation reduces to the solution of a cubic equation with respect to $z = \cosh x$ (one should only look for real solutions that satisfy the inequality $|z| \geq 1$).

10. $\sum_{k=0}^{n-1} \sinh(kx + a) = 0.$

We convert the sum on the left-hand side into a product by the following formula (Gradshteyn & Ryzhik, 2000):

$$\sum_{k=0}^{n-1} \sinh(kx + a) = \sinh\left(\frac{n-1}{2}x + a\right) \frac{\sinh(nx/2)}{\sinh(x/2)}.$$

It follows that $x = -\frac{2a}{n-1}$ is a root of the equation.

11. $a \sinh^2 x + b \sinh x \cosh x + c \cosh^2 x = 0.$

$1°$. For $a = 0$ and considering that $\cosh x > 0$, the solution of this equation reduces to solving a simpler equation, $b \sinh x + c \cosh x = 0$, which was discussed earlier (see equation 1.3.2.7).

$2°$. For $a \neq 0$, after dividing by $\cosh^2 x$ and substituting $z = \tanh x$, the equation reduces to a quadratic equation of the form 1.1.1.2:

$$az^2 + bz + c = 0.$$

One should only look for real solutions that satisfy the inequality $|z| < 1$.

12. $a_1 \sinh^2 x + a_2 \sinh x \cosh x + a_3 \cosh^2 x + b_1 \sinh(2x) + b_2 \cosh(2x) = c.$

Using the identities

$$\sinh(2x) = 2 \sinh x \cosh x, \quad \cosh(2x) = \cosh^2 x + \sinh^2 x, \quad c = c(\cosh^2 x - \sinh^2 x),$$

one can reduce this equation to an equation of the form 1.3.2.11:

$$(a_1 + b_2 + c) \sinh^2 x + (a_2 + 2b_1) \sinh x \cosh x + (a_3 + b_2 - c) \cosh^2 x = 0.$$

13. $a \sinh x + b \sinh^2 x + c \cosh^2 x + d \cosh(2x) + k = 0.$

Expressing $\cosh^2 x$ and $\cosh(2x)$ in terms of hyperbolic sine, we arrive at a quadratic equation for $z = \sinh x$:

$$(b + c + 2d)z^2 + az + c + d + k = 0.$$

14. $a \cosh x + b \cosh^2 x + c \sinh^2 x + d \cosh(2x) + k = 0.$

Expressing $\sinh^2 x$ and $\cosh(2x)$ in terms of hyperbolic cosine, we arrive at a quadratic equation for $z = \cosh x$:

$$(b + c + 2d)z^2 + az - c - d + k = 0.$$

One should only consider the real roots that satisfy the condition $z \geq 1$.

15. $a\cosh(2x) + b\cosh^2(2x) + c\cosh(4x) + d\cosh^4 x + k = 0.$

Using the hyperbolic identities

$$\cosh(4x) = 2\cosh^2(2x) - 1,$$
$$\cosh^4 x = (\cosh^2 x)^2 = \left[\tfrac{1}{2} + \tfrac{1}{2}\cosh(2x)\right]^2 = \tfrac{1}{4} + \tfrac{1}{2}\cosh(2x) + \tfrac{1}{4}\cosh^2(2x),$$

we transform the original equation into a quadratic equation with respect to $z = \cosh(2x)$:

$$(b + 2c + \tfrac{1}{4}d)z^2 + (a + \tfrac{1}{2}d)z + k - c + \tfrac{1}{4}d = 0.$$

One should only consider the real roots that satisfy the condition $z \geq 1$.

▶ **Hyperbolic equations of the general form.**

16. $\displaystyle\sum_{k=1}^{n} a_k \sinh(kx) = 0.$

Using the multiple-argument hyperbolic formula

$$\sinh(kx) = \sinh x \left[\sum_{j=0}^{[(k-1)/2]} \frac{(-1)^j 2^{k-2j-1}(k-j-1)!}{j!(k-2j-1)!} \cosh^{k-2j-1} x\right],$$

where $[A]$ stands for the integer part of the number A, the equation in question can be reduced to the simple hyperbolic equation $\sinh x = 0$, which has the trivial solution $x = 0$, and an algebraic equation of degree $n-1$ with respect to $z = \cosh x$ (one should only look for real solutions that satisfy the inequality $z \geq 1$).

17. $\displaystyle\sum_{k=0}^{n} a_k \cosh(kx) = 0.$

Using the multiple-argument hyperbolic formula

$$\cosh(kx) = k\sum_{j=0}^{[k/2]} \frac{(-1)^j 2^{k-2j-1}(k-j-1)!}{j!(k-2j)!} \cosh^{k-2j} x,$$

where $[A]$ stands for the integer part of the number A, the equation in question can be reduced to an algebraic equation of degree n with respect to $z = \cosh x$ (one should only look for real solutions that satisfy the inequality $z \geq 1$).

18. $\displaystyle\sum_{k=0}^{n} a_k \sinh^k x \cosh^{n-k} x = 0.$

After dividing by $\cosh^n x \neq 0$ and substituting $z = \tanh x$, the equation reduces to an algebraic equation of the form

$$\sum_{k=0}^{n} a_k z^k = 0 \qquad (z = \tanh x).$$

One should only consider the roots that satisfy the condition $|z| < 1$.

19. $P(\sinh x, \cosh x) = 0$.

Here $P(y_1, y_2)$ is a polynomial in two variables, y_1 and y_2, of total degree m. The substitution $z = e^x$ leads to an algebraic equation for the unknown z:

$$z^m P\left(\frac{z^2 - 1}{2z}, \frac{z^2 + 1}{2z}\right) = 0.$$

One should only consider the roots that satisfy the condition $z > 0$.

20. $P(\tanh x, \coth x) = 0$.

Here $P(y, z)$ is a polynomial in two variables of maximum degree m in z. The substitution $y = \tanh x$ leads to an algebraic equation for the unknown y:

$$y^m P(y, 1/y) = 0.$$

One should only consider the roots that satisfy the condition $|z| < 1$.

1.3.3. Equations Containing Logarithmic Functions

1. $\log_a(bx + c) = d$.

Solution: $x = \dfrac{a^d - c}{b}$.

2. $\log_a(b_1 x + c_1) + \log_a(b_2 x + c_2) = \log_a(b_3 x + c_3) + \log_a(b_4 x + c_4) + d$.

The real roots of this equation are determined from the quadratic equation (written in non-canonical form):

$$(b_1 x + c_1)(b_2 x + c_2) = a^d (b_3 x + c_3)(b_4 x + c_4),$$

and must satisfy the four inequalities $b_k x + c_k > 0$ for $k = 1, 2, 3, 4$.

3. $a \log_c^2(\beta x) + b \log_c(\beta x) + c = 0$.

The substitution $z = \log_c(\beta x)$ leads to the quadratic equation $az^2 + bz + c = 0$.

4. $b_1 \log_a x + b_2 \log_{a^2} x + \cdots + b_n \log_{a^n} x = c$.

Taking into account the formula $\log_{a^k} x = \frac{1}{k} \log_a x$, we get

$$x = a^{c/S}, \quad S = b_1 + \frac{1}{2} b_2 + \cdots + \frac{1}{n} b_n.$$

5. $b_1 \log_{a_1} x + b_2 \log_{a_2} x + \cdots + b_n \log_{a_n} x = c$.

Taking into account the formula $\log_a x = \log_k x / \log_k a$ ($k > 0$ and $k \neq 1$), we get

$$x = k^{c/S}, \quad S = \frac{b_1}{\log_k a_1} + \frac{b_2}{\log_k a_2} + \cdots + \frac{b_n}{\log_k a_n}.$$

As k one can choose, for example, Euler's number e.

6. $\sum_{k=0}^{n} a_k \log_c^k(\beta x) = 0$.

The substitution $z = \log_c(\beta x)$ leads to the algebraic equation $\sum_{k=0}^{n} a_k z^k = 0$.

7. $x\log_a(bx) = c$.

Here $a > 0$ and $bx > 0$. Solution:
$$x = \frac{1}{b}\exp\bigl[W(bc\ln a)\bigr],$$

where $W(z)$ is the Lambert W function, which is implicitly defined by the relation $We^W = z$. For real $z = t$, the function $W(t)$ is single-valued for $t \geq -1/e$ and $W \geq -1$ (see equation 10.1.1.2 in Section 10.1.1 for details on the Lambert W function).

8. $\log_a x + bx^k + c = 0$.

Solution:
$$x = a^{-c}\exp\left[-\frac{1}{k}W(bka^{-ck}\ln a)\right],$$

where $W(z)$ is the Lambert W function.

9. $\log_a(ba^x + c) = px + q$.

Using the properties of the logarithm, we obtain the exponential equation $ba^x + c = a^{px+b}$, or
$$bz + c = a^b z^p, \quad z = a^x.$$

For $p = 1$, this is a linear equation, and for $p = 2$ or $p = -1$, it is a quadratic equation (one only considers the roots that satisfy the conditions $z > 0$ and $bz + c > 0$).

References

Bailey, W.N., *Generalised Hypergeometric Series*, Cambridge Univ. Press, Cambridge, 1935.
Botta, V. and da Silva, J.V., On the behavior of roots of trinomial equations, *Acta Math. Hungarica*, Vol. 157, No. 1, pp. 54–62, 2019.
Bronshtein, I.N. and Semendyayev, K.A., *Handbook of Mathematics*, 6th ed., Springer, Berlin, 2015.
Cockle, J., Sketch of a theory of transcendental roots, *Phil. Mag.*, Vol. 20, pp. 145–148, 1860.
Cubic Equation, from *Wikipedia*, https://en.wikipedia.org/wiki/Cubic_equation
Glasser, M.L., Hypergeometric functions and the trinomial equation, *J. Comput. Applied Math.*, Vol. 118, No. 1–2, pp. 169–173, 2000.
Gradshteyn, I.S. and Ryzhik, I.M., *Table of Integrals, Series, and Products*, 6th ed., Academic Press, New York, 2000.
Harley, R., On the solution of the transcendental solution of algebraic equations, *Quart. J. Pure Appl. Math.*, Vol. 5, pp. 337–361, 1862.
Hermite, C., Sulla risoluzione delle equazioni del quinto grado, *Annali di Math. Pura ed Appl.*, Vol. 1, pp. 256–259, 1858.
Kato, M. and Noumi, M., Monodromy groups of hypergeometric functions satisfying algebraic equations, *Tohoku Math. J., Second Series*, Vol. 55, No. 2, pp. 189–205, 2003.
King, R.B., *Beyond the Quartic Equation*, Birkhäuser, Boston, 1996.
Korn, G.A. and Korn, T.M., *Mathematical Handbook for Scientists and Engineers*, 2nd ed., Dover Publ., New York, 2000.
Lee, P.D. and Spearman, B.K., The factorization of $x^5 + ax^m + 1$, *Scientiae Math. Japonicae*, Vol. 73, No. 2–3, pp. 171–174, 2011.

Mellin, H.J., Résolution de l'équation algébrique générale à l'aide de la fonction gamma, *C. R. Acad. Sci. Paris Sér. I Math.,* Vol. 172, pp. 658–661, 1921.

Mikhalkin, E.N., Certain formulas for solutions to trinomial and tetranomial algebraic equations, *J. Sib. Federal Univ. Math. Phys.,* Vol. 5, No. 2, pp. 230–223, 2012.

Mikhalkin, E.N., On solving general algebraic equations by integrals of elementary functions, *Sib. Math. J.,* Vol. 47, pp. 301–306, 2006.

Passare, M. and Tsikh, A., Algebraic equations and hypergeometric series, In: *The Legacy of Niels Henrik Abel* (eds. Laudal, O.A., Piene, R.), Springer, Berlin-Heidelberg, 2004.

Polyanin, A.D. and Manzhirov, A.V., *Handbook of Mathematics for Engineers and Scientists,* Chapman & Hall/CRC Press, Boca Raton–London, 2007.

Quartic Equation, from *Wikipedia,* https://en.wikipedia.org/wiki/Quartic_equation

Ritelli, D. and Spaletta, G., Trinomial equation: The hypergeometric way, *Open J. Math. Sci.,* Vol. 5, pp. 236–247, 2021.

Spearman, B.K. and Williams, K.S., Characterization of solvable quintics $x^5 + ax + b$, *Am. Math. Month.,* Vol. 101, No. 10, pp. 986–992, 1994.

Spearman, B.K. and Williams, K.S., On solvable quintics $x^5 + ax + b$ and $x^5 + ax^2 + b$, *Rocky Mt. J. Math.,* Vol. 28, No. 2, pp. 753–772, 1996.

Sturmfels, B., Solving algebraic equations in terms of $\mathscr{A}$-hypergeometric series, *Discrete Math.,* Vol. 210, pp. 171–181, 2000.

Umemura, H., Resolution of algebraic equations by theta constants, In: *Tata Lectures on Theta II* (ed. Mumford, D.), Birkhäuser, Boston, pp. 3.261–3.272, 1984.

Weisstein, E.W., *CRC Concise Encyclopedia of Mathematics, 2nd ed.,* Chapman & Hall/CRC Press, Boca Raton, London, 2002.

Chapter 2

Ordinary Differential Equations

▶ **Preliminary remarks.** Ordinary differential equations (ODEs) are mathematical equations containing an unknown function of one argument and its derivatives with respect to this argument. The order of an ordinary differential equation is the maximum order of the derivative of the unknown function included in the equation.

The integration of ordinary differential equations in closed form is the representation of solutions to these equations by analytical formulas that are written using an a priori specified set of admissible functions and a pre-listed set of mathematical operations. A solution is said to be obtained by quadrature if elementary functions and functions included in the equation are used as admissible functions (this is necessary when the equation depends on arbitrary functions), and the admissible operations are a finite set of arithmetic operations, superposition operations (to form composite functions), operations of differentiation, and operations of taking the indefinite integral. A solution can be written in an explicit, implicit, or parametric form.

This chapter provides a brief description of closed-form solutions (usually via quadratures) to various linear and nonlinear ordinary differential equations. It also gives some transformations, integrals, and reductions, leading to simpler ODEs. The chapter includes some ODEs whose solutions are expressed in terms of special functions. No stationary solutions of the form $y = \text{const}$, which are easy to find without integrating differential equations, are discussed here.

In solutions, integration constants (arbitrary constants not included in the ODEs under consideration) are denoted by $C, C_0, C_1, \ldots, C_n$.

2.1. First-Order Ordinary Differential Equations

2.1.1. Simplest First-Order ODEs

1. $y'_x = f(y)$.

First-order autonomous ODE. It is a special case of the separable ODE 2.1.1.2.

General solution: $x = \int \dfrac{dy}{f(y)} + C$.

2. $y'_x = f(x)g(y)$.

Separable ODE.

General solution: $\int \dfrac{dy}{g(y)} = \int f(x)\,dx + C$.

3. $g(x)y'_x = f_1(x)y + f_0(x)$.

First-order linear ODE.

General solution:
$$y = Ce^F + e^F \int e^{-F} \frac{f_0(x)}{g(x)} \, dx, \quad \text{where} \quad F(x) = \int \frac{f_1(x)}{g(x)} \, dx.$$

4. $g(x) y'_x = f_1(x) y + f_0(x) y^k$.

Bernoulli equation. Here k is an arbitrary number. For $k \neq 1$, the substitution $w(x) = y^{1-k}$ leads to a linear ODE:
$$g(x) w'_x = (1-k) f_1(x) w + (1-k) f_0(x).$$

General solution:
$$y = \left[Ce^F + (1-k) e^F \int e^{-F} \frac{f_0(x)}{g(x)} \, dx \right]^{\frac{1}{1-k}}, \quad \text{where} \quad F(x) = (1-k) \int \frac{f_1(x)}{g(x)} \, dx.$$

5. $y'_x = f(y/x)$.

First-order homogeneous ODE. The substitution $u(x) = y/x$ leads to a separable ODE: $x u'_x = f(u) - u$.

General solution:
$$\int \frac{du}{f(u) - u} = \ln|x| + C, \quad u = \frac{y}{x}.$$

Particular solutions: $y = A_k x$, where the A_k are roots of the algebraic (transcendental) equation $A_k - f(A_k) = 0$.

2.1.2. Riccati Equations

1. $y'_x = ay^2 + bx^k$.

Special Riccati equation, k is an arbitrary number.

1°. The substitution $y = -u'_x/(au)$ leads to a second-order linear ODE of the form 2.2.2.4: $u''_{xx} + abx^k u = 0$.

2°. General solution for $k \neq -2$:
$$y = -\frac{1}{a} \frac{w'_x}{w}, \quad w(x) = \begin{cases} \sqrt{x} \left[C_1 J_{\frac{1}{2q}} \left(\frac{1}{q} \sqrt{ab} \, x^q \right) + C_2 Y_{\frac{1}{2q}} \left(\frac{1}{q} \sqrt{ab} \, x^q \right) \right] & \text{if } ab > 0, \\ \sqrt{x} \left[C_1 I_{\frac{1}{2q}} \left(\frac{1}{q} \sqrt{|ab|} \, x^q \right) + C_2 K_{\frac{1}{2q}} \left(\frac{1}{q} \sqrt{|ab|} \, x^q \right) \right] & \text{if } ab < 0, \end{cases}$$

where $q = \frac{1}{2}(k+2)$; $J_m(z)$ and $Y_m(z)$ are Bessel functions and $I_m(z)$ and $K_m(z)$ are modified Bessel functions (see equations 2.2.2.22 and 2.2.2.23).

3°. Solution for $k = -2$:
$$y = \frac{\lambda}{x} - x^{2a\lambda} \left(\frac{ax}{2a\lambda + 1} x^{2a\lambda} + C \right)^{-1},$$

where λ is a root of the quadratic equation $a\lambda^2 + \lambda + b = 0$.

2. $y'_x = ay^2 + be^{\lambda x}$.

The substitution $y = -u'_x/(au)$ leads to a second-order linear ODE of the form 2.2.3.1: $u''_{xx} + abe^{\lambda x}u = 0$.

3. $y'_x = y^2 + f(x)y - a^2 - af(x)$.

Particular solution: $y_0 = a$. The general solution can be obtained with the formulas in Item 1° of equation 2.1.2.19.

4. $y'_x = f(x)y^2 + ay - ab - b^2 f(x)$.

Particular solution: $y_0 = b$. The general solution can be obtained with the formulas in Item 1° of equation 2.1.2.19.

5. $y'_x = y^2 + xf(x)y + f(x)$.

Particular solution: $y_0 = -1/x$. The general solution can be obtained with the formulas in Item 1° of equation 2.1.2.19.

6. $y'_x = f(x)y^2 - ax^k f(x)y + akx^{k-1}$.

Particular solution: $y_0 = ax^k$. The general solution can be obtained with the formulas in Item 1° of equation 2.1.2.19.

7. $y'_x = f(x)y^2 + akx^{k-1} - a^2 x^{2k} f(x)$.

Particular solution: $y_0 = ax^k$. The general solution can be obtained with the formulas in Item 1° of equation 2.1.2.19.

8. $y'_x = -(k+1)x^k y^2 + x^{k+1} f(x)y - f(x)$.

Particular solution: $y_0 = x^{-k-1}$. The general solution can be obtained with the formulas in Item 1° of equation 2.1.2.19.

9. $xy'_x = f(x)y^2 + ky + ax^{2k} f(x)$.

General solution: $y = \begin{cases} \sqrt{a}\, x^k \tan\left[\sqrt{a} \int x^{k-1} f(x)\, dx + C\right] & \text{if } a > 0, \\ \sqrt{|a|}\, x^k \tanh\left[-\sqrt{|a|} \int x^{k-1} f(x)\, dx + C\right] & \text{if } a < 0. \end{cases}$

10. $xy'_x = x^{2k} f(x)y^2 + [ax^k f(x) - k]y + bf(x)$.

The substitution $z = x^k y$ leads to a separable ODE: $z'_x = x^{k-1} f(x)(z^2 + az + b)$.

11. $y'_x = f(x)y^2 + g(x)y - a^2 f(x) - ag(x)$.

Particular solution: $y_0 = a$. The general solution can be obtained with the formulas in Item 1° of equation 2.1.2.19.

12. $y'_x = f(x)y^2 + g(x)y + akx^{k-1} - a^2 x^{2k} f(x) - ax^k g(x)$.

Particular solution: $y_0 = ax^k$. The general solution can be obtained with the formulas in Item 1° of equation 2.1.2.19.

13. $y'_x = ae^{\lambda x}y^2 + ae^{\lambda x}f(x)y + \lambda f(x)$.

Particular solution: $y_0 = -\dfrac{\lambda}{a}e^{-\lambda x}$. The general solution can be obtained with the formulas in Item 1° of equation 2.1.2.19.

14. $y'_x = f(x)y^2 - ae^{\lambda x}f(x)y + a\lambda e^{\lambda x}$.

Particular solution: $y_0 = ae^{\lambda x}$. The general solution can be obtained with the formulas in Item 1° of equation 2.1.2.19.

15. $y'_x = f(x)y^2 + a\lambda e^{\lambda x} - a^2 e^{2\lambda x}f(x)$.

Particular solution: $y_0 = ae^{\lambda x}$. The general solution can be obtained with the formulas in Item 1° of equation 2.1.2.19.

16. $y'_x = f(x)y^2 + \lambda y + ae^{2\lambda x}f(x)$.

General solution: $y = \begin{cases} \sqrt{a}\, e^{\lambda x} \tan\left[\sqrt{a} \int e^{\lambda x} f(x)\, dx + C\right] & \text{if } a > 0, \\ \sqrt{|a|}\, e^{\lambda x} \tanh\left[-\sqrt{|a|} \int e^{\lambda x} f(x)\, dx + C\right] & \text{if } a < 0. \end{cases}$

17. $y'_x = y^2 - f^2(x) + f'_x(x)$.

Particular solution: $y_0 = f(x)$. The general solution can be obtained with the formulas in Item 1° of equation 2.1.2.19.

18. $y'_x = f(x)y^2 - f(x)g(x)y + g'_x(x)$.

Particular solution: $y_0 = g(x)$. The general solution can be obtained with the formulas in Item 1° of equation 2.1.2.19.

19. $y'_x = f_2(x)y^2 + f_1(x)y + f_0(x)$.

General Riccati equation.

1°. Given a particular solution $y_0 = y_0(x)$ of the Riccati equation, the general solution can be written as:
$$y = y_0(x) + \Phi(x)\left[C - \int \Phi(x) f_2(x)\, dx\right]^{-1},$$
where
$$\Phi(x) = \exp\left\{\int [2f_2(x)y_0(x) + f_1(x)]\, dx\right\}.$$
To the particular solution $y_0(x)$ there corresponds $C = \infty$.

2°. The substitution
$$u(x) = \exp\left(-\int f_2 y\, dx\right)$$
reduces the general Riccati equation to a second-order linear ODE:
$$f_2 u''_{xx} - [(f_2)'_x + f_1 f_2] u'_x + f_0 f_2^2 u = 0,$$
which often may be easier to solve than the original Riccati equation.

3°. Many solvable Riccati equations can be found in the handbooks by Kamke (1977) and Polyanin & Zaitsev (2003, 2018).

2.1.3. Abel Equations

1. $y'_x = f_3(x)y^3 + f_2(x)y^2 + f_1(x)y + f_0(x), \qquad f_3(x) \not\equiv 0.$

Abel equation of the first kind (general form). The Abel equation of the first kind is not integrable in closed form for arbitrary $f_n(x)$.

Given below are some special cases where this ODE is integrable by quadrature.

1°. In the degenerate case $f_2(x) = f_0(x) = 0$, we have the Bernoulli equation 2.1.1.4 with $k = 3$.

2°. If the functions $f_n(x)$ ($n = 0, 1, 2, 3$) are all proportional, i.e., $f_n(x) = a_n g(x)$, then the original equation is a separable ODE of the form 2.1.1.2.

3°. The Abel equation is homogeneous:
$$y'_x = a\frac{y^3}{x^3} + b\frac{y^2}{x^2} + c\frac{y}{x} + d.$$

See ODE 2.1.1.5 with $f(u) = au^3 + bu^2 + cu + d$.

4°. The Abel equation is generalized homogeneous:
$$y'_x = ax^{2n+1}y^3 + bx^n y^2 + \frac{c}{x}y + dx^{-n-2}.$$

See ODE 2.1.4.4 with $k = n+1$. The substitution $w = x^{n+1}y$ leads to a separable equation:
$$xw'_x = aw^3 + bw^2 + (c+n+1)w + d.$$

5°. The Abel equation has the form
$$y'_x = ax^{3k-m}y^3 + bx^{2k}y^2 + \frac{m-k}{x}y + dx^{2m}.$$

Then it is reducible with the substitution $y = x^{m-k}z$ to a separable ODE of the form 2.1.1.2:
$$z'_x = x^{k+m}(az^3 + bz^2 + c).$$

6°. Let $f_0 \equiv 0$, $f_1 \equiv 0$, and $(f_3/f_2)'_x = af_2$ for some constant a. Then the substitution $y = f_2 f_3^{-1} u$ leads to a separable equation:
$$u'_x = f_2^2 f_3^{-1}(u^3 + u^2 + au).$$

7°. If
$$f_0 = \frac{f_1 f_2}{3 f_3} - \frac{2 f_2^3}{27 f_3^2} - \frac{1}{3}\frac{d}{dx}\frac{f_2}{f_3}, \qquad f_n = f_n(x),$$

then the solution of the original ODE is given by
$$y(x) = E\left(C - 2\int f_3 E^2\, dx\right)^{-1/2} - \frac{f_2}{3 f_3}, \quad \text{where} \quad E = \exp\left[\int\left(f_1 - \frac{f_2^2}{3 f_3}\right)dx\right].$$

2. $yy'_x = y + f(x).$

Abel equation of the second kind in the canonical form. This ODE is not integrable for an arbitrary $f(x)$. In the handbooks by Polyanin & Zaitsev (2003, 2018), general solutions of this equation with the right-hand sides of the following forms are described:

$$f(x) = kx + Ax^m,$$
$$f(x) = kx + A\alpha x^p + A^2\beta x^q,$$
$$f(x) = kx + A\alpha x^{1/2} + A^2\beta + A^3\gamma x^{-1/2},$$
$$f(x) = Ax^2 - \tfrac{9}{625}A^{-1},$$
$$f(x) = kx + A\alpha x^{1/3} + A^2\beta x^{-1/3} + A^4\gamma x^{-5/3},$$
$$f(x) = kx + \alpha x^{1/3} + \beta + \gamma x^{-1/3} + \delta x^{-2/3},$$
$$f(x) = kx + A^2\alpha x^{-1/7} + A^3\beta x^{-5/7} + A^4\delta x^{-9/7},$$
$$f(x) = \pm\frac{1}{\sqrt{Ax^2 + Bx + C}},$$
$$f(x) = kx + \frac{\alpha x^2 + \beta}{\sqrt{x^2 + \gamma}},$$
$$f(x) = A + B\exp(-2x/A),$$

where A, B, and C are arbitrary constants such that the above functions make sense, and k, m, p, q, α, β, γ, and δ are some given numbers.

The illustrative examples below are several special cases where the ODE in question is integrated by quadrature.

1°. The Abel equation

$$yy'_x - y = a,$$

which can be written as a separable ODE, has the general solution in implicit form

$$x = y - a\ln|y + a| + C.$$

2°. The Abel equation

$$yy'_x - y = ax + b, \qquad a \neq 0,$$

has the general solution in parametric form

$$x = C\exp\left(-\int \frac{\tau\,d\tau}{\tau^2 - \tau - a}\right) - \frac{b}{a}, \quad y = C\tau\exp\left(-\int \frac{\tau\,d\tau}{\tau^2 - \tau - a}\right).$$

3°. The Abel equation

$$yy'_x - y = a(e^{2x/a} - 1)$$

has the general solution in parametric form

$$x = a\ln\left|\frac{\tau^2 + 1}{\tau}(\arctan\tau - C)\right|, \quad y = \frac{a}{\tau}\left[\tau + (\tau^2 - 1)(\arctan\tau - C)\right].$$

4°. The Abel equation

$$yy'_x - y = -\frac{2(m+1)}{(m+3)^2}x + Ax^m$$

has the general solution in parametric form

$$x = \frac{m+3}{m-1}a\tau\Phi^{\frac{2}{m-1}}, \quad y = a\Phi^{\frac{2}{m-1}}\left(\Phi\Psi + \frac{2}{m-1}\tau\right),$$
$$\Phi = \int(1 \pm \tau^{m+1})^{-1/2}\,d\tau - C, \quad \Psi = \sqrt{1 \pm \tau^{m+1}},$$

where the free parameter of the equation A is expressed in terms of the solution parameter a as $A = \pm \dfrac{m+1}{2} \left(\dfrac{m-1}{m+3} \right)^{m+1} a^{1-m}$.

5°. The Abel equation
$$yy'_x - y = a + be^{-2x/a},$$
for $ab > 0$, has the general solution in parametric form
$$x = a \ln \left| \dfrac{\sqrt{\tau^2 + ab}}{a \ln|\tau + \sqrt{\tau^2 + ab}| + C} \right|, \quad y = \tau \dfrac{a \ln|\tau + \sqrt{\tau^2 + ab}| + C}{\sqrt{\tau^2 + ab}} - a.$$

Remark 2.1. The transformation $y = a\hat{y}$, $x = a\hat{x} + b$ brings the original ODE to a similar equation,
$$\hat{y}\hat{y}'_{\hat{x}} - \hat{y} = a^{-1} f(a\hat{x} + b).$$
Therefore the function $f(x)$ on the right-hand side of the Abel equation of the second kind in the canonical form can be identified with the two-parameter family of functions $a^{-1} f(ax + b)$.

3. $yy'_x = f(x)y + g(x).$

Abel equation of the second kind. With the aid of the substitution $z = \displaystyle\int f(x)\, dx$, this ODE is reducible to the canonical form 2.1.3.2:
$$yy'_z = y + \Phi(z).$$
Here the function $\Phi(z)$ is defined parametrically (x is the parameter) by the relations
$$\Phi = \dfrac{g(x)}{f(x)}, \quad z = \int f(x)\, dx.$$

Given below are some special cases where the original ODE is integrable by quadrature or admits a simplification.

1°. The Abel equation
$$yy'_x = x^{k-1}[(1+2k)x + ak]y - kx^{2k}(x+a)$$
is reduced by the transformation
$$x = \dfrac{w}{z}, \quad y = -\dfrac{1}{z^k} + x^{k+1} + ax^k,$$
to a separable ODE: $w'_z = w^{-k} - a$.

2°. The Abel equation
$$yy'_x = [a(2n+k)x^k + b]x^{n-1}y + (-a^2 n x^{2k} - abx^k + c)x^{2n-1}$$
is reduced by the substitution $y = x^n(z + ax^k)$ to a Bernoulli equation with respect to $x = x(z)$:
$$(nz^2 - bz - c)x'_z = -zx - ax^{k+1}.$$

3°. The Abel equation

$$yy'_x = [(3-m)x - 1]y + (m-1)(x^3 - x^2 - ax)$$

is reduced by the transformation

$$x = \frac{w}{z}, \quad y = -z^{m-1} + x^2 - x - a,$$

to an Abel equation in the canonical form 2.1.3.2:

$$ww'_z = w + az + z^m.$$

Many other solvable Abel equations of this form can be found in the handbooks by Polyanin & Zaitsev (2003, 2018).

4. $[y + g(x)]y'_x = f_2(x)y^2 + f_1(x)y + f_0(x).$

Abel equation of the second kind (general form). The Abel equation of the second kind is not integrable in closed form for arbitrary $f_n(x)$ and $g(x)$.

Special cases of the integrable Abel equation of the second kind. Given below are some special cases where this ODE is integrable by quadrature.

1°. If $g(x) = $ const and the functions $f_n(x)$ ($n = 0, 1, 2$) are all proportional, i.e., $f_n(x) = a_n f(x)$, then the Abel equation is a separable ODE of the form 2.1.1.2.

2°. The Abel equation is homogeneous:

$$(y + sx)y'_x = \frac{a}{x}y^2 + by + cx.$$

See equation 2.1.1.5. The substitution $u = y/x$ leads the original ODE to a separable ODE.

3°. The Abel equation is generalized homogeneous:

$$(y + sx^n)y'_x = \frac{a}{x}y^2 + bx^{n-1}y + cx^{2n-1}.$$

See ODE 2.1.4.4 with $k = -n$. The substitution $w = yx^{-n}$ leads to a separable equation:

$$x(w + s)w'_x = (a - n)w^2 + (b - ns)w + c.$$

4°. The Abel equation

$$(y + a_2 x + c_2)y'_x = b_1 y + a_1 x + c_1$$

is a special case of ODE 2.1.4.3 with $f(w) = w$ and $b_2 = 1$.

5°. The unnormalized Abel equation

$$[(a_1 x + a_2 x^k)y + b_1 x + b_2 x^k]y'_x = c_2 y^2 + c_1 y + c_0$$

can be reduced to the form 2.1.3.4 by dividing by $(a_1 x + a_2 x^k)$. Taking y to be the independent variable and $x = x(y)$ to be the dependent one, we obtain the Bernoulli equation

$$(c_2 y^2 + c_1 y + c_0)x'_y = (a_1 y + b_1)x + (a_2 y + b_2)x^k.$$

See ODE 2.1.1.4.

6°. The general solution of the Abel equation
$$(y + Ax^n + a)y'_x + Anx^{n-1}y + kx^m + b = 0$$
is given by
$$y^2 + \frac{2k}{m+1}x^{m+1} + 2(Ax^n y + ay + bx) = C.$$

7°. The Abel equation
$$(y + ax^{k+1} + bx^k)y'_x = (akx^k + cx^{k-1})y$$
is reducible with the substitution $y = x^k(z-b)$ to a Bernoulli equation with respect to $x = x(z)$:
$$[-kz^2 + (bk+c)z - bc]x'_z = zx + ax^2.$$

8°. The general solution of the Abel equation
$$(y+g)y'_x = f_2 y^2 + f_1 y + f_1 g - f_2 g^2, \qquad f_n = f_n(x), \; g = g(x),$$
is given by
$$y = -g + CE + E\int (f_1 + g'_x - 2f_2 g)E^{-1}\, dx, \quad \text{where} \quad E = \exp\left(\int f_2\, dx\right).$$

9°. The general solution of the Abel equation
$$(y+g)y'_x = f_2 y^2 + (2f_2 g - g'_x)y + f_0, \qquad f_n = f_n(x), \; g = g(x),$$
has the form
$$y = -g \pm E\left[2\int (f_0 + gg'_x - f_2 g^2)E^{-2}\, dx + C\right]^{1/2}, \quad \text{where} \quad E = \exp\left(\int f_2\, dx\right).$$

Remark 2.2. *Many other solvable Abel equations of this form can be found in the handbooks by Polyanin & Zaitsev (2003, 2018).*

Some transformations of the Abel equation of the second kind.

1°. The substitution
$$w = (y+g)E, \quad \text{where} \quad E = \exp\left(-\int f_2\, dx\right), \tag{1}$$
brings the original ODE to the simpler form 2.1.3.3:
$$ww'_x = F_1(x)w + F_0(x), \tag{2}$$
where
$$F_1 = (f_1 - 2f_2 g + g'_x)E, \quad F_0 = (f_0 - f_1 g + f_2 g^2)E^2.$$

2°. In turn, ODE (2) can be reduced, by the introduction of the new independent variable
$$z = \int F_1(x)\, dx, \tag{3}$$
to the canonical form 2.1.3.2:
$$ww'_z - w = \Phi(z). \tag{4}$$
Here the function $\Phi(z)$ is defined parametrically (x is the parameter) by the relations
$$\Phi = \frac{F_0(x)}{F_1(x)}, \quad z = \int F_1(x)\, dx.$$
Substitutions (1) and (3), which take the Abel equation to the canonical form, are called *canonical*.

Remark 2.3. Any Abel equations of the second kind related by linear (in y) transformations of the form
$$\tilde{x} = \psi_1(x), \quad \tilde{y} = \psi_2(x)y + \psi_3(x)$$
have identical canonical forms, up to the two-parameter family of functions specified in Remark 2.1.

3°. The substitution $y + g = 1/u$ reduces the Abel equation of the second kind to an Abel equation of the first kind of a special form:
$$u'_x + (f_0 - f_1 g + f_2 g^2)u^3 + (f_1 - 2f_2 g + g'_x)u^2 + f_2 u = 0.$$

2.1.4. Other First-Order ODEs Solved for the Derivative

▶ In equations 2.1.4.1–2.1.4.29, the functions f, g, and h are arbitrary functions whose arguments can depend on both x and y.

1. $y'_x = f(ax + by)$.

If $b \neq 0$, the substitution $u(x) = ax + by$ leads to an autonomous ODE of the form 2.1.1.1: $u'_x = bf(u) + a$.

2. $y'_x = f(y + ax^k + b) - akx^{k-1}$.

The substitution $u = y + ax^k + b$ leads to an autonomous ODE of the form 2.1.1.1: $u'_x = f(u)$.

3. $y'_x = f\left(\dfrac{a_1 x + b_1 y + c_1}{a_2 x + b_2 y + c_2}\right)$.

This ODE can be reduced to a homogeneous ODE of the form 2.1.1.5. To this end, for $a_1 x + b_1 y \neq k(a_2 x + b_2 y)$, one should use the changes of variables $\xi = x - x_0$ and $\eta = y - y_0$, where the constants x_0 and y_0 are determined by solving the linear algebraic system
$$a_1 x_0 + b_1 y_0 + c_1 = 0,$$
$$a_2 x_0 + b_2 y_0 + c_2 = 0.$$

As a result, one arrives at the following equation for $\eta = \eta(\xi)$:
$$\eta'_\xi = f\left(\frac{a_1 \xi + b_1 \eta}{a_2 \xi + b_2 \eta}\right).$$

On dividing the numerator and denominator of the argument of f by ξ, one obtains a homogeneous equation whose right-hand side is dependent on the ratio η/ξ only:
$$\eta'_\xi = f\left(\frac{a_1 + b_1 \eta/\xi}{a_2 + b_2 \eta/\xi}\right).$$

For $a_1 x + b_1 y = k(a_2 x + b_2 y)$, we have an equation of the type 2.1.4.1.

4. $y'_x = x^{-k-1} f(x^k y)$.

Generalized homogeneous ODE. The substitution $z = x^k y$ leads to a separable ODE: $xz'_x = kz + f(z)$.

5. $y'_x = \dfrac{y}{x} f(x^n y^m)$.

Generalized homogeneous ODE. The substitution $z = x^n y^m$ leads to a separable ODE: $x z'_x = nz + mz f(z)$.

6. $y'_x = -\dfrac{n}{m}\dfrac{y}{x} + y^k f(x) g(x^n y^m)$.

The substitution $z = x^n y^m$ leads to a separable ODE:

$$z'_x = m x^{\frac{n-nk}{m}} f(x) z^{\frac{k+m-1}{m}} g(z).$$

7. $y'_x = x^{n-1} y^{1-m} f(ax^n + by^m)$.

The substitution $w = ax^n + by^m$ leads to a separable ODE: $w'_x = x^{n-1}[an + bm f(w)]$.

8. $[x^k f(y) + x g(y)] y'_x = h(y)$.

This is a Bernoulli equation with respect to $x = x(y)$ (see equation 2.1.1.4).

9. $x[f(x^n y^m) + m x^k g(x^n y^m)] y'_x = y[h(x^n y^m) - n x^k g(x^n y^m)]$.

The transformation $t = x^n y^m$, $z = x^{-k}$ leads to a linear ODE with respect to $z = z(t)$:

$$t[nf(t) + mh(t)] z'_t = -kf(t) z - km g(t).$$

10. $x[f(x^n y^m) + m y^k g(x^n y^m)] y'_x = y[h(x^n y^m) - n y^k g(x^n y^m)]$.

The transformation $t = x^n y^m$, $z = y^{-k}$ leads to a linear ODE with respect to $z = z(t)$:

$$t[nf(t) + mh(t)] z'_t = -kh(t) z + kn g(t).$$

11. $x[sf(x^n y^m) - m g(x^k y^s)] y'_x = y[n g(x^k y^s) - k f(x^n y^m)]$.

The transformation $t = x^n y^m$, $w = x^k y^s$ leads to a separable ODE: $tf(t) w'_t = w g(w)$.

12. $[f(y) + a m x^n y^{m-1}] y'_x + g(x) + a n x^{n-1} y^m = 0$.

General solution:

$$\int f(y)\, dy + \int g(x)\, dx + a x^n y^m = C.$$

13. $y'_x = f(x) e^{\lambda y} + g(x)$.

The substitution $u = e^{-\lambda y}$ leads to a linear ODE: $u'_x = -\lambda g(x) u - \lambda f(x)$.

14. $y'_x = f(x) e^{\lambda y} + g(x) + h(x) e^{-\lambda y}$.

The substitution $u = e^{-\lambda y}$ leads to the general Riccati equation:
$u'_x = -\lambda h(x) u^2 - \lambda g(x) u - \lambda f(x)$.

15. $y'_x = e^{-\lambda x} f(e^{\lambda x} y)$.

The substitution $u = e^{\lambda x} y$ leads to a separable ODE: $u'_x = f(u) + \lambda u$.

16. $y'_x = e^{\lambda y} f(e^{\lambda y} x)$.

The substitution $u = e^{\lambda y} x$ leads to a separable ODE: $x u'_x = \lambda u^2 f(u) + u$.

17. $y'_x = y f(e^{\alpha x} y^m)$.

The substitution $z = e^{\alpha x} y^m$ leads to a separable ODE: $z'_x = \alpha z + m z f(z)$.

18. $y'_x = \dfrac{1}{x} f(x^n e^{\alpha y})$.

The substitution $z = x^n e^{\alpha y}$ leads to a separable ODE: $xz'_x = nz + \alpha z f(z)$.

19. $y'_x = -\dfrac{n}{x} + f(x) g(x^n e^y)$.

The substitution $z = x^n e^y$ leads to a separable ODE: $z'_x = f(x) z g(z)$.

20. $y'_x = -\dfrac{\alpha}{m} y + y^k f(x) g(e^{\alpha x} y^m)$.

The substitution $z = e^{\alpha x} y^m$ leads to a separable ODE:
$$z'_x = m \exp\left[\dfrac{\alpha}{m}(1-k)x\right] f(x) z^{\frac{k+m-1}{m}} g(z).$$

21. $y'_x = e^{\alpha x - \beta y} f(ae^{\alpha x} + be^{\beta y})$.

The substitution $w = ae^{\alpha x} + be^{\beta y}$ leads to a separable ODE: $w'_x = e^{\alpha x}[a\alpha + b\beta f(w)]$.

22. $[e^{\alpha x} f(y) + a\beta] y'_x + e^{\beta y} g(x) + a\alpha = 0$.

Solution:
$$\int e^{-\beta y} f(y)\, dy + \int e^{-\alpha x} g(x)\, dx - ae^{-\alpha x - \beta y} = C.$$

23. $x[f(x^n e^{\alpha y}) + \alpha y g(x^n e^{\alpha y})] y'_x = h(x^n e^{\alpha y}) - n y g(x^n e^{\alpha y})$.

The substitution $t = x^n e^{\alpha y}$ leads to a linear ODE with respect to $y = y(t)$:
$$t[nf(t) + \alpha h(t)] y'_t = -n g(t) y + h(t).$$

24. $[f(e^{\alpha x} y^m) + mx g(e^{\alpha x} y^m)] y'_x = y[h(e^{\alpha x} y^m) - \alpha x g(e^{\alpha x} y^m)]$.

The substitution $t = e^{\alpha x} y^m$ leads to a linear ODE with respect to $x = x(t)$:
$$t[\alpha f(t) + mh(t)] x'_t = mg(t) x + f(t).$$

25. $y'_x = f(x) \cosh(\lambda y) + g(x) \sinh(\lambda y) + h(x)$.

The substitution $w = e^{\lambda y}$ leads to the general Riccati equation:
$$w'_x = \tfrac{1}{2}\lambda[f(x) + g(x)] w^2 + \lambda h(x) w + \tfrac{1}{2}\lambda[f(x) - g(x)].$$

26. $y'_x = f(x) y \ln y + g(x) y$.

The substitution $y = e^u$ leads to a linear ODE: $u'_x = f(x) u + g(x)$.

27. $y'_x = f(x) y \ln^2 y + g(x) y \ln y + h(x) y$.

The substitution $y = e^u$ leads to the general Riccati equation:
$$u'_x = f(x) u^2 + g(x) u + h(x).$$

28. $y'_x = f(x) \cos(ay) + g(x) \sin(ay) + h(x)$.

The substitution $u = \tan(\tfrac{1}{2} ay)$ leads to the general Riccati equation:
$$u'_x = \tfrac{1}{2} a[h(x) - f(x)] u^2 + a g(x) u + \tfrac{1}{2} a[f(x) + h(x)].$$

29. $y'_x = f(y + a \tan x) - a \tan^2 x$.

The substitution $u = y + a \tan x$ leads to a separable ODE: $u'_x = a + f(u)$.

2.1.5. ODEs Not Solved for the Derivative and ODEs Defined Parametrically

1. $x = f(y'_x)$.

General solution in parametric form:
$$x = f(t), \quad y = \int t f'_t(t)\,dt + C.$$

2. $y = f(y'_x)$.

General solution in parametric form:
$$x = \int f'_t(t)\frac{dt}{t} + C, \quad y = f(t).$$

3. $f(y'_x) + ax + by + s = 0$.

Solution in parametric form:
$$x = C - \int \frac{f'_t(t)\,dt}{a + bt}, \quad by = -ax - s - f(t).$$

In addition, there is a particular solution $y = \alpha x + \beta$, where α and β determined by solving the system of two algebraic equations:
$$a + b\alpha = 0, \quad f(\alpha) + b\beta + s = 0.$$

4. $y = xy'_x + f(y'_x)$.

Clairaut's equation. General solution: $y = Cx + f(C)$.

In addition, there is a singular solution, which may be written in parametric form as (Kamke, 1977):
$$x = -f'_t(t), \quad y = -tf'_t(t) + f(t).$$

5. $y = xf(y'_x) + g(y'_x)$.

Lagrange–d'Alembert equation. For the case $f(t) = t$, see equation 2.1.5.4.

We set $t = y'_x$ and then differentiate the ODE with respect to x. Taking into account the relations $y''_{xx} = t'_x = 1/x'_t$, we arrive at a linear ODE for the function $x = x(t)$ (Kamke, 1977):
$$[t - f(t)]x'_t = f'_t(t)x + g'_t(t).$$

6. $xf(y'_x) + yg(y'_x) + h(y'_x) = 0$.

The Legendre transformation $X = y'_x$, $Y = xy'_x - y$, $Y'_X = x$ leads to a linear ODE:
$$[f(X) + Xg(X)]Y'_X - g(X)Y + h(X) = 0.$$

Inverse transformation: $x = Y'_X$, $y = XY'_X - Y$, $y'_x = X$.

7. $y = x^k f(y'_x) + xy'_x$.

We set $t = y'_x$ and then differentiate the ODE with respect to x. Taking into account that $y''_{xx} = t'_x = 1/x'_t$, we arrive at a Bernoulli equation for the function $x = x(t)$ (Kamke, 1977):
$$kf(t)x'_t - f'_t(t)x - x^{2-k} = 0.$$

8. $(xy'_x - y)^k f(y'_x) + yg(y'_x) + xh(y'_x) = 0.$

The Legendre transformation $x = u'_t$, $y = tu'_t - u$ ($y'_x = t$) leads to a Bernoulli equation (Kamke, 1977):
$$[tg(t) + h(t)]u'_t = g(t)u - f(t)u^k.$$

9. $x = f(t), \quad y'_x = g(t).$

An ODE defined parametrically by two equations (t is a parameter).
General solution in parametric form (Polyanin & Zhurov, 2016):
$$x = f(t), \quad y = \int f'_t(t)g(t)\, dt + C.$$

10. $x = f(t)y + g(t), \quad y'_x = h(t).$

An ODE defined parametrically by two equations (t is a parameter).
General solution in parametric form (Polyanin & Zhurov, 2017):
$$x = fy + g, \quad y = CE + E \int \frac{hg'_t\, dt}{(1 - fh)E},$$

where C is an arbitrary constant, $E = \exp\left(\int \frac{hf'_t\, dt}{1 - fh}\right)$.

2.2. Second-Order Linear Ordinary Differential Equations

2.2.1. Preliminary Remarks and Some Formulas

▶ **Linear homogeneous ODEs.**

A *second-order linear homogeneous ordinary differential equation* has the general form
$$f_2(x)y''_{xx} + f_1(x)y'_x + f_0(x)y = 0. \tag{1}$$

1°. The *trivial solution*, $y = 0$, is a particular solution of the homogeneous linear ODE.

2°. Let $y_1(x)$, $y_2(x)$ be a fundamental system of solutions (nontrivial linearly independent particular solutions) of equation (1). Then its general solution is given by
$$y = C_1 y_1(x) + C_2 y_2(x),$$
where C_1 and C_2 are arbitrary constants.

3°. Let $y_1 = y_1(x)$ be any nontrivial particular solution of equation (1). Then its general solution can be represented as
$$y = y_1\left(C_1 + C_2 \int \frac{e^{-F}}{y_1^2}\, dx\right), \quad \text{where} \quad F = \int \frac{f_1(x)}{f_2(x)}\, dx. \tag{2}$$

4°. The substitution $u = y'_x/y$ brings the second-order homogeneous linear ODE (1) to a first-order Riccati ODE of the form 2.1.2.19:
$$f_2(x)u'_x + f_2(x)u^2 + f_1(x)u + f_0(x) = 0.$$

Ordinary Differential Equations

▶ **Linear nonhomogeneous ODEs.**

A *second-order linear nonhomogeneous ordinary differential equation* has the general form

$$f_2(x)y''_{xx} + f_1(x)y'_x + f_0(x)y = g(x). \tag{3}$$

1°. The general solution of the nonhomogeneous linear equation (3) is the sum of the general solution of the corresponding homogeneous equation (1) and any particular solution of the nonhomogeneous equation (3).

2°. Let $y_1 = y_1(x)$, $y_2 = y_2(x)$ be a fundamental system of solutions of the corresponding linear homogeneous ODE (1). Then the general solution of the linear nonhomogeneous equation (3) can be represented as

$$y = C_1 y_1 + C_2 y_2 + y_2 \int y_1 \frac{g(x)}{f_2(x)} \frac{dx}{W(x)} - y_1 \int y_2 \frac{g(x)}{f_2(x)} \frac{dx}{W(x)}, \tag{4}$$

where $W(x) = y_1(y_2)'_x - y_2(y_1)'_x$ is the *Wronskian determinant*.

Liouville's formula holds:

$$W(x) = W(x_0) \exp\left[-\int_{x_0}^{x} \frac{f_1(t)}{f_2(t)} dt\right].$$

3°. Given a nontrivial particular solution $y_1 = y_1(x)$ of the homogeneous ODE (1), a second particular solution $y_2 = y_2(x)$ can be calculated from the formula

$$y_2 = y_1 \int \frac{e^{-F}}{y_1^2} dx, \quad \text{where} \quad F = \int \frac{f_1(x)}{f_2(x)} dx.$$

Then the general solution of the nonhomogeneous ODE (3) can be constructed by formula (4) with $W(x) = e^{-F}$.

2.2.2. Equations Involving Power Functions

▶ **ODEs of the form $y''_{xx} + f(x)y = 0$.**

1. $y''_{xx} + ay = 0$.

Equation of free oscillations.

General solution: $y = \begin{cases} C_1 \sinh(x\sqrt{|a|}) + C_2 \cosh(x\sqrt{|a|}) & \text{if } a < 0, \\ C_1 + C_2 x & \text{if } a = 0, \\ C_1 \sin(x\sqrt{a}) + C_2 \cos(x\sqrt{a}) & \text{if } a > 0. \end{cases}$

2. $y''_{xx} - (ax + b)y = 0, \quad a \neq 0$.

The substitution $\xi = a^{-2/3}(ax + b)$ leads to the *Airy equation*

$$y''_{\xi\xi} - \xi y = 0, \tag{$*$}$$

which often arises in various applications. The general solution of equation ($*$) can be written as:

$$y = C_1 \operatorname{Ai}(\xi) + C_2 \operatorname{Bi}(\xi),$$

where $\operatorname{Ai}(\xi)$ and $\operatorname{Bi}(\xi)$ are the *Airy functions* of the first and second kind, respectively.

The Airy functions admit the following integral representation:
$$\operatorname{Ai}(\xi) = \frac{1}{\pi} \int_0^\infty \cos\left(\tfrac{1}{3}t^3 + \xi t\right) dt,$$
$$\operatorname{Bi}(\xi) = \frac{1}{\pi} \int_0^\infty \left[\exp\left(-\tfrac{1}{3}t^3 + \xi t\right) + \sin\left(\tfrac{1}{3}t^3 + \xi t\right)\right] dt.$$

The Airy functions can be expressed in terms of the Bessel functions and the modified Bessel functions of order $1/3$ by the relations:
$$\operatorname{Ai}(\xi) = \tfrac{1}{3}\sqrt{\xi}\left[I_{-1/3}(z) - I_{1/3}(z)\right], \quad \operatorname{Ai}(-\xi) = \tfrac{1}{3}\sqrt{\xi}\left[J_{-1/3}(z) + J_{1/3}(z)\right],$$
$$\operatorname{Bi}(\xi) = \sqrt{\tfrac{1}{3}\xi}\left[I_{-1/3}(z) + I_{1/3}(z)\right], \quad \operatorname{Bi}(-\xi) = \sqrt{\tfrac{1}{3}\xi}\left[J_{-1/3}(z) - J_{1/3}(z)\right],$$
where $z = \tfrac{2}{3}\xi^{3/2}$.

For large values of ξ, the leading terms of the asymptotic expansions of the Airy functions are:
$$\operatorname{Ai}(\xi) = \frac{1}{2\sqrt{\pi}}\xi^{-1/4}\exp(-z), \quad \operatorname{Ai}(-\xi) = \frac{1}{\sqrt{\pi}}\xi^{-1/4}\sin\left(z + \frac{\pi}{4}\right),$$
$$\operatorname{Bi}(\xi) = \frac{1}{\sqrt{\pi}}\xi^{-1/4}\exp(z), \quad \operatorname{Bi}(-\xi) = \frac{1}{\sqrt{\pi}}\xi^{-1/4}\cos\left(z + \frac{\pi}{4}\right).$$

The Airy equation $(*)$ is a special case of equation 2.2.2.4 with $a = n = 1$.

3. $y''_{xx} - (ax^2 + b)y = 0$.

Weber equation (two canonical forms of the equation correspond to $a = \pm\tfrac{1}{4}$).

1°. The transformation $z = x^2\sqrt{a}$, $u = e^{z/2}y$ leads to the degenerate hypergeometric equation 2.2.2.16:
$$zu''_{zz} + \left(\tfrac{1}{2} - z\right)u'_z - \tfrac{1}{4}\left(\frac{b}{\sqrt{a}} + 1\right)u = 0.$$

2°. For $b = a$, there is a particular solution: $y = \exp\left(\tfrac{1}{2}ax^2\right)$.

3°. For $a = k^2 > 0$, $b = -(2n+1)k$, where $n = 1, 2, \ldots$, there is a solution of the form:
$$y = \exp\left(-\tfrac{1}{2}kx^2\right)H_n\left(\sqrt{k}\,x\right), \quad k > 0,$$
where $H_n(z) = (-1)^n \exp(z^2)\dfrac{d^n}{dz^n}\exp(-z^2)$ is the Hermite polynomial of order n.

4. $y''_{xx} - ax^k y = 0$.

1°. For $k = -2$, this is the Euler equation 2.2.2.21 (the solution is expressed in terms of elementary functions).

2°. Assume that $2/(k+2) = 2m+1$, where m is an integer. Then the general solution is
$$y = \begin{cases} x(x^{1-2q}D)^{m+1}\left[C_1\exp\left(\dfrac{\sqrt{a}}{q}x^q\right) + C_2\exp\left(-\dfrac{\sqrt{a}}{q}x^q\right)\right] & \text{if } m \geq 0, \\ (x^{1-2q}D)^{-m}\left[C_1\exp\left(\dfrac{\sqrt{a}}{q}x^q\right) + C_2\exp\left(-\dfrac{\sqrt{a}}{q}x^q\right)\right] & \text{if } m < 0, \end{cases}$$
where $D = \dfrac{d}{dx}$, $q = \dfrac{k+2}{2} = \dfrac{1}{2m+1}$.

3°. For any k, the general solution is expressed in terms of the Bessel functions and modified Bessel functions:

$$y = \begin{cases} C_1\sqrt{x}\, J_{\frac{1}{2q}}\left(\frac{\sqrt{-a}}{q}x^q\right) + C_2\sqrt{x}\, Y_{\frac{1}{2q}}\left(\frac{\sqrt{-a}}{q}x^q\right) & \text{if } a < 0, \\ C_1\sqrt{x}\, I_{\frac{1}{2q}}\left(\frac{\sqrt{a}}{q}x^q\right) + C_2\sqrt{x}\, K_{\frac{1}{2q}}\left(\frac{\sqrt{a}}{q}x^q\right) & \text{if } a > 0, \end{cases}$$

where $q = \frac{1}{2}(k+2)$. For details about the functions $J_\nu(z)$, $Y_\nu(z)$ and $I_\nu(z)$, $K_\nu(z)$, see equations 2.2.2.22 and 2.2.2.23.

5. $y''_{xx} + (ax^{2n} + bx^{n-1})y = 0$.

The substitution $\xi = x^{n+1}$ leads to a linear equation of the form 2.2.2.17:

$$(n+1)^2 \xi y''_{\xi\xi} + n(n+1) y'_\xi + (a\xi + b) y = 0.$$

▶ **ODEs of the form** $y''_{xx} + (ax + b) y'_x + f(x) y = 0$.

6. $y''_{xx} + ay'_x + by = 0$.

Second-order constant-coefficient linear equation. In physics this equation is called an *equation of damped vibrations*.

General solution:

$$y = \begin{cases} \exp\left(-\frac{1}{2}ax\right)\left[C_1 \exp\left(\frac{1}{2}\lambda x\right) + C_2 \exp\left(-\frac{1}{2}\lambda x\right)\right] & \text{if } \lambda^2 = a^2 - 4b > 0, \\ \exp\left(-\frac{1}{2}ax\right)\left[C_1 \sin\left(\frac{1}{2}\lambda x\right) + C_2 \cos\left(\frac{1}{2}\lambda x\right)\right] & \text{if } \lambda^2 = 4b - a^2 > 0, \\ \exp\left(-\frac{1}{2}ax\right)\left(C_1 x + C_2\right) & \text{if } a^2 = 4b. \end{cases}$$

7. $y''_{xx} + ay'_x + (bx + c)y = 0$.

1°. General solution with $b > 0$:

$$y = \exp\left(-\tfrac{1}{2}ax\right)\sqrt{\xi}\left[C_1 J_{1/3}\left(\tfrac{2}{3}\sqrt{b}\,\xi^{3/2}\right) + C_2 Y_{1/3}\left(\tfrac{2}{3}\sqrt{b}\,\xi^{3/2}\right)\right], \quad \xi = x + \frac{4c - a^2}{4b},$$

where $J_{1/3}(z)$ and $Y_{1/3}(z)$ are Bessel functions.

2°. General solution with $b < 0$:

$$y = \exp\left(-\tfrac{1}{2}ax\right)\sqrt{\xi}\left[C_1 J_{1/3}\left(\tfrac{2}{3}\sqrt{|b|}\,\xi^{3/2}\right) + C_2 Y_{1/3}\left(\tfrac{2}{3}\sqrt{|b|}\,\xi^{3/2}\right)\right], \quad \xi = x + \frac{4c - a^2}{4b},$$

where $I_{1/3}(z)$ and $K_{1/3}(z)$ are modified Bessel functions.

3°. For $b = 0$, see equation 2.2.2.6.

8. $y''_{xx} + ay'_x - (bx^2 + c)y = 0$.

The substitution $y = w\exp\left(\frac{1}{2}x^2\sqrt{b}\right)$ leads to a linear equation of the form 2.2.2.17:

$$w''_{xx} + (2\sqrt{b}\,x + a)w'_x + (a\sqrt{b}\,x - c + \sqrt{b}\,)w = 0.$$

9. $y''_{xx} + axy'_x + by = 0$.

General solution:

$$y = C_1 \Phi\left(\tfrac{1}{2}a^{-1}b, \tfrac{1}{2}, -\tfrac{1}{2}ax^2\right) + C_2 \Psi\left(\tfrac{1}{2}a^{-1}b, \tfrac{1}{2}, -\tfrac{1}{2}ax^2\right),$$

where $\Phi(a, b; x)$ and $\Psi(a, b; x)$ are the degenerate hypergeometric functions (see equation 2.2.2.16).

10. $y''_{xx} + axy'_x + bxy = 0$.

General solution:

$$y = e^{-bx/a}\left[C_1 \Phi\left(\tfrac{1}{2}a^{-3}b^2, \tfrac{1}{2}, -\tfrac{1}{2}a\xi^2\right) + C_2 \Psi\left(\tfrac{1}{2}a^{-3}b^2, \tfrac{1}{2}, -\tfrac{1}{2}a\xi^2\right)\right], \quad \xi = x - 2a^{-2}b,$$

where $\Phi(a,b;x)$ and $\Psi(a,b;x)$ are the degenerate hypergeometric functions (see equation 2.2.2.16).

11. $y''_{xx} + (ax+b)y'_x + (\alpha x^2 + \beta x + \gamma)y = 0$.

The substitution $y = u\exp(sx^2)$, where s is a root of the quadratic equation $4s^2 + 2as + \alpha = 0$, leads to an ODE of the form 2.2.2.17:

$$u''_{xx} + [(a+4s)x + b]u'_x + [(\beta + 2bs)x + \gamma + 2s]u = 0.$$

▶ **ODEs of the form** $(a_2 x + b_2)y''_{xx} + (a_1 x + b_1)y'_x + f(x)y = 0$.

12. $xy''_{xx} + ay'_x + by = 0$.

1°. The general solution is expressed in terms of Bessel functions and modified Bessel functions:

$$y = \begin{cases} x^{\frac{1-a}{2}}\left[C_1 J_\nu(2\sqrt{bx}) + C_2 Y_\nu(2\sqrt{bx})\right] & \text{if } bx > 0, \\ x^{\frac{1-a}{2}}\left[C_1 I_\nu(2\sqrt{|bx|}) + C_2 K_\nu(2\sqrt{|bx|})\right] & \text{if } bx < 0, \end{cases}$$

where $\nu = |1-a|$. For details about the functions $J_\nu(z)$, $Y_\nu(z)$ and $I_\nu(z)$, $K_\nu(z)$, see equations 2.2.2.22 and 2.2.2.23.

2°. For $a = \tfrac{1}{2}(2n+1)$, where $n = 0, 1, \ldots$, the general solution is

$$y = \begin{cases} C_1 \dfrac{d^n}{dx^n}\cos\sqrt{4bx} + C_2 \dfrac{d^n}{dx^n}\sin\sqrt{4bx} & \text{if } bx > 0, \\ C_1 \dfrac{d^n}{dx^n}\cosh\sqrt{4|bx|} + C_2 \dfrac{d^n}{dx^n}\sinh\sqrt{4|bx|} & \text{if } bx < 0. \end{cases}$$

13. $xy''_{xx} + ay'_x + bxy = 0$.

1°. The general solution is expressed in terms of Bessel functions and modified Bessel functions:

$$y = \begin{cases} x^{\frac{1-a}{2}}\left[C_1 J_\nu(\sqrt{b}\,x) + C_2 Y_\nu(\sqrt{b}\,x)\right] & \text{if } b > 0, \\ x^{\frac{1-a}{2}}\left[C_1 I_\nu(\sqrt{|b|}\,x) + C_2 K_\nu(\sqrt{|b|}\,x)\right] & \text{if } b < 0, \end{cases}$$

where $\nu = \tfrac{1}{2}|1-a|$.

2°. For $a = 2n$, where $n = 1, 2, \ldots$, the general solution is

$$y = \begin{cases} C_1\left(\dfrac{1}{x}\dfrac{d}{dx}\right)^n\cos(x\sqrt{b}) + C_2\left(\dfrac{1}{x}\dfrac{d}{dx}\right)^n\sin(x\sqrt{b}) & \text{if } b > 0, \\ C_1\left(\dfrac{1}{x}\dfrac{d}{dx}\right)^n\cosh(x\sqrt{-b}) + C_2\left(\dfrac{1}{x}\dfrac{d}{dx}\right)^n\sinh(x\sqrt{-b}) & \text{if } b < 0. \end{cases}$$

14. $xy''_{xx} + ky'_x + bx^{1-2k}y = 0.$

For $k = 1$, this is the Euler equation 2.2.2.21. For $k \neq 1$, the general solution is

$$y = \begin{cases} C_1 \sin\left(\dfrac{\sqrt{b}}{k-1}x^{1-k}\right) + C_2 \cos\left(\dfrac{\sqrt{b}}{k-1}x^{1-k}\right) & \text{if } b > 0, \\ C_1 \exp\left(\dfrac{\sqrt{-b}}{k-1}x^{1-k}\right) + C_2 \exp\left(\dfrac{-\sqrt{-b}}{k-1}x^{1-k}\right) & \text{if } b < 0. \end{cases}$$

15. $xy''_{xx} + ay'_x + bx^k y = 0.$

If $k = -1$ or $b = 0$, we have the Euler equation 2.2.2.21.

1°. For $b > 0$ and $k \neq -1$, the general solution is expressed in terms of the Bessel functions:

$$y = x^{\frac{1-a}{2}}\left[C_1 J_\nu\left(\frac{2\sqrt{b}}{k+1}x^{\frac{k+1}{2}}\right) + C_2 Y_\nu\left(\frac{2\sqrt{b}}{k+1}x^{\frac{k+1}{2}}\right)\right], \quad \text{where} \quad \nu = \frac{|1-a|}{k+1}.$$

2°. For $b < 0$ and $k \neq -1$, the general solution is expressed in terms of the modified Bessel functions:

$$y = x^{\frac{1-a}{2}}\left[C_1 I_\nu\left(\frac{2\sqrt{-b}}{k+1}x^{\frac{k+1}{2}}\right) + C_2 K_\nu\left(\frac{2\sqrt{-b}}{k+1}x^{\frac{k+1}{2}}\right)\right], \quad \text{where} \quad \nu = \frac{|1-a|}{k+1}.$$

16. $xy''_{xx} + (b-x)y'_x - ay = 0.$

Degenerate hypergeometric equation.

1°. If $b \neq 0, -1, -2, -3, \ldots$, Kummer's series is a particular solution:

$$\Phi(a, b; x) = 1 + \sum_{k=1}^{\infty} \frac{(a)_k}{(b)_k} \frac{x^k}{k!},$$

where $(a)_k = a(a+1)\ldots(a+k-1)$, $(a)_0 = 1$. If $b > a > 0$, this solution can be written in terms of a definite integral:

$$\Phi(a, b; x) = \frac{\Gamma(b)}{\Gamma(a)\Gamma(b-a)} \int_0^1 e^{xt} t^{a-1}(1-t)^{b-a-1}\, dt,$$

where $\Gamma(z) = \displaystyle\int_0^\infty e^{-t} t^{z-1}\, dt$ is the gamma function.

Table 2.1 presents particular cases when the Kummer function $\Phi(a, b; z)$ is expressed in terms of elementary functions or simpler special functions.

If b is not an integer, then the general solution has the form:

$$y = C_1 \Phi(a, b; x) + C_2 x^{1-b}\Phi(a-b+1, 2-b; x).$$

2°. The following function is a solution of the degenerate hypergeometric equation:

$$\Psi(a, b; x) = \frac{\Gamma(1-b)}{\Gamma(a-b+1)}\Phi(a, b; x) + \frac{\Gamma(b-1)}{\Gamma(a)}x^{1-b}\Phi(a-b+1, 2-b; x).$$

TABLE 2.1. Special cases of the Kummer function function $\Phi = \Phi(a, b; z)$.

a	b	z	Φ	Conventional notation
a	a	x	e^x	
1	2	$2x$	$\dfrac{1}{x} e^x \sinh x$	
a	$a+1$	$-x$	$ax^{-a}\gamma(a,x)$	Incomplete gamma function $\gamma(a,x) = \displaystyle\int_0^x e^{-t} t^{a-1}\, dt$
$\dfrac{1}{2}$	$\dfrac{3}{2}$	$-x^2$	$\dfrac{\sqrt{\pi}}{2}\,\mathrm{erf}\, x$	Error function $\mathrm{erf}\, x = \dfrac{2}{\sqrt{\pi}} \displaystyle\int_0^x \exp(-t^2)\, dt$
$-n$	$\dfrac{1}{2}$	$\dfrac{x^2}{2}$	$\dfrac{n!}{(2n)!}\left(-\dfrac{1}{2}\right)^{-n} H_{2n}(x)$	Hermite polynomials $H_n(x) = (-1)^n e^{x^2} \dfrac{d^n}{dx^n}\left(e^{-x^2}\right),$
$-n$	$\dfrac{3}{2}$	$\dfrac{x^2}{2}$	$\dfrac{n!}{(2n+1)!}\left(-\dfrac{1}{2}\right)^{-n} H_{2n+1}(x)$	$n = 0, 1, 2, \ldots$
$-n$	b	x	$\dfrac{n!}{(b)_n} L_n^{(b-1)}(x)$	Laguerre polynomials $L_n^{(\alpha)}(x) = \dfrac{e^x x^{-\alpha}}{n!} \dfrac{d^n}{dx^n}\left(e^{-x} x^{n+\alpha}\right),$ $\alpha = b-1,$ $(b)_n = b(b+1)\ldots(b+n-1)$
$\nu + \dfrac{1}{2}$	$2\nu + 1$	$2x$	$\Gamma(1+\nu) e^x \left(\dfrac{x}{2}\right)^{-\nu} I_\nu(x)$	Modified Bessel functions $I_\nu(x)$
$n+1$	$2n+2$	$2x$	$\Gamma\!\left(n+\dfrac{3}{2}\right) e^x \left(\dfrac{x}{2}\right)^{-n-\frac{1}{2}} I_{n+\frac{1}{2}}(x)$	

Calculate the limit as $b \to n$ (n is an integer) to obtain

$$\Psi(a, n+1; x) = \frac{(-1)^{n-1}}{n!\, \Gamma(a-n)} \Bigg\{ \Phi(a, n+1; x) \ln x + \sum_{r=0}^{\infty} \frac{(a)_r}{(n+1)_r} \big[\psi(a+r) - \psi(1+r) - \psi(1+n+r)\big] \frac{x^r}{r!} \Bigg\}$$
$$+ \frac{(n-1)!}{\Gamma(a)} \sum_{r=0}^{n-1} \frac{(a-n)_r}{(1-n)_r} \frac{x^{r-n}}{r!},$$

where $n = 0, 1, 2, \ldots$ (the last sum must be omitted for $n = 0$), $\psi(z) = [\ln \Gamma(z)]'_z$ is the logarithmic derivative of the gamma function:

$$\psi(1) = -\gamma, \quad \psi(n) = -\gamma + \sum_{k=1}^{n-1} k^{-1}, \quad \gamma = 0.5772\ldots \text{ (Euler's constant)}.$$

If b is a negative number, then the function Ψ can be expressed in terms of Ψ a positive second argument using the relation

$$\Psi(a, b; x) = x^{1-b} \Psi(a - b + 1, 2 - b; x),$$

which holds for any value of x.

Table 2.2 presents particular cases when the Kummer function $\Psi(a,b;z)$ is expressed in terms of simpler special functions.

TABLE 2.2. Special cases of the Kummer function function $\Psi = \Psi(a,b;z)$.

a	b	z	Ψ	Conventional notation
$1-a$	$1-a$	x	$e^x \Gamma(a,x)$	Incomplete gamma function $\Gamma(a,x) = \int_x^\infty e^{-t} t^{a-1}\, dt$
$\dfrac{1}{2}$	$\dfrac{1}{2}$	x^2	$\sqrt{\pi}\exp(x^2)\operatorname{erfc} x$	Complementary error function $\operatorname{erfc} x = \dfrac{2}{\sqrt{\pi}}\int_x^\infty \exp(-t^2)\,dt$
1	1	$-x$	$-e^{-x}\operatorname{Ei}(x)$	Exponential integral $\operatorname{Ei}(x) = \int_{-\infty}^x \dfrac{e^t}{t}\,dt$
1	1	$-\ln x$	$-x^{-1}\operatorname{li} x$	Logarithmic integral $\operatorname{li} x = \int_0^x \dfrac{dt}{t}$
$\dfrac{1}{2}-\dfrac{n}{2}$	$\dfrac{3}{2}$	x^2	$2^{-n}x^{-1}H_n(x)$	Hermite polynomials $H_n(x) = (-1)^n e^{x^2}\dfrac{d^n}{dx^n}(e^{-x^2})$, $n=0,1,2,\ldots$
$\nu+\dfrac{1}{2}$	$2\nu+1$	$2x$	$\pi^{-1/2}(2x)^{-\nu}e^x K_\nu(x)$	Modified Bessel functions $K_\nu(x)$

3°. For $b \neq 0, -1, -2, -3, \ldots$, the general solution of the degenerate hypergeometric equation can be written in the form

$$y = C_1 \Phi(a,b;x) + C_2 \Psi(a,b;x),$$

while for $b = 0, -1, -2, -3, \ldots$, it can be represented as

$$y = x^{1-b}\bigl[C_1 \Phi(a-b+1,\, 2-b;\, x) + C_2 \Psi(a-b+1,\, 2-b;\, x)\bigr].$$

4°. The degenerate hypergeometric functions $\Phi(a,b;x)$ and $\Psi(a,b;x)$ are discussed in detail in the books by Bateman & Erdélyi (1953, Vol. 1), Abramowitz & Stegun (1964), Andrews et al. (2001), Olver et al. (2010), and Beals & Wong (2016).

17. $(a_2 x + b_2)y''_{xx} + (a_1 x + b_1)y'_x + (a_0 x + b_0)y = 0.$

Let $\mathcal{J}(a,b;x)$ denote an arbitrary solution of the degenerate hypergeometric equation

$$xy''_{xx} + (b-x)y'_x - ay = 0 \quad \text{(see ODE 2.2.2.16)}$$

and let $Z_\nu(x)$ stand for an arbitrary solution of the Bessel equation

$$x^2 y''_{xx} + xy'_x + (x^2 - \nu^2)y = 0 \quad \text{(see ODE 2.2.2.22)}.$$

The results of solving the original equation are presented in Table 2.3.

TABLE 2.3. Solutions of equation 2.2.2.17 for different values of the determining parameters.

Solution: $y = e^{kx} w(z)$, where $z = \dfrac{x-\mu}{\lambda}$					
Constraints	k	λ	μ	w	Parameters
$a_2 \neq 0$, $a_1^2 \neq 4a_0 a_2$	$\dfrac{\sqrt{D} - a_1}{2a_2}$	$-\dfrac{a_2}{2a_2 k + a_1}$	$-\dfrac{b_2}{a_2}$	$\mathcal{J}(a, b; z)$	$a = B(k)/(2a_2 k + a_1)$, $b = (a_2 b_1 - a_1 b_2) a_2^{-2}$
$a_2 = 0$, $a_1 \neq 0$	$-\dfrac{a_0}{a_1}$	1	$-\dfrac{2b_2 k + b_1}{a_1}$	$\mathcal{J}(a, \tfrac{1}{2}; \beta z^2)$	$a = B(k)/(2a_1)$, $\beta = -a_1/(2b_2)$
$a_2 \neq 0$, $a_1^2 = 4a_0 a_2$	$-\dfrac{a_1}{2a_2}$	a_2	$-\dfrac{b_2}{a_2}$	$z^{\nu/2} Z_\nu(\beta \sqrt{z})$	$\nu = 1 - (2b_2 k + b_1) a_2^{-1}$, $\beta = 2\sqrt{B(k)}$
$a_2 = a_1 = 0$, $a_0 \neq 0$	$-\dfrac{b_1}{2b_2}$	1	$\dfrac{b_1^2 - 4b_0 b_2}{4a_0 b_2}$	$z^{1/2} Z_{1/3}(\beta z^{3/2})$ see also A14.1.2.12	$\beta = \dfrac{2}{3}\left(\dfrac{a_0}{b_2}\right)^{1/2}$
Notation: $D = a_1^2 - 4a_0 a_2$, $B(k) = b_2 k^2 + b_1 k + b_0$					

▶ **ODEs of the form** $x^2 y''_{xx} + (ax + b) y'_x + f(x) y = 0$.

18. $x^2 y''_{xx} + [a^2 x^2 - n(n+1)] y = 0$, $\quad n = 0, 1, 2, \ldots$
General solution:
$$y x^{n+1} = (x^3 D)^n \left(\frac{C_1 \cos ax + C_2 \sin ax}{x^{2n-1}} \right), \quad \text{where} \quad D = \frac{d}{dx}.$$

19. $x^2 y''_{xx} - [a^2 x^2 + n(n+1)] y = 0$, $\quad n = 0, 1, 2, \ldots$
General solution:
$$y x^{n+1} = (x^3 D)^n \left(\frac{C_1 e^{ax} + C_2 e^{-ax}}{x^{2n-1}} \right), \quad \text{where} \quad D = \frac{d}{dx}.$$

20. $x^2 y''_{xx} + (ax^2 + bx + c) y = 0$.
The substitution $y = x^\lambda u$, where λ is a root of the quadratic equation $\lambda^2 - \lambda + c = 0$, leads to an equation of the form 2.2.2.17:
$$x u''_{xx} + 2\lambda u'_x + (ax + b) u = 0.$$
For $a = -\tfrac{1}{4}$, $b = k$, and $c = \tfrac{1}{4} - m^2$, the original equation is referred to as *Whittaker's equation*.

21. $x^2 y''_{xx} + axy'_x + by = 0$.
Euler equation. General solution:
$$y = \begin{cases} |x|^{\frac{1-a}{2}} (C_1 |x|^\mu + C_2 |x|^{-\mu}) & \text{if } (1-a)^2 > 4b, \\ |x|^{\frac{1-a}{2}} (C_1 + C_2 \ln |x|) & \text{if } (1-a)^2 = 4b, \\ |x|^{\frac{1-a}{2}} [C_1 \sin(\mu \ln |x|) + C_2 \cos(\mu \ln |x|)] & \text{if } (1-a)^2 < 4b, \end{cases}$$
where $\mu = \tfrac{1}{2} |(1-a)^2 - 4b|^{1/2}$.

22. $x^2 y''_{xx} + x y'_x + (x^2 - \nu^2) y = 0.$

Bessel equation.

1°. Let ν be an arbitrary noninteger. Then the general solution is given by

$$y = C_1 J_\nu(x) + C_2 Y_\nu(x), \qquad (1)$$

where $J_\nu(x)$ and $Y_\nu(x)$ are the Bessel functions of the first and second kind:

$$J_\nu(x) = \sum_{k=0}^{\infty} \frac{(-1)^k (x/2)^{\nu+2k}}{k!\, \Gamma(\nu+k+1)}, \qquad Y_\nu(x) = \frac{J_\nu(x) \cos \pi\nu - J_{-\nu}(x)}{\sin \pi\nu}. \qquad (2)$$

Solution (1) is denoted by $y = Z_\nu(x)$, which is referred to as the cylindrical function.

2°. Integral representations with $x > 0$:

$$J_\nu(x) = \frac{1}{\pi} \int_0^\pi \cos(x \sin\theta - \nu\theta)\, d\theta - \frac{1}{\pi} \sin(\pi\nu) \int_0^\infty \exp(-x \sinh t - \nu t)\, dt,$$

$$Y_\nu(x) = \frac{1}{\pi} \int_0^\pi \sin(x \sin\theta - \nu\theta)\, d\theta - \frac{1}{\pi} \int_0^\infty [e^{\nu t} + e^{-\nu t} \cos(\pi\nu)] e^{-x \sinh t}\, dt.$$

3°. In the case $\nu = n + \tfrac{1}{2}$, where $n = 0, 1, 2, \ldots$, the Bessel functions are expressed in terms of elementary functions:

$$J_{n+\frac{1}{2}}(x) = \sqrt{\frac{2}{\pi}}\, x^{n+\frac{1}{2}} \left(-\frac{1}{x}\frac{d}{dx}\right)^n \frac{\sin x}{x}, \quad J_{-n-\frac{1}{2}}(x) = \sqrt{\frac{2}{\pi}}\, x^{n+\frac{1}{2}} \left(\frac{1}{x}\frac{d}{dx}\right)^n \frac{\cos x}{x},$$

$$Y_{n+\frac{1}{2}}(x) = (-1)^{n+1} J_{-n-\frac{1}{2}}(x).$$

4°. Let $\nu = n$ be an arbitrary integer. The following relations hold:

$$J_{-n}(x) = (-1)^n J_n(x), \quad Y_{-n}(x) = (-1)^n Y_n(x).$$

The solution is given by formula (1) in which the function $J_n(x)$ is obtained by substituting $\nu = n$ into formula (2), while $Y_n(x)$ is found by taking the limit as $\nu \to n$ and for nonnegative n becomes

$$Y_n(x) = \frac{2}{\pi} J_n(x) \ln \frac{x}{2} - \frac{1}{\pi} \sum_{k=0}^{n-1} \frac{(n-k-1)!}{k!} \left(\frac{2}{x}\right)^{n-2k}$$

$$- \frac{1}{\pi} \sum_{k=0}^{\infty} (-1)^k \left(\frac{x}{2}\right)^{n+2k} \frac{\psi(k+1) + \psi(n+k+1)}{k!\,(n+k)!},$$

where $\psi(1) = -C$, $\psi(n) = -C + \sum_{k=1}^{n-1} k^{-1}$, $C = 0.5772\ldots$ is Euler's constant, $\psi(x) = [\ln \Gamma(x)]'_x$ is the logarithmic derivative of the gamma function.

5°. The Bessel functions are discussed in detail in the books by Bateman & Erdélyi (1953, Vol. 2), McLachlan (1955), Abramowitz & Stegun (1964), Magnus et al. (1966), Andrews et al. (2001), Weisstein (2003), Olver et al. (2010), and Beals & Wong (2016).

23. $x^2 y''_{xx} + x y'_x - (x^2 + \nu^2) y = 0.$

Modified Bessel equation. It can be reduced to equation 2.2.2.22 by means of the substitution $x = i\bar{x}$ ($i^2 = -1$).

1°. General solution:
$$y = C_1 I_\nu(x) + C_2 K_\nu(x),$$
where $I_\nu(x)$ and $K_\nu(x)$ are modified Bessel functions of the first and second kind:
$$I_\nu(x) = \sum_{k=0}^{\infty} \frac{(x/2)^{2k+\nu}}{k!\,\Gamma(\nu+k+1)}, \qquad K_\nu(x) = \frac{\pi}{2} \frac{I_{-\nu}(x) - I_\nu(x)}{\sin \pi\nu}.$$

The modified Bessel function $I_\nu(x)$ can be expressed in terms of the Bessel function J_ν as:
$$I_\nu(x) = e^{-\pi\nu i/2} J_\nu(x e^{\pi i/2}), \qquad i^2 = -1.$$

2°. Integral representations with $x > 0$:
$$I_\nu(x) = \frac{x^\nu}{\pi^{1/2} 2^\nu \Gamma(\nu + \tfrac{1}{2})} \int_{-1}^{1} \exp(-xt)(1-t^2)^{\nu-1/2}\,dt, \quad (\nu > -\tfrac{1}{2}),$$
$$K_\nu(x) = \int_0^\infty \exp(-x \cosh t) \cosh(\nu t)\,dt.$$

For integer $\nu = n$,
$$I_n(x) = \frac{1}{\pi} \int_0^\pi \exp(x\cos t) \cos(nt)\,dt \qquad (n = 0, 1, 2, \ldots),$$
$$K_0(x) = \int_0^\infty \cos(x \sinh t)\,dt = \int_0^\infty \frac{\cos(xt)}{\sqrt{t^2+1}}\,dt \qquad (x > 0).$$

3°. In the case $\nu = \pm n \pm \tfrac{1}{2}$, where $n = 0, 1, 2, \ldots$, the modified Bessel functions are expressed in terms of elementary functions:
$$I_{n+1/2}(x) = \frac{1}{\sqrt{2\pi x}} \left[e^x \sum_{k=0}^{n} \frac{(-1)^k (n+k)!}{k!\,(n-k)!\,(2x)^k} - (-1)^n e^{-x} \sum_{k=0}^{n} \frac{(n+k)!}{k!\,(n-k)!\,(2x)^k} \right],$$
$$I_{-n-1/2}(x) = \frac{1}{\sqrt{2\pi x}} \left[e^x \sum_{k=0}^{n} \frac{(-1)^k (n+k)!}{k!\,(n-k)!\,(2x)^k} + (-1)^n e^{-x} \sum_{k=0}^{n} \frac{(n+k)!}{k!\,(n-k)!\,(2x)^k} \right],$$
$$K_{n+1/2}(x) = K_{-n-1/2}(x) = \sqrt{\frac{\pi}{2x}}\, e^{-x} \sum_{k=0}^{n} \frac{(n+k)!}{k!\,(n-k)!\,(2x)^k}.$$

4°. If $\nu = n$ is a nonnegative integer, then
$$K_n(x) = (-1)^{n+1} I_n(x) \ln \frac{x}{2} + \frac{1}{2} \sum_{m=0}^{n-1} (-1)^m \left(\frac{x}{2}\right)^{2m-n} \frac{(n-m-1)!}{m!}$$
$$+ \frac{1}{2}(-1)^n \sum_{m=0}^{\infty} \left(\frac{x}{2}\right)^{n+2m} \frac{\psi(n+m+1) + \psi(m+1)}{m!\,(n+m)!},$$

where $\psi(z)$ is the logarithmic derivative of the gamma function; for $n = 0$, the first sum must be dropped.

5°. The modified Bessel functions are discussed in detail in the books by Bateman & Erdélyi (1953, Vol. 2), McLachlan (1955), Abramowitz & Stegun (1964), Magnus et al. (1966), Andrews et al. (2001), Weisstein (2003), Olver et al. (2010), and Beals & Wong (2016).

24. $x^2 y''_{xx} + axy'_x + (bx^k + c)y = 0$, $\quad k \neq 0$.

The special case $b = 0$ corresponds to the Euler equation 2.2.2.21.

1°. For $b > 0$, the general solution is

$$y = x^{\frac{1-a}{2}} \left[C_1 J_\nu \left(\frac{2}{k} \sqrt{b} \, x^{\frac{k}{2}} \right) + C_2 Y_\nu \left(\frac{2}{k} \sqrt{b} \, x^{\frac{k}{2}} \right) \right],$$

where $\nu = \frac{1}{k}\sqrt{(1-a)^2 - 4c}$; $J_\nu(z)$ and $Y_\nu(z)$ are the Bessel functions of the first and second kind.

2°. For $b < 0$, the general solution is

$$y = x^{\frac{1-a}{2}} \left[C_1 I_\nu \left(\frac{2}{k} \sqrt{|b|} \, x^{\frac{k}{2}} \right) + C_2 K_\nu \left(\frac{2}{k} \sqrt{|b|} \, x^{\frac{k}{2}} \right) \right],$$

where $\nu = \frac{1}{k}\sqrt{(1-a)^2 - 4c}$; $I_\nu(z)$ and $K_\nu(z)$ are the modified Bessel functions of the first and second kind.

25. $x^2 y''_{xx} + axy'_x + x^n(bx^n + c)y = 0$.

The substitution $\xi = x^n$ leads to an equation of the form 2.2.2.17:

$$n^2 \xi y''_{\xi\xi} + n(n - 1 + a) y'_\xi + (b\xi + c) y = 0.$$

26. $x^2 y''_{xx} + (ax + b) y'_x + cy = 0$.

The transformation $x = z^{-1}$, $y = z^k e^z w$, where k is a root of the quadratic equation $k^2 + (1 - a)k + c = 0$, leads to an equation of the form 2.2.2.17:

$$z w''_{zz} + [(2 - b)z + 2k + 2 - a] w'_z + [(1 - b)z + 2k + 2 - a - bk] w = 0.$$

▶ **ODEs of the form** $(a_2 x^2 + b_2 x + c_2) y''_{xx} + (b_1 x + c_1) y'_x + c_0 y = 0$.

27. $(1 - x^2) y''_{xx} + n(n - 1) y = 0$, $\quad n = 0, 1, 2, \ldots$

This equation arises in hydrodynamics when describing axially symmetric Stokes flows.

1°. For $n \geq 2$, the general solution is given by:

$$y = C_1 \mathcal{J}_n(x) + C_2 \mathcal{H}_n(x),$$

where $\mathcal{J}_n(x)$ and $\mathcal{H}_n(x)$ are the *Gegenbauer functions* which can be expressed in terms of the Legendre polynomials $P_n(x)$ and the Legendre functions of the second kind $Q_n(x)$ (see equation 2.2.2.29) as follows:

$$\mathcal{J}_n(x) = \frac{P_{n-2}(x) - P_n(x)}{2n - 1}, \quad \mathcal{H}_n(x) = \frac{Q_{n-2}(x) - Q_n(x)}{2n - 1}.$$

2°. For $n = 0$ and $n = 1$, the general solution is: $y = C_1 + C_2 x$.

28. $(1-x^2)y''_{xx} - xy'_x - ay = 0$.

$1°$. For $a = k^2 > 0$, the general solution is:

$$y = \begin{cases} C_1 \cos(k \operatorname{arcosh} |x|) + C_2 \sin(k \operatorname{arcosh} |x|) & \text{if } |x| > 1, \\ C_1 \exp(k \arccos x) + C_2 \exp(-k \arccos x) & \text{if } |x| < 1, \end{cases}$$

where $\operatorname{arcosh} x = \ln(x + \sqrt{x^2-1})$.

$2°$. For $a = -k^2 < 0$, the general solution is:

$$y = \begin{cases} C_1 \exp(k \operatorname{arcosh} |x|) + C_2 \exp(-k \operatorname{arcosh} |x|) & \text{if } |x| > 1, \\ C_1 \cos(k \arccos x) + C_2 \sin(k \arccos x) & \text{if } |x| < 1. \end{cases}$$

$3°$. For $a = -n^2$, where n is a nonnegative integer, particular solutions are

$$y_0 = T_n(x) = \cos(n \arccos x) = \frac{(-1)^n}{2^n \left(\frac{1}{2}\right)_n} \sqrt{1-x^2}\, \frac{d^n}{dx^n}\left[(1-x^2)^{n-\frac{1}{2}}\right]$$

$$= \frac{n}{2} \sum_{m=0}^{[n/2]} (-1)^m \frac{(n-m-1)!}{m!\,(n-2m)!} (2x)^{n-2m},$$

where $T_n(x)$ is the *Chebyshev polynomial of the first kind*, and $[b]$ stands for the integer part of the number b.

29. $(1-x^2)y''_{xx} - 2xy'_x + n(n+1)y = 0$, $\quad n = 0, 1, 2, \ldots$

Legendre equation (special case).

The general solution is given by

$$y = C_1 P_n(x) + C_2 Q_n(x),$$

where the *Legendre polynomials* $P_n(x)$ and the *Legendre functions of the second kind* $Q_n(x)$ are defined as

$$P_n(x) = \frac{1}{n!\, 2^n} \frac{d^n}{dx^n}(x^2-1)^n, \quad Q_n(x) = \frac{1}{2} P_n(x) \ln \frac{1+x}{1-x} - \sum_{m=1}^{n} \frac{1}{m} P_{m-1}(x) P_{n-m}(x).$$

The functions $P_n = P_n(x)$ can be conveniently calculated using the recurrence relations

$$P_0(x) = 1, \quad P_1(x) = x, \quad P_2(x) = \frac{1}{2}(3x^2-1),$$

$$P_{n+1}(x) = \frac{2n+1}{n+1} x P_n(x) - \frac{n}{n+1} P_{n-1}(x), \quad n = 2, 3, \ldots$$

Three leading functions $Q_n = Q_n(x)$ are

$$Q_0(x) = \frac{1}{2} \ln \frac{1+x}{1-x}, \quad Q_1(x) = \frac{x}{2} \ln \frac{1+x}{1-x} - 1, \quad Q_2(x) = \frac{3x^2-1}{4} \ln \frac{1+x}{1-x} - \frac{3}{2} x.$$

All n zeros of the polynomial $P_n(x)$ are real and lie on the interval $-1 < x < 1$; the functions $P_n(x)$ form an orthogonal system on the closed interval $-1 \le x \le 1$, with the following relations taking place:

$$\int_{-1}^{1} P_n(x) P_m(x)\, dx = \begin{cases} 0 & \text{if } n \ne m, \\ \dfrac{2}{2n+1} & \text{if } n = m. \end{cases}$$

30. $(1 - x^2) y''_{xx} - 2x y'_x + \nu(\nu + 1) y = 0.$

Legendre equation (special case), where ν is an arbitrary number. The case $\nu = n$ where n is a nonnegative integer is considered in 2.2.2.29.

The substitution $z = x^2$ leads to the hypergeometric equation 2.2.2.33. Therefore, with $|x| < 1$ the general solution can be written as

$$y = C_1 F\left(-\frac{\nu}{2}, \frac{1+\nu}{2}, \frac{1}{2}; x^2\right) + C_2 x F\left(\frac{1-\nu}{2}, 1+\frac{\nu}{2}, \frac{3}{2}; x^2\right),$$

where $F(\alpha, \beta, \gamma; x)$ is the hypergeometric series.

The associated Legendre functions, which are solutions of the Legendre equation, are discussed in detail in the books by Bateman & Erdélyi (1953, Vol. 1), Abramowitz & Stegun (1964), Andrews et al. (2001), Olver et al. (2010), and Beals & Wong (2016).

31. $(ax^2 + b) y''_{xx} + a x y'_x + c y = 0.$

The substitution $z = \displaystyle\int \frac{dx}{\sqrt{ax^2 + b}}$ leads to a constant coefficient linear ODE:

$$y''_{zz} + c y = 0.$$

32. $(1 - x^2) y''_{xx} + (ax + b) y'_x + c y = 0.$

The substitution $2z = 1 + x$ leads to the hypergeometric equation 2.2.2.33:

$$z(1 - z) y''_{zz} + [a z + \tfrac{1}{2}(b - a)] y'_z + c y = 0.$$

33. $x(x - 1) y''_{xx} + [(\alpha + \beta + 1) x - \gamma] y'_x + \alpha \beta y = 0.$

Gaussian hypergeometric equation.

1°. For $\gamma \ne 0, -1, -2, -3, \ldots$, a solution can be expressed in terms of the hypergeometric series:

$$F(\alpha, \beta, \gamma; x) = 1 + \sum_{k=1}^{\infty} \frac{(\alpha)_k (\beta)_k}{(\gamma)_k} \frac{x^k}{k!}, \quad (\alpha)_k = \alpha(\alpha + 1) \ldots (\alpha + k - 1),$$

which, *a fortiori*, is convergent for $|x| < 1$.

2°. For $\gamma > \beta > 0$, the solution from Item 1° can be expressed in terms of a definite integral:

$$F(\alpha, \beta, \gamma; x) = \frac{\Gamma(\gamma)}{\Gamma(\beta)\Gamma(\gamma - \beta)} \int_0^1 t^{\beta - 1} (1 - t)^{\gamma - \beta - 1} (1 - tx)^{-\alpha}\, dt,$$

where $\Gamma(\beta)$ is the gamma function.

3°. If γ is not an integer, the general solution of the hypergeometric equation has the form:

$$y = C_1 F(\alpha, \beta, \gamma; x) + C_2 x^{1-\gamma} F(\alpha - \gamma + 1,\ \beta - \gamma + 1,\ 2 - \gamma;\ x).$$

In the degenerate cases $\gamma = 0, -1, -2, -3, \ldots$, a particular solution of the hypergeometric equation corresponds to $C_1 = 0$ and $C_2 = 1$. If γ is a positive integer, another particular solution corresponds to $C_1 = 1$ and $C_2 = 0$. In both these cases, the general solution can be constructed by means of formula (2) given in Section 2.2.1.

Table 2.4 presents some special cases when the hypergeometric function $F(\alpha, \beta, \gamma; x)$ is expressed in terms of elementary functions.

Table 2.5 gives the general solutions of the hypergeometric equation for some values of the determining parameters.

4°. The function F possesses the following properties:

$$F(\alpha, \beta, \gamma; x) = F(\beta, \alpha, \gamma; x),$$
$$F(\alpha, \beta, \gamma; x) = (1-x)^{\gamma-\alpha-\beta} F(\gamma - \alpha,\ \gamma - \beta,\ \gamma;\ x),$$
$$F(\alpha, \beta, \gamma; x) = (1-x)^{-\alpha} F\left(\alpha,\ \gamma - \beta,\ \gamma;\ \frac{x}{x-1}\right),$$
$$\frac{d^n}{dx^n} F(\alpha, \beta, \gamma; x) = \frac{(\alpha)_n (\beta)_n}{(\gamma)_n} F(\alpha + n,\ \beta + n,\ \gamma + n;\ x).$$

5°. The hypergeometric functions are discussed in detail in the books by Bateman & Erdélyi (1953, Vol. 1), Abramowitz & Stegun (1964), Magnus et al. (1966), Slavyanov & Lay (2000), Andrews et al. (2001), Olver et al. (2010), and Beals & Wong (2016).

▶ **Other linear ODEs.**

34. $x^4 y''_{xx} + ay = 0$.

The transformation $z = 1/x$, $u = y/x$ leads to a constant coefficient linear equation of the form 2.2.2.1: $u''_{zz} + au = 0$.

35. $x^4 y''_{xx} + (ax^2 + bx + c)y = 0$.

The transformation $z = 1/x$, $u = y/x$ leads to a linear equation of the form 2.2.2.20:

$$z^2 u''_{zz} + (cz^2 + bz + a)u = 0.$$

36. $x^2(x-a)^2 y''_{xx} + by = 0$.

General solution:

$$y = C_1 |x|^m |x-a|^{1-m} + C_2 |x|^{1-m} |x-a|^m,$$

where m is a root of the quadratic equation $m(m-1)a^2 = -b$.

37. $ax^2(x-1)^2 y''_{xx} + (bx^2 + cx + d)y = 0$.

Let p and q be roots of the quadratic equations

$$ap(p-1) + d = 0, \qquad aq(q-1) + b + c + d = 0.$$

TABLE 2.4. Some special cases where the hypergeometric function $F(\alpha, \beta, \gamma; z)$ is expressed in terms of elementary functions.

α	β	γ	z	F
$-n$	β	γ	x	$\sum_{k=0}^{n} \frac{(-n)_k (\beta)_k}{(\gamma)_k} \frac{x^k}{k!}$, where $n = 1, 2, \ldots$
$-n$	β	$-n-m$	x	$\sum_{k=0}^{n} \frac{(-n)_k (\beta)_k}{(-n-m)_k} \frac{x^k}{k!}$, where $n = 1, 2, \ldots$
α	β	β	x	$(1-x)^{-\alpha}$
α	$\frac{1}{2}\alpha + 1$	$\frac{1}{2}\alpha$	x	$(1+x)(1-x)^{-\alpha-1}$
α	$\alpha + \frac{1}{2}$	$2\alpha + 1$	x	$\left(\frac{1+\sqrt{1-x}}{2}\right)^{-2\alpha}$
α	$\alpha + \frac{1}{2}$	2α	x	$\frac{1}{\sqrt{1-x}} \left(\frac{1+\sqrt{1-x}}{2}\right)^{1-2\alpha}$
α	$\alpha + \frac{1}{2}$	$\frac{3}{2}$	x^2	$\frac{(1+x)^{1-2\alpha} - (1-x)^{1-2\alpha}}{2x(1-2\alpha)}$
α	$\alpha + \frac{1}{2}$	$\frac{1}{2}$	$-\tan^2 x$	$\cos^{2\alpha} x \cos(2\alpha x)$
α	$\alpha + \frac{1}{2}$	$\frac{1}{2}$	x^2	$\frac{1}{2}\left[(1+x)^{-2\alpha} + (1-x)^{-2\alpha}\right]$
α	$\alpha - \frac{1}{2}$	2α	x	$2^{2\alpha-1}\left(1+\sqrt{1-x}\right)^{1-2\alpha}$
α	$2-\alpha$	$\frac{3}{2}$	$\sin^2 x$	$\frac{\sin[(2\alpha-2)x]}{(\alpha-1)\sin(2x)}$
α	$1-\alpha$	$\frac{1}{2}$	$-x^2$	$\frac{\left(\sqrt{1+x^2}+x\right)^{2\alpha-1} + \left(\sqrt{1+x^2}-x\right)^{2\alpha-1}}{2\sqrt{1+x^2}}$
α	$1-\alpha$	$\frac{3}{2}$	$\sin^2 x$	$\frac{\sin[(2\alpha-1)x]}{(\alpha-1)\sin(2x)}$
α	$1-\alpha$	$\frac{1}{2}$	$\sin^2 x$	$\frac{\cos[(2\alpha-1)x]}{\cos x}$
α	$-\alpha$	$\frac{1}{2}$	$-x^2$	$\frac{1}{2}\left[\left(\sqrt{1+x^2}+x\right)^{2\alpha} + \left(\sqrt{1+x^2}-x\right)^{2\alpha}\right]$
α	$-\alpha$	$\frac{1}{2}$	$\sin^2 x$	$\cos(2\alpha x)$
1	1	2	$-x$	$\frac{1}{x}\ln(x+1)$
$\frac{1}{2}$	1	$\frac{3}{2}$	x^2	$\frac{1}{2x}\ln\frac{1+x}{1-x}$
$\frac{1}{2}$	1	$\frac{3}{2}$	$-x^2$	$\frac{1}{x}\arctan x$
$\frac{1}{2}$	$\frac{1}{2}$	$\frac{3}{2}$	x^2	$\frac{1}{x}\arcsin x$
$\frac{1}{2}$	$\frac{1}{2}$	$\frac{3}{2}$	$-x^2$	$\frac{1}{x}\operatorname{arsinh} x$

TABLE 2.5. General solutions of the hypergeometric equation 2.2.2.33 for some values of the determining parameters.

α	β	γ	Solution: $y = y(x)$
0	β	γ	$C_1 + C_2 \int |x|^{-\gamma} |x-1|^{\gamma-\beta-1}\, dx$
α	$\alpha + \frac{1}{2}$	$2\alpha + 1$	$C_1 (1+\sqrt{1-x})^{-2\alpha} + C_2 x^{-2\alpha} (1+\sqrt{1-x})^{2\alpha}$
α	$\alpha - \frac{1}{2}$	$\frac{1}{2}$	$C_1 (1+\sqrt{x})^{1-2\alpha} + C_2 (1-\sqrt{x})^{1-2\alpha}$
α	$\alpha + \frac{1}{2}$	$\frac{3}{2}$	$\dfrac{1}{\sqrt{x}} \left[C_1 (1+\sqrt{x})^{1-2\alpha} + C_2 (1-\sqrt{x})^{1-2\alpha} \right]$
1	β	γ	$|x|^{1-\gamma} |x-1|^{\gamma-\beta-1} \left(C_1 + C_2 \int |x|^{\gamma-2} |x-1|^{\beta-\gamma}\, dx \right)$
α	β	α	$|x-1|^{-\beta} \left(C_1 + C_2 \int |x|^{-\alpha} |x-1|^{\beta-1}\, dx \right)$
α	β	$\alpha+1$	$|x|^{-\alpha} \left(C_1 + C_2 \int |x|^{\alpha-1} |x-1|^{-\beta}\, dx \right)$

The substitution $y = x^p (x-1)^q w$ leads to the hypergeometric equation of the form 2.2.2.33:

$$ax(x-1)w''_{xx} + 2a[(p+q)x - p]w'_x + (2apq - c - 2d)w = 0.$$

38. $(x^2 + 1)^2 y''_{xx} + ay = 0.$

Halm's equation. General solution:

$$y = \begin{cases} \sqrt{x^2+1}\,[C_1 \cos(\beta \arctan x) + C_2 \sin(\beta \arctan x)] & \text{if } a+1 = \beta^2 > 0, \\ \sqrt{x^2+1}\,[C_1 \cosh(\beta \arctan x) + C_2 \sinh(\beta \arctan x)] & \text{if } a+1 = -\beta^2 < 0, \\ \sqrt{x^2+1}\,(C_1 + C_2 \arctan x) & \text{if } a = -1. \end{cases}$$

39. $(x^2 - 1)^2 y''_{xx} + ay = 0.$

General solution:

$$y = \begin{cases} \sqrt{|x^2-1|}\,[C_1 \cos(\beta \ln |z|) + C_2 \sin(\beta \ln |z|)] & \text{if } a-1 = 4\beta^2 > 0, \\ (x+1)(C_1 |z|^{(2\beta-1)/2} + C_2 |z|^{-(2\beta+1)/2}) & \text{if } a-1 = -4\beta^2 < 0, \\ \sqrt{|x^2-1|}\,(C_1 + C_2 \ln |z|) & \text{if } a = 1, \end{cases}$$

where $z = (x+1)/(x-1)$.

40. $(x^2 \pm a^2)^2 y''_{xx} + b^2 y = 0.$

This equation describes bending of a double-walled compressed bar with a parabolic cross-section.

1°. For the upper sign (constricted bar), the general solution is as follows:

$$y = \sqrt{x^2 + a^2}\,(C_1 \cos u + C_2 \sin u), \quad \text{where } u = \sqrt{1 + (b/a)^2}\,\arctan(x/a).$$

2°. For the lower sign (bar with salients), the general solution is given by:

$$y = \sqrt{a^2 - x^2}\,(C_1 \cos u + C_2 \sin u), \quad \text{where } u = \frac{\sqrt{b^2 - a^2}}{2a} \ln \frac{a+x}{a-x}; \quad |x| < a.$$

41. $(1 - x^2)^2 y''_{xx} - 2x(1 - x^2)y'_x + [\nu(\nu + 1)(1 - x^2) - \mu^2]y = 0.$

Legendre equation, ν and μ are arbitrary parameters.

The transformation $x = 1 - 2\xi$, $y = |x^2 - 1|^{\mu/2} w$ leads to the hypergeometric equation 2.2.2.33:

$$\xi(\xi - 1)w''_{\xi\xi} + (\mu + 1)(1 - 2\xi)w'_\xi + (\nu - \mu)(\nu + \mu + 1)w = 0$$

with parameters $\alpha = \mu - \nu$, $\beta = \mu + \nu + 1$, $\gamma = \mu + 1$.

In particular, the original equation is integrable by quadrature if $\nu = \mu$ or $\nu = -\mu - 1$.

The *associated Legendre functions* of the first and the second kind, $P^\mu_\nu(x)$ and $Q^\mu_\nu(x)$, which are solutions of the Legendre equation, are discussed in detail in the books by Bateman & Erdélyi (1953, Vol. 1), Abramowitz & Stegun (1964), Andrews et al. (2001), Olver et al. (2010), and Beals & Wong (2016).

42. $(x - a)^2(x - b)^2 y''_{xx} - cy = 0, \quad a \neq b.$

The transformation $\xi = \ln\left|\dfrac{x-a}{x-b}\right|$, $y = (x - b)\eta$ leads to a constant coefficient linear ODE: $(a - b)^2(\eta''_{\xi\xi} - \eta'_\xi) - c\eta = 0$. Therefore, the general solution is as follows:

$$y = C_1 |x - a|^{(1+\lambda)/2} |x - b|^{(1-\lambda)/2} + C_2 |x - a|^{(1-\lambda)/2} |x - b|^{(1+\lambda)/2},$$

where $\lambda^2 = 4c(a - b)^{-2} + 1 \neq 0$.

43. $(ax^2 + bx + c)^2 y''_{xx} + Ay = 0.$

The transformation

$$\xi = \int \frac{dx}{ax^2 + bx + c}, \quad w = \frac{y}{\sqrt{|ax^2 + bx + c|}}$$

leads to a constant coefficient linear ODE of the form 2.2.2.1:

$$w''_{\xi\xi} + (A + ac - \tfrac{1}{4}b^2)w = 0.$$

44. $(x^2 - 1)^2 y''_{xx} + 2x(x^2 - 1)y'_x + [(x^2 - 1)(a^2 x^2 - \lambda) - m^2]y = 0.$

Equation for prolate spheroidal wave functions, $m = 0, 1, \ldots$ It arises when separating variables in the wave equation written in the system of prolate spheroidal coordinates.

1°. In applications, one usually looks for eigenvalues $\lambda = \lambda_{mn}$ and eigenfunctions $y = y_{mn}(x)$ that assume finite values at $x = \pm 1$. The following functions are solutions of the eigenvalue problem:

$$S^{(1)}_{mn}(a, x) = \sum_{r=0,1}^{\infty} d^{mn}_r(a) P^m_{m+r}(x) \quad \text{(prolate angular functions of the first kind)},$$

$$S^{(2)}_{mn}(a, x) = \sum_{r=-\infty}^{\infty} d^{mn}_r(a) Q^m_{m+r}(x) \quad \text{(prolate angular functions of the second kind)},$$

where $P_n^m(x)$ and $Q_n^m(x)$ are the associated Legendre functions of the first and second kind. For $-1 \le x \le 1$, we have

$$P_n^m(x) = (1-x^2)^{m/2}\frac{d^m}{dx^m}P_n(x).$$

The summation is performed over either even or odd values of r, depending on whether $|n-m|$ is even or odd, respectively.

$2°$. The following recurrence relations for the coefficients $d_k = d_k^{mn}(a)$ hold:

$$\alpha_k d_{k+2} + (\beta_k - \lambda_{mn})d_k + \gamma_k d_{k-2} = 0,$$

where

$$\alpha_k = \frac{a^2(2m+k+1)(2m+k+2)}{(2m+2k+3)(2m+2k+5)},$$

$$\beta_k = (m+k)(m+k+1) + a^2\frac{2(m+k)(m+k+1) - 2m^2 - 1}{(2m+2k-1)(2m+2k+3)},$$

$$\gamma_k = \frac{a^2 k(k-1)}{(2m+2k-3)(2m+2k-1)}.$$

$3°$. For $a \to 0$, the eigenvalues are defined by:

$$\lambda_{mn} = n(n+1) + \frac{1}{2}\left[1 - \frac{(2m-1)(2m+1)}{(2n-1)(2n+3)}\right]a^2 + O(a^4).$$

$4°$. For $a \to \infty$, we have:

$$\lambda_{mn} = aq + m^2 - \tfrac{1}{8}(q^2+5) - \tfrac{1}{64}q(q^2+11-32m^2)a^{-1} + O(a^{-2}), \quad q = 2(n-m)+1.$$

$5°$. References for the ODE 2.2.2.44: Bateman & Erdélyi (1955, Vol. 3), Abramowitz & Stegun (1964), and Olver et al. (2010).

45. $(x^2+1)^2 y''_{xx} + 2x(x^2+1)y'_x + [(x^2+1)(a^2x^2 - \lambda) + m^2]y = 0.$

Equation of oblate spheroidal wave functions, $m = 0, 1, \ldots$ The transformations $x = \pm i\tilde{x}$, $a = \mp i\tilde{a}$ lead to equation 2.2.2.44.

See the books by Bateman & Erdélyi (1955, Vol. 3), Abramowitz & Stegun (1964), and Olver et al. (2010) for more information on this equation.

46. $x^2(ax^n - 1)y''_{xx} + x(apx^n + q)y'_x + (arx^n + s)y = 0.$

Let us find the roots A_1, A_2 and B_1, B_2 of the quadratic equations

$$A^2 - (q+1)A - s = 0, \quad B^2 - (p-1)B + r = 0$$

and use parameters $c, \alpha, \beta,$ and γ defined as

$$c = A_1, \quad \alpha = (A_1 + B_1)n^{-1}, \quad \beta = (A_1 + B_2)n^{-1}, \quad \gamma = 1 + (A_1 - A_2)n^{-1}.$$

Then the general solution of the original equation has the form $y = x^c u(ax^n)$, where $u = u(z)$ is the general solution of the hypergeometric equation 2.2.2.33: $z(z-1)u''_{zz} + [(\alpha+\beta+1)z - \gamma]u'_z + \alpha\beta u = 0.$

2.2.3. Equations Involving Exponential and Other Elementary Functions

1. $y''_{xx} + ae^{\lambda x} y = 0,\qquad \lambda \ne 0.$

General solution:

$$y = \begin{cases} C_1 J_0\!\left(2\lambda^{-1}\sqrt{a}\,e^{\lambda x/2}\right) + C_2 Y_0\!\left(2\lambda^{-1}\sqrt{a}\,e^{\lambda x/2}\right) & \text{if } a > 0, \\ C_1 I_0\!\left(2\lambda^{-1}\sqrt{|a|}\,e^{\lambda x/2}\right) + C_2 K_0\!\left(2\lambda^{-1}\sqrt{|a|}\,e^{\lambda x/2}\right) & \text{if } a < 0, \end{cases}$$

where $J_0(z)$ and $Y_0(z)$ are Bessel functions, and $I_0(z)$ and $K_0(z)$ are modified Bessel functions.

2. $y''_{xx} + (ae^x - b)y = 0.$

General solution:

$$y = \begin{cases} C_1 J_{2\sqrt{b}}\!\left(2\sqrt{a}\,e^{x/2}\right) + C_2 Y_{2\sqrt{b}}\!\left(2\sqrt{a}\,e^{x/2}\right) & \text{if } a > 0, \\ C_1 I_{2\sqrt{b}}\!\left(2\sqrt{|a|}\,e^{x/2}\right) + C_2 K_{2\sqrt{b}}\!\left(2\sqrt{|a|}\,e^{x/2}\right) & \text{if } a < 0, \end{cases}$$

where $J_\nu(z)$ and $Y_\nu(z)$ are Bessel functions, and $I_\nu(z)$ and $K_\nu(z)$ are modified Bessel functions.

3. $y''_{xx} - (ae^{2\lambda x} + be^{\lambda x} + c)y = 0.$

The transformation $z = e^{\lambda x}$, $w = z^{-k}y$, where $k = \sqrt{c}/\lambda$, leads to an equation of the form 2.2.2.17:

$$\lambda^2 z w''_{zz} + \lambda^2 (2k+1) w'_z - (az + b)w = 0.$$

4. $y''_{xx} + ay'_x + be^{2ax} y = 0.$

The transformation $\xi = e^{ax}$, $u = ye^{ax}$ leads to a constant coefficient linear ODE of the form 2.2.2.1: $u''_{\xi\xi} + ba^{-2} u = 0.$

5. $y''_{xx} - ay'_x + be^{2ax} y = 0.$

The substitution $\xi = e^{ax}$ leads to a constant coefficient linear equation of the form 2.2.2.1: $y''_{\xi\xi} + ba^{-2} y = 0.$

6. $y''_{xx} + ay'_x + (be^{\lambda x} + c)y = 0.$

General solution:

$$y = \begin{cases} e^{-ax/2}\left[C_1 J_\nu\!\left(2\lambda^{-1}\sqrt{b}\,e^{\lambda x/2}\right) + C_2 Y_\nu\!\left(2\lambda^{-1}\sqrt{b}\,e^{\lambda x/2}\right)\right] & \text{if } b > 0, \\ e^{-ax/2}\left[C_1 I_\nu\!\left(2\lambda^{-1}\sqrt{|b|}\,e^{\lambda x/2}\right) + C_2 K_\nu\!\left(2\lambda^{-1}\sqrt{|b|}\,e^{\lambda x/2}\right)\right] & \text{if } b < 0, \end{cases}$$

where $\nu = \frac{1}{\lambda}\sqrt{a^2 - 4c}$; $J_\nu(z)$ and $Y_\nu(z)$ are Bessel functions, and $I_\nu(z)$ and $K_\nu(z)$ are modified Bessel functions.

7. $y''_{xx} - (a - 2q \cosh 2x)y = 0.$

Modified Mathieu equation. The substitution $x = i\xi$ leads to the Mathieu equation 2.2.3.8:

$$y''_{\xi\xi} + (a - 2q\cos 2\xi)y = 0.$$

For eigenvalues $a = a_n(q)$ and $a = b_n(q)$, the corresponding solutions of the modified Mathieu equation are

$$\text{Ce}_{2n+p}(x, q) = \text{ce}_{2n+p}(ix, q) = \sum_{k=0}^{\infty} A_{2k+p}^{2n+p} \cosh[(2k+p)x],$$

$$\text{Se}_{2n+p}(x, q) = -i\,\text{se}_{2n+p}(ix, q) = \sum_{k=0}^{\infty} B_{2k+p}^{2n+p} \sinh[(2k+p)x],$$

where p can be either 0 or 1, and the coefficients A_{2k+p}^{2n+p} and B_{2k+p}^{2n+p} are specified in Item 2° of equation 2.2.3.8.

8. $y''_{xx} + (a - 2q\cos 2x)y = 0.$

Mathieu equation.

1°. Given numbers a and q, there exists a solution $y(x)$ and a characteristic index μ such that

$$y(x + \pi) = e^{2\pi\mu} y(x).$$

For small q, an approximate value of μ can be found from the equation

$$\cosh(\pi\mu) = 1 + 2\sin^2\left(\tfrac{1}{2}\pi\sqrt{a}\right) + \frac{\pi q^2}{(1-a)\sqrt{a}} \sin(\pi\sqrt{a}) + O(q^4).$$

If $y_1(x)$ is the solution of the Mathieu equation satisfying the initial conditions $y_1(0) = 1$ and $y'_1(0) = 0$, the characteristic index can be determined from the relation

$$\cosh(2\pi\mu) = y_1(\pi).$$

The solution $y_1(x)$, and hence μ, can be determined with a required degree of accuracy by means of numerical or approximate methods.

The general solution differs depending on the value of $y_1(\pi)$ and can be expressed in terms of two auxiliary periodic functions $\varphi_1(x)$ and $\varphi_2(x)$ (see Table 2.6).

TABLE 2.6. The general solution of the Mathieu equation 2.2.3.8 expressed in terms of auxiliary periodical functions $\varphi_1(x)$ and $\varphi_2(x)$.

Constraint	General solution $y = y(x)$	Period of φ_1 and φ_2	Index
$y_1(\pi) > 1$	$C_1 e^{2\mu x}\varphi_1(x) + C_2 e^{-2\mu x}\varphi_2(x)$	π	μ is a real number
$y_1(\pi) < -1$	$C_1 e^{2\rho x}\varphi_1(x) + C_2 e^{-2\rho x}\varphi_2(x)$	2π	$\mu = \rho + \tfrac{1}{2}i$, $i^2 = -1$, ρ is the real part of μ
$\lvert y_1(\pi)\rvert < 1$	$(C_1\cos\nu x + C_2\sin\nu x)\varphi_1(x)$ $+ (C_1\cos\nu x - C_2\sin\nu x)\varphi_2(x)$	π	$\mu = i\nu$ is a pure imaginary number, $\cos(2\pi\nu) = y_1(\pi)$
$y_1(\pi) = \pm 1$	$C_1\varphi_1(x) + C_2 x\varphi_2(x)$	π	$\mu = 0$

$2°$. In applications, of major interest are periodical solutions of the Mathieu equation that exist for certain values of the parameters a and q (those values of a are referred to as eigenvalues). The most important periodic solutions of the Mathieu equation are called *Mathieu functions* and are denoted by $\mathrm{ce}_n(x,q)$ and $\mathrm{se}_n(x,q)$. The Mathieu functions are listed in Table 2.7.

$3°$. Listed below are two leading terms of the asymptotic expansions of the Mathieu functions $\mathrm{ce}_n(x,q)$ and $\mathrm{se}_n(x,q)$, as well as of the corresponding eigenvalues $a_n(q)$ and $b_n(q)$, as $q \to 0$:

$$\mathrm{ce}_0(x,q) = \frac{1}{\sqrt{2}}\left(1 - \frac{q}{2}\cos 2x\right), \quad a_0(q) = -\frac{q^2}{2} + \frac{7q^4}{128};$$

$$\mathrm{ce}_1(x,q) = \cos x - \frac{q}{8}\cos 3x, \quad a_1(q) = 1 + q;$$

$$\mathrm{ce}_2(x,q) = \cos 2x + \frac{q}{4}\left(1 - \frac{\cos 4x}{3}\right), \quad a_2(q) = 4 + \frac{5q^2}{12};$$

$$\mathrm{ce}_n(x,q) = \cos nx + \frac{q}{4}\left[\frac{\cos(n+2)x}{n+1} - \frac{\cos(n-2)x}{n-1}\right], \quad a_n(q) = n^2 + \frac{q^2}{2(n^2-1)} \quad (n \geq 3);$$

$$\mathrm{se}_1(x,q) = \sin x - \frac{q}{8}\sin 3x, \quad b_1(q) = 1 - q;$$

$$\mathrm{se}_2(x,q) = \sin 2x - q\frac{\sin 4x}{12}, \quad b_2(q) = 4 - \frac{q^2}{12};$$

$$\mathrm{se}_n(x,q) = \sin nx - \frac{q}{4}\left[\frac{\sin(n+2)x}{n+1} - \frac{\sin(n-2)x}{n-1}\right], \quad b_n(q) = n^2 + \frac{q^2}{2(n^2-1)} \quad (n \geq 3).$$

TABLE 2.7. The Mathieu functions $\mathrm{ce}_n = \mathrm{ce}_n(x,q)$ and $\mathrm{se}_n = \mathrm{se}_n(x,q)$ (for odd n, the functions ce_n and se_n are 2π-periodic, and for even n, they are π-periodic); definite eigenvalues $a = a_n(q)$ and $a = b_n(q)$ correspond to each value of parameter q.

Mathieu functions	Recurrence relations for coefficients	Normalization conditions
$\mathrm{ce}_{2n}(x,q) = \sum\limits_{m=0}^{\infty} A_{2m}^{2n} \cos(2mx)$	$qA_2^{2n} = a_{2n} A_0^{2n}$; $qA_4^{2n} = (a_{2n}-4)A_2^{2n} - 2qA_0^{2n}$; $qA_{2m+2}^{2n} = (a_{2n}-4m^2)A_{2m}^{2n}$ $\quad - qA_{2m-2}^{2n}, \quad m \geq 2$	$(A_0^{2n})^2 + \sum\limits_{m=0}^{\infty}(A_{2m}^{2n})^2$ $= \begin{cases} 2 & \text{if } n=0, \\ 1 & \text{if } n \geq 1 \end{cases}$
$\mathrm{ce}_{2n+1}(x,q) = \sum\limits_{m=0}^{\infty} A_{2m+1}^{2n+1} \cos[(2m+1)x]$	$qA_3^{2n+1} = (a_{2n+1}-1-q)A_1^{2n+1}$; $qA_{2m+3}^{2n+1} = [a_{2n+1}-(2m+1)^2]A_{2m+1}^{2n+1}$ $\quad - qA_{2m-1}^{2n+1}, \quad m \geq 1$	$\sum\limits_{m=0}^{\infty}(A_{2m+1}^{2n+1})^2 = 1$
$\mathrm{se}_{2n}(x,q) = \sum\limits_{m=0}^{\infty} B_{2m}^{2n} \sin(2mx)$, $\mathrm{se}_0(x,q) = 0$	$qB_4^{2n} = (b_{2n}-4)B_2^{2n}$; $qB_{2m+2}^{2n} = (b_{2n}-4m^2)B_{2m}^{2n}$ $\quad - qB_{2m-2}^{2n}, \quad m \geq 2$	$\sum\limits_{m=0}^{\infty}(B_{2m}^{2n})^2 = 1$
$\mathrm{se}_{2n+1}(x,q) = \sum\limits_{m=0}^{\infty} B_{2m+1}^{2n+1} \sin[(2m+1)x]$	$qB_3^{2n+1} = (b_{2n+1}-1-q)B_1^{2n+1}$; $qB_{2m+3}^{2n+1} = [b_{2n+1}-(2m+1)^2]B_{2m+1}^{2n+1}$ $\quad - qB_{2m-1}^{2n+1}, \quad m \geq 1$	$\sum\limits_{m=0}^{\infty}(B_{2m+1}^{2n+1})^2 = 1$

4°. The Mathieu functions are discussed in the books by McLachlan (1947), Whittaker & Watson (1952), Bateman & Erdélyi (1955, vol. 3), and Abramowitz & Stegun (1964), Olver et al. (2010), and Beals & Wong (2016) in more detail.

9. $y''_{xx} + a \tan x \, y'_x + by = 0.$

1°. The substitution $\xi = \sin x$ leads to a linear ODE of the form 2.2.2.32:
$$(\xi^2 - 1)y''_{\xi\xi} + (1 - a)\xi y'_\xi - by = 0.$$

2°. General solution for $a = -2$:
$$y \cos x = \begin{cases} C_1 \sin(kx) + C_2 \cos(kx) & \text{if } b+1 = k^2 > 0, \\ C_1 \sinh(kx) + C_2 \cosh(kx) & \text{if } b+1 = -k^2 < 0. \end{cases}$$

3°. General solution for $a = 2$ and $b = 3$:
$$y = C_1 \cos^3 x + C_2 \sin x \, (1 + 2\cos^2 x).$$

2.2.4. Equations Involving Arbitrary Functions

▶ Notation: $f = f(x)$ and $g = g(x)$ are arbitrary functions; a, b, and λ are arbitrary parameters.

1. $y''_{xx} + ay = f.$

Equation of forced oscillations without friction.
General solution:
$$y = \begin{cases} C_1 \cos(kx) + C_2 \sin(kx) + k^{-1} \int_{x_0}^{x} f(\xi) \sin[k(x-\xi)] \, d\xi & \text{if } a = k^2 > 0, \\ C_1 \cosh(kx) + C_2 \sinh(kx) + k^{-1} \int_{x_0}^{x} f(\xi) \sinh[k(x-\xi)] \, d\xi & \text{if } a = -k^2 < 0, \\ C_1 x + C_2 + \int_{x_0}^{x} (x-\xi) f(\xi) \, d\xi & \text{if } a = 0, \end{cases}$$

where x_0 is an arbitrary number.

2. $y''_{xx} + ay'_x + by = f.$

Equation of forced oscillations with friction. The substitution $y = w \exp(-\tfrac{1}{2}ax)$ leads to an equation of the form 2.2.4.1:
$$w''_{xx} + (b - \tfrac{1}{4}a^2)w = f \exp(\tfrac{1}{2}ax).$$

3. $y''_{xx} + fy'_x = g.$

General solution:
$$y = C_1 + \int e^{-F} \left(C_2 + \int e^{F} g \, dx \right) dx, \quad \text{where} \quad F = \int f \, dx.$$

4. $y''_{xx} + fy'_x + a(f - a)y = 0.$

Particular solution: $y_0 = e^{-ax}$.

5. $y''_{xx} + xfy'_x - fy = 0.$

Particular solution: $y_0 = x$.

6. $xy''_{xx} + (xf + a)y'_x + (a-1)fy = 0.$
Particular solution: $y_0 = x^{1-a}$.

7. $xy''_{xx} + [(ax+1)f + ax - 1]y'_x + a^2xfy = 0.$
Particular solution: $y_0 = (ax+1)e^{-ax}$.

8. $xy''_{xx} + [(ax^2 + bx)f + 2]y'_x + bfy = 0.$
Particular solution: $y_0 = a + b/x$.

9. $x^2 y''_{xx} + xfy'_x + a(f - a - 1)y = 0.$
Particular solution: $y_0 = x^{-a}$.

10. $y''_{xx} + (f + ae^{\lambda x})y'_x + ae^{\lambda x}(f + \lambda)y = 0.$
Particular solution: $y_0 = \exp\left(-\dfrac{a}{\lambda}e^{\lambda x}\right)$.

11. $y''_{xx} - (f^2 + f'_x)y = 0.$
Particular solution: $y_0 = \exp\left(\int f\, dx\right)$.

12. $y''_{xx} + 2fy'_x + (f^2 + f'_x)y = 0.$
General solution:
$$y = (C_2 x + C_1)\exp\left(-\int f\, dx\right).$$

13. $y''_{xx} + (1-a)fy'_x - a(f^2 + f'_x)y = 0.$
Particular solution: $y_0 = \exp\left(a\int f\, dx\right)$.

14. $y''_{xx} + fy'_x + (fg - g^2 + g'_x)y = 0.$
Particular solution: $y_0 = \exp\left(-\int g\, dx\right)$.

15. $fy''_{xx} - af'_x y'_x - bf^{2a+1}y = 0.$
General solution:
$$y = C_1 \exp\left(\sqrt{b}\int f^a\, dx\right) + C_2 \exp\left(-\sqrt{b}\int f^a\, dx\right).$$

16. $f^2 y''_{xx} + f(f'_x + a)y'_x + by = 0.$
The substitution $\xi = \int f^{-1}dx$ leads to a constant coefficient linear ODE: $y''_{\xi\xi} + ay'_\xi + by = 0$.

17. $y''_{xx} - f'_x y'_x + a^2 e^{2f} y = 0.$
General solution:
$$y = C_1 \sin\left(a\int e^f\, dx\right) + C_2 \cos\left(a\int e^f\, dx\right).$$

18. $y''_{xx} - f'_x y'_x - a^2 e^{2f} y = 0.$
General solution:
$$y = C_1 \exp\left(a\int e^f\, dx\right) + C_2 \exp\left(-a\int e^f\, dx\right).$$

2.3. Second-Order Nonlinear Ordinary Differential Equations

2.3.1. Equations of the Form $y''_{xx} = f(x, y)$

1. $y''_{xx} = f(y)$.

A two-terms autonomous second-order ODE. The substitution $u(y) = y'_x$ leads to a first-order separated ODE: $uu'_y = f(y)$.

General solution: $\int \left[C_1 + 2 \int f(y)\, dy \right]^{-1/2} dy = C_2 \pm x$.

2. $y''_{xx} = Ax^n y^m$.

Emden–Fowler equation.

$1°$. The transformation $z = x^{n+2} y^{m-1}$, $w = xy'_x/y$ leads to a first-order ODE (an Abel equation of the second kind):

$$z[(m-1)w + n + 2]w'_z = -w^2 + w + Az.$$

$2°$. The transformation $y = w/t$, $x = 1/t$ leads to the Emden–Fowler equation with the independent variable raised to a different power: $w''_{tt} = At^{-n-m-3} w^m$.

$3°$. With $m \neq 1$, the Emden–Fowler equation has a particular solution:

$$y = \lambda x^{\frac{n+2}{1-m}}, \quad \text{where } \lambda = \left[\frac{(n+2)(n+m+1)}{A(m-1)^2} \right]^{\frac{1}{m-1}}.$$

$4°$. Table 2.8 presents all solvable Emden–Fowler equations whose general solutions are outlined in the handbooks by Polyanin & Zaitsev (2003, 2018). The one-parameter families (in the space of the parameters n and m) and isolated points are presented in a consecutive fashion. Equations are arranged in order of increasing m and increasing n (for identical m).

As illustrative examples, below are two special cases where this ODE is integrated by quadrature.

1. Emden–Fowler equation

$$y''_{xx} = Ax^{-m-3} y^m$$

for $m \neq -1$ admits the general solution in parametric form

$$x = aC_1^{m-1} \left[\int (1 \pm \tau^{m+1})^{-1/2}\, d\tau + C_2 \right]^{-1}, \quad y = bC_1^{m+1} \tau \left[\int (1 \pm \tau^{m+1})^{-1/2}\, d\tau + C_2 \right]^{-1},$$

where the free parameter of the equation A is expressed in terms of the solution parameters a and b as $A = \pm \frac{1}{2}(m+1) a^{m+1} b^{1-m}$.

For $m = -1$, the general solution in parametric form is

$$x = C_1 \left[\int \exp(\mp \tau^2)\, d\tau + C_2 \right]^{-1}, \quad y = b\exp(\mp \tau^2) \left[\int \exp(\mp \tau^2)\, d\tau + C_2 \right]^{-1},$$

where $A = \mp 2b^2$.

2. Emden–Fowler equation

$$y''_{xx} = Ax^{-\frac{m+3}{2}} y^m$$

TABLE 2.8. Solvable cases of the Emden–Fowler equation $y''_{xx} = Ax^n y^m$.

No	m	n	No	m	n
\multicolumn{3}{c|}{*One-parameter families*}	13	$-5/3$	$-5/6$
			14	$-5/3$	$-1/2$
1	arbitrary	0	15	$-5/3$	1
2	arbitrary	$-m-3$	16	$-5/3$	2
3	arbitrary	$-\frac{1}{2}(m+3)$	17	$-7/5$	$-13/5$
4	0	arbitrary	18	$-7/5$	1
5	1	arbitrary	19	$-1/2$	$-7/2$
			20	$-1/2$	$-5/2$
\multicolumn{3}{c|}{*Isolated points*}	21	$-1/2$	-2
6	-7	1	22	$-1/2$	$-4/3$
7	-7	3	23	$-1/2$	$-7/6$
8	$-5/2$	$-1/2$	24	$-1/2$	$-1/2$
9	-2	-2	25	$-1/2$	1
10	-2	1	26	2	-5
11	$-5/3$	$-10/3$	27	2	$-20/7$
12	$-5/3$	$-7/3$	28	2	$-15/7$

for $m \neq -1$ admits the general solution in parametric form

$$x = aC_2^2 \exp\left[2\int\left(\frac{8}{m+1}\tau^{m+1} + \tau^2 + C_1\right)^{-1/2} d\tau\right],$$

$$y = bC_2\tau \exp\left[\int\left(\frac{8}{m+1}\tau^{m+1} + \tau^2 + C_1\right)^{-1/2} d\tau\right],$$

where the free parameter of the equation A is expressed in terms of the solution parameters a and b as $A = (a/b^2)^{(m-1)/2}$.

For $m = -1$, the general solution in parametric form is

$$x = aC_2^2 \exp\left[2\int\left(8\ln|\tau| + \tau^2 + C_1\right)^{-1/2} d\tau\right],$$

$$y = bC_2\tau \exp\left[\int\left(8\ln|\tau| + \tau^2 + C_1\right)^{-1/2} d\tau\right],$$

where $A = b^2/a$.

3. $y''_{xx} + f(x)y = ay^{-3}$.

Ermakov's equation. Let $w = w(x)$ be a nontrivial solution of the second-order linear ODE: $w''_{xx} + f(x)w = 0$. The transformation $\xi = \int \frac{dx}{w^2}$, $z = \frac{y}{w}$ takes the original ODE to an autonomous equation of the form 2.3.1.1: $z''_{\xi\xi} = az^{-3}$.

General solution: $C_1 y^2 = aw^2 + w^2 \left(C_2 + C_1 \int \frac{dx}{w^2}\right)^2$.

▶ In what follows, f, g, h, and ψ are arbitrary composite functions of their arguments indicated in parentheses after the function name (the arguments can depend on x and y).

4. $y''_{xx} = f(ay + bx + c)$.

The substitution $w = ay + bx + c$ leads to an ODE of the form 2.3.1.1: $w''_{xx} = af(w)$.

5. $y''_{xx} = f(y + ax^2 + bx + c)$.

The substitution $w = y + ax^2 + bx + c$ leads to an ODE of the form 2.3.1.1: $w''_{xx} = f(w) + 2a$.

6. $y''_{xx} = x^{-1} f(yx^{-1})$.

Homogeneous second-order ODE. The transformation $t = -\ln|x|$, $z = y/x$ leads to an autonomous ODE of the form 2.3.2.1: $z''_{tt} - z'_t = f(z)$.

7. $y''_{xx} = x^{-3} f(yx^{-1})$.

The transformation $\xi = 1/x$, $w = y/x$ leads to an ODE of the form 2.3.1.1: $w''_{\xi\xi} = f(w)$.

8. $y''_{xx} = x^{-3/2} f(yx^{-1/2})$.

Having set $w = yx^{-1/2}$, we obtain the equation

$$\frac{d}{dx}(xw'_x)^2 = \frac{1}{2} ww'_x + 2f(w)w'_x,$$

the integration of which results in a first-order separable ODE.

General solution:

$$\int \left[C_1 + \tfrac{1}{4} w^2 + 2 \int f(w)\, dw \right]^{-1/2} dw = C_2 \pm \ln x.$$

9. $y''_{xx} = x^{k-2} f(x^{-k} y)$.

Generalized homogeneous ODE. The transformation $z = x^{-k} y$, $w = xy'_x/y$ leads to a first-order ODE:

$$z(w - k)w'_z = z^{-1} f(z) + w - w^2.$$

10. $y''_{xx} = yx^{-2} f(x^n y^m)$.

Generalized homogeneous ODE. The transformation $z = x^n y^m$, $w = xy'_x/y$ leads to a first-order ODE:

$$z(mw + n)w'_z = f(z) + w - w^2.$$

11. $y''_{xx} = y^{-3} f\left(\dfrac{y}{\sqrt{ax^2 + bx + c}} \right)$.

Setting $u(x) = y(ax^2 + bx + c)^{-1/2}$ and integrating the equation, we obtain a first-order separable ODE:

$$(ax^2 + bx + c)^2 (u'_x)^2 = (\tfrac{1}{4} b^2 - ac) u^2 + 2 \int u^{-3} f(u)\, du + C_1.$$

12. $y''_{xx} = e^{-ax} f(e^{ax} y)$.

The transformation $z = e^{ax} y$, $w = y'_x/y$ leads to a first-order ODE:

$$z(w + a)w'_z = z^{-1} f(z) - w^2.$$

13. $y''_{xx} = y f(e^{ax} y^m)$.

The transformation $z = e^{ax} y^m$, $w = y'_x/y$ leads to a first-order ODE:

$$z(mw + a)w'_z = f(z) - w^2.$$

14. $y''_{xx} = x^{-2} f(x^n e^{ay})$.

The transformation $z = x^n e^{ay}$, $w = xy'_x$ leads to a first-order ODE:
$$z(aw + n)w'_z = f(z) + w.$$

15. $y''_{xx} = \dfrac{\psi''_{xx}}{\psi} y + \psi^{-3} f\!\left(\dfrac{y}{\psi}\right)$, $\quad \psi = \psi(x)$.

The transformation $\xi = \int \dfrac{dx}{\psi^2}$, $w = \dfrac{y}{\psi}$ leads to an autonomous ODE of the form 2.3.1.1: $w''_{\xi\xi} = f(w)$.

General solution:
$$\int \left[C_1 + 2 \int f(w)\, dw \right]^{-1/2} dw = C_2 \pm \int \dfrac{dx}{\psi^2(x)}.$$

2.3.2. Equations of the Form $f(x, y) y''_{xx} = g(x, y, y'_x)$

1. $y''_{xx} - y'_x = f(y)$.

Autonomous ODE. It occurs in various problems of mass and heat transfer.

The substitution $w(y) = y'_x$ leads to a first-order Abel ODE of the form 2.1.3.2: $ww'_y - w = f(y)$. For solvable equations of the form in question, see handbooks by Polyanin & Zaitsev (2003, 2018).

Below are a few illustrative examples representing special cases where this ODE is integrable by quadrature.

1°. The equation
$$y''_{xx} - y'_x = -\dfrac{2(m+1)}{(m+3)^2} y \pm \dfrac{m+1}{2a^2} y^m, \qquad m \neq \pm 1, \quad m \neq -3,$$

has the general solution in parametric form
$$x = \dfrac{m+3}{m-1} \ln\!\left(aC_1^{1-m} \dfrac{m-1}{m+3} \int \dfrac{d\tau}{\sqrt{1 \pm \tau^{m+1}}} + C_2 \right),$$
$$y = C_1^2 \tau \left(aC_1^{1-m} \dfrac{m-1}{m+3} \int \dfrac{d\tau}{\sqrt{1 \pm \tau^{m+1}}} + C_2 \right)^{\frac{2}{m-1}}.$$

2°. The equation
$$y''_{xx} - y'_x = \pm 2a^2 y^{-1}$$

has the general solution in parametric form
$$x = -\ln\!\left[C_1 \int \exp(\pm \tau^2)\, d\tau + C_2 \right],$$
$$y = aC_1 \exp(\pm \tau^2) \left[C_1 \int \exp(\pm \tau^2)\, d\tau + C_2 \right]^{-1}.$$

3°. The equation
$$y''_{xx} - y'_x = -\tfrac{2}{9} y + \tfrac{16}{9} a^{3/2} y^{-1/2}$$

has the general solution in parametric form

$$x = -3\ln\{C_1 \exp(-\tau)[\exp(3\tau) + C_2 \sin(\sqrt{3}\tau)]\},$$

$$y = a\exp(2\tau)\frac{[2\exp(3\tau) - C_2 \sin(\sqrt{3}\tau) + \sqrt{3}C_2 \cos(\sqrt{3}\tau)]^2}{[\exp(3\tau) + C_2 \sin(\sqrt{3}\tau)]^2}.$$

4°. The equation

$$y''_{xx} - y'_x = -\tfrac{9}{100}y \pm \tfrac{9}{100}a^{8/3}y^{-5/3}$$

has the general solution in parametric form

$$x = -\tfrac{5}{4}\ln[\pm(\tau^4 - 6\tau^2 + 4C_1\tau - 3)] + C_2,$$

$$y = a(\tau^3 - 3\tau + C_1)^{3/2}[\pm(\tau^4 - 6\tau^2 + 4C_1\tau - 3)]^{-9/8}.$$

5°. The equation

$$y''_{xx} - y'_x = -\tfrac{3}{16}y - \tfrac{3}{64}a^{8/3}y^{-5/3}$$

has the general solution in parametric form

$$x = C_1 - 2\ln[\sin\tau \cosh(\tau + C_2) + \cos\tau \sinh(\tau + C_2)],$$

$$y = a[\tan\tau + \tanh(\tau + C_2)]^{-3/2}.$$

2. $y''_{xx} + f(y)y'_x + g(y) = 0.$

Liénard's equation. The substitution $w(y) = y'_x$ leads to a first-order Abel ODE of the form 2.1.3.3:

$$ww'_y + f(y)w + g(y) = 0.$$

For solvable ODEs of the form in question, see handbooks by Polyanin & Zaitsev (2003, 2018).

3. $y''_{xx} + [ay + f(x)]y'_x + f'_x(x)y = 0.$

Integrating yields a first-order ODE (a Riccati equation):

$$y'_x + f(x)y + \tfrac{1}{2}ay^2 = C.$$

4. $y''_{xx} + [2ay + f(x)]y'_x + af(x)y^2 = g(x).$

On setting $u = y'_x + ay^2$, we obtain a first-order linear ODE: $u'_x + f(x)u = g(x).$

5. $y''_{xx} = ay'_x + e^{2ax}f(y).$

For $a = 0$, this is an ODE of the form 2.3.1.1.

General solution for $a \neq 0$:

$$\int \Big[C_1 + 2\int f(y)\,dy\Big]^{-1/2} dy = C_2 \pm \frac{1}{a}e^{ax}.$$

6. $y''_{xx} = f(y)y'_x.$

General solution:

$$\int \frac{dy}{F(y) + C_1} = C_2 + x, \quad \text{where} \quad F(y) = \int f(y)\,dy.$$

7. $y''_{xx} = [e^{\alpha x} f(y) + \alpha] y'_x$.

The substitution $w(y) = e^{-\alpha x} y'_x$ leads to a first-order separable ODE: $w'_y = f(y)$.

General solution:
$$\int \frac{dy}{F(y) + C_1} = C_2 + \frac{1}{\alpha} e^{\alpha x}, \quad \text{where} \quad F(y) = \int f(y)\, dy.$$

8. $xy''_{xx} = ky'_x + x^{2k+1} f(y)$.

$1°$. General solution for $k \neq -1$:
$$\int \left[C_1 + 2 \int f(y)\, dy \right]^{-1/2} dy = \pm \frac{x^{k+1}}{k+1} + C_2.$$

$2°$. General solution for $k = -1$:
$$\int \left[C_1 + 2 \int f(y)\, dy \right]^{-1/2} dy = \pm \ln|x| + C_2.$$

9. $xy''_{xx} = f(y) y'_x$.

The substitution $w(y) = xy'_x/y$ leads to a first-order linear ODE: $yw'_y = -w + 1 + f(y)$.

10. $xy''_{xx} = [x^k f(y) + k - 1] y'_x$.

General solution:
$$\int \frac{dy}{F(y) + C_1} = C_2 + \frac{1}{k} x^k, \quad \text{where} \quad F(y) = \int f(y)\, dy.$$

11. $x^2 y''_{xx} + xy'_x = f(y)$.

The substitution $x = \pm e^t$ leads to an autonomous ODE of the form 2.3.1.1: $y''_{tt} = f(y)$.

12. $(ax^2 + b) y''_{xx} + axy'_x + f(y) = 0$.

The substitution $\xi = \int \frac{dx}{\sqrt{ax^2 + b}}$ leads to an autonomous ODE of the form 2.3.1.1: $y''_{\xi\xi} + f(y) = 0$.

13. $y''_{xx} = f(y) y'_x + g(x)$.

Integrating yields a first-order ODE:
$$y'_x = \int f(y)\, dy + \int g(x)\, dx + C.$$

14. $xy''_{xx} + (k+1) y'_x = x^{k-1} f(yx^k)$.

The transformation $\xi = x^k$, $w = yx^k$ leads to an autonomous ODE of the form 2.3.1.1: $w''_{\xi\xi} = k^{-2} f(w)$.

15. $gy''_{xx} + \frac{1}{2} g'_x y'_x = f(y)$, $\quad g = g(x)$.

Integrating yields a first-order separable ODE: $g(x)(y'_x)^2 = 2 \int f(y)\, dy + C_1$.

General solution for $g(x) \geq 0$:
$$\int \left[C_1 + 2 \int f(y)\, dy \right]^{-1/2} dy = C_2 \pm \int \frac{dx}{\sqrt{g(x)}}.$$

16. $y''_{xx} = -ay'_x + e^{ax}f(ye^{ax})$.

The transformation $\xi = e^{ax}$, $w = ye^{ax}$ leads to an autonomous ODE of the form 2.3.1.1: $w''_{\xi\xi} = a^{-2}f(w)$.

17. $xy''_{xx} = f(x^k e^{ay})y'_x$.

The transformation $z = x^k e^{ay}$, $w = xy'_x$ leads to the following first-order separable ODE: $z(aw + k)w'_z = [f(z) + 1]w$.

18. $x^2 y''_{xx} + xy'_x = f(x^k e^{ay})$.

The transformation $z = x^k e^{ay}$, $w = xy'_x$ leads to the following first-order separable ODE: $z(aw + k)w'_z = f(z)$.

19. $yy''_{xx} + (y'_x)^2 = f(x)$.

General solution:
$$y^2 = C_1 x + C_2 + 2\int_0^x (x-t)f(t)\,dt,$$

where C_1 and C_2 are arbitrary constants.

20. $yy''_{xx} + (y'_x)^2 + f(x)yy'_x + g(x) = 0$.

The substitution $u = y^2$ leads to a linear ODE,
$$u''_{xx} + f(x)u'_x + 2g(x) = 0,$$

which can be reduced by the change of variable $w(x) = u'_x$ to a first-order linear ODE.

21. $yy''_{xx} - (y'_x)^2 + f(x)yy'_x + g(x)y^2 = 0$.

The substitution $u = y'_x/y$ leads to a first-order linear ODE:
$$u'_x + f(x)u + g(x) = 0.$$

22. $yy''_{xx} + a(y'_x)^2 + f(x)yy'_x + g(x)y^2 = 0$.

The substitution $w = y^{a+1}$ leads to a linear ODE:
$$w''_{xx} + f(x)w'_x + (a+1)g(x)w = 0.$$

23. $yy''_{xx} + (1-k)(y'_x)^2 = f(x)y^k$.

For $k = 2$, this is an ODE of the form 2.3.2.21.

General solution for $k \neq 2$:
$$y^{2-k} = C_1 x + C_2 + (2-k)\int_0^x (x-t)f(t)\,dt,$$

where C_1 and C_2 are arbitrary constants.

24. $yy''_{xx} - k(y'_x)^2 + f(x)y^2 + ay^{4k-2} = 0$.

1°. For $k = 1$, this is an ODE of the form 2.3.2.22.

$2°$. For $k \neq 1$, the substitution $w = y^{1-k}$ leads to Ermakov's equation 2.3.1.3:

$$w''_{xx} + (1-k)f(x)w + a(1-k)w^{-3} = 0.$$

25. $yy''_{xx} - k(y'_x)^2 + f(x)y^2 + g(x)y^{k+1} = 0.$

The substitution $w = y^{1-k}$ leads to a nonhomogeneous linear ODE:

$$w''_{xx} + (1-k)f(x)w + (1-k)g(x) = 0.$$

26. $yy''_{xx} = f(x)(y'_x)^2.$

The substitution $w(x) = xy'_x/y$ leads to a Bernoulli equation 2.1.1.4:

$$xw'_x = w + [f(x) - 1]w^2.$$

27. $y''_{xx} - a(y'_x)^2 = f(x)e^{ay}.$

General solution in implicit form for $a \neq 0$:

$$e^{-ay} = C_1 x + C_2 - a \int_0^x (x-t)f(t)\,dt,$$

where C_1 and C_2 are arbitrary constants.

28. $y''_{xx} - a(y'_x)^2 + f(x)e^{ay} + g(x) = 0.$

The substitution $w = e^{-ay}$ leads to a nonhomogeneous linear ODE:

$$w''_{xx} - ag(x)w = af(x).$$

29. $y''_{xx} - a(y'_x)^2 + be^{4ay} + f(x) = 0.$

The substitution $w = e^{-ay}$ leads to Ermakov's equation 2.3.1.3:

$$w''_{xx} - af(x)w = abw^{-3}.$$

30. $y''_{xx} + a(y'_x)^2 - \frac{1}{2}y'_x = e^x f(y).$

The substitution $w(y) = e^{-x}(y'_x)^2$ leads to a first-order linear ODE: $w'_y + 2aw = 2f(y).$

31. $y''_{xx} + \alpha(y'_x)^2 = [e^{\beta x}f(y) + \beta]y'_x.$

General solution:

$$\int \frac{e^{\alpha y}\,dy}{F(y) + C_1} = C_2 + \frac{1}{\beta}e^{\beta x}, \quad \text{where} \quad F(y) = \int e^{\alpha y}f(y)\,dy.$$

32. $y''_{xx} + f(y)(y'_x)^2 + g(y) = 0.$

The substitution $z(y) = (y'_x)^2$ leads to a first-order linear ODE:

$$z'_y + 2f(y)z + 2g(y) = 0.$$

33. $y''_{xx} + f(y)(y'_x)^2 - \frac{1}{2}y'_x = e^x g(y).$

The substitution $w(y) = e^{-x}(y'_x)^2$ leads to a first-order linear ODE:

$$w'_y + 2f(y)w = 2g(y).$$

34. $y''_{xx} = f(y)(y'_x)^2 + g(x)y'_x$.

Dividing by y'_x, we obtain an exact differential equation. Its solution follows from the first-order ODE:
$$\ln|y'_x| = \int f(y)\,dy + \int g(x)\,dx + C.$$

Solving this equation for y'_x, we arrive at a separable ODE. In addition, $y = C_1$ is a singular solution, with C_1 being an arbitrary constant.

35. $y''_{xx} = xf(y)(y'_x)^3$.

Taking y to be the independent variable, we obtain a linear ODE with respect to $x = x(y)$: $x''_{yy} = -f(y)x$.

36. $y''_{xx} = f(x)g(y)h(y'_x)$.

1°. In the special case of power-law functions $f(x)$, $g(y)$, and $g(u)$, this is the generalized Emden–Fowler equation, $y''_{xx} = ax^k y^m (y'_x)^n$.

2°. Solutions of many nonlinear ODEs of this type are given in handbooks by Polyanin & Zaitsev (2003, 2018).

37. $y''_{xx} = \dfrac{y}{x^2} f\!\left(\dfrac{xy'_x}{y}\right)$.

The substitution $w(x) = xy'_x/y$ leads to a first-order separable ODE:
$$xw'_x = f(w) + w - w^2.$$

38. $gy''_{xx} + \tfrac{1}{2}g'_x y'_x = f(y)h(y'_x\sqrt{g})$, $\quad g = g(x)$.

The substitution $w(y) = y'_x\sqrt{g}$ leads to a first-order separable ODE: $ww'_y = f(y)h(w)$.

39. $y''_{xx} = f(y'^2_x + ay)$.

The substitution $w(y) = (y'_x)^2 + ay$ leads to a first-order separable ODE: $w'_y = 2f(w) + a$.

2.3.3. ODEs of General Form Containing Arbitrary Functions of Two Arguments

1. $y''_{xx} = F(x, y'_x)$.

The right-hand side of this ODE does not explicitly depend on the desired function y.

The substitution $u(x) = y'_x$ leads to a first-order ODE: $u'_x = F(x, u)$.

2. $y''_{xx} = F(y, y'_x)$.

Autonomous ODE. The right-hand side of this equation does not explicitly depend on the independent variable x.

The substitution $u(y) = y'_x$ leads to a first-order ODE: $uu'_y = F(y, u)$.

3. $y''_{xx} = F(ax + by, y'_x)$.

The substitution $bw = ax + by$ leads to an autonomous ODE of the form 2.3.3.2:
$$w''_{xx} = F(bw, w'_x - a/b).$$

4. $y''_{xx} = yF(x, y'_x/y)$.

ODE homogeneous in the dependent variable y. This ODE remains unchanged under scaling of the dependent variable, $y \Longrightarrow cy$, where c is an arbitrary nonzero number.

The substitution $u(x) = y'_x/y$ leads to a first-order ODE: $u'_x + u^2 = F(x, u)$.

5. $y''_{xx} = x^{-2}F(y, xy'_x)$.

ODE homogeneous in the independent variable x. This ODE remains unchanged under scaling of the independent variable, $x \Longrightarrow cx$, where c is an arbitrary nonzero number.

The substitution $t = \ln|x|$ leads to an autonomous equation of the form 2.3.3.2 and the substitution $u(y) = xy'_x$ leads to a first-order ODE: $uu'_y - u = F(y, u)$.

6. $y''_{xx} = \frac{1}{x}F\left(\frac{y}{x}, y'_x\right)$.

ODE homogeneous in both variables. This ODE is invariant under simultaneous same scaling (dilatation) of the independent and dependent variables, $x \Longrightarrow cx$ and $y \Longrightarrow cy$, where c is an arbitrary nonzero number.

The transformation
$$z = y/x, \quad w = y'_x$$
leads to a first-order ODE of the form $(w - z)w'_z = F(z, w)$.

7. $y''_{xx} = \frac{1}{ax + by + c}F\left(\frac{ax + by + c}{\alpha x + \beta y + \gamma}, y'_x\right)$.

1°. For $a\beta - b\alpha \neq 0$, the transformation
$$z = x - x_0, \quad w = y - y_0,$$
where x_0 and y_0 are the constants determined by the linear algebraic system of equations
$$ax_0 + by_0 + c = 0, \quad \alpha x_0 + \beta y_0 + \gamma = 0,$$
leads to a homogeneous ODE of the form 2.3.3.6:
$$w''_{zz} = \frac{1}{z}\Phi\left(\frac{w}{z}, w'_z\right), \quad \text{where} \quad \Phi(\xi, u) = \frac{1}{a + b\xi}F\left(\frac{a + b\xi}{\alpha + \beta\xi}, u\right).$$

2°. For $a\beta - b\alpha = 0$, the substitution $bw = ax + by + c$ leads to an autonomous ODE of the form 2.3.3.2.

8. $y''_{xx} = x^{-k-2}F(x^k y, x^{k+1} y'_x)$.

Generalized homogeneous ODE. This ODE remains unchanged under simultaneous scaling of the independent and dependent variables in accordance with the rules $x \Longrightarrow cx$ and $y \Longrightarrow c^{-k}y$, where c is an arbitrary nonzero constant.

1°. The transformation $t = \ln x$, $z = x^k y$ leads to an ODE of the form 2.3.3.2:
$$z''_{tt} - (2k + 1)z'_t + k(k + 1)z = F(z, z'_t - kz).$$

2°. The transformation $z = x^k y$, $u = x^{k+1} y'_x$ leads to a first-order ODE:
$$(u + kz)u'_z = (k + 1)u + F(z, u).$$

3°. The ODE admits a particular solution of the form:
$$y = ax^{-k},$$
where a is root of the algebraic (transcendental) equation $ak(k+1) - F(a, -ak) = 0$.

9. $y''_{xx} = \dfrac{y}{x^2} F\!\left(x^p y^q,\ \dfrac{x}{y} y'_x\right).$

Generalized homogeneous ODE. For $q \neq 0$, this ODE is an alternative form of equation 2.3.3.8 with $k = p/q$.

1°. The transformation $z = x^p y^q$, $w = xy'_x/y$ leads to a first-order ODE:
$$z(qw + p)w'_z = F(z, w) + w - w^2.$$

2°. The original ODE admits a particular solution of the form:
$$y = ax^k, \quad k = -p/q,$$
where a is root of the algebraic (transcendental) equation $k(k-1) - F(a^q, k) = 0$.

10. $y''_{xx} = a^2 y + F(x, y'_x + ay).$

1°. The substitution $w = y'_x + ay$ leads to a first-order ODE: $w'_x = aw + F(x, w)$.

2°. For $F(x, w) = f(w)$, the equation admits a particular solution of the form:
$$y = Ce^{-ax} + k,$$
where C is an arbitrary constant, and k is a root of the algebraic (transcendental) equation $a^2 k + f(ak) = 0$.

11. $y''_{xx} = (a^2 x^2 + a)y + F(x, y'_x - axy).$

The substitution $w = y'_x - axy$ leads to a first-order ODE: $w'_x = -axw + F(x, w)$.

12. $y''_{xx} = F\!\left(x,\ y'_x - \dfrac{y}{x}\right).$

The substitution $w(x) = y'_x - \dfrac{y}{x}$ leads to a first-order ODE: $xw'_x = -w + xF(x, w)$.

13. $y''_{xx} = F(x,\ xy'_x - y).$

The substitution $w(x) = xy'_x - y$ leads to a first-order ODE: $w'_x = xF(x, w)$.

14. $y''_{xx} = x^{-2} F(y,\ xy'_x - y).$

The substitution $w(y) = xy'_x - y$ leads to a first-order ODE: $(y + w)w'_y = F(y, w)$.

15. $xy''_{xx} + (a+1)y'_x = F(x,\ xy'_x + ay).$

The substitution $w = xy'_x + ay$ leads to a first-order ODE: $w'_x = F(x, w)$.

16. $x^2 y''_{xx} = 2y + F(x,\ xy'_x + y).$

The substitution $w = xy'_x + y$ leads to a first-order ODE: $xw'_x = 2w + F(x, w)$.

17. $x^2 y''_{xx} = a(a+1)y + F(x,\ xy'_x + ay).$

The substitution $w = xy'_x + ay$ leads to a first-order ODE: $xw'_x = (a+1)w + F(x, w)$.

18. $y''_{xx} = 2ayy'_x + F(x, y'_x - ay^2)$.

The substitution $w = y'_x - ay^2$ leads to a first-order ODE: $w'_x = F(x, w)$.

19. $y''_{xx} = e^{-\alpha x} F(e^{\alpha x} y, e^{\alpha x} y'_x)$.

ODE invariant under "translation–scaling" transformation. This ODE remains unchanged under the simultaneous translation and scaling of the variables, $x \to x + b$ and $y \Longrightarrow cy$, where $c = e^{-\alpha b}$ and b is an arbitrary constant.

1°. The substitution $z(x) = e^{\alpha x} y$ leads to a second-order autonomous ODE of the form 2.3.3.2:
$$z''_{xx} - 2\alpha z'_x + \alpha^2 z = F(z, z'_x - \alpha z).$$

2°. The transformation $z = e^{\alpha x} y$, $u = e^{\alpha x} y'_x$ leads to a first-order ODE of the form:
$$(u + \alpha z) u'_z = \alpha u + F(z, u).$$

3°. The original ODE admits a particular solution of the form $y = Ae^{-\alpha x}$, where A is root of the algebraic (transcendental) equation $A\alpha^2 - F(A, -A\alpha) = 0$.

20. $y''_{xx} = y F(e^{\beta x} y^m, y'_x/y)$.

ODE invariant under "translation–scaling" transformation. The transformation $z = e^{\beta x} y^m$, $w = y'_x/y$ leads to a first-order ODE:
$$(mw + \beta) w'_z = F(z, w) - w^2.$$

For $m \neq 0$, this ODE is an alternative form of equation 2.3.3.19 with $\alpha = \beta/m$.

The original ODE admits a particular solution of the form $y = Ae^{-\beta x/m}$, where A is root of the algebraic (transcendental) equation $(\beta/m)^2 - F(A^m, -\beta/m) = 0$.

21. $y''_{xx} = x^{-2} F(x^m e^{\alpha y}, xy'_x)$.

ODE invariant under "scaling–translation" transformation. This ODE remains unchanged under simultaneous scaling and translation of variables, $x \Longrightarrow bx$ and $y \Longrightarrow y + c$, where $b = e^{-c\alpha/m}$ and c is an arbitrary constant.

The transformation $z = x^m e^{\alpha y}$, $w = xy'_x$ leads to a first-order ODE:
$$z(\alpha w + m) w'_z = F(z, w) + w.$$

The original ODE admits a particular solution of the form $y = -(m/\alpha) \ln(Ax)$, where A is a constant.

22. $y''_{xx} = e^{2ay} F(xe^{ay}, e^{-ay} y'_x)$.

The transformation $z = xe^{ay}$, $w = e^{-ay} y'_x$ leads to a first-order ODE:
$$(azw + 1) w'_z = F(z, w) - aw^2.$$

23. $y''_{xx} = ae^y y'_x + F(x, y'_x - ae^y)$.

The substitution $w = y'_x - ae^y$ leads to a first-order ODE: $w'_x = F(x, w)$.

24. $y''_{xx} = (e^{2x} + e^x)y + F(x, y'_x - e^x y)$.

The substitution $w = y'_x - e^x y$ leads to a first-order ODE: $w'_x = -e^x w + F(x, w)$.

25. $y''_{xx} = F(x, y'_x \sinh x - y \cosh x) + y.$

The substitution $w = y'_x \sinh x - y \cosh x$ leads to a first-order ODE of the form: $w'_x = F(x, w) \sinh x.$

26. $y''_{xx} = F(x, y'_x \cosh x - y \sinh x) + y.$

The substitution $w = y'_x \cosh x - y \sinh x$ leads to a first-order ODE of the form: $w'_x = F(x, w) \cosh x.$

27. $y''_{xx} = x^{-2} F(ay + b \ln x, xy'_x).$

The transformation $z = ay + b \ln x$, $w = xy'_x$ leads to a first-order ODE: $(aw + b)w'_z = F(z, w) + w.$

28. $y''_{xx} = yF(ax + b \ln y, y'_x/y).$

The transformation $z = ax + b \ln y$, $w = y'_x/y$ leads to a first-order ODE: $(bw + a)w'_z = F(z, w) - w^2.$

29. $y''_{xx} = F(x, y'_x \sin x - y \cos x) - y.$

The substitution $w = y'_x \sin x - y \cos x$ leads to a first-order ODE: $w'_x = F(x, w) \sin x.$

30. $y''_{xx} = F(x, y'_x \cos x + y \sin x) - y.$

The substitution $w = y'_x \cos x + y \sin x$ leads to a first-order ODE: $w'_x = F(x, w) \cos x.$

31. $y''_{xx} = (\varphi^2 + \varphi'_x)y + F(x, y'_x - \varphi y), \qquad \varphi = \varphi(x).$

The substitution $w = y'_x - \varphi y$ leads to a first-order ODE: $w'_x = -\varphi w + F(x, w).$

32. $y''_{xx} = \dfrac{\varphi''_{xx}}{\varphi} y + F\!\left(x, y'_x - \dfrac{\varphi'_x}{\varphi} y\right), \qquad \varphi = \varphi(x).$

The substitution $w = y'_x - \dfrac{\varphi'_x}{\varphi} y$ leads to a first-order ODE: $w'_x = -\dfrac{\varphi'_x}{\varphi} w + F(x, w).$

33. $f^2 y''_{xx} + f f'_x y'_x = \Phi(y, f y'_x), \qquad f = f(x).$

The substitution $w(y) = f y'_x$ leads to a first-order ODE: $w w'_y = \Phi(y, w).$

2.4. Higher-Order Ordinary Differential Equations

2.4.1. Higher-Order Linear Ordinary Differential Equations

▶ **Preliminary remarks.**

Throughout this section, we denote higher derivatives by $y_x^{(n)}$ to mean $d^n y/dx^n$.

$1°.$ The *trivial solution* $y = 0$ is a particular solution of any linear homogeneous ODE.

$2°.$ The general solution of a homogeneous linear equation of the nth-order

$$f_n(x)y_x^{(n)} + f_{n-1}(x)y_x^{(n-1)} + \cdots + f_1(x)y'_x + f_0(x)y = 0 \tag{1}$$

has the form

$$y = C_1 y_1(x) + C_2 y_2(x) + \cdots + C_n y_n(x). \tag{2}$$

Here $y_1(x), y_2(x), \ldots, y_n(x)$ make up a fundamental set of solutions (the y_k are linearly independent solutions; $y_k \not\equiv 0$); $C_1, C_2, \ldots, C_n$ are arbitrary constants.

$3°$. Let $y_0 = y_0(x)$ be a nontrivial particular solution of equation (1). Then the substitution

$$y = y_0(x) \int z(x)\, dx \qquad (3)$$

leads to a linear $(n-1)$st-order ODE for $z(x)$.

Note that instead of substitution (3), one can use the substitution

$$z(x) = \varphi(x)\big[y_0(x)y'_x - y'_0(x)y\big],$$

where $\varphi(x)$ is any given function.

$4°$. Let $y_1 = y_1(x)$ and $y_2 = y_2(x)$ be two nontrivial linearly independent particular solutions of equation (1). Then the substitution

$$y = y_1 \int y_2 w\, dx - y_2 \int y_1 w\, dx \qquad (4)$$

leads to a linear $(n-2)$nd-order ODE for $w = w(x)$.

▶ **Linear equations with constant coefficients.**

1. $y'''_{xxx} + ay = 0.$

General solution:

$$y = \begin{cases} C_1 + C_2 x + C_3 x^2 & \text{if } a = 0, \\ C_1 \exp(-kx) + C_2 \exp\left(\tfrac{1}{2}kx\right)\cos\left(\tfrac{\sqrt{3}}{2}kx\right) + C_3 \exp\left(\tfrac{1}{2}kx\right)\sin\left(\tfrac{\sqrt{3}}{2}kx\right) & \text{if } a \neq 0, \end{cases}$$

where C_1, C_2, and C_3 are arbitrary constants, and $k = a^{1/3}$.

2. $y'''_{xxx} + a_2 y''_{xx} + a_1 y'_x + a_0 y = 0.$

Third-order linear homogeneous ODE with constant coefficients.

Denote $P(\lambda) = \lambda^3 + a_2 \lambda^2 + a_1 \lambda + a_0$.

$1°$. Let the characteristic polynomial $P(\lambda)$ factorize as

$$P(\lambda) = (\lambda - \lambda_1)(\lambda - \lambda_2)(\lambda - \lambda_3),$$

where λ_1, λ_2, and λ_3 are real numbers.

General solution:

$$y = \begin{cases} C_1 e^{\lambda_1 x} + C_2 e^{\lambda_2 x} + C_3 e^{\lambda_3 x} & \text{if all roots } \lambda_k \text{ are different}, \\ (C_1 + C_2 x)e^{\lambda_1 x} + C_3 e^{\lambda_3 x} & \text{if } \lambda_1 = \lambda_2 \neq \lambda_3, \\ (C_1 + C_2 x + C_3 x^2)e^{\lambda_1 x} & \text{if } \lambda_1 = \lambda_2 = \lambda_3. \end{cases}$$

$2°$. Let $P(\lambda) = (\lambda - \lambda_1)(\lambda^2 + 2b_1 \lambda + b_0)$, where $b_1^2 < b_0$.

General solution:

$$y = C_1 e^{\lambda_1 x} + e^{-b_1 x}(C_2 \cos \mu x + C_3 \sin \mu x), \text{ where } \mu = \sqrt{b_0 - b_1^2}.$$

3. $y''''_{xxxx} + ay = 0$.

   1°. General solution for $a = 0$:
   $$y = C_1 + C_2 x + C_3 x^2 + C_4 x^3.$$

   2°. General solution for $a = 4k^4 > 0$:
   $$y = C_1 \cosh kx \cos kx + C_2 \cosh kx \sin kx + C_3 \sinh kx \cos kx + C_4 \sinh kx \sin kx.$$

   3°. General solution for $a = -s^4 < 0$:
   $$y = C_1 \cos sx + C_2 \sin sx + C_3 \cosh sx + C_4 \sinh sx.$$

4. $y''''_{xxxx} + (a+b)y''_{xx} + aby = 0$.

   1°. General solution for $a = b$:
   $$y = \begin{cases} (C_1 + C_2 x)\cos(\sqrt{a}\,x) + (C_3 + C_4 x)\sin(\sqrt{a}\,x) & \text{if } a > 0, \\ (C_1 + C_2 x)\exp(\sqrt{|a|}\,x) + (C_3 + C_4 x)\exp(-\sqrt{|a|}\,x) & \text{if } a < 0. \end{cases}$$

   2°. General solution for $a > 0$, $b > 0$, and $a \neq b$:
   $$y = C_1 \cos(\sqrt{a}\,x) + C_2 \sin(\sqrt{a}\,x) + C_3 \cos(\sqrt{b}\,x) + C_4 \sin(\sqrt{b}\,x).$$

   3°. General solution for $a > 0$, $b < 0$, and $a \neq b$:
   $$y = C_1 \cos(\sqrt{a}\,x) + C_2 \sin(\sqrt{a}\,x) + C_3 \exp(\sqrt{|b|}\,x) + C_4 \exp(-\sqrt{|b|}\,x).$$

   4°. General solution for $a < 0$, $b > 0$, and $a \neq b$:
   $$y = C_1 \exp(\sqrt{|a|}\,x) + C_2 \exp(-\sqrt{|a|}\,x) + C_3 \cos(\sqrt{b}\,x) + C_4 \sin(\sqrt{b}\,x).$$

   5°. General solution for $a < 0$, $b < 0$, and $a \neq b$:
   $$y = C_1 \exp(\sqrt{|a|}\,x) + C_2 \exp(-\sqrt{|a|}\,x) + C_3 \exp(\sqrt{|b|}\,x) + C_4 \exp(-\sqrt{|b|}\,x).$$

5. $y''''_{xxxx} + a_3 y'''_{xxx} + a_2 y''_{xx} + a_1 y'_x + a_0 y = 0$.

   Fourth-order linear homogeneous ODE with constant coefficients. For $a_0 = 0$, the substitution $w(x) = y'_x$ leads to a third-order ODE. Let
   $$P(\lambda) = \lambda^4 + a_3 \lambda^3 + a_2 \lambda^2 + a_1 \lambda + a_0$$
   denote the characteristic polynomial of the equation and let $a_0 \neq 0$.

   1°. Let P be factorizable, so that $P(\lambda) = (\lambda - \lambda_1)(\lambda - \lambda_2)(\lambda - \lambda_3)(\lambda - \lambda_4)$, where $\lambda_1, \lambda_2, \lambda_3$, and λ_4 are real numbers. The following cases are possible:

   a) $\lambda_1, \ldots, \lambda_4$ are all different, then the general solution is
   $$y = C_1 e^{\lambda_1 x} + C_2 e^{\lambda_2 x} + C_3 e^{\lambda_3 x} + C_4 e^{\lambda_4 x};$$

   b) $\lambda_1 = \lambda_2$; λ_3 and λ_4 are different and not equal to λ_1, then the general solution is
   $$y = (C_1 + C_2 x)e^{\lambda_1 x} + C_3 e^{\lambda_3 x} + C_4 e^{\lambda_4 x};$$

   c) $\lambda_1 = \lambda_2 = \lambda_3 \neq \lambda_4$, then the general solution is
   $$y = (C_1 + C_2 x + C_3 x^2)e^{\lambda_1 x} + C_4 e^{\lambda_4 x};$$

   d) $\lambda_1 = \lambda_2 = \lambda_3 = \lambda_4$, then the general solution is
   $$y = (C_1 + C_2 x + C_3 x^2 + C_4 x^3)e^{\lambda_1 x}.$$

2°. Suppose that $P(\lambda) = (\lambda - \lambda_1)(\lambda - \lambda_2)(\lambda^2 + 2b_1\lambda + b_0)$, where λ_1 and λ_2 are real numbers, and $b_1^2 - b_0 < 0$. The following cases are possible:

a) $\lambda_1 \neq \lambda_2$, then the general solution is

$$y = C_1 e^{\lambda_1 x} + C_2 e^{\lambda_2 x} + e^{-b_1 x}[C_3 \cos(\mu x) + C_4 \sin(\mu x)], \quad \mu = \sqrt{b_0 - b_1^2};$$

b) $\lambda_1 = \lambda_2$, then the general solution is

$$y = (C_1 + C_2 x)e^{\lambda_1 x} + e^{-b_1 x}[C_3 \cos(\mu x) + C_4 \sin(\mu x)], \quad \mu = \sqrt{b_0 - b_1^2}.$$

3°. Suppose that $P(\lambda) = (\lambda^2 + 2b_1\lambda + b_0)(\lambda^2 + 2\beta_1\lambda + \beta_0)$, where $b_1^2 - b_0 < 0$ and $\beta_1^2 - \beta_0 < 0$. The following cases are possible:

a) $(b_1 - \beta_1)^2 + (b_0 - \beta_0)^2 \neq 0$, then the general solution is

$$y = e^{-b_1 x}[C_1 \cos(\mu x) + C_2 \sin(\mu x)] + e^{-\beta_1 x}[C_3 \cos(\nu x) + C_4 \sin(\nu x)],$$

where $\mu = \sqrt{b_0 - b_1^2}$, $\nu = \sqrt{\beta_0 - \beta_1^2}$;

b) $b_1 = \beta_1$ and $b_0 = \beta_0$, then the general solution is

$$y = e^{-b_1 x}[(C_1 + C_2 x) \cos(\mu x) + (C_3 + C_4 x) \sin(\mu x)], \quad \mu = \sqrt{b_0 - b_1^2}.$$

6. $y_x^{(n)} = f(x)$.

Simplest linear nonhomogeneous nth-order ordinary differential equation.

General solution:

$$y = \sum_{k=0}^{n-1} C_k x^k + \int_{x_0}^{x} \frac{(x-t)^{n-1}}{(n-1)!} f(t)\, dt,$$

where C_k are arbitrary constants and x_0 is any number for which the integral on the right-hand side makes sense.

7. $y_x^{(n)} = ay$.

1°. General solution for $a = 0$:

$$y = \sum_{k=1}^{n} C_{k-1} x^{k-1},$$

where C_{k-1} ($k = 1, 2, \ldots, n$) are arbitrary constants.

2°. General solution for $a > 0$:

$$y = \sum_{k=1}^{k \leq \frac{1}{2}(n+2)} \exp(\alpha_k x)\left[A_k \cos(\beta_k x) + B_k \sin(\beta_k x)\right],$$

$$\alpha_k = a^{1/n} \cos \frac{2(k-1)\pi}{n}, \quad \beta_k = a^{1/n} \sin \frac{2(k-1)\pi}{n},$$

where A_k and B_k are arbitrary constants.

3°. General solution for $a < 0$:

$$y = \sum_{k=1}^{k \leq \frac{1}{2}(n+1)} \exp(\alpha_k x)\big[A_k \cos(\beta_k x) + B_k \sin(\beta_k x)\big],$$

$$\alpha_k = |a|^{1/n} \cos \frac{(2k-1)\pi}{n}, \qquad \beta_k = |a|^{1/n} \sin \frac{(2k-1)\pi}{n},$$

where A_k and B_k are arbitrary constants.

8. $y_x^{(n)} + a_{n-1} y_x^{(n-1)} + \cdots + a_1 y_x' + a_0 y = 0.$

Linear homogeneous nth-order ODE of general form with constant coefficients.

The general solution of this equation is determined by the roots of the characteristic equation

$$P(\lambda) = 0, \qquad \text{where} \quad P(\lambda) = \lambda^n + a_{n-1} \lambda^{n-1} + \cdots + a_1 \lambda + a_0. \tag{1}$$

The following cases are possible:

1°. All roots $\lambda_1, \lambda_2, \ldots, \lambda_n$ of the characteristic equation (1) are real and distinct. Then the general solution of the homogeneous linear ODE has the form

$$y = C_1 \exp(\lambda_1 x) + C_2 \exp(\lambda_2 x) + \cdots + C_n \exp(\lambda_n x). \tag{2}$$

2°. There are m equal real roots $\lambda_1 = \lambda_2 = \cdots = \lambda_m$ ($m \leq n$), and the other roots are real and distinct. In this case, the general solution is given by

$$y = \exp(\lambda_1 x)(C_1 + C_2 x + \cdots + C_m x^{m-1})$$
$$+ C_{m+1} \exp(\lambda_{m+1} x) + C_{m+2} \exp(\lambda_{m+2} x) + \cdots + C_n \exp(\lambda_n x).$$

3°. There are m equal complex conjugate roots $\lambda = \alpha \pm i\beta$ ($2m \leq n$), and the other roots are real and distinct. In this case, the general solution is

$$y = \exp(\alpha x) \cos(\beta x)(A_1 + A_2 x + \cdots + A_m x^{m-1})$$
$$+ \exp(\alpha x) \sin(\beta x)(B_1 + B_2 x + \cdots + B_m x^{m-1})$$
$$+ C_{2m+1} \exp(\lambda_{2m+1} x) + C_{2m+2} \exp(\lambda_{2m+2} x) + \cdots + C_n \exp(\lambda_n x),$$

where $A_1, \ldots, A_m, B_1, \ldots, B_m, C_{2m+1}, \ldots, C_n$ are arbitrary constants.

4°. In the general case where there are r different roots $\lambda_1, \lambda_2, \ldots, \lambda_r$ of multiplicities $m_1, m_2, \ldots, m_r$, respectively, the left-hand side of the characteristic equation (1) can be represented as the product

$$P(\lambda) = (\lambda - \lambda_1)^{m_1} (\lambda - \lambda_2)^{m_2} \ldots (\lambda - \lambda_r)^{m_r},$$

where $m_1 + m_2 + \cdots + m_r = n$. The general solution of the original equation is given by the formula

$$y = \sum_{k=1}^{r} \exp(\lambda_k x)(C_{k,0} + C_{k,1} x + \cdots + C_{k,m_k-1} x^{m_k-1}),$$

where $C_{k,l}$ are arbitrary constants.

If the characteristic equation (1) has complex conjugate roots, then in the above solution, one should extract the real part on the basis of the relation $\exp(\alpha \pm i\beta) = e^\alpha(\cos\beta \pm i\sin\beta)$.

9. $y_x^{(n)} + a_{n-1} y_x^{(n-1)} + \cdots + a_1 y_x' + a_0 y = f(x).$

Linear nonhomogeneous nth-order ODE of general form with constant coefficients.

1°. The general solution of this equation is the sum of the general solution to the corresponding homogeneous ODE with $f(x) \equiv 0$ (see the previous equation) and any particular solution of the nonhomogeneous linear ODE.

If all the roots $\lambda_1, \lambda_2, \ldots, \lambda_n$ of the characteristic equation

$$P(\lambda) = 0, \quad \text{where} \quad P(\lambda) = \lambda^n + a_{n-1}\lambda^{n-1} + \cdots + a_1 \lambda + a_0 = 0,$$

are different, the original ODE has the general solution:

$$y = \sum_{\nu=1}^{n} C_\nu e^{\lambda_\nu x} + \sum_{\nu=1}^{n} \frac{e^{\lambda_\nu x}}{P_\lambda'(\lambda_\nu)} \int f(x) e^{-\lambda_\nu x} \, dx, \tag{1}$$

where $P_\lambda'(\lambda) = n\lambda^{n-1} + a_{n-1}(n-1)\lambda^{n-2} + \cdots + a_1$. For complex roots, the real part in (1) should be taken.

2°. Table 2.9 lists the forms of particular solutions corresponding to some special forms of functions on the right-hand side of the linear nonhomogeneous equation.

3°. Consider the Cauchy problem for the original ODE subject to the homogeneous initial conditions

$$y(0) = y_x'(0) = \cdots = y_x^{(n-1)}(0) = 0. \tag{2}$$

Let $y(x)$ be a solution to the original ODE with arbitrary $f(x)$ subject to conditions (2) and let $u(x)$ be a solution of the auxiliary, simpler problem with $f(x) \equiv 1$, so that $u(x) = y(x)|_{f(x)\equiv 1}$. Then the formula

$$y(x) = \int_0^x f(t) u_x'(x-t) \, dt$$

holds. It is called the *Duhamel integral*.

▶ **Linear ODEs containing power functions.**

10. $y_{xxxx}'''' - 2a^2 y_{xx}'' + a^4 y - \lambda(ax-b)(y_{xx}'' - a^2 y) = 0.$

This equation arises in turbulence theory. Setting $z(x) = y_{xx}'' - a^2 y$, one obtains a second-order linear equation of the form 2.2.2.7:

$$z_{xx}'' - a^2 z - \lambda(ax-b)z = 0. \tag{1}$$

Let the original equation be subjected to the boundary conditions

$$y(0) = y_x'(0) = 0, \quad y(1) = y_x'(1) = 0. \tag{2}$$

Then the solution satisfying the first two conditions in (2) can represented as:

$$2ay = e^{ax} \int_0^x e^{-ax} z \, dx - e^{-ax} \int_0^x e^{ax} z \, dx.$$

To meet the last two conditions in (2), one should take the solution of (1) that satisfies the integral relations $\int_0^1 e^{-ax} z \, dx = \int_0^1 e^{ax} z \, dx = 0$.

TABLE 2.9. Forms of particular solutions of the constant-coefficient nonhomogeneous linear equation $y_x^{(n)} + a_{n-1} y_x^{(n-1)} + \cdots + a_1 y_x' + a_0 y = f(x)$ that correspond to some special forms of the function $f(x)$.

Form of the function $f(x)$	Roots of the characteristic equation $\lambda^n + a_{n-1}\lambda^{n-1} + \cdots + a_1\lambda + a_0 = 0$	Form of a particular solution $y = \widetilde{y}(x)$
$P_m(x)$	Zero is not a root of the characteristic equation (i.e., $a_0 \neq 0$)	$\widetilde{P}_m(x)$
	Zero is a root of the characteristic equation (multiplicity r)	$x^r \widetilde{P}_m(x)$
$P_m(x) e^{\alpha x}$ (α is a real constant)	α is not a root of the characteristic equation	$\widetilde{P}_m(x) e^{\alpha x}$
	α is a root of the characteristic equation (multiplicity r)	$x^r \widetilde{P}_m(x) e^{\alpha x}$
$P_m(x) \cos \beta x$ $+ Q_n(x) \sin \beta x$	$i\beta$ is not a root of the characteristic equation	$\widetilde{P}_\nu(x) \cos \beta x$ $+ \widetilde{Q}_\nu(x) \sin \beta x$
	$i\beta$ is a root of the characteristic equation (multiplicity r)	$x^r [\widetilde{P}_\nu(x) \cos \beta x$ $+ \widetilde{Q}_\nu(x) \sin \beta x]$
$[P_m(x) \cos \beta x$ $+ Q_n(x) \sin \beta x] e^{\alpha x}$	$\alpha + i\beta$ is not a root of the characteristic equation	$[\widetilde{P}_\nu(x) \cos \beta x$ $+ \widetilde{Q}_\nu(x) \sin \beta x] e^{\alpha x}$
	$\alpha + i\beta$ is a root of the characteristic equation (multiplicity r)	$x^r [\widetilde{P}_\nu(x) \cos \beta x$ $+ \widetilde{Q}_\nu(x) \sin \beta x] e^{\alpha x}$

Notation: P_m and Q_n are polynomials of degrees m and n with given coefficients; $\widetilde{P}_m$, $\widetilde{P}_\nu$, and $\widetilde{Q}_\nu$ are polynomials of degrees m and ν whose coefficients are determined by substituting the particular solution into the original ODE; $\nu = \max(m, n)$; and α and β are real numbers, $i^2 = -1$.

11. $x^2 y_{xxxx}'''' + 6x y_{xxx}''' + 6 y_{xx}'' - a^2 y = 0$.

Equation of transverse vibrations of a thin rod.

General solution:

$$y = \frac{1}{\sqrt{x}} \left[C_1 J_1 \left(2\sqrt{ax} \right) + C_2 Y_1 \left(2\sqrt{ax} \right) + C_3 I_1 \left(2\sqrt{ax} \right) + C_4 K_1 \left(2\sqrt{ax} \right) \right],$$

where $J_1(z)$ and $Y_1(z)$ are Bessel functions, and $I_1(z)$ and $K_1(z)$ are modified Bessel functions.

12. $y_x^{(n)} = axy + b$, $\quad a > 0$.

General solution (Kamke, 1977):

$$y = \sum_{\nu=0}^{n} C_\nu \varepsilon_\nu \int_0^\infty \exp\left[\varepsilon_\nu xt - \frac{t^{n+1}}{a(n+1)} \right] dt, \quad \varepsilon_\nu = \exp\left(\frac{2\pi \nu i}{n+1} \right),$$

where $\sum_{\nu=0}^{n} C_\nu = \dfrac{b}{a}$ and $i^2 = -1$.

13. $x^{2n} y_x^{(n)} = ay$.

The transformation $x = t^{-1}$, $y = wt^{1-n}$ leads to a constant coefficient linear ODE of the form 2.4.1.7: $w_t^{(n)} = (-1)^n aw$.

14. $x^n y_x^{(2n)} = ay$.

General solution:
$$y = x^{n/2} \sum_{k=1}^{n} \left[C_{k1} I_n(2\beta_k \sqrt{x}) + C_{k2} K_n(2\beta_k \sqrt{x}) \right],$$

where $I_n(z)$ and $K_n(z)$ are modified Bessel functions; $\beta_1, \ldots, \beta_n$ are roots of the equation $\beta^n = \sqrt{a}$.

15. $x^{3n} y_x^{(2n)} = ay$.

The transformation $x = t^{-1}$, $y = wt^{1-2n}$ leads to an equation of the form 2.4.1.14: $t^n w_t^{(2n)} = aw$.

16. $x^{n+1/2} y_x^{(2n+1)} = ay$.

General solution:
$$y = x^{(2n+1)/4} \sum_{k=0}^{2n} C_k \left[J_{-n-1/2}(2\beta_k \sqrt{x}) + i J_{n+1/2}(2\beta_k \sqrt{x}) \right],$$

where $J_m(z)$ are Bessel functions; $\beta_0, \ldots, \beta_{2n}$ are roots of the equation $\beta^{2n+1} = -ai$; $i^2 = -1$.

17. $x^{3n+3/2} y_x^{(2n+1)} = ay$.

The transformation $x = t^{-1}$, $y = wt^{-2n}$ leads to a linear ODE of the form 2.4.1.16: $t^{n+1/2} w_t^{(2n+1)} = -aw$.

18. $y_x^{(n)} = ax^\beta y$.

For specific β, see equations 2.4.1.7, 2.4.1.12 (with $b=0$), 2.4.1.13 to 2.4.1.17, and 2.4.1.19 (with $b=0$), and 2.4.1.45 (with $a_1 = \cdots = a_{n-1} = 0$).

1°. Let $n \geq 2$, $\beta > -n$, and $(n+\beta)(s+1) \neq 1, 2, \ldots, n-1$, where $s = 0, 1, \ldots$ Then the equation has n solutions that can be represented as:

$$y_j(x) = x^{j-1} E_{n, 1+\beta/n, (\beta+j-1)/n}(ax^{\beta+n}), \qquad j = 1, 2, \ldots, n. \tag{1}$$

Here $E_{n,m,l}(z)$ is a Mittag-Leffler type special function defined by:

$$E_{n,m,l}(z) = 1 + \sum_{k=1}^{\infty} b_k z^k,$$
$$b_k = \prod_{s=0}^{k-1} \frac{\Gamma(n(ms+l)+1)}{\Gamma(n(ms+l+1)+1)} = \prod_{s=0}^{k-1} \frac{1}{[n(ms+l)+1] \ldots [n(ms+l)+n]}, \tag{2}$$

where $\Gamma(\xi)$ is the gamma function, l is an arbitrary number, and $m > 0$.

If $\beta \geq 0$, solutions (1) are linearly independent. Series expansions of (2) are convenient for small x.

2°. Let $n \geq 2$, $\beta < -n$, and $(n+\beta)(s+1) \neq -1, -2, \ldots, -(n-1)$, where $s = 0, 1, \ldots$ Then the equation in question has n solutions that can be represented as:

$$y_j(x) = x^{j-1} E_{n,-1-\beta/n,-1-(\beta+j)/n}\bigl(a(-1)^n x^{\beta+n}\bigr), \qquad j = 1, 2, \ldots, n, \qquad (3)$$

where $E_{n,m,l}(z)$ is the Mittag-Leffler type special function defined by (2). If $\beta \leq -2n$, solutions (3) are linearly independent. Series expansions of (3) are convenient for large x.

3°. The transformation $x = t^{-1}$, $y = w t^{1-n}$ leads to an equation of similar form:

$$w_t^{(n)} = a(-1)^{n+1} t^{-2n-\beta} w.$$

For details on Items 1° and 2°, see Saigo & Kilbas (2000).

19. $x^{2n+1} y_x^{(n)} = ay + bx^n$.

The transformation $x = t^{-1}$, $y = w t^{1-n}$ leads to a linear ODE of the form 2.4.1.12: $w_t^{(n)} = (-1)^n (atw + b)$.

20. $(ax+b)^{2n+1} y_x^{(n)} = (cx+d) y$.

The transformation

$$\xi = \frac{cx+d}{ax+b}, \qquad w = \frac{y}{(ax+b)^{n-1}}$$

leads to an equation of the form 2.4.1.12: $w_\xi^{(n)} = \Delta^{-n} \xi w$, where $\Delta = bc - ad$.

21. $(ax+b)^n (cx+d)^n y_x^{(n)} = ky$.

1°. The transformation $\xi = \ln \dfrac{ax+b}{cx+d}$, $w = \dfrac{y}{(cx+d)^{n-1}}$ leads to a constant coefficient linear ODE.

2°. The transformation $\zeta = \dfrac{ax+b}{cx+d}$, $w = \dfrac{y}{(cx+d)^{n-1}}$ leads to the Euler equation 2.4.1.45: $\zeta^n w_\zeta^{(n)} = k \Delta^{-n} w$, where $\Delta = ad - bc$.

22. $(ax^2 + bx + c)^n y_x^{(n)} = ky$.

The transformation $\xi = \displaystyle\int \frac{dx}{ax^2 + bx + c}$, $w = y(ax^2 + bx + c)^{\frac{1-n}{2}}$ leads to a constant coefficient linear ODE.

23. $(ax+b)^n (cx+d)^{3n} y_x^{(2n)} = ky$.

The transformation $\xi = \dfrac{ax+b}{cx+d}$, $w = \dfrac{y}{(cx+d)^{2n-1}}$ leads to an equation of the form 2.4.1.14: $\xi^n w_\xi^{(2n)} = k \Delta^{-2n} w$, where $\Delta = ad - bc$.

24. $(ax+b)^{n+1/2} (cx+d)^{3n+3/2} y_x^{(2n+1)} = ky$.

The transformation $\xi = \dfrac{ax+b}{cx+d}$, $w = \dfrac{y}{(cx+d)^{2n}}$ leads to an equation of the form 2.4.1.16: $\xi^{n+1/2} w_\xi^{(2n+1)} = k \Delta^{-2n-1} w$, where $\Delta = ad - bc$.

25. $y_x^{(n)} + ax^k y_x' + akx^{k-1} y = 0.$

Integrating yields an $(n-1)$st-order linear ODE: $y_x^{(n-1)} + ax^k y = C.$

26. $y_x^{(n)} + ax^{k+1} y_x' - a(n-1) x^k y = 0.$

The substitution $z = xy_x' - (n-1)y$ leads to an $(n-1)$st-order linear ODE: $z_x^{(n-1)} + ax^{k+1} z = 0.$

27. $y_x^{(n)} + ax^{k+1} y_x' + a(k+n) x^k y = 0.$

The transformation $x = t^{-1}$, $y = wt^{1-n}$ leads to an equation of the form 2.4.1.25:

$$w_t^{(n)} + bt^\nu w_t' + b\nu t^{\nu-1} w = 0,$$

where $b = a(-1)^{n+1}$, $\nu = 1 - k - 2n.$

28. $y_x^{(n)} + (ax + b) x^k y_x' - ax^k y = 0.$

Particular solution: $y_0 = ax + b.$

29. $y_x^{(n)} + (ax + b) x^k y_x' - 2ax^k y = 0.$

Particular solution: $y_0 = (ax + b)^2.$

30. $y_x^{(n)} + (ax + b) x^k y_x' - 3ax^k y = 0.$

Particular solution: $y_0 = (ax + b)^3.$

31. $y_x^{(n)} + (ax + b) x^k y_x' - a(n-1) x^k y = 0.$

Particular solution: $y_0 = (ax + b)^{n-1}$. The substitution

$$z = (ax + b) y_x' - a(n-1) y$$

leads to an $(n-1)$st-order linear ODE:

$$z_x^{(n-1)} + (ax + b) x^k z = 0.$$

32. $y_x^{(n)} + ax^{k+1} y_x' - amx^k y = 0, \quad m = 1, 2, \ldots, n-1.$

Particular solution: $y_0 = x^m$. The substitution $z = xy_x' - my$ leads to an $(n-1)$st-order linear ODE:

$$D^{n-m-1} \left(\frac{z_x^{(m)}}{x} \right) + ax^k z = 0, \quad \text{where} \quad D = \frac{d}{dx}.$$

33. $y_x^{(2n)} = a^n y + bx^k (y_{xx}'' - ay).$

This is a special case of equation 2.4.1.65 with $f(x) = bx^k$. The substitution $w = y_{xx}'' - ay$ leads to a $(2n-2)$nd-order linear ODE:

$$w_x^{(2n-2)} + aw_x^{(2n-4)} + \cdots + a^{n-1} w = bx^k w.$$

34. $y_x^{(n)} + ax^k y_x^{(m)} - (ab^m x^k + b^n) y = 0.$

Particular solution: $y_0 = e^{bx}.$

35. $y_x^{(n)} + (ax^k - b^{n-m})y_x^{(m)} - ab^m x^k y = 0.$

Particular solution: $y_0 = e^{bx}$.

36. $y_x^{(n)} + ay_x^{(n-1)} + bx^m y_x' + abx^m y = 0.$

Particular solution: $y_0 = e^{-ax}$.

37. $xy_x^{(n)} - nmy_x^{(n-1)} + axy = 0,$ $\quad n = 2, 3, 4, \ldots, \quad m = 1, 2, 3, \ldots$

General solution:
$$y = x^{(m+1)n-1} \left(x^{1-n} \frac{d}{dx} \right)^m (x^{1-n} w),$$

where w is the general solution of the constant coefficient linear ODE: $w_x^{(n)} + aw = 0$.

38. $xy_x^{(n)} + ny_x^{(n-1)} = axy + b.$

The substitution $w = xy$ leads to a constant coefficient linear ODE: $w_x^{(n)} = aw + b$.

39. $xy_x^{(n)} + ny_x^{(n-1)} = ax^2 y + b.$

The substitution $w = xy$ leads to an equation of the form 2.4.1.12: $w_x^{(n)} = axw + b$.

40. $xy_x^{(n)} + (n - m - 1)y_x^{(n-1)} + ax^k y_x' - amx^{k-1} y = 0.$

Particular solution: $y_0 = x^m$.

41. $xy_x^{(n)} + ax^k y_x^{(m)} - (ax^k + amx^{k-1} + x + n)y = 0.$

Particular solution: $y_0 = xe^x$.

42. $xy_x^{(n)} = \sum_{\nu=0}^{n-1} [(aA_{\nu+1} - A_\nu)x + A_{\nu+1}] y_x^{(\nu)}.$

Here $A_n = 1$, $A_0 = 0$; a and A_ν are arbitrary numbers ($\nu = 1, \ldots, n-1$).

Denote $P(\lambda) = \sum_{\nu=0}^{n-1} A_{\nu+1} \lambda^\nu$. Let the roots $\lambda_1, \ldots, \lambda_{n-1}$ of the algebraic equation $P(\lambda) = 0$ be all different and let $P(a) \neq 0$. Then the general solution of the ODE in question is

$$y = C_1 e^{\lambda_1 x} + C_2 e^{\lambda_2 x} + \cdots + C_{n-1} e^{\lambda_{n-1} x} + C_n e^{ax} \left[x - \frac{P'_\lambda(a)}{P(a)} \right].$$

43. $x^2 y_x^{(n)} + 2nxy_x^{(n-1)} + n(n-1)y_x^{(n-2)} = ax^2 y + b.$

The substitution $w = x^2 y$ leads to a constant coefficient linear ODE: $w_x^{(n)} = aw + b$.

44. $x^2 y_x^{(n)} + 2nxy_x^{(n-1)} + n(n-1)y_x^{(n-2)} = ax^3 y + b.$

The substitution $w = x^2 y$ leads to an equation of the form 2.4.1.12: $w_x^{(n)} = axw + b$.

45. $a_n x^n y_x^{(n)} + a_{n-1} x^{n-1} y_x^{(n-1)} + \cdots + a_1 xy_x' + a_0 y = 0.$

Euler equation. If all roots λ_k ($k = 1, \ldots, n$) of the algebraic equation

$$\sum_{\nu=1}^{n} a_\nu \lambda(\lambda - 1) \ldots (\lambda - \nu + 1) = -a_0$$

are different, the general solution of the original differential equation is given by:
$$y = C_1|x|^{\lambda_1} + \cdots + C_n|x|^{\lambda_n}.$$

In the general case, the substitution $t = \ln|x|$ leads to a constant coefficient linear ODE of the form 2.4.1.8:
$$\sum_{\nu=1}^{n} a_\nu D(D-1)\ldots(D-\nu+1)y = -a_0 y, \qquad \text{where} \quad D = \frac{d}{dx}.$$

46. $x^{2n+1}y_x^{(n)} + nx^{2n}y_x^{(n-1)} = axy.$

The substitution $w = xy$ leads to an ODE of the form 2.4.1.13: $x^{2n}w_x^{(n)} = aw$.

47. $x^{2n+1}y_x^{(n)} + nx^{2n}y_x^{(n-1)} = ay.$

The substitution $w = xy$ leads to an equation of the form 2.4.1.19 (with $b = 0$): $x^{2n+1}w_x^{(n)} = aw$.

48. $x^n y_x^{(2n)} + 2nx^{n-1}y_x^{(2n-1)} = ay.$

The substitution $w = xy$ leads to an equation of the form 2.4.1.14: $x^n w_x^{(2n)} = aw$.

49. $x^{3n} y_x^{(2n)} + 2nx^{3n-1}y_x^{(2n-1)} = ay.$

The substitution $w = xy$ leads to an ODE of the form 2.4.1.15: $x^{3n} w_x^{(2n)} = aw$.

50. $x^{n+1} y_x^{(2n+1)} + (2n+1)x^n y_x^{(2n)} = a\sqrt{x}\, y.$

The substitution $w = xy$ leads to an ODE of the form 2.4.1.16: $x^{n+1/2} w_x^{(2n+1)} = aw$.

51. $x^{3n+3/2} y_x^{(2n+1)} + (2n+1)x^{3n+1/2} y_x^{(2n)} = ay.$

The substitution $w = xy$ leads to an ODE of the form 2.4.1.17: $x^{3n+3/2} w_x^{(2n+1)} = aw$.

52. $\delta(\delta+\beta_1-1)\ldots(\delta+\beta_q-1)y = x(\delta+\alpha_1)\ldots(\delta+\alpha_p)y, \quad \delta = x(d/dx).$

One of the solutions to this ODE is expressed in terms of the *generalized hypergeometric series* (Bailey, 1935):
$$y = {}_pF_q(\alpha_1,\ldots,\alpha_p; \beta_1,\ldots,\beta_q; x) = \sum_{k=0}^{\infty} \frac{(\alpha_1)_k \ldots (\alpha_p)_k}{(\beta_1)_k \ldots (\beta_q)_k} \frac{x^k}{k!},$$
$$(\alpha)_k = \alpha(\alpha+1)\ldots(\alpha+k-1).$$

For $p = q+1$, the generalized hypergeometric series converges for $|x| < 1$ and diverges for $|x| > 1$.

▶ **Equations containing arbitrary functions.**

53. $fy_x^{(n)} - f_x^{(n)} y = 0, \qquad f = f(x).$

Particular solution: $y_0 = f(x)$.

54. $fy_x^{(2n+1)} + f_x^{(2n+1)}y = g(x)$, $\quad f = f(x)$.

First integral:
$$\sum_{k=0}^{2n} (-1)^k f_x^{(2n-k)} y_x^{(k)} = \int g(x)\, dx + C.$$

55. $y_x^{(n)} + (ax+b)f(x)y_x' - af(x)y = 0.$

Particular solution: $y_0 = ax + b$.

56. $y_x^{(n)} + (ax+b)f(x)y_x' - 2af(x)y = 0.$

Particular solution: $y_0 = (ax+b)^2$.

57. $y_x^{(n)} + (ax+b)f(x)y_x' - (n-1)af(x)y = 0.$

Particular solution: $y_0 = (ax+b)^{n-1}$. The substitution
$$z = (ax+b)y_x' - a(n-1)y$$
leads to an $(n-1)$st-order linear ODE:
$$z_x^{(n-1)} + (ax+b)f(x)z = 0.$$

58. $y_x^{(n)} + xf(x)y_x' - mf(x)y = 0,$ $\quad m = 1, 2, \ldots, n-1.$

Particular solution: $y_0 = x^m$. The substitution $z = xy_x' - my$ leads to an $(n-1)$st-order ODE:
$$D^{n-m-1}\left(\frac{z_x^{(m)}}{x}\right) + f(x)z = 0, \quad \text{where} \quad D = \frac{d}{dx}.$$

59. $y_x^{(n)} + f(x)y_x' + f_x'(x)y = g(x).$

Integrating yields an $(n-1)$st-order linear ODE:
$$y_x^{(n-1)} + f(x)y = \int g(x)\, dx + C.$$

60. $y_x^{(2n)} = y + f(x)(y_x' \cosh x - y \sinh x).$

The substitution $w = y_x' \cosh x - y \sinh x$ leads to a $(2n-1)$st-order linear ODE.

61. $y_x^{(2n)} = y + f(x)(y_x' \sinh x - y \cosh x).$

The substitution $w = y_x' \sinh x - y \cosh x$ leads to a $(2n-1)$st-order linear ODE.

62. $y_x^{(2n)} = (-1)^n y + f(x)(y_x' \sin x - y \cos x).$

The substitution $w = y_x' \sin x - y \cos x$ leads to a $(2n-1)$st-order linear ODE.

63. $y_x^{(2n)} = (-1)^n y + f(x)(y_x' \cos x + y \sin x).$

The substitution $w = y_x' \cos x + y \sin x$ leads to a $(2n-1)$st-order linear ODE.

64. $y_x^{(n)} = \dfrac{\varphi_x^{(n)}}{\varphi} y + f(x)\left(y_x' - \dfrac{\varphi_x'}{\varphi}y\right),$ $\quad \varphi = \varphi(x)$.

Particular solution: $y_0 = \varphi(x)$. The substitution $w = y_x' - \dfrac{\varphi_x'}{\varphi} y$ leads to an $(n-1)$st-order linear ODE.

65. $y_x^{(2n)} = a^n y + f(x)(y_{xx}'' - ay)$.

The substitution $w = y_{xx}'' - ay$ leads to a $(2n-2)$nd-order linear ODE:
$$w_x^{(2n-2)} + a w_x^{(2n-4)} + \cdots + a^{n-1} w = f(x) w.$$

66. $y_x^{(2n)} = a^2 y + f(x)[y_x^{(n)} + ay]$.

The substitution $w = y_x^{(n)} + ay$ leads to an nth-order linear ODE: $w_x^{(n)} = [f(x) + a]w$.

67. $y_x^{(n)} + f(x) y_x^{(m)} - [a^n + a^m f(x)] y = 0$.

Particular solution: $y_0 = e^{ax}$.

68. $y_x^{(n)} + (f - a^{n-m}) y_x^{(m)} - a^m f y = 0$, $\quad f = f(x)$.

Particular solution: $y_0 = e^{ax}$.

69. $y_x^{(n)} + a y_x^{(n-1)} + f y_x' + a f y = 0$, $\quad f = f(x)$.

Particular solution: $y_0 = e^{-ax}$.

70. $y_x^{(n)} = \sum_{k=0}^{n-1} (a_{k+1} f - a_k) y_x^{(k)}$.

Here $f = f(x)$; $a_n = 1$, $a_0 = 0$; a_k are arbitrary numbers ($k = 1, 2, \ldots, n-1$).

1°. Particular solutions: $y_k = e^{\lambda_k x}$ ($k = 1, 2, \ldots, n-1$), where the λ_k are roots of the polynomial equation $\sum_{k=0}^{n-1} a_{k+1} \lambda^k = 0$.

2°. The equation in question admits the first integral
$$\sum_{k=0}^{n-1} a_{k+1} y_x^{(k)} = C \exp\left(\int f\, dx\right),$$
which is a linear nonhomogeneous $(n-1)$st-order ODE with constant coefficients of the form 2.4.1.9.

71. $x y_x^{(n)} + (a + n - 1) y_x^{(n-1)} = f(x)(x y_x' + ay)$.

The substitution $w = x y_x' + ay$ leads to an $(n-1)$st-order linear ODE: $w_x^{(n-1)} = f(x) w$.

72. $x^n y_x^{(n)} + b_{n-1} x^{n-1} y_x^{(n-1)} + \cdots + b_1 x y_x' + b_0 y = f(x)$.

Nonhomogeneous Euler equation. The substitution $x = ae^t$ ($a \neq 0$) leads to a constant coefficient nonhomogeneous linear ODE of the form 2.4.1.9.

73. $x^m y_x^{(n)} = \sum_{k=0}^{n-1} [x^m(a_{k+1} f - a_k) + a_{k+1}] y_x^{(k)}$.

Here $f = f(x)$; $a_n = 1$, $a_0 = 0$; m and a_k are arbitrary numbers ($k = 1, 2, \ldots, n-1$).

Particular solutions: $y_k = e^{\lambda_k x}$ ($k = 1, 2, \ldots, n-1$), where the λ_k are roots of the polynomial equation $\sum_{k=0}^{n-1} a_{k+1} \lambda^k = 0$.

74. $\sum_{k=2}^{n} f_k(x) y_x^{(k)} = g(x)(xy_x' - y).$

Particular solution: $y_0 = x$. The substitution $w(x) = xy_x' - y$ leads to an $(n-1)$st-order linear ODE.

75. $\sum_{k=m+1}^{n} f_k(x) y_x^{(k)} = g(x)(xy_x' - my),$ $\quad m = 1, 2, \ldots, n-1.$

Particular solution: $y_0 = x^m$. The substitution $w(x) = xy_x' - my$ leads to an $(n-1)$st-order linear ODE.

76. $\sum_{k=0}^{n} (f_k - a f_{k+1}) y_x^{(k)} = 0.$

Here $f_k = f_k(x)$ $(k = 1, 2, \ldots, n)$; $f_{n+1} \equiv f_0 \equiv 0$.
  Particular solution: $y_0 = e^{ax}$.

77. $\sum_{k=0}^{n} x^k [f_k + (k-m) f_{k+1}] y_x^{(k)} = 0.$

Here $f_k = f_k(x)$ $(k = 1, 2, \ldots, n)$; $f_{n+1} \equiv f_0 \equiv 0$.
  Particular solution: $y_0 = x^m$.

2.4.2. Third- and Fourth-Order Nonlinear Ordinary Differential Equations

1. $y_{xxx}''' = f(y).$

The substitution $w(y) = (y_x')^2$ leads to a second-order ODE: $w_{yy}'' = \pm 2 f(y) w^{-1/2}$. In particular, for $f(y) = ay^n$, the obtained equation is an Emden–Fowler equation of the form 2.3.1.2 with $m = -1/2$.

2. $yy_{xxx}''' = f(x).$

1°. On integrating the equation, we have
$$yy_{xx}'' - \frac{1}{2}(y_x')^2 = \int f(x)\, dx + C.$$

The substitution $y = w^2$ reduces the latter ODE to the form
$$w_{xx}'' = \frac{1}{2}\left[\int f(x)\, dx + C\right] w^{-3}.$$

2°. The transformation $x = 1/t$, $y = u/t^2$ leads to an equation of the same form: $uu_{ttt}''' = -t^{-2} f(1/t)$.

3. $(y_x')^2 - yy_{xx}'' - ay_{xxx}''' = 0.$

This autonomous ODE occurs in the theory of the hydrodynamic boundary layer.
  Particular solutions:
$$y = \frac{6a}{x+C},$$
$$y = Ce^{\lambda x} - a\lambda,$$

where C and λ are arbitrary constants.

4. $yy'''_{xxx} + 3y'_x y''_{xx} = f(x)$.

General solution:
$$y^2 = C_1 x^2 + C_2 x + C_3 + \int_0^x (x-t)^2 f(t)\, dt,$$

where C_1, C_2, and C_3 are arbitrary constants.

5. $y'''_{xxx} = (f-a) y''_{xx} + (af - b) y'_x + bfy, \qquad f = f(y)$.

Particular solution:
$$y = C_1 e^{\lambda_1 x} + C_2 e^{\lambda_2 x},$$

where C_1 and C_2 are arbitrary constants, and λ_1 and λ_2 are the roots of the quadratic equation $\lambda^2 + a\lambda + b = 0$.

Note that in the original ODE, the function f can depend on two arguments x and y.

6. $y''''_{xxxx} = a y^{-5/3}$.

Multiply both sides of the equation by $y^{5/3}$ and differentiate the resulting expression with respect to x to obtain
$$3 y y_x^{(5)} + 5 y'_x y''''_{xxxx} = 0.$$

Integrating this equation three times, we arrive at the chain of equations
$$3 y y''''_{xxxx} + 2 y'_x y'''_{xxx} - (y''_{xx})^2 = 2 C_2, \tag{1}$$
$$3 y y'''_{xxx} - y'_x y''_{xx} = 2 C_2 x + C_1, \tag{2}$$
$$3 y y''_{xx} - 2 (y'_x)^2 = C_2 x^2 + C_1 x + C_0, \tag{3}$$

where C_0, C_1, and C_2 are arbitrary constants. By eliminating the highest derivatives from (1)–(3) with the help of the original equation, we obtain a first-order ODE:
$$(2 P y'_x - 3 P'_x y)^2 = 9(C_1^2 - 4 C_0 C_2) y^2 - 2 P^3 + 54 a P y^{4/3},$$

where $P = C_2 x^2 + C_1 x + C_0$. The substitution $y = (P/w)^{3/2}$ leads to a separable ODE, the integration of which finally yields:
$$\int [9(C_1^2 - 4 C_0 C_2) + 54 a w - 2 w^3]^{-1/2} \frac{dw}{w} \pm \int \frac{dx}{3P} = C_3.$$

7. $y''''_{xxxx} = a y^m$.

This is a special case of ODE 2.4.2.9 with $f(w) = a y^m$.

1°. Integrating yields
$$2 y'_x y'''_{xxx} - (y''_{xx})^2 = \frac{2a}{m+1} y^{m+1} + \frac{4}{3} C,$$

where C is an arbitrary constant, and $m \neq -1$. The substitution $w(y) = (y'_x)^{3/2}$ leads to a second-order ODE:
$$w''_{yy} = \left(\frac{3a}{2m+2} y^{m+1} + C \right) w^{-5/3}.$$

The value $C = 0$ corresponds to an Emden–Fowler equation of the form 2.3.1.2.

2°. Particular solution: $y = \left[\dfrac{8(m+1)(m+3)(3m+1)}{a(m-1)^4}\right]^{\frac{1}{m-1}} (x+C)^{\frac{4}{1-m}}$.

8. $y''''_{xxxx} = ax^{-3m-5} y^m$.

The transformation $x = t^{-1}$, $y = t^{-3} w(t)$ leads to an equation of the form 2.4.2.7: $w''''_{tttt} = aw^m$.

9. $y''''_{xxxx} = f(y)$.

Autonomous equation. Integrating yields

$$2 y'_x y'''_{xxx} - (y''_{xx})^2 = 2 \int f(y)\, dy + 2C.$$

The substitution $w(y) = |y'_x|^{3/2}$ leads to a second-order ODE:

$$w''_{yy} = \tfrac{3}{2} \left[\int f(y)\, dy + C\right] w^{-5/3}.$$

10. $y''''_{xxxx} + a y''_{xx} = f(y)$.

Integrating yields

$$2 y'_x y'''_{xxx} - (y''_{xx})^2 + a(y'_x)^2 = 2 \int f(y)\, dy + 2C,$$

where C is an arbitrary constant. The substitution $w(y) = |y'_x|^{3/2}$ leads to a second-order ODE:

$$w''_{yy} = -\tfrac{3}{4} a w^{-1/3} + \tfrac{3}{2} \left[\int f(y)\, dy + C\right] w^{-5/3}.$$

2.4.3. Higher-Order Nonlinear Ordinary Differential Equations

1. $y_x^{(2n)} = a y^{\frac{1+2n}{1-2n}}$.

Let us multiply both sides by $y^{\frac{2n+1}{2n-1}}$ and differentiate with respect to x to obtain

$$(2n-1) y y_x^{(2n+1)} + (2n+1) y'_x y_x^{(2n)} = 0.$$

Three integrals of this ODE containing arbitrary constants C_0, C_1, and C_2 are presented in 2.4.3.23, where one should set $f \equiv 0$. Eliminating the highest derivatives from those integrals and the original equation, one can always obtain a $(2n-3)$rd-order ODE. With the aid of the transformation

$$t = \int \frac{dx}{P}, \quad w = y P^{\frac{1-2n}{2}}, \qquad \text{where} \quad P = C_2 x^2 + C_1 x + C_0,$$

this equation can be reduced to an autonomous form. Therefore, the substitution $z(w) = w'_t$ finally leads to a $(2n-4)$th-order ODE with respect to $z = z(w)$.

2. $y_x^{(n)} = f(y)$.

Autonomous equation. This is a special case of ODE 2.4.3.31.

1°. The substitution $w(y) = y'_x$ leads to an $(n-1)$st-order ODE.

2°. For even $n = 2m$, the first integral of the equation is:

$$\sum_{k=1}^{m-1}(-1)^k y_x^{(k)} y_x^{(2m-k)} + \tfrac{1}{2}(-1)^m \left[y_x^{(m)}\right]^2 + \int f(y)\,dy = C.$$

Furthermore, the order of the resulting equation can be reduced by one with the substitution $w(y) = y_x'$.

3. $y_x^{(n)} = x^{-n} f(y)$.

The substitution $t = \ln|x|$ leads to an autonomous equation of the form 2.4.3.31.

4. $y_x^{(n)} = x^{1-n} f(y/x)$.

Homogeneous equation. This is a special case of ODE 2.4.3.36. The transformation $t = \ln x$, $w = y/x$ leads to an autonomous equation of the form 2.4.3.31.

5. $y_x^{(n)} = x^{-n-1} f(x^{1-n} y)$.

The transformation $x = t^{-1}$, $y = t^{1-n} w$ leads to an autonomous equation of the form 2.4.3.2: $w_t^{(n)} = (-1)^n f(w)$.

6. $y_x^{(2n)} = x^{-\frac{2n+1}{2}} f\left(x^{\frac{1-2n}{2}} y\right)$.

The transformation $x = e^t$, $y = x^{\frac{2n-1}{2}} w(t)$ leads to an autonomous equation of the form 2.4.3.26, whose order can be reduced by two.

7. $y_x^{(n)} = x^{-n-k} f(y x^k)$.

This is a special case of ODE 2.4.3.37.

 1°. The transformation $t = \ln x$, $z = y x^k$ leads to an autonomous equation of the form 2.4.3.31.

 2°. The transformation $z = y x^k$, $w = x y_x'/y$ leads to an $(n-1)$st-order ODE.

8. $y_x^{(n)} = y x^{-n} f(x^p y^q)$.

This is a special case of ODE 2.4.3.38. The transformation $t = x^p y^q$, $w = x y_x'/y$ leads to an $(n-1)$st-order ODE.

9. $y y_x^{(2n+1)} = f(x)$.

Integrating yields a $2n$th-order ODE:

$$2\sum_{m=0}^{n-1}(-1)^m y_x^{(m)} y_x^{(2n-m)} + (-1)^n \left[y_x^{(n)}\right]^2 = 2\int f(x)\,dx + C,$$

where the notation $y_x^{(0)} \equiv y$ is used.

10. $y_x^{(n)} = (ax + by + c)^{1-n} f\left(\dfrac{ax + by + c}{\alpha x + \beta y + \gamma}\right)$.

 1°. For $a\beta - b\alpha = 0$, the substitution $bw = ax + by + c$ leads to an autonomous ODE of the form 2.4.3.31.

2°. For $a\beta - b\alpha \neq 0$, the transformation

$$z = x - x_0, \quad w = y - y_0,$$

where x_0 and y_0 are the constants determined by solving the linear algebraic system

$$ax_0 + by_0 + c = 0,$$
$$\alpha x_0 + \beta y_0 + \gamma = 0,$$

leads to a homogeneous ODE of the form 2.4.3.36:

$$w_z^{(n)} = z^{1-n} F\left(\frac{w}{z}\right), \quad \text{where} \quad F(\xi) = (a + b\xi)^{1-n} f\left(\frac{a + b\xi}{\alpha + \beta \xi}\right).$$

11. $(ax + b)^n (cx + d) y_x^{(n)} = f\left(\dfrac{y}{(cx + d)^{n-1}}\right)$.

The transformation $\xi = \ln\left|\dfrac{ax + b}{cx + d}\right|$, $w = \dfrac{y}{(cx + d)^{n-1}}$ leads to an autonomous equation of the form 2.4.3.31.

12. $y_x^{(n)} = (ax^2 + bx + c)^{-\frac{1+n}{2}} f\left(y(ax^2 + bx + c)^{\frac{1-n}{2}}\right)$.

1°. The transformation

$$t = \int \frac{dx}{ax^2 + bx + c}, \quad w = y(ax^2 + bx + c)^{\frac{1-n}{2}} \qquad (1)$$

leads to an autonomous equation with respect to $w = w(t)$, which admits reduction of order by the substitution $z(w) = w_t'$.

2°. Let $n = 2m$ be an even integer ($m = 1, 2, 3, \dots$). In this case, transformation (1) yields an equation of the form 2.4.3.26, whose order can be reduced by two.

Setting

$$P = ax^2 + bx + c, \quad y = wP^{\frac{2m-1}{2}}$$

and multiplying both sides of the equation by $w_x' = P^{-\frac{1+2m}{2}}\left(Py_x' + \frac{1-2m}{2} P_x' y\right)$, we obtain

$$\left(Py_x' + \frac{1 - 2m}{2} P_x' y\right) y_x^{(2m)} = f(w) w_x'.$$

Integrating both sides of this relation with respect to x (the left-hand side is integrated by parts), we get

$$\sum_{k=0}^{m-2} (-1)^k \psi_x^{(k)} y_x^{(2m-1-k)} + (-1)^{m-1} \int \psi_x^{(m-1)} y_x^{(m+1)} \, dx = \int f(w) \, dw + C, \qquad (2)$$

where

$$\psi_x^{(k)} = \frac{d^k}{dx^k}\left(Py_x' + \frac{1 - 2m}{2} P_x' y\right) = P y_x^{(k+1)} + \left(k - m + \frac{1}{2}\right) P_x' y_x^{(k)} + ak(k - 2m) y_x^{(k-1)}$$

(recall that $n = 2m$). It can be shown that the integrand on the left-hand side of (2) is a total differential. Finally, we arrive at the first integral

$$\sum_{k=0}^{m-2}(-1)^k\left[Py_x^{(k+1)}+\left(k-m+\tfrac{1}{2}\right)P'_xy_x^{(k)}+ak(k-2m)y_x^{(k-1)}\right]y_x^{(2m-1-k)}$$
$$+(-1)^{m-1}\left\{\tfrac{1}{2}P\left[y_x^{(m)}\right]^2-\tfrac{1}{2}P'_xy_x^{(m-1)}y_x^{(m)}+a(1-m^2)y_x^{(m-2)}y_x^{(m)}+\tfrac{1}{2}am^2\left[y_x^{(m-1)}\right]^2\right\}$$
$$=\int f(w)\,dw+C.$$

13. $y_x^{(n)} = e^{\alpha x}f(ye^{-\alpha x}).$

The substitution $w(x) = ye^{-\alpha x}$ leads to an autonomous equation of the form 2.4.3.31.

14. $y_x^{(n)} = yf(e^{\alpha x}y^m).$

The transformation $z = e^{\alpha x}y^m$, $w(z) = y'_x/y$ leads to an $(n-1)$st-order ODE.

15. $y_x^{(n)} = x^{-n}f(x^m e^{\alpha y}).$

The transformation $z = x^m e^{\alpha y}$, $w(z) = xy'_x$ leads to an $(n-1)$st-order ODE.

16. $y_x^{(n)} = a^n y + F(x, y'_x - ay).$

1°. The substitution $u = y'_x - ay$ leads to an $(n-1)$st-order ODE:

$$u_x^{(n-1)} + au_x^{(n-2)} + \cdots + a^{n-1}u = F(x, u).$$

2°. For $F(x, u) = f(u)$, the equation admits a particular solution of the form:

$$y = Ce^{ax} + k,$$

where C is an arbitrary constant, and k is a root of the algebraic (transcendental) equation $a^n k + f(-ak) = 0$.

17. $y_x^{(n)} = F(x, xy'_x - y).$

The substitution $u = xy'_x - y$ leads to an $(n-1)$st-order ODE: $\dfrac{d^{n-2}}{dx^{n-2}}\left(\dfrac{u'_x}{x}\right) = F(x, u).$

18. $y_x^{(n)} = F(x, xy'_x - my).$

Here m is a positive integer and $n \geq m + 1$. The substitution $u = xy'_x - my$ leads to an $(n-1)$st-order ODE: $\zeta_x^{(n-m-1)} = F(x, u)$, where $\zeta = u_x^{(m)}/x$.

19. $y_x^{(2n)} = a^n y + F(x, y''_{xx} - ay).$

1°. The substitution $u(x) = y''_{xx} - ay$ leads to a $(2n-2)$nd-order ODE:

$$u_x^{(2n-2)} + au_x^{(2n-4)} + \cdots + a^{n-1}u = F(x, u).$$

In particular, for $n = 2$, we get

$$u''_{xx} + au = F(x, u).$$

2°. For $F(x, u) = f(u)$, the original equation admits a particular solution of the form:

$$y = \begin{cases} C_1 \exp(-x\sqrt{a}) + C_2 \exp(x\sqrt{a}) + k & \text{if } a > 0, \\ C_1 \cos(x\sqrt{-a}) + C_2 \sin(x\sqrt{-a}) + k & \text{if } a < 0, \end{cases}$$

where C_1 and C_2 are arbitrary constants, and k is a root of the algebraic (transcendental) equation $a^n k + f(-ak) = 0$.

20. $y_x^{(2n)} = a^2 y + F(x, y_x^{(n)} + ay)$.

The substitution $u = y_x^{(n)} + ay$ leads to an nth-order ODE: $u_x^{(n)} = au + F(x, u)$.

21. $xy_x^{(n)} + (a + n - 1) y_x^{(n-1)} = F(x, xy_x' + ay)$.

The substitution $u = xy_x' + ay$ leads to an $(n-1)$st-order ODE: $u_x^{(n-1)} = F(x, u)$.

22. $y_x^{(n)} = \dfrac{\varphi_x^{(n)}}{\varphi} y + F\!\left(x, y_x' - \dfrac{\varphi_x'}{\varphi} y\right)$, $\varphi = \varphi(x)$.

The substitution $w = y_x' - \dfrac{\varphi_x'}{\varphi} y$ leads to an $(n-1)$st-order ODE.

23. $(2n - 1) y y_x^{(2n+1)} + (2n + 1) y_x' y_x^{(2n)} = f(x)$.

Having integrated the equation once, we obtain

$$(2n - 1) y y_x^{(2n)} + 2\sum_{k=1}^{n-1} (-1)^{k+1} y_x^{(k)} y_x^{(2n-k)} + (-1)^{n+1} \left[y_x^{(n)}\right]^2 = \int f(x)\,dx + 2C_2.$$

The second integration leads to a $(2n - 1)$st-order ODE:

$$\sum_{k=0}^{n-1} (2n - 1 - 2k)(-1)^k y_x^{(k)} y_x^{(2n-1-k)} = 2C_2 x + C_1 + \int_{x_0}^{x} (x - t) f(t)\,dt.$$

The third integration leads to a $(2n - 2)$nd-order ODE:

$$\sum_{k=0}^{n-2} (k + 1)(2n - k - 1)(-1)^k y_x^{(k)} y_x^{(2n-2-k)} + \tfrac{1}{2}(-1)^{n-1} n^2 \left[y_x^{(n-1)}\right]^2$$
$$= C_2 x^2 + C_1 x + C_0 + \frac{1}{2}\int_{x_0}^{x} (x - t)^2 f(t)\,dt.$$

24. $y y_x^{(n)} - y_x' y_x^{(n-1)} = f(x) y^2$.

Integrating yields an $(n-1)$st-order linear ODE: $y_x^{(n-1)} = \left[\int f(x)\,dx + C\right] y$.

25. $y y_x^{(n)} = y_x' y_x^{(n-1)} + f(x) y y_x^{(n-1)}$.

Integrating yields an $(n-1)$st-order linear ODE: $y_x^{(n-1)} = C \exp\!\left[\int f(x)\,dx\right] y$.

26. $\sum_{m=1}^{n} a_m y_x^{(2m)} = f(y).$

The first integral has the form:

$$\sum_{m=1}^{n} a_m \left\{ \sum_{\nu=1}^{m-1} (-1)^{\nu} y_x^{(\nu)} y_x^{(2m-\nu)} + \frac{1}{2}(-1)^m \left[y_x^{(m)}\right]^2 \right\} + \int f(y)\,dy = C,$$

where C is an arbitrary constant. Furthermore, the order of the resulting equation can be reduced by one by the substitution $w(y) = y_x'$.

27. $\sum_{m=1}^{n} a_m x^m y_x^{(m)} = f(y).$

The substitution $t = \ln|x|$ leads to an autonomous equation of the form 2.4.3.31.

28. $y \sum_{m=0}^{n} a_m y_x^{(2m+1)} = f(x).$

Integrating yields a $2n$th-order ODE:

$$\sum_{m=0}^{n} a_m \left\{ 2 \sum_{\nu=0}^{m-1} (-1)^{\nu} y_x^{(\nu)} y_x^{(2m-\nu)} + (-1)^m \left[y_x^{(m)}\right]^2 \right\} = 2 \int f(x)\,dx + C,$$

where $y_x^{(0)}$ stands for y.

29. $\sum_{m=0}^{n} a_m y_x^{(m)} y_x^{(2n+1-m)} = f(x).$

The first integral has the form:

$$2 \sum_{m=0}^{n-1} A_m y_x^{(m)} y_x^{(2n-m)} + A_n \left[y_x^{(n)}\right]^2 = 2 \int f(x)\,dx + C,$$

where

$$A_m = \sum_{k=0}^{m} (-1)^{m+k} a_k = a_m - a_{m-1} + a_{m-2} - \cdots.$$

If the condition

$$A_n = 2 \sum_{m=0}^{n-1} (-1)^{n-1+m} A_m$$

is satisfied, the obtained ODE can be integrated twice more (in particular, see equation 2.4.3.23).

30. $y_x^{(n)} = F(x, y_x', y_{xx}'', \ldots, y_x^{(n-1)}).$

The right-hand side of this equation does not explicitly depend on the desired function y.

The substitution $u(x) = y_x'$ leads to an $(n-1)$st-order ODE:

$$u_x^{(n-1)} = F(x, u, u_x', \ldots, u_x^{(n-2)}).$$

31. $y_x^{(n)} = F(y, y_x', y_{xx}'', \ldots, y_x^{(n-1)}).$

Autonomous nth-order ODE. The right-hand side of this equation does not explicitly depend on the independent variable x.

The substitution $u(y) = y_x'$ leads to an $(n-1)$st-order ODE.

32. $y_x^{(n)} = yF(y_x'/y, y_{xx}''/y, \ldots, y_x^{(n-1)}/y).$

This is a special case of ODE 2.4.3.33.

Particular solution:
$$y = Ce^{\lambda x},$$
where C is an arbitrary constant, and λ is a root of algebraic (transcendental) equation: $\lambda^n = F(\lambda, \lambda^2, \ldots, \lambda^{n-1}).$

33. $y_x^{(n)} = yF(x, y_x'/y, y_{xx}''/y, \ldots, y_x^{(n-1)}/y).$

ODE homogeneous in the dependent variable y. This ODE remains unchanged under scaling of the dependent variable, $y \Longrightarrow cy$, where c is an arbitrary nonzero number.

The substitution $u(x) = y_x'/y$ leads to an $(n-1)$st-order ODE.

34. $y_x^{(n)} = x^{-n}F(y, xy_x', x^2 y_{xx}'', \ldots, x^{n-1}y_x^{(n-1)}).$

ODE homogeneous in the independent variable x. This ODE remains unchanged under scaling of the independent variable, $x \Longrightarrow cx$, where c is an arbitrary nonzero number.

1°. The substitution $t = \ln|x|$ leads to an autonomous ODE of the form 2.4.3.31.

2°. The substitution $u(y) = xy_x'$ leads to an $(n-1)$st-order ODE.

35. $y_x^{(n)} = F(x, xy_x' - y, y_{xx}'', y_{xxx}''', \ldots, y_x^{(n-1)}).$

The substitution $u(x) = xy_x' - y$ leads to an $(n-1)$st-order ODE:
$$\zeta_x^{(n-2)} = F(x, u, \zeta, \zeta_x', \ldots, \zeta_x^{(n-3)}), \quad \text{where} \quad \zeta = u_x'/x.$$

36. $y_x^{(n)} = x^{1-n}F(y/x, y_x', xy_{xx}'', \ldots, x^{n-2}y_x^{(n-1)}).$

ODE homogeneous in both variables. This ODE is invariant under simultaneous same scaling (dilatation) of the independent and dependent variables, $x \Longrightarrow cx$ and $y \Longrightarrow cy$, where c is an arbitrary nonzero number.

1°. The transformation $t = \ln x$, $z = y/x$ leads to an autonomous ODE of the form 2.4.3.31.

2°. The transformation $z = y/x$, $w = xy_x'/y$ leads to an $(n-1)$st-order ODE.

37. $y_x^{(n)} = x^{-k-n}F(x^k y, x^{k+1}y_x', \ldots, x^{k+n-1}y_x^{(n-1)}).$

Generalized homogeneous ODE. This ODE remains unchanged under simultaneous scaling of the dependent and independent variables in accordance with the rules $x \Longrightarrow cx$ and $y \Longrightarrow c^{-k}y$, where c is an arbitrary nonzero constant.

1°. The transformation $t = \ln x$, $u = x^k y$ leads to an autonomous equation of the form 2.4.3.31.

2°. This ODE admits a particular solution of the form $y = Ax^{-k}$, where A is a constant.

38. $y_x^{(n)} = yx^{-n}F(x^p y^q, xy_x'/y, x^2 y_{xx}''/y, \ldots, x^{n-1}y_x^{(n-1)}/y).$

Generalized homogeneous equation. For $q \neq 0$, this ODE is an alternative form of equation 2.4.3.37 with $k = p/q$.

1°. The transformation $t = x^p y^q$, $z = xy_x'/y$ leads to an $(n-1)$st-order ODE.

2°. This ODE admits a particular solution of the form $y = Ax^{-p/q}$, where A is a constant.

39. $y_x^{(n)} = e^{-\alpha x} F\big(e^{\alpha x} y,\ e^{\alpha x} y_x',\ e^{\alpha x} y_{xx}'',\ \ldots,\ e^{\alpha x} y_x^{(n-1)}\big).$

ODE invariant under "translation–scaling" transformation. This ODE remains unchanged under simultaneous translation and scaling of the variables, $x \Longrightarrow x + b$ and $y \Longrightarrow cy$, where $c = e^{-b\alpha}$ and b is an arbitrary constant.

1°. The substitution $z = e^{\alpha x} y$ leads to an autonomous equation of the form 2.4.3.31 and the transformation $z = e^{\alpha x} y$, $w = y_x'/y$ leads to an $(n-1)$st-order ODE.

2°. This ODE admits a particular solution of the form $y = A e^{-\alpha x}$, where A is a constant.

40. $y_x^{(n)} = y F\big(e^{\beta x} y^m,\ y_x'/y,\ y_{xx}''/y,\ \ldots,\ y_x^{(n-1)}/y\big).$

ODE invariant under "translation–scaling" transformation. For $m \ne 0$, this ODE is an alternative form of equation 2.4.3.39 with $\alpha = \beta/m$.

1°. The transformation $z = e^{\beta x} y^m$, $w = y_x'/y$ leads to an $(n-1)$st-order ODE.

2°. This ODE admits a particular solution of the form $y = A e^{-\beta x/m}$, where A is a constant.

41. $y_x^{(n)} = x^{-n} F\big(x^m e^{\alpha y},\ xy_x',\ x^2 y_{xx}'',\ \ldots,\ x^{n-1} y_x^{(n-1)}\big).$

ODE invariant under "scaling–translation" transformation. This ODE remains unchanged under simultaneous scaling and translation of the variables, $x \Longrightarrow bx$ and $y \Longrightarrow y + c$, where $b = e^{-c\alpha/m}$ and c is an arbitrary constant.

1°. The transformation $z = x^m e^{\alpha y}$, $w = xy_x'$ leads to an $(n-1)$st-order ODE.

2°. This ODE admits a particular solution of the form $y = -(m/\alpha)\ln(Ax)$, where A is a constant.

▶ Note that the handbooks by Kamke (1977) and Zaitsev & Polyanin (2003, 2018) were extensively used in compiling this chapter; references to these sources are usually omitted. In these books (see also Murphy, 1960), one can also find exact solutions of other linear and nonlinear ordinary differential equations. The main analytical methods for finding solutions to ODEs are described, for example, in the books by Ince (1956), Murphy (1960), Kamke (1977), Zaitsev & Polyanin (2018), and Zwillinger (2022). For numerical methods of integrating such equations, see, for example, Butcher (1987 and 2016), Fox & Mayers (1987), Hairer et al. (1993), Shampine (1994), Hairer & Wanner (1996), and Ascher & Petzold (1998).

References

Abramowitz, M. and Stegun, I.A. (Editors), *Handbook of Mathematical Functions with Formulas, Graphs and Mathematical Tables*, National Bureau of Standards Applied Mathematics, Washington, 1964.

Andrews, G.E., Askey, R., and Roy, R., *Special Functions*, Cambridge Univ. Press, Cambridge, 2001.

Ascher, U. and Petzold, L., *Computer Methods for Ordinary Differential Equations and Differential-Algebraic Equations*, SIAM, Philadelphia, 1998.

Bailey, W.N., *Generalised Hypergeometric Series*, Cambridge Univ. Press, Cambridge, 1935.
Bateman, H. and Erdélyi, A., *Higher Transcendental Functions*, Vols. 1 and 2, McGraw-Hill, New York, 1953.
Bateman, H. and Erdélyi, A., *Higher Transcendental Functions*, Vol. 3, McGraw-Hill, New York, 1955.
Beals, R. and Wong, R., *Special Functions and Orthogonal Polynomials*, Cambridge Univ. Press, Cambridge, 2016.
Butcher, J.C., *The Numerical Analysis of Ordinary Differential Equations: Runge-Kutta and General Linear Methods*, Wiley-Interscience, New York, 1987.
Butcher, J.C., *Numerical Methods for Ordinary Differential Equations, 2nd ed.*, Wiley, New York, 2016.
Fox, L. and Mayers, D.F., *Numerical Solution of Ordinary Differential Equations for Scientists and Engineers*, Chapman & Hall, 1987.
Hairer, E., Nørsett, S.P., and Wanner, G., *Solving Ordinary Differential Equations I: Nonstiff Problems, 2nd ed.*, Springer, Berlin, 1993.
Hairer, E. and Wanner, G., *Solving Ordinary Differential Equations II. Stiff and Differential-Algebraic Problems, 2nd ed.*, Springer, New York, 1996.
Ince, E.L., *Ordinary Differential Equations*, Dover Publ., New York, 1956.
Kamke, E., *Differentialgleichungen: Lösungsmethoden und Lösungen, I, Gewöhnliche Differentialgleichungen*, B. G. Teubner, Leipzig, 1977.
Magnus, W., Oberhettinger, F., and Soni, R.P., *Formulas and Theorems for the Special Functions of Mathematical Physics, 3rd Edition*, Springer, Berlin, 1966.
McLachlan, N.W., *Bessel Functions for Engineers*, Clarendon Press, Oxford, 1955.
McLachlan, N.W., *Theory and Application of Mathieu Functions*, Clarendon Press, Oxford, 1947.
Murphy, G.M., *Ordinary Differential Equations and Their Solutions*, D. Van Nostrand, New York, 1960.
Olver, F.W.J., Lozier, D.W., Boisvert, R.F., and Clark, C.W. (Editors), *NIST Handbook of Mathematical Functions*, NIST and Cambridge Univ. Press, Cambridge, 2010.
Polyanin, A.D. and Zaitsev, V.F., *Handbook of Exact Solutions for Ordinary Differential Equations, 2nd ed.*, Chapman & Hall/CRC Press, Boca Raton–London, 2003.
Polyanin, A.D. and Zaitsev, V.F., *Handbook of Ordinary Differential Equations: Exact Solutions, Methods, and Problems*, CRC Press, Boca Raton–London, 2018.
Polyanin, A.D. and Zhurov, A.I. Parametrically defined nonlinear differential equations and their solutions: Applications in fluid dynamics, *Appl. Math. Lett.*, Vol. 55, pp. 72–80, 2016.
Polyanin, A.D. and Zhurov, A.I. Parametrically defined nonlinear differential equations, differential-algebraic equations, and implicit ODEs: Transformations, general solutions, and integration methods, *Appl. Math. Lett.*, Vol. 64, pp. 59–66, 2017.
Saigo, M. and Kilbas, A.A., Solution of one class of linear differential equations in terms of Mittag-Leffler type functions [in Russian], *Dif. Uravneniya*, Vol. 38, No. 2, pp. 168–176, 2000.
Shampine, L.F., *Numerical Solution of Ordinary Differential Equations*, Chapman & Hall/CRC Press, Boca Raton, 1994.
Slavyanov, S.Yu. and Lay, W., *Special Functions: A Unified Theory Based on Singularities*, Oxford Univ. Press, Oxford, 2000.
Weisstein, E.W., *CRC Concise Encyclopedia of Mathematics, 2nd ed.*, Chapman and Hall/CRC Press, Boca Raton–London, 2003.
Whittaker, E.T. and Watson, G.N., *A Course of Modern Analysis*, Vols. 1–2, Cambridge Univ. Press, Cambridge, 1952.
Zwillinger, D., *Handbook of Differential Equations, 4th ed.*, CRC Press, Boca Raton, 2022.

Chapter 3

Systems of Ordinary Differential Equations

▶ **Preliminary remarks.** This chapter describes exact solutions to some linear and nonlinear systems consisting of two or more coupled ordinary differential equations of the first or second order. It also presents some transformations, first integrals, and reductions leading to simpler separate (uncoupled) ODEs. Special attention is paid to nonlinear systems of a reasonably general form with arbitrary functions.

3.1. Linear Systems of ODEs

3.1.1. Systems of Two First-Order ODEs

1. $x'_t = ax + by, \quad y'_t = cx + dy.$

System of two constant-coefficient first-order linear homogeneous ODEs.

Let us write out the characteristic equation

$$\lambda^2 - (a+d)\lambda + ad - bc = 0 \tag{1}$$

and find its discriminant

$$D = (a-d)^2 + 4bc. \tag{2}$$

1°. Case $ad - bc \neq 0$. The origin of coordinates $x = y = 0$ is the only one stationary point; it is

a node if $D = 0$;
a node if $D > 0$ and $ad - bc > 0$;
a saddle if $D > 0$ and $ad - bc < 0$;
a focus if $D < 0$ and $a + d \neq 0$;
a center if $D < 0$ and $a + d = 0$.

1.1. Suppose $D > 0$. The characteristic equation (1) has two distinct real roots:

$$\lambda_1 = \tfrac{1}{2}(a+d+\sqrt{D}), \quad \lambda_2 = \tfrac{1}{2}(a+d-\sqrt{D}).$$

The general solution of the original system of differential equations is expressed as

$$x = C_1 b e^{\lambda_1 t} + C_2 b e^{\lambda_2 t},$$
$$y = C_1(\lambda_1 - a)e^{\lambda_1 t} + C_2(\lambda_2 - a)e^{\lambda_2 t},$$

where C_1 and C_2 are arbitrary constants.

DOI: 10.1201/9781003051329-3

1.2. Suppose $D < 0$. The characteristic equation (1) has two complex conjugate roots,
$$\lambda_{1,2} = \sigma \pm i\beta, \quad \sigma = \tfrac{1}{2}(a+d), \quad \beta = \tfrac{1}{2}\sqrt{|D|}, \quad i^2 = -1.$$
The general solution of the original system of differential equations is given by
$$x = be^{\sigma t}[C_1 \sin(\beta t) + C_2 \cos(\beta t)],$$
$$y = e^{\sigma t}\{[(\sigma - a)C_1 - \beta C_2]\sin(\beta t) + [\beta C_1 + (\sigma - a)C_2]\cos(\beta t)\},$$
where C_1 and C_2 are arbitrary constants.

1.3. Suppose $D = 0$ and $a \neq d$. The characteristic equation (1) has two equal real roots, $\lambda_1 = \lambda_2$. The general solution of the original system of differential equations is
$$x = 2b\left(C_1 + \frac{C_2}{a-d} + C_2 t\right)\exp\left(\frac{a+d}{2}t\right),$$
$$y = [(d-a)C_1 + C_2 + (d-a)C_2 t]\exp\left(\frac{a+d}{2}t\right),$$
where C_1 and C_2 are arbitrary constants.

1.4. Suppose $a = d \neq 0$ and $b = 0$. General solution:
$$x = C_1 e^{at}, \quad y = (cC_1 t + C_2)e^{at}.$$

1.5. Suppose $a = d \neq 0$ and $c = 0$. General solution:
$$x = (bC_1 t + C_2)e^{at}, \quad y = C_1 e^{at}.$$

2°. Case $ad - bc = 0$ and $a^2 + b^2 > 0$. The entire straight line $ax + by = 0$ consists of singular points. The system in question may be rewritten in the form
$$x'_t = ax + by, \quad y'_t = k(ax + by).$$

2.1. Suppose $a + bk \neq 0$. General solution:
$$x = bC_1 + C_2 e^{(a+bk)t}, \quad y = -aC_1 + kC_2 e^{(a+bk)t}.$$

2.2. Suppose $a + bk = 0$. General solution:
$$x = C_1(bkt - 1) + bC_2 t, \quad y = k^2 b C_1 t + (bk^2 t + 1)C_2.$$

2. $x'_t = a_1 x + b_1 y + c_1, \quad y'_t = a_2 x + b_2 y + c_2.$

The general solution of this system is given by the sum of any one of its particular solutions and the general solution of the corresponding homogeneous system of ODEs (see system 3.1.1.1).

1°. Suppose $a_1 b_2 - a_2 b_1 \neq 0$. A particular solution:
$$x = x_0, \quad y = y_0,$$
where the constants x_0 and y_0 are determined by solving the linear algebraic system of equations
$$a_1 x_0 + b_1 y_0 + c_1 = 0, \quad a_2 x_0 + b_2 y_0 + c_2 = 0.$$

$2°$. Suppose $a_1b_2 - a_2b_1 = 0$ and $a_1^2 + b_1^2 > 0$. Then the original system can be rewritten as
$$x'_t = ax + by + c_1, \quad y'_t = k(ax + by) + c_2.$$

2.1. If $\sigma = a + bk \neq 0$, the original system has a particular solution of the form
$$x = b\sigma^{-1}(c_1k - c_2)t - \sigma^{-2}(ac_1 + bc_2), \quad y = kx + (c_2 - c_1k)t.$$

2.2. If $\sigma = a + bk = 0$, the original system has a particular solution of the form
$$x = \tfrac{1}{2}b(c_2 - c_1k)t^2 + c_1t, \quad y = kx + (c_2 - c_1k)t.$$

3. $x'_t = f(t)x + g(t)y, \quad y'_t = g(t)x + f(t)y.$

General solution:
$$x = e^F(C_1 e^G + C_2 e^{-G}), \quad y = e^F(C_1 e^G - C_2 e^{-G}),$$
where C_1 and C_2 are arbitrary constants, and
$$F = \int f(t)\,dt, \quad G = \int g(t)\,dt.$$

4. $x'_t = f(t)x + g(t)y, \quad y'_t = -g(t)x + f(t)y.$

General solution (Kamke, 1977):
$$x = F(C_1 \cos G + C_2 \sin G), \quad y = F(-C_1 \sin G + C_2 \cos G),$$
where C_1 and C_2 are arbitrary constants, and
$$F = \exp\left[\int f(t)\,dt\right], \quad G = \int g(t)\,dt.$$

5. $x'_t = f(t)x + g(t)y, \quad y'_t = ag(t)x + [f(t) + bg(t)]y.$

The transformation
$$x = \exp\left[\int f(t)\,dt\right] u, \quad y = \exp\left[\int f(t)\,dt\right] v, \quad \tau = \int g(t)\,dt$$
leads to a system of constant coefficient linear ODEs of the form 3.1.1.1:
$$u'_\tau = v, \quad v'_\tau = au + bv.$$

6. $x'_t = f(t)x + g(t)y, \quad y'_t = a[f(t) + ah(t)]x + a[g(t) - h(t)]y.$

Let us multiply the first equation by $-a$ and add it to the second equation to obtain
$$y'_t - ax'_t = -ah(t)(y - ax).$$
By setting $U = y - ax$ and then integrating, one obtains
$$y - ax = C_1 \exp\left[-a \int h(t)\,dt\right], \qquad (*)$$
where C_1 is an arbitrary constant. On solving $(*)$ for y and on substituting the resulting expression into the first equation of the system, one arrives at a first-order linear ODE for x.

7. $x'_t = f(t)x + g(t)y$, $y'_t = h(t)x + p(t)y$.

1°. Let us express y from the first equation and substitute into the second one to obtain a second-order linear ODE:

$$gx''_{tt} - (fg + gp + g'_t)x'_t + (fgp - g^2h + fg'_t - f'_tg)x = 0. \tag{1}$$

This equation is easy to integrate if, for example, the following conditions are met:

1) $fgp - g^2h + fg'_t - f'_tg = 0$;
2) $fgp - g^2h + fg'_t - f'_tg = ag$, $fg + gp + g'_t = bg$.

In the first case, equation (1) has a particular solution $u = C = \text{const}$. In the second case, it is a constant-coefficient linear ODE.

A considerable number of other solvable cases of equation (1) can be found in Section 2.2.

2°. Suppose a particular solution of the system in question is known:

$$x = x_0(t), \quad y = y_0(t).$$

Then the general solution can be written out in the form

$$x(t) = C_1 x_0(t) + C_2 x_0(t) \int \frac{g(t)F(t)P(t)}{x_0^2(t)}\, dt,$$

$$y(t) = C_1 y_0(t) + C_2 \left[\frac{F(t)P(t)}{x_0(t)} + y_0(t) \int \frac{g(t)F(t)P(t)}{x_0^2(t)}\, dt \right],$$

where C_1 and C_2 are arbitrary constants, and

$$F(t) = \exp\left[\int f(t)\, dt\right], \quad P(t) = \exp\left[\int p(t)\, dt\right].$$

3.1.2. Systems of Two Second-Order ODEs

1. $x''_{tt} = ax + by$, $y''_{tt} = cx + dy$.

System of two constant-coefficient second-order linear homogeneous ODEs (a special case of system 3.1.3.2 with $n = 2$).

The characteristic equation is a biquadratic algebraic equation of the form

$$\lambda^4 - (a+d)\lambda^2 + ad - bc = 0.$$

1°. Case $ad - bc \neq 0$.

1.1. Suppose $(a - d)^2 + 4bc \neq 0$. The characteristic equation has four distinct roots $\lambda_1, \ldots, \lambda_4$. The general solution of the system in question is written as

$$x = C_1 b e^{\lambda_1 t} + C_2 b e^{\lambda_2 t} + C_3 b e^{\lambda_3 t} + C_4 b e^{\lambda_4 t},$$
$$y = C_1(\lambda_1^2 - a)e^{\lambda_1 t} + C_2(\lambda_2^2 - a)e^{\lambda_2 t} + C_3(\lambda_3^2 - a)e^{\lambda_3 t} + C_4(\lambda_4^2 - a)e^{\lambda_4 t},$$

where $C_1, \ldots, C_4$ are arbitrary constants.

1.2. General solution with $(a-d)^2 + 4bc = 0$ and $a \neq d$:

$$x = 2C_1\left(bt + \frac{2bk}{a-d}\right)e^{kt/2} + 2C_2\left(bt - \frac{2bk}{a-d}\right)e^{-kt/2} + 2bC_3 te^{kt/2} + 2bC_4 te^{-kt/2},$$

$$y = C_1(d-a)te^{kt/2} + C_2(d-a)te^{-kt/2} + C_3[(d-a)t + 2k]e^{kt/2}$$
$$+ C_4[(d-a)t - 2k]e^{-kt/2},$$

where $C_1, \ldots, C_4$ are arbitrary constants and $k = \sqrt{2(a+d)}$.

1.3. General solution with $a = d \neq 0$ and $b = 0$:

$$x = 2\sqrt{a}\, C_1 e^{\sqrt{a}\,t} + 2\sqrt{a}\, C_2 e^{-\sqrt{a}\,t},$$
$$y = cC_1 t e^{\sqrt{a}\,t} - cC_2 t e^{-\sqrt{a}\,t} + C_3 e^{\sqrt{a}\,t} + C_4 e^{-\sqrt{a}\,t}.$$

1.4. General solution with $a = d \neq 0$ and $c = 0$:

$$x = bC_1 t e^{\sqrt{a}\,t} - bC_2 t e^{-\sqrt{a}\,t} + C_3 e^{\sqrt{a}\,t} + C_4 e^{-\sqrt{a}\,t},$$
$$y = 2\sqrt{a}\, C_1 e^{\sqrt{a}\,t} + 2\sqrt{a}\, C_2 e^{-\sqrt{a}\,t}.$$

2°. Case $ad - bc = 0$ and $a^2 + b^2 > 0$. The original system can be rewritten in the form

$$x''_{tt} = ax + by, \quad y''_{tt} = k(ax + by).$$

2.1. General solution with $a + bk \neq 0$:

$$x = C_1 \exp(t\sqrt{a+bk}) + C_2 \exp(-t\sqrt{a+bk}) + C_3 bt + C_4 b,$$
$$y = C_1 k \exp(t\sqrt{a+bk}) + C_2 k \exp(-t\sqrt{a+bk}) - C_3 at - C_4 a.$$

2.2. General solution with $a + bk = 0$:

$$x = C_1 bt^3 + C_2 bt^2 + C_3 t + C_4,$$
$$y = kx + 6C_1 t + 2C_2.$$

2. $x''_{tt} = a_1 x + b_1 y + c_1, \quad y''_{tt} = a_2 x + b_2 y + c_2.$

The general solution of this system is expressed as the sum of any one of its particular solutions and the general solution of the corresponding homogeneous system (see system 3.1.2.1).

1°. Suppose $a_1 b_2 - a_2 b_1 \neq 0$. A particular solution:

$$x = x_0, \quad y = y_0,$$

where the constants x_0 and y_0 are determined by solving the linear algebraic system of equations

$$a_1 x_0 + b_1 y_0 + c_1 = 0, \quad a_2 x_0 + b_2 y_0 + c_2 = 0.$$

2°. Suppose $a_1 b_2 - a_2 b_1 = 0$ and $a_1^2 + b_1^2 > 0$. Then the system can be rewritten as

$$x''_{tt} = ax + by + c_1, \quad y''_{tt} = k(ax + by) + c_2.$$

2.1. If $\sigma = a + bk \neq 0$, the original system has a particular solution

$$x = \tfrac{1}{2} b\sigma^{-1}(c_1 k - c_2) t^2 - \sigma^{-2}(ac_1 + bc_2), \quad y = kx + \tfrac{1}{2}(c_2 - c_1 k) t^2.$$

2.2. If $\sigma = a + bk = 0$, the system has a particular solution
$$x = \tfrac{1}{24}b(c_2 - c_1 k)t^4 + \tfrac{1}{2}c_1 t^2, \quad y = kx + \tfrac{1}{2}(c_2 - c_1 k)t^2.$$

3. $x''_{tt} - ay'_t + bx = 0, \quad y''_{tt} + ax'_t + by = 0.$

This system is used to describe the horizontal motion of a pendulum taking into account Earth's rotation.

General solution with $a^2 + 4b > 0$:
$$x = C_1 \cos(\alpha t) + C_2 \sin(\alpha t) + C_3 \cos(\beta t) + C_4 \sin(\beta t),$$
$$y = -C_1 \sin(\alpha t) + C_2 \cos(\alpha t) - C_3 \sin(\beta t) + C_4 \cos(\beta t),$$

where $C_1, \ldots, C_4$ are arbitrary constants and
$$\alpha = \tfrac{1}{2}a + \tfrac{1}{2}\sqrt{a^2 + 4b}, \quad \beta = \tfrac{1}{2}a - \tfrac{1}{2}\sqrt{a^2 + 4b}.$$

4. $x''_{tt} + a_1 x'_t + b_1 y'_t + c_1 x + d_1 y = k_1 e^{i\omega t},$
$y''_{tt} + a_2 x'_t + b_2 y'_t + c_2 x + d_2 y = k_2 e^{i\omega t}.$

Systems of this type often arise in oscillation theory (e.g., oscillations of a ship and a ship gyroscope). The general solution of this constant-coefficient linear nonhomogeneous system of ODEs is expressed as the sum of any one of its particular solutions and the general solution of the corresponding homogeneous system (with $k_1 = k_2 = 0$).

1°. A particular solution is sought by the method of undetermined coefficients in the form
$$x = A_* e^{i\omega t}, \quad y = B_* e^{i\omega t}.$$

On substituting these expressions into the system of differential equations in question, one arrives at a linear nonhomogeneous system of algebraic equations for the coefficients A_* and B_*.

2°. The general solution of a homogeneous system of differential equations is determined by a linear combination of its linearly independent particular solutions, which are sought using the method of undetermined coefficients in the form of exponential functions,
$$x = Ae^{\lambda t}, \quad y = Be^{\lambda t}.$$

On substituting these expressions into the system and on collecting the coefficients of the unknowns A and B, one obtains
$$(\lambda^2 + a_1 \lambda + c_1)A + (b_1 \lambda + d_1)B = 0,$$
$$(a_2 \lambda + c_2)A + (\lambda^2 + b_2 \lambda + d_2)B = 0.$$

For a nontrivial solution to exist, the determinant of this system must vanish. This requirement results in the characteristic equation
$$(\lambda^2 + a_1 \lambda + c_1)(\lambda^2 + b_2 \lambda + d_2) - (b_1 \lambda + d_1)(a_2 \lambda + c_2) = 0,$$

which is used to determine λ. If the roots of this equation, $\lambda_1, \ldots, \lambda_4$, are all distinct, then the general solution of the original system of differential equations has the form
$$x = -C_1(b_1\lambda_1 + d_1)e^{\lambda_1 t} - C_2(b_1\lambda_2 + d_1)e^{\lambda_2 t} - C_3(b_1\lambda_1 + d_1)e^{\lambda_3 t} - C_4(b_1\lambda_4 + d_1)e^{\lambda_4 t},$$
$$y = C_1(\lambda_1^2 + a_1\lambda_1 + c_1)e^{\lambda_1 t} + C_2(\lambda_2^2 + a_1\lambda_2 + c_1)e^{\lambda_2 t}$$
$$+ C_3(\lambda_3^2 + a_1\lambda_3 + c_1)e^{\lambda_3 t} + C_4(\lambda_4^2 + a_1\lambda_4 + c_1)e^{\lambda_4 t},$$

where $C_1, \ldots, C_4$ are arbitrary constants.

5. $x''_{tt} = a(ty'_t - y)$, $y''_{tt} = b(tx'_t - x)$.

The transformation
$$u = tx_t - x, \quad v = ty'_t - y \tag{1}$$

leads to a first-order linear system of ODEs:
$$u'_t = atv, \quad v'_t = btu.$$

The general solution of this system is expressed as

with $ab > 0$: $\begin{cases} u(t) = C_1 a \exp(\frac{1}{2}\sqrt{ab}\,t^2) + C_2 a \exp(-\frac{1}{2}\sqrt{ab}\,t^2), \\ v(t) = C_1 \sqrt{ab} \exp(\frac{1}{2}\sqrt{ab}\,t^2) - C_2 \sqrt{ab} \exp(-\frac{1}{2}\sqrt{ab}\,t^2); \end{cases}$

with $ab < 0$: $\begin{cases} u(t) = C_1 a \cos(\frac{1}{2}\sqrt{|ab|}\,t^2) + C_2 a \sin(\frac{1}{2}\sqrt{|ab|}\,t^2), \\ v(t) = -C_1 \sqrt{|ab|} \sin(\frac{1}{2}\sqrt{|ab|}\,t^2) + C_2 \sqrt{|ab|} \cos(\frac{1}{2}\sqrt{|ab|}\,t^2), \end{cases}$ (2)

where C_1 and C_2 are arbitrary constants. On substituting (2) into (1) and integrating, one arrives at the general solution of the original system in the form

$$x = C_3 t + t \int \frac{u(t)}{t^2}\,dt, \quad y = C_4 t + t \int \frac{v(t)}{t^2}\,dt,$$

where C_3 and C_4 are arbitrary constants.

6. $x''_{tt} = f(t)(a_1 x + b_1 y) + g(t)$, $y''_{tt} = f(t)(a_2 x + b_2 y) + h(t)$.

Let k_1 and k_2 be roots of the quadratic equation
$$k^2 - (a_1 + b_2)k + a_1 b_2 - a_2 b_1 = 0.$$

Then, on multiplying the equations of the system by the constants a_2 and $k - a_1$, respectively, and on adding them together, one can rewrite the system in the form of two independent (uncoupled) linear ODEs:

$z''_1 = k_1 f(t) z_1 + a_2 g(t) + (k_1 - a_1) h(t)$, $z_1 = a_2 x + (k_1 - a_1) y$ (for $k = k_1$);
$z''_2 = k_2 f(t) z_2 + a_2 g(t) + (k_2 - a_1) h(t)$, $z_2 = a_2 x + (k_2 - a_1) y$ (for $k = k_2$).

Here, a prime stands for a derivative with respect to t.

7. $x''_{tt} + f(t)x + g(t)(x-y) + h(t) = 0$, $y''_{tt} + f(t)y - g(t)(x-y) + p(t) = 0$.

First, we add both ODEs together term by term, and then subtract the second from the first ODE. Introducing further new variables according to the formulas

$$u = x + y, \quad v = x - y \quad \left(\text{or } x = \tfrac{1}{2}(u+v), \quad y = \tfrac{1}{2}(u-v)\right),$$

we obtain two uncoupled second-order linear ODEs:
$$u''_{tt} + f(x)u + h(t) + p(t) = 0,$$
$$v''_{tt} + [f(t) + 2g(t)]v + h(t) - p(t) = 0.$$

For a simple special case $f(t) = a^2 = \text{const}$, $g(t) = b^2 = \text{const}$, and $h(t) = p(t) = 0$, we have

$$u = C_1 \cos(at) + C_2 \sin(at), \quad v = C_3 \cos(\sqrt{a^2 + 2b^2}\,t) + C_4 \sin(\sqrt{a^2 + 2b^2}\,t),$$

where $C_1, \ldots, C_4$ are arbitrary constants.

8. $x_{tt}'' = f(t)(a_1 x_t' + b_1 y_t') + g(t), \quad y_{tt}'' = f(t)(a_2 x_t' + b_2 y_t') + h(t).$

$1°$. Let k_1 and k_2 be roots of the quadratic equation

$$k^2 - (a_1 + b_2)k + a_1 b_2 - a_2 b_1 = 0.$$

Then, on multiplying the equations of the system by the constants a_2 and $k - a_1$, respectively, and on adding them together, one can rewrite the system in the form of two independent (uncoupled) linear ODEs:

$$z_1'' = k_1 f(t) z_1' + a_2 g(t) + (k_1 - a_1) h(t), \quad z_1 = a_2 x + (k_1 - a_1) y \quad \text{(for } k = k_1\text{)};$$
$$z_2'' = k_2 f(t) z_2' + a_2 g(t) + (k_2 - a_1) h(t), \quad z_2 = a_2 x + (k_2 - a_1) y \quad \text{(for } k = k_2\text{)}.$$

$2°$. Let $g(t) = h(t) \equiv 0$. Integrating the ODEs for z_1 and z_2 from Item $1°$ and returning to the original variables, one arrives at a linear algebraic system for the unknowns x and y:

$$a_2 x + (k_1 - a_1) y = C_1 \int \exp[k_1 F(t)] \, dt + C_2,$$
$$a_2 x + (k_2 - a_1) y = C_3 \int \exp[k_2 F(t)] \, dt + C_4,$$

where $C_1, \ldots, C_4$ are arbitrary constants and $F(t) = \int f(t) \, dt$.

9. $x_{tt}'' = af(t)(ty_t' - y), \quad y_{tt}'' = bf(t)(tx_t' - x).$

The transformation

$$u = tx_t' - x, \quad v = ty_t' - y \tag{1}$$

leads to a system of first-order equations:

$$u_t' = atf(t)v, \quad v_t' = btf(t)u.$$

The general solution of this system is expressed as

$$\text{if } ab > 0, \begin{cases} u(t) = C_1 a \exp\left(\sqrt{ab} \int tf(t) \, dt\right) + C_2 a \exp\left(-\sqrt{ab} \int tf(t) \, dt\right), \\ v(t) = C_1 \sqrt{ab} \exp\left(\sqrt{ab} \int tf(t) \, dt\right) - C_2 \sqrt{ab} \exp\left(-\sqrt{ab} \int tf(t) \, dt\right); \end{cases}$$

$$\text{if } ab < 0, \begin{cases} u(t) = C_1 a \cos\left(\sqrt{|ab|} \int tf(t) \, dt\right) + C_2 a \sin\left(\sqrt{|ab|} \int tf(t) \, dt\right), \\ v(t) = -C_1 \sqrt{|ab|} \sin\left(\sqrt{|ab|} \int tf(t) \, dt\right) + C_2 \sqrt{|ab|} \cos\left(\sqrt{|ab|} \int tf(t) \, dt\right), \end{cases} \tag{2}$$

where C_1 and C_2 are arbitrary constants. On substituting (2) into (1) and integrating, one obtains the general solution of the original system

$$x = C_3 t + t \int \frac{u(t)}{t^2} \, dt, \quad y = C_4 t + t \int \frac{v(t)}{t^2} \, dt,$$

where C_3 and C_4 are arbitrary constants.

10. $t^2 x_{tt}'' + a_1 t x_t' + b_1 t y_t' + c_1 x + d_1 y = 0,$
$\quad\; t^2 y_{tt}'' + a_2 t x_t' + b_2 t y_t' + c_2 x + d_2 y = 0.$

Linear ODE system homogeneous in the independent variable (an Euler-type system of ODEs).

1°. The general solution is determined by a linear combination of linearly independent particular solutions that are sought by the method of undetermined coefficients in the form of power-law functions
$$x = A|t|^k, \quad y = B|t|^k.$$
On substituting these expressions into the system and on collecting the coefficients of the unknowns A and B, one obtains
$$[k^2 + (a_1 - 1)k + c_1]A + (b_1 k + d_1)B = 0,$$
$$(a_2 k + c_2)A + [k^2 + (b_2 - 1)k + d_2]B = 0.$$
For a nontrivial solution to exist, the determinant of this system must vanish. This requirement results in the characteristic equation
$$[k^2 + (a_1 - 1)k + c_1][k^2 + (b_2 - 1)k + d_2] - (b_1 k + d_1)(a_2 k + c_2) = 0,$$
which is used to determine k. If the roots of this equation, $k_1, \ldots, k_4$, are all distinct, then the general solution of the system of differential equations in question has the form
$$x = -C_1(b_1 k_1 + d_1)|t|^{k_1} - C_2(b_1 k_2 + d_1)|t|^{k_2} - C_3(b_1 k_3 + d_1)|t|^{k_3} - C_4(b_1 k_4 + d_1)|t|^{k_4},$$
$$y = C_1[k_1^2 + (a_1 - 1)k_1 + c_1]|t|^{k_1} + C_2[k_2^2 + (a_1 - 1)k_2 + c_1]|t|^{k_2}$$
$$+ C_3[k_3^2 + (a_1 - 1)k_3 + c_1]|t|^{k_3} + C_4[k_4^2 + (a_1 - 1)k_4 + c_1]|t|^{k_4},$$
where $C_1, \ldots, C_4$ are arbitrary constants.

2°. The substitution $t = \sigma e^\tau$ ($\sigma \neq 0$) leads to a system of constant-coefficient linear differential equations:
$$x''_{\tau\tau} + (a_1 - 1)x'_\tau + b_1 y'_\tau + c_1 x + d_1 y = 0,$$
$$y''_{\tau\tau} + a_2 x'_\tau + (b_2 - 1)y'_\tau + c_2 x + d_2 y = 0.$$

11. $(\alpha t^2 + \beta t + \gamma)^2 x''_{tt} = ax + by, \quad (\alpha t^2 + \beta t + \gamma)^2 y''_{tt} = cx + dy.$

The transformation
$$\tau = \int \frac{dt}{\alpha t^2 + \beta t + \gamma}, \quad u = \frac{x}{\sqrt{|\alpha t^2 + \beta t + \gamma|}}, \quad v = \frac{y}{\sqrt{|\alpha t^2 + \beta t + \gamma|}}$$
leads to a constant-coefficient linear system of ODEs of the form 3.1.2.1:
$$u''_{\tau\tau} = (a - \alpha\gamma + \tfrac{1}{4}\beta^2)u + bv,$$
$$v''_{\tau\tau} = cu + (d - \alpha\gamma + \tfrac{1}{4}\beta^2)v.$$

12. $x''_{tt} = f(t)(tx'_t - x) + g(t)(ty'_t - y), \quad y''_{tt} = h(t)(tx'_t - x) + p(t)(ty'_t - y).$

The transformation
$$u = tx_t - x, \quad v = ty'_t - y \tag{1}$$
leads to a linear system of first-order ODEs:
$$u'_t = tf(t)u + tg(t)v, \quad v'_t = th(t)u + tp(t)v. \tag{2}$$

In order to find the general solution of this system, it suffices to know any one of its particular solutions (see system 3.1.1.7).

For solutions of some systems of the form (2), see systems 3.1.1.3–3.1.1.6.

If all functions in (2) are proportional, that is,

$$f(t) = a\varphi(t), \quad g(t) = b\varphi(t), \quad h(t) = c\varphi(t), \quad p(t) = d\varphi(t),$$

then the introduction of the new independent variable $\tau = \int t\varphi(t)\,dt$ leads to a constant-coefficient system of the form 3.1.1.1.

2°. Suppose the solution of system (2) has been found in the form

$$u = u(t, C_1, C_2), \quad v = v(t, C_1, C_2), \tag{3}$$

where C_1 and C_2 are arbitrary constants. Then, on substituting (3) into (1) and integrating, one obtains the solution of the original system:

$$x = C_3 t + t\int \frac{u(t, C_1, C_2)}{t^2}\,dt, \quad y = C_4 t + t\int \frac{v(t, C_1, C_2)}{t^2}\,dt,$$

where C_3 and C_4 are arbitrary constants.

3.1.3. Other Systems of Two ODEs

1. $x'_t + ay'_t + by = f(t), \quad x''_{tt} + ay''_{tt} + by'_t + x + ay = g(t).$

The unique solution:

$$x = g + \frac{a}{b}g'_t - f'_t - \frac{a}{b}(f + f''_{tt}), \quad y = \frac{1}{b}(f + f''_{tt} - g'_t).$$

We would like to draw attention to the fact that the solution of this system of ODEs does not contain arbitrary constants. There are no other solutions.

Note that for $a = b = 1$, the solution of this ODE system is given in Kamke (1977).

2. $x^{(n)}_t = a_1 x + b_1 y + f_1(t), \quad y^{(n)}_t = a_2 x + b_2 y + f_2(t).$

Let us multiply the second ODE of the system by k and add it termwise to the first ODE to obtain, after rearrangement,

$$(x + ky)^{(n)}_t = (a_1 + ka_2)\left(x + \frac{b_1 + kb_2}{a_1 + ka_2}y\right) + f_1(t) + kf_2(t). \tag{1}$$

Let us take the constant k so that $\dfrac{b_1 + kb_2}{a_1 + ka_2} = k$, which results in a quadratic equation for k:

$$a_2 k^2 + (a_1 - b_2)k - b_1 = 0. \tag{2}$$

In this case, (1) is a nonhomogeneous linear constant-coefficient nth-other ODE for $z = x + ky$:

$$z^{(n)}_t = (a_1 + ka_2)z + f_1(t) + kf_2(t). \tag{3}$$

Integrating this equation yields

$$x + ky = C_1\varphi_1(t, k) + \cdots + C_n\varphi_n(t, k) + \psi(t, k),$$

where $\varphi_m(t, k)$ are linearly independent solutions of the corresponding homogeneous ODE (3) for $f_1(t) = f_2(t) \equiv 0$ (for solutions to this equation, see equation 2.4.1.7). It follows that if the roots of the quadratic equation (2) are distinct, we have two relations,

$$x + k_1 y = C_1 \varphi_1(t, k_1) + \cdots + C_n \varphi_n(t, k_1) + \psi(t, k_1),$$
$$x + k_2 y = C_{n+1} \varphi_1(t, k_2) + \cdots + C_{2n} \varphi_n(t, k_2) + \psi(t, k_2),$$

which represent a linear algebraic system of equations for the functions x and y.

Remark 3.1. *The above method for the solution of the original system of ODEs (3.1.3.2) is known as D'Alembert's method.*

3.1.4. Systems of Three and More ODEs

1. $x'_t = ax, \quad y'_t = bx + cy, \quad z'_t = dx + ky + pz.$

General solution:

$$x = C_1 e^{at}, \quad y = \frac{bC_1}{a-c} e^{at} + C_2 e^{ct},$$

$$z = \frac{C_1}{a-p}\left(d + \frac{bk}{a-c}\right) e^{at} + \frac{kC_2}{c-p} e^{ct} + C_3 e^{pt},$$

where C_1, C_2, and C_3 are arbitrary constants.

2. $x'_t = cy - bz, \quad y'_t = az - cx, \quad z'_t = bx - ay.$

   1°. First integrals:

$$ax + by + cz = A, \tag{1}$$
$$x^2 + y^2 + z^2 = B^2, \tag{2}$$

where A and B are arbitrary constants. It follows that the integral curves are circles formed by the intersection of planes (1) and spheres (2).

   2°. General solution (Kamke, 1977):

$$x = aC_0 + kC_1 \cos(kt) + (cC_2 - bC_3) \sin(kt),$$
$$y = bC_0 + kC_2 \cos(kt) + (aC_3 - cC_1) \sin(kt),$$
$$z = cC_0 + kC_3 \cos(kt) + (bC_1 - aC_2) \sin(kt),$$

where $k = \sqrt{a^2 + b^2 + c^2}$ and the three of four constants of integration $C_0, \ldots, C_3$ are related by one constraint

$$aC_1 + bC_2 + cC_3 = 0.$$

3. $ax'_t = bc(y - z), \quad by'_t = ac(z - x), \quad cz'_t = ab(x - y).$

   1°. First integral:

$$a^2 x + b^2 y + c^2 z = A,$$

where A is an arbitrary constant. It follows that the integral curves are plane curves.

2°. General solution (Kamke, 1977):

$$x = C_0 + kC_1 \cos(kt) + a^{-1}bc(C_2 - C_3)\sin(kt),$$
$$y = C_0 + kC_2 \cos(kt) + ab^{-1}c(C_3 - C_1)\sin(kt),$$
$$z = C_0 + kC_3 \cos(kt) + abc^{-1}(C_1 - C_2)\sin(kt),$$

where $k = \sqrt{a^2 + b^2 + c^2}$ and three of the four constants of integration $C_0, \ldots, C_3$ are related by one constraint

$$a^2 C_1 + b^2 C_2 + c^2 C_3 = 0.$$

4. $x'_t = (a_1 f + g)x + a_2 f y + a_3 f z,$
$y'_t = b_1 f x + (b_2 f + g)y + b_3 f z, \quad z'_t = c_1 f x + c_2 f y + (c_3 f + g)z.$

Here, $f = f(t)$ and $g = g(t)$.

The transformation

$$x = \exp\left[\int g(t)\,dt\right] u, \quad y = \exp\left[\int g(t)\,dt\right] v, \quad z = \exp\left[\int g(t)\,dt\right] w, \quad \tau = \int f(t)\,dt$$

leads to the system of constant coefficient linear ODEs

$$u'_\tau = a_1 u + a_2 v + a_3 w, \quad v'_\tau = b_1 u + b_2 v + b_3 w, \quad w'_\tau = c_1 u + c_2 v + c_3 w.$$

5. $x'_t = h(t)y - g(t)z, \quad y'_t = f(t)z - h(t)x, \quad z'_t = g(t)x - f(t)y.$

1°. First integral:

$$x^2 + y^2 + z^2 = C^2,$$

where C is an arbitrary constant.

2°. The system concerned can be reduced to a Riccati equation (see Kamke, 1977).

6. $x'_k = a_{k1}x_1 + a_{k2}x_2 + \cdots + a_{kn}x_n; \quad k = 1, 2, \ldots, n.$

System of n constant-coefficient first-order linear homogeneous ODEs.

The general solution of a homogeneous system of ODEs is determined by a linear combination of linearly independent particular solutions, which are sought by the method of undetermined coefficients in the form of exponential functions,

$$x_k = A_k e^{\lambda t}; \quad k = 1, 2, \ldots, n.$$

On substituting these expressions into the system and on collecting the coefficients of the unknowns A_k, one obtains a linear homogeneous system of algebraic equations:

$$a_{k1} A_1 + a_{k2} A_2 + \cdots + (a_{kk} - \lambda) A_k + \cdots + a_{kn} A_n = 0; \quad k = 1, 2, \ldots, n.$$

For a nontrivial solution to exist, the determinant of this system must vanish. This requirement results in a characteristic equation that serves to determine λ.

7. $\sum_{k=1}^{n} \left(a_{mk} t^2 \dfrac{d^2 x_k}{dt^2} + b_{mk} t \dfrac{dx_k}{dt} + c_{mk} x_k \right) = f_m(t), \quad m = 1, 2, \ldots, n.$

Linear nonhomogeneous Euler system of second-order ODEs.

1°. The substitutions $t = \pm e^\xi$ bring this system to a constant-coefficient linear system of ODEs:

$$\sum_{k=1}^{n}\left[a_{mk}\frac{d^2 x_k}{d\xi^2} + (b_{mk} - a_{mk})\frac{dx_k}{d\xi} + c_{mk}x_k\right] = f_m(\pm e^\xi), \qquad m = 1, 2, \ldots, n,$$

which can be solved using, for example, the Laplace transform.

2°. Particular solutions to a homogeneous Euler system (for the system, corresponding to $f_m(t) \equiv 0$) are sought in the form of power functions:

$$x_1 = A_1 t^\sigma, \quad x_2 = A_2 t^\sigma, \ldots, \quad y_n = A_n t^\sigma, \tag{1}$$

where the coefficients $A_1, A_2, \ldots, A_n$ are determined by solving the associated homogeneous system of algebraic equations obtained by substituting expressions (1) into the differential equations of the system in question and dividing by t^σ. Since the system is homogeneous, for it to have nontrivial solutions, its determinant must vanish. This results in a dispersion equation for the exponent σ.

3.2. Nonlinear Systems of Two ODEs

3.2.1. Systems of First-Order ODEs

▶ **Autonomous ODE systems.**

1. $x'_t = ax^2 + by^2$, $\quad y'_t = 2axy \qquad (ab > 0)$.

Let us multiply the first equation by a and the second one by $\lambda = \pm\sqrt{ab}$ and add them together to obtain the two independent first-order separable ODEs:

$$u'_t = u^2, \qquad u = ax + \sqrt{ab}\, y;$$
$$v'_t = v^2, \qquad v = ax - \sqrt{ab}\, y.$$

Integrating these equations, we obtain the general solution to the original ODE system:

$$x = -\frac{1}{2a}\left(\frac{1}{t+C_1} + \frac{1}{t+C_2}\right), \quad y = -\frac{1}{2\sqrt{ab}}\left(\frac{1}{t+C_2} - \frac{1}{t+C_1}\right),$$

where C_1 and C_2 are arbitrary constants.

2. $x'_t = xf(ax - by) + g(ax - by)$, $\quad y'_t = yf(ax - by) + h(ax - by)$.

Here $f(z)$, $g(z)$, and $h(z)$ are arbitrary functions.

1°. Particular solution:

$$x = x_0 + Cb\exp\bigl[f(ax_0 - by_0)t\bigr], \quad y = y_0 + Ca\exp\bigl[f(ax_0 - by_0)t\bigr],$$

where C is an arbitrary constant, and x_0, y_0 is any solution of the transcendental (algebraic) system of equations

$$x_0 f(ax_0 - by_0) + g(ax_0 - by_0) = 0, \quad y_0 f(ax_0 - by_0) + h(ax_0 - by_0) = 0.$$

$2°$. Let us multiply the first equation by a and the second one by $-b$ and add them together to obtain the first-order separable ODE

$$z'_t = zf(z) + ag(z) - bh(z), \qquad z = ax - by. \qquad (1)$$

We will consider this equation in conjunction with the first equation of the system,

$$x'_t = xf(z) + g(z). \qquad (2)$$

Autonomous equation (1) can be treated separately; its general solution can be written out in implicit form (see ODE 2.1.1.1). The function $x = x(t)$ can be determined by solving the first-order linear equation (2), and the function $y = y(t)$ is found as $y = (ax - z)/b$.

3. $x'_t = b_2 f(a_1 x + b_1 y) + b_1 g(a_2 x + b_2 y),$
   $y'_t = -a_2 f(a_1 x + b_1 y) - a_1 g(a_2 x + b_2 y).$

It is assumed that $a_1 b_2 - a_2 b_1 \neq 0$.

Multiplying the ODEs by suitable constants and adding together, one obtains two independent first-order separable ODEs:

$$u'_t = (a_1 b_2 - a_2 b_1) f(u), \qquad u = a_1 x + b_1 y;$$
$$v'_t = -(a_1 b_2 - a_2 b_1) g(v), \qquad v = a_2 x + b_2 y.$$

4. $x'_t = e^{-ax} f(y - x), \qquad y'_t = e^{-ay} g(y - x).$

$1°$. Particular solution for $a \neq 0$:

$$x = \frac{1}{a} \ln[af(\lambda)t + C], \qquad y = \frac{1}{a} \ln[af(\lambda)t + C] + \lambda,$$

where C is an arbitrary constant, and λ is a root of the transcendental (algebraic) equation $f(\lambda) = e^{-a\lambda} g(\lambda)$.

$2°$. First, in the original system, we exclude the variable t and then make the substitution $y = x + z$. As a result, we obtain a first-order separable ODE of the form 2.1.1.1:

$$z'_x = e^{-az} \frac{g(z)}{f(z)} - 1.$$

5. $x'_t = e^{-ax} f(y - x) + g(y - x), \qquad y'_t = e^{-ay} h(y - x) + g(y - x).$

For $a \neq 0$, the transformation

$$x = \frac{1}{a} \ln u, \qquad y = \frac{1}{a} \ln v$$

leads to a ODE system of the form 3.2.1.8:

$$u'_t = aug\left(\frac{1}{a} \ln \frac{v}{u}\right) + af\left(\frac{1}{a} \ln \frac{v}{u}\right), \qquad v'_t = avg\left(\frac{1}{a} \ln \frac{v}{u}\right) + ah\left(\frac{1}{a} \ln \frac{v}{u}\right).$$

6. $x'_t = xf(y/x), \qquad y'_t = yg(y/x).$

First-order ODE system homogeneous in both unknown variables, where $f(z)$ and $g(z)$ are arbitrary functions. This ODE system is invariant under the same scaling of the unknowns, $x \Longrightarrow cx$ and $y \Longrightarrow cy$, where c is an arbitrary nonzero number.

1°. Particular solution:

$$x = C\exp(st), \quad y = C\lambda\exp(st), \quad s = f(\lambda),$$

where C is an arbitrary constant, and λ is a root of the transcendental (algebraic) equation $f(\lambda) = g(\lambda)$.

2°. Excluding t from the original system, we arrive at a first-order homogeneous ODE, $y'_x = g(y/x)/f(y/x)$, which is easy to integrate by substituting $u = y/x$.

7. $x'_t = x^k f(y/x), \quad y'_t = y^k g(y/x).$

First-order generalized homogeneous ODE system. This ODE system remains unchanged under simultaneous scaling of the dependent and independent variables in accordance with the rule $x \Longrightarrow cx$, $y \Longrightarrow cy$, and $t \Longrightarrow c^{1-k}t$, where c is an arbitrary positive constant.

1°. Particular solution for $k \neq 1$:

$$x = s(t+C)^{1/(1-k)}, \quad y = s\lambda(t+C)^{1/(1-k)}, \quad s = [(1-k)f(\lambda)]^{1/(1-k)},$$

where C is an arbitrary constant, and λ is a root of the transcendental (algebraic) equation $f(\lambda) = \lambda^{k-1} g(\lambda)$.

Particular solution for $k = 1$:

$$x = Ce^{\beta t}, \quad y = \lambda Ce^{\beta t}, \quad \beta = f(\lambda),$$

where C is an arbitrary constant, and λ is a root of the transcendental (algebraic) equation $f(\lambda) = g(\lambda)$.

2°. Excluding t from the original system, we arrive at a first-order homogeneous ODE, $y'_x = (y/x)^k g(y/x)/f(y/x)$, which is easy to integrate by substituting $u = y/x$.

8. $x'_t = xf(y/x) + g(y/x), \quad y'_t = yf(y/x) + h(y/x).$

Here $f(z)$, $g(z)$, and $h(z)$ are arbitrary functions.

1°. Let λ be a root of the algebraic (transcendental) equation

$$\lambda g(\lambda) = h(\lambda).$$

Particular solution if $f(\lambda) \neq 0$:

$$x = C\exp[f(\lambda)t] - \frac{g(\lambda)}{f(\lambda)}, \quad y = \lambda\left\{C\exp[f(\lambda)t] - \frac{g(\lambda)}{f(\lambda)}\right\},$$

where C is an arbitrary constant.

Particular solution if $f(\lambda) = 0$:

$$x = g(\lambda)t + C, \quad y = \lambda[g(\lambda)t + C].$$

2°. The substitution $y = xz$ leads to the system

$$x'_t = xf(z) + g(z), \quad xz'_t = h(z) - zg(z).$$

Dividing the first ODE by the second ODE term by term, we eliminate the variable t. As a result, we arrive at the Bernoulli equation

$$x'_z = \frac{g(z)}{h(z) - zg(z)} x + \frac{f(z)}{h(z) - zg(z)} x^2.$$

9. $x'_t = xf(y/x) + h(y/x), \quad y'_t = yg(y/x) + (y/x)h(y/x).$

General solution:

$$x = F(t)\left[\int \frac{h(\varphi)}{F(t)} dt + C\right], \quad y = \varphi(t)F(t)\left[\int \frac{h(\varphi)}{F(t)} dt + C\right],$$

$$F(t) = \exp\left[\int f(\varphi)\, dt\right],$$

where C is an arbitrary constant, and the function $\varphi = \varphi(t)$ is described by a first-order separable ODE of the form 2.1.1.1:

$$\varphi'_t = \varphi[g(\varphi) - f(\varphi)].$$

10. $x'_t = x^k f(y/x) + xg(y/x), \quad y'_t = y^k h(y/x) + yg(y/x).$

For $k = 1$, see system 3.2.1.6.

For $k \neq 1$, the transformation

$$u = x^{1-k}, \quad v = y^{1-k}$$

leads to an ODE system of the form 3.2.1.8:

$$u'_t = (1-k)ug((v/u)^{1/(1-k)}) + (1-k)f((v/u)^{1/(1-k)}),$$
$$v'_t = (1-k)vg((v/u)^{1/(1-k)}) + (1-k)h((v/u)^{1/(1-k)}).$$

11. $x'_t = xf(y/x)\ln x + xg(y/x), \quad y'_t = yf(y/x)\ln y + yh(y/x).$

The transformation $x = e^u$, $y = e^v$ leads to a system of the form 3.2.1.2:

$$u'_t = uf(e^{v-u}) + g(e^{v-u}), \quad v'_t = vf(e^{v-u}) + h(e^{v-u}).$$

12. $x'_t = xf\left(\dfrac{y}{x}\right) - yg\left(\dfrac{y}{x}\right) + \dfrac{x}{\sqrt{x^2+y^2}} h\left(\dfrac{y}{x}\right),$

$y'_t = yf\left(\dfrac{y}{x}\right) + xg\left(\dfrac{y}{x}\right) + \dfrac{y}{\sqrt{x^2+y^2}} h\left(\dfrac{y}{x}\right).$

Solution:
$$x = r(t)\cos\varphi(t), \quad y = r(t)\sin\varphi(t),$$

where the function $\varphi = \varphi(t)$ is determined from the first-order separable ODE:

$$\varphi'_t = g(\tan\varphi),$$

and the function $r = r(t)$ satisfies the first-order linear ODE:

$$r'_t = rf(\tan\varphi) + h(\tan\varphi).$$

13. $x'_t = xf\left(\dfrac{y}{x}\right) + yg\left(\dfrac{y}{x}\right) + \dfrac{x}{\sqrt{x^2 - y^2}} h\left(\dfrac{y}{x}\right),$

$y'_t = yf\left(\dfrac{y}{x}\right) + xg\left(\dfrac{y}{x}\right) + \dfrac{y}{\sqrt{x^2 - y^2}} h\left(\dfrac{y}{x}\right).$

Solution:
$$x = r(t)\cosh\psi(t), \quad y = r(t)\sinh\psi(t),$$
where the function $\psi = \psi(t)$ is determined from the first-order separable ODE:
$$\psi'_t = g(\tanh\psi),$$
and the function $r = r(t)$ satisfies the first-order linear ODE:
$$r'_t = rf(\tanh\psi) + h(\tanh\psi).$$

14. $x'_t = xf(yx^a), \quad y'_t = yg(yx^a).$

First-order generalized homogeneous ODE system in both unknown variables. This ODE system remains unchanged under scaling of the dependent variables in accordance with the rule $x \Longrightarrow cx$ and $y \Longrightarrow c^{-a}y$, where c is an arbitrary positive constant.

1°. Particular solution:
$$x = Ce^{\lambda t}, \quad y = kC^{-a}e^{-a\lambda t}, \quad \lambda = f(k),$$
where C is an arbitrary constant, and k is a root of the algebraic (transcendental) equation $af(k) + g(k) = 0$.

2°. General solution:
$$x = C\exp\left[\int f(\varphi(t))\,dt\right], \quad y = C^{-a}\varphi(t)\exp\left[-a\int f(\varphi(t))\,dt\right],$$
where C is an arbitrary constant, and the function $\varphi = \varphi(t)$ is described by the first-order separable ODE
$$\varphi'_t = \varphi[g(\varphi) + af(\varphi)].$$

15. $x'_t = x^{1-ab} f(yx^a), \quad y'_t = y^{1+b} g(yx^a).$

First-order generalized homogeneous ODE system. This ODE system remains unchanged under simultaneous scaling of the dependent and independent variables in accordance with the rule $x \Longrightarrow cx$ and $y \Longrightarrow c^{-a}y$, and $t \Longrightarrow c^{ab}t$, where c is an arbitrary positive constant.

1°. Particular solution for $ab \ne 0$:
$$x = A(t+C)^{\frac{1}{ab}}, \quad y = kA^{-a}(t+C)^{-\frac{1}{b}}, \quad A = [abf(k)]^{\frac{1}{ab}},$$
where C is an arbitrary constant, and k is a root of the algebraic (transcendental) equation $af(k) + k^b g(k) = 0$.

2°. Excluding t from the original system, we arrive at a first-order generalized homogeneous ODE of the form 2.1.4.4:
$$y'_x = x^{-a-1}(yx^a)^{b+1}\dfrac{g(yx^a)}{f(yx^a)},$$
which is easy to integrate by substituting $u = yx^a$.

16. $x'_t = xf(ax+by) - byg(y/x), \quad y'_t = yf(ax+by) + ayg(y/x).$

1°. Solution:
$$x = br(t)\cos^2\varphi(t), \quad y = ar(t)\sin^2\varphi(t),$$
where the functions $\varphi = \varphi(t)$ and $r = r(t)$ are determined from uncoupled first-order separable ODEs:
$$r'_t = rf(abr), \quad \varphi'_t = \frac{1}{2}a\tan\varphi\, g\!\left(\frac{a}{b}\tan^2\varphi\right).$$
Their general solutions can be represented in an implicit form (see equation 2.1.1.1).

2°. Solution:
$$x = br(t)\cosh^2\psi(t), \quad y = -ar(t)\sinh^2\psi(t),$$
where the function $\psi = \psi(t)$ and $r = r(t)$ are determined from uncoupled first-order separable ODEs:
$$r'_t = rf(abr), \quad \psi'_t = \frac{1}{2}a\tanh\psi\, g\!\left(-\frac{a}{b}\tanh^2\psi\right).$$

3°. Multiplying the first ODE by a and the second one by b adding together, one obtains the first-order separable ODE:
$$z'_t = zf(z), \quad z = ax + by.$$

17. $x'_t = xf(x^2+y^2) - yg(y/x), \quad y'_t = yf(x^2+y^2) + xg(y/x).$

Solution:
$$x = r(t)\cos\varphi(t), \quad y = r(t)\sin\varphi(t),$$
where the functions $\varphi = \varphi(t)$ and $r = r(t)$ are determined from uncoupled first-order separable ODEs:
$$r'_t = rf(r^2), \quad \varphi'_t = g(\tan\varphi).$$

18. $x'_t = xf(x^2-y^2) + yg(y/x), \quad y'_t = yf(x^2-y^2) + xg(y/x).$

Solution:
$$x = r(t)\cosh\psi(t), \quad y = r(t)\sinh\psi(t),$$
where the functions $\psi = \psi(t)$ and $r = r(t)$ are determined from uncoupled first-order separable ODEs:
$$r'_t = rf(r^2), \quad \psi'_t = g(\tanh\psi).$$

19. $x'_t = xf(x^2+y^2, y/x), \quad y'_t = yf(x^2+y^2, y/x).$

This is a special case of ODE system 3.2.1.21.

General solution:
$$x = r(t)\cos C, \quad y = r(t)\sin C,$$
where C is an arbitrary constant, and the function $r = r(t)$ satisfies a separable first-order ODE: $r'_t = rf(r^2, \tan C).$

20. $x'_t = x^k f(x^2+y^2, y/x) g(y/x), \quad y'_t = y^k f(x^2+y^2, y/x) h(y/x).$

This is a special case of ODE system 3.2.1.23.

Particular solution:
$$x = r(t)\cos\beta, \quad y = r(t)\sin\beta,$$
where the constant β is determined by the transcendental equation
$$g(\tan\beta) = (\tan\beta)^{k-1} h(\tan\beta),$$
where the function $r = r(t)$ satisfies a separable first-order ODE:
$$r'_t = (\cos\beta)^{k-1} g(\tan\beta) r^k f(r^2, \tan\beta).$$

21. $x'_t = x^k F(x,y), \quad y'_t = g(y) F(x,y).$

General solution in implicit form:
$$x = \varphi(y), \quad \int \frac{dy}{g(y) F(\varphi(y), y)} = t + C_2,$$
where
$$\varphi(y) = \begin{cases} \left[C_1 + (1-k) \int \dfrac{dy}{g(y)} \right]^{\frac{1}{1-k}} & \text{if } k \neq 1, \\ C_1 \exp\left[\int \dfrac{dy}{g(y)} \right] & \text{if } k = 1, \end{cases}$$
and C_1 and C_2 are arbitrary constants.

22. $x'_t = e^{\lambda x} F(x,y), \quad y'_t = g(y) F(x,y).$

General solution in implicit form:
$$x = \varphi(y), \quad \int \frac{dy}{g(y) F(\varphi(y), y)} = t + C_2,$$
where
$$\varphi(y) = \begin{cases} -\dfrac{1}{\lambda} \ln\left[C_1 - \lambda \int \dfrac{dy}{g(y)} \right] & \text{if } \lambda \neq 0, \\ C_1 + \int \dfrac{dy}{g(y)} & \text{if } \lambda = 0, \end{cases}$$
and C_1 and C_2 are arbitrary constants.

23. $x'_t = F(x,y), \quad y'_t = G(x,y).$

Autonomous system of two ODEs of general form.

Suppose
$$y = y(x, C_1),$$
where C_1 is an arbitrary constant, is the general solution of the first-order ODE:
$$F(x,y) y'_x = G(x,y).$$
Then the general solution of the system in question results in the following dependence for the variable x:
$$\int \frac{dx}{F(x, y(x, C_1))} = t + C_2.$$

▶ **Non-autonomous ODE systems.**

24. $x'_t = a_1(t)x + b_1(t)x^k + c_1(t)x^k y^{1-m}$,
    $y'_t = a_2(t)y + b_2(t)y^m + c_2(t)x^{1-k}y^m$.

$1°$. For $k \neq 1$ and $m \neq 1$, the transformation

$$x = u^{1/(1-k)}, \quad y = v^{1/(1-k)}$$

leads to the linear ODE system

$$u'_t = (1-k)a_1(t)u + (1-k)b_1(t) + (1-k)c_1(t)v,$$
$$v'_t = (1-m)a_2(t)v + (1-m)b_2(t) + (1-m)c_2(t)u.$$

For $a_n = \text{const}$, $b_n = \text{const}$, and $c_n = \text{const}$ ($n = 1, 2$), see system 3.1.1.2.

$2°$. For $k = 1$, the original system is easy to integrate since the second ODE is the Bernoulli equation (see equation 2.1.1.4), and the first ODE is a linear with respect to x.

25. $x'_t = a_1(t) + b_1(t)e^{\beta x} + c_1(t)e^{\beta x - \lambda y}$,
    $y'_t = a_2(t) + b_2(t)e^{\lambda y} + c_2(t)e^{\lambda y - \beta x}$.

For $\beta\lambda \neq 0$, the transformation

$$x = -\frac{1}{\beta}\ln u, \quad y = -\frac{1}{\lambda}\ln v$$

leads to the linear ODE system

$$u'_t = -\beta a_1(t)u + -\beta b_1(t) - \beta c_1(t)v,$$
$$v'_t = -\lambda a_2(t)v - \lambda b_2(t) - \lambda c_2(t)u.$$

For $a_n = \text{const}$, $b_n = \text{const}$, and $c_n = \text{const}$ ($n = 1, 2$), see system 3.1.1.2.

26. $x'_t + a(t)x = x^k f(y/x), \quad y'_t + a(t)y = y^k g(y/x)$.

The transformation

$$x = \varphi(t)u, \quad y = \varphi(t)v, \quad \tau = \int [\varphi(t)]^{k-1} dt, \quad \varphi(t) = \exp\left(-\int a(t)\,dt\right)$$

leads to an autonomous ODE system of the form 3.2.1.7:

$$u'_\tau = u^k f(v/u), \quad v'_\tau = v^k g(v/u).$$

27. $x'_t = a(t)x^k f_1(y/x) + b(t)xg(y/x), \quad y'_t = a(t)y^k f_2(y/x) + b(t)yg(y/x)$.

Here $a(t)$, $b(t)$, $f_1(z)$, $f_2(z)$, and $g(z)$ are arbitrary functions.

Let s be any solution of the transcendental (algebraic) equation

$$f_1(s) = s^{k-1} f_2(s).$$

Then the original ODE system admits an exact solution of the form

$$x = \varphi(t)\psi(t), \quad y = s\varphi(t)\psi(t),$$

where the functions $\varphi = \varphi(t)$ and $\psi = \psi(t)$ satisfy two ODEs:

$$\varphi'_t = \lambda a(t)\varphi^k \psi^{k-1}, \quad \psi'_t = \beta b(t)\psi,$$
$$\lambda = f(s), \quad \beta = g(s).$$

For $k \neq 1$, the general solution of these ODE is determined by the formulas

$$\varphi(t) = \left[(1-k)\lambda \int a(t)\psi^{k-1}(t)\, dt + C_1\right]^{1/(1-k)},$$

$$\psi(t) = C_2 \exp\left[\beta \int b(t)\, dt\right],$$

where C_1 and C_2 are arbitrary constants.

In the special case where $a(t) = b(t) = 1$ and $g(y/x) = \beta = \text{const}$ ($\beta \neq 0$), we have

$$\varphi(t) = \left(C_1 - \frac{C_2^{k-1}\lambda}{\beta} e^{\beta(k-1)t}\right)^{1/(1-k)}, \quad \psi(t) = C_2 e^{\beta t}.$$

28. $x'_t = a(t)e^{\lambda x} f_1(y-x) + b(t)g(y-x),$
    $y'_t = a(t)e^{\lambda y} f_2(y-x) + b(t)g(y-x).$

Here $a(t)$, $b(t)$, $f_1(z)$, $f_2(z)$, and $g(z)$ are arbitrary functions.

The transformation
$$x = \ln X, \quad y = \ln Y$$

leads to a ODE system of the form 3.2.1.27 with $k = \lambda + 1$:

$$X'_t = a(t) X^{\lambda+1} F_1(Y/X) + b(t) X G(Y/X),$$
$$Y'_t = a(t) Y^{\lambda+1} F_2(Y/X) + b(t) Y G(Y/X),$$

where $F_1(z) = f_1(\ln z)$, $F_2(z) = f_2(\ln z)$, and $G(z) = g(\ln z)$.

29. $x'_t = e^{-ax} f(y-x) + h(t), \quad y'_t = e^{-ay} g(y-x) + h(t).$

The transformation
$$x = u + \varphi(t), \quad y = v + \varphi(t), \quad \tau = \int \exp[-a\varphi(t)]\, dt, \quad \varphi(t) = \int h(t)\, dt,$$

leads to an autonomous ODE system of the form 3.2.1.4:

$$u'_\tau = e^{-au} f(v-u), \quad v'_\tau = e^{-av} g(v-u).$$

30. $x'_t = f(t, y - ax), \quad y'_t = g(t, y - ax).$

   1°. Solution:

$$x = \int f(t, \varphi(t))\, dt + C, \quad y = \varphi(t) + a \int f(t, \varphi(t))\, dt + aC,$$

where C is an arbitrary constant, and the function $\varphi = \varphi(t)$ is described by the ODE:

$$\varphi'_t = g(t, \varphi) - a f(t, \varphi). \tag{1}$$

2°. If the functions f and g do not explicitly depend on t, then equation (1) is a separable ODE of the form 2.1.1.1. In this case, the original system has a particular solution linear in t:
$$x = f(k)t + C, \quad y = af(k)t + aC + k,$$
where C is an arbitrary constant, and k is a root of the algebraic (transcendental) equation $af(k) = g(k)$.

31. $x'_t = xf(t, y/x), \quad y'_t = yg(t, y/x).$

1°. Solution:
$$x = C \exp\left[\int f(t, \varphi(t))\, dt\right], \quad y = C\varphi(t) \exp\left[\int f(t, \varphi(t))\, dt\right],$$
where C is an arbitrary constant, and the function $\varphi = \varphi(t)$ is described by the ODE:
$$\varphi'_t = \varphi[g(t, \varphi) - f(t, \varphi)]. \tag{1}$$

2°. If the functions f and g do not explicitly depend on t, then equation (1) is a separable ODE of the form 2.1.1.1. In this case, the original system has a particular solution of the exponential form
$$x = Ce^{\lambda t}, \quad y = Cke^{\lambda t}, \quad \lambda = f(k),$$
where C is an arbitrary constant, and k is a root of the algebraic (transcendental) equation $f(k) = g(k)$.

32. $x'_t = xf(t, yx^a), \quad y'_t = yg(t, yx^a).$

This is a generalization of the previous system of ODEs.

1°. Solution:
$$x = C \exp\left[\int f(t, \varphi(t))\, dt\right], \quad y = C^{-a}\varphi(t) \exp\left[-a \int f(t, \varphi(t))\, dt\right],$$
where C is an arbitrary constant, and the function $\varphi = \varphi(t)$ is described by the ODE:
$$\varphi'_t = \varphi[g(t, \varphi) + af(t, \varphi)]. \tag{1}$$

2°. If the functions f and g do not explicitly depend on t, then equation (1) is a separable ODE of the form 2.1.1.1. In this case, the original system has a particular solution of the exponential form
$$x = Ce^{\lambda t}, \quad y = kC^{-a}e^{-a\lambda t}, \quad \lambda = f(k),$$
where C is an arbitrary constant, and k is a root of the algebraic (transcendental) equation $af(k) + g(k) = 0$.

33. $x'_t = f_1(x)g_1(y)\Phi(t, x, y), \quad y'_t = f_2(x)g_2(y)\Phi(t, x, y).$

First integral:
$$\int \frac{f_2(x)}{f_1(x)}\, dx - \int \frac{g_1(y)}{g_2(y)}\, dy = C, \tag{*}$$
where C is an arbitrary constant.

On solving (*) for x (or y) and on substituting the resulting expression into one of the equations of the system concerned, one arrives at a first-order equation for y (or x).

34. $x = tx'_t + F(x'_t, y'_t), \quad y = ty'_t + G(x'_t, y'_t).$

Clairaut system of ODEs.

The following are solutions of the system (Kamke, 1977):

(i) straight lines

$$x = C_1 t + F(C_1, C_2), \quad y = C_2 t + G(C_1, C_2),$$

where C_1 and C_2 are arbitrary constants;

(ii) envelopes of these lines;

(iii) continuously differentiable curves that are formed by segments of curves (i) and (ii).

3.2.2. Systems of Second- and Third-Order ODEs

▶ **ODE systems without first derivatives.**

1. $x''_{tt} = ax^2 + by^2, \quad y''_{tt} = 2axy \quad (ab > 0).$

Let us multiply the first equation by a and the second one by $\lambda = \pm\sqrt{ab}$ and add them together to obtain two independent autonomous second-order ODEs:

$$u''_{tt} = u^2, \quad u = ax + \sqrt{ab}\, y;$$
$$v''_{tt} = v^2, \quad v = ax - \sqrt{ab}\, y.$$

The general solutions of these equations can be expressed in terms of the elliptic Weierstrass function (see, for example, Polyanin & Zaitsev, 2003).

2. $x''_{tt} = xf(ax - by) + g(ax - by), \quad y''_{tt} = yf(ax - by) + h(ax - by).$

Here $f(z)$, $g(z)$, and $h(z)$ are arbitrary functions.

$1°$. Let x_0, y_0 be a solution of the transcendental (algebraic) system of equations

$$x_0 f(ax_0 - by_0) + g(ax_0 - by_0) = 0, \quad y_0 f(ax_0 - by_0) + h(ax_0 - by_0) = 0.$$

We introduce the notation

$$\lambda = f(ax_0 - by_0).$$

Particular solution for $\lambda > 0$:

$$x = x_0 + b[C_1 \exp(-\sqrt{\lambda}\, t) + C_2 \exp(\sqrt{\lambda}\, t)],$$
$$y = y_0 + a[C_1 \exp(-\sqrt{\lambda}\, t) + C_2 \exp(\sqrt{\lambda}\, t)],$$

where C_1 and C_2 are arbitrary constants.

Particular periodic solution for $\lambda < 0$:

$$x = x_0 + b[C_1 \cos(\sqrt{|\lambda|}\, t) + C_2 \sin(\sqrt{|\lambda|}\, t)],$$
$$y = y_0 + a[C_1 \cos(\sqrt{|\lambda|}\, t) + C_2 \sin(\sqrt{|\lambda|}\, t)].$$

2°. Let us multiply the first equation by a and the second one by $-b$ and add them together to obtain the second-order autonomous ODE:

$$z''_{tt} = zf(z) + ag(z) - bh(z), \qquad z = ax - by. \tag{1}$$

We will consider this equation in conjunction with the first equation of the system,

$$x''_{tt} = xf(z) + g(z). \tag{2}$$

Autonomous equation (1) can be treated separately; its general solution can be written out in implicit form (see ODE 2.3.1.1). The function $x = x(t)$ can be determined by solving the linear equation (2), and the function $y = y(t)$ is found as $y = (ax - z)/b$.

3. $x''_{tt} = b_2 f(a_1 x + b_1 y) + b_1 g(a_2 x + b_2 y),$
$y''_{tt} = -a_2 f(a_1 x + b_1 y) - a_1 g(a_2 x + b_2 y).$

It is assumed that $a_1 b_2 - a_2 b_1 \neq 0$.

Multiplying the ODEs by suitable constants and adding together, one obtains two independent second-order autonomous ODEs:

$$u''_{tt} = (a_1 b_2 - a_2 b_1) f(u), \qquad u = a_1 x + b_1 y;$$
$$v''_{tt} = -(a_1 b_2 - a_2 b_1) g(v), \qquad v = a_2 x + b_2 y.$$

Their general solutions can be represented in an implicit form (see ODE 2.3.1.1).

4. $x''_{tt} = xf(y/x), \quad y''_{tt} = yg(y/x).$

Second-order ODE system homogeneous in both unknown variables, where $f(z)$ and $g(z)$ are arbitrary functions. This ODE system is invariant under the same scaling of the unknowns, $x \Longrightarrow cx$ and $y \Longrightarrow cy$, where c is an arbitrary nonzero number.

1°. A particular periodic solution (Polyanin & Manzhirov, 2007):

$$x = C_1 \sin(kt) + C_2 \cos(kt), \quad k = \sqrt{-f(\lambda)},$$
$$y = \lambda[C_1 \sin(kt) + C_2 \cos(kt)],$$

where C_1 and C_2 are arbitrary constants and λ is a root of the transcendental (algebraic) equation

$$f(\lambda) = g(\lambda). \tag{*}$$

For the special case $f(z) = g(z)$, the constant λ is arbitrary.

2°. Particular solution (Polyanin & Manzhirov, 2007):

$$x = C_1 \exp(kt) + C_2 \exp(-kt), \quad k = \sqrt{f(\lambda)},$$
$$y = \lambda[C_1 \exp(kt) + C_2 \exp(-kt)],$$

where C_1 and C_2 are arbitrary constants and λ is a root of the transcendental (algebraic) equation (*).

5. $x_{tt}'' = xf(y/x) + g(y/x)$, $y_{tt}'' = yf(y/x) + h(y/x)$.

Let λ be a root of the algebraic (transcendental) equation

$$\lambda g(\lambda) = h(\lambda).$$

1°. Particular solution if $f(\lambda) > 0$:

$$x = C_1 \exp(-\sqrt{f(\lambda)}\,t) + C_2 \exp(\sqrt{f(\lambda)}\,t) - \frac{g(\lambda)}{f(\lambda)},$$

$$y = \lambda\left[C_1 \exp(-\sqrt{f(\lambda)}\,t) + C_2 \exp(\sqrt{f(\lambda)}\,t) - \frac{g(\lambda)}{f(\lambda)}\right],$$

where C_1 and C_2 are arbitrary constants.

2°. Particular periodic solution if $f(\lambda) < 0$:

$$x = C_1 \cos(\sqrt{|f(\lambda)|}\,t) + C_2 \sin(\sqrt{|f(\lambda)|}\,t) - \frac{g(\lambda)}{f(\lambda)},$$

$$y = \lambda\left[C_1 \cos(\sqrt{|f(\lambda)|}\,t) + C_2 \sin(\sqrt{|f(\lambda)|}\,t) - \frac{g(\lambda)}{f(\lambda)}\right].$$

3°. Particular solution if $f(\lambda) = 0$:

$$x = \tfrac{1}{2}g(\lambda)t^2 + C_1 t + C_2, \quad y = \lambda[\tfrac{1}{2}g(\lambda)t^2 + C_1 t + C_2].$$

6. $x_{tt}'' = x^k f(y/x)$, $y_{tt}'' = y^k g(y/x)$.

Second-order generalized homogeneous ODE system, where $f(z)$ and $g(z)$ are arbitrary functions. This ODE system remains unchanged under simultaneous scaling of the dependent and independent variables in accordance with the rules $x \Longrightarrow cx$, $y \Longrightarrow cy$, and $t \Longrightarrow c^{(1-k)/2}t$, where c is an arbitrary positive constant.

1°. Particular solution for $k \neq 1$:

$$x = s(t+C)^{\frac{2}{1-k}}, \quad y = s\lambda(t+C)^{\frac{2}{1-k}}, \quad s = \left[\frac{(1-k)^2}{2(1+k)} f(\lambda)\right]^{\frac{2}{1-k}},$$

where C is an arbitrary constant, and λ is a root of the transcendental (algebraic) equation $f(\lambda) = \lambda^{k-1} g(\lambda)$.

2°. Particular solution (generalizes the solution from Item 1°):

$$x = x(t), \quad y = \lambda x(t),$$

where $x = x(t)$ is a solution of the second-order autonomous ODE $x_{tt}'' = f(\lambda)x^k$, and λ is a root of the transcendental (algebraic) equation $f(\lambda) = \lambda^{k-1} g(\lambda)$.

7. $x_{tt}'' = xf(yx^a)$, $y_{tt}'' = yg(yx^a)$.

Second-order generalized homogeneous ODE system in both unknown variables. This ODE system remains unchanged under scaling of the dependent variables in accordance with the rules $x \Longrightarrow cx$ and $y \Longrightarrow c^{-a}y$, where c is an arbitrary positive constant.

Particular solutions:
$$x = Ce^{\lambda t}, \quad y = kC^{-a}e^{-a\lambda t}, \quad \lambda = \pm\sqrt{f(k)},$$
where C is an arbitrary constant, and k is a root of the algebraic (transcendental) equation $a^2 f(k) = g(k)$.

8. $x''_{tt} = e^{ax}f(y-x), \quad y''_{tt} = e^{ay}g(y-x).$

Second-order ODE system invariant under "translation–scaling" transformation, where $f(z)$ and $g(z)$ are arbitrary functions. This ODE system remains unchanged under simultaneous translation and scaling of variables in accordance with the rules $x \Longrightarrow x + b$, $y \Longrightarrow y + b$ and $t \Longrightarrow ct$, where $c = e^{-ab/2}$ and b is an arbitrary constant.

Particular solution:
$$x = x(t), \quad y = x(t) + \lambda,$$
where $x = x(t)$ is a solution of the second-order autonomous ODE $x''_{tt} = f(\lambda)e^{ax}$ and λ is a root of the transcendental (algebraic) equation $f(\lambda) = e^{a\lambda}g(\lambda)$.

9. $x''_{tt} = kxr^{-3}, \quad y''_{tt} = kyr^{-3}, \quad$ where $\quad r = \sqrt{x^2 + y^2}.$

Equations of motion of a point mass in the xy-plane under gravity.

By changing to the polar coordinates
$$x = r\cos\varphi, \quad y = r\sin\varphi, \quad r = r(t), \quad \varphi = \varphi(t),$$
one can obtain two first integrals (Kamke, 1977):
$$r^2\varphi'_t = C_1, \quad (r'_t)^2 + r^2(\varphi'_t)^2 = -2kr^{-1} + C_2, \tag{1}$$
where C_1 and C_2 are arbitrary constants. Assuming that $C_1 \neq 0$ and integrating further, one finds that
$$r[C\cos(\varphi - \varphi_0) - k] = C_1^2, \quad C^2 = C_1^2 C_2 + k^2.$$
This is an equation of a conic section. The function $\varphi(t)$ can be found from the first equation in (1).

10. $x''_{tt} = xf(r), \quad y''_{tt} = yf(r), \quad$ where $\quad r = \sqrt{x^2 + y^2}.$

Equations of motion of a point mass in the xy-plane under a central force.

By changing to the polar coordinates
$$x = r\cos\varphi, \quad y = r\sin\varphi, \quad r = r(t), \quad \varphi = \varphi(t),$$
one can obtain two first integrals
$$r^2\varphi'_t = C_1, \quad (r'_t)^2 + r^2(\varphi'_t)^2 = 2\int rf(r)\,dr + C_2,$$
where C_1 and C_2 are arbitrary constants. Integrating further, one finds that
$$t + C_3 = \pm\int \frac{r\,dr}{\sqrt{2r^2 F(r) + r^2 C_2 - C_1^2}}, \quad \varphi = C_1\int \frac{dt}{r^2} + C_4, \tag{*}$$

where C_3 and C_4 are arbitrary constants and

$$F(r) = \int rf(r)\,dr.$$

It is assumed in the second relation in $(*)$ that the function $r = r(t)$ is obtained by solving the first equation in $(*)$ for $r(t)$.

11. $x''_{tt} + a(t)x = x^{-3}f(y/x)$, $\quad y''_{tt} + a(t)y = y^{-3}g(y/x)$.

Generalized Ermakov system of ODEs.

$1°$. First integral:

$$\frac{1}{2}(xy'_t - yx'_t)^2 + \int^{y/x}[uf(u) - u^{-3}g(u)]\,du = C,$$

where C is an arbitrary constant.

$2°$. Suppose $\varphi = \varphi(t)$ is a nontrivial solution of the second-order linear ODE:

$$\varphi''_{tt} + a(t)\varphi = 0. \tag{1}$$

Then the transformation

$$\tau = \int \frac{dt}{\varphi^2(t)}, \quad u = \frac{x}{\varphi(t)}, \quad v = \frac{y}{\varphi(t)} \tag{2}$$

leads to the autonomous system of ODEs

$$u''_{\tau\tau} = u^{-3}f(v/u), \quad v''_{\tau\tau} = v^{-3}g(v/u). \tag{3}$$

$3°$. A particular solution of system (3) is

$$u = A\sqrt{C_2\tau^2 + C_1\tau + C_0}, \quad v = Ak\sqrt{C_2\tau^2 + C_1\tau + C_0}, \quad A = \left[\frac{f(k)}{C_0C_2 - \frac{1}{4}C_1^2}\right]^{1/4},$$

where C_0, C_1, and C_2 are arbitrary constants, and k is a root of the algebraic (transcendental) equation

$$k^4 f(k) = g(k).$$

$4°$. References for the system of ODEs 3.2.2.11: Ray & Reid (1979), Athorne (1991), Govinger et al. (1993), Berkovich (2002), and Polyanin & Manzhirov (2007).

12. $x''_{tt} = xf(x^2 + y^2, y/x)$, $\quad y''_{tt} = yf(x^2 + y^2, y/x)$.

Particular solution:

$$x = r(t)\cos C, \quad y = r(t)\sin C,$$

where C is an arbitrary constant, and the function $r = r(t)$ satisfies the solvable autonomous second-order ODE of the form 2.3.1.1: $r''_{tt} = rf(r^2, \tan C)$.

13. $x''_{tt} = x^k f(x^2 + y^2, y/x)g(y/x)$, $\quad y''_{tt} = y^k f(x^2 + y^2, y/x)h(y/x)$.

Here $f(z_1, z_2)$, $g(z_2)$, and $h(z_2)$ are arbitrary functions, k is a free parameter.

Particular solution:

$$x = r(t)\cos\beta, \quad y = r(t)\sin\beta,$$

where the constant β is determined by the transcendental equation
$$g(\tan\beta) = (\tan\beta)^{k-1} h(\tan\beta),$$
where the function $r = r(t)$ satisfies the solvable autonomous second-order ODE of the form 2.3.1.1: $r''_{tt} = (\cos\beta)^{k-1} g(\tan\beta) r^k f(r^2, \tan\beta)$.

14. $x''_{tt} = \dfrac{1}{x^3} F\left(\dfrac{x}{\varphi(t)}, \dfrac{y}{\varphi(t)}\right)$, $y''_{tt} = \dfrac{1}{y^3} F\left(\dfrac{x}{\varphi(t)}, \dfrac{y}{\varphi(t)}\right)$,

$\varphi(t) = \sqrt{at^2 + bt + c}$.

1°. The transformation
$$\tau = \int \frac{dt}{\varphi(t)}, \quad u = \frac{x}{\varphi(t)}, \quad v = \frac{y}{\varphi(t)}$$
leads to the autonomous ODE system
$$u''_{\tau\tau} + (ac - \tfrac{1}{4}b^2)u = u^{-3} F(u, v),$$
$$v''_{\tau\tau} + (ac - \tfrac{1}{4}b^2)v = v^{-3} G(u, v).$$

2°. Particular solutions:
$$x = A\sqrt{at^2 + bt + c}, \quad y = B\sqrt{at^2 + bt + c},$$
where A and B are the roots of the system of algebraic (transcendental) equations
$$(ac - \tfrac{1}{4}b^2) A^4 = F(A, B), \quad (ac - \tfrac{1}{4}b^2) B^4 = G(A, B).$$

▶ ODE systems with first derivatives.

15. $x''_{tt} = f(y'_t/x'_t)$, $y''_{tt} = g(y'_t/x'_t)$.

1°. The transformation
$$u = x'_t, \quad w = y'_t \tag{1}$$
leads to a system of the first-order ODEs:
$$u'_t = f(w/u), \quad w'_t = g(w/u). \tag{2}$$
Eliminating t yields a first-order homogeneous ODE, whose solution is given by
$$\int \frac{f(\xi)\, d\xi}{g(\xi) - \xi f(\xi)} = \ln|u| + C, \quad \xi = \frac{w}{u}, \tag{3}$$
where C is an arbitrary constant. On solving (3) for w, one obtains $w = w(u, C)$. On substituting this expression into the first equation of (2), one can find $u = u(t)$ and then $w = w(t)$. Finally, one can determine $x = x(t)$ and $y = y(t)$ from (1) by simple integration.

2°. *The Suslov problem.* The problem of a point particle sliding down an inclined rough plane is described by the equations
$$x''_{tt} = 1 - \frac{kx'_t}{\sqrt{(x'_t)^2 + (y'_t)^2}}, \quad y''_{tt} = -\frac{ky'_t}{\sqrt{(x'_t)^2 + (y'_t)^2}},$$

which correspond to a special case of the system in question with

$$f(z) = 1 - \frac{k}{\sqrt{1+z^2}}, \quad g(z) = -\frac{kz}{\sqrt{1+z^2}}.$$

The solution of the corresponding Cauchy problem under the initial conditions

$$x(0) = y(0) = x'_t(0) = 0, \quad y'_t(0) = 1$$

leads, for the case $k = 1$, to the following dependences $x(t)$ and $y(t)$ written in parametric form (Klimov & Zhuravlev, 2002):

$$x = -\tfrac{1}{16} + \tfrac{1}{16}\xi^4 - \tfrac{1}{4}\ln\xi, \quad y = \tfrac{2}{3} - \tfrac{1}{2}\xi - \tfrac{1}{6}\xi^3, \quad t = \tfrac{1}{4} - \tfrac{1}{4}\xi^2 - \tfrac{1}{2}\ln\xi \quad (0 \leq \xi \leq 1).$$

16. $x''_{tt} = xf(y/x, y'_t/x'_t), \quad y''_{tt} = yg(y/x, y'_t/x'_t).$

Second-order ODE system homogeneous in both unknown variables, where $f(z_1, z_2)$ and $g(z_1, z_2)$ are arbitrary functions. This ODE system is invariant under the same scaling of the unknowns, $x \Longrightarrow cx$ and $y \Longrightarrow cy$, where c is an arbitrary nonzero number.

Let k be the root of the transcendental (algebraic) equation

$$f(k,k) = g(k,k).$$

1. Particular solution for $\lambda = f(k,k) > 0$:

$$x = C_1 \exp(-\sqrt{\lambda}\,t) + C_2 \exp(\sqrt{\lambda}\,t), \quad y = k[C_1 \exp(-\sqrt{\lambda}\,t) + C_2 \exp(\sqrt{\lambda}\,t)],$$

where C_1 and C_2 are arbitrary constants.

2. Particular periodic solution for $\lambda = f(k,k) < 0$:

$$x = C_1 \cos(\sqrt{|\lambda|}\,t) + C_2 \sin(\sqrt{|\lambda|}\,t), \quad y = k[\cos(\sqrt{|\lambda|}\,t) + C_2 \sin(\sqrt{|\lambda|}\,t)].$$

17. $x''_{tt} = x^{-3}f(y/x, y'_t/x'_t), \quad y''_{tt} = y^{-3}g(y/x, y'_t/x'_t).$

Solution:

$$x = A\sqrt{C_2 t^2 + C_1 t + C_0}, \quad y = B\sqrt{C_2 t^2 + C_1 t + C_0},$$

where C_0, C_1, and C_2 are arbitrary constants, and A and B are the roots of the system of algebraic (transcendental) equations

$$A^4(C_0 C_2 - \tfrac{1}{4}C_1^2) = f(B/A, B/A),$$
$$B^4(C_0 C_2 - \tfrac{1}{4}C_1^2) = g(B/A, B/A).$$

18. $x''_{tt} = x\Phi(x, y, t, x'_t, y'_t), \quad y''_{tt} = y\Phi(x, y, t, x'_t, y'_t).$

1°. First integral:

$$xy'_t - yx'_t = C,$$

where C is an arbitrary constant.

Remark 3.2. *In the equations, the arbitrary function Φ can also be dependent on the second and higher derivatives with respect to t.*

2°. Particular solution: $y = C_1 x$, where C_1 is an arbitrary constant, and the function $x = x(t)$ is determined by the second-order ODE

$$x_{tt}'' = x\Phi(x, C_1 x, t, x_t', C_1 x_t').$$

19. $x_{tt}'' + x^{-3} f(y/x) = x\Phi(x, y, t, x_t', y_t'),$
$y_{tt}'' + y^{-3} g(y/x) = y\Phi(x, y, t, x_t', y_t').$

First integral:

$$\frac{1}{2}(xy_t' - yx_t')^2 + \int^{y/x} [u^{-3} g(u) - u f(u)]\, du = C,$$

where C is an arbitrary constant.

Remark 3.3. *The function Φ can also be dependent on the second and higher derivatives with respect to t.*

20. $x_{tt}'' = x_t' \Phi(x, y, t, x_t', y_t') + f(y), \quad y_{tt}'' = -y_t' \Phi(x, y, t, x_t', y_t') + g(x).$

First integral:

$$x_t' y_t' - \int f(y)\, dy - \int g(x)\, dx = C.$$

21. $x_{tt}'' = a y_t' \Phi(x, y, t, x_t', y_t') + f(x), \quad y_{tt}'' = b x_t' \Phi(x, y, t, x_t', y_t') + g(y).$

First integral:

$$b(x_t')^2 - a(y_t')^2 - 2 \int f(x)\, dx + a \int g(y)\, dy = C.$$

22. $x_{tt}'' = F(t, tx_t' - x, ty_t' - y), \quad y_{tt}'' = G(t, tx_t' - x, ty_t' - y).$

1°. The transformation

$$u = tx_t - x, \quad v = ty_t' - y \qquad (1)$$

leads to a system of first-order ODEs:

$$u_t' = tF(t, u, v), \quad v_t' = tG(t, u, v). \qquad (2)$$

2°. Suppose a solution of system (2) has been found in the form

$$u = u(t, C_1, C_2), \quad v = v(t, C_1, C_2), \qquad (3)$$

where C_1 and C_2 are arbitrary constants. Then, substituting (3) into (1) and integrating, one obtains a solution of the original system:

$$x = C_3 t + t \int \frac{u(t, C_1, C_2)}{t^2}\, dt, \quad y = C_4 t + t \int \frac{v(t, C_1, C_2)}{t^2}\, dt.$$

3°. If the functions F and G are independent of t, then, on eliminating t from system (2), one arrives at a first-order ODE:

$$g(u, v) u_v' = F(u, v).$$

23. $x''_{tt} = xf(x, x'_t) + x'_t g(x, x'_t), \quad y''_{tt} = yf(x, x'_t) + y'_t g(x, x'_t).$

Let $x = x(t)$ be a solution to the first ODE of the original system. Then the second equation of the system is a linear second-order ODE in y; this ODE has a particular solution: $y_1 = x(t)$. In this case, the general solution of the second ODE can be found using formula (2) from Section 2.2.1.

24. $(x'_t)^2 - xx''_{tt} - ax'''_{ttt} = 0, \quad x'_t y'_t - xy''_{tt} - ay'''_{ttt} = 0.$

This system of ODEs occurs in the theory of the hydrodynamic boundary layer.

1°. Particular solution:
$$x = \frac{6a}{t+C_1}, \quad y = \frac{C_2}{t+C_1} + \frac{C_3}{(t+C_1)^2} + C_4,$$

where $C_1, \ldots, C_4$ are arbitrary constants.

2°. Particular solution:
$$x = C_1 e^{\lambda t} - a\lambda, \quad y = C_2(C_1 e^{\lambda t} - a\lambda),$$

where C_1, C_2, and λ are arbitrary constants.

3°. Let $x = x(t)$ be a solution to the first ODE of the original system. Then the second equation of the system is a third-order linear ODE in y; this ODE has two particular solutions: $y_1 = 1$ and $y_2 = x(t)$. Therefore, the second equation can be reduced to a first-order linear ODE (see formulae (4) in Section 2.4.1), which is easy to integrate.

25. $x'''_{ttt} = x''_{tt} f(x, x'_t, x''_{tt}) + x'_t g(x, x'_t, x''_{tt}),$
$y'''_{ttt} = y''_{tt} f(x, x'_t, x''_{tt}) + y'_t g(x, x'_t, x''_{tt}).$

Let $x = x(t)$ be a solution to the first ODE of the original system. Then the second equation of the system is a third-order linear ODE in y; this ODE has two particular solutions: $y_1 = 1$ and $y_2 = x(t)$. Therefore, the second equation can be reduced to a first-order linear ODE (see formulae (4) in Section 2.4.1), which is easy to integrate.

3.3. Nonlinear Systems of Three or More ODEs

3.3.1. Systems of Three ODEs

1. $ax'_t = (b-c)yz, \quad by'_t = (c-a)zx, \quad cz'_t = (a-b)xy.$

First integrals (Kamke, 1977):
$$ax^2 + by^2 + cz^2 = C_1,$$
$$a^2 x^2 + b^2 y^2 + c^2 z^2 = C_2,$$

where C_1 and C_2 are arbitrary constants. On solving the first integrals for y and z and on substituting the resulting expressions into the first equation of the system, one arrives at a separable first-order ODE.

2. $x'_t = cg(y) - bh(z)$, $\quad y'_t = ah(z) - cf(x)$, $\quad z'_t = bf(x) - ag(y)$.

A special case of system 3.3.1.4 with $F_1 = f(x)$, $F_2 = g(y)$, and $F_3 = h(z)$.

First integrals:
$$ax + by + cz = C_1,$$
$$\int f(x)\,dx + \int g(y)\,dy + \int h(z)\,dz = C_2,$$

where C_1 and C_2 are arbitrary constants.

3. $ax'_t = (b-c)yzF(x,y,z,t)$,
$by'_t = (c-a)zxF(x,y,z,t)$, $\quad cz'_t = (a-b)xyF(x,y,z,t)$.

First integrals:
$$ax^2 + by^2 + cz^2 = C_1,$$
$$a^2x^2 + b^2y^2 + c^2z^2 = C_2,$$

where C_1 and C_2 are arbitrary constants. On solving the first integrals for y and z and on substituting the resulting expressions into the first equation of the system, one arrives at a first-order ODE; if F is independent of t, this ODE will be separable.

4. $x'_t = cF_2 - bF_3$, $\quad y'_t = aF_3 - cF_1$, $\quad z'_t = bF_1 - aF_2$.

Here, $F_n = F_n(x,y,z)$ are arbitrary functions ($n = 1, 2, 3$).

First integral:
$$ax + by + cz = C_1,$$

where C_1 is an arbitrary constant. On eliminating t and z from the first two equations of the system (using the above first integral), one arrives at the first-order ODE

$$\frac{dy}{dx} = \frac{aF_3(x,y,z) - cF_1(x,y,z)}{cF_2(x,y,z) - bF_3(x,y,z)}, \quad \text{where} \quad z = \frac{1}{c}(C_1 - ax - by).$$

5. $x'_t = czF_2 - byF_3$, $\quad y'_t = axF_3 - czF_1$, $\quad z'_t = byF_1 - axF_2$.

Here, $F_n = F_n(x,y,z)$ are arbitrary functions ($n = 1, 2, 3$).

First integral:
$$ax^2 + by^2 + cz^2 = C_1,$$

where C_1 is an arbitrary constant. On eliminating t and z from the first two equations of the system (using the above first integral), one arrives at the first-order ODE

$$\frac{dy}{dx} = \frac{axF_3(x,y,z) - czF_1(x,y,z)}{czF_2(x,y,z) - byF_3(x,y,z)}, \quad \text{where} \quad z = \pm\sqrt{\frac{1}{c}(C_1 - ax^2 - by^2)}.$$

6. $x'_t = x(cF_2 - bF_3)$, $\quad y'_t = y(aF_3 - cF_1)$, $\quad z'_t = z(bF_1 - aF_2)$.

Here, $F_n = F_n(x,y,z)$ are arbitrary functions ($n = 1, 2, 3$).

First integral:
$$|x|^a |y|^b |z|^c = C_1,$$

where C_1 is an arbitrary constant. On eliminating t and z from the first two equations of the system (using the above first integral), one obtains a first-order ODE.

7. $x'_t = h(z)F_2 - g(y)F_3$, $y'_t = f(x)F_3 - h(z)F_1$, $z'_t = g(y)F_1 - f(x)F_2$.

Here, $F_n = F_n(x, y, z)$ are arbitrary functions ($n = 1, 2, 3$).

First integral:
$$\int f(x)\,dx + \int g(y)\,dy + \int h(z)\,dz = C_1,$$

where C_1 is an arbitrary constant. On eliminating t and z from the first two equations of the system (using the above first integral), one obtains a first-order ODE.

8. $x''_{tt} = \dfrac{\partial F}{\partial x}$, $y''_{tt} = \dfrac{\partial F}{\partial y}$, $z''_{tt} = \dfrac{\partial F}{\partial z}$,

where $F = F(r)$, $r = \sqrt{x^2 + y^2 + z^2}$.

Equations of motion of a point particle under gravity.

The system can be rewritten as a single vector equation:
$$\mathbf{r}''_{tt} = \operatorname{grad} F \quad \text{or} \quad \mathbf{r}''_{tt} = \frac{F'(r)}{r}\mathbf{r},$$

where $\mathbf{r} = (x, y, z)$.

1°. First integrals (Kamke, 1977):

$(\mathbf{r}'_t)^2 = 2F(r) + C_1$ (law of conservation of energy),

$[\mathbf{r} \times \mathbf{r}'_t] = \mathbf{C}$ (law of conservation of areas),

$(\mathbf{r} \cdot \mathbf{C}) = 0$ (all trajectories are plane curves).

2°. Solution:
$$\mathbf{r} = \mathbf{a}\, r \cos\varphi + \mathbf{b}\, r \sin\varphi.$$

Here the constant vectors $\mathbf{a}$ and $\mathbf{b}$ must satisfy the conditions
$$|\mathbf{a}| = |\mathbf{b}| = 1, \quad (\mathbf{a} \cdot \mathbf{b}) = 0,$$

and the functions $r = r(t)$ and $\varphi = \varphi(t)$ are given by
$$t = \int \frac{r\,dr}{\sqrt{2r^2 F(r) + C_1 r^2 - C_3^2}} + C_2,$$
$$\varphi = C_3 \int \frac{dr}{r\sqrt{2r^2 F(r) + C_1 r^2 - C_3^2}}, \quad C_3 = |\mathbf{C}|.$$

9. $x''_{tt} = xF$, $y''_{tt} = yF$, $z''_{tt} = zF$, where $F = F(x, y, z, t, x'_t, y'_t, z'_t)$.

First integrals (laws of conservation of areas):
$$zy'_t - yz'_t = C_1, \quad xz'_t - zx'_t = C_2, \quad yx'_t - xy'_t = C_3,$$

where C_1, C_2, and C_3 are arbitrary constants.

Corollary of the conservation laws:
$$C_1 x + C_2 y + C_3 z = 0.$$

This implies that all integral curves are plane curves.

Remark 3.4. *The function F can also be dependent on the second and higher derivatives with respect to t.*

10. $x''_{tt} = F_1$, $y''_{tt} = F_2$, $z''_{tt} = F_3$,

where $F_n = F_n(t, tx'_t - x, ty'_t - y, tz'_t - z)$.

1°. The transformation
$$u = tx_t' - x, \quad v = ty_t' - y, \quad w = tz_t' - z \tag{1}$$

leads to the system of first-order ODEs:

$$u_t' = tF_1(t,u,v,w), \quad v_t' = tF_2(t,u,v,w), \quad w_t' = tF_3(t,u,v,w). \tag{2}$$

2°. Suppose a solution of system (2) has been found in the form

$$u(t) = u(t, C_1, C_2, C_3), \quad v(t) = v(t, C_1, C_2, C_3), \quad w(t) = w(t, C_1, C_2, C_3), \tag{3}$$

where C_1, C_2, and C_3 are arbitrary constants. Then, substituting (3) into (1) and integrating, one obtains a solution of the original system:

$$x = C_4 t + t\int \frac{u(t)}{t^2}\,dt, \quad y = C_5 t + t\int \frac{v(t)}{t^2}\,dt, \quad z = C_6 t + t\int \frac{w(t)}{t^2}\,dt,$$

where C_4, C_5, and C_6 are arbitrary constants.

3.3.2. Equations of Dynamics of a Rigid Body with a Fixed Point*

▶ **Kinematic and dynamic Euler equations.**

The motion (rotation) of a rigid about a fixed point under the action of external forces is governed by a system of six first-order coupled nonlinear ODEs:

$$Ap_t' + (C-B)qr = M_1, \tag{1}$$
$$Bq_t' + (A-C)pr = M_2, \tag{2}$$
$$Cr_t' + (B-A)pq = M_3, \tag{3}$$
$$p = \psi_t' \sin\theta \sin\varphi + \theta_t' \cos\varphi, \tag{4}$$
$$q = \psi_t' \sin\theta \cos\varphi - \theta_t' \sin\varphi, \tag{5}$$
$$r = \psi_t' \cos\theta + \varphi_t', \tag{6}$$

where p, q, and r are the components of the body's angular velocity in a moving orthonormal reference frame, $\xi\eta\zeta$, rigidly connected with the body and formed by the principal axes of inertia (the origin placed at the fixed point); A, B, and C are the moments of inertia about the principal axes; and M_1, M_2, and M_3 are the components of the moment of external forces in the frame $\xi\eta\zeta$, which usually depend of the Euler angles ψ, θ, and φ defining the position of the moving frame relative to the fixed one.

It is required to determine p, q, r, ψ, θ, and φ as functions of time t from system (1)–(6).

From now on, the following quantities will be used in this section: m is the mass of the body, $\mathbf{r}$ is the position vector of the center of mass, $\mathbf{K} = (K_1, K_2, K_3)^{\mathrm{T}} = (Ap, Bq, Cr)^{\mathrm{T}}$ is the angular momentum of the body (in the frame $\xi\eta\zeta$), $\boldsymbol{\gamma} = (\gamma_1, \gamma_2, \gamma_3)$ is a vertical unit vector ($\gamma_1^2 + \gamma_2^2 + \gamma_3^2 = 1$), which is introduced when the body is in a homogeneous gravitational field so that the direction of $\boldsymbol{\gamma}$ is opposite to the gravitational acceleration $\mathbf{g}$, with $g = |\mathbf{g}|$.

---

*This section was written by Alexander Fomichev.

Equations (1)–(3) are known as *Euler's dynamic equations* and (4)–(6) as *Euler's kinematic equations*. In general, system (1)–(6) cannot be solved by quadrature. However, there are three special cases where the system is reduced to quadratures for any initial conditions; this is due to the availability of first integrals, which do not exist in the general case. The three solvable cases are discussed below.

▶ **Euler's case.**

The Euler's case takes place when the body has an arbitrary shape and the external moments are all zero:
$$M_1 = M_2 = M_3 = 0. \tag{7}$$
With formulas (7), the dynamic equations (1)–(3) can be solved independently of the kinematic equations.

To be specific, we assume that $A \geq B \geq C$ and $A > C$ (the special case $A = B = C$ is trivial). ODE system (1)–(3) with condition (7) has the following first integrals:
$$Ap^2 + Bq^2 + Cr^2 = 2T \quad \text{(conservation of energy)},$$
$$A^2p^2 + B^2q^2 + C^2r^2 = K^2 \quad \text{(conservation of angular momentum)},$$
where $T > 0$ and K are arbitrary constants.

For $A > C$, p and r can always be expressed via q:
$$p = \pm\sqrt{a - bq^2}, \quad r = \pm\sqrt{c - dq^2}, \tag{8}$$
with the constants a, b, c, and d expressible in terms of the initial parameters of the problem. Substituting (8) into the equation for q yields
$$Bq'_t \pm (A - C)\sqrt{(a - bq^2)(c - dq^2)} = 0.$$
Integrating gives the solution in implicit form
$$t - t_0 = \pm \frac{B}{A - C} \int_0^q \frac{dq}{\sqrt{(a - bq^2)(c - dq^2)}}.$$

Effectively, the problem is reduced to the inversion of an elliptic integral, resulting in expressions of $p(t)$, $q(t)$, and $r(t)$ in terms of elliptic functions of time.

To solve the kinematic equations, it is convenient to direct the z-axis of the fixed frame along the constant angular momentum **K**, in which case we obtain
$$K_1 = Ap = K \sin\theta \sin\varphi, \quad K_2 = Bq = K \sin\theta \cos\varphi, \quad K_3 = Cr = K \cos\theta. \tag{9}$$
From the second and third equations (9), as well as equations (4)–(5), follow the formulas for the Euler angles
$$\cos\theta(t) = \frac{Cr(t)}{K}, \quad \cos\varphi(t) = \frac{Bq(t)}{K\sin\theta(t)},$$
$$\psi(t) = \psi_0 + \int_0^t \frac{p(t)\sin\varphi(t) + q(t)\cos\varphi(t)}{\sin\theta(t)} dt.$$

This solution is known to have geometric interpretations suggested by Poinsot & MacCullagh (see, for example, Zhuravlev, 1996 and Borisov & Mamaev, 2001).

▶ **Lagrange's case.**

The body, which is in a homogeneous gravitational field, is dynamically symmetric and its center of mass lies on the dynamic symmetry axis (the ζ-axis). Then, in equations (1)–(3), one should set

$$A = B, \quad \mathbf{M} = (M_1, M_2, M_3)^T = mg(\mathbf{r} \times \boldsymbol{\gamma}). \tag{10}$$

System (1)–(6) with (10) admits the first three integrals, which are given below.

1. Conservation of the angular momentum projection onto the ζ-axis:

$$K_3 = \text{const.}$$

2. Conservation of the angular momentum projection onto the direction of $\boldsymbol{\gamma}$:

$$(\mathbf{K} \cdot \boldsymbol{\gamma}) = K_1\gamma_1 + K_2\gamma_2 + K_3\gamma_3 = C_1 = \text{const.}$$

3. Energy integral:

$$\frac{h}{2}(\theta_t')^2 + \frac{K_3^2}{2C} + \frac{(C_1 - K_3\cos\theta)^2}{2A\sin\theta} + mgl\cos\theta = h = \text{const.}$$

The availability of these integrals reduces the problem to the equation

$$(\theta_t')^2 = 2h - \frac{K_3^2}{C} - \frac{(C_1 - K_3\cos\theta)^2}{\sin\theta} - 2\cos\theta,$$

which is obtained if one sets $A = mgl = 1$ (without loss of generality). With the change of variable $u = \cos\theta$, this equation can be reduced to the elliptic quadrature

$$u_t' = \sqrt{R(u)},$$

$$R(u) = 2(h_1 - u)(1 - u^2) - (C_1 - K_3 u)^2, \quad h_1 = h - \frac{K_3^2}{2C}.$$

To determine the full motion of the system, one has to integrate the following two equations:

$$\psi_t' = \frac{C_1 - K_3 u}{1 - u^2}, \quad \varphi_t' = \left(\frac{1}{C} - 1\right)K_3 + \frac{C_1 - K_3 u}{1 - u^2}.$$

Depending on the initial data and specific parameters of the problem, the solution defines four types of motion, in one of which the axis of the top asymptotically tends to a vertical positions.

▶ **Sofia Kovalevskaya's case.**

The body is dynamically symmetric with $A = B$ and, in addition, the condition $A = 2C$ holds. The center of mass lies in the equatorial plane of the inertia ellipsoid (its center at the fixed point) and its position in the frame $\xi\eta\zeta$ is $\mathbf{r} = (L, 0, 0)^T$. The system is in a homogeneous gravitational field, so that $\mathbf{M} = mg(\mathbf{r} \times \boldsymbol{\gamma})$. For simplicity, we assume that $A = 1$, $mg = 1$, and $L = 1$.

This case is much more complex than the previous two, both in the way how the equations are integrated and from the viewpoint of the qualitative analysis of the motion. The Euler equations (1)–(6) allow the first three integrals, which are given below.

1. Conservation of the angular momentum projection onto the vertical:

$$(\mathbf{K} \cdot \boldsymbol{\gamma}) = K_1\gamma_1 + K_2\gamma_2 + K_3\gamma_3 = c = \text{const.}$$

2. Energy integral:

$$\tfrac{1}{2}(K_1^2 + K_2^2 + K_3^2) - L\gamma_1 = h = \text{const.}$$

3. An integral that has no clear physical meaning:

$$\left(\frac{K_1^2 + K_2^2}{2} + \gamma_1 x\right)^2 + (K_1 K_2 + \gamma_2 x)^2 = k = \text{const.}$$

The equations of motion are integrated using Kovalevskaya's variables (s_1, s_2), which are defined as follows:

$$s_1 = \frac{R - \sqrt{R_1 R_2}}{2(z_1 - z_2)^2}, \quad s_2 = \frac{R + \sqrt{R_1 R_2}}{2(z_1 - z_2)^2},$$

$$z_1 = K_\xi + iK_\eta, \quad z_2 = K_\xi - iK_\eta, \quad i^2 = -1,$$

$$R = R(z_1, z_2) = \tfrac{1}{4}z_1^2 z_2^2 - \tfrac{1}{2}h(z_1^2 + z_2^2) + c(z_1 + z_2) + \tfrac{1}{4}k^2 - 1,$$

$$R_1 = R(z_1, z_1), \quad R_2 = R(z_2, z_2).$$

In these variables, the equations of motion become

$$\frac{ds_1}{dt} = \frac{\sqrt{P(s_1)}}{s_1 - s_2}, \quad \frac{ds_2}{dt} = \frac{\sqrt{P(s_2)}}{s_2 - s_1}, \tag{11}$$

where

$$P(s) = \left[(2s + \tfrac{1}{2}h)^2 - \tfrac{1}{16}k^2\right]\left[4s^3 + 2hs^2 + \tfrac{1}{16}(4h^2 - k^2 + 4)s + \tfrac{1}{16}c^2\right].$$

By eliminating t, system (11) can be reduced to a separable equation, which is easy to integrate. As a results, equations (11) also convert to separable equations.

When writing Section 3.3.2, the following literature was used: Jacobi (1884), Lagrange (1889), Kowalewsky (1889, 1890), Klein & Sommerfeld (1965), Zhuravlev (1996, 2001), Borisov & Mamaev (2001), Klimov & Zhuravlev (2002), and Polyanin & Zaitsev (2018).

References

Athorne, C., Rational Ermakov systems of Fuchsian type, *J. Phys. A: Math. Gen.*, Vol. 24, No. 5, 945, 1991.

Berkovich, L.M., *Factorization and Transformations of Ordinary Differential Equations* [in Russian], Regulyarnaya i Khaoticheskaya Dinamika, Moscow, 2002.

Borisov, A.V. and Mamaev, I.S., *Rigid Body Dynamics* [in Russian], Regular and Chaotic Dynamics, Izhevsk, 2001.

Govinger, K.S., Athorne, C., and Leach, P.J.L., The algebraic structure of generalized Ermakov systems of three dimensions, *J. Phys. A: Math. Gen.*, Vol. 26, pp. 4035–4046, 1993.

Jacobi, C.G.J., *Vorlesungen über Dynamik*, G. Reimer, Berlin, 1884.

Klein, F. and Sommerfeld, A., *Über die Theorie des Kreisels*, Johnson Reprint corp., New York, 1965.

Kowalewsky, S., Sur le problème de la rotation d'un corps solide autour d'un point fixe, *Acta. Math.,* Vol. 12, No. 2, pp. 177–232, 1889.

Kowalewsky, S., Mémoires sur un cas particuliès du problème de la rotation d'un point fixe, où l'intégration s'effectue à l'aide de fonctions ultraelliptiques du tems, *Mémoires présentés par divers savants à l'Académie des seiences de l'Institut national de France,* Paris, Vol. 31, pp. 1–62, 1890.

Lagrange, J.L., *Mécanique Analytique. Œuvres de Lagrange,* Vol. 12, Gauthier–Villars, Paris, 1889.

Kamke, E., *Differentialgleichungen: Lösungsmethoden und Lösungen, I, Gewöhnliche Differentialgleichungen,* B. G. Teubner, Leipzig, 1977.

Klein, F. and Sommerfeld, A., *Über die Theorie des Kreisels,* Johnson Reprint Corp., New York, 1965.

Klimov, D.M. and Zhuravlev, V.Ph., *Group-Theoretic Methods in Mechanics and Applied Mathematics,* Taylor & Francis, London, 2002.

Polyanin, A.D. and Manzhirov, A.V., *Handbook of Mathematics for Engineers and Scientists,* Chapman & Hall/CRC Press, Boca Raton–London, 2007.

Polyanin, A.D. and Zaitsev, V.F., *Handbook of Exact Solutions for Ordinary Differential Equations,* 2nd ed., Chapman & Hall/CRC Press, Boca Raton–London, 2003.

Polyanin, A.D. and Zaitsev, V.F., *Handbook of Ordinary Differential Equations: Exact Solutions, Methods, and Problems,* CRC Press, Boca Raton–London, 2018.

Polyanin, A.D. and Zhurov, A.I., Exact solutions to homogeneous and quasi-homogeneous systems of nonlinear ODEs, *arXiv:2107.10759 [nlin.SI],* 2021.

Ray, J.R. and Reid, J.L., More exact invariants for the time dependent harmonic oscillator, *Phys. Letters, Ser. A,* Vol. 71, pp. 317–319, 1979.

Zhuravlev, V.Ph., The solid angle theorem in rigid body dynamics, *J. Appl. Math. Mech.,* Vol. 60, No. 2, pp. 319–322, 1996.

Zhuravlev, V.Ph., *Foundations of Theoretical Mechanics* [in Russian], Fizmatlit, Moscow, 2001.

Chapter 4

First-Order Partial Differential Equations

▶ **Preliminary remarks.** First-order partial differential equations (PDEs) are mathematical equations that contain two or more first-order partial derivatives of the unknown function.

This chapter presents exact solutions to linear and nonlinear PDEs of the first order in two independent variables. It briefly describes the most common methods for solving such linear and quasilinear PDEs based on integrating systems of ordinary differential equations (ODEs). Degenerate solutions of first-order PDEs that depend on only one of the independent variables are not considered here.

4.1. Linear Partial Differential Equations in Two Independent Variables

4.1.1. Preliminary Remarks. Solution Methods

▶ **The representation of the general solution via particular solutions.**

In the general case, a *first-order linear nonhomogeneous PDE in two independent variables* has the form

$$f(x,y)u_x + g(x,y)u_y = h_1(x,y)u + h_0(x,y). \qquad (1)$$

The general solution of the linear nonhomogeneous PDE (1) can be represented as the sum of any particular solution to this equation and the general solution of the corresponding homogeneous equation (with $h_0 \equiv 0$). In what follows, we give a more detailed statement about the representation of the general solution via particular solutions.

The general solution of PDE (1) can be represented in the form

$$u = u_2 + u_1 \Phi(u_0),$$

where $u_0 = u_0(x,y)$ is any nonconstant particular solution of the "truncated" homogeneous PDE (1) with $h_0 = h_1 = 0$; $u_1 = u_1(x,y)$ is a nontrivial particular solution of the truncated homogeneous PDE (1) with $h_0 = 0$; $u_2 = u_2(x,y)$ is a particular solution of the nonhomogeneous PDE (1); and $\Phi = \Phi(u_0)$ is an arbitrary function.

▶ **Method of characteristics (solution by using the characteristic ODE system).**

Given two distinct (functionally independent) integrals,

$$z_1(x,y) = C_1, \qquad z_2(x,y,u) = C_2 \qquad (2)$$

DOI: 10.1201/9781003051329-4

of the *characteristic system* of ODEs

$$\frac{dx}{f(x,y)} = \frac{dy}{g(x,y)} = \frac{du}{h_1(x,y)u + h_0(x,y)}, \qquad (3)$$

the general solution of the nonhomogeneous PDE (1) is defined by

$$\Psi(z_1, z_2) = 0, \qquad (4)$$

where Ψ is an arbitrary function of two variables. With formula (4) solved for z_1 or z_2, we often specify the general solution in the form

$$z_k = \Phi(z_{3-k}),$$

where $k = 1, 2$ and $\Phi(z)$ is an arbitrary function of one variable.

Remark 4.1. The degenerate case $h_0 = h_1 = 0$ in (3) corresponds to the simplest second integral $z_2 = C_2$ in (2). In this case, the general solution of PDE (1) is $u = \Phi(z_1)$, where $\Phi(z_1)$ is an arbitrary function, and z_1 is called a *principal integral*.

Remark 4.2. In the general case, the second integral (2) can be represented in the form $p_1(x,y)u + p_0(x,y) = C_2$, where $p_0(x,y)$ and $p_1(x,y)$ are some functions.

▶ **Solution based on a change of variables.**

Suppose a particular solution $z = z(x, y)$ of the truncated linear homogeneous PDE

$$f(x,y)z_x + g(x,y)z_y = 0 \qquad (z \not\equiv \text{const}) \qquad (5)$$

is known. By switching from x and y to the new variables x and $z = z(x, y)$ in PDE (1), we obtain

$$\bar{f}(x,z)u_x = \bar{h}_1(x,z)u + \bar{h}_0(x,z), \qquad (6)$$

where $\bar{f}(x,z) = f(x,y)$, $\bar{h}_1(x,z) = h_1(x,y)$, and $\bar{h}_0(x,z) = h_0(x,y)$ are the coefficients of the original equation (1) rewritten in term of x and z.

Equation (6) can be treated as an ordinary differential equation for $u = u(x)$ with parameter z. The general solution of equation (6) has the form

$$u = E\left[\int \frac{\bar{h}_0(x,z)}{\bar{f}(x,z)} \frac{dx}{E} + \Phi(z)\right], \qquad E = \exp\left[\int \frac{\bar{h}_1(x,z)}{\bar{f}(x,z)} dx\right],$$

where Φ is an arbitrary function; in the integration, z is treated as a parameter. To find a general integral of PDE (1), one should compute the integrals in the last relation and then return to the original variables x and y.

4.1.2. Equations of the Form $f(x,y)u_x + g(x,y)u_y = 0$

▶ In this section, for brevity, only the principal integral $z = z(x, y)$ will often be given. In such cases, the general solution to the PDE under consideration is determined by the formula $u = \Phi(z)$, where $\Phi(z)$ is an arbitrary function.

1. $au_x + bu_y = 0.$

General solution: $u = \Phi(bx - ay)$, where $\Phi(z)$ is an arbitrary function.

2. $axu_x + byu_y = 0$.

For $a = b$, this is a *conoidal equation*. General solution: $u = \Phi(|x|^b|y|^{-a})$, where $\Phi(z)$ is an arbitrary function.

3. $ayu_x + bxu_y = 0$.

General solution: $u = \Phi(bx^2 - ay^2)$, where $\Phi(z)$ is an arbitrary function.

4. $(a_1x + b_1y + c_1)u_x + (a_2x + b_2y + c_2)u_y = 0$.

The principal integral is determined by solutions of the following auxiliary system of algebraic equations for the parameters s, λ, μ, α, β, and γ:

$$(a_1 - s)(b_2 - s) = a_2 b_1, \tag{1}$$

$$a_1\lambda + a_2\mu = s\lambda, \qquad b_1\lambda + b_2\mu = s\mu, \tag{2}$$

$$c_1\alpha + c_2\beta - s\gamma = c_1\lambda + c_2\mu, \tag{3}$$

$$(a_1 - s)\alpha + a_2\beta = \lambda s, \qquad b_1\alpha + (b_2 - s)\beta = \mu s. \tag{4}$$

Case 1: $(a_1 - b_2)^2 + 4a_2b_1 \neq 0$. Equation (1) has two different roots s_1 and s_2. To these roots there correspond two sets of solutions, λ_1, μ_1 and λ_2, μ_2, of system (2).

1.1. If $a_1b_2 - a_2b_1 \neq 0$, then $s_1 \neq 0$ and $s_2 \neq 0$. Hence, the principal integral has the form

$$z = \frac{|s_1(\lambda_1 x + \mu_1 y) + \lambda_1 c_1 + \mu_1 c_2|^{s_2}}{|s_2(\lambda_2 x + \mu_2 y) + \lambda_2 c_1 + \mu_2 c_2|^{s_1}}.$$

1.2. If $a_1b_2 - a_2b_1 = 0$, then $s_1 = s = a_1 + b_2$ and $s_2 = 0$.
Principal integral for $\lambda_2 c_1 + \mu_2 c_2 \neq 0$:

$$z = s\frac{\lambda_2 x + \mu_2 y}{\lambda_2 c_1 + \mu_2 c_2} - \ln|s_1(\lambda_1 x + \mu_1 y) + \lambda_1 c_1 + \mu_1 c_2|.$$

Principal integral for $\lambda_2 c_1 + \mu_2 c_2 = 0$:

$$z = \lambda_2 x + \mu_2 y.$$

Case 2: $(a_1 - b_2)^2 + 4a_2b_1 = 0$. Equation (1) has a double root $s = \frac{1}{2}(a_1 + b_2)$. System (2) gives λ and μ not equal to zero simultaneously.

2.1. If $s \neq 0$, then we find γ from (3) and take nonzero α and β that satisfy relations (4). This leads to the principal integral

$$z = \ln|s(\lambda x + \mu y) + c_1\lambda + c_2\mu| - \frac{s(\alpha x + \beta y + \gamma)}{s(\lambda x + \mu y) + c_1\lambda + c_2\mu}.$$

2.2. If $s = 0$, then $b_2 = -a_1$. We have

$$z = a_2 x^2 - 2a_1 xy - b_1 y^2 + 2c_2 x - 2c_1 y.$$

5. $(a_1y^2 + b_1xy + c_1x^2)u_x + (a_2y^2 + b_2xy + c_2x^2)u_y = 0$.

Principal integral:

$$z = \int \frac{(a_1 v^2 + b_1 v + c_1)\,dv}{a_1 v^3 + (b_1 - a_2)v^2 + (c_1 - b_2)v - c_2} + \ln|x|, \qquad v = \frac{y}{x}.$$

6. $u_x + (ay + bx^k)u_y = 0.$

Principal integral: $z = ye^{-ax} - b\int x^k e^{-ax}\,dx.$

7. $u_x + (ax^k y + bx^n)u_y = 0.$

Principal integral: $z = y\exp\!\left(-\dfrac{a}{k+1}x^{k+1}\right) - b\int x^n \exp\!\left(-\dfrac{a}{k+1}x^{k+1}\right)dx.$

8. $u_x + (ae^{\lambda x} + b)u_y = 0.$

Principal integral: $z = \lambda(bx - y) + ae^{\lambda x}.$

9. $u_x + (ae^{\lambda y} + b)u_y = 0.$

Principal integral: $z = \lambda(bx - y) + \ln|b + ae^{\lambda y}|.$

10. $ae^{\lambda x}u_x + be^{\beta y}u_y = 0.$

Principal integral: $z = \dfrac{1}{\beta b}e^{-\beta y} - \dfrac{1}{\lambda a}e^{-\lambda x}.$

11. $u_x + [f(x)y + g(x)]u_y = 0.$

Principal integral: $z = e^{-F}y - \int e^{-F}g(x)\,dx,$ where $F = \int f(x)\,dx.$

12. $u_x + [f(x)y + g(x)y^k]u_y = 0.$

Principal integral: $z = e^{-F}y^{1-k} - (1-k)\int e^{-F}g(x)\,dx,$ where $F = (1-k)\int f(x)\,dx.$

13. $u_x + [f(x)e^{\lambda y} + g(x)]u_y = 0.$

Principal integral: $z = e^{-\lambda y}E + \lambda\int f(x)E\,dx,$ where $E = \exp\!\left[\lambda\int g(x)\,dx\right].$

14. $f(x)u_x + g(y)u_y = 0.$

Principal integral: $z = \int \dfrac{dx}{f(x)} - \int \dfrac{dy}{g(y)}.$

15. $[f(x) + g(y)]u_x + f'_x(x)u_y = 0.$

Principal integral: $z = f(x)e^{-y} - \int e^{-y}g(y)\,dy.$

16. $[x^k f(y) + xg(y)]u_x + h(y)u_y = 0.$

Principal integral:

$$z = x^{1-k}E + (k-1)\int \dfrac{f(y)E}{h(y)}\,dy, \quad \text{where}\quad E = \exp\!\left[(k-1)\int \dfrac{g(y)}{h(y)}\,dy\right].$$

17. $[f(y) + amx^k y^{m-1}]u_x - [g(x) + akx^{k-1}y^m]u_y = 0.$

Principal integral: $z = \int f(y)\,dy + \int g(x)\,dx + ax^k y^m.$

18. $[e^{\alpha x}f(y) + c\beta]u_x - [e^{\beta y}g(x) + c\alpha]u_y = 0.$

Principal integral: $z = \displaystyle\int e^{-\beta y}f(y)\,dy + \int e^{-\alpha x}g(x)\,dx - ce^{-\alpha x - \beta y}.$

19. $u_x + f(ax + by + c)u_y = 0, \quad b \neq 0.$

Principal integral: $z = \displaystyle\int \frac{dv}{a + bf(v)} - x,$ where $v = ax + by + c.$

20. $u_x + f(y/x)u_y = 0.$

Principal integral: $z = \displaystyle\int \frac{dv}{f(v) - v} - \ln|x|,$ where $v = \dfrac{y}{x}.$

21. $xu_x + yf(x^n y^m)u_y = 0.$

Principal integral: $z = \displaystyle\int \frac{dv}{v[mf(v) + n]} - \ln|x|,$ where $v = x^n y^m.$

22. $u_x + yf(e^{\alpha x}y^m)u_y = 0.$

Principal integral: $z = \displaystyle\int \frac{dv}{v[\alpha + mf(v)]} - x,$ where $v = e^{\alpha x}y^m.$

23. $xu_x + f(x^n e^{\alpha y})u_y = 0.$

Principal integral: $z = \displaystyle\int \frac{dv}{v[n + \alpha f(v)]} - \ln|x|,$ where $v = x^n e^{\alpha y}.$

4.1.3. Equations of the Form $f(x, y)u_x + g(x, y)u_y = h(x, y)$

▶ In the solutions of equations 4.1.3.1–4.1.3.18, $\Phi(z)$ is an arbitrary composite function whose argument z can depend on both x and y.

1. $au_x + bu_y = c.$

Equation of a cylindrical surface. Two forms of representation of the general solution:

$$u = \frac{c}{a}x + \Phi(bx - ay), \qquad u = \frac{c}{b}y + \Phi(bx - ay).$$

2. $au_x + bu_y = f(x).$

General solution: $u = \dfrac{1}{a}\displaystyle\int f(x)\,dx + \Phi(bx - ay).$

3. $u_x + au_y = f(x)y^k.$

General solution: $u = \displaystyle\int_{x_0}^{x}(y - ax + at)^k f(t)\,dt + \Phi(y - ax),$ where x_0 can be taken arbitrarily.

4. $u_x + au_y = f(x)e^{\lambda y}.$

General solution: $u = e^{\lambda(y - ax)}\displaystyle\int f(x)e^{a\lambda x}\,dx + \Phi(y - ax).$

5. $au_x + bu_y = f(x) + g(y)$.

General solution: $u = \dfrac{1}{a} \int f(x)\, dx + \dfrac{1}{b} \int g(y)\, dy + \Phi(bx - ay)$.

6. $u_x + au_y = f(x)g(y)$.

General solution: $u = \int_{x_0}^{x} f(t)g(y - ax + at)\, dt + \Phi(y - ax)$, where x_0 can be taken arbitrarily.

7. $u_x + au_y = f(x, y)$.

General solution: $u = \int_{x_0}^{x} f(t, y - ax + at)\, dt + \Phi(y - ax)$, where x_0 can be taken arbitrarily.

8. $u_x + [ay + f(x)]u_y = g(x)$.

General solution: $u = \int g(x)\, dx + \Phi(z)$, where $z = e^{-ax} y - \int f(x) e^{-ax}\, dx$.

9. $u_x + [ay + f(x)]u_y = g(x)h(y)$.

General solution:

$$u = \int g(x)\, h\!\left(e^{ax} z + e^{ax} \int f(x) e^{-ax}\, dx\right) dx + \Phi(z), \quad \text{where } z = e^{-ax} y - \int f(x) e^{-ax}\, dx.$$

In the integration, z is treated as a parameter.

10. $u_x + [f(x)y + g(x)y^k]u_y = h(x)$.

General solution: $u = \int h(x)\, dx + \Phi(z)$, where

$$z = e^{-F} y^{1-k} - (1-k) \int e^{-F} g(x)\, dx, \quad F = (1-k) \int f(x)\, dx.$$

11. $u_x + [f(x) + g(x) e^{\lambda y}] u_y = h(x)$.

General solution: $u = \int h(x)\, dx + \Phi(z)$, where

$$z = e^{-\lambda y} F(x) + \lambda \int g(x) F(x)\, dx, \quad F(x) = \exp\!\left[\lambda \int f(x)\, dx\right].$$

12. $axu_x + byu_y = f(x, y)$.

General solution:

$$u = \dfrac{1}{a} \int \dfrac{1}{x} f\!\left(x, z^{1/a} x^{b/a}\right) dx + \Phi(z), \quad \text{where } z = y^a x^{-b}.$$

In the integration, z is treated as a parameter.

13. $f(x) u_x + g(y) u_y = h_1(x) + h_2(y)$.

General solution: $u = \int \dfrac{h_1(x)}{f(x)}\, dx + \int \dfrac{h_2(y)}{g(y)}\, dy + \Phi\!\left(\int \dfrac{dx}{f(x)} - \int \dfrac{dy}{g(y)}\right)$.

14. $f(x)u_x + g(y)u_y = h(x,y)$.

The transformation $\xi = \int \dfrac{dx}{f(x)}$, $\eta = \int \dfrac{dy}{g(y)}$ leads to an equation of the form 4.1.3.7 for $u = u(\xi, \eta)$.

15. $f(y)u_x + g(x)u_y = h(x,y)$.

The transformation $\xi = \int g(x)\,dx$, $\eta = \int f(y)\,dy$ leads to an equation of the form 4.1.3.7 for $u = u(\xi, \eta)$.

16. $f(x)u_x + [g_1(x)y + g_0(x)]u_y = h(x,y)$.

General solution:
$$u = \Phi(z) + \int \frac{h(x,\, zG + Q)}{f}\,dx, \qquad z = \frac{y - Q}{G},$$

where $G = \exp\left(\int \dfrac{g_1}{f}\,dx\right)$ and $Q = G\int \dfrac{g_0\,dx}{fG}$. In the integration, z is treated as a parameter.

17. $f(x)u_x + [g_1(x)y + g_0(x)y^k]u_y = h(x,y)$.

For $k = 1$ see equation 4.1.3.16. For $k \neq 1$, the substitution $\xi = y^{1-k}$ leads to an equation of the form 4.1.3.16:
$$f(x)u_x + (1-k)[g_1(x)\xi + g_0(x)]u_\xi = h\!\left(x,\, \xi^{\frac{1}{1-k}}\right).$$

18. $f(x)u_x + [g_1(x) + g_0(x)e^{\lambda y}]u_y = h(x,y)$.

The substitution $\xi = e^{-\lambda y}$ leads to an equation of the form 4.1.3.16:
$$f(x)u_x - \lambda[g_1(x)\xi + g_0(x)]u_\xi = h\!\left(x,\, -\frac{1}{\lambda}\ln\xi\right).$$

4.1.4. Equations of the Form $f(x,y)u_x + g(x,y)u_y = h(x,y)u + r(x,y)$

▶ In the solutions of equations 4.1.4.1–4.1.3.12 and 4.1.3.15, $\Phi(z)$ is an arbitrary composite function whose argument z can depend on both x and y.

1. $au_x + bu_y = cu$.

Two forms of representation of the general solution:
$$u = \exp\!\left(\frac{c}{a}x\right)\Phi(bx - ay), \qquad u = \exp\!\left(\frac{c}{b}y\right)\Phi(bx - ay).$$

2. $au_x + bu_y = f(x)u$.

General solution: $u = \exp\!\left[\dfrac{1}{a}\int f(x)\,dx\right]\Phi(bx - ay)$.

3. $au_x + bu_y = f(x)u + g(x)$.

General solution:
$$u = \exp\left[\frac{1}{a}\int f(x)\,dx\right]\left\{\Phi(bx - ay) + \frac{1}{a}\int g(x)\exp\left[-\frac{1}{a}\int f(x)\,dx\right]dx\right\}.$$

4. $au_x + bu_y = [f(x) + g(y)]u$.

General solution: $u = \exp\left[\dfrac{1}{a}\int f(x)\,dx + \dfrac{1}{b}\int g(y)\,dy\right]\Phi(bx - ay)$.

5. $u_x + au_y = f(x,y)u$.

General solution: $u = \exp\left[\displaystyle\int_{x_0}^{x} f(t, y - ax + at)\,dt\right]\Phi(y - ax)$, where x_0 can be taken arbitrarily.

6. $u_x + au_y = f(x,y)u + g(x,y)$.

General solution:
$$u = F(x,z)\left[\Phi(z) + \int \frac{g(x, z + ax)}{F(x,z)}\,dx\right], \quad F(x,z) = \exp\left[\int f(x, z + ax)\,dx\right],$$

where $z = y - ax$. In the integration, z is treated as a parameter.

7. $axu_x + byu_y = f(x)u + g(x)$.

General solution:
$$u = \exp\left[\frac{1}{a}\int \frac{f(x)\,dx}{x}\right]\left\{\Phi(x^{-b/a}y) + \frac{1}{a}\int \frac{g(x)}{x}\exp\left[-\frac{1}{a}\int \frac{f(x)\,dx}{x}\right]dx\right\}.$$

8. $axu_x + byu_y = f(x,y)u$.

General solution:
$$u = \exp\left[\frac{1}{a}\int \frac{1}{x}f(x, z^{1/a}x^{b/a})\,dx\right]\Phi(z), \quad \text{where} \quad z = y^a x^{-b}.$$

In the integration, z is treated as a parameter.

9. $xu_x + ayu_y = f(x,y)u + g(x,y)$.

General solution:
$$u = F(x,z)\left[\Phi(z) + \int \frac{g(x, zx^a)}{xF(x,z)}\,dx\right], \quad F(x,z) = \exp\left[\int \frac{1}{x}f(x, zx^a)\,dx\right],$$

where $z = yx^{-a}$. In the integration, z is treated as a parameter.

10. $(x - a)u_x + (y - b)u_y = u$.

Differential equation of a conic surface with vertex at the point $(a, b, 0)$.

General solution: $u = (x - a)\Phi\left(\dfrac{y - b}{x - a}\right)$.

11. $f(x)u_x + g(y)u_y = [h_1(x) + h_2(y)]u.$

General solution:
$$u = \exp\left[\int \frac{h_1(x)}{f(x)} dx + \int \frac{h_2(y)}{g(y)} dy\right] \Phi\left(\int \frac{dx}{f(x)} dx - \int \frac{dy}{g(y)} dy\right).$$

12. $f_1(x)u_x + f_2(y)u_y = au + g_1(x) + g_2(y).$

General solution:
$$u = E_1(x)\Phi(z) + E_1(x)\int \frac{g_1(x)\,dx}{f_1(x)E_1(x)} + E_2(y)\int \frac{g_2(y)\,dy}{f_2(y)E_2(y)},$$

where
$$E_1(x) = \exp\left[a\int \frac{dx}{f_1(x)}\right], \quad E_2(y) = \exp\left[a\int \frac{dy}{f_2(y)}\right], \quad z = \int \frac{dx}{f_1(x)} - \int \frac{dy}{f_2(y)}.$$

13. $f(x)u_x + g(y)u_y = h(x,y)u + r(x,y).$

The transformation $\xi = \int \frac{dx}{f(x)}, \; \eta = \int \frac{dy}{g(y)}$ leads to an equation of the form 4.1.4.6 for $u = u(\xi, \eta)$.

14. $f(y)u_x + g(x)u_y = h(x,y)u + r(x,y).$

The transformation $\xi = \int g(x)\,dx, \; \eta = \int f(y)\,dy$ leads to an equation of the form 4.1.4.6 for $u = u(\xi, \eta)$.

15. $f(x)u_x + [g_1(x)y + g_0(x)]u_y = h(x,y)u + r(x,y).$

General solution:
$$u = H(x,z)\left[\Phi(z) + \int \frac{r(x, zG+Q)}{f(x)H(x,z)} dx\right],$$

where
$$z = \frac{y-Q}{G}, \quad H(x,z) = \exp\left[\int \frac{h(x, zG+Q)}{f(x)} dx\right],$$
$$G = G(x) = \exp\left[\int \frac{g_1(x)}{f(x)} dx\right], \quad Q = Q(x) = G(x)\int \frac{g_0(x)\,dx}{f(x)G(x)}.$$

In the integration, z is treated as a parameter.

16. $f(x)u_x + [g_1(x)y + g_0(x)y^k]u_y = h(x,y)u + r(x,y).$

For $k = 0$ and $k = 1$, see equation 4.1.4.15. For $k \neq 1$, the substitution $\xi = y^{1-k}$ leads to an equation of the form 4.1.4.15:
$$f(x)u_x + (1-k)[g_1(x)\xi + g_0(x)]u_\xi = h\left(x, \xi^{\frac{1}{1-k}}\right)u + r\left(x, \xi^{\frac{1}{1-k}}\right).$$

17. $f(x)u_x + [g_1(x) + g_0(x)e^{\lambda y}]u_y = h(x,y)u + r(x,y).$

The substitution $\xi = e^{-\lambda y}$ leads to the equation of the form 4.1.4.15:
$$f(x)u_x - \lambda[g_1(x)\xi + g_0(x)]u_\xi = h\left(x, -\frac{1}{\lambda}\ln\xi\right)u + r\left(x, -\frac{1}{\lambda}\ln\xi\right).$$

4.2. Quasilinear Partial Differential Equations in Two Independent Variables

4.2.1. Preliminary Remarks. Solution Methods

▶ **Method of characteristics (solution by using the characteristic ODE system).**

A *first-order quasilinear PDE in two independent variables* has the general form

$$f(x,y,u)u_x + g(x,y,u)u_y = h(x,y,u). \tag{1}$$

Such equations are used to describe various phenomena and processes in continuous mechanics, gas dynamics, fluid dynamics, heat and mass transfer theory, wave theory, acoustics, and many other disciplines.

If two independent integrals,

$$z_1(x,y,u) = C_1, \quad z_2(x,y,u) = C_2, \tag{2}$$

of the *characteristic system of ODEs*

$$\frac{dx}{f(x,y,u)} = \frac{dy}{g(x,y,u)} = \frac{du}{h(x,y,u)} \tag{3}$$

are known, then the *general solution of the first-order PDE* (1) is given by

$$\Psi(z_1, z_2) = 0, \tag{4}$$

where Ψ is an arbitrary function of two variables. With formula (4) solved for z_1 or z_2, we often specify the general solution in the form

$$z_k = \Phi(z_{3-k}),$$

where $k = 1, 2$, and $\Phi(z)$ is an arbitrary function of one variable.

Remark 4.3. As a rule, the general solution of equation (1) cannot be represented in an explicit form, solved with respect to the desired function u.

Remark 4.4. The degenerate case $h = 0$ in (3) corresponds to the simplest second integral $u = C_2$ in (2).

Remark 4.5. Methods for solving first-order quasilinear PDEs of the form (1) are described in the books by Kamke (1965), Rhee, Aris and Amundson (1986), Polyanin, Zaitsev, and Moussiaux (2002), Polyanin & Zaitsev (2012).

▶ **First-order quasilinear PDEs of special form.**

1°. Consider the *first-order quasilinear PDE of a special form*

$$f(x,y)u_x + g(x,y)u_y = h(x,y)\varphi(u).$$

The substitution $w = \int \dfrac{du}{\varphi(u)}$ takes this equation to a simpler first-order linear PDE

$$f(x,y)w_x + g(x,y)w_y = h(x,y).$$

For such linear equations, see Sections 4.1.1 and 4.1.3.

$2°$. The *first-order PDE with power-law nonlinearity*

$$f(x,y)u_x + g(x,y)u_y = h_1(x,y)u + h_2(x,y)u^k \quad (k \neq 0, 1)$$

is reduced by substituting $w = u^{1-k}$ to equation (1) from Section 4.1.1:

$$f(x,y)w_x + g(x,y)w_y = (1-k)h_1(x,y)w + (1-k)h_2(x,y).$$

$3°$. The *first-order PDE with exponential nonlinearity*

$$f(x,y)u_x + g(x,y)u_y = h_1(x,y) + h_2(x,y)e^{\lambda u} \quad (\lambda \neq 0)$$

is reduced by substituting $w = e^{-\lambda u}$ to equation (1) from Section 4.1.1:

$$f(x,y)w_x + g(x,y)w_y = -\lambda h_1(x,y)w - \lambda h_2(x,y).$$

4.2.2. Equations of the Form $f(x,y)u_x + g(x,y)u_y = h(x,y,u)$

▶ In the solutions of equations 4.2.2.1–4.2.2.18, $\Phi(z)$ is an arbitrary composite function whose argument z can depend on both x and y.

1. $u_x + au_y = bu^k$.

General solution for $k \neq 1$: $u = \bigl[\Phi(y-ax) + b(1-k)x\bigr]^{\frac{1}{1-k}}$. For $k=1$, see equation 4.1.4.1.

2. $u_x + au_y = be^{-\lambda u}$.

General solution: $u = \dfrac{1}{\lambda}\ln\bigl[\Phi(y-ax) + b\lambda x\bigr]$.

3. $u_x + au_y = f(x)u + g(x)u^k$.

General solution:

$$u^{1-k} = F(x)\Phi(y-ax) + (1-k)F(x)\int \frac{g(x)}{F(x)}\,dx, \quad \text{where } F(x) = \exp\left[(1-k)\int f(x)\,dx\right].$$

4. $u_x + au_y = f(x) + g(x)e^{-\lambda u}$.

General solution:

$$e^{\lambda u} = F(x)\Phi(y-ax) + \lambda F(x)\int \frac{g(x)}{F(x)}\,dx, \quad \text{where } F(x) = \exp\left[\lambda \int f(x)\,dx\right].$$

5. $au_x + bu_y = f(u)$.

General solution:

$$\int \frac{du}{f(u)} = \frac{x}{a} + \Phi(bx-ay).$$

6. $au_x + bu_y = f(x)g(u)$.

General solution:

$$\int \frac{du}{g(u)} = \frac{1}{a}\int f(x)\,dx + \Phi(bx-ay).$$

7. $u_x + au_y = f(x)g(y)h(u)$.

General solution: $\displaystyle\int \frac{du}{h(u)} = \int_{x_0}^{x} f(t)g(y - ax + at)\,dt + \Phi(y - ax)$, where x_0 can be taken arbitrarily.

8. $axu_x + bxu_y = cu^k$.

General solution:
$$\frac{c}{a}\ln|x| + \Phi(bx - ay) = \begin{cases} \frac{1}{1-k}u^{1-k} & \text{if } k \neq 1, \\ \ln|u| & \text{if } k = 1. \end{cases}$$

9. $axu_x + byu_y = cu^k$.

General solution:
$$\frac{c}{a}\ln|x| + \Phi\big(|x|^b|y|^{-a}\big) = \begin{cases} \frac{1}{1-k}u^{1-k} & \text{if } k \neq 1, \\ \ln|u| & \text{if } k = 1. \end{cases}$$

10. $ayu_x + bxu_y = cu^k$.

General solution for $ab > 0$:
$$\frac{c}{\sqrt{ab}}\ln\big|\sqrt{ab}\,x + ay\big| + \Phi(ay^2 - bx^2) = \begin{cases} \frac{1}{1-k}u^{1-k} & \text{if } k \neq 1, \\ \ln|u| & \text{if } k = 1. \end{cases}$$

11. $axu_x + byu_y = f(u)$.

General solution:
$$\int \frac{du}{f(u)} = \frac{1}{a}\ln|x| + \Phi\big(|x|^b|y|^{-a}\big).$$

12. $ayu_x + bxu_y = f(u)$.

General solution for $ab > 0$:
$$\int \frac{du}{f(u)} = \frac{1}{\sqrt{ab}}\ln\big|\sqrt{ab}\,x + ay\big| + \Phi(ay^2 - bx^2).$$

13. $ax^n u_x + by^k u_y = f(u)$.

General solution:
$$\int \frac{du}{f(u)} = \frac{1}{a(1-n)}x^{1-n} + \Phi(z), \quad \text{where } z = \frac{1}{a(1-n)}x^{1-n} - \frac{1}{b(1-k)}y^{1-k}.$$

14. $ay^n u_x + bx^k u_y = f(u)$.

General solution:
$$a\int \frac{du}{f(u)} = \int \left(\frac{b}{a}\frac{n+1}{k+1}x^{k+1} - z\right)^{-\frac{n}{n+1}} dx, \quad \text{where } z = \frac{b}{a}\frac{n+1}{k+1}x^{k+1} - y^{n+1}.$$

In the integration, z is treated as a parameter.

15. $ae^{\lambda x}u_x + be^{\beta y}u_y = f(u)$.

General solution: $\displaystyle\int \frac{du}{f(u)} = -\frac{1}{a\lambda}e^{-\lambda x} + \Phi(z)$, where $z = a\lambda e^{-\beta y} - b\beta e^{-\lambda x}$.

16. $ae^{\lambda y}u_x + be^{\beta x}u_y = f(u)$.

General solution: $\displaystyle\int \frac{du}{f(u)} = \frac{c(\beta x - \lambda y)}{z} + \Phi(z)$, where $z = a\beta e^{\lambda y} - b\lambda e^{\beta x}$.

17. $f(x)u_x + g(y)u_y = h(u)$.

General solution: $\displaystyle\int \frac{du}{h(u)} = \int \frac{dx}{f(x)} + \Phi(z)$, where $z = \int \frac{dx}{f(x)} - \int \frac{dy}{g(y)}$.

18. $f(y)u_x + g(x)u_y = h(u)$.

The transformation $\xi = \displaystyle\int g(x)\,dx$, $\eta = \displaystyle\int f(y)\,dy$ leads to an equation of the form 4.2.2.7:

$$u_\xi + u_\eta = F(\xi)G(\eta)h(u), \quad \text{where} \quad F(\xi) = \frac{1}{g(x)}, \quad G(\eta) = \frac{1}{f(y)}.$$

4.2.3. Equations of the Form $u_x + f(x, y, u)u_y = 0$

▶ In the solutions of equations 4.2.3.1–4.2.3.17, $\Phi(u)$ is an arbitrary function.

1. $u_x + auu_y = 0$.

Hopf's equation. It is used as a model equation of nonlinear wave theory and gas dynamics, where the independent variables x and y play the role of time and the spatial coordinate, respectively.

1°. General solution:

$$\Phi(axu - y, u) = 0 \quad \text{or} \quad y = axu + \tilde{\Phi}(u),$$

where Φ and $\tilde{\Phi}$ are arbitrary functions.

2°. The solution of the Cauchy problem with the initial condition

$$u = \varphi(y) \quad \text{at} \quad x = 0$$

can be represented in parametric form as

$$y = \xi + a\varphi(\xi)x, \quad u = \varphi(\xi).$$

For $a > 0$ and $\varphi'(\xi) > 0$, these relations describe the classical single-valued solution.

3°. Consider the Cauchy problem for the Hopf's equation with the discontinuous piecewise-constant initial condition

$$u(0, y) = \begin{cases} u_1 & \text{for } y < 0, \\ u_2 & \text{for } y > 0. \end{cases}$$

It is assumed that $a > 0$, $u_1 > 0$, and $u_2 > 0$.

For $u_1 < u_2$, the generalized solution has the form

$$u(x,y) = \begin{cases} u_1 & \text{for } y/x < V_1, \\ y/(ax) & \text{for } V_1 \le y/x \le V_2, \\ u_2 & \text{for } y/x > V_2, \end{cases} \quad \text{where} \quad V_1 = au_1, \; V_2 = au_2.$$

This solution is continuous in the half-plane $x > 0$ and describes a *rarefaction wave*.

For $u_1 > u_2$, the generalized solution has the form

$$u(x,y) = \begin{cases} u_1 & \text{for } y/x < V, \\ u_2 & \text{for } y/x > V, \end{cases} \quad \text{where} \quad V = \tfrac{1}{2}a(u_1+u_2).$$

This solution experiences a discontinuity along the line $y = Vx$ and describes a *shock wave*.

2. $u_x + (au + bx)u_y = 0.$

General solution: $y = axu + \tfrac{1}{2}bx^2 + \Phi(u)$.

3. $u_x + (au + by)u_y = 0.$

General solution: $x = \dfrac{1}{b}\ln|au + by| + \Phi(u)$.

4. $u_x + [au + yf(x)]u_y = 0.$

General solution: $yF(x) - au\displaystyle\int F(x)\,dx = \Phi(u)$, where $F(x) = \exp\left[-\displaystyle\int f(x)\,dx\right]$.

5. $u_x + [au + f(y)]u_y = 0.$

General solution: $x = \displaystyle\int_{y_0}^{y} \dfrac{dt}{f(t) + au} + \Phi(u)$. In the integration, u is treated as a parameter.

6. $u_x + (au^k + b)u_y = 0.$

General solution: $y = axu^k + bx + \Phi(u)$.

7. $u_x + (au^k + bx)u_y = 0.$

General solution: $y = axu^k + \tfrac{1}{2}bx^2 + \Phi(u)$.

8. $u_x + (ae^{\lambda u} + b)u_y = 0.$

General solution: $y = x(ae^{\lambda u} + b) + \Phi(u)$.

9. $u_x + (ae^{\lambda u} + bx)u_y = 0.$

General solution: $y = axe^{\lambda u} + \tfrac{1}{2}bx^2 + \Phi(u)$.

10. $u_x + f(u)u_y = 0.$

A model equation of gas dynamics. This equation is also encountered in hydrodynamics, multiphase flows, wave theory, acoustics, chemical engineering, and other applications.

1°. General solution:
$$y = xf(u) + \Phi(u),$$
where Φ is an arbitrary function.

2°. The solution of the Cauchy problem with the initial condition
$$u = \varphi(y) \quad \text{at} \quad x = 0$$
can be represented in the parametric form
$$y = \xi + \mathcal{F}(\xi)x, \quad u = \varphi(\xi),$$
where $\mathcal{F}(\xi) = f(\varphi(\xi))$.

3°. Consider the Cauchy problem with the discontinuous piecewise-constant initial condition
$$u(0, y) = \begin{cases} u_1 & \text{for } y < 0, \\ u_2 & \text{for } y > 0. \end{cases}$$

It is assumed that $x \geq 0$, $f > 0$, and $f' > 0$ for $u > 0$, $u_1 > 0$, and $u_2 > 0$.

Generalized solution for $u_1 < u_2$:
$$u(x, y) = \begin{cases} u_1 & \text{for } y/x < V_1, \\ f^{-1}(y/x) & \text{for } V_1 \leq y/x \leq V_2, \\ u_2 & \text{for } y/x > V_2, \end{cases} \quad \text{where} \quad V_1 = f(u_1), \; V_2 = f(u_2).$$

Here f^{-1} is the inverse of the function f, i.e., $f^{-1}(f(u)) \equiv u$. This solution is continuous in the half-plane $x > 0$ and describes a rarefaction wave.

Generalized solution for $u_1 > u_2$:
$$u(x, y) = \begin{cases} u_1 & \text{for } y/x < V, \\ u_2 & \text{for } y/x > V, \end{cases} \quad \text{where} \quad V = \frac{1}{u_2 - u_1} \int_{u_1}^{u_2} f(u)\, du.$$

This solution undergoes a discontinuity along the line $y = Vx$ and describes a shock wave.

11. $u_x + [f(u) + ax]u_y = 0.$

General solution: $y = xf(u) + \frac{1}{2}ax^2 + \Phi(u)$.

12. $u_x + [f(u) + ay]u_y = 0.$

General solution for $a \neq 0$: $x = \dfrac{1}{a} \ln|ay + f(u)| + \Phi(u)$.

13. $u_x + [f(u) + g(x)]u_y = 0.$

General solution: $y = xf(u) + \displaystyle\int g(x)\, dx + \Phi(u)$.

14. $u_x + [f(u) + g(y)]u_y = 0.$

General solution: $x = \displaystyle\int_{y_0}^{y} \frac{dt}{g(t) + f(u)} + \Phi(u)$. In the integration, u is treated as a parameter.

15. $u_x + [yf(u) + g(x)]u_y = 0.$

General solution: $y\exp[-xf(u)] - \int_{x_0}^{x} g(t)\exp[-tf(u)]\,dt = \Phi(u),$ where x_0 can be taken arbitrarily.

16. $u_x + [xf(u) + yg(u) + h(u)]u_y = 0.$
General solution:
$$y + \frac{xf(u) + h(u)}{g(u)} + \frac{f(u)}{g^2(u)} = \exp[g(u)x]\Phi(u).$$

17. $u_x + f(x)g(y)h(u)u_y = 0.$
General solution:
$$\int \frac{dy}{g(y)} - h(u)\int f(x)\,dx = \Phi(u).$$

4.2.4. Equations of the Form $u_x + f(x,y,u)u_y = g(x,y,u)$

▶ In the solutions of equations 4.2.4.1–4.2.4.18, $\Phi(z)$ is an arbitrary composite function whose argument z can depend on x, y, and u.

1. $u_x + auu_y = b.$
General solution: $\Phi(u - bx,\ au^2 - 2by) = 0.$

2. $u_x + auu_y = bx.$
General solution: $y = axu - \tfrac{1}{3}abx^3 + \Phi(u - \tfrac{1}{2}bx^2).$

3. $u_x + auu_y = by.$
General solution:
$$x = \int_{y_0}^{y} \frac{dt}{\sqrt{ab(t^2 - y^2) + a^2u^2}} + \Phi(au^2 - by^2),$$
where y_0 may be taken arbitrary.

4. $u_x + auu_y + bu = 0.$
For $a, b > 0$, this is a *model equation of nonlinear waves with damping* (Whitham, 1974).

1°. General solution: $\Phi(au + by,\ ue^{bx}) = 0.$

2°. The solution of the Cauchy problem with the initial condition $u(0, y) = f(y)$ can be represented in parametric form as
$$y = \xi + \frac{a}{b}(1 - e^{-bx})f(\xi), \quad u = e^{-bx}f(\xi).$$

5. $u_x + auu_y = f(x).$
General solution:
$$y = ax[u - F(x)] + a\int F(x)\,dx + \Phi(u - F(x)), \quad \text{where } F(x) = \int f(x)\,dx.$$

6. $u_x + auu_y = f(y)$.

General solution:
$$x = \pm \int_{y_0}^{y} \frac{dt}{\sqrt{2aF(t) - 2az}} + \Phi(z), \qquad z = F(y) - \frac{1}{2}au^2,$$

where $F(y) = \int f(y)\,dy$. In the integration, z is treated as a parameter.

7. $u_x + auu_y = f(y - bx)$.

This is a model equation describing nonlinear waves issuing from a moving source (the variables x and y play the role of time and the spatial coordinate, respectively, and b is the source velocity).

1°. General solution:
$$x = \pm \int_{t_0}^{y-bx} \frac{dt}{\sqrt{b^2 + 2aF(t) - 2az}} + \Phi(z), \qquad z = F(y - bx) - \frac{1}{2}au^2 + bu,$$

where $F(t) = \int f(t)\,dt$. In the integration, z is treated as a parameter; t_0 may be taken arbitrary.

2°. Solution with a steady profile:
$$u = b - \left[(b - u_0)^2 - 2\int_{\xi}^{\infty} f(t)\,dt\right]^{1/2}, \qquad \xi = y - bx,$$

where u_0 is the constant of integration.

8. $u_x + [au + f(x)]u_y = g(x)$.

General solution:
$$y = ax[u - G(x)] + a\int G(x)\,dx + F(x) + \Phi(u - G(x)),$$

where
$$F(x) = \int f(x)\,dx, \qquad G(x) = \int g(x)\,dx.$$

9. $u_x + f(u)u_y = g(x)$.

General solution:
$$y = \int_{x_0}^{x} f(G(t) - G(x) + u)\,dt + \Phi(u - G(x)), \quad \text{where} \quad G(x) = \int g(x)\,dx.$$

10. $u_x + f(u)u_y = g(y)$.

General solution:
$$x = \int_{y_0}^{y} \psi(G(t) - G(y) + F(u))\,dt + \Phi(F(u) - G(y)),$$

where $G(y) = \int g(y)\,dy$ and $F(u) = \int f(u)\,du$. The function $\psi = \psi(z)$ is defined parametrically by $\psi = \dfrac{1}{f(u)}$, $z = F(u)$.

11. $u_x + f(u)u_y = g(u)$.

General solution:
$$y = \int \frac{f(u)}{g(u)} du + \Phi\left(x - \int \frac{du}{g(u)}\right).$$

12. $u_x + [f(u) + g(x)]u_y = h(x)$.

General solution:
$$y = \int_{x_0}^{x} f(H(t) - H(x) + u) \, dt + G(x) + \Phi(u - H(x)),$$

where
$$G(x) = \int g(x) \, dx, \quad H(x) = \int h(x) \, dx.$$

13. $u_x + [f(u) + g(x)]u_y = h(u)$.

General solution:
$$y = \int \frac{f(u)}{h(u)} du + \int_{u_0}^{u} \frac{g(H(t) - H(u) + x)}{h(t)} dt + \Phi(x - H(u)), \quad \text{where} \quad H(u) = \int \frac{du}{h(u)}.$$

14. $u_x + [f(u) + yg(x)]u_y = h(x)$.

General solution:
$$yG(x) - \int_{x_0}^{x} G(t) f(H(t) - H(x) + u) \, dt = \Phi(u - H(x)),$$

where $G(x) = \exp\left[-\int g(x) \, dx\right]$, $H(x) = \int h(x) \, dx$, and x_0 may be taken arbitrary. In the integration, u is treated as a parameter.

15. $u_x + f(x)g(y)h(u)u_y = p(x)$.

General solution:
$$\int \frac{dy}{g(y)} = \int_{x_0}^{x} f(t) h(P(t) - P(x) + u) \, dt + \Phi(u - P(x)), \quad \text{where} \quad P(x) = \int p(x) \, dx.$$

In the integration, u is treated as a parameter; x_0 may be taken arbitrary.

16. $u_x + f(x)g(y)h(u)u_y = p(u)$.

General solution:
$$\int \frac{dy}{g(y)} = \int_{u_0}^{u} \frac{h(t)}{p(t)} f(P(t) - P(u) + x) \, dt + \Phi(x - P(u)), \quad \text{where} \quad P(u) = \int \frac{du}{p(u)}.$$

In the integration, u is treated as a parameter; u_0 may be taken arbitrary.

17. $u_x + f(x, u)u_y = g(x)$.

General solution:
$$y = \int_{x_0}^{x} f(t, G(t) - G(x) + u) \, dt + \Phi(u - G(x)), \quad \text{where} \quad G(x) = \int g(x) \, dx.$$

In the integration, u is treated as a parameter; x_0 may be taken arbitrary.

18. $u_x + f(x, u)u_y = g(u)$.

General solution:
$$y = \int_{u_0}^{u} \frac{f(G(t) - G(u) + x, t)}{g(t)} \, dt + \Phi(x - G(u)), \quad \text{where} \quad G(u) = \int \frac{du}{g(u)}.$$

In the integration, u is treated as a parameter; u_0 may be taken arbitrary.

4.3. Nonlinear Partial Differential Equations in Two Independent Variables

4.3.1. Preliminary Remarks. A Complete Integral

A *nonlinear first-order partial differential equation* in two independent variables has the general form
$$F(x, y, u, u_x, u_y) = 0. \tag{1}$$

Such equations are encountered in analytical mechanics, calculus of variations, optimal control, differential games, dynamic programming, geometric optics, differential geometry, and other fields.

$1°$. Suppose a particular solution of PDE (1),
$$u = \Theta(x, y, C_1, C_2), \tag{2}$$

dependent on two free parameters, C_1 and C_2, that are not included in the original equation, is known. The two-parameter family of solutions (2) is called a *complete integral* of PDE (1) if the rank of the matrix
$$M = \begin{pmatrix} \Theta_1 & \Theta_{x1} & \Theta_{y1} \\ \Theta_2 & \Theta_{x2} & \Theta_{y2} \end{pmatrix} \tag{3}$$

is equal to two in the domain under consideration (for example, this is valid if $\Theta_{x1}\Theta_{y2} - \Theta_{x2}\Theta_{y1} \neq 0$). In equation (3), Θ_n denotes the partial derivative of Θ with respect to C_n ($n = 1, 2$), Θ_{xn} is the second partial derivative with respect to x and C_n, and Θ_{yn} is the second partial derivative with respect to y and C_n.

In some cases, a complete integral can be found using the method of undetermined coefficients by presetting an appropriate structure of the particular solution sought. (The complete integral is determined by the differential equation nonuniquely.)

A complete integral of equation (1) is often written in implicit form:
$$\Theta(x, y, u, C_1, C_2) = 0. \tag{4}$$

$2°$. The *general integral* of PDE (1) can be represented in parametric form by using the complete integral (2) [or (4)] and the two equations
$$\begin{aligned} C_2 &= \Phi(C_1), \\ \frac{\partial \Theta}{\partial C_1} &+ \frac{\partial \Theta}{\partial C_2} \Phi'(C_1) = 0, \end{aligned} \tag{5}$$

where Φ is an arbitrary function and the prime stands for the derivative. In a sense, the general integral plays the role of the general solution dependent on an arbitrary function (the question whether it describes all solutions calls for further analysis).

3°. *Singular integrals* of PDE (1) can be found without invoking a complete integral by eliminating p and q from the following system of three algebraic (or transcendental) equations:
$$F = 0, \quad \frac{\partial F}{\partial u_x} = 0, \quad \frac{\partial F}{\partial u_y} = 0,$$
where the first equation coincides with equation (1).

Remark 4.6. *Methods for solving first-order nonlinear PDEs of the form (1) are described in the books by Kamke (1965), Courant & Hilbert (1989), Tran et al. (1999), Polyanin et al. (2002), Polyanin & Zaitsev (2012).*

4.3.2. Equations Quadratic in One Derivative

▶ *In this section and the two subsequent sections, as a rule, only complete integrals are presented. In order to construct the corresponding general solution, one should use the formulas of Section 4.3.1.*

1. $u_x + au_y^2 = by$.

This equation governs the free vertical drop of a point body near the Earth's surface (y is the vertical coordinate measured downward, x time, $m = \frac{1}{2a}$ the mass of the body, and $g = 2ab$ the gravitational acceleration).

Complete integral (Markeev, 1990): $u = -C_1 x \pm \dfrac{2a}{3b}\left(\dfrac{by + C_1}{a}\right)^{3/2} + C_2$.

2. $u_x + au_y^2 + by^2 = 0$.

This equation governs free oscillations of a point body of mass $m = 1/(2a)$ in an elastic field with elastic coefficient $k = 2b$ (x is time and y is the displacement from the equilibrium).

Complete integral (Gantmakher, 1966): $u = -C_1 x + C_2 \pm \displaystyle\int \sqrt{\dfrac{C_1 - by^2}{a}}\, dx + C_2$.

3. $u_x + au_y^2 = f(x) + g(y)$.

Complete integral: $u = -C_1 x + \displaystyle\int f(x)\, dx + \int \sqrt{\dfrac{g(y) + C_1}{a}}\, dy + C_2$.

4. $u_x + au_y^2 = f(x)y + g(x)$.

Complete integral:
$$u = \varphi(x)y + \int [g(x) - a\varphi^2(x)]\, dx + C_1, \quad \text{where } \varphi(x) = \int f(x)\, dx + C_2.$$

5. $u_x + au_y^2 = f(x)u + g(x)$.

Complete integral:
$$u = F(x)(C_1 + C_2 y) + F(x)\int [g(x) - aC_2^2 F^2(x)]\,\dfrac{dx}{F(x)}, \quad \text{where } F(x) = \exp\left[\int f(x)\, dx\right].$$

6. $u_x + au_y^2 + bu_y = f(x) + g(y)$.
Complete integral:
$$u = -C_1 x + C_2 + \int f(x)\, dx - \frac{b}{2a} y \pm \frac{1}{2a} \int \sqrt{4ag(y) + b^2 + 4aC_1}\, dy.$$

7. $u_x + au_y^2 + bu_y = f(x)y + g(x)$.
Complete integral:
$$u = \varphi(x) y + \int \left[g(x) - a\varphi^2(x) - b\varphi(x) \right] dx + C_1, \quad \text{where } \varphi(x) = \int f(x)\, dx + C_2.$$

8. $u_x + au_y^2 + bu_y = f(x)u + g(x)$.
Complete integral:
$$u = (C_1 y + C_2) F(x) + F(x) \int \left[g(x) - aC_1^2 F^2(x) - bC_1 F(x) \right] \frac{dx}{F(x)},$$
where $F(x) = \exp\left[\int f(x)\, dx \right]$.

9. $u_x - f(x) u_y^2 = 0$.
Complete integral:
$$u = C_1^2 \int f(x)\, dx + C_1 y + C_2.$$

10. $u_x + f(x) u_y^2 + g(x) u_y = h(x)$.
Complete integral:
$$u = C_1 y + C_2 + \int \left[h(x) - C_1^2 f(x) - C_1 g(x) \right] dx.$$

11. $u_x + f(x) u_y^2 + g(x) u_y = h(x) u + p(x) y + s(x)$.
Complete integral (Polyanin et al., 2002):
$$u = y\varphi(x) + \psi(x),$$
where
$$\varphi(x) = C_1 H(x) + H(x) \int \frac{p(x)}{H(x)}\, dx, \quad H(x) = \exp\left[\int h(x)\, dx \right],$$
$$\psi(x) = C_2 H(x) + H(x) \int \left[s(x) - f(x)\varphi^2(x) - g(x)\varphi(x) \right] \frac{dx}{H(x)}.$$

12. $u_x + \left[f(x) y + g(x) \right] u_y^2 = 0$.
Complete integral:
$$u = \varphi(x) y - \int g(x) \varphi^2(x)\, dx + C_1, \qquad \varphi(x) = \left[C_2 + \int f(x)\, dx \right]^{-1}.$$

13. $u_x - f(y) u_y^2 = 0$.
Complete integral:
$$u = \pm C_1^2 x + C_1 \int \frac{dy}{\sqrt{|f(y)|}} + C_2.$$
Here the plus sign is taken if $f(y) > 0$, and the minus sign if $f(y) < 0$.

14. $u_x + f(y)u_y^2 + g(y)u_y = h(x) + r(y)$.

Complete integral (Polyanin et al., 2002):

$$u = -C_1 x + C_2 + \int h(x)\,dx + \int \frac{-g(y) \pm \sqrt{g^2(y) + 4f(y)r(y) + 4C_1 f(y)}}{2f(y)}\,dy.$$

15. $u_x - f(u)u_y^2 = 0$.

Complete integral in implicit form: $\int f(u)\,du = C_1^2 x + C_1 y + C_2$.

16. $u_x + f(u)u_y^2 + g(u)u_y = h(u)$.

Complete integral in implicit form:

$$C_1 x + C_2 y + \int \frac{2C_2^2 f(u)\,du}{C_1 + C_2 g(u) \pm \sqrt{[C_1 + C_2 g(u)]^2 + 4C_2^2 f(u)h(u)}} = C_3.$$

One of the constants C_1, C_2, or C_3 can be set equal to ± 1.

17. $u_x - f(u)u_y^2 - [yg(x) + h(x)]u_y = 0$.

The transformation

$$t = \int \varphi^2(x)\,dx, \quad z = \varphi(x)y + \int h(x)\varphi(x)\,dx, \quad \varphi(x) = \exp\left[\int g(x)\,dx\right],$$

leads to a simpler equation of the form 4.3.2.15:

$$u_t - f(u)u_z^2 = 0.$$

18. $u_x + f(y)g(u)u_y^2 = h(u)$.

Complete integral in implicit form for $f(y) > 0$:

$$x + C_1 \int \frac{dy}{\sqrt{f(y)}} + \int \frac{2C_1^2 g(u)\,du}{1 + \sqrt{1 + 4C_1^2 g(u)h(u)}} = C_2.$$

19. $u_x - f(x)g(y)h(u)u_y^2 = 0$.

Complete integral in implicit form:

$$\int h(u)\,du = \pm C_1^2 \int f(x)\,dx + C_1 \int \frac{dy}{\sqrt{|g(y)|}} + C_2.$$

Here the plus sign is taken if $g(y) > 0$, and the minus sign if $g(y) < 0$.

20. $f_1(x)u_x + f_2(y)u_y^2 = g_1(x) + g_2(y)$.

Complete integral: $u = \int \frac{g_1(x) - C_1}{f_1(x)}\,dx + \int \sqrt{\frac{g_2(y) + C_1}{f_2(y)}}\,dy + C_2$.

4.3.3. Equations Quadratic in Two Derivatives

1. $au_x^2 + bu_y^2 = c.$

For $a = b$, this is a *differential equation of light rays*.
Complete integral: $u = C_1 x + C_2 y + C_3$, where $aC_1^2 + bC_2^2 = c$.
An alternative form of the complete integral: $\dfrac{u^2}{c} = \dfrac{(x - C_1)^2}{a} + \dfrac{(y - C_2)^2}{b}$.

2. $u_x^2 + u_y^2 = a - 2by.$

This equation governs parabolic motion of a point mass in vacuum (the coordinate x is measured along the Earth's surface, the coordinate y is measured vertically upward from the Earth's surface, and a is the gravitational acceleration)

Complete integral (Appell, 1953): $u = C_1 x \pm \dfrac{1}{3b}(a - C_1^2 - 2by)^{3/2} + C_2$.

3. $u_x^2 + u_y^2 = (a/u)^2 - 1.$

This equation describes a family of spherical surfaces of radius a with centers at points of the xy-plane.

Complete integral in implicit form (Courant & Hilbert, 1989):
$$(x - C_1)^2 + (y - C_2)^2 + u^2 = a^2.$$

Another complete integral: $\dfrac{(y - C_1 x - C_2)^2}{1 + C_1^2} + u^2 = a^2$.

4. $u_x^2 + u_y^2 = \dfrac{a}{\sqrt{x^2 + y^2}} + b.$

This equation arises from the solution of the two-body problem in celestial mechanics.
Complete integral (Appell, 1953 and Kamke, 1965):
$$u = \pm \int \sqrt{b + \frac{a}{r} - \frac{C_1^2}{r^2}}\, dr + C_1 \arctan \frac{y}{x} + C_2, \quad \text{where } r = \sqrt{x^2 + y^2}.$$

5. $u_x^2 + u_y^2 = f(x).$

Complete integral (Kamke, 1965): $u = C_1 y + C_2 \pm \displaystyle\int \sqrt{f(x) - C_1^2}\, dx$.

6. $u_x^2 + u_y^2 = f(x) + g(y).$

Complete integral:
$$u = \pm \int \sqrt{f(x) + C_1}\, dx \pm \int \sqrt{g_2(y) - C_1}\, dy + C_2.$$

The signs before each of the integrals can be chosen independently of each other.

7. $u_x^2 + u_y^2 = f(x^2 + y^2).$

Hamilton's equation for the plane motion of a point mass under the action of a central force.
Complete integral (Kamke, 1965):
$$u = C_1 \arctan \frac{x}{y} + C_2 \pm \frac{1}{2} \int \sqrt{zf(z) - C_1^2}\, \frac{dz}{z}, \quad z = x^2 + y^2.$$

8. $u_x^2 + u_y^2 = f(u)$.

Complete integral in implicit form: $\displaystyle\int \frac{du}{\sqrt{f(u)}} = \pm\sqrt{(x+C_1)^2 + (y+C_2)^2}$.

9. $u_x^2 + \dfrac{1}{x^2} u_y^2 = f(x)$.

This equation governs the plane motion of a point mass in a central force field, with x and y being polar coordinates.

Complete integral (Appel, 1953): $u = C_1 y \pm \displaystyle\int \sqrt{f(x) - \dfrac{C_1^2}{x^2}}\, dx + C_2$.

10. $u_x^2 + f(x) u_y^2 = g(x)$.

Complete integral: $u = C_1 y + C_2 + \displaystyle\int \sqrt{g(x) - C_1^2 f(x)}\, dx$.

11. $u_x^2 + f(y) u_y^2 = g(y)$.

Complete integral: $u = C_1 x + C_2 + \displaystyle\int \sqrt{\dfrac{g(y) - C_1^2}{f(y)}}\, dy$.

12. $u_x^2 + f(u) u_y^2 = g(u)$.

Complete integral in implicit form: $\displaystyle\int \sqrt{\dfrac{C_1^2 + C_2^2 f(u)}{g(u)}}\, du = C_1 x + C_2 y + C_3$.

One of the constants C_1 or C_2 can be set equal to ± 1.

13. $f_1(x) u_x^2 + f_2(y) u_y^2 = g_1(x) + g_2(y)$.

A *separable PDE*. This equation is encountered in differential geometry in studying geodesic lines of Liouville surfaces.

Complete integral (Appel, 1953 and Kamke, 1965):

$$u = \pm \int \sqrt{\frac{g_1(x) + C_1}{f_1(x)}}\, dx \pm \int \sqrt{\frac{g_2(y) - C_1}{f_2(y)}}\, dy + C_2.$$

The signs before each of the integrals can be chosen independently of each other.

14. $f_1(x,y) u_x^2 + f_2(x,y) u_y^2 = g(x,y) h(u)$.

The substitution $w = \displaystyle\int \dfrac{du}{\sqrt{h(u)}}$ leads to a simpler equation:

$$f_1(x,y) w_x^2 + f_2(x,y) w_y^2 = g(x,y).$$

For solutions of this equation for some types of the right-hand side, see 4.3.3.1, 4.3.3.2, 4.3.3.4–4.3.3.7, 4.3.3.9–4.3.3.11, and 4.3.3.13.

4.3.4. Equations with Arbitrary Nonlinearities in Derivatives

1. $u_x + f(u_y) = 0.$

This equation is encountered in optimal control and differential games (Subbotin, 1995).

$1°$. Complete integral: $u = C_1 y - f(C_1) x + C_2$.

$2°$. On differentiating the equation with respect to y, we arrive at a quasilinear equation of the form 4.2.3.10:
$$w_x + f'(w) w_y = 0, \qquad w = u_y.$$

$3°$. The solution of the Cauchy problem with the initial condition $u(0, y) = \varphi(y)$ can be written in parametric form as
$$y = f'(\zeta) x + \xi, \quad u = \big[\zeta f'(\zeta) - f(\zeta)\big] x + \varphi(\xi), \qquad \text{where} \quad \zeta = \varphi'(\xi).$$

2. $u_x + f(u_y) = g(x).$

Complete integral: $u = C_1 y - f(C_1) x + \int g(x)\, dx + C_2$.

3. $u_x + f(u_y) = g(x) y + h(x).$

Complete integral:
$$u = \varphi(x) y + \int \big[h(x) - f(\varphi(x))\big] dx + C_1, \quad \text{where } \varphi(x) = \int g(x)\, dx + C_2.$$

4. $u_x + f(u_y) = g(x) u + h(x).$

Complete integral:
$$u = (C_1 y + C_2) \varphi(x) + \varphi(x) \int \big[h(x) - f(C_1 \varphi(x))\big] \frac{dx}{\varphi(x)}, \quad \text{where } \varphi(x) = \exp\!\left[\int g(x)\, dx\right].$$

5. $u_x - [g(x) y + h(x)] f(u_y) = 0.$

Complete integral:
$$u = \varphi(x) y + \int f(\varphi(x)) h(x)\, dx + C_1,$$

where the function $\varphi(x)$ is defined implicitly by the relation $\int \dfrac{d\varphi}{f(\varphi)} = \int g(x)\, dx + C_2.$

6. $u_x + [g_1(x) y + g_0(x)] f(u_y) + h_2(x) u + h_1(x) y + h_0(x) = 0.$

$1°$. Complete integral (Polyanin et al., 2022):
$$u = \varphi(x) y + \psi(x),$$

where the functions $\varphi = \varphi(x)$ and $\psi = \psi(x)$ are determined by solving the ODEs
$$\varphi'_x + g_1(x) f(\varphi) + h_2(x) \varphi + h_1(x) = 0, \tag{1}$$
$$\psi'_x + g_0(x) f(\varphi) + h_2(x) \psi + h_0(x) = 0. \tag{2}$$

2°. For $g_1(x) \equiv 0$, the general solutions of equations (1) and (2) are

$$\varphi(x) = C_1 H(x) - H(x) \int \frac{h_1(x)}{H(x)} dx, \quad H(x) = \exp\left[-\int h_2(x) dx\right],$$

$$\psi(x) = C_2 H(x) - H(x) \int \frac{h_0(x) + g_0(x) f(\varphi(x))}{H(x)} dx.$$

7. $u_x - u f(u_y) = 0.$

Complete integral:
$$u = (y + C_1)\varphi(x),$$

where the function $\varphi(x)$ is defined implicitly by the relation $\displaystyle\int \frac{d\varphi}{\varphi f(\varphi)} = x + C_2.$

8. $u_x - u f(u_y/u) = 0.$

Complete integral: $u = C_1 \exp[C_2 y + f(C_2) x].$

9. $u_x - u f(u^\beta u_y) = 0.$

For $\beta = -1$, see equation 4.3.4.8.

Complete integral for $\beta \neq -1$ (Polyanin et al., 2002):

$$u = \left[(1+\beta)y + C_1\right]^{\frac{1}{1+\beta}} \varphi(x + C_2),$$

where the function $\varphi(x)$ is defined implicitly by the relation $\displaystyle x = \int \frac{d\varphi}{\varphi f(\varphi^{\beta+1})}.$

10. $u_x - f(e^{\beta u} u_y) = 0.$

Complete integral:
$$u = \frac{1}{\beta} \ln(\beta y + C_1) + \varphi(x + C_2),$$

where the function $\varphi(x)$ is defined implicitly by the relation $\displaystyle x = \int \frac{d\varphi}{f(e^{\beta\varphi})}.$

11. $u_x - g(x) u^\beta f(u_y/u) = h(x) u.$

The transformation

$$u(x,y) = H(x) z(t,y), \quad t = \int g(x) H^{\beta-1}(x) dx, \quad H(x) = \exp\left[\int h(x) dx\right],$$

leads to a simpler equation, $z_t - z^\beta f(z_y/z) = 0$, which, by solving for z_y, may be rewritten in the form of equation 4.3.4.9.

12. $u_x - g(x) e^{\beta u} f(u_y) = h(x).$

The transformation

$$u(x,y) = z(t,y) + H(x), \quad t = \int g(x) \exp[\beta H(x)] dx, \quad H(x) = \int h(x) dx,$$

leads to a simpler equation, $z_t - e^{\beta z} f(z_y) = 0$, which, by solving for z_y, may be rewritten in the form of equation 4.3.4.10.

13. $u_x - F(x, u_y) = 0$.

Complete integral (Kamke, 1965): $u = \int F(x, C_1)\, dx + C_1 y + C_2$.

14. $u_x + F(x, u_y) = au$.

Complete integral: $u = e^{ax}(C_1 y + C_2) - e^{ax} \int e^{-ax} F(x, C_1 e^{ax})\, dx$.

15. $u_x + F(x, u_y) = g(x) u$.

Complete integral (Polyanin et al., 2002):
$$u = \varphi(x)(C_1 y + C_2) - \varphi(x) \int F(x, C_1 \varphi(x)) \frac{dx}{\varphi(x)}, \quad \text{where } \varphi(x) = \exp\left[\int g(x)\, dx\right].$$

16. $F(u_x, u_y) = 0$.

Complete integral (Kamke, 1965 and Courant & Hilbert, 1989):
$$u = C_1 x + C_2 y + C_3,$$
where C_1 and C_3 are arbitrary constants and the constant C_2 is related to C_1 by the algebraic equation $F(C_1, C_2) = 0$.

17. $u = x u_x + y u_y + F(u_x, u_y)$.

Clairaut's equation.

Complete integral (Courant & Hilbert, 1989): $u = C_1 x + C_2 y + F(C_1, C_2)$.

18. $F_1(x, u_x) = F_2(y, u_y)$.

A separable first-order PDE.

Complete integral (Kamke, 1965):
$$u = \varphi(x) + \psi(y) + C_1,$$
where the functions $\varphi = \varphi(x)$ and $\psi = \psi(y)$ are determined from the ODEs
$$F_1(x, \varphi'_x) = C_2, \qquad F_2(y, \psi'_y) = C_2.$$

19. $F_1(x, u_x) + F_2(y, u_y) + au = 0$.

A separable first-order PDE.

Complete integral (Polyanin et al., 2002):
$$u = \varphi(x) + \psi(y),$$
where the functions $\varphi = \varphi(x)$ and $\psi = \psi(y)$ are determined from the ODEs
$$F_1(x, \varphi'_x) + a\varphi = C_1, \qquad F_2(y, \psi'_y) + a\psi = -C_1,$$
where C_1 is an arbitrary constant. If $a \neq 0$, one can set $C_1 = 0$ in these equations.

20. $F_1\!\left(x, \frac{1}{u} u_x\right) + u^k F_2\!\left(y, \frac{1}{u} u_y\right) = 0$.

A separable first-order PDE. Complete integral:
$$u(x, y) = \varphi(x) \psi(y).$$
The functions $\varphi = \varphi(x)$ and $\psi = \psi(y)$ are determined by solving the ODEs
$$\varphi^{-k} F_1(x, \varphi'_x/\varphi) = C, \qquad \psi^k F_2(y, \psi'_y/\psi) = -C,$$
where C is an arbitrary constant.

21. $F_1(x, u_x) + e^{\lambda u} F_2(y, u_y) = 0.$

A *separable first-order PDE.*

Complete integral (Polyanin et al., 2002):
$$u(x, y) = \varphi(x) + \psi(y).$$

The functions $\varphi = \varphi(x)$ and $\psi = \psi(y)$ are determined by solving the ODEs
$$e^{-\lambda \varphi} F_1(x, \varphi'_x) = C, \quad e^{\lambda \psi} F_2(y, \psi'_y) = -C,$$

where C is an arbitrary constant.

22. $F_1\left(x, \dfrac{1}{u} u_x\right) + F_2\left(y, \dfrac{1}{u} u_y\right) = k \ln u.$

A *separable first-order PDE.*

Complete integral (Polyanin et al., 2002):
$$u(x, y) = \varphi(x) \psi(y).$$

The functions $\varphi = \varphi(x)$ and $\psi = \psi(y)$ are determined by solving the ODEs
$$F_1(x, \varphi'_x/\varphi) - k \ln \varphi = C, \quad F_2(y, \psi'_y/\psi) - k \ln \psi = -C,$$

where C is an arbitrary constant.

23. $u_x + y F_1(x, u_y) + F_2(x, u_y) = 0.$

Complete integral:
$$u = \varphi(x) y - \int F_2(x, \varphi(x)) \, dx + C_1,$$

where the function $\varphi(x)$ is determined by solving the ODE $\varphi'_x + F_1(x, \varphi) = 0$.

24. $F(u_x + ay, u_y + ax) = 0.$

Complete integral: $u = -axy + C_1 x + C_2 y + C_3$, where $F(C_1, C_2) = 0$.

25. $u_x^2 + u_y^2 = F(x^2 + y^2, y u_x - x u_y).$

Complete integral (Kamke, 1965):
$$u = -C_1 \arctan \frac{y}{x} + \frac{1}{2} \int \sqrt{\xi F(\xi, C_1) - C_1^2} \, \frac{d\xi}{\xi} + C_2, \quad \text{where } \xi = x^2 + y^2.$$

26. $F(x, u_x, u_y) = 0.$

Complete integral: $u = C_1 y + \varphi(x, C_1) + C_2$, where the function $\varphi = \varphi(x, C_1)$ is determined from the ordinary differential equation $F(x, \varphi'_x, C_1) = 0$.

27. $F(ax + by, u_x, u_y) = 0.$

For $b = 0$, see equation 4.3.4.26. Complete integral for $b \neq 0$:
$$u = C_1 x + \varphi(z, C_1) + C_2, \quad z = ax + by,$$

where the function $\varphi = \varphi(z)$ is determined from the nonlinear ordinary differential equation $F(z, a\varphi'_z + C_1, b\varphi'_z) = 0$.

28. $F(u, u_x, u_y) = 0.$

Complete integral (Kamke, 1965):
$$u = u(z), \quad z = C_1 x + C_2 y,$$

where C_1 and C_2 are arbitrary constants and $u = u(z)$ is described by the autonomous ODE $F(u, C_1 u'_z, C_2 u'_z) = 0$.

29. $F(ax + by + cu, u_x, u_y) = 0.$

For $c = 0$, see equation 4.3.4.27. If $c \neq 0$, then the substitution $cu = ax + by + cu$ leads to an equation of the form 4.3.4.28: $F(cu, u_x - a/c, u_y - b/c) = 0$.

30. $F(x, u_x + ay, u_y + ax) = 0.$

Complete integral:
$$u = -axy + C_1 y + \varphi(x) + C_2,$$

where the function $\varphi(x)$ is determined by solving the first-order ordinary differential equation $F(x, \varphi'_x, C_1) = 0$.

31. $F(x, u_x, u_y, u - yu_y) = 0.$

Complete integral: $u = C_1 y + \varphi(x)$, where the function $\varphi(x)$ is determined from the ordinary differential equation $F(x, \varphi'_x, C_1, \varphi) = 0$.

32. $F(u, u_x, u_y, xu_x + yu_y) = 0.$

Complete integral (Polyanin et al., 2022):
$$u = \varphi(\xi), \quad \xi = C_1 x + C_2 y,$$

where the function $\varphi(\xi)$ is determined by solving the nonlinear ordinary differential equation $F(\varphi, C_1 \varphi'_\xi, C_2 \varphi'_\xi, \xi \varphi'_\xi) = 0$.

33. $F(ax + by, u_x, u_y, u - xu_x - yu_y) = 0.$

Complete integral:
$$u = C_1 x + C_2 y + \varphi(\xi), \quad \xi = ax + by,$$

where the function $\varphi(\xi)$ is determined by solving the nonlinear ordinary differential equation $F(\xi, a\varphi'_\xi + C_1, b\varphi'_\xi + C_2, \varphi - \xi\varphi'_\xi) = 0$.

34. $F(x, u_x, G(y, u_y)) = 0.$

Complete integral:
$$u = \varphi(x, C_1) + \psi(y, C_1) + C_2,$$

where the functions φ and ψ are determined by the ODEs
$$F(x, \varphi'_x, C_1) = 0, \quad G(y, \psi'_y) = C_1.$$

On solving these equations for the derivatives, we obtain linear separable equations, which are easy to integrate.

▶ More exact solutions of linear and nonlinear PDEs of the first order can be found in specialized reference books by Kamke (1965) and Polyanin et al. (2002) (as a rule, these books are not cited in the text); see also Polyanin & Manzhirov (2007), Polyanin & Nazaikinskii (2016), Polyanin & Zaitsev (2012). These books, as well as those by Courant & Hilbert (1989) and Tran et al. (1999), describe the basic analytical methods for solving such equations.

Applications of first-order PDEs in various fields of mechanics, physics, chemical engineering sciences, and other disciplines are discussed in the books by Appell (1953), Lopez (2012), Rhee et al. (1986), Subbotin (1995), and Whitham (1974).

References

Appell, P., *Traité de Mécanique Rationnelle, T. 1: Statique. Dinamyque du Point (Ed. 6)*, Gauthier-Villars, Paris, 1953.
Courant, R. and Hilbert, D., *Methods of Mathematical Physics, Vol. 2*, Wiley-Interscience, New York, 1989.
Gantmakher, F.R., *Lectures on Analytical Mechanics* [in Russian], Fizmatlit, Moscow, 1966.
Kamke, E., *Differentialgleichungen: Lösungsmethoden und Lösungen, II, Partielle Differentialgleichungen Erster Ordnung für eine gesuchte Funktion*, Akad. Verlagsgesellschaft Geest & Portig, Leipzig, 1965.
Lopez, G., *Partial Differential Equations of First Order and Their Applications to Physics, 2nd ed.*,, World Scientific Publ., Singapore, 2012.
Markeev, A.P., *Theoretical Mechanics* [in Russian], Nauka, Moscow, 1990.
Polyanin, A.D. and Manzhirov, A.V., *Handbook of Mathematics for Engineers and Scientists* (Chapters 13 and T17), Chapman & Hall/CRC Press, Boca Raton–London, 2007.
Polyanin, A.D. and Nazaikinskii, V.E., *Handbook of Linear Partial Differential Equations for Engineers and Scientists, 2nd ed.* (Chapters 2 and 13), Chapman & Hall/CRC Press, Boca Raton–London, 2016.
Polyanin, A.D. and Zaitsev, V.F., *Handbook of Nonlinear Partial Differential Equations, 2nd ed.* (Chapters 1 and 24), Chapman & Hall/CRC Press, Boca Raton–London, 2012.
Polyanin, A.D., Zaitsev, V.F., and Moussiaux, A., *Handbook of First Order Partial Differential Equations*, Taylor & Francis, London, 2002.
Rhee, H., Aris, R., and Amundson, N.R., *First Order Partial Differential Equations, Vol. 1*, Prentice Hall, Englewood Cliffs, New Jersey, 1986.
Subbotin, A.I., *Generalized Solutions of First Order PDEs: the Dynamical Optimization Perspective*, Birkhäuser, Boston, 1995.
Tran, D.V., Tsuji, M., and Nguyen, D.T.S., *The Characteristic Method and Its Generalizations for First-Order Nonlinear Partial Differential Equations*, Chapman & Hall, London, 1999.
Whitham, G.B., *Linear and Nonlinear Waves*, Wiley, New York, 1974.

Chapter 5

Linear Equations and Problems of Mathematical Physics

▶ **Preliminary remarks.** Linear equations of mathematical physics are usually called linear partial differential equations (PDEs) of the second or higher orders; these are used to describe various natural phenomena or processes.

Linear homogeneous PDEs with constant coefficients, including linear equations of mathematical physics, and some linear PDEs with variable coefficients allow particular *multiplicative separable solutions* in the form of a product of functions of different independent variables. For arbitrary linear homogeneous PDEs, the *principle of linear superposition* is valid: any linear combination of its particular (exact) solutions is also a solution of the equation under consideration. The most common methods for solving linear PDEs are the *method of separation of variables* and the *method of integral transforms*; these make it possible to represent solutions to many problems of mathematical physics in the form of infinite series and definite integrals.

This chapter describes exact solutions to the most common linear equations and problems of mathematical physics in two independent variables and other linear PDEs of the second or higher orders.

5.1. Parabolic Equations

5.1.1. Heat (Diffusion) Equation $u_t = a u_{xx}$

▶ **Particular solutions:**

$$u = Ax + B,$$
$$u = A(x^2 + 2at) + B,$$
$$u = A(x^3 + 6atx) + B,$$
$$u = A(x^4 + 12atx^2 + 12a^2t^2) + B,$$
$$u = x^{2n} + \sum_{k=1}^{n} \frac{(2n)(2n-1)\ldots(2n-2k+1)}{k!}(at)^k x^{2n-2k},$$
$$u = x^{2n+1} + \sum_{k=1}^{n} \frac{(2n+1)(2n)\ldots(2n-2k+2)}{k!}(at)^k x^{2n-2k+1},$$
$$u = A\exp(a\mu^2 t \pm \mu x) + B,$$
$$u = A\frac{1}{\sqrt{t}}\exp\left(-\frac{x^2}{4at}\right) + B,$$

DOI: 10.1201/9781003051329-5

$$u = A\frac{x}{t^{3/2}} \exp\left(-\frac{x^2}{4at}\right) + B,$$

$$u = A\exp(-a\mu^2 t)\cos(\mu x + B) + C,$$

$$u = A\exp(-\mu x)\cos(\mu x - 2a\mu^2 t + B) + C,$$

$$u = A\operatorname{erf}\left(\frac{x}{2\sqrt{at}}\right) + B,$$

where A, B, C, and μ are arbitrary constants, n is a positive integer, and $\operatorname{erf} z = \frac{2}{\sqrt{\pi}}\int_0^z \exp(-\xi^2)\,d\xi$ is the error function (probability integral).

▶ **Formulas allowing the construction of particular solutions.**

Suppose $u = u(x,t)$ is a solution of the heat equation. Then the functions (Miller, 1977 and Polyanin & Nazaikinskii, 2016):

$$u_1 = Au(\pm\lambda x + C_1,\ \lambda^2 t + C_2) + B,$$

$$u_2 = A\exp(\lambda x + a\lambda^2 t)u(x + 2a\lambda t + C_1,\ t + C_2),$$

$$u_3 = \frac{A}{\sqrt{|\delta + \beta t|}} \exp\left[-\frac{\beta x^2}{4a(\delta + \beta t)}\right] u\left(\pm\frac{x}{\delta + \beta t},\ \frac{\gamma + \lambda t}{\delta + \beta t}\right), \qquad \lambda\delta - \beta\gamma = 1,$$

where A, B, C_1, C_2, β, δ, and λ are arbitrary constants, are also solutions of this equation. The last formula is usually encountered with $\beta = 1$, $\gamma = -1$, and $\delta = \lambda = 0$.

▶ **Particular solutions in the form of functional series.**

A solution containing an arbitrary function of the spatial variable:

$$u(x,t) = f(x) + \sum_{n=1}^{\infty} \frac{(at)^n}{n!} f_x^{(2n)}(x), \qquad f_x^{(m)}(x) = \frac{d^m}{dx^m}f(x),$$

where $f(x)$ is any infinitely differentiable function. This solution satisfies the initial condition $u(x,0) = f(x)$. The sum will be finite if $f(x)$ is a polynomial.

Solutions containing arbitrary functions of the time:

$$u(x,t) = g(t) + \sum_{n=1}^{\infty} \frac{1}{a^n(2n)!} x^{2n} g_t^{(n)}(t),$$

$$u(x,t) = xh(t) + x\sum_{n=1}^{\infty} \frac{1}{a^n(2n+1)!} x^{2n} h_t^{(n)}(t),$$

where $g(t)$ and $h(t)$ are any infinitely differentiable functions. The sums will be finite when $g(t)$ and $h(t)$ are polynomials. The first solution satisfies the boundary condition of the first kind $u|_{x=0} = g(t)$, and the second one satisfies the boundary condition of the second kind $u_x|_{x=0} = h(t)$.

▶ **Cauchy problem and boundary value problems.**

For solutions of the Cauchy problem and various boundary value problems for the heat equation, see Section 5.1.2 with $\Phi(x,t) \equiv 0$.

5.1.2. Nonhomogeneous Heat Equation $u_t = au_{xx} + \Phi(x,t)$

▶ **Domain:** $-\infty < x < \infty$. **Cauchy problem.**

An initial condition is prescribed:
$$u = f(x) \quad \text{at} \quad t = 0.$$

The solution of the Cauchy problem for the heat equation is
$$u = \int_{-\infty}^{\infty} f(\xi) G(x,\xi,t)\,d\xi + \int_0^t \int_{-\infty}^{\infty} \Phi(\xi,\tau) G(x,\xi,t-\tau)\,d\xi\,d\tau,$$

where
$$G(x,\xi,t) = \frac{1}{2\sqrt{\pi at}} \exp\left[-\frac{(x-\xi)^2}{4at}\right].$$

▶ **Solutions of boundary value problems in terms of the Green's function.**

We consider boundary value problems for the nonhomogeneous heat equation on an interval $0 \le x \le l$ with the general initial condition
$$u = f(x) \quad \text{at} \quad t = 0$$
and various homogeneous boundary conditions. The solution of these problems can be represented in terms of the Green's function as
$$u = \int_0^l f(\xi) G(x,\xi,t)\,d\xi + \int_0^t \int_0^l \Phi(\xi,\tau) G(x,\xi,t-\tau)\,d\xi\,d\tau.$$

Here, the upper limit l can be finite or infinite; if $l = \infty$, there is no boundary condition corresponding to it.

Below are Green's functions for the heat equation for various types of homogeneous boundary conditions.

▶ **Domain:** $0 \le x < \infty$. **First boundary value problem.**

A boundary condition is prescribed:
$$u = 0 \quad \text{at} \quad x = 0.$$

Green's function:
$$G(x,\xi,t) = \frac{1}{2\sqrt{\pi at}}\left\{\exp\left[-\frac{(x-\xi)^2}{4at}\right] - \exp\left[-\frac{(x+\xi)^2}{4at}\right]\right\}.$$

▶ **Domain:** $0 \le x < \infty$. **Second boundary value problem.**

A boundary condition is prescribed:
$$u_x = 0 \quad \text{at} \quad x = 0.$$

Green's function:
$$G(x,\xi,t) = \frac{1}{2\sqrt{\pi at}}\left\{\exp\left[-\frac{(x-\xi)^2}{4at}\right] + \exp\left[-\frac{(x+\xi)^2}{4at}\right]\right\}.$$

▶ **Domain: $0 \leq x < \infty$. Third boundary value problem.**

A boundary condition is prescribed:
$$u_x - ku = 0 \quad \text{at} \quad x = 0.$$

Green's function:
$$G(x, \xi, t) = \frac{1}{2\sqrt{\pi a t}} \left\{ \exp\left[-\frac{(x-\xi)^2}{4at}\right] + \exp\left[-\frac{(x+\xi)^2}{4at}\right] \right.$$
$$\left. - 2k \int_0^\infty \exp\left[-\frac{(x+\xi+\eta)^2}{4at} - k\eta\right] d\eta \right\}.$$

▶ **Domain: $0 \leq x \leq l$. First boundary value problem.**

Boundary conditions are prescribed:
$$u = 0 \quad \text{at} \quad x = 0, \qquad u = 0 \quad \text{at} \quad x = l.$$

Two forms of representation of the Green's function:
$$G(x, \xi, t) = \frac{2}{l} \sum_{n=1}^\infty \sin\left(\frac{n\pi x}{l}\right) \sin\left(\frac{n\pi \xi}{l}\right) \exp\left(-\frac{an^2\pi^2 t}{l^2}\right)$$
$$= \frac{1}{2\sqrt{\pi at}} \sum_{n=-\infty}^\infty \left\{ \exp\left[-\frac{(x-\xi+2nl)^2}{4at}\right] - \exp\left[-\frac{(x+\xi+2nl)^2}{4at}\right] \right\}.$$

The first series converges rapidly at large t and the second series at small t.

▶ **Domain: $0 \leq x \leq l$. Second boundary value problem.**

Boundary conditions are prescribed:
$$u_x = 0 \quad \text{at} \quad x = 0, \qquad u_x = 0 \quad \text{at} \quad x = l.$$

Two forms of representation of the Green's function:
$$G(x, \xi, t) = \frac{1}{l} + \frac{2}{l} \sum_{n=1}^\infty \cos\left(\frac{n\pi x}{l}\right) \cos\left(\frac{n\pi \xi}{l}\right) \exp\left(-\frac{an^2\pi^2 t}{l^2}\right)$$
$$= \frac{1}{2\sqrt{\pi at}} \sum_{n=-\infty}^\infty \left\{ \exp\left[-\frac{(x-\xi+2nl)^2}{4at}\right] + \exp\left[-\frac{(x+\xi+2nl)^2}{4at}\right] \right\}.$$

The first series converges rapidly at large t and the second series at small t.

▶ **Domain: $0 \leq x \leq l$. Third boundary value problem ($k_1 > 0$, $k_2 > 0$).**

Boundary conditions are prescribed:
$$u_x - k_1 u = 0 \quad \text{at} \quad x = 0, \qquad u_x + k_2 u = 0 \quad \text{at} \quad x = l.$$

Green's function:

$$G(x,\xi,t) = \sum_{n=1}^{\infty} \frac{1}{\|y_n\|^2} y_n(x) y_n(\xi) \exp(-a\mu_n^2 t),$$

$$y_n(x) = \cos(\mu_n x) + \frac{k_1}{\mu_n} \sin(\mu_n x), \quad \|y_n\|^2 = \frac{k_2}{2\mu_n^2} \frac{\mu_n^2 + k_1^2}{\mu_n^2 + k_2^2} + \frac{k_1}{2\mu_n^2} + \frac{l}{2}\left(1 + \frac{k_1^2}{\mu_n^2}\right),$$

where μ_n are positive roots of the transcendental equation $\dfrac{\tan(\mu l)}{\mu} = \dfrac{k_1 + k_2}{\mu^2 - k_1 k_2}$.

▶ **Solutions of boundary value problems with nonhomogeneous boundary conditions.**

Any linear problem with arbitrary nonhomogeneous boundary conditions can be reduced to a linear problem with homogeneous boundary conditions. To this end, one should perform the change of variable

$$u(x,t) = v(x,t) + \varphi(x,t),$$

where v is the new unknown function and φ is any function that satisfies the nonhomogeneous boundary conditions.

Table 5.1 gives examples of such transformations for linear initial-boundary value problems with one space variable for parabolic and hyperbolic equations. Functions $g_1(t)$ and $g_2(t)$, which are included in nonhomogeneous boundary conditions, can be chosen arbitrarily. In the third boundary value problem, it is assumed that $k_1 > 0$ and $k_2 > 0$.

Note that the selection of the function φ is of a purely algebraic nature and is not connected with the equation in question; there are infinitely many suitable functions φ that satisfy nonhomogeneous boundary conditions.

5.1.3. Heat Type Equation of the Form $u_t = au_{xx} + bu_x + cu + \Phi(x,t)$

The substitution

$$u = \exp(\beta t + \mu x) v(x,t), \qquad \beta = c - \frac{b^2}{4a}, \qquad \mu = -\frac{b}{2a}$$

leads to the nonhomogeneous heat equation

$$v_t = av_{xx} + \exp(-\beta t - \mu x) \Phi(x,t),$$

which is considered in Sections 5.1.1 and 5.1.2.

5.1.4. Heat Equation with Axial Symmetry $u_t = a(u_{rr} + r^{-1} u_r)$

This is the *two-dimensional heat (diffusion) equation with axial symmetry*, where $r = \sqrt{x^2 + y^2}$ is the radial coordinate.

TABLE 5.1. Simple transformations of the form $u(x,t) = v(x,t) + \varphi(x,t)$ that lead to homogeneous boundary conditions in initial-boundary value problems with one space variable ($0 \leq x \leq l$).

No.	Initial-boundary value problem	Boundary conditions	Function $\varphi = \varphi(x,t)$ satisfying the boundary conditions
1	First initial-boundary value problem	Dirichlet boundary conditions: $u = g_1(t)$ at $x = 0$ $u = g_2(t)$ at $x = l$	$\varphi = g_1(t) + \dfrac{x}{l}[g_2(t) - g_1(t)]$
2	Second initial-boundary value problem	Neumann boundary conditions: $u_x = g_1(t)$ at $x = 0$ $u_x = g_2(t)$ at $x = l$	$\varphi = xg_1(t) + \dfrac{x^2}{2l}[g_2(t) - g_1(t)]$
3	Third initial-boundary value problem	Robin boundary conditions: $u_x - k_1 u = g_1(t)$ at $x = 0$ $u_x + k_2 u = g_2(t)$ at $x = l$	$\varphi = \dfrac{(k_2 x - 1 - k_2 l)g_1(t) + (1 + k_1 x)g_2(t)}{k_2 + k_1 + k_1 k_2 l}$
4	Mixed initial-boundary value problem	Mixed boundary conditions: $u = g_1(t)$ at $x = 0$ $u_x = g_2(t)$ at $x = l$	$\varphi = g_1(t) + xg_2(t)$
5	Mixed initial-boundary value problem	Mixed boundary conditions: $u_x = g_1(t)$ at $x = 0$ $u = g_2(t)$ at $x = l$	$\varphi = (x - l)g_1(t) + g_2(t)$

▶ **Particular solutions:**

$$u = A + B \ln r,$$

$$u = A + B(r^2 + 4at),$$

$$u = A + B(r^4 + 16atr^2 + 32a^2 t^2),$$

$$u = A + B\left(r^{2n} + \sum_{k=1}^{n} \frac{4^k [n(n-1)\ldots(n-k+1)]^2}{k!}(at)^k r^{2n-2k}\right),$$

$$u = A + B(4at \ln r + r^2 \ln r - r^2),$$

$$u = A + \frac{B}{t}\exp\left(-\frac{r^2}{4at}\right),$$

$$u = A + B \int_1^{\zeta} e^{-z}\frac{dz}{z}, \quad \zeta = \frac{r^2}{4at},$$

$$u = A + B \exp(-a\mu^2 t) J_0(\mu r),$$

$$u = A + B \exp(-a\mu^2 t) Y_0(\mu r),$$

$$u = A + \frac{B}{t}\exp\left(-\frac{r^2 + \mu^2}{4t}\right) I_0\left(\frac{\mu r}{2t}\right),$$

$$u = A + \frac{B}{t}\exp\left(-\frac{r^2 + \mu^2}{4t}\right) K_0\left(\frac{\mu r}{2t}\right),$$

where A, B, and μ are arbitrary constants, n is an arbitrary positive integer, $J_0(z)$ and $Y_0(z)$ are Bessel functions, and $I_0(z)$ and $K_0(z)$ are modified Bessel functions.

▶ **Formulas allowing the construction of particular solutions.**

Suppose $u = u(r,t)$ is a solution of the original equation. Then the functions

$$u_1 = Au(\pm\lambda r,\ \lambda^2 t + C) + B,$$

$$u_2 = \frac{A}{\delta + \beta t}\exp\left[-\frac{\beta r^2}{4a(\delta + \beta t)}\right]u\left(\pm\frac{r}{\delta + \beta t},\ \frac{\gamma + \lambda t}{\delta + \beta t}\right),\qquad \lambda\delta - \beta\gamma = 1,$$

where A, B, C, β, δ, and λ are arbitrary constants, are also solutions of this equation. The second formula is usually encountered with $\beta = 1$, $\gamma = -1$, and $\delta = \lambda = 0$.

▶ **Boundary value problems.**

For solutions of various boundary value problems, see Section 5.1.5 with $\Phi(r,t) \equiv 0$.

5.1.5. Nonhomogeneous Heat Equation with Axial Symmetry
$$u_t = a(u_{rr} + r^{-1}u_r) + \Phi(r,t)$$

▶ **Solutions of boundary value problems in terms of the Green's function.**

We consider boundary value problems for the nonhomogeneous heat equation with axial symmetry in the domain $0 \leq r \leq R$ with the general initial condition

$$u = f(r) \quad \text{at} \quad t = 0$$

and various homogeneous boundary conditions (the solutions bounded at $r = 0$ are sought). The solution of these problems can be represented in terms of the Green's function as

$$u = \int_0^R f(\xi)G(r,\xi,t)\,d\xi + \int_0^t\int_0^R \Phi(\xi,\tau)G(r,\xi,t-\tau)\,d\xi\,d\tau.$$

Below are Green's functions for the nonhomogeneous heat equation with axial symmetry for various types of homogeneous boundary conditions.

▶ **Domain: $0 \leq r \leq R$. First boundary value problem.**

A boundary condition is prescribed:

$$u = 0 \quad \text{at} \quad r = R.$$

Green's function:

$$G(r,\xi,t) = \frac{2}{R^2}\sum_{n=1}^{\infty}\frac{\xi}{J_1^2(\mu_n)}J_0\left(\mu_n\frac{r}{R}\right)J_0\left(\mu_n\frac{\xi}{R}\right)\exp\left(-\frac{a\mu_n^2 t}{R^2}\right),$$

where μ_n are positive zeros of the Bessel function, $J_0(\mu) = 0$. Below are the numerical values of the first ten roots:

$$\mu_1 = 2.4048, \quad \mu_2 = 5.5201, \quad \mu_3 = 8.6537, \quad \mu_4 = 11.7915, \quad \mu_5 = 14.9309,$$
$$\mu_6 = 18.0711, \quad \mu_7 = 21.2116, \quad \mu_8 = 24.3525, \quad \mu_9 = 27.4935, \quad \mu_{10} = 30.6346.$$

The zeros of the Bessel function $J_0(\mu)$ may be approximated by the formula

$$\mu_n = 2.4 + 3.13(n-1) \quad (n = 1, 2, 3, \ldots),$$

which is accurate to within 0.3%. As $n \to \infty$, we have $\mu_{n+1} - \mu_n \to \pi$.

▶ **Domain: $0 \leq r \leq R$. Second boundary value problem.**

A boundary condition is prescribed:

$$u_r = 0 \quad \text{at} \quad r = R.$$

Green's function:

$$G(r, \xi, t) = \frac{2}{R^2}\xi + \frac{2}{R^2} \sum_{n=1}^{\infty} \frac{\xi}{J_0^2(\mu_n)} J_0\left(\frac{\mu_n r}{R}\right) J_0\left(\frac{\mu_n \xi}{R}\right) \exp\left(-\frac{a\mu_n^2 t}{R^2}\right),$$

where μ_n are positive zeros of the first-order Bessel function, $J_1(\mu) = 0$. Below are the numerical values of the first ten roots:

$$\mu_1 = 3.8317, \quad \mu_2 = 7.0156, \quad \mu_3 = 10.1735, \quad \mu_4 = 13.3237, \quad \mu_5 = 16.4706,$$
$$\mu_6 = 19.6159, \quad \mu_7 = 22.7601, \quad \mu_8 = 25.9037, \quad \mu_9 = 29.0468, \quad \mu_{10} = 32.1897.$$

As $n \to \infty$, we have $\mu_{n+1} - \mu_n \to \pi$.

▶ **Domain: $0 \leq r \leq R$. Third boundary value problem.**

A boundary condition is prescribed:

$$u_r + ku = g(t) \quad \text{at} \quad r = R.$$

Green's function:

$$G(r, \xi, t) = \frac{2}{R^2} \sum_{n=1}^{\infty} \frac{\mu_n^2 \xi}{(k^2 R^2 + \mu_n^2) J_0^2(\mu_n)} J_0\left(\frac{\mu_n r}{R}\right) J_0\left(\frac{\mu_n \xi}{R}\right) \exp\left(-\frac{a\mu_n^2 t}{R^2}\right),$$

where the μ_n are positive roots of the transcendental equation

$$\mu J_1(\mu) - kR J_0(\mu) = 0.$$

The numerical values of the first six roots μ_n can be found in Carslaw & Jaeger (1984).

5.1.6. Heat Equation with Central Symmetry $u_t = a(u_{rr} + 2r^{-1}u_r)$

This is the *three-dimensional heat (diffusion) equation with central symmetry*, where $r = \sqrt{x^2 + y^2 + z^2}$ is the radial coordinate.

▶ **Particular solutions:**

$$u = A + Br^{-1},$$
$$u = A + B(r^2 + 6at),$$
$$u = A + B(r^4 + 20atr^2 + 60a^2t^2),$$
$$u = A + B\left[r^{2n} + \sum_{k=1}^{n} \frac{(2n+1)(2n)\dots(2n-2k+2)}{k!}(at)^k r^{2n-2k}\right],$$
$$u = A + 2aBtr^{-1} + Br,$$
$$u = Ar^{-1}\exp(a\mu^2 t \pm \mu r) + B,$$
$$u = A + \frac{B}{t^{3/2}}\exp\left(-\frac{r^2}{4at}\right),$$
$$u = A + \frac{B}{r\sqrt{t}}\exp\left(-\frac{r^2}{4at}\right),$$
$$u = Ar^{-1}\exp(-a\mu^2 t)\cos(\mu r + B) + C,$$
$$u = Ar^{-1}\exp(-\mu r)\cos(\mu r - 2a\mu^2 t + B) + C,$$
$$u = \frac{A}{r}\mathrm{erf}\left(\frac{r}{2\sqrt{at}}\right) + B,$$

where A, B, C, and μ are arbitrary constants, n is an arbitrary positive integer, and erf z is the error function.

▶ **Formulas allowing the construction of particular solutions.**

Suppose $u = u(r, t)$ is a solution of the original equation. Then the functions

$$u_1 = Au(\pm\lambda r, \lambda^2 t + C) + B,$$
$$u_2 = \frac{A}{|\delta + \beta t|^{3/2}}\exp\left[-\frac{\beta r^2}{4a(\delta + \beta t)}\right]u\left(\pm\frac{r}{\delta + \beta t}, \frac{\gamma + \lambda t}{\delta + \beta t}\right), \qquad \lambda\delta - \beta\gamma = 1,$$

where A, B, C, β, δ, and λ are arbitrary constants, are also solutions of this equation. The second formula is usually encountered with $\beta = 1$, $\gamma = -1$, and $\delta = \lambda = 0$.

▶ **Reduction to a constant coefficient heat equation.**

The substitution

$$u = v(r, t)/r$$

brings the original PDE with variable coefficients to the constant coefficient heat equation

$$v_t = av_{rr},$$

which is discussed in Section 5.1.1.

▶ **Boundary value problems.**

For solutions of various boundary value problems, see Section 5.1.7 with $\Phi(r, t) \equiv 0$.

5.1.7. Nonhomogeneous Heat Equation with Central Symmetry
$$u_t = a(u_{rr} + 2r^{-1}u_r) + \Phi(r,t)$$

▶ **Solutions of boundary value problems in terms of the Green's function.**

We consider boundary value problems for the nonhomogeneous heat equation with central symmetry in the domain $0 \leq r \leq R$ with the general initial condition

$$u = f(r) \quad \text{at} \quad t = 0$$

and various homogeneous boundary conditions (the solutions bounded at $r = 0$ are sought). The solution of these problems can be represented in terms of the Green's function as

$$u = \int_0^R f(\xi) G(r, \xi, t) \, d\xi + \int_0^t \int_0^R \Phi(\xi, \tau) G(r, \xi, t - \tau) \, d\xi \, d\tau.$$

Below are the Green's functions for the nonhomogeneous heat equation with central symmetry for various types of homogeneous boundary conditions.

▶ **Domain: $0 \leq r \leq R$. First boundary value problem.**

A boundary condition is prescribed:

$$u = 0 \quad \text{at} \quad r = R.$$

Green's function:

$$G(r, \xi, t) = \frac{2\xi}{Rr} \sum_{n=1}^{\infty} \sin\left(\frac{n\pi r}{R}\right) \sin\left(\frac{n\pi \xi}{R}\right) \exp\left(-\frac{an^2\pi^2 t}{R^2}\right).$$

▶ **Domain: $0 \leq r \leq R$. Second boundary value problem.**

A boundary condition is prescribed:

$$u_r = 0 \quad \text{at} \quad r = R.$$

Green's function:

$$G(r, \xi, t) = \frac{3\xi^2}{R^3} + \frac{2\xi}{Rr} \sum_{n=1}^{\infty} \frac{\mu_n^2 + 1}{\mu_n^2} \sin\left(\frac{\mu_n r}{R}\right) \sin\left(\frac{\mu_n \xi}{R}\right) \exp\left(-\frac{a\mu_n^2 t}{R^2}\right),$$

where μ_n are positive roots of the transcendental equation $\tan \mu - \mu = 0$. The first five roots are

$$\mu_1 = 4.4934, \quad \mu_2 = 7.7253, \quad \mu_3 = 10.9041, \quad \mu_4 = 14.0662, \quad \mu_5 = 17.2208.$$

▶ **Domain: $0 \leq r \leq R$. Third boundary value problem.**

A boundary condition is prescribed:

$$u_r + ku = g(t) \quad \text{at} \quad r = R.$$

Green's function:

$$G(r,\xi,t) = \frac{2\xi}{Rr} \sum_{n=1}^{\infty} \frac{\mu_n^2 + (kR-1)^2}{\mu_n^2 + kR(kR-1)} \sin\left(\frac{\mu_n r}{R}\right) \sin\left(\frac{\mu_n \xi}{R}\right) \exp\left(-\frac{a\mu_n^2 t}{R^2}\right),$$

where the μ_n are positive roots of the transcendental equation

$$\mu \cot \mu + kR - 1 = 0.$$

The numerical values of the first six roots μ_n can be found in Carslaw & Jaeger (1984).

5.1.8. Heat Type Equation of the Form $u_t = u_{xx} + (1-2\beta)x^{-1}u_x$

This dimensionless heat type equation is encountered in problems of the diffusion boundary layer. For $\beta = 0$, $\beta = \frac{1}{2}$, or $\beta = -\frac{1}{2}$, see the equations in Sections 5.1.4, 5.1.1, or 5.1.6, respectively.

▶ **Particular solutions:**

$$u = A + Bx^{2\beta},$$
$$u = A + 4(1-\beta)Bt + Bx^2,$$
$$u = A + 16(2-\beta)(1-\beta)Bt^2 + 8(2-\beta)Btx^2 + Bx^4,$$
$$u = x^{2n} + \sum_{p=1}^{n} \frac{4^p}{p!} s_{n,p} s_{n-\beta,p} t^p x^{2(n-p)}, \quad s_{q,p} = q(q-1)\ldots(q-p+1),$$
$$u = A + 4(1+\beta)Btx^{2\beta} + Bx^{2\beta+2},$$
$$u = A + Bt^{\beta-1} \exp\left(-\frac{x^2}{4t}\right),$$
$$u = A + B\frac{x^{2\beta}}{t^{\beta+1}} \exp\left(-\frac{x^2}{4t}\right),$$
$$u = A + B\gamma\left(\beta, \frac{x^2}{4t}\right),$$
$$u = A + B\exp(-\mu^2 t)x^\beta J_\beta(\mu x),$$
$$u = A + B\exp(-\mu^2 t)x^\beta Y_\beta(\mu x),$$
$$u = A + B\frac{x^\beta}{t} \exp\left(-\frac{x^2+\mu^2}{4t}\right) I_{\pm\beta}\left(\frac{\mu x}{2t}\right),$$
$$u = A + B\frac{x^\beta}{t} \exp\left(-\frac{x^2+\mu^2}{4t}\right) K_\beta\left(\frac{\mu x}{2t}\right),$$

where A, B, and μ are arbitrary constants, n is an arbitrary positive number, $\gamma(\beta, z) = \int_0^z e^{-\xi} \xi^{\beta-1} d\xi$ is the incomplete gamma function, $\Gamma(\beta) = \gamma(\beta, \infty)$ is the gamma function, $J_\beta(z)$ and $Y_\beta(z)$ are Bessel functions, and $I_\beta(z)$ and $K_\beta(z)$ are modified Bessel functions.

▶ **Formulas and transformations for constructing particular solutions.**

1°. Suppose $u = u(x,t)$ is a solution of the original equation. Then the functions

$$u_1 = Au(\pm \lambda x, \lambda^2 t + C),$$

$$u_2 = A|a+bt|^{\beta-1} \exp\left[-\frac{bx^2}{4(a+bt)}\right] u\left(\pm\frac{x}{a+bt}, \frac{c+kt}{a+bt}\right), \qquad ak - bc = 1,$$

where A, C, a, b, and c are arbitrary constants, are also solutions of this equation. The second formula is usually encountered with $a = k = 0$, $b = 1$, and $c = -1$.

2°. The substitution $u = x^{2\beta} v(x,t)$ brings the equation with parameter β to an equation of the same type with parameter $-\beta$:

$$v_t = v_{xx} + (1+2\beta)x^{-1} v_x.$$

▶ **Domain: $0 \leq x < \infty$. First boundary value problem.**

The following initial and boundary conditions are prescribed:

$$u = f(x) \quad \text{at} \quad t = 0, \qquad u = g(t) \quad \text{at} \quad x = 0.$$

Solution for $0 < \beta < 1$ (Sutton, 1943):

$$u = \frac{x^\beta}{2t} \int_0^\infty f(\xi) \xi^{1-\beta} \exp\left(-\frac{x^2+\xi^2}{4t}\right) I_\beta\left(\frac{\xi x}{2t}\right) d\xi$$

$$+ \frac{x^{2\beta}}{2^{2\beta+1}\Gamma(\beta+1)} \int_0^t g(\tau) \exp\left[-\frac{x^2}{4(t-\tau)}\right] \frac{d\tau}{(t-\tau)^{1+\beta}}.$$

▶ **Domain: $0 \leq x < \infty$. Second boundary value problem.**

The following initial and boundary conditions are prescribed:

$$u = f(x) \quad \text{at} \quad t = 0, \qquad (x^{1-2\beta} u_x) = g(t) \quad \text{at} \quad x = 0.$$

Solution for $0 < \beta < 1$ (Sutton, 1943):

$$u = \frac{x^\beta}{2t} \int_0^\infty f(\xi) \xi^{1-\beta} \exp\left(-\frac{x^2+\xi^2}{4t}\right) I_{-\beta}\left(\frac{\xi x}{2t}\right) d\xi$$

$$- \frac{2^{2\beta-1}}{\Gamma(1-\beta)} \int_0^t g(\tau) \exp\left[-\frac{x^2}{4(t-\tau)}\right] \frac{d\tau}{(t-\tau)^{1-\beta}}.$$

5.1.9. Heat Type Equation of the Form $u_t = [f(x)u_x]_x$

The equation describes heat transfer in a quiescent medium (solid body) in the case where the thermal diffusivity $f(x)$ is a coordinate dependent function.

1°. For any $f(x)$, the original PDE admits particular solutions of polynomial form in t:

$$u_n(x,t) = \sum_{k=0}^n t^k \varphi_{n,k}(x),$$

where n is a positive integer.

Below are examples of particular solutions of this type (Polyanin, 2002):

$$u = A + B \int \frac{dx}{f(x)},$$

$$u = At + A \int \frac{x\,dx}{f(x)} + B,$$

$$u = At\varphi(x) + A \int \left(\int \varphi(x)\,dx\right) \frac{dx}{f(x)} + B, \quad \varphi(x) = \int \frac{dx}{f(x)},$$

$$u = At^2 + 2At\psi(x) + 2A \int \left(\int \psi(x)\,dx\right) \frac{dx}{f(x)} + B, \quad \psi(x) = \int \frac{x\,dx}{f(x)},$$

$$u = At^2\varphi(x) + 2AtI(x) + 2A \int \left(\int I(x)\,dx\right) \frac{dx}{f(x)} + B, \quad I(x) = \int \left(\int \varphi(x)\,dx\right) \frac{dx}{f(x)},$$

where A and B are arbitrary constants.

$2°$. There are particular solutions of the multiplicative form

$$u = e^{-\lambda t} w(x),$$

where the function $w(x)$ is identified by solving the following linear ODE with free parameter λ:

$$[f(x) w'_x]'_x + \lambda w = 0.$$

$3°$. A solution in the form of an infinite series:

$$u = \Theta(x) + \sum_{n=1}^{\infty} \frac{1}{n!} t^n L^n [\Theta(x)], \qquad L \equiv \frac{d}{dx}\left[f(x)\frac{d}{dx}\right].$$

It contains an arbitrary function of the space variable, $\Theta = \Theta(x)$. This solution satisfies the initial condition $u(x, 0) = \Theta(x)$.

5.1.10. Equations of the Form
$$s(x)u_t = [p(x)u_x]_x - q(x)u + \Phi(x, t)$$

Equations of this form are often arise in heat and mass transfer theory and chemical engineering sciences. Throughout this subsection, we assume that the functions s, p, p'_x, and q are continuous and $s > 0$, $p > 0$, and $x_1 \leq x \leq x_2$.

▶ **General formulas for solving linear nonhomogeneous boundary value problems.**

The solution of the equation in question subjected to the initial condition

$$u = f(x) \quad \text{at} \quad t = 0$$

and the arbitrary linear nonhomogeneous boundary conditions

$$a_1 u_x + b_1 u = g_1(t) \quad \text{at} \quad x = x_1,$$
$$a_2 u_x + b_2 u = g_2(t) \quad \text{at} \quad x = x_2,$$

can be represented as the sum

$$u(x,t) = \int_0^t \int_{x_1}^{x_2} \Phi(\xi,\tau)\mathcal{G}(x,\xi,t-\tau)\,d\xi\,d\tau + \int_{x_1}^{x_2} s(\xi)f(\xi)\mathcal{G}(x,\xi,t)\,d\xi$$
$$+ p(x_1)\int_0^t g_1(\tau)\Lambda_1(x,t-\tau)\,d\tau + p(x_2)\int_0^t g_2(\tau)\Lambda_2(x,t-\tau)\,d\tau. \quad (1)$$

Here, the modified Green's function is given by

$$\mathcal{G}(x,\xi,t) = \sum_{n=1}^{\infty} \frac{y_n(x)y_n(\xi)}{\|y_n\|^2} \exp(-\lambda_n t), \qquad \|y_n\|^2 = \int_{x_1}^{x_2} s(x)y_n^2(x)\,dx, \quad (2)$$

where the λ_n and $y_n(x)$ are the eigenvalues and corresponding eigenfunctions of the following Sturm–Liouville problem for a second-order linear ODE:

$$[p(x)y'_x]'_x + [\lambda s(x) - q(x)]y = 0,$$
$$a_1 y'_x + b_1 y = 0 \quad \text{at} \quad x = x_1, \quad (3)$$
$$a_2 y'_x + b_2 y = 0 \quad \text{at} \quad x = x_2.$$

The functions $\Lambda_1(x,t)$ and $\Lambda_2(x,t)$ that occur in the integrands of the last two terms in solution (1) are expressed via the Green's function (2). The corresponding formulas for $\Lambda_m(x,t)$ are given in Table 5.2 for the basic types of boundary value problems.

TABLE 5.2. Expressions of the functions $\Lambda_1(x,t)$ and $\Lambda_2(x,t)$ involved in the integrands of the last two terms in solution (1).

Type of problem	Form of boundary conditions	Functions $\Lambda_m(x,t)$
First boundary value problem $(a_1 = a_2 = 0,\ b_1 = b_2 = 1)$	$u = g_1(t)$ at $x = x_1$ $u = g_2(t)$ at $x = x_2$	$\Lambda_1(x,t) = \frac{\partial}{\partial \xi}\mathcal{G}(x,\xi,t)\big|_{\xi=x_1}$ $\Lambda_2(x,t) = -\frac{\partial}{\partial \xi}\mathcal{G}(x,\xi,t)\big|_{\xi=x_2}$
Second boundary value problem $(a_1 = a_2 = 1,\ b_1 = b_2 = 0)$	$u_x = g_1(t)$ at $x = x_1$ $u_x = g_2(t)$ at $x = x_2$	$\Lambda_1(x,t) = -\mathcal{G}(x,x_1,t)$ $\Lambda_2(x,t) = \mathcal{G}(x,x_2,t)$
Third boundary value problem $(a_1 = a_2 = 1,\ b_1 < 0,\ b_2 > 0)$	$u_x + b_1 u = g_1(t)$ at $x = x_1$ $u_x + b_2 u = g_2(t)$ at $x = x_2$	$\Lambda_1(x,t) = -\mathcal{G}(x,x_1,t)$ $\Lambda_2(x,t) = \mathcal{G}(x,x_2,t)$
Mixed boundary value problem $(a_1 = b_2 = 0,\ a_2 = b_1 = 1)$	$u = g_1(t)$ at $x = x_1$ $u_x = g_2(t)$ at $x = x_2$	$\Lambda_1(x,t) = \frac{\partial}{\partial \xi}\mathcal{G}(x,\xi,t)\big|_{\xi=x_1}$ $\Lambda_2(x,t) = \mathcal{G}(x,x_2,t)$
Mixed boundary value problem $(a_1 = b_2 = 1,\ a_2 = b_1 = 0)$	$u_x = g_1(t)$ at $x = x_1$ $u = g_2(t)$ at $x = x_2$	$\Lambda_1(x,t) = -\mathcal{G}(x,x_1,t)$ $\Lambda_2(x,t) = -\frac{\partial}{\partial \xi}\mathcal{G}(x,\xi,t)\big|_{\xi=x_2}$

▶ **General properties of the Sturm–Liouville problem (3).**

$1°$. There are infinitely many eigenvalues. All eigenvalues are real and different and can be ordered so that $\lambda_1 < \lambda_2 < \lambda_3 < \ldots$, with $\lambda_n \to \infty$ as $n \to \infty$ (therefore, there can exist only finitely many negative eigenvalues). Each eigenvalue is of multiplicity 1.

$2°$. The eigenfunctions are determined up to a constant multiplier. Each eigenfunction $y_n(x)$ has exactly $n-1$ zeros in the open interval (x_1, x_2).

3°. Eigenfunctions $y_n(x)$ and $y_m(x)$, $n \ne m$, are orthogonal with weight $s(x)$ on the interval $x_1 \le x \le x_2$:

$$\int_{x_1}^{x_2} s(x) y_n(x) y_m(x)\, dx = 0 \quad \text{for} \quad n \ne m.$$

4°. An arbitrary function $F(x)$ that has a continuous derivative and satisfies the boundary conditions of the Sturm–Liouville problem can be expanded into an absolutely and uniformly convergent series in eigenfunctions:

$$F(x) = \sum_{n=1}^{\infty} F_n y_n(x), \quad F_n = \frac{1}{\|y_n\|^2} \int_{x_1}^{x_2} s(x) F(x) y_n(x)\, dx,$$

where the norm $\|y_n\|^2$ is defined in (2).

5°. If the conditions

$$q(x) \ge 0, \quad a_1 b_1 \le 0, \quad a_2 b_2 \ge 0 \tag{4}$$

are satisfied, there are no negative eigenvalues. If $q \equiv 0$ and $b_1 = b_2 = 0$, then $\lambda_1 = 0$ is the least eigenvalue, to which there corresponds the eigenfunction $\varphi_1 = \text{const}$. Otherwise, all eigenvalues are positive, provided that conditions (4) are satisfied.

6°. The following asymptotic relation holds for large eigenvalues as $n \to \infty$:

$$\lambda_n = \frac{\pi^2 n^2}{\Delta^2} + O(1), \quad \Delta = \int_{x_1}^{x_2} \sqrt{\frac{s(x)}{p(x)}}\, dx.$$

5.1.11. Liquid-Film Mass Transfer Equation $(1 - y^2) u_x = a u_{yy}$

This equation describes steady-state heat and mass transfer in a fluid film with a parabolic velocity profile. The variables have the following physical meanings: u is a dimensionless temperature (concentration); x and y are dimensionless coordinates measured, respectively, along and across the film ($y = 0$ corresponds to the free surface of the film and $y = 1$ to the solid surface the film flows down), and $\text{Pe} = 1/a$ is the Peclet number. Mixed boundary conditions are usually encountered in practical applications.

▶ **Particular solutions:**

$$u(x, y) = kx - \frac{k}{12a} y^4 + \frac{k}{2a} y^2 + Ay + B,$$
$$u(x, y) = A \exp(-a\lambda^2 x) \exp\left(-\tfrac{1}{2}\lambda y^2\right) \Phi\left(\tfrac{1}{4} - \tfrac{1}{4}\lambda, \tfrac{1}{2}; \lambda y^2\right),$$
$$u(x, y) = A \exp(-a\lambda^2 x)\, y \exp\left(-\tfrac{1}{2}\lambda y^2\right) \Phi\left(\tfrac{3}{4} - \tfrac{1}{4}\lambda, \tfrac{3}{2}; \lambda y^2\right),$$
$$\Phi(\alpha, \beta; z) = 1 + \sum_{m=1}^{\infty} \frac{\alpha(\alpha+1)\ldots(\alpha+m-1)}{\beta(\beta+1)\ldots(\beta+m-1)} \frac{z^m}{m!},$$

where A, B, k, and λ are arbitrary constants, and $\Phi(\alpha, \beta; z)$ is the degenerate hypergeometric function.

▶ **Mass exchange between fluid film and gas.**

The mass exchange between a fluid film and the gas above the free surface, provided that the admixture concentration at the film surface is constant and there is no mass transfer through the solid surface, meets the boundary conditions

$$u = 0 \quad \text{at} \quad x = 0 \quad (0 < y < 1),$$
$$u = 1 \quad \text{at} \quad y = 0 \quad (x > 0),$$
$$u_y = 0 \quad \text{at} \quad y = 1 \quad (x > 0).$$

The solution of the original equation subject to these boundary conditions is given by (see Davis, 1973):

$$u(x, y) = 1 - \sum_{m=1}^{\infty} A_m \exp(-a\lambda_m^2 x) F_m(y), \tag{1}$$
$$F_m(y) = y \exp(-\tfrac{1}{2}\lambda_m y^2) \Phi(\tfrac{3}{4} - \tfrac{1}{4}\lambda_m, \tfrac{3}{2}; \lambda_m y^2),$$

where the function F_m and the coefficients A_m and λ_m are independent of the parameter a. The eigenvalues λ_m are solutions of the transcendental equation

$$\lambda_m \Phi(\tfrac{3}{4} - \tfrac{1}{4}\lambda_m, \tfrac{3}{2}; \lambda_m) - \Phi(\tfrac{3}{4} - \tfrac{1}{4}\lambda_m, \tfrac{1}{2}; \lambda_m) = 0.$$

The series coefficients A_m are calculated from

$$A_m = \frac{\int_0^1 (1 - y^2) F_m(y)\, dy}{\int_0^1 (1 - y^2)[F_m(y)]^2\, dy}, \quad \text{where} \quad m = 1, 2, \ldots$$

Table 5.3 shows the first ten eigenvalues λ_m and coefficients A_m calculated in Rotem & Neilson (1966).

TABLE 5.3. Eigenvalues λ_m and coefficients A_m in solution (1).

m	λ_m	A_m	m	λ_m	A_m
1	2.2631	1.3382	6	22.3181	−0.1873
2	6.2977	−0.5455	7	26.3197	0.1631
3	10.3077	0.3589	8	30.3209	−0.1449
4	14.3128	−0.2721	9	34.3219	0.1306
5	18.3159	0.2211	10	38.3227	−0.1191

The solution asymptotics as $ax \to 0$ is given by

$$u = \operatorname{erfc}\left(\frac{y}{2\sqrt{ax}}\right),$$

where $\operatorname{erfc} z = \int_z^{\infty} \exp(-\xi^2)\, d\xi$ is the complementary error function.

▶ **Dissolution of a plate by a laminar fluid film.**

The dissolution of a plate by a laminar fluid film, provided that the concentration at the solid surface is constant and there is no mass flux from the film into the gas, satisfies the boundary conditions

$$u = 0 \quad \text{at} \quad x = 0 \quad (0 < y < 1),$$
$$u_y = 0 \quad \text{at} \quad y = 0 \quad (x > 0),$$
$$u = 1 \quad \text{at} \quad y = 1 \quad (x > 0).$$

The solution of the original equation subject to these boundary conditions is given by

$$u(x,y) = 1 - \sum_{m=0}^{\infty} A_m \exp(-a\lambda_m^2 x) G_m(y), \tag{2}$$
$$G_m(y) = \exp\left(-\tfrac{1}{2}\lambda_m y^2\right) \Phi\left(\tfrac{1}{4} - \tfrac{1}{4}\lambda_m, \tfrac{1}{2}; \lambda_m y^2\right),$$

where the functions G_m and the constants A_m and λ_m are independent of the parameter a.

The eigenvalues λ_m are solutions of the transcendental equation

$$\Phi\left(\tfrac{1}{4} - \tfrac{1}{4}\lambda_m, \tfrac{1}{2}; \lambda_m\right) = 0.$$

The following approximate relation is convenient to calculate λ_m:

$$\lambda_m = 4m + 1.68 \qquad (m = 0, 1, 2, \dots). \tag{3}$$

The maximum error of this formula is less than 0.2%.

The coefficients A_m are approximated by

$$A_0 = 1.2, \quad A_m = (-1)^m 2.27\, \lambda_m^{-7/6} \quad \text{for} \quad m = 1, 2, 3, \dots,$$

where the eigenvalues λ_m are defined by (3). The maximum error of the expressions for A_m is less than 0.1%.

The solution asymptotics as $ax \to 0$ is given by

$$u = \frac{1}{\Gamma\left(\tfrac{1}{3}\right)} \Gamma\left(\tfrac{1}{3}, \tfrac{2}{9}\zeta\right), \quad \zeta = \frac{(1-y)^3}{ax},$$

where $\Gamma(\alpha, z) = \displaystyle\int_z^\infty e^{-\xi} \xi^{\alpha - 1}\, d\xi$ is the incomplete gamma function, $\Gamma(\alpha) = \Gamma(\alpha, 0)$ is the gamma function, and $\Gamma\left(\tfrac{1}{3}\right) \approx 2.679$.

5.1.12. Equations of the Diffusion (Thermal) Boundary Layer

1. $f(x)u_x + g(x)y u_y = u_{yy}.$

This equation is encountered in diffusion boundary layer problems (mass exchange of drops and bubbles with a flow); see Levich (1962).

The transformation (A and B are any numbers)

$$t = \int \frac{h^2(x)}{f(x)}\, dx + A, \quad z = yh(x), \quad \text{where} \quad h(x) = B \exp\left[-\int \frac{g(x)}{f(x)}\, dx\right],$$

leads to a constant coefficient heat equation, $u_t = u_{zz}$, which is considered in Section 5.1.1.

2. $f(x)y^{n-1}u_x + g(x)y^n u_y = u_{yy}$.

For $n = 2$, this equation is encountered in diffusion boundary layer problems (mass exchange of solid particles with a flow); see Levich (1962).

The transformation

$$t = \frac{1}{4}(n+1)^2 \int \frac{h^2(x)}{f(x)} dx, \quad z = h(x)y^{\frac{n+1}{2}}, \quad \text{where} \quad h(x) = \exp\left[-\frac{n+1}{2}\int \frac{g(x)}{f(x)} dx\right],$$

leads to the simpler equation

$$u_t = u_{zz} + \frac{1-2k}{z} u_z, \quad k = \frac{1}{n+1},$$

which is considered in Section 5.1.8.

5.1.13. Schrödinger Equation $i\hbar u_t = -\dfrac{\hbar^2}{2m} u_{xx} + U(x)u$

▶ **Eigenvalue problem. Cauchy problem.**

Schrödinger's equation is the basic equation of quantum mechanics; u is the wave function, $i^2 = -1$, $\hbar$ is Planck's constant, m is the mass of the particle, and $U(x)$ is the potential energy of the particle in the force field.

1°. In discrete spectrum problems, the particular solutions are sought in the form

$$u = \exp\left(-\frac{iE_n}{\hbar} t\right)\psi_n(x),$$

where the eigenfunctions ψ_n and the respective energies E_n have to be determined by solving the ODE eigenvalue problem

$$\frac{d^2\psi_n}{dx^2} + \frac{2m}{\hbar^2}[E_n - U(x)]\psi_n = 0, \tag{1}$$
$$\psi_n \to 0 \text{ as } x \to \pm\infty, \quad \int_{-\infty}^{\infty} |\psi_n|^2 \, dx = 1.$$

The last relation is the normalizing condition for ψ_n.

2°. In the cases where the eigenfunctions $\psi_n(x)$ form an orthonormal basis in $L_2(\mathbb{R})$, the solution of the Cauchy problem for Schrödinger's equation with the initial condition

$$u = f(x) \quad \text{at} \quad t = 0 \tag{2}$$

is given by

$$u = \int_{-\infty}^{\infty} G(x,\xi,t)f(\xi)\,d\xi, \quad G(x,\xi,t) = \sum_{n=0}^{\infty} \psi_n(x)\psi_n(\xi)\exp\left(-\frac{iE_n}{\hbar}t\right).$$

Various potentials $U(x)$ are considered below and particular solutions of the boundary value problem (1) or the Cauchy problem for Schrödinger's equation are presented.

▶ **Free particle:** $U(x) = 0$.

The solution of the Cauchy problem with the initial condition (2) is given by

$$u = \frac{1}{2\sqrt{i\pi\tau}} \int_{-\infty}^{\infty} \exp\left[-\frac{(x-\xi)^2}{4i\tau}\right] f(\xi)\,d\xi,$$

$$\tau = \frac{\hbar t}{2m}, \quad \sqrt{ia} = \begin{cases} e^{\pi i/4}\sqrt{|a|} & \text{if } a > 0, \\ e^{-\pi i/4}\sqrt{|a|} & \text{if } a < 0. \end{cases}$$

▶ **Linear potential (motion in a uniform external field):** $U(x) = ax$.

Solution of the Cauchy problem with the initial condition (2):

$$u = \frac{1}{2\sqrt{i\pi\tau}} \exp\left(-ib\tau x - \tfrac{1}{3}ib^2\tau^3\right) \int_{-\infty}^{\infty} \exp\left[-\frac{(x+b\tau^2-\xi)^2}{4i\tau}\right] f(\xi)\,d\xi,$$

$$\tau = \frac{\hbar t}{2m}, \quad b = \frac{2am}{\hbar^2}.$$

▶ **Linear harmonic oscillator:** $U(x) = \tfrac{1}{2}m\omega^2 x^2$.

Eigenvalues:
$$E_n = \hbar\omega\left(n + \tfrac{1}{2}\right), \qquad n = 0, 1, \ldots$$

Normalized eigenfunctions:

$$\psi_n(x) = \frac{1}{\pi^{1/4}\sqrt{2^n n!\, x_0}} \exp\left(-\tfrac{1}{2}\xi^2\right) H_n(\xi), \qquad \xi = \frac{x}{x_0}, \quad x_0 = \sqrt{\frac{\hbar}{m\omega}},$$

where $H_n(\xi) = (-1)^n \exp(\xi^2)\dfrac{d^n}{d\xi^n}\exp(-\xi^2)$ are the Hermite polynomials, $n = 0, 1, \ldots$.
The functions $\psi_n(x)$ form an orthonormal basis in $L_2(\mathbb{R})$.

▶ **Isotropic free particle:** $U(x) = a/x^2$.

Here the variable $x \geq 0$ plays the role of the radial coordinate, and $a > 0$. The equation with $U(x) = a/x^2$ results from Schrödinger's equation for a free particle with n space coordinates if one passes to spherical (cylindrical) coordinates and separates the angular variables.

The solution of Schrödinger's equation satisfying the initial condition (2) has the form

$$u = \frac{\exp\left[-\tfrac{1}{2}i\pi(\mu+1)\operatorname{sign} t\right]}{2|\tau|} \int_0^{\infty} \sqrt{xy} \exp\left(i\frac{x^2+y^2}{4\tau}\right) J_\mu\left(\frac{xy}{2|\tau|}\right) f(y)\,dy,$$

$$\tau = \frac{\hbar t}{2m}, \quad \mu = \sqrt{\frac{2am}{\hbar^2} + \frac{1}{4}} \geq 1,$$

where $J_\mu(\xi)$ is the Bessel function.

▶ **Morse potential:** $U(x) = U_0(e^{-2x/a} - 2e^{-x/a})$.

Eigenvalues:
$$E_n = -U_0\left[1 - \frac{1}{\beta}(n + \tfrac{1}{2})\right]^2, \quad \beta = \frac{a\sqrt{2mU_0}}{\hbar}, \quad 0 \le n < \beta - 2.$$

Eigenfunctions:
$$\psi_n(x) = \xi^s e^{-\xi/2} \Phi(-n, 2s+1, \xi), \quad \xi = 2\beta e^{-x/a}, \quad s = \frac{a\sqrt{-2mE_n}}{\hbar},$$

where $\Phi(a, b, \xi)$ is the degenerate hypergeometric function.

In this case, the number of eigenvalues (energy levels) E_n and eigenfunctions ψ_n is finite: $n = 0, 1, \ldots, n_{\max}$.

5.2. Hyperbolic Equations

5.2.1. Wave Equation $u_{tt} = a^2 u_{xx}$

This wave equation is also known as the *equation of vibration of a string*. It is often encountered in elasticity, aerodynamics, acoustics, and electrodynamics.

▶ **General solution. Some formulas.**

1°. General solution:
$$u = \varphi(x + at) + \psi(x - at),$$
where $\varphi(x)$ and $\psi(x)$ are arbitrary functions.

2°. If $u(x,t)$ is a solution of the wave equation, then the functions
$$u_1 = Au(\pm \lambda x + C_1, \pm \lambda t + C_2) + B,$$
$$u_2 = Au\left(\frac{x - vt}{\sqrt{1 - (v/a)^2}}, \frac{t - va^{-2}x}{\sqrt{1 - (v/a)^2}}\right),$$
$$u_3 = Au\left(\frac{x}{x^2 - a^2 t^2}, \frac{t}{x^2 - a^2 t^2}\right)$$

are also solutions of the equation everywhere these functions are defined (A, B, C_1, C_2, v, and λ are arbitrary constants). The signs at λ's in the formula for u_1 are taken arbitrarily. The function u_2 results from the invariance of the wave equation under the *Lorentz transformations*.

▶ **Domain:** $-\infty < x < \infty$. **Cauchy problem.**

Initial conditions are prescribed:
$$u = f(x) \quad \text{at} \quad t = 0, \quad u_t = g(x) \quad \text{at} \quad t = 0.$$

Solution (*D'Alembert's formula*):
$$u = \frac{1}{2}[f(x + at) + f(x - at)] + \frac{1}{2a}\int_{x-at}^{x+at} g(\xi)\, d\xi.$$

▶ **Domain: $0 \leq x < \infty$. First boundary value problem.**

The following two initial and one boundary conditions are prescribed:

$$u = f(x) \quad \text{at} \quad t = 0, \qquad u_t = g(x) \quad \text{at} \quad t = 0, \qquad u = h(t) \quad \text{at} \quad x = 0.$$

Solution:

$$u = \begin{cases} \dfrac{1}{2}[f(x+at) + f(x-at)] + \dfrac{1}{2a}\displaystyle\int_{x-at}^{x+at} g(\xi)\,d\xi & \text{for } t < \dfrac{x}{a}, \\ \dfrac{1}{2}[f(x+at) - f(at-x)] + \dfrac{1}{2a}\displaystyle\int_{at-x}^{x+at} g(\xi)\,d\xi + h\!\left(t - \dfrac{x}{a}\right) & \text{for } t > \dfrac{x}{a}. \end{cases}$$

In the domain $t < x/a$ the boundary conditions have no effect on the solution and the expression of u coincides with D'Alembert's solution for an infinite line (see the previous paragraph).

▶ **Domain: $0 \leq x < \infty$. Second boundary value problem.**

The following two initial and one boundary conditions are prescribed:

$$u = f(x) \quad \text{at} \quad t = 0, \qquad u_t = g(x) \quad \text{at} \quad t = 0, \qquad u_x = h(t) \quad \text{at} \quad x = 0.$$

Solution:

$$u = \begin{cases} \dfrac{1}{2}[f(x+at) + f(x-at)] + \dfrac{1}{2a}[G(x+at) - G(x-at)] & \text{for } t < \dfrac{x}{a}, \\ \dfrac{1}{2}[f(x+at) + f(at-x)] + \dfrac{1}{2a}[G(x+at) + G(at-x)] - aH\!\left(t - \dfrac{x}{a}\right) & \text{for } t > \dfrac{x}{a}, \end{cases}$$

where $G(z) = \displaystyle\int_0^z g(\xi)\,d\xi$ and $H(z) = \displaystyle\int_0^z h(\xi)\,d\xi$.

▶ **Domain: $0 \leq x \leq l$. Boundary value problems.**

For solutions of various boundary value problems, see Section 5.2.2 for $\Phi(x,t) \equiv 0$.

5.2.2. Nonhomogeneous Wave Equation $u_{tt} = a^2 u_{xx} + \Phi(x,t)$

▶ **Solutions of boundary value problems in terms of the Green's function.**

We consider boundary value problems for the nonhomogeneous wave equation on an interval $0 \leq x \leq l$ with the general initial conditions

$$u = f(x) \quad \text{at} \quad t = 0, \qquad u_t = g(x) \quad \text{at} \quad t = 0$$

and various homogeneous boundary conditions. The solution of these problems can be represented in terms of the Green's function as

$$u = \frac{\partial}{\partial t}\int_0^l f(\xi) G(x,\xi,t)\,d\xi + \int_0^l g(\xi) G(x,\xi,t)\,d\xi + \int_0^t\!\!\int_0^l \Phi(\xi,\tau) G(x,\xi,t-\tau)\,d\xi\,d\tau.$$

Here, the upper limit l can assume any finite values.

Below are Green's functions for the wave equation for various types of homogeneous boundary conditions.

▶ **Domain: $0 \leq x \leq l$. First boundary value problem.**

Boundary conditions are prescribed:
$$u = 0 \quad \text{at} \quad x = 0, \qquad u = 0 \quad \text{at} \quad x = l.$$

Green's function:
$$G(x, \xi, t) = \frac{2}{a\pi} \sum_{n=1}^{\infty} \frac{1}{n} \sin\left(\frac{n\pi x}{l}\right) \sin\left(\frac{n\pi \xi}{l}\right) \sin\left(\frac{n\pi a t}{l}\right).$$

▶ **Domain: $0 \leq x \leq l$. Second boundary value problem.**

Boundary conditions are prescribed:
$$u_x = 0 \quad \text{at} \quad x = 0, \qquad u_x = 0 \quad \text{at} \quad x = l.$$

Green's function:
$$G(x, \xi, t) = \frac{t}{l} + \frac{2}{a\pi} \sum_{n=1}^{\infty} \frac{1}{n} \cos\left(\frac{n\pi x}{l}\right) \cos\left(\frac{n\pi \xi}{l}\right) \sin\left(\frac{n\pi a t}{l}\right).$$

▶ **Domain: $0 \leq x \leq l$. Third boundary value problem ($k_1 > 0$, $k_2 > 0$).**

Boundary conditions are prescribed:
$$u_x - k_1 u = 0 \quad \text{at} \quad x = 0, \qquad u_x + k_2 u = 0 \quad \text{at} \quad x = l.$$

Green's function:
$$G(x, \xi, t) = \frac{1}{a} \sum_{n=1}^{\infty} \frac{1}{\lambda_n \|u_n\|^2} \sin(\lambda_n x + \varphi_n) \sin(\lambda_n \xi + \varphi_n) \sin(\lambda_n a t),$$
$$\varphi_n = \arctan\frac{\lambda_n}{k_1}, \quad \|u_n\|^2 = \frac{l}{2} + \frac{(\lambda_n^2 + k_1 k_2)(k_1 + k_2)}{2(\lambda_n^2 + k_1^2)(\lambda_n^2 + k_2^2)};$$

the λ_n are positive roots of the transcendental equation $\cot(\lambda l) = \dfrac{\lambda^2 - k_1 k_2}{\lambda(k_1 + k_2)}$.

▶ **Solutions of boundary value problems with nonhomogeneous boundary conditions.**

Any linear problem for the wave equation with arbitrary nonhomogeneous boundary conditions can be reduced to a linear problem with homogeneous boundary conditions (see Table 5.1 and the explanatory text to it).

5.2.3. Klein–Gordon Equation $u_{tt} = a^2 u_{xx} - bu$

This equation is encountered in quantum field theory and a number of physical applications.

▶ **Particular solutions:**

$$u = \cos(\lambda x)[A\cos(\mu t) + B\sin(\mu t)], \quad b = -a^2\lambda^2 + \mu^2,$$
$$u = \sin(\lambda x)[A\cos(\mu t) + B\sin(\mu t)], \quad b = -a^2\lambda^2 + \mu^2,$$
$$u = \exp(\pm\mu t)[A\cos(\lambda x) + B\sin(\lambda x)], \quad b = -a^2\lambda^2 - \mu^2,$$
$$u = \exp(\pm\lambda x)[A\cos(\mu t) + B\sin(\mu t)], \quad b = a^2\lambda^2 + \mu^2,$$
$$u = \exp(\pm\lambda x)[A\exp(\mu t) + B\exp(-\mu t)], \quad b = a^2\lambda^2 - \mu^2,$$
$$u = AJ_0(\xi) + BY_0(\xi), \quad \xi = \frac{\sqrt{b}}{a}\sqrt{a^2(t+C_1)^2 - (x+C_2)^2}, \quad b > 0,$$
$$u = AI_0(\xi) + BK_0(\xi), \quad \xi = \frac{\sqrt{-b}}{a}\sqrt{a^2(t+C_1)^2 - (x+C_2)^2}, \quad b < 0,$$

where A, B, C_1, and C_2 are arbitrary constants, $J_0(\xi)$ and $Y_0(\xi)$ are Bessel functions, and $I_0(\xi)$ and $K_0(\xi)$ are modified Bessel functions.

▶ **Formulas allowing the construction of particular solutions.**

Suppose $u = u(x,t)$ is a solution of the Klein–Gordon equation. Then the functions

$$u_1 = Au(\pm x + C_1, \pm t + C_2) + B,$$
$$u_2 = Au\left(\frac{x - vt}{\sqrt{1-(v/a)^2}}, \frac{t - va^{-2}x}{\sqrt{1-(v/a)^2}}\right),$$

where A, B, C_1, C_2, and v are arbitrary constants, are also solutions of this equation. The signs in the formula for u_1 are taken arbitrarily.

▶ **Domain: $0 \leq x \leq l$. Boundary value problems.**

For solutions of boundary value problems, see Section 5.2.4 for $\Phi(x,t) \equiv 0$.

5.2.4. Nonhomogeneous Klein–Gordon Equation
$$u_{tt} = a^2 u_{xx} - bu + \Phi(x,t)$$

▶ **Solutions of boundary value problems in terms of the Green's function.**

We consider boundary value problems for the nonhomogeneous Klein–Gordon equation on an interval $0 \leq x \leq l$ with the general initial conditions

$$u = f(x) \quad \text{at} \quad t = 0, \qquad u_t = g(x) \quad \text{at} \quad t = 0$$

and various homogeneous boundary conditions. The solution of these problems can be represented in terms of the Green's function as

$$u = \frac{\partial}{\partial t}\int_0^l f(\xi)G(x,\xi,t)\,d\xi + \int_0^l g(\xi)G(x,\xi,t)\,d\xi + \int_0^t\int_0^l \Phi(\xi,\tau)G(x,\xi,t-\tau)\,d\xi\,d\tau.$$

Here, the upper limit l can assume any finite values.

Below are the Green's functions for the Klein–Gordon equation subject to homogeneous boundary conditions.

▶ **Domain: $0 \leq x \leq l$. First boundary value problem.**

Boundary conditions are prescribed:
$$u = 0 \quad \text{at} \quad x = 0, \qquad u = 0 \quad \text{at} \quad x = l.$$

Green's function for $b > 0$:
$$G(x, \xi, t) = \frac{2}{l} \sum_{n=1}^{\infty} \sin(\lambda_n x) \sin(\lambda_n \xi) \frac{\sin\bigl(t\sqrt{a^2 \lambda_n^2 + b}\bigr)}{\sqrt{a^2 \lambda_n^2 + b}}, \qquad \lambda_n = \frac{\pi n}{l}.$$

▶ **Domain: $0 \leq x \leq l$. Second boundary value problem.**

Boundary conditions are prescribed:
$$u_x = 0 \quad \text{at} \quad x = 0, \qquad u_x = 0 \quad \text{at} \quad x = l.$$

Green's function for $b > 0$:
$$G(x, \xi, t) = \frac{1}{l\sqrt{b}} \sin\bigl(t\sqrt{b}\bigr) + \frac{2}{l} \sum_{n=1}^{\infty} \cos(\lambda_n x) \cos(\lambda_n \xi) \frac{\sin\bigl(t\sqrt{a^2 \lambda_n^2 + b}\bigr)}{\sqrt{a^2 \lambda_n^2 + b}}, \qquad \lambda_n = \frac{\pi n}{l}.$$

▶ **Domain: $0 \leq x \leq l$. Third boundary value problem.**

Boundary conditions are prescribed:
$$u_x - k_1 u = 0 \quad \text{at} \quad x = 0, \qquad u_x + k_2 u = 0 \quad \text{at} \quad x = l.$$

Green's function:
$$G(x, \xi, t) = \sum_{n=1}^{\infty} \frac{y_n(x) y_n(\xi) \sin\bigl(t\sqrt{a^2 \lambda_n^2 + b}\bigr)}{\|y_n\|^2 \sqrt{a^2 \lambda_n^2 + b}},$$

$$y_n(x) = \cos(\lambda_n x) + \frac{k_1}{\lambda_n} \sin(\lambda_n x), \qquad \|y_n\|^2 = \frac{k_2}{2\lambda_n^2} \frac{\lambda_n^2 + k_1^2}{\lambda_n^2 + k_2^2} + \frac{k_1}{2\lambda_n^2} + \frac{l}{2}\left(1 + \frac{k_1^2}{\lambda_n^2}\right).$$

Here the λ_n are positive roots of the transcendental equation $\dfrac{\tan(\lambda l)}{\lambda} = \dfrac{k_1 + k_2}{\lambda^2 - k_1 k_2}$.

5.2.5. Wave Equation with Axial Symmetry
$$u_{tt} = a^2(u_{rr} + r^{-1} u_r) + \Phi(r, t)$$

This is the *two-dimensional nonhomogeneous wave equation with axial symmetry*, where $r = \sqrt{x^2 + y^2}$ is the radial coordinate.

▶ **Solutions of boundary value problems in terms of the Green's function.**

We consider boundary value problems for the nonhomogeneous wave equation with axial symmetry in the domain $0 \leq r \leq R$ with the general initial conditions
$$u = f(r) \quad \text{at} \quad t = 0, \qquad u_t = g(r) \quad \text{at} \quad t = 0,$$

and various homogeneous boundary conditions at $r = R$ (the solutions bounded at $r = 0$ are sought). The solution of these problems can be represented in terms of the Green's function as

$$u = \frac{\partial}{\partial t}\int_0^R f(\xi)G(r,\xi,t)\,d\xi + \int_0^R g(\xi)G(r,\xi,t)\,d\xi + \int_0^t\int_0^R \Phi(\xi,\tau)G(r,\xi,t-\tau)\,d\xi\,d\tau.$$

Below are Green's functions for the wave equation with axial symmetry for three types of homogeneous boundary conditions.

▶ **Domain: $0 \leq r \leq R$. First boundary value problem.**

A boundary condition is prescribed:
$$u = 0 \quad \text{at} \quad r = R.$$

Green's function:
$$G(r,\xi,t) = \frac{2\xi}{aR}\sum_{n=1}^{\infty}\frac{1}{\lambda_n J_1^2(\lambda_n)}J_0\left(\frac{\lambda_n r}{R}\right)J_0\left(\frac{\lambda_n \xi}{R}\right)\sin\left(\frac{\lambda_n at}{R}\right),$$

where λ_n are positive zeros of the Bessel function, $J_0(\lambda) = 0$. The numerical values of the first ten λ_n are specified in Section 5.1.5 (see the paragraph "Domain: $0 \leq r \leq R$. First boundary value problem.").

▶ **Domain: $0 \leq r \leq R$. Second boundary value problem.**

A boundary condition is prescribed:
$$u_r = 0 \quad \text{at} \quad r = R.$$

Green's function:
$$G(r,\xi,t) = \frac{2t\xi}{R^2} + \frac{2\xi}{aR}\sum_{n=1}^{\infty}\frac{1}{\lambda_n J_0^2(\lambda_n)}J_0\left(\frac{\lambda_n r}{R}\right)J_0\left(\frac{\lambda_n \xi}{R}\right)\sin\left(\frac{\lambda_n at}{R}\right),$$

where λ_n are positive zeros of the first-order Bessel function, $J_1(\lambda) = 0$. The numerical values of the first ten roots λ_n are specified in Section 5.1.5 (see the paragraph "Domain: $0 \leq r \leq R$. Second boundary value problem.").

▶ **Domain: $0 \leq r \leq R$. Third boundary value problem.**

A boundary condition is prescribed:
$$u_r + ku = 0 \quad \text{at} \quad r = R.$$

Green's function:
$$G(r,\xi,t) = \frac{2\xi}{aR}\sum_{n=1}^{\infty}\frac{\lambda_n}{(k^2R^2 + \lambda_n^2)J_0^2(\lambda_n)}J_0\left(\frac{\lambda_n r}{R}\right)J_0\left(\frac{\lambda_n \xi}{R}\right)\sin\left(\frac{\lambda_n at}{R}\right),$$

where the λ_n are positive roots of the transcendental equation
$$\lambda J_1(\lambda) - kRJ_0(\lambda) = 0.$$

5.2.6. Wave Equation with Central Symmetry
$$u_{tt} = a^2(u_{rr} + 2r^{-1}u_r) + \Phi(r,t)$$

This is the *equation of vibration of a gas with central symmetry*, where $r = \sqrt{x^2 + y^2 + z^2}$ is the radial coordinate.

▶ **General solution for $\Phi(r,t) \equiv 0$:**
$$u(t,r) = \frac{\varphi(r+at) + \psi(r-at)}{r},$$
where $\varphi(r_1)$ and $\psi(r_2)$ are arbitrary functions.

▶ **Reduction to a constant coefficient wave equation.**

The substitution $v(r,t) = ru(r,t)$ leads to the nonhomogeneous constant coefficient wave equation
$$v_{tt} = a^2 v_{rr} + r\Phi(r,t),$$
which is discussed in Section 5.2.2.

▶ **Domain: $0 \leq r < \infty$. Cauchy problem.**

Initial conditions are prescribed:
$$u = f(r) \quad \text{at} \quad t=0, \qquad u_t = g(r) \quad \text{at} \quad t=0, \tag{1}$$

Solution
$$u = \frac{1}{2r}\big[(r-at)f(|r-at|) + (r+at)f(|r+at|)\big] + \frac{1}{2ar}\int_{r-at}^{r+at} \xi g(|\xi|)\,d\xi$$
$$+ \frac{1}{2ar}\int_0^t d\tau \int_{r-a(t-\tau)}^{r+a(t-\tau)} \xi \Phi(|\xi|,\tau)\,d\xi.$$

▶ **Solutions of boundary value problems in terms of the Green's function.**

We consider boundary value problems for the nonhomogeneous wave equation with central symmetry in the domain $0 \leq r \leq R$ with the general initial conditions (1) and various homogeneous boundary conditions at $r = R$ (we look for solutions bounded at $r = 0$). The solution of these problems can be represented in terms of the Green's function as
$$u = \frac{\partial}{\partial t}\int_0^R f(\xi)G(r,\xi,t)\,d\xi + \int_0^R g(\xi)G(r,\xi,t)\,d\xi + \int_0^t\int_0^R \Phi(\xi,\tau)G(r,\xi,t-\tau)\,d\xi\,d\tau.$$

Below are the Green's functions for the wave equation with central symmetry for three types of homogeneous boundary conditions.

▶ **Domain: $0 \leq r \leq R$. First boundary value problem.**

A boundary condition is prescribed:
$$u = 0 \quad \text{at} \quad r = R.$$

Green's function:
$$G(r,\xi,t) = \frac{2\xi}{\pi a r} \sum_{n=1}^{\infty} \frac{1}{n} \sin\left(\frac{n\pi r}{R}\right) \sin\left(\frac{n\pi \xi}{R}\right) \sin\left(\frac{a n \pi t}{R}\right).$$

▶ **Domain: $0 \leq r \leq R$. Second boundary value problem.**

A boundary condition is prescribed:
$$u_r = 0 \quad \text{at} \quad r = R.$$

Green's function:
$$G(r,\xi,t) = \frac{3t\xi^2}{R^3} + \frac{2\xi}{ar} \sum_{n=1}^{\infty} \frac{\mu_n^2 + 1}{\mu_n^3} \sin\left(\frac{\mu_n r}{R}\right) \sin\left(\frac{\mu_n \xi}{R}\right) \sin\left(\frac{\mu_n a t}{R}\right),$$

where μ_n are positive roots of the transcendental equation $\tan \mu - \mu = 0$. The first five roots are

$$\mu_1 = 4.4934, \quad \mu_2 = 7.7253, \quad \mu_3 = 10.9041, \quad \mu_4 = 14.0662, \quad \mu_5 = 17.2208.$$

▶ **Domain: $0 \leq r \leq R$. Third boundary value problem.**

A boundary condition is prescribed:
$$u_r + ku = 0 \quad \text{at} \quad r = R.$$

Green's function:
$$G(r,\xi,t) = \frac{2\xi}{ar} \sum_{n=1}^{\infty} \frac{\mu_n^2 + (kR-1)^2}{\mu_n[\mu_n^2 + kR(kR-1)]} \sin\left(\frac{\mu_n r}{R}\right) \sin\left(\frac{\mu_n \xi}{R}\right) \sin\left(\frac{\mu_n a t}{R}\right).$$

Here the μ_n are positive roots of the transcendental equation
$$\mu \cot \mu + kR - 1 = 0.$$

5.2.7. Equations of the Form $s(x)u_{tt} = [p(x)u_x]_x - q(x)u + \Phi(x,t)$

It is assumed that the functions s, p, p'_x, and q are continuous and the inequalities $s > 0$, $p > 0$ hold for $x_1 \leq x \leq x_2$.

The solution of the equation in question under the general initial conditions
$$u = f_0(x) \quad \text{at} \quad t = 0, \qquad u_t = f_1(x) \quad \text{at} \quad t = 0$$

and the arbitrary linear nonhomogeneous boundary conditions
$$a_1 u_x + b_1 u = g_1(t) \quad \text{at} \quad x = x_1,$$
$$a_2 u_x + b_2 u = g_2(t) \quad \text{at} \quad x = x_2$$

can be represented as the sum

$$u(x,t) = \int_0^t \int_{x_1}^{x_2} \Phi(\xi,\tau)\mathcal{G}(x,\xi,t-\tau)\,d\xi\,d\tau$$
$$+ \frac{\partial}{\partial t}\int_{x_1}^{x_2} s(\xi)f_0(\xi)\mathcal{G}(x,\xi,t)\,d\xi + \int_{x_1}^{x_2} s(\xi)f_1(\xi)\mathcal{G}(x,\xi,t)\,d\xi$$
$$+ p(x_1)\int_0^t g_1(\tau)\Lambda_1(x,t-\tau)\,d\tau + p(x_2)\int_0^t g_2(\tau)\Lambda_2(x,t-\tau)\,d\tau. \quad (1)$$

Here, the modified Green's function is determined by

$$\mathcal{G}(x,\xi,t) = \sum_{n=1}^{\infty} \frac{y_n(x)y_n(\xi)\sin(t\sqrt{\lambda_n})}{\|y_n\|^2\sqrt{\lambda_n}}, \qquad \|y_n\|^2 = \int_{x_1}^{x_2} s(x)y_n^2(x)\,dx, \quad (2)$$

where the λ_n and $y_n(x)$ are the eigenvalues and corresponding eigenfunctions of the Sturm–Liouville problem for the second-order linear ODE

$$[p(x)y_x']_x' + [\lambda s(x) - q(x)]y = 0,$$
$$a_1 y_x' + b_1 y = 0 \quad \text{at} \quad x = x_1, \quad (3)$$
$$a_2 y_x' + b_2 y = 0 \quad \text{at} \quad x = x_2.$$

The functions $\Lambda_1(x,t)$ and $\Lambda_2(x,t)$ that occur in the integrands of the last two terms in solution (1) are expressed in terms of the Green's function of (2). For the corresponding formulas for specific boundary value problems, see Table 5.2.

General properties of the Sturm–Liouville problem (3) are described in Section 5.1.10, where exactly the same eigenvalue problem was considered.

5.2.8. Telegraph Type Equations
$$u_{tt} + ku_t = a^2 u_{xx} + bu_x + cu + \Phi(x,t)$$

1. $u_{tt} + ku_t = a^2 u_{xx} + bu.$

For $k > 0$ and $b < 0$, it is a *telegraph equation*.

The substitution $u = \exp\bigl(-\tfrac{1}{2}kt\bigr)v(x,t)$ leads to the Klein–Gordon equation

$$v_{tt} = a^2 v_{xx} + \bigl(b + \tfrac{1}{4}k^2\bigr)v,$$

which is discussed in Section 5.2.3.

2. $u_{tt} + ku_t = a^2 u_{xx} + bu_x + cu + \Phi(x,t).$

The substitution

$$u = \exp\bigl(-\tfrac{1}{2}a^{-2}bx - \tfrac{1}{2}kt\bigr)v(x,t)$$

leads to the equation

$$v_{tt} = a^2 v_{xx} + \bigl(c + \tfrac{1}{4}k^2 - \tfrac{1}{4}a^{-2}b^2\bigr)v + \exp\bigl(\tfrac{1}{2}a^{-2}bx + \tfrac{1}{2}kt\bigr)\Phi(x,t),$$

which is discussed in Section 5.2.4.

5.3. Elliptic Equations

5.3.1. Laplace Equation $\Delta u = 0$

The Laplace equation is often encountered in heat and mass transfer theory, fluid mechanics, elasticity, electrostatics, and other areas of mechanics and physics.

The two-dimensional Laplace equation has the following form:

$$u_{xx} + u_{yy} = 0 \quad \text{in the Cartesian coordinates,}$$
$$r^{-1}(ru_r)_r + r^{-2}u_{\varphi\varphi} = 0 \quad \text{in the polar coordinates,}$$

where $x = r\cos\varphi$, $y = r\sin\varphi$, and $r = \sqrt{x^2 + y^2}$.

▶ **Particular solutions.**

$1°$. Particular solutions in the Cartesian coordinates:

$$u = Ax + By + C,$$
$$u = A(x^2 - y^2) + Bxy,$$
$$u = A(x^3 - 3xy^2) + B(3x^2y - y^3),$$
$$u = \frac{Ax + By}{x^2 + y^2} + C,$$
$$u = \exp(\pm\mu x)(A\cos\mu y + B\sin\mu y),$$
$$u = (A\cos\mu x + B\sin\mu x)\exp(\pm\mu y),$$
$$u = (A\sinh\mu x + B\cosh\mu x)(C\cos\mu y + D\sin\mu y),$$
$$u = (A\cos\mu x + B\sin\mu x)(C\sinh\mu y + D\cosh\mu y),$$

where A, B, C, D, and μ are arbitrary constants.

$2°$. Particular solutions in the polar coordinates:

$$u = A\ln r + B,$$
$$u = (Ar^m + Br^{-m})(C\cos m\varphi + D\sin m\varphi),$$

where A, B, C, and D are arbitrary constants, and $m = 1, 2, \ldots$

▶ **Formulas allowing the construction of particular solutions.**

If $u(x, y)$ is a solution of the Laplace equation, then the functions

$$u_1 = Au(\pm\lambda x + C_1, \pm\lambda y + C_2) + B,$$
$$u_2 = Au(x\cos\beta + y\sin\beta, -x\sin\beta + y\cos\beta),$$
$$u_3 = Au\left(\frac{x}{x^2 + y^2}, \frac{y}{x^2 + y^2}\right)$$

are also solutions everywhere they are defined; A, B, C_1, C_2, β, and λ are arbitrary constants. The signs at λ's in the formula for u_1 are taken arbitrarily.

▶ **Method for constructing particular solutions.**

A fairly general method for constructing particular solutions involves the following. Let $f(z) = u(x,y) + iv(x,y)$ be any analytic function of the complex variable $z = x + iy$ (u and v are real functions of the real variables x and y; $i^2 = -1$). Then the real and imaginary parts of f both satisfy the two-dimensional Laplace equation:

$$\Delta u = 0, \quad \Delta v = 0.$$

Thus, by specifying analytic functions $f(z)$ and taking their real and imaginary parts, one obtains various solutions of the two-dimensional Laplace equation.

Below are solutions of boundary value problems for the two-dimensional Laplace equation with homogeneous boundary conditions of various types.

Remark 5.1. *The first boundary value problem for the Laplace equation and other elliptic equations is often called the Dirichlet problem, and the second boundary value problem is called the Neumann problem.*

▶ **Domain:** $-\infty < x < \infty, 0 \leq y < \infty$. **First boundary value problem.**

A half-plane is considered. A boundary condition is prescribed:

$$u = f(x) \quad \text{at} \quad y = 0.$$

Solution:

$$u(x,y) = \frac{1}{\pi}\int_{-\infty}^{\infty} \frac{y f(\xi)\, d\xi}{(x-\xi)^2 + y^2} = \frac{1}{\pi}\int_{-\pi/2}^{\pi/2} f(x + y \tan\theta)\, d\theta.$$

▶ **Domain:** $-\infty < x < \infty, 0 \leq y < \infty$. **Second boundary value problem.**

A half-plane is considered. A boundary condition is prescribed:

$$u_y = f(x) \quad \text{at} \quad y = 0.$$

Solution:

$$u(x,y) = \frac{1}{\pi}\int_{-\infty}^{\infty} f(\xi) \ln\sqrt{(x-\xi)^2 + y^2}\, d\xi + C,$$

where C is an arbitrary constant.

▶ **Domain:** $0 \leq x < \infty, 0 \leq y < \infty$. **First boundary value problem.**

A quadrant of the plane is considered. Boundary conditions are prescribed:

$$u = f_1(y) \quad \text{at} \quad x = 0, \qquad u = f_2(x) \quad \text{at} \quad y = 0.$$

Solution:

$$u(x,y) = \frac{4}{\pi}xy \int_0^\infty \frac{f_1(\eta)\eta\, d\eta}{[x^2 + (y-\eta)^2][x^2 + (y+\eta)^2]}$$
$$+ \frac{4}{\pi}xy \int_0^\infty \frac{f_2(\xi)\xi\, d\xi}{[(x-\xi)^2 + y^2][(x+\xi)^2 + y^2]}.$$

▶ **Domain:** $-\infty < x < \infty$, $0 \leq y \leq a$. **First boundary value problem.**

An infinite strip is considered. Boundary conditions are prescribed:

$$u = f_1(x) \quad \text{at} \quad y = 0, \qquad u = f_2(x) \quad \text{at} \quad y = a.$$

Solution:

$$u(x,y) = \frac{1}{2a}\sin\left(\frac{\pi y}{a}\right)\int_{-\infty}^{\infty}\frac{f_1(\xi)\,d\xi}{\cosh[\pi(x-\xi)/a] - \cos(\pi y/a)}$$
$$+ \frac{1}{2a}\sin\left(\frac{\pi y}{a}\right)\int_{-\infty}^{\infty}\frac{f_2(\xi)\,d\xi}{\cosh[\pi(x-\xi)/a] + \cos(\pi y/a)}.$$

▶ **Domain:** $-\infty < x < \infty$, $0 \leq y \leq a$. **Second boundary value problem.**

An infinite strip is considered. Boundary conditions are prescribed:

$$u_y = f_1(x) \quad \text{at} \quad y = 0, \qquad u_y = f_2(x) \quad \text{at} \quad y = a.$$

Solution:

$$u(x,y) = \frac{1}{2\pi}\int_{-\infty}^{\infty} f_1(\xi)\ln\{\cosh[\pi(x-\xi)/a] - \cos(\pi y/a)\}\,d\xi$$
$$- \frac{1}{2\pi}\int_{-\infty}^{\infty} f_2(\xi)\ln\{\cosh[\pi(x-\xi)/a] + \cos(\pi y/a)\}\,d\xi + C,$$

where C is an arbitrary constant.

▶ **Domain:** $0 \leq x \leq a$, $0 \leq y \leq b$. **First boundary value problem.**

A rectangle is considered. Boundary conditions are prescribed:

$$u = f_1(y) \quad \text{at} \quad x = 0, \qquad u = f_2(y) \quad \text{at} \quad x = a,$$
$$u = f_3(x) \quad \text{at} \quad y = 0, \qquad u = f_4(x) \quad \text{at} \quad y = b.$$

Solution:

$$u(x,y) = \sum_{n=1}^{\infty} A_n \sinh\left[\frac{n\pi}{b}(a-x)\right]\sin\left(\frac{n\pi}{b}y\right) + \sum_{n=1}^{\infty} B_n \sinh\left(\frac{n\pi}{b}x\right)\sin\left(\frac{n\pi}{b}y\right)$$
$$+ \sum_{n=1}^{\infty} C_n \sin\left(\frac{n\pi}{a}x\right)\sinh\left[\frac{n\pi}{a}(b-y)\right] + \sum_{n=1}^{\infty} D_n \sin\left(\frac{n\pi}{a}x\right)\sinh\left(\frac{n\pi}{a}y\right),$$

where the coefficients A_n, B_n, C_n, and D_n are expressed as

$$A_n = \frac{2}{\lambda_n}\int_0^b f_1(\xi)\sin\left(\frac{n\pi\xi}{b}\right)d\xi, \qquad B_n = \frac{2}{\lambda_n}\int_0^b f_2(\xi)\sin\left(\frac{n\pi\xi}{b}\right)d\xi,$$
$$C_n = \frac{2}{\mu_n}\int_0^a f_3(\xi)\sin\left(\frac{n\pi\xi}{a}\right)d\xi, \qquad D_n = \frac{2}{\mu_n}\int_0^a f_4(\xi)\sin\left(\frac{n\pi\xi}{a}\right)d\xi,$$
$$\lambda_n = b\sinh\left(\frac{n\pi a}{b}\right), \qquad \mu_n = a\sinh\left(\frac{n\pi b}{a}\right).$$

▶ **Domain:** $0 \leq r \leq R$. **First boundary value problem.**

A circle is considered. A boundary condition is prescribed:

$$u = f(\varphi) \quad \text{at} \quad r = R.$$

Solution in the polar coordinates:

$$u(r, \varphi) = \frac{1}{2\pi} \int_0^{2\pi} f(\psi) \frac{R^2 - r^2}{r^2 - 2Rr\cos(\varphi - \psi) + R^2} \, d\psi.$$

This formula is conventionally referred to as the *Poisson integral*.

▶ **Domain:** $0 \leq r \leq R$. **Second boundary value problem.**

A circle is considered. A boundary condition is prescribed:

$$u_r = f(\varphi) \quad \text{at} \quad r = R,$$

where the function $f(\varphi)$ must satisfy the solvability condition $\int_0^{2\pi} f(\varphi) \, d\varphi = 0$.

Solution in the polar coordinates:

$$u(r, \varphi) = \frac{R}{2\pi} \int_0^{2\pi} f(\psi) \ln \frac{r^2 - 2Rr\cos(\varphi - \psi) + R^2}{R^2} \, d\psi + C,$$

where C is an arbitrary constant; this formula is known as the *Dini integral*.

5.3.2. Poisson Equation $\Delta u + \Phi(x, y) = 0$

The two-dimensional Poisson equation has the following form:

$$u_{xx} + u_{yy} + \Phi(x, y) = 0 \quad \text{in the Cartesian coordinates,}$$
$$r^{-1}(ru_r)_r + r^{-2}u_{\varphi\varphi} + \Phi(r, \varphi) = 0 \quad \text{in the polar coordinates.}$$

Below are solutions of the two-dimensional Poisson equation for various types of homogeneous boundary conditions.

▶ **Domain:** $-\infty < x < \infty$, $-\infty < y < \infty$.

Solution:

$$u(x, y) = \frac{1}{2\pi} \int_{-\infty}^{\infty} \int_{-\infty}^{\infty} \Phi(\xi, \eta) \ln \frac{1}{\sqrt{(x-\xi)^2 + (y-\eta)^2}} \, d\xi \, d\eta.$$

▶ **Domain:** $-\infty < x < \infty$, $0 \leq y < \infty$. **First boundary value problem.**

A half-plane is considered. A boundary condition is prescribed:

$$u = f(x) \quad \text{at} \quad y = 0.$$

Solution:

$$u(x, y) = \frac{1}{\pi} \int_{-\infty}^{\infty} \frac{y f(\xi) \, d\xi}{(x-\xi)^2 + y^2} + \frac{1}{2\pi} \int_0^{\infty} \int_{-\infty}^{\infty} \Phi(\xi, \eta) \ln \frac{\sqrt{(x-\xi)^2 + (y+\eta)^2}}{\sqrt{(x-\xi)^2 + (y-\eta)^2}} \, d\xi \, d\eta.$$

▶ **Domain: $0 \leq x < \infty$, $0 \leq y < \infty$. First boundary value problem.**

A quadrant of the plane is considered. Boundary conditions are prescribed:
$$u = f_1(y) \quad \text{at} \quad x = 0, \qquad u = f_2(x) \quad \text{at} \quad y = 0.$$

Solution:
$$u(x,y) = \frac{4}{\pi} xy \int_0^\infty \frac{f_1(\eta)\eta \, d\eta}{[x^2+(y-\eta)^2][x^2+(y+\eta)^2]} + \frac{4}{\pi} xy \int_0^\infty \frac{f_2(\xi)\xi \, d\xi}{[(x-\xi)^2+y^2][(x+\xi)^2+y^2]}$$
$$+ \frac{1}{2\pi} \int_0^\infty \int_0^\infty \Phi(\xi,\eta) \ln \frac{\sqrt{(x-\xi)^2+(y+\eta)^2}\sqrt{(x+\xi)^2+(y-\eta)^2}}{\sqrt{(x-\xi)^2+(y-\eta)^2}\sqrt{(x+\xi)^2+(y+\eta)^2}} \, d\xi \, d\eta.$$

▶ **Domain: $0 \leq x \leq a$, $0 \leq y \leq b$. First boundary value problem.**

A rectangle is considered. Boundary conditions are prescribed:
$$u = f_1(y) \quad \text{at} \quad x = 0, \qquad u = f_2(y) \quad \text{at} \quad x = a,$$
$$u = f_3(x) \quad \text{at} \quad y = 0, \qquad u = f_4(x) \quad \text{at} \quad y = b.$$

Solution:
$$u(x,y) = \int_0^a \int_0^b \Phi(\xi,\eta) G(x,y,\xi,\eta) \, d\eta \, d\xi$$
$$+ \int_0^b f_1(\eta) \left[\frac{\partial}{\partial \xi} G(x,y,\xi,\eta)\right]_{\xi=0} d\eta - \int_0^b f_2(\eta) \left[\frac{\partial}{\partial \xi} G(x,y,\xi,\eta)\right]_{\xi=a} d\eta$$
$$+ \int_0^a f_3(\xi) \left[\frac{\partial}{\partial \eta} G(x,y,\xi,\eta)\right]_{\eta=0} d\xi - \int_0^a f_4(\xi) \left[\frac{\partial}{\partial \eta} G(x,y,\xi,\eta)\right]_{\eta=b} d\xi.$$

Two forms of representation of the Green's function:
$$G(x,y,\xi,\eta) = \frac{2}{a} \sum_{n=1}^\infty \frac{\sin(p_n x)\sin(p_n \xi)}{p_n \sinh(p_n b)} H_n(y,\eta) = \frac{2}{b} \sum_{m=1}^\infty \frac{\sin(q_m y)\sin(q_m \eta)}{q_m \sinh(q_m a)} Q_m(x,\xi),$$

where
$$p_n = \frac{\pi n}{a}, \quad H_n(y,\eta) = \begin{cases} \sinh(p_n \eta) \sinh[p_n(b-y)] & \text{for } b \geq y > \eta \geq 0, \\ \sinh(p_n y) \sinh[p_n(b-\eta)] & \text{for } b \geq \eta > y \geq 0; \end{cases}$$
$$q_m = \frac{\pi m}{b}, \quad Q_m(x,\xi) = \begin{cases} \sinh(q_m \xi) \sinh[q_m(a-x)] & \text{for } a \geq x > \xi \geq 0, \\ \sinh(q_m x) \sinh[q_m(a-\xi)] & \text{for } a \geq \xi > x \geq 0. \end{cases}$$

▶ **Domain: $0 \leq r \leq R$, $0 \leq \varphi \leq 2\pi$. First boundary value problem.**

A circle is considered. A boundary condition is prescribed:
$$u = f(\varphi) \quad \text{at} \quad r = R.$$

Solution in the polar coordinates:
$$u(r,\varphi) = \frac{1}{2\pi} \int_0^{2\pi} f(\eta) \frac{R^2-r^2}{r^2-2Rr\cos(\varphi-\eta)+R^2} \, d\eta + \int_0^{2\pi} \int_0^R \Phi(\xi,\eta) G(r,\varphi,\xi,\eta) \xi \, d\xi \, d\eta,$$

where
$$G(r,\varphi,\xi,\eta) = \frac{1}{4\pi} \ln \frac{r^2 \xi^2 - 2R^2 r\xi \cos(\varphi-\eta) + R^4}{R^2[r^2 - 2r\xi \cos(\varphi-\eta) + \xi^2]}.$$

5.3.3. Helmholtz Equation $\Delta u + \lambda u = -\Phi(x, y)$

Many problems related to steady-state oscillations (mechanical, acoustical, thermal, electromagnetic) lead to the two-dimensional Helmholtz equation. For $\lambda < 0$, this equation describes mass transfer processes with volume chemical reactions of the first order.

The two-dimensional Helmholtz equation has the following form:

$$u_{xx} + u_{yy} + \lambda u = -\Phi(x, y) \quad \text{in the Cartesian coordinates,}$$
$$r^{-1}(r u_r)_r + r^{-2} u_{\varphi\varphi} + \lambda u = -\Phi(r, \varphi) \quad \text{in the polar coordinates.}$$

▶ **Particular solutions of the homogeneous Helmholtz equation with $\Phi \equiv 0$.**

1°. Particular solutions of the homogeneous equation in the Cartesian coordinates:

$$u = (Ax + B)(C \cos \mu y + D \sin \mu y), \quad \lambda = \mu^2,$$
$$u = (Ax + B)(C \cosh \mu y + D \sinh \mu y), \quad \lambda = -\mu^2,$$
$$u = (A \cos \mu x + B \sin \mu x)(Cy + D), \quad \lambda = \mu^2,$$
$$u = (A \cosh \mu x + B \sinh \mu x)(Cy + D), \quad \lambda = -\mu^2,$$
$$u = (A \cos \mu_1 x + B \sin \mu_1 x)(C \cos \mu_2 y + D \sin \mu_2 y), \quad \lambda = \mu_1^2 + \mu_2^2,$$
$$u = (A \cos \mu_1 x + B \sin \mu_1 x)(C \cosh \mu_2 y + D \sinh \mu_2 y), \quad \lambda = \mu_1^2 - \mu_2^2,$$
$$u = (A \cosh \mu_1 x + B \sinh \mu_1 x)(C \cos \mu_2 y + D \sin \mu_2 y), \quad \lambda = -\mu_1^2 + \mu_2^2,$$
$$u = (A \cosh \mu_1 x + B \sinh \mu_1 x)(C \cosh \mu_2 y + D \sinh \mu_2 y), \quad \lambda = -\mu_1^2 - \mu_2^2,$$

where A, B, C, and D are arbitrary constants.

2°. Particular solutions of the homogeneous equation in the polar coordinates:

$$u = [A J_0(\mu r) + B Y_0(\mu r)](C \varphi + D), \quad \lambda = \mu^2,$$
$$u = [A I_0(\mu r) + B K_0(\mu r)](C \varphi + D), \quad \lambda = -\mu^2,$$
$$u = [A J_m(\mu r) + B Y_m(\mu r)](C \cos m\varphi + D \sin m\varphi), \quad \lambda = \mu^2,$$
$$u = [A I_m(\mu r) + B K_m(\mu r)](C \cos m\varphi + D \sin m\varphi), \quad \lambda = -\mu^2,$$

where $m = 1, 2, \ldots$; A, B, C, D are arbitrary constants; $J_m(\mu)$ and $Y_m(\mu)$ are Bessel functions; and $I_m(\mu)$ and $K_m(\mu)$ are modified Bessel functions.

▶ **Formulas allowing the construction of particular solutions.**

Suppose $u = u(x, y)$ is a solution of the homogeneous Helmholtz equation. Then the functions
$$u_1 = u(\pm x + C_1, \pm y + C_2),$$
$$u_2 = u(x \cos \theta + y \sin \theta + C_1, -x \sin \theta + y \cos \theta + C_2),$$

where C_1, C_2, and θ are arbitrary constants, are also solutions of the equation. The signs in the formula for u_1 are taken arbitrarily.

Below are solutions of the two-dimensional Helmholtz equation for various types of homogeneous boundary conditions.

▶ **Domain:** $-\infty < x < \infty$, $-\infty < y < \infty$.

1°. Solution for $\lambda = -s^2 < 0$:

$$u(x,y) = \frac{1}{2\pi} \int_{-\infty}^{\infty} \int_{-\infty}^{\infty} \Phi(\xi,\eta) K_0(s\varrho) \, d\xi \, d\eta, \qquad \varrho = \sqrt{(x-\xi)^2 + (y-\eta)^2}.$$

Here and in what follows, $K_0(z) = \int_0^{\infty} \frac{\cos(zt)}{\sqrt{1+t^2}} \, dt$ is the modified Bessel function of the second kind.

2°. Solution for $\lambda = k^2 > 0$:

$$u(x,y) = -\frac{i}{4} \int_{-\infty}^{\infty} \int_{-\infty}^{\infty} \Phi(\xi,\eta) H_0^{(2)}(k\varrho) \, d\xi \, d\eta, \qquad \varrho = \sqrt{(x-\xi)^2 + (y-\eta)^2}.$$

Here and in what follows, $H_0^{(2)}(z) = J_0(z) - iY_0(z)$ is the Hankel function of the second kind of order 0, where $J_0(x)$ and $Y_0(z)$ are the Bessel functions, and $i^2 = -1$.

Remark 5.2. *The radiation Sommerfeld conditions at infinity were used to obtain the solution with $\lambda > 0$; see Tikhonov & Samarskii (1990) and Polyanin & Nazaikinskii (2016).*

▶ **Domain:** $-\infty < x < \infty$, $0 \leq y < \infty$. **First boundary value problem.**

A half-plane is considered. A boundary condition is prescribed:

$$u = f(x) \quad \text{at} \quad y = 0.$$

Solution:

$$u(x,y) = \int_{-\infty}^{\infty} f(\xi) \left[\frac{\partial}{\partial \eta} G(x,y,\xi,\eta) \right]_{\eta=0} d\xi + \int_0^{\infty} \int_{-\infty}^{\infty} \Phi(\xi,\eta) G(x,y,\xi,\eta) \, d\xi \, d\eta.$$

1°. The Green's function for $\lambda = -s^2 < 0$:

$$G(x,y,\xi,\eta) = \frac{1}{2\pi} \big[K_0(s\varrho_1) - K_0(s\varrho_2) \big],$$

$$\varrho_1 = \sqrt{(x-\xi)^2 + (y-\eta)^2}, \qquad \varrho_2 = \sqrt{(x-\xi)^2 + (y+\eta)^2}.$$

2°. The Green's function for $\lambda = k^2 > 0$:

$$G(x,y,\xi,\eta) = -\frac{i}{4} \big[H_0^{(2)}(k\varrho_1) - H_0^{(2)}(k\varrho_2) \big].$$

Remark 5.3. *The Sommerfeld radiation conditions at infinity were used to obtain the solution with $\lambda > 0$; see Tikhonov & Samarskii (1990) and Polyanin & Nazaikinskii (2016).*

▶ **Domain:** $-\infty < x < \infty$, $0 \leq y < \infty$. **Second boundary value problem.**

A half-plane is considered. A boundary condition is prescribed:

$$u_y = f(x) \quad \text{at} \quad y = 0.$$

Solution:

$$u(x,y) = -\int_{-\infty}^{\infty} f(\xi) G(x,y,\xi,0) \, d\xi + \int_0^{\infty} \int_{-\infty}^{\infty} \Phi(\xi,\eta) G(x,y,\xi,\eta) \, d\xi \, d\eta.$$

1°. The Green's function for $\lambda = -s^2 < 0$:
$$G(x,y,\xi,\eta) = \frac{1}{2\pi}\big[K_0(s\varrho_1) + K_0(s\varrho_2)\big],$$
$$\varrho_1 = \sqrt{(x-\xi)^2 + (y-\eta)^2}, \quad \varrho_2 = \sqrt{(x-\xi)^2 + (y+\eta)^2}.$$

2°. The Green's function for $\lambda = k^2 > 0$:
$$G(x,y,\xi,\eta) = -\frac{i}{4}\big[H_0^{(2)}(k\varrho_1) + H_0^{(2)}(k\varrho_2)\big].$$

Remark 5.4. *The Sommerfeld radiation conditions at infinity were used to obtain the solution with $\lambda > 0$; see Tikhonov & Samarskii (1990) and Polyanin & Nazaikinskii (2016).*

▶ **Domain: $0 \le x < \infty$, $0 \le y < \infty$. First boundary value problem.**

A quadrant of the plane is considered. Boundary conditions are prescribed:
$$u = f_1(y) \quad \text{at} \quad x = 0, \qquad u = f_2(x) \quad \text{at} \quad y = 0.$$

Solution:
$$u(x,y) = \int_0^\infty f_1(\eta)\left[\frac{\partial}{\partial \xi}G(x,y,\xi,\eta)\right]_{\xi=0} d\eta + \int_0^\infty f_2(\xi)\left[\frac{\partial}{\partial \eta}G(x,y,\xi,\eta)\right]_{\eta=0} d\xi$$
$$+ \int_0^\infty \int_0^\infty \Phi(\xi,\eta) G(x,y,\xi,\eta)\, d\xi\, d\eta.$$

1°. The Green's function for $\lambda = -s^2 < 0$:
$$G(x,y,\xi,\eta) = \frac{1}{2\pi}\big[K_0(s\varrho_1) - K_0(s\varrho_2) - K_0(s\varrho_3) + K_0(s\varrho_4)\big],$$
$$\varrho_1 = \sqrt{(x-\xi)^2 + (y-\eta)^2}, \quad \varrho_2 = \sqrt{(x-\xi)^2 + (y+\eta)^2},$$
$$\varrho_3 = \sqrt{(x+\xi)^2 + (y-\eta)^2}, \quad \varrho_4 = \sqrt{(x+\xi)^2 + (y+\eta)^2}.$$

2°. The Green's function for $\lambda = k^2 > 0$:
$$G(x,y,\xi,\eta) = -\frac{i}{4}\big[H_0^{(2)}(k\varrho_1) - H_0^{(2)}(k\varrho_2) - H_0^{(2)}(k\varrho_3) + H_0^{(2)}(k\varrho_4)\big].$$

▶ **Domain: $0 \le x < \infty$, $0 \le y < \infty$. Second boundary value problem.**

A quadrant of the plane is considered. Boundary conditions are prescribed:
$$u_x = f_1(y) \quad \text{at} \quad x = 0, \qquad u_y = f_2(x) \quad \text{at} \quad y = 0.$$

Solution:
$$u(x,y) = -\int_0^\infty f_1(\eta) G(x,y,0,\eta)\, d\eta - \int_0^\infty f_2(\xi) G(x,y,\xi,0)\, d\xi$$
$$+ \int_0^\infty \int_0^\infty \Phi(\xi,\eta) G(x,y,\xi,\eta)\, d\xi\, d\eta.$$

1°. The Green's function for $\lambda = -s^2 < 0$:

$$G(x,y,\xi,\eta) = \frac{1}{2\pi}\big[K_0(s\varrho_1) + K_0(s\varrho_2) + K_0(s\varrho_3) + K_0(s\varrho_4)\big],$$

$$\varrho_1 = \sqrt{(x-\xi)^2 + (y-\eta)^2}, \quad \varrho_2 = \sqrt{(x-\xi)^2 + (y+\eta)^2},$$
$$\varrho_3 = \sqrt{(x+\xi)^2 + (y-\eta)^2}, \quad \varrho_4 = \sqrt{(x+\xi)^2 + (y+\eta)^2}.$$

2°. The Green's function for $\lambda = k^2 > 0$:

$$G(x,y,\xi,\eta) = -\frac{i}{4}\big[H_0^{(2)}(k\varrho_1) + H_0^{(2)}(k\varrho_2) + H_0^{(2)}(k\varrho_3) + H_0^{(2)}(k\varrho_4)\big].$$

▶ **Domain: $0 \leq x \leq a$, $0 \leq y \leq b$. First boundary value problem.**

A rectangle is considered. Boundary conditions are prescribed:

$$u = f_1(y) \text{ at } x = 0, \quad u = f_2(y) \text{ at } x = a,$$
$$u = f_3(x) \text{ at } y = 0, \quad u = f_4(x) \text{ at } y = b.$$

1°. Eigenvalues of the homogeneous problem with $\Phi \equiv 0$ (it is convenient to label them with a double subscript):

$$\lambda_{nm} = \pi^2\left(\frac{n^2}{a^2} + \frac{m^2}{b^2}\right); \quad n = 1, 2, \ldots; \quad m = 1, 2, \ldots$$

Eigenfunctions and the norm squared:

$$u_{nm} = \sin\left(\frac{n\pi x}{a}\right)\sin\left(\frac{m\pi y}{b}\right), \quad \|u_{nm}\|^2 = \frac{ab}{4}.$$

2°. Solution for $\lambda \neq \lambda_{nm}$:

$$u(x,y) = \int_0^a\int_0^b \Phi(\xi,\eta)G(x,y,\xi,\eta)\,d\eta\,d\xi$$
$$+ \int_0^b f_1(\eta)\left[\frac{\partial}{\partial\xi}G(x,y,\xi,\eta)\right]_{\xi=0}d\eta - \int_0^b f_2(\eta)\left[\frac{\partial}{\partial\xi}G(x,y,\xi,\eta)\right]_{\xi=a}d\eta$$
$$+ \int_0^a f_3(\xi)\left[\frac{\partial}{\partial\eta}G(x,y,\xi,\eta)\right]_{\eta=0}d\xi - \int_0^a f_4(\xi)\left[\frac{\partial}{\partial\eta}G(x,y,\xi,\eta)\right]_{\eta=b}d\xi.$$

Two forms of representation of the Green's function:

$$G(x,y,\xi,\eta) = \frac{2}{a}\sum_{n=1}^\infty \frac{\sin(p_n x)\sin(p_n \xi)}{\beta_n \sinh(\beta_n b)}H_n(y,\eta) = \frac{2}{b}\sum_{k=1}^\infty \frac{\sin(q_k y)\sin(q_k \eta)}{\mu_k \sinh(\mu_k a)}Q_k(x,\xi),$$

where

$$p_n = \frac{\pi n}{a}, \quad \beta_n = \sqrt{p_n^2 - \lambda}, \quad H_n(y,\eta) = \begin{cases}\sinh(\beta_n \eta)\sinh[\beta_n(b-y)] & \text{if } b \geq y > \eta \geq 0, \\ \sinh(\beta_n y)\sinh[\beta_n(b-\eta)] & \text{if } b \geq \eta > y \geq 0;\end{cases}$$

$$q_k = \frac{\pi k}{b}, \quad \mu_k = \sqrt{q_k^2 - \lambda}, \quad Q_k(x,\xi) = \begin{cases}\sinh(\mu_k \xi)\sinh[\mu_k(a-x)] & \text{if } a \geq x > \xi \geq 0, \\ \sinh(\mu_k x)\sinh[\mu_k(a-\xi)] & \text{if } a \geq \xi > x \geq 0.\end{cases}$$

▶ **Domain:** $0 \leq r \leq R$. **First boundary value problem.**

A circle is considered. A boundary condition is prescribed:
$$u = 0 \quad \text{at} \quad r = R.$$

Eigenvalues of the homogeneous boundary value problem with $\Phi \equiv 0$:
$$\lambda_{nm} = \frac{\mu_{nm}^2}{R^2}; \quad n = 0, 1, 2, \ldots; \; m = 1, 2, 3, \ldots$$

Here μ_{nm} are positive zeros of the Bessel functions, $J_n(\mu) = 0$.

Eigenfunctions:
$$u_{nm}^{(1)} = J_n(r\sqrt{\lambda_{nm}})\cos n\varphi, \quad u_{nm}^{(2)} = J_n(r\sqrt{\lambda_{nm}})\sin n\varphi.$$

Eigenfunctions possessing the axial symmetry property: $u_{0m}^{(1)} = J_0(r\sqrt{\lambda_{0m}})$.

▶ **Domain:** $0 \leq r \leq R$. **Second boundary value problem.**

A circle is considered. A boundary condition is prescribed:
$$u_r = 0 \quad \text{at} \quad r = R.$$

Eigenvalues of the homogeneous boundary value problem with $\Phi \equiv 0$:
$$\lambda_{nm} = \frac{\mu_{nm}^2}{R^2},$$

where μ_{nm} are roots of the transcendental equations $J_n'(\mu) = 0$.

Eigenfunctions:
$$u_{nm}^{(1)} = J_n(r\sqrt{\lambda_{nm}})\cos n\varphi, \quad u_{nm}^{(2)} = J_n(r\sqrt{\lambda_{nm}})\sin n\varphi.$$

Here, $n = 0, 1, 2, \ldots$; for $n \neq 0$, the parameter m assumes the values $m = 1, 2, 3, \ldots$; for $n = 0$, the root μ_{00} is zero (the corresponding eigenfunction is $u_{00} = 1$).

Eigenfunctions possessing the axial symmetry property: $u_{0m}^{(1)} = J_0(r\sqrt{\lambda_{0m}})$.

5.3.4. Convective Heat and Mass Transfer Equations

1. $u_{xx} + u_{yy} = au_x + bu_y + cu.$

This equation describes a stationary temperature field in a medium moving with a constant velocity, provided there is volume heat release (or absorption) proportional to temperature. For $b = c = 0$, it describes a temperature field in a continuous medium moving with a constant velocity along the x-axis (this occurs, for example, if a liquid-metal coolant flows past a flat plate or if a plate is in a seepage flow through a granular medium).

1°. The substitution
$$u(x, y) = \exp\left[\tfrac{1}{2}(ax + by)\right] w(x, y)$$

brings the original equation to the Helmholtz equation
$$w_{xx} + w_{yy} = \left(c + \tfrac{1}{4}a^2 + \tfrac{1}{4}b^2\right)w,$$

which is discussed in Section 5.3.3.

2°. Let $b = c = 0$. Consider the second boundary value problem in the upper half-plane ($-\infty < x < \infty$, $0 \leq y < \infty$). We assume that a thermal flux is prescribed on the surface of a plate of finite length and the medium has a constant temperature far away from the plate:

$$u_y = f(x) \quad \text{for} \quad y = 0, \ |x| < 1,$$
$$u_y = 0 \quad \text{for} \quad y = 0, \ |x| > 1,$$
$$u \to u_\infty \quad \text{as} \quad x^2 + y^2 \to \infty.$$

The solution of this problem in the Cartesian coordinates has the form

$$u(x,y) = u_\infty - \frac{1}{\pi} \int_{-1}^{1} f(\xi) \exp\left[\tfrac{1}{2} a(x - \xi)\right] K_0\left(\tfrac{1}{2} a \sqrt{(x-\xi)^2 + y^2}\right) d\xi,$$

where $K_0(z)$ is the modified Bessel function of the second kind.

2. $u_{xx} + u_{yy} = \text{Pe}\,(1 - y^2)u_x$.

Graetz–Nusselt equation. It governs steady-state heat exchange in a laminar fluid flow with a parabolic velocity profile in a plane channel. The equation for the temperature u is written in terms of dimensionless Cartesian coordinates x, y related to the channel half-width h; $\text{Pe} = Uh/a$ is the Peclet number, a is the thermal diffusivity, and U is the fluid velocity at the channel axis ($y = 0$). The walls of the channel correspond to $y = \pm 1$.

1°. Particular solutions:

$$u(y) = A + By,$$
$$u(x,y) = 12Ax + A\,\text{Pe}\,(6y^2 - y^4) + B,$$
$$u(x,y) = \sum_{n=1}^{m} A_n \exp\left(-\frac{\lambda_n^2}{\text{Pe}} x\right) f_n(y).$$

Here, A, B, A_n, and λ_n are arbitrary constants, and the functions f_n are defined by

$$f_n(y) = \exp\left(-\tfrac{1}{2}\lambda_n y^2\right) \Phi\left(\alpha_n, \tfrac{1}{2}; \lambda_n y^2\right), \quad \alpha_n = \tfrac{1}{4} - \tfrac{1}{4}\lambda_n - \tfrac{1}{4}\lambda_n^3 \,\text{Pe}^{-2}, \tag{1}$$

where $\Phi(\alpha, \beta; \xi) = 1 + \sum_{k=1}^{\infty} \frac{\alpha(\alpha+1)\ldots(\alpha+k-1)}{\beta(\beta+1)\ldots(\beta+k-1)} \frac{\xi^k}{k!}$ is the degenerate hypergeometric function.

2°. Let the walls of the channel be maintained at a constant temperature, $u = 0$ for $x < 0$ and $u = u_0$ for $x > 0$. Due to the symmetry of the problem about the x-axis, it suffices to consider only half of the domain, $0 \leq y \leq 1$. The boundary conditions are written as

$$y = 0, \quad u_y = 0; \quad y = 1, \quad u = \begin{cases} 0 & \text{for } x < 0, \\ u_0 & \text{for } x > 0; \end{cases}$$
$$x \to -\infty, \quad u \to 0; \quad x \to \infty, \quad u \to u_0.$$

The solution of the original equation under these boundary conditions is sought in the form

$$u(x,y) = u_0 \sum_{n=1}^{\infty} B_n \exp\left(\frac{\mu_n^2}{\text{Pe}} x\right) g_n(y) \qquad \text{for} \quad x < 0,$$

$$u(x,y) = u_0 \left[1 - \sum_{n=1}^{\infty} A_n \exp\left(-\frac{\lambda_n^2}{\text{Pe}} x\right) f_n(y)\right] \qquad \text{for} \quad x > 0.$$

The series coefficients must satisfy the matching conditions at the boundary:
$$u(x,y)|_{x\to 0,\ x<0} - u(x,y)|_{x\to 0,\ x>0} = 0,$$
$$u_x(x,y)|_{x\to 0,\ x<0} - u_x(x,y)|_{x\to 0,\ x>0} = 0.$$

For $x > 0$, the function $f_n(y)$ is defined by relation (1), where the eigenvalues λ_n are roots of the transcendental equation
$$\Phi(\alpha_n, \tfrac{1}{2}; \lambda_n) = 0, \quad \text{where} \quad \alpha_n = \tfrac{1}{4} - \tfrac{1}{4}\lambda_n - \tfrac{1}{4}\lambda_n^3\,\mathsf{Pe}^{-2}.$$

For $\mathsf{Pe} \to \infty$, it is convenient to use the following approximate relation to identify the λ_n:
$$\lambda_n = 4(n-1) + 1.68 \qquad (n = 1, 2, 3, \ldots). \tag{2}$$

The error of this formula does not exceed 0.2%. The corresponding numerical values of the coefficients A_n are well approximated by the relations
$$A_1 = 1.2, \quad A_n = 2.27\,(-1)^{n-1}\lambda_n^{-7/6} \qquad \text{for} \quad n = 2, 3, 4, \ldots,$$
whose maximum error is less than 0.1%, provided that the λ_n are calculated by (2).

For $\mathsf{Pe} \to 0$, the following asymptotic relations hold:
$$\lambda_n = \sqrt{\pi(n-\tfrac{1}{2})\,\mathsf{Pe}}, \quad A_n = \frac{4(-1)^{n-1}}{\pi^2(2n-1)^2}, \quad f_n(y) = \cos\big[\pi(n-\tfrac{1}{2})y\big] \quad n=1,2,3,\ldots$$

No results for $x < 0$ are given here, because they are of secondary importance in applications.

3°. Let a constant thermal flux be prescribed at the walls for $x > 0$ and let, for $x < 0$, the walls be insulated from heat and the temperature vanishes as $x \to -\infty$. Then the boundary conditions have the form
$$y = 0, \quad u_y = 0; \qquad y = 1, \quad u_y = \begin{cases} 0 & \text{for } x < 0, \\ q & \text{for } x > 0; \end{cases} \qquad x \to -\infty, \quad u \to 0.$$

In the domain of thermal stabilization, the asymptotic behavior of the solution (as $x \to \infty$) is as follows:
$$u(x,y) = q\left(\frac{3}{2}\frac{x}{\mathsf{Pe}} + \frac{3}{4}y^2 - \frac{1}{8}y^4 + \frac{9}{4\,\mathsf{Pe}^2} - \frac{39}{280}\right).$$

4°. See also Graetz (1883), Nusselt (1910), Deavours (1974), and Polyanin et al. (2002).

3. $u_{xx} + u_{yy} = f(y)u_x.$

This equation describes steady-state heat exchange in a laminar fluid flow with an arbitrary velocity profile $f = f(y)$ in a plane channel.

1°. Particular solutions:
$$u(x,y) = Ax + A\int_{y_0}^{y}(y-\xi)f(\xi)\,d\xi + By + C,$$
$$u(x,y) = B + \sum_{n=1}^{m} A_n \exp(-\beta_n x)w_n(y).$$

Here, A, B, C, y_0, A_n, and β_n are arbitrary constants, and the functions $w_n = w_n(y)$ are determined by the second-order linear ODE

$$w_n'' + \left[\beta_n f(y) + \beta_n^2\right] w_n = 0.$$

2°. The first solution in Item 1° describes the temperature distribution far away from the inlet section of the tube, in the domain of thermal stabilization, provided that a constant thermal flux is prescribed at the channel walls.

4. $u_{rr} + r^{-1} u_r + u_{zz} = \text{Pe}\,(1 - r^2) u_z.$

This equation governs steady-state heat exchange in a laminar fluid flow with parabolic (Poiseuille's) velocity profile in a circular tube. The equation for the temperature u is written in terms of dimensionless cylindrical coordinates r, z related to the tube radius R; $\text{Pe} = UR/a$ is the Peclet number, a is the thermal diffusivity, and U is the fluid velocity at the tube axis ($r = 0$). The walls of the tube correspond to $r = 1$.

1°. Particular solutions:

$$u(r) = A + B \ln r,$$
$$u(r, z) = 16 A z + A\,\text{Pe}\,(4r^2 - r^4) + B,$$
$$u(r, z) = \sum_{n=1}^{m} A_n \exp\!\left(-\frac{\lambda_n^2}{\text{Pe}} z\right) f_n(r).$$

Here, A, B, A_n, and λ_n are arbitrary constants, and the functions f_n are defined by

$$f_n(r) = \exp\!\left(-\tfrac{1}{2}\lambda_n r^2\right) \Phi(\alpha_n, 1;\,\lambda_n r^2), \qquad \alpha_n = \tfrac{1}{2} - \tfrac{1}{4}\lambda_n - \tfrac{1}{4}\lambda_n^3\,\text{Pe}^{-2}, \qquad (1)$$

where $\Phi(\alpha, \beta; \xi)$ is the degenerate hypergeometric function.

2°. Let the tube wall be maintained at a constant temperature such that $u = 0$ for $z < 0$ and $u = u_0$ for $z > 0$. The boundary conditions are written as

$$r = 0, \quad u_r = 0; \qquad r = 1, \quad u = \begin{cases} 0 & \text{for } z < 0, \\ u_0 & \text{for } z > 0; \end{cases}$$
$$z \to -\infty, \quad u \to 0; \qquad z \to \infty, \quad u \to u_0.$$

The solution of the original equation under these boundary conditions is sought in the form

$$u(r, z) = u_0 \sum_{n=1}^{\infty} B_n \exp\!\left(\frac{\mu_n^2}{\text{Pe}} z\right) g_n(r) \qquad \text{for } z < 0,$$

$$u(r, z) = u_0 \left[1 - \sum_{n=1}^{\infty} A_n \exp\!\left(-\frac{\lambda_n^2}{\text{Pe}} z\right) f_n(r)\right] \qquad \text{for } z > 0.$$

The series coefficients must satisfy the matching conditions at the boundary:

$$u(r, z)\big|_{z \to 0,\, z<0} - u(r, z)\big|_{z \to 0,\, z>0} = 0,$$
$$u_z(r, z)\big|_{z \to 0,\, z<0} - u_z(r, z)\big|_{z \to 0,\, z>0} = 0.$$

For $z > 0$, the functions $f_n(r)$ are defined by relations (1), where the eigenvalues λ_n are roots of the transcendental equation

$$\Phi(\alpha_n, 1; \lambda_n) = 0, \quad \text{where} \quad \alpha_n = \tfrac{1}{2} - \tfrac{1}{4}\lambda_n - \tfrac{1}{4}\lambda_n^3 \, \text{Pe}^{-2}.$$

For $\text{Pe} \to \infty$, it is convenient to use the following approximate relation to identify the λ_n:

$$\lambda_n = 4(n-1) + 2.7 \qquad (n = 1, 2, 3, \dots). \tag{2}$$

The error of this formula does not exceed 0.3%. The corresponding numerical values of the coefficients A_n are well approximated by the relations

$$A_n = 2.85\,(-1)^{n-1}\lambda_n^{-2/3} \qquad \text{for} \quad n = 1, 2, 3, \dots,$$

whose maximum error is 0.5%,

No results for $z < 0$ are given here, since they are of secondary importance in applications.

3°. Let a constant thermal flux be prescribed at the wall for $z > 0$ and let, for $z < 0$, the tube surface be insulated from heat and the temperature vanishes as $z \to -\infty$. Then the boundary conditions have the form

$$r = 0, \quad u_r = 0; \qquad r = 1, \quad u_r = \begin{cases} 0 & \text{for } z < 0, \\ q & \text{for } z > 0; \end{cases} \qquad z \to -\infty, \quad u \to 0.$$

In the domain of thermal stabilization, the asymptotic behavior of the solution (as $z \to \infty$) is as follows:

$$u(r,z) = q\left(4\frac{z}{\text{Pe}} + r^2 - \frac{1}{4}r^4 + \frac{8}{\text{Pe}^2} - \frac{7}{24}\right).$$

4°. See also Davis (1973), Cebeci & Bradshaw (1984), Polyanin et al. (2002).

5. $a(u_{xx} + u_{yy}) = v_1(x,y)u_x + v_2(x,y)u_y.$

This is an equation of steady-state convective heat and mass transfer in the Cartesian coordinate system. Here $v_1 = v_1(x,y)$ and $v_2 = v_2(x,y)$ are the components of the fluid velocity that are assumed to be known from the solution of the hydrodynamic problem.

1°. In plane problems of convective heat exchange in liquid metals modeled by an ideal fluid, as well as in describing seepage (filtration) flows employing the model of potential flows, the fluid velocity components $v_1(x,y)$ and $v_2(x,y)$ can be expressed in terms of the potential $\varphi = \varphi(x,y)$ and stream function $\psi = \psi(x,y)$ as follows:

$$v_1 = \varphi_x = -\psi_y, \qquad v_2 = \varphi_y = \psi_x. \tag{1}$$

The function φ is determined by solving the Laplace equation $\Delta\varphi = 0$.

2°. By switching in the convective heat exchange equation from x, y to the new variables φ, ψ (the *Boussinesq transformation*) and taking into account (1), we arrive a simpler PDE with constant coefficients,

$$u_{\varphi\varphi} + u_{\psi\psi} = \frac{1}{a}u_\varphi. \tag{2}$$

The Boussinesq transformation maps any plane contour in a potential flow onto a cut in the φ-axis, simultaneously with the reduction of the original equation to the form (2). Consequently, the heat transfer problem of a potential flow about this contour is reduced to the heat exchange problem of a longitudinal flow of an ideal fluid past a flat plate (see linear PDE 5.3.4.1, in which one should rename a to $1/a$ and put $b = c = 0$).

6. $\dfrac{1}{r^2}(r^2 u_r)_r + \dfrac{1}{r^2 \sin\theta}(\sin\theta\, u_\theta)_\theta = \cos\theta\, u_r - \dfrac{\sin\theta}{r} u_\theta.$

This PDE is obtained from the equation $u_{xx} + u_{yy} + u_{zz} = u_x$ by switching to the spherical coordinate system in the axisymmetric case.

The general solution satisfying the decay condition ($u \to 0$ as $r \to \infty$) is expressed as

$$u(r,\theta) = \left(\frac{\pi}{r}\right)^{1/2} \exp\left(\frac{r\cos\theta}{2}\right) \sum_{n=0}^{\infty} A_n K_{n+\frac{1}{2}}\left(\frac{r}{2}\right) P_n(\cos\theta),$$

where A_n are arbitrary constants. The Legendre polynomials $P_n(\xi)$ and the modified Bessel functions $K_{n+\frac{1}{2}}(z)$ are given by

$$P_n(\xi) = \frac{1}{n!\,2^n}\frac{d^n}{d\xi^n}(\xi^2-1)^n, \quad K_{n+\frac{1}{2}}\left(\frac{r}{2}\right) = \left(\frac{\pi}{r}\right)^{1/2} \exp\left(-\frac{r}{2}\right) \sum_{m=0}^{n}\frac{(n+m)!}{(n-m)!\,m!\,r^m}.$$

5.3.5. Equations of Heat and Mass Transfer in Anisotropic Media

1. $(ax^n u_x)_x + (by^m u_y)_y = 0.$

This is a two-dimensional equation of heat and mass transfer theory in a inhomogeneous anisotropic medium. Here $a_1(x) = ax^n$ and $a_2(y) = by^m$ are the principal thermal diffusivities.

1°. Particular solutions (A, B, C are arbitrary constants):

$$u = Ax^{1-n} + By^{1-m} + C,$$
$$u = A\left[\frac{x^{2-n}}{a(2-n)} - \frac{y^{2-m}}{b(2-m)}\right] + B,$$
$$u = Ax^{1-n}y^{1-m} + B.$$

2°. For $n \neq 2$ and $m \neq 2$, there are particular solutions of the form

$$u = u(\xi), \qquad \xi = \left[b(2-m)^2 x^{2-n} + a(2-n)^2 y^{2-m}\right]^{1/2}.$$

The function $u = u(\xi)$ satisfies the ODE

$$u''_{\xi\xi} + \frac{A}{\xi} u'_\xi = 0, \qquad A = \frac{4-nm}{(2-n)(2-m)}. \tag{1}$$

The general solution of ODE (1) is given by

$$u(\xi) = \begin{cases} C_1 \xi^{1-A} + C_2 & \text{for } A \neq 1, \\ C_1 \ln\xi + C_2 & \text{for } A = 1, \end{cases}$$

where C_1 and C_2 are arbitrary constants.

3°. There are multiplicative separable particular solutions in the form

$$u = \varphi(x)\psi(y), \tag{2}$$

where $\varphi(x)$ and $\psi(y)$ are determined by the following second-order linear ODEs (A_1 is an arbitrary constant):

$$(ax^n \varphi'_x)'_x = -A_1 \varphi, \tag{3}$$
$$(by^m \psi'_y)'_y = A_1 \psi. \tag{4}$$

The solution of ODE (3) is given by

$$\varphi(x) = \begin{cases} x^{\frac{1-n}{2}}\left[C_1 J_\nu\left(\beta x^{\frac{2-n}{2}}\right) + C_2 Y_\nu\left(\beta x^{\frac{2-n}{2}}\right)\right] & \text{for } A_1 > 0, \\ x^{\frac{1-n}{2}}\left[C_1 I_\nu\left(\beta x^{\frac{2-n}{2}}\right) + C_2 K_\nu\left(\beta x^{\frac{2-n}{2}}\right)\right] & \text{for } A_1 < 0, \end{cases}$$

$$\nu = \frac{|1-n|}{2-n}, \qquad \beta = \frac{2}{2-n}\sqrt{\frac{|A_1|}{a}},$$

where C_1 and C_2 are arbitrary constants, $J_\nu(z)$ and $Y_\nu(z)$ are Bessel functions, and $I_\nu(z)$ and $K_\nu(z)$ are modified Bessel functions.

The solution of ODE (4) is expressed as

$$\psi(y) = \begin{cases} y^{\frac{1-m}{2}}\left[C_1 J_\sigma\left(\mu y^{\frac{2-m}{2}}\right) + C_2 Y_\sigma\left(\mu y^{\frac{2-m}{2}}\right)\right] & \text{for } A_1 < 0, \\ y^{\frac{1-m}{2}}\left[C_1 I_\sigma\left(\mu y^{\frac{2-m}{2}}\right) + C_2 K_\sigma\left(\mu y^{\frac{2-m}{2}}\right)\right] & \text{for } A_1 > 0, \end{cases}$$

$$\sigma = \frac{|1-m|}{2-m}, \qquad \mu = \frac{2}{2-m}\sqrt{\frac{|A_1|}{b}},$$

where C_1 and C_2 are arbitrary constants.

The sum of solutions of the form (2) corresponding to different values of the parameter A_1 is also a solution of the original equation; the solutions of some boundary value problems may be obtained by separation of variables.

4°. See also equation 5.3.5.3, Item 4°, for $c = 0$.

2. $(ax^n u_x)_x + (by^m u_y)_y = c.$

The substitution

$$u = w(x, y) + \frac{c}{a(2-n)} x^{2-n}$$

leads to a homogeneous equation of the form 5.3.5.1:

$$(ax^n w_x)_x + (by^m w_y)_y = 0.$$

3. $(ax^n u_x)_x + (by^m u_y)_y = cu.$

This is a two-dimensional equation of heat and mass transfer theory with a linear source in an inhomogeneous anisotropic medium.

1°. For $n \neq 2$ and $m \neq 2$, there are particular solutions of the form
$$u = u(\xi), \qquad \xi = \left[b(2-m)^2 x^{2-n} + a(2-n)^2 y^{2-m}\right]^{1/2}.$$
The function $u = u(\xi)$ satisfies the ordinary differential equation
$$u''_{\xi\xi} + \frac{A}{\xi} u'_\xi = Bu, \tag{1}$$
where
$$A = \frac{4-nm}{(2-n)(2-m)}, \qquad B = \frac{4c}{ab(2-n)^2(2-m)^2}.$$
The general solution of equation (1) is given by
$$u(\xi) = \xi^{\frac{1-A}{2}}\left[C_1 J_\nu\!\left(\xi\sqrt{|B|}\right) + C_2 Y_\nu\!\left(\xi\sqrt{|B|}\right)\right] \quad \text{for } B < 0,$$
$$u(\xi) = \xi^{\frac{1-A}{2}}\left[C_1 I_\nu\!\left(\xi\sqrt{B}\right) + C_2 K_\nu\!\left(\xi\sqrt{B}\right)\right] \quad \text{for } B > 0,$$
where $\nu = \frac{1}{2}|1-A|$; C_1 and C_2 are arbitrary constants; $J_\nu(z)$ and $Y_\nu(z)$ are Bessel functions; and $I_\nu(z)$ and $K_\nu(z)$ are modified Bessel functions.

2°. There are multiplicative separable particular solutions of the form
$$u = \varphi(x)\psi(y),$$
where $\varphi(x)$ and $\psi(y)$ are determined by the following second-order linear ordinary differential equations (A_1 is an arbitrary constant):
$$(ax^n \varphi'_x)'_x = A_1 \varphi, \qquad (by^m \psi'_y)'_y = (c - A_1)\psi. \tag{2}$$
The solutions of equations (2) are expressed in terms of the Bessel functions (or modified Bessel functions); see equation 5.3.5.1, Item 3°.

3°. There are additive separable particular solutions of the form
$$u = f(x) + g(y),$$
where $f(x)$ and $g(y)$ are determined by the following second-order linear ordinary differential equations (A_2 is an arbitrary constant):
$$(ax^n f'_x)'_x - cf = A_2, \qquad (by^m g'_y)'_y - cg = -A_2. \tag{3}$$
The solutions of equations (3) are expressed in terms of the Bessel functions (or modified Bessel functions).

4°. The transformation
$$x^{\frac{2-n}{2}} = Ar\cos\theta, \qquad y^{\frac{2-m}{2}} = Br\sin\theta,$$
where $A^2 = a(2-n)^2$ and $B^2 = b(2-m)^2$, leads to the PDE
$$u_{rr} + \frac{4-nm}{(2-n)(2-m)}\frac{1}{r}u_r + \frac{1}{r^2}u_{\theta\theta} - \frac{2}{r^2}\frac{(nm-n-m)\cos 2\theta + (n-m)}{(2-n)(2-m)\sin 2\theta}u_\theta = 4cu,$$
which admits separable solutions of the form $u = F_1(r)F_2(\theta)$.

4. $(ae^{\beta x}u_x)_x + (be^{\mu y}u_y)_y = 0.$

This is a two-dimensional equation of heat and mass transfer theory in an inhomogeneous anisotropic medium. Here $a_1(x) = ae^{\beta x}$ and $a_2(y) = be^{\mu y}$ are the principal thermal diffusivities.

1°. Particular solutions (A, B, C are arbitrary constants):
$$u = Ae^{-\beta x} + Be^{-\mu y} + C,$$
$$u = \frac{A}{a\beta^2}(\beta x + 1)e^{-\beta x} - \frac{A}{b\mu^2}(\mu y + 1)e^{-\mu y} + B,$$
$$u = Ae^{-\beta x - \mu y} + B.$$

2°. There are multiplicative separable particular solutions of the form
$$u = \varphi(x)\psi(y), \tag{1}$$
where $\varphi(x)$ and $\psi(y)$ are determined by the following second-order linear ordinary differential equations (A_1 is an arbitrary constant):
$$(ae^{\beta x}\varphi'_x)'_x = -A_1\varphi, \tag{2}$$
$$(be^{\mu y}\psi'_y)'_y = A_1\psi. \tag{3}$$

The solution of equation (2) is given by
$$\varphi(x) = \begin{cases} e^{-\beta x/2}\left[C_1 J_1(ke^{-\beta x/2}) + C_2 Y_1(ke^{-\beta x/2})\right] & \text{for } A_1 > 0, \\ e^{-\beta x/2}\left[C_1 I_1(ke^{-\beta x/2}) + C_2 K_1(ke^{-\beta x/2})\right] & \text{for } A_1 < 0, \end{cases}$$
where $k = -(2/\beta)\sqrt{|A_1|/a}$; C_1 and C_2 are arbitrary constants; $J_1(z)$ and $Y_1(z)$ are Bessel functions; and $I_1(z)$ and $K_1(z)$ are modified Bessel functions.

The solution of equation (3) is given by
$$\psi(y) = \begin{cases} e^{-\mu y/2}\left[C_1 J_1(se^{-\mu y/2}) + C_2 Y_1(se^{-\mu y/2})\right] & \text{for } A_1 < 0, \\ e^{-\mu y/2}\left[C_1 I_1(se^{-\mu y/2}) + C_2 K_1(se^{-\mu y/2})\right] & \text{for } A_1 > 0, \end{cases}$$
where $s = -(2/\mu)\sqrt{|A_1|/b}$; C_1 and C_2 are arbitrary constants.

The sum of solutions of the form (1) corresponding to different values of the parameter A_1 is also a solution of the original equation.

3°. See also equation 5.3.5.6, Item 3°, for $c = 0$.

5. $(ae^{\beta x} u_x)_x + (be^{\mu y} u_y)_y = c.$

The substitution
$$u = w(x, y) - \frac{c}{a\beta^2}(\beta x + 1)e^{-\beta x}$$
leads to a homogeneous equation of the form 5.3.5.4:
$$(ae^{\beta x} w_x)_x + (be^{\mu y} w_y)_y = 0.$$

6. $(ae^{\beta x} u_x)_x + (be^{\mu y} u_y)_y = cu.$

This is a two-dimensional equation of heat and mass transfer theory with a linear source in an inhomogeneous anisotropic medium.

1°. For $\beta\mu \neq 0$, there are particular solutions of the form
$$u = u(\xi), \qquad \xi = (b\mu^2 e^{-\beta x} + a\beta^2 e^{-\mu y})^{1/2}.$$
The function $u = u(\xi)$ satisfies the ordinary differential equation
$$u''_{\xi\xi} - \frac{1}{\xi}u'_\xi = Bu, \qquad B = \frac{4c}{ab\beta^2\mu^2}.$$
For the solution of this equation, see 5.3.5.3 (Item 1° for $A = -1$).

2°. The original equation admits multiplicative (and additive) separable solutions. See equation 5.3.5.11 with $f(x) = ae^{\beta x}$ and $g(y) = be^{\mu y}$.

3°. The transformation
$$e^{-\beta x/2} = Ar\cos\theta, \quad e^{-\mu y/2} = Br\sin\theta,$$
where $A^2 = a\beta^2$ and $B^2 = b\mu^2$, leads to the PDE
$$u_{rr} - \frac{1}{r}u_r + \frac{1}{r^2}u_{\theta\theta} - \frac{2}{r^2}\cot 2\theta\, u_\theta = 4cu,$$
which admits separable solutions of the form $u(r,\theta) = F_1(r)F_2(\theta)$.

7. $(ax^n u_x)_x + (be^{\beta y} u_y)_y = cu.$

1°. For $n \neq 2$ and $\beta \neq 0$, there are particular solutions of the form
$$u = u(r), \quad r^2 = \frac{x^{2-n}}{a(2-n)^2} + \frac{e^{-\beta y}}{b\beta^2}.$$

The function $u = u(r)$ satisfies the ordinary differential equation
$$u''_{rr} + \frac{n}{2-n}\frac{1}{r}u'_r = 4cu.$$

For the solution of this equation, see equation 5.3.5.3 (Item 1°).

2°. The original equation admits multiplicative (and additive) separable solutions. See equation 5.3.5.11 with $f(x) = ax^n$ and $g(y) = be^{\beta y}$.

3°. The transformation
$$x^{1-\frac{1}{2}n} = Ar\cos\theta, \quad e^{-\frac{1}{2}\beta y} = Br\sin\theta,$$
where $A^2 = a(2-n)^2$ and $B^2 = b\beta^2$, leads to the PDE
$$u_{rr} + \frac{n}{2-n}\frac{1}{r}u_r + \frac{1}{r^2}u_{\theta\theta} - \frac{2}{r^2}\frac{(1-n)\cos 2\theta + 1}{(2-n)\sin 2\theta}u_\theta = 4cu,$$
which admits separable solutions of the form $u = F_1(r)F_2(\theta)$.

8. $[f(x)u_x]_x + u_{yy} = 0.$

1°. Particular solutions:
$$u = C_1 y^2 + C_2 y - 2\int \frac{C_1 x + C_3}{f(x)}\,dx + C_4,$$
$$u = C_1 y^3 + C_2 y - 6y\int \frac{C_1 x + C_3}{f(x)}\,dx + C_4,$$
$$u = [C_1\Phi(x) + C_2]y + C_3\Phi(x) + C_4, \quad \Phi(x) = \int \frac{dx}{f(x)},$$
$$u = [C_1\Phi(x) + C_2]y^2 + C_3\Phi(x) + C_4 - 2\int\left\{\frac{1}{f(x)}\int[C_1\Phi(x) + C_2]\,dx\right\}dx,$$
where $C_1, \ldots, C_5$ are arbitrary constants.

2°. Separable particular solution:
$$u = (C_1 e^{\lambda y} + C_2 e^{-\lambda y}) H(x),$$
where C_1, C_2, and λ are arbitrary constants, and the function $H = H(x)$ satisfies the ordinary differential equation $[f(x) H'_x]'_x + \lambda^2 H = 0$.

3°. Separable particular solution:
$$u = [C_1 \sin(\lambda y) + C_2 \cos(\lambda y)] Z(x),$$
where C_1, C_2, and λ are arbitrary constants, and the function $Z = Z(x)$ satisfies the ordinary differential equation $[f(x) Z'_x]'_x - \lambda^2 Z = 0$.

4°. Particular solutions with even powers of y:
$$u = \sum_{k=0}^{n} \zeta_k(x) y^{2k},$$
where the functions $\zeta_k = \zeta_k(x)$ are defined by the recurrence relations
$$\zeta_n(x) = A_n \Phi(x) + B_n, \qquad \Phi(x) = \int \frac{dx}{f(x)},$$
$$\zeta_{k-1}(x) = A_k \Phi(x) + B_k - 2k(2k-1) \int \frac{1}{f(x)} \left\{ \int \zeta_k(x)\, dx \right\} dx,$$
where A_k and B_k are arbitrary constants ($k = n, \ldots, 1$).

5°. Particular solutions with odd powers of y:
$$u = \sum_{k=0}^{n} \eta_k(x) y^{2k+1},$$
where the functions $\eta_k = \eta_k(x)$ are defined by the recurrence relations
$$\eta_n(x) = A_n \Phi(x) + B_n, \qquad \Phi(x) = \int \frac{dx}{f(x)},$$
$$\eta_{k-1}(x) = A_k \Phi(x) + B_k - 2k(2k+1) \int \frac{1}{f(x)} \left\{ \int \eta_k(x)\, dx \right\} dx,$$
where A_k and B_k are arbitrary constants ($k = n, \ldots, 1$).

9. $[f(x) u_x]_x + [g(y) u_y]_y = 0.$

This is a two-dimensional sourceless equation of heat and mass transfer theory in an inhomogeneous anisotropic medium. The functions $f = f(x)$ and $g = g(y)$ are the principal thermal diffusivities.

1°. Particular solutions:
$$u = A_1 \int \frac{dx}{f(x)} + B_1 \int \frac{dy}{g(y)} + C_1,$$
$$u = A_2 \int \frac{x\, dx}{f(x)} - A_2 \int \frac{y\, dy}{g(y)} + B_2,$$
$$u = A_3 \int \frac{dx}{f(x)} \int \frac{dy}{g(y)} + B_3,$$
where the A_k, B_k, and C_1 are arbitrary constants. A linear combination of these solutions is also a solution of the original equation.

2°. There are multiplicative separable particular solutions of the form
$$u = \varphi(x)\psi(y), \qquad (1)$$
where $\varphi(x)$ and $\psi(y)$ are determined by the following second-order linear ordinary differential equations (A is an arbitrary constant):
$$\begin{aligned}(f\varphi'_x)'_x &= A\varphi, & f &= f(x), \\ (g\psi'_y)'_y &= -A\psi, & g &= g(y).\end{aligned} \qquad (2)$$
The sum of solutions of the form (1) corresponding to different values of the parameter A in (2) is also a solution of the original equation (the solutions of some boundary value problems may be obtained by separation of variables).

10. $[f(x)u_x]_x + [g(y)u_y]_y = h(x).$

The substitution
$$u = w(x,y) + \int \left(\int h(x)\,dx\right) \frac{dx}{f(x)}$$
leads to a homogeneous equation of the form 5.3.5.9:
$$[f(x)w_x]_x + [g(y)w_y]_y = 0.$$

11. $[f(x)u_x]_x + [g(y)u_y]_y = \beta u.$

This is a two-dimensional equation of heat and mass transfer theory with a linear source in an inhomogeneous anisotropic medium. The functions $f = f(x)$ and $g = g(y)$ are the principal thermal diffusivities.

1°. There are multiplicative separable particular solutions of the form
$$u = \varphi(x)\psi(y), \qquad (1)$$
where $\varphi(x)$ and $\psi(y)$ are determined by the following second-order linear ordinary differential equations (A is an arbitrary constant):
$$\begin{aligned}(f\varphi'_x)'_x &= A\varphi, & f &= f(x), \\ (g\psi'_y)'_y &= (\beta - A)\psi, & g &= g(y).\end{aligned} \qquad (2)$$
The sum of solutions of the form (1) corresponding to different values of the parameter A in (2) is also a solution of the original equation; the solutions of some boundary value problems may be obtained by separation of variables.

2°. There are additive separable particular solutions of the form
$$u = \Phi(x) + \Psi(y),$$
where $\Phi(x)$ and $\Psi(y)$ are determined by the following second-order linear ordinary differential equations (C is an arbitrary constant):
$$\begin{aligned}(f\Phi'_x)'_x - \beta\Phi &= C, & f &= f(x), \\ (g\Psi'_y)'_y - \beta\Psi &= -C, & g &= g(y).\end{aligned}$$
In the special case $\beta = 0$, the solutions of these equations can be represented as
$$\Phi(x) = C\int \frac{x\,dx}{f(x)} + A_1 \int \frac{dx}{f(x)} + B_1,$$
$$\Psi(y) = -C\int \frac{y\,dy}{g(y)} + A_2 \int \frac{dy}{g(y)} + B_2,$$
where A_1, A_2, B_1, and B_2 are arbitrary constants.

5.3.6. Tricomi and Related Equations

1. $yu_{xx} + u_{yy} = 0.$

Tricomi equation. It is used to describe transonic gas flows.

1°. Particular solutions:
$$u = Axy + Bx + Cy + D,$$
$$u = A(3x^2 - y^3) + B(x^3 - xy^3) + C(6yx^2 - y^4),$$

where A, B, C, and D are arbitrary constants.

2°. Particular solutions with even powers of x:
$$u = \sum_{k=0}^{n} \varphi_k(y) x^{2k},$$

where the functions $\varphi_k = \varphi_k(y)$ are defined by the recurrence relations
$$\varphi_n(y) = A_n y + B_n, \quad \varphi_{k-1}(y) = A_k y + B_k - 2k(2k-1) \int_0^y (y-t) t \varphi_k(t) \, dt,$$

where A_k and B_k are arbitrary constants ($k = n, \ldots, 1$).

3°. Particular solutions with odd powers of x:
$$u = \sum_{k=0}^{n} \psi_k(y) x^{2k+1},$$

where the functions $\psi_k = \psi_k(y)$ are defined by the recurrence relations
$$\psi_n(y) = A_n y + B_n, \quad \psi_{k-1}(y) = A_k y + B_k - 2k(2k+1) \int_0^y (y-t) t \psi_k(t) \, dt,$$

where A_k and B_k are arbitrary constants ($k = n, \ldots, 1$).

4°. Separable particular solutions:
$$u = \left[A \sinh(3\lambda x) + B \cosh(3\lambda x) \right] \sqrt{y} \left[C J_{1/3}(2\lambda y^{3/2}) + D Y_{1/3}(2\lambda y^{3/2}) \right],$$
$$u = \left[A \sin(3\lambda x) + B \cos(3\lambda x) \right] \sqrt{y} \left[C I_{1/3}(2\lambda y^{3/2}) + D K_{1/3}(2\lambda y^{3/2}) \right],$$

where A, B, C, D, and λ are arbitrary constants, $J_{1/3}(z)$ and $Y_{1/3}(z)$ are Bessel functions, and $I_{1/3}(z)$ and $K_{1/3}(z)$ are modified Bessel functions.

5°. For $y > 0$, see also equation 5.3.6.2 with $n = 1$.

2. $y^n u_{xx} + u_{yy} = 0.$

1°. Particular solutions:
$$u = Axy + Bx + Cy + D,$$
$$u = Ax^2 - \frac{2A}{(n+1)(n+2)} y^{n+2},$$
$$u = Ax^3 - \frac{6A}{(n+1)(n+2)} xy^{n+2},$$
$$u = Ayx^2 - \frac{2A}{(n+2)(n+3)} y^{n+3},$$

where A, B, C, and D are arbitrary constants.

$2°$. Particular solutions with even powers of x:

$$u = \sum_{k=0}^{m} \varphi_k(y) x^{2k}.$$

The functions $\varphi_k = \varphi_k(y)$ are defined by the recurrence relations

$$\varphi_m(y) = A_m y + B_m, \quad \varphi_{k-1}(y) = A_k y + B_k - 2k(2k-1)\int_a^y (y-t) t^n \varphi_k(t)\,dt,$$

where A_k and B_k are arbitrary constants ($k = m, \ldots, 1$), a is any number.

$3°$. Particular solutions with odd powers of x:

$$u = \sum_{k=0}^{m} \psi_k(y) x^{2k+1}.$$

The functions $\psi_k = \psi_k(y)$ are defined by the recurrence relations

$$\psi_m(y) = A_m y + B_m, \quad \psi_{k-1}(y) = A_k y + B_k - 2k(2k+1)\int_a^y (y-t) t^n \psi_k(t)\,dt,$$

where A_k and B_k are arbitrary constants ($k = m, \ldots, 1$), a is any number.

$4°$. Separable particular solutions:

$$u = \left[A\sinh(\lambda q x) + B\cosh(\lambda q x)\right] \sqrt{y}\left[CJ_{\frac{1}{2q}}(\lambda y^q) + DY_{\frac{1}{2q}}(\lambda y^q)\right], \quad q = \tfrac{1}{2}(n+2),$$

$$u = \left[A\sin(\lambda q x) + B\cos(\lambda q x)\right] \sqrt{y}\left[CI_{\frac{1}{2q}}(\lambda y^q) + DK_{\frac{1}{2q}}(\lambda y^q)\right],$$

where A, B, C, D, and λ are arbitrary constants, $J_\nu(z)$ and $Y_\nu(z)$ are Bessel functions, and $I_\nu(z)$ and $K_\nu(z)$ are modified Bessel functions.

$5°$. Fundamental solutions (for $y > 0$):

$$u_1(x, y, x_0, y_0) = k_1 (r_1^2)^{-\beta} F(\beta, \beta, 2\beta; 1-\xi), \quad \beta = \frac{n}{2(n+2)}, \quad \xi = \frac{r_2^2}{r_1^2},$$

$$u_2(x, y, x_0, y_0) = k_2 (r_1^2)^{-\beta}(1-\xi)^{1-2\beta} F(1-\beta, 1-\beta, 2-2\beta; 1-\xi).$$

Here $F(a, b, c; \xi)$ is the hypergeometric function and

$$r_1^2 = (x - x_0)^2 + \frac{4}{(n+2)^2}\left(y^{\frac{n+2}{2}} + y_0^{\frac{n+2}{2}}\right), \quad k_1 = \frac{1}{4\pi}\left(\frac{4}{n+2}\right)^{2\beta} \frac{\Gamma^2(\beta)}{\Gamma(2\beta)},$$

$$r_2^2 = (x - x_0)^2 + \frac{4}{(n+2)^2}\left(y^{\frac{n+2}{2}} - y_0^{\frac{n+2}{2}}\right), \quad k_2 = \frac{1}{4\pi}\left(\frac{4}{n+2}\right)^{2\beta} \frac{\Gamma^2(1-\beta)}{\Gamma(2-2\beta)},$$

where $\Gamma(\beta)$ is the gamma function; x_0 and y_0 are arbitrary constants.

The fundamental solutions satisfy the conditions

$$\left.\frac{\partial u_1}{\partial y}\right|_{y=0} = 0, \quad \left. u_2 \right|_{y=0} = 0 \quad (x \text{ and } x_0 \text{ are any}, y_0 > 0).$$

3. $u_{xx} + f(x) u_{yy} = 0.$

Generalized Tricomi equation.

1°. Particular solutions:

$$u = C_1 xy + C_2 y + C_3 x + C_4,$$

$$u = C_1 y^2 + C_2 xy + C_3 y + C_4 x - 2C_1 \int_a^x (x-t) f(t)\, dt + C_5,$$

$$u = C_1 y^3 + C_2 xy + C_3 y + C_4 x - 6C_1 y \int_a^x (x-t) f(t)\, dt + C_5,$$

$$u = (C_1 x + C_2) y^2 + C_3 xy + C_4 y + C_5 x - 2 \int_a^x (x-t)(C_1 t + C_2) f(t)\, dt + C_6,$$

where C_1, C_2, C_3, C_4, C_5, and C_6 are arbitrary constants, a is any number.

2°. Separable particular solution:

$$u = (C_1 e^{\lambda y} + C_2 e^{-\lambda y}) H(x),$$

where C_1, C_2, and λ are arbitrary constants, and the function $H = H(x)$ satisfies the ordinary differential equation $H''_{xx} + \lambda^2 f(x) H = 0$.

3°. Separable particular solution:

$$u = [C_1 \sin(\lambda y) + C_2 \cos(\lambda y)] Z(x),$$

where C_1, C_2, and λ are arbitrary constants, and the function $Z = Z(x)$ satisfies the ordinary differential equation $Z''_{xx} - \lambda^2 f(x) Z = 0$.

4°. Particular solutions with even powers of y:

$$u = \sum_{k=0}^{n} \varphi_k(x) y^{2k}.$$

The functions $\varphi_k = \varphi_k(x)$ are defined by the recurrence relations

$$\varphi_n(x) = A_n x + B_n, \quad \varphi_{k-1}(x) = A_k x + B_k - 2k(2k-1) \int_a^x (x-t) f(t) \varphi_k(t)\, dt,$$

where A_k and B_k are arbitrary constants ($k = n, \ldots, 1$) and a is any number.

5°. Particular solutions with odd powers of y:

$$u = \sum_{k=0}^{n} \psi_k(x) y^{2k+1}.$$

The functions $\psi_k = \psi_k(x)$ are defined by the recurrence relations

$$\psi_n(x) = A_n x + B_n, \quad \psi_{k-1}(x) = A_k x + B_k - 2k(2k+1) \int_a^x (x-t) f(t) \psi_k(t)\, dt,$$

where A_k and B_k are arbitrary constants ($k = n, \ldots, 1$) and a is any number.

4. $[f_1(x) u_x]_x + [f_2(y) u_y]_y + \lambda [g_1(x) + g_2(y)] u = 0.$

This equation is encountered in the theory of vibration of inhomogeneous membranes. Its separable solutions are sought in the form $u = \varphi(x) \psi(y)$, where $\varphi(x)$ and $\psi(y)$ are determined by the following second-order linear ODEs (C is an arbitrary constant):

$$[f_1(x) \varphi'_x]'_x + [\lambda g_1(x) + C] \varphi = 0,$$
$$[f_2(y) \psi'_y]'_y + [\lambda g_2(y) - C] \psi = 0.$$

5.4. Simplifications of Second-Order Linear Partial Differential Equations

5.4.1. Reduction of PDEs in Two Independent Variables to Canonical Forms

▶ **Characteristic ODEs.**

Consider a second-order partial differential equation in two independent variables of the general form

$$a(x,y)u_{xx} + 2b(x,y)u_{xy} + c(x,y)u_{yy} = F(x,y,u,u_x,u_y), \qquad (1)$$

where a, b, c are some functions of x and y that have continuous derivatives up to the second order inclusive.*

The left-side of PDE (1) can be simplified by using suitable transformations of independent variables. These transformations depend on the functional coefficients a, b, and c of the equation and are described below.

Given a point (x,y), PDE (1) is said to be

$$\begin{array}{ll} \textit{parabolic} & \text{if } b^2 - ac = 0, \\ \textit{hyperbolic} & \text{if } b^2 - ac > 0, \\ \textit{elliptic} & \text{if } b^2 - ac < 0 \end{array}$$

at this point.

To reduce PDE (1) to a simpler canonical form, one should write out the characteristic ODE

$$a\,(dy)^2 - 2b\,dx\,dy + c\,(dx)^2 = 0,$$

which splits into two ODEs

$$a\,dy - \left(b + \sqrt{b^2 - ac}\right) dx = 0 \qquad (2)$$

and

$$a\,dy - \left(b - \sqrt{b^2 - ac}\right) dx = 0, \qquad (3)$$

and then find their general integrals.

Remark 5.5. *The characteristic equations (2) and (3) can be used if $a \not\equiv 0$. If $a \equiv 0$, the simpler equations*

$$dx = 0,$$
$$2b\,dy - c\,dx = 0$$

should be used; the first equation has the obvious general solution $x = C$.

---

*The right-hand side of PDE (1) may be nonlinear. The classification and the procedure of reducing such equations to a canonical form are only determined by the left-hand side of the PDE.

▶ **Canonical form of parabolic equations** (case $b^2 - ac = 0$).

In this case, ODEs (2) and (3) coincide and have a common general integral

$$\varphi(x, y) = C.$$

By switching from x, y to new independent variables ξ, η in accordance with the relations

$$\xi = \varphi(x, y), \qquad \eta = \eta(x, y),$$

where $\eta = \eta(x, y)$ is any twice differentiable function that satisfies the condition of nondegeneracy of the Jacobian $\frac{D(\xi, \eta)}{D(x, y)}$ in the given domain, we reduce PDE (1) to the canonical form

$$u_{\eta\eta} = F_1(\xi, \eta, u, u_\xi, u_\eta). \tag{4}$$

For η one can take $\eta = x$ or $\eta = y$.

It is obvious that the transformed PDE (4) has only one highest-derivative term, just as the heat equation (see Section 5.1.1).

Remark 5.6. *In the degenerate case where the function F_1 does not depend on the derivative u_ξ, equation (4) is an ordinary differential equation in the variable η, and ξ serves as a parameter.*

▶ **Canonical forms of hyperbolic equations** (case $b^2 - ac > 0$).

The general integrals

$$\varphi(x, y) = C_1, \qquad \psi(x, y) = C_2$$

of ODEs (2) and (3) are real and distinct. These integrals determine two distinct families of real characteristics.

$1°$. *First canonical form.* By switching from x, y to new independent variables ξ, η in accordance with the relations

$$\xi = \varphi(x, y), \qquad \eta = \psi(x, y),$$

we reduce equation (1) to

$$u_{\xi\eta} = F_2(\xi, \eta, u, u_\xi, u_\eta). \tag{5}$$

This is the so-called *first canonical form of a hyperbolic equation*.

$2°$. *Second canonical form.* The transformation

$$\xi = t + z, \qquad \eta = t - z$$

brings PDE (5) to another canonical form,

$$u_{tt} - u_{zz} = F_3(t, z, u, u_t, u_z), \tag{6}$$

where $F_3 = 4F_2$. This is the so-called *second canonical form of a hyperbolic equation*. Apart from notation, the left-hand side of the last equation coincides with that of the wave equation (see Section 5.2.1).

▶ **Canonical form of elliptic equations (case $b^2 - ac < 0$).**

In this case, the general integrals of ODEs (2) and (3) are complex conjugate; these determine two families of complex characteristics.

Let the general integral of ODE (2) have the form

$$\varphi(x, y) + i\psi(x, y) = C, \qquad i^2 = -1,$$

where $\varphi(x, y)$ and $\psi(x, y)$ are real-valued functions.

By switching from x, y to new independent variables ξ, η in accordance with the relations

$$\xi = \varphi(x, y), \qquad \eta = \psi(x, y),$$

we reduce PDE (1) to the canonical form

$$u_{\xi\xi} + u_{\eta\eta} = F_4(\xi, \eta, u, u_\xi, u_\eta).$$

Apart from notation, the left-hand side of the last equation coincides with that of the Laplace equation (see Section 5.3.1).

5.4.2. Simplifications of Linear Constant-Coefficient Partial Differential Equations

▶ **Summary table of transformations simplifying linear constant-coefficient PDEs.**

1°. When reduced to a canonical form, linear homogeneous constant-coefficient PDEs

$$au_{xx} + 2bu_{xy} + cu_{yy} + pu_x + qu_y + su = 0 \tag{7}$$

admit further simplifications. In general, the transformation

$$u(x, y) = \exp(\beta_1 \xi + \beta_2 \eta) w(\xi, \eta) \tag{8}$$

can be used. Here ξ and η are new variables used to reduce PDE (7) to the canonical form (see above); the coefficients β_1 and β_2 in (8) are chosen so that there is only one first derivative remaining in a parabolic equation or both first derivatives vanish in a hyperbolic or an elliptic equation. For final results, see Table 5.4.

2°. The coefficients k and k_1 in the reduced hyperbolic and elliptic equations (see the last column in Table 5.4) are expressed as

$$k = \frac{2bpq - aq^2 - cp^2}{16a(b^2 - ac)^2} - \frac{s}{4a(b^2 - ac)}, \qquad k_1 = \frac{s}{4b^2} + \frac{cp^2 - 2bpq}{16b^4}. \tag{9}$$

▶ **Special linear PDEs admitting a general solution in elementary functions.**

1°. If the coefficients in PDE (7) satisfy the relation

$$2bpq - aq^2 - cp^2 - 4s(b^2 - ac) = 0,$$

TABLE 5.4. Reduction of linear homogeneous constant coefficient PDEs (7) using the transformation (8); the constants k and k_1 are given by formulas (9).

Type of equation, conditions on coefficients	Variables ξ and η in transformation (8)	Coefficients β_1 and β_2 in transformation (8)	Reduced equation
Parabolic equation, $a=b=0$, $c\neq 0$, $p\neq 0$	$\xi = -\dfrac{c}{p}x$, $\eta = y$	$\beta_1 = \dfrac{4cs-q^2}{4c^2}$, $\beta_2 = -\dfrac{q}{2c}$	$w_\xi - w_{\eta\eta} = 0$
Parabolic equation, $b^2 - ac = 0$ $(aq-bp\neq 0, |a|+|b|\neq 0)$	$\xi = \dfrac{a(ay-bx)}{bp-aq}$, $\eta = x$	$\beta_1 = \dfrac{4as-p^2}{4a^2}$, $\beta_2 = -\dfrac{p}{2a}$	$w_\xi - w_{\eta\eta} = 0$
Hyperbolic equation, $a\neq 0$, $D = b^2 - ac > 0$	$\xi = ay - (b+\sqrt{D})x$, $\eta = ay - (b-\sqrt{D})x$	$\beta_{1,2} = \dfrac{aq-bp}{4aD} \pm \dfrac{p}{4a\sqrt{D}}$	$w_{\xi\eta} + kw = 0$
Hyperbolic equation, $a = 0$, $b\neq 0$	$\xi = x$, $\eta = 2by - cx$	$\beta_1 = \dfrac{cp - 2bq}{4b^2}$, $\beta_2 = -\dfrac{p}{4b^2}$	$w_{\xi\eta} + k_1 w = 0$
Elliptic equation, $D = b^2 - ac < 0$	$\xi = ay - bx$, $\eta = \sqrt{|D|}\, x$	$\beta_1 = \dfrac{aq-bp}{2aD}$, $\beta_2 = -\dfrac{p}{2a\sqrt{|D|}}$	$w_{\xi\xi} + w_{\eta\eta} + 4kw = 0$
Ordinary differential equation, $b^2 - ac = 0$, $aq - bp = 0$	$\xi = ay - bx$, $\eta = x$	$\beta_1 = \beta_2 = 0$	$aw_{\eta\eta} + pw_\eta + sw = 0$

then $k = 0$. In this case with $a \neq 0$, the general solution of the corresponding hyperbolic equation has the form

$$u(x,y) = \exp(\beta_1 \xi + \beta_2 \eta)[f(\xi) + g(\eta)], \quad D = b^2 - ac > 0,$$

$$\xi = ay - (b+\sqrt{D})x, \quad \eta = ay - (b-\sqrt{D})x,$$

$$\beta_1 = \frac{aq-bp}{4aD} + \frac{p}{4a\sqrt{D}}, \quad \beta_2 = \frac{aq-bp}{4aD} - \frac{p}{4a\sqrt{D}},$$

where $f(\xi)$ and $g(\xi)$ are arbitrary functions.

$2°$. In the degenerate case $b^2 - ac = 0$, $aq - bp = 0$ (where the original equation reduces to an ordinary differential equation; see the last row in Table 5.4), the general solution of PDE (7) is expressed as

$$u = \exp\left(-\frac{px}{2a}\right)\left[f(ay-bx)\exp\left(\frac{x\sqrt{\lambda}}{2a}\right) + g(ay-bx)\exp\left(-\frac{x\sqrt{\lambda}}{2a}\right)\right]$$
$$\text{if } \lambda = p^2 - 4as > 0,$$

$$u = \exp\left(-\frac{px}{2a}\right)\left[f(ay-bx)\sin\left(\frac{x\sqrt{|\lambda|}}{2a}\right) + g(ay-bx)\cos\left(\frac{x\sqrt{|\lambda|}}{2a}\right)\right]$$
$$\text{if } \lambda = p^2 - 4as < 0,$$

$$u = \exp\left(-\frac{px}{2a}\right)[f(ay-bx) + xg(ay-bx)] \quad \text{if } p^2 - 4as = 0,$$

where $f(z)$ and $g(z)$ are arbitrary functions.

5.5. Third-Order Linear Partial Differential Equations

5.5.1. Equations Containing the First Derivative in t and the Third Derivative in x

1. $u_t + u_{xxx} = \Phi(x,t)$.

For $\Phi \equiv 0$, this is the *linearized Korteweg–de Vries equation*.

$1°$. Particular solutions of the homogeneous equation with $\Phi(x,t) = 0$:

$$u(x,t) = a(x^3 - 6t) + bx^2 + cx + k,$$
$$u(x,t) = a(x^5 - 60x^2 t) + b(x^4 - 24xt),$$
$$u(x,t) = a\sin(\lambda x + \lambda^3 t) + b\cos(\lambda x + \lambda^3 t) + c,$$
$$u(x,t) = a\sinh(\lambda x - \lambda^3 t) + b\cosh(\lambda x - \lambda^3 t) + c,$$
$$u(x,t) = \exp(-\lambda^3 t)\big[a\exp(\lambda x) + b\exp(-\tfrac{1}{2}\lambda x)\sin\big(\tfrac{\sqrt{3}}{2}\lambda x + c\big)\big],$$

where a, b, c, k, and λ are arbitrary constants.

$2°$. Fundamental solution:

$$\mathscr{E}(x,t) = \frac{1}{\pi}\int_0^\infty \cos(t\xi^3 + x\xi)\,d\xi = \frac{1}{(3t)^{1/3}}\operatorname{Ai}(z), \quad z = \frac{x}{(3t)^{1/3}},$$

where $\operatorname{Ai}(z) = \frac{1}{\pi}\int_0^\infty \cos(\tfrac{1}{3}\xi^3 + z\xi)\,d\xi$ is the Airy function.

$3°$. Domain: $-\infty < x < \infty$. Cauchy problem.
An initial condition is prescribed:

$$u = f(x) \quad \text{at} \quad t = 0.$$

Solution:

$$u(x,t) = \int_{-\infty}^\infty \mathscr{E}(x-y,t)f(y)\,dy + \int_0^t \int_{-\infty}^\infty \mathscr{E}(x-y,t-\tau)\Phi(y,\tau)\,dy\,d\tau.$$

$4°$. Domain: $0 \le x < \infty$. First boundary value problem.
Initial and boundary conditions are prescribed:

$$u = 0 \quad \text{at} \quad t = 0, \qquad u = f(t) \quad \text{at} \quad x = 0, \qquad u \to 0 \quad \text{as} \quad x \to \infty.$$

Solution (Faminskii, 1999):

$$u(x,t) = 3\int_0^t \operatorname{Ai}''\!\left(\frac{x}{(t-\tau)^{1/3}}\right)\frac{f(\tau)}{t-\tau}\,d\tau,$$

where $\operatorname{Ai}''(z)$ is the second derivative of the Airy function.

$5°$. Domain: $-\infty < x \le 0$. First boundary value problem.
Initial and boundary conditions are prescribed:

$$u = 0 \quad \text{at} \quad t = 0, \qquad u = f(t) \quad \text{at} \quad x = 0, \qquad u \to 0 \quad \text{as} \quad x \to -\infty.$$

Solution (Faminskii, 1999):

$$u(x,t) = -\frac{3}{2}\int_0^t \operatorname{Ai}''\!\left(\frac{x}{(t-\tau)^{1/3}}\right)\frac{f(\tau)}{t-\tau}\,d\tau.$$

2. $u_t + au_{xxx} - bu_{xx} = \Phi(x,t)$.

Linearized Burgers–Korteweg–de Vries equation.

1°. Fundamental solution:
$$\mathscr{E}(x,t) = \frac{1}{2\pi}\int_{-\infty}^{\infty}\exp[(-bu^2+iau^3)t]e^{ixu}\,du$$
$$= \frac{1}{\pi}\int_{0}^{\infty}\exp(-bu^2)\cos(au^3 t + xu)\,du.$$

2°. Domain: $-\infty < x < \infty$. Cauchy problem.
An initial condition is prescribed:
$$w = f(x) \quad \text{at} \quad t = 0.$$

Solution:
$$w(x,t) = \int_{-\infty}^{\infty}\mathscr{E}(x-y,t)f(y)\,dy + \int_{0}^{t}\int_{-\infty}^{\infty}\mathscr{E}(x-y,t-\tau)\Phi(y,\tau)\,dy\,d\tau.$$

5.5.2. Equations Containing the First Derivative in t and a Mixed Third Derivative

1. $u_t - au_{xx} - bu_{txx} = 0$.

Equation of seepage of a compressible fluid in a cracked porous medium (e.g., see Barenblatt et al., 1960 and Barenblatt, 1963). This equation also describes one-dimensional unsteady motions of second-grade non-Newtonian fluids (e.g., see Puri, 1984 and Christov, 2010).

1°. Particular solutions:
$$u(x,t) = Ax^4 + (12aAt+B)x^2 + 12a^2At^2 + 2a(12A+B)t + C,$$
$$u(x,t) = \exp\left(\frac{a\beta^2 t}{1-b\beta^2}\right)\left[A\exp(-\beta x) + B\exp(\beta x)\right],$$
$$u(x,t) = \exp\left(-\frac{a\beta^2 t}{1+b\beta^2}\right)\left[A\cos(\beta x) + B\sin(\beta x)\right],$$

where A, B, C, and β are arbitrary constants. The last solution is periodic in x.

2°. Solutions periodic in t:
$$u(x,t) = e^{-\lambda x}\left[A\cos(\omega t - \mu x) + B\sin(\omega t - \mu x)\right],$$
$$\lambda = \pm\left[\frac{\omega\sqrt{a^2+b^2}+b\omega^2}{2(a^2+b^2\omega^2)}\right]^{1/2}, \quad \mu = \pm\left[\frac{\omega\sqrt{a^2+b^2}-b\omega^2}{2(a^2+b^2\omega^2)}\right]^{1/2}.$$

Here, one only takes the upper or lower signs simultaneously; A and B are arbitrary constants.

For $\lambda > 0$, these formulas give the solution of Stokes' second problem for a second-grade fluid with the boundary conditions

$$u = A\cos(\omega t) + B\sin(\omega t) \quad \text{at} \quad x = 0, \qquad u \to 0 \quad \text{as} \quad x \to \infty.$$

This is a problem without initial conditions, $-\infty < t < \infty$; see also Item 12°.

3°. Fundamental solution:

$$\begin{aligned}
\mathscr{E}_e(x,t) &= \frac{1}{2\pi}\int_{-\infty}^{\infty} \frac{1}{1+b\xi^2}\exp\left(-\frac{a\xi^2 t}{1+b\xi^2} + ix\xi\right) d\xi \\
&= \frac{1}{\pi}\int_0^{\infty} \exp\left(-\frac{a\xi^2 t}{1+b\xi^2}\right)\frac{\cos(x\xi)}{1+b\xi^2}\,d\xi.
\end{aligned} \qquad (1)$$

4°. Domain: $-\infty < x < \infty$. Cauchy problem.
An initial condition is prescribed:

$$u = f(x) \quad \text{at} \quad t = 0. \qquad (2)$$

One also assumes that $u \to 0$ and $u_x \to 0$ as $|x| \to \infty$.

Solution:

$$u(x,t) = \int_{-\infty}^{\infty} \mathscr{E}_e(x-y,t)[f(y) - bf''(y)]\,dy,$$

where the function $\mathscr{E}_e(x,t)$ is defined in (1).

5°. Domain: $0 \leq x < \infty$. First boundary value problem with initial condition (2).
A boundary condition is prescribed:

$$u = 0 \quad \text{at} \quad x = 0.$$

Solution:

$$u(x,t) = \int_0^{\infty} [\mathscr{E}_e(x-y,t) - \mathscr{E}_e(x+y,t)][f(y) - bf''(y)]\,dy,$$

where the function $\mathscr{E}_e(x,t)$ is defined in (1).

6°. Domain: $0 \leq x < \infty$. Second boundary value problem with initial condition (2).
A boundary condition is prescribed:

$$u_x = 0 \quad \text{at} \quad x = 0.$$

Solution:

$$u(x,t) = \int_0^{\infty} [\mathscr{E}_e(x-y,t) + \mathscr{E}_e(x+y,t)][f(y) - bf''(y)]\,dy.$$

7°. Domain: $0 \leq x \leq l$. First boundary value problem with initial condition (2).
Boundary conditions are prescribed:

$$u = 0 \quad \text{at} \quad x = 0, \qquad u = 0 \quad \text{at} \quad x = l.$$

Solution:
$$u(x,t) = \sum_{n=1}^{\infty} A_n \sin(\beta_n x) \exp\left(-\frac{a\beta_n^2 t}{1+b\beta_n^2}\right),$$
$$A_n = \frac{2}{l}\int_0^l f(x)\sin(\beta_n x)\,dx, \quad \beta_n = \frac{\pi n}{l}.$$

8°. Domain: $0 \leq x \leq l$. Second boundary value problem with initial condition (2). Boundary conditions are prescribed:
$$u_x = 0 \quad \text{at} \quad x = 0, \qquad u_x = 0 \quad \text{at} \quad x = l.$$

Solution:
$$u(x,t) = A_0 + \sum_{n=1}^{\infty} A_n \cos(\beta_n x) \exp\left(-\frac{a\beta_n^2 t}{1+b\beta_n^2}\right),$$
$$A_0 = \frac{1}{l}\int_0^l f(x)\,dx, \quad A_n = \frac{2}{l}\int_0^l f(x)\cos(\beta_n x)\,dx, \quad \beta_n = \frac{\pi n}{l}.$$

9°. Domain: $0 \leq x \leq l$. Mixed boundary value problem with initial condition (2). Boundary conditions are prescribed:
$$u = 0 \quad \text{at} \quad x = 0, \qquad u_x = 0 \quad \text{at} \quad x = l.$$

Solution:
$$u(x,t) = \sum_{n=1}^{\infty} A_n \sin(\beta_n x) \exp\left(-\frac{a\beta_n^2 t}{1+b\beta_n^2}\right),$$
$$A_n = \frac{2}{l}\int_0^l f(x)\sin(\beta_n x)\,dx, \quad \beta_n = \frac{\pi(2n+1)}{2l}.$$

10°. Domain: $0 \leq x \leq l$. Mixed boundary value problem with initial condition (2). Boundary conditions are prescribed:
$$u_x = 0 \quad \text{at} \quad x = 0, \qquad u = 0 \quad \text{at} \quad x = l.$$

Solution:
$$u(x,t) = \sum_{n=1}^{\infty} A_n \cos(\beta_n x) \exp\left(-\frac{a\beta_n^2 t}{1+b\beta_n^2}\right),$$
$$A_n = \frac{2}{l}\int_0^l f(x)\cos(\beta_n x)\,dx, \quad \beta_n = \frac{\pi(2n+1)}{2l}.$$

11°. Domain: $0 \leq x < \infty$. First boundary value problem. The following conditions are prescribed:

$u = 0$ at $t = 0$ (initial condition),
$u = f(t)$ at $x = 0$ (boundary condition),
$u \to 0$ as $x \to \infty$ (boundary condition).

It is assumed that the initial and boundary conditions are consistent; i.e., $f(0) = 0$.

Solution:
$$u(x,t) = \frac{2}{\pi} \int_0^\infty U(\xi, t) \sin(x\xi)\, d\xi,$$

$$U(\xi, t) = \frac{\xi}{1+b\xi^2} \int_0^t \varphi(\tau) \exp\left[-\frac{a\xi^2(t-\tau)}{1+b\xi^2}\right] d\tau, \quad \varphi(\tau) = af(\tau) + bf'(\tau).$$

An alternative representation of the solution:

$$u(x,t) = \int_0^t \varphi(\tau) G(x, t-\tau)\, d\tau, \quad \varphi(\tau) = af(\tau) + bf'(\tau),$$

$$G(x,t) = \frac{2}{\pi} \int_0^\infty \frac{\xi \sin(x\xi)}{1+b\xi^2} \exp\left(-\frac{a\xi^2 t}{1+b\xi^2}\right) d\xi.$$

12°. *Domain*: $0 \leq x < \infty$. *Problems without initial conditions* ($-\infty < t < \infty$).

In applications, there are problems in which the process is studied at a time instant significantly remote from the initial instant. In this case, the initial conditions essentially do not affect the distribution of the unknown at the observation time. Such problems do not require an initial condition, and the boundary conditions are assumed to be prescribed for all preceding time instants, $-\infty < t$. However, in addition, a boundedness condition in the entire domain is imposed on the solution.

As an example, consider the first boundary value problem for the half-space $0 \leq x < \infty$ with the boundary conditions

$$u = f(t) \quad \text{at} \quad x = 0, \quad u \to 0 \quad \text{as} \quad x \to \infty.$$

The estimate $|f(t)| < C\exp(-\lambda|t|)$ with $C > 0$ and $\lambda > a/b$ is assumed to hold as $t \to -\infty$.

Solution:

$$u(x,t) = \int_{-\infty}^t \varphi(\tau) G(x, t-\tau)\, d\tau, \quad \varphi(\tau) = af(\tau) + bf'(\tau),$$

$$G(x,t) = \frac{2}{\pi} \int_0^\infty \frac{\xi \sin(x\xi)}{1+b\xi^2} \exp\left(-\frac{a\xi^2 t}{1+b\xi^2}\right) d\xi.$$

13°. *Stokes' first problem* (a special case of Item 12°). The problem of unidirectional plane flow of a second-grade fluid in a half-plane caused by an impulsive motion of a plate is characterized by the boundary conditions

$$u = U_0 \vartheta(t) \quad \text{at} \quad x = 0, \quad u \to 0 \quad \text{as} \quad x \to \infty,$$

where $\vartheta(t)$ is the Heaviside step function and $-\infty < t < \infty$.

Three representations of the solution:

$$u(x,t) = U_0 \vartheta(t)\left[1 - \frac{2}{\pi}\int_0^\infty \frac{\sin(x\xi)}{\xi(1+b\xi^2)} \exp\left(-\frac{a\xi^2 t}{1+b\xi^2}\right) d\xi\right]$$

$$= U_0 \vartheta(t)\left[1 - \frac{1}{\pi}\int_0^{1/b} \exp(-at\eta) \sin\left(x\sqrt{\frac{\eta}{1-b\eta}}\right) \frac{d\eta}{\eta}\right]$$

$$= U_0 \vartheta(t) e^{-at/b} \int_0^\infty e^{-\zeta} \operatorname{erfc}\left(\frac{x}{2\sqrt{b\zeta}}\right) I_0\left(2\sqrt{\frac{at\zeta}{b}}\right) d\zeta,$$

where erfc z is the complementary error function and $I_0(y)$ is the modified Bessel function of the first kind of order zero.

14°. *A modified Stokes' second problem* (a special case of Item 12°). The transient version of Stokes's second problem for a second-grade fluid is characterized by the boundary conditions

$$u = U_0 \vartheta(t) \cos(\omega t) \quad \text{at} \quad x = 0, \quad u \to 0 \quad \text{as} \quad x \to \infty,$$

where $\vartheta(t)$ is the Heaviside step function and $-\infty < t < \infty$.

Solution:

$$u(x,t) = U_0 \vartheta(t) \frac{2}{\pi} \int_0^\infty \frac{\xi \sin(x\xi)}{a^2 \xi^4 + \omega^2(1+b\xi^2)^2} \bigg\{ -\frac{a^2 \xi^2}{1+b\xi^2} \exp\left(-\frac{a\xi^2 t}{1+b\xi^2}\right)$$
$$+ a\omega \sin(\omega t) + [a^2 \xi^2 + b\omega^2(1+b\xi^2)] \cos(\omega t) \bigg\} d\xi.$$

15°. Domain: $0 \leq x < \infty$. *Transient filtration of a fluid in an adit* (u is the pressure). The following conditions are prescribed:

$$u = u_0 \quad \text{at} \quad t = 0, \quad u = u_1 + (u_0 - u_1)e^{-at/b} \quad \text{at} \quad x = 0,$$

where u_1 is the pressure in the stratum to the left of the boundary $x = 0$.

Solution:

$$u(x,t) = u_1 + (u_0 - u_1) \exp\left(-\frac{at}{b}\right)$$
$$+ (u_0 - u_1) \frac{2}{\pi} \int_0^\infty \frac{\sin(x\xi)}{\xi} \left[\exp\left(-\frac{a\xi^2 t}{1+b\xi^2}\right) - \exp\left(-\frac{at}{b}\right)\right] d\xi.$$

Remark 5.7. The problems and their solutions given in Items 13°–15° are discussed in detail in the articles by Barenblatt et al. (1960), Puri (1984), Bandelli et al. (1995), Christov (2010), Christov & Christov (2012), Christov & Jordan (2012).

2. $u_t - au_{xx} - bu_{txx} = \Phi(x,t)$.

1°. Domain: $-\infty < x < \infty$. Cauchy problem.
An initial condition is prescribed:

$$u = f(x) \quad \text{at} \quad t = 0.$$

Solution:

$$u(x,t) = \int_0^t \int_{-\infty}^\infty \mathscr{E}_e(x-y, t-\tau) \Phi(y,\tau) \, dy \, d\tau + \int_{-\infty}^\infty \mathscr{E}_e(x-y, t)[f(y) - bf''(y)] \, dy,$$

where

$$\mathscr{E}_e(x,t) = \frac{1}{2\pi} \int_{-\infty}^\infty \frac{1}{1+b\xi^2} \exp\left(-\frac{a\xi^2 t}{1+b\xi^2} + ix\xi\right) d\xi$$
$$= \frac{1}{\pi} \int_0^\infty \exp\left(-\frac{a\xi^2 t}{1+b\xi^2}\right) \frac{\cos(x\xi)}{1+b\xi^2} \, d\xi.$$

2°. In Items 3°–6°, we consider problems on an interval $0 \leq x \leq l$ with the general initial condition

$$u = f(x) \quad \text{at} \quad t = 0$$

and various homogeneous boundary conditions. The solution can be represented in terms of the Green's function as

$$u(x,t) = \int_0^t \int_0^l G(x,\xi,t-\tau)\Phi(\xi,\tau)\,d\xi\,d\tau + \int_0^l G(x,\xi,t)[f(\xi) - bf''(\xi)]\,d\xi.$$

The function $f(x)$ is assumed to be consistent with the homogeneous boundary conditions for u at $x = 0$ and $x = l$.

3°. Domain: $0 \leq x \leq l$. First boundary value problem.
Boundary conditions are prescribed:

$$u = 0 \quad \text{at} \quad x = 0, \qquad u = 0 \quad \text{at} \quad x = l.$$

Green's function:

$$G(x,\xi,t) = \frac{2}{l}\sum_{n=1}^{\infty} \frac{\sin(\beta_n x)\sin(\beta_n \xi)}{1+b\beta_n^2} \exp\left(-\frac{a\beta_n^2 t}{1+b\beta_n^2}\right), \quad \beta_n = \frac{\pi n}{l}.$$

4°. Domain: $0 \leq x \leq l$. Second boundary value problem.
Boundary conditions are prescribed:

$$u_x = 0 \quad \text{at} \quad x = 0, \qquad u_x = 0 \quad \text{at} \quad x = l.$$

Green's function:

$$G(x,\xi,t) = \frac{1}{l} + \frac{2}{l}\sum_{n=1}^{\infty} \frac{\cos(\beta_n x)\cos(\beta_n \xi)}{1+b\beta_n^2} \exp\left(-\frac{a\beta_n^2 t}{1+b\beta_n^2}\right), \quad \beta_n = \frac{\pi n}{l}.$$

5°. Domain: $0 \leq x \leq l$. Mixed boundary value problem.
Boundary conditions are prescribed:

$$u = 0 \quad \text{at} \quad x = 0, \qquad u_x = 0 \quad \text{at} \quad x = l.$$

Green's function:

$$G(x,\xi,t) = \frac{2}{l}\sum_{n=0}^{\infty} \frac{\sin(\beta_n x)\sin(\beta_n \xi)}{1+b\beta_n^2} \exp\left(-\frac{a\beta_n^2 t}{1+b\beta_n^2}\right), \quad \beta_n = \frac{\pi(2n+1)}{2l}.$$

6°. Domain: $0 \leq x \leq l$. Mixed boundary value problem.
Boundary conditions are prescribed:

$$u_x = 0 \quad \text{at} \quad x = 0, \qquad u = 0 \quad \text{at} \quad x = l.$$

Green's function:

$$G(x,\xi,t) = \frac{2}{l}\sum_{n=0}^{\infty} \frac{\cos(\beta_n x)\cos(\beta_n \xi)}{1+b\beta_n^2} \exp\left(-\frac{a\beta_n^2 t}{1+b\beta_n^2}\right), \quad \beta_n = \frac{\pi(2n+1)}{2l}.$$

3. $u_t - au_{xx} - bu_{txx} + cu = \Phi(x,t)$.

The substitution $u = e^{-ct}w$ leads to an equation of the form 5.5.2.2:

$$w_t - (a-bc)w_{xx} - bw_{txx} = e^{ct}\Phi(x,t).$$

5.5.3. Equations Containing the Second Derivative in t and a Mixed Third Derivative

1. $u_{tt} - au_{xx} - bu_{txx} = 0.$

This equation describes one-dimensional unsteady motions of viscous compressible barotropic fluids.

1°. Particular solutions:
$$u(x,t) = (A_1 t + A_2)x^2 + (B_1 t + B_2)x + \tfrac{1}{3}aA_1 t^3 + (aA_2 + bA_1)t^2 + C_1 t + C_2;$$
$$u(x,t) = e^{\beta t}\left[A_1 \exp\left(-\frac{\beta x}{\sqrt{a + b\beta}}\right) + A_2 \exp\left(\frac{\beta x}{\sqrt{a + b\beta}}\right)\right], \quad a + b\beta > 0;$$
$$u(x,t) = e^{\beta t}\left[A_1 \cos\left(\frac{\beta x}{\sqrt{|a + b\beta|}}\right) + A_2 \sin\left(\frac{\beta x}{\sqrt{|a + b\beta|}}\right)\right], \quad a + b\beta < 0,$$

where A_n, B_n, C_n, and β are arbitrary constants. The last solution is periodic in x.

2°. Solutions periodic in t:
$$u(x,t) = e^{-\lambda x}\left[A \cos(\omega t - \mu x) + B \sin(\omega t - \mu x)\right],$$
$$\lambda = \pm\omega\left[\frac{\sqrt{a^2 + b^2\omega^2} - a}{2(a^2 + b^2\omega^2)}\right]^{1/2}, \quad \mu = \pm\omega\left[\frac{\sqrt{a^2 + b^2\omega^2} + a}{2(a^2 + b^2\omega^2)}\right]^{1/2}.$$

Here, one only takes either the upper or lower signs simultaneously; A and B are arbitrary constants.

3°. Domain: $0 \leq x \leq l$. First boundary value problem.
The following conditions are prescribed:

$$\begin{aligned} u &= f(x) & \text{at} \quad t &= 0 & &\text{(initial condition)}, \\ u_t &= g(x) & \text{at} \quad t &= 0 & &\text{(initial condition)}, \\ u &= 0 & \text{at} \quad x &= 0 & &\text{(boundary condition)}, \\ u &= 0 & \text{at} \quad x &= l & &\text{(boundary condition)}. \end{aligned}$$

Solution:
$$u(x,t) = \sum_{n=1}^{\infty}\left[A_n \psi_{n1}(t) + B_n \psi_{n2}(t)\right]\sin(\beta_n x), \quad \beta_n = \frac{\pi n}{l},$$

where
$$\psi_{n1}(t) = \frac{\lambda_{n1}\exp(-\lambda_{n2}t) - \lambda_{n2}\exp(-\lambda_{n1}t)}{\lambda_{n1} - \lambda_{n2}}, \quad \psi_{n2}(t) = \frac{\exp(-\lambda_{n2}t) - \exp(-\lambda_{n1}t)}{\lambda_{n1} - \lambda_{n2}},$$
$$\lambda_{n1} = \frac{b\beta_n^2 + \beta_n\sqrt{b^2\beta_n^2 - 4a}}{2}, \quad \lambda_{n2} = \frac{b\beta_n^2 - \beta_n\sqrt{b^2\beta_n^2 - 4a}}{2},$$
$$A_n = \frac{2}{l}\int_0^l f(x)\sin(\beta_n x)\,dx, \quad B_n = \frac{2}{l}\int_0^l g(x)\sin(\beta_n x)\,dx.$$

Remark 5.8. This solution holds for $b^2\beta_n^2 - 4a > 0$ as well as for $b^2\beta_n^2 - 4a < 0$. In the latter case, one should write $\sqrt{b^2\beta_n^2 - 4a} = i\sqrt{4a - b^2\beta_n^2}$, where $i^2 = -1$, and transform the exponentials occurring in the functions $\psi_{n1}(t)$ and $\psi_{n2}(t)$ by using de Moivre's formulas.

4°. Domain: $0 \leq x \leq l$. Second boundary value problem.
The following conditions are prescribed:

$$u = f(x) \quad \text{at} \quad t = 0 \quad \text{(initial condition)},$$
$$u_t = g(x) \quad \text{at} \quad t = 0 \quad \text{(initial condition)},$$
$$u_x = 0 \quad \text{at} \quad x = 0 \quad \text{(boundary condition)},$$
$$u_x = 0 \quad \text{at} \quad x = l \quad \text{(boundary condition)}.$$

Solution:

$$u(x,t) = A_0 + B_0 t + \sum_{n=1}^{\infty} \left[A_n \psi_{n1}(t) + B_n \psi_{n2}(t) \right] \cos(\beta_n x), \quad \beta_n = \frac{\pi n}{l},$$

where

$$\psi_{n1}(t) = \frac{\lambda_{n1} \exp(-\lambda_{n2} t) - \lambda_{n2} \exp(-\lambda_{n1} t)}{\lambda_{n1} - \lambda_{n2}}, \quad \psi_{n2}(t) = \frac{\exp(-\lambda_{n2} t) - \exp(-\lambda_{n1} t)}{\lambda_{n1} - \lambda_{n2}},$$

$$\lambda_{n1} = \frac{b\beta_n^2 + \beta_n \sqrt{b^2 \beta_n^2 - 4a}}{2}, \quad \lambda_{n2} = \frac{b\beta_n^2 - \beta_n \sqrt{b^2 \beta_n^2 - 4a}}{2},$$

$$A_0 = \frac{1}{l} \int_0^l f(x)\, dx, \quad B_0 = \frac{1}{l} \int_0^l g(x)\, dx,$$

$$A_n = \frac{2}{l} \int_0^l f(x) \cos(\beta_n x)\, dx, \quad B_n = \frac{2}{l} \int_0^l g(x) \cos(\beta_n x)\, dx, \quad n = 1, 2, \ldots.$$

2. $cu_{tt} + u_t = a u_{xx} + b u_{txx}$.

This equation describes one-dimensional unsteady motions of a viscoelastic incompressible Oldroyd-B fluid ($a > 0$, $b > 0$, and $c > 0$).

1°. Solutions periodic in x:

$$u(x,t) = e^{-\gamma t} \left[A \cos(\beta x) + B \sin(\beta x) \right],$$
$$\gamma_{1,2} = \frac{1}{2c} \left[b\beta^2 + 1 \pm \sqrt{(b\beta^2 + 1)^2 - 4ac\beta^2} \right],$$

where A, B, and β are arbitrary constants.

2°. Solutions periodic in t:

$$u(x,t) = e^{-\lambda x} \left[A \cos(\omega t - \mu x) + B \sin(\omega t - \mu x) \right],$$

$$\lambda = \pm \left[\frac{\omega \sqrt{(bc\omega^2 + a)^2 + (ac - b)^2 \omega^2} - (ac - b)\omega^2}{2(a^2 + b^2 \omega^2)} \right]^{1/2},$$

$$\mu = \pm \left[\frac{\omega \sqrt{(bc\omega^2 + a)^2 + (ac - b)^2 \omega^2} + (ac - b)\omega^2}{2(a^2 + b^2 \omega^2)} \right]^{1/2}.$$

Here, one only takes either the upper or lower signs simultaneously; A and B are arbitrary constants.

3°. *Stokes' first problem* (Tanner, 1962). The problem of unidirectional plane flow of an *Oldroyd-B fluid* in a half-plane due to the impulsive motion of the plate is characterized by the boundary conditions

$$u = U_0 \vartheta(t) \quad \text{at} \quad x = 0, \quad u \to 0 \quad \text{as} \quad x \to \infty,$$

where $\vartheta(t)$ is the Heaviside step function and $-\infty < t < \infty$.

Solution (Christov, 2010):

$$u(x,t) = U_0 \vartheta(t) \left\{ \frac{1}{2} + \frac{1}{\pi} \int_0^\infty \exp\left(-xg(\eta)[\cos h(\eta) - \sin h(\eta)]\right) \right.$$
$$\left. \times \sin\left(\frac{t\eta}{c} - xg(\eta)[\cos h(\eta) + \sin h(\eta)]\right) \frac{d\eta}{\eta} \right\},$$

where

$$g(\eta) = \left(\frac{\eta}{2ac}\right)^{1/2} \left(\frac{1+\eta^2}{1+k^2\eta^2}\right)^{1/4}, \quad h(\eta) = \frac{1}{2}[\tan^{-1}\eta - \tan^{-1}(k\eta)], \quad k = \frac{b}{ac}.$$

4°. For the special case of $b = ac$, the equation admits the factorization

$$(c\,\partial_t + 1)(\partial_t - a\,\partial_{xx})[u] = 0.$$

Hence, its general solution has the form

$$u = f(x)\exp(-t/c) + u(x,t),$$

where $f(x)$ is an arbitrary function, and the function $u = u(x,t)$ is an arbitrary solution of the heat equation $\partial_t u - a\,\partial_{xx} u = 0$.

The solution of Stokes' first problem (see Item 3°) with $b = ac$ can be expressed via the complementary error function:

$$u(x,t) = U_0 \vartheta(t)\,\mathrm{erfc}\left(\frac{x}{2\sqrt{at}}\right).$$

5.6. Fourth-Order Linear Partial Differential Equations

5.6.1. Equation of Transverse Vibration of an Elastic Rod
$$u_{tt} + a^2 u_{xxxx} = 0$$

This equation is encountered in studying transverse vibration of elastic rods.

▶ **Particular solutions:**

$$u = (C_1 x^3 + C_2 x^2 + C_3 x + C_4)t + C_5 x^3 + C_6 x^2 + C_7 x + C_8,$$
$$u = 12a^2 C_1 t^2 + C_2 t - C_1 x^4 + C_3 x^3 + C_4 x^2 + C_5 x + C_6,$$
$$u = \left[C_1 \sin(\lambda x) + C_2 \cos(\lambda x) + C_3 \sinh(\lambda x) + C_4 \cosh(\lambda x)\right] \sin(\lambda^2 a t),$$
$$u = \left[C_1 \sin(\lambda x) + C_2 \cos(\lambda x) + C_3 \sinh(\lambda x) + C_4 \cosh(\lambda x)\right] \cos(\lambda^2 a t),$$

where $C_1, \ldots, C_8$ and λ are arbitrary constants. Here, the first solution is degenerate, the second solution is additive separable, and the last two solutions are multiplicative separable.

▶ **Domain:** $-\infty < x < \infty$. **Cauchy problem.**

Initial conditions are prescribed:
$$u = f(x) \quad \text{at} \quad t = 0, \qquad u_t = ag''(x) \quad \text{at} \quad t = 0,$$
where the prime denotes the derivative with respect to x.

Boussinesq solution:
$$u = \frac{1}{\sqrt{2\pi}} \int_{-\infty}^{\infty} f\left(x - 2\xi\sqrt{at}\right)\left(\cos\xi^2 + \sin\xi^2\right) d\xi$$
$$+ \frac{1}{a\sqrt{2\pi}} \int_{-\infty}^{\infty} g\left(x - 2\xi\sqrt{at}\right)\left(\cos\xi^2 - \sin\xi^2\right) d\xi.$$

▶ **Domain:** $0 \leq x < \infty$. **Free vibration of a semi-infinite rod.**

The following conditions are prescribed:
$$u = 0 \quad \text{at} \quad t = 0, \qquad u_t = 0 \quad \text{at} \quad t = 0 \qquad \text{(initial conditions)},$$
$$u = f(t) \quad \text{at} \quad x = 0, \qquad u_{xx} = 0 \quad \text{at} \quad x = 0 \qquad \text{(boundary conditions)}.$$

Boussinesq solution:
$$u = \frac{1}{\sqrt{\pi}} \int_{x/\sqrt{2at}}^{\infty} f\left(t - \frac{x^2}{2a\xi^2}\right)\left(\sin\frac{\xi^2}{2} + \cos\frac{\xi^2}{2}\right) d\xi.$$

▶ **Domain:** $0 \leq x \leq l$. **Boundary value problems.**

For solutions of the equation of transverse vibration of elastic rods for various boundary value problems, see Section 5.6.2 for $\Phi \equiv 0$.

5.6.2. Nonhomogeneous Equation of the Form $u_{tt} + a^2 u_{xxxx} = \Phi(x, t)$

▶ **Domain:** $0 \leq x \leq l$. **Solution in terms of the Green's function.**

We consider boundary value problems for the nonhomogeneous equation of transverse vibration of an elastic rod on an interval $0 \leq x \leq l$ with the general initial conditions
$$u = f(x) \quad \text{at} \quad t = 0, \qquad u_t = g(x) \quad \text{at} \quad t = 0$$
and various homogeneous boundary conditions. The solution of these problems can be represented in terms of the Green's function as
$$u = \frac{\partial}{\partial t} \int_0^l f(\xi) G(x, \xi, t)\, d\xi + \int_0^l g(\xi) G(x, \xi, t)\, d\xi + \int_0^t \int_0^l \Phi(\xi, \tau) G(x, \xi, t - \tau)\, d\xi\, d\tau.$$

Below are Green's functions for the equation of transverse vibration of elastic bars for various types of homogeneous boundary conditions.

▶ **Both ends of the rod are clamped.**

Boundary conditions are prescribed:
$$u = u_x = 0 \quad \text{at} \quad x = 0, \qquad u = u_x = 0 \quad \text{at} \quad x = l.$$

Green's function:
$$G(x,\xi,t) = \frac{4}{al}\sum_{n=1}^{\infty}\frac{\lambda_n^2}{[\varphi_n''(l)]^2}\varphi_n(x)\varphi_n(\xi)\sin(\lambda_n^2 at),$$

where
$$\varphi_n(x) = [\sinh(\lambda_n l) - \sin(\lambda_n l)][\cosh(\lambda_n x) - \cos(\lambda_n x)]$$
$$- [\cosh(\lambda_n l) - \cos(\lambda_n l)][\sinh(\lambda_n x) - \sin(\lambda_n x)]$$

and the λ_n are positive roots of the transcendental equation $\cosh(\lambda l)\cos(\lambda l) = 1$. The numerical values of the roots can be calculated from the formulas

$$\lambda_n = \frac{\mu_n}{l}, \quad \text{where} \quad \mu_1 = 4.730, \quad \mu_2 = 7.859, \quad \mu_n = \frac{\pi}{2}(2n+1) \quad \text{for} \quad n \geq 3.$$

▶ **Both ends of the rod are hinged.**

Boundary conditions are prescribed:
$$u = u_{xx} = 0 \quad \text{at} \quad x = 0, \qquad u = u_{xx} = 0 \quad \text{at} \quad x = l.$$

Green's function:
$$G(x,\xi,t) = \frac{2l}{a\pi^2}\sum_{n=1}^{\infty}\frac{1}{n^2}\sin(\lambda_n x)\sin(\lambda_n \xi)\sin(\lambda_n^2 at), \qquad \lambda_n = \frac{\pi n}{l}.$$

▶ **One end of the rod is clamped and the other is hinged.**

Boundary conditions are prescribed:
$$u = u_x = 0 \quad \text{at} \quad x = 0, \qquad u = u_{xx} = 0 \quad \text{at} \quad x = l.$$

Green's function:
$$G(x,\xi,t) = \frac{2}{al}\sum_{n=1}^{\infty}\lambda_n^2\frac{\varphi_n(x)\varphi_n(\xi)}{|\varphi_n'(l)\varphi_n'''(l)|}\sin(\lambda_n^2 at),$$

where
$$\varphi_n(x) = [\sinh(\lambda_n l) - \sin(\lambda_n l)][\cosh(\lambda_n x) - \cos(\lambda_n x)]$$
$$- [\cosh(\lambda_n l) - \cos(\lambda_n l)][\sinh(\lambda_n x) - \sin(\lambda_n x)]$$

and λ_n are positive roots of the transcendental equation $\tan(\lambda l) - \tanh(\lambda l) = 0$.

▶ **One end of the rod is clamped and the other is free.**

Boundary conditions are prescribed:
$$u = u_x = 0 \quad \text{at} \quad x = 0, \qquad u_{xx} = u_{xxx} = 0 \quad \text{at} \quad x = l.$$

Green's function:
$$G(x,\xi,t) = \frac{4}{al}\sum_{n=1}^{\infty}\frac{\varphi_n(x)\varphi_n(\xi)}{\lambda_n^2\varphi_n^2(l)}\sin(\lambda_n^2 at),$$

where
$$\varphi_n(x) = [\sinh(\lambda_n l) + \sin(\lambda_n l)][\cosh(\lambda_n x) - \cos(\lambda_n x)]$$
$$- [\cosh(\lambda_n l) + \cos(\lambda_n l)][\sinh(\lambda_n x) - \sin(\lambda_n x)]$$

and λ_n are positive roots of the transcendental equation $\cosh(\lambda l)\cos(\lambda l) = -1$.

▶ **One end of the rod is hinged and the other is free.**

Boundary conditions are prescribed:
$$u = u_{xx} = 0 \quad \text{at} \quad x = 0, \qquad u_{xx} = u_{xxx} = 0 \quad \text{at} \quad x = l.$$

Green's function:
$$G(x, \xi, t) = \frac{4}{al} \sum_{n=1}^{\infty} \frac{\varphi_n(x)\varphi_n(\xi)}{\lambda_n^2 \varphi_n^2(l)} \sin(\lambda_n^2 a t),$$

where
$$\varphi_n(x) = \sin(\lambda_n l)\sinh(\lambda_n x) + \sinh(\lambda_n l)\sin(\lambda_n x)$$

and λ_n are positive roots of the transcendental equation $\tan(\lambda l) - \tanh(\lambda l) = 0$.

5.6.3. Biharmonic Equation $\Delta\Delta u = 0$

The biharmonic equation is encountered in plane problems of elasticity (u is the *Airy stress function*). It is also used to describe slow flows of viscous incompressible fluids (u is the stream function).

In the rectangular Cartesian system of coordinates, the biharmonic operator has the form
$$\Delta\Delta \equiv \Delta^2 = \frac{\partial^4}{\partial x^4} + 2\frac{\partial^4}{\partial x^2 \partial y^2} + \frac{\partial^4}{\partial y^4}.$$

▶ **Particular solutions:**

$$u = (A\cosh\beta x + B\sinh\beta x + Cx\cosh\beta x + Dx\sinh\beta x)(a\cos\beta y + b\sin\beta y),$$
$$u = (A\cos\beta x + B\sin\beta x + Cx\cos\beta x + Dx\sin\beta x)(a\cosh\beta y + b\sinh\beta y),$$
$$u = Ar^2 \ln r + Br^2 + C\ln r + D, \quad r = \sqrt{(x-a)^2 + (y-b)^2},$$

where A, B, C, D, a, b, and β are arbitrary constants.

▶ **Various representations of the general solution.**

$1°$. Various representations of the general solution in terms of harmonic functions:
$$u = xu_1 + u_2,$$
$$u = yu_1 + u_2,$$
$$u = (x^2 + y^2)u_1 + u_2,$$

where $u_1 = u_1(x, y)$ and $u_2 = u_2(x, y)$ are arbitrary functions satisfying the Laplace equation $\Delta u_k = 0$ ($k = 1, 2$).

$2°$. Complex form of representation of the general solution:
$$u = \text{Re}[\bar{z}f(z) + g(z)],$$

where $f(z)$ and $g(z)$ are arbitrary analytic functions of the complex variable $z = x + iy$; $\bar{z} = x - iy$, $i^2 = -1$. The symbol $\text{Re}[A]$ stands for the real part of the complex quantity A.

▶ **Boundary value problems for the upper half-plane.**

1°. Domain: $-\infty < x < \infty, 0 \le y < \infty$. The desired function and its derivative along the normal are prescribed at the boundary:
$$u = 0 \quad \text{at} \quad y = 0, \qquad u_y = f(x) \quad \text{at} \quad y = 0.$$

Solution:
$$u(x,y) = \int_{-\infty}^{\infty} f(\xi) G(x-\xi, y)\, d\xi, \qquad G(x,y) = \frac{1}{\pi} \frac{y^2}{x^2 + y^2}.$$

2°. Domain: $-\infty < x < \infty, 0 \le y < \infty$. The derivatives of the desired function are prescribed at the boundary:
$$u_x = f(x) \quad \text{at} \quad y = 0, \qquad u_y = g(x) \quad \text{at} \quad y = 0.$$

Solution:
$$u = \frac{1}{\pi} \int_{-\infty}^{\infty} f(\xi) \left[\arctan\left(\frac{x-\xi}{y}\right) + \frac{y(x-\xi)}{(x-\xi)^2 + y^2} \right] d\xi + \frac{y^2}{\pi} \int_{-\infty}^{\infty} \frac{g(\xi)\, d\xi}{(x-\xi)^2 + y^2} + C,$$

where C is an arbitrary constant.

▶ **Boundary value problem for a circle.**

Domain: $0 \le r \le a, 0 \le \varphi \le 2\pi$. Boundary conditions in the polar coordinates:
$$u = f(\varphi) \quad \text{at} \quad r = a, \qquad u_r = g(\varphi) \quad \text{at} \quad r = a.$$

Solution:
$$u = \frac{1}{2\pi a}(r^2 - a^2)^2 \left[\int_0^{2\pi} \frac{[a - r\cos(\eta - \varphi)] f(\eta)\, d\eta}{[r^2 + a^2 - 2ar\cos(\eta - \varphi)]^2} - \frac{1}{2} \int_0^{2\pi} \frac{g(\eta)\, d\eta}{r^2 + a^2 - 2ar\cos(\eta - \varphi)} \right].$$

5.6.4. Nonhomogeneous Biharmonic Equation $\Delta\Delta u = \Phi(x, y)$

▶ **Domain:** $-\infty < x < \infty, -\infty < y < \infty$.

Solution:
$$u = \int_{-\infty}^{\infty} \int_{-\infty}^{\infty} \Phi(\xi, \eta) \mathcal{E}(x-\xi, y-\eta)\, d\xi\, d\eta, \qquad \mathcal{E}(x,y) = \frac{1}{8\pi}(x^2 + y^2) \ln\sqrt{x^2 + y^2}.$$

▶ **Domain:** $-\infty < x < \infty, 0 \le y < \infty$. **Boundary value problem.**

The upper half-plane is considered. The derivatives are prescribed at the boundary:
$$u_x = f(x) \quad \text{at} \quad y = 0, \qquad u_y = g(x) \quad \text{at} \quad y = 0.$$

Solution:
$$u = \frac{1}{\pi} \int_{-\infty}^{\infty} f(\xi) \left[\arctan\left(\frac{x-\xi}{y}\right) + \frac{y(x-\xi)}{(x-\xi)^2 + y^2} \right] d\xi + \frac{y^2}{\pi} \int_{-\infty}^{\infty} \frac{g(\xi)\, d\xi}{(x-\xi)^2 + y^2}$$
$$+ \frac{1}{8\pi} \int_{-\infty}^{\infty} d\xi \int_0^{\infty} \left[\frac{1}{2}(R_+^2 - R_-^2) - R_-^2 \ln\frac{R_+}{R_-} \right] \Phi(\xi, \eta)\, d\eta + C,$$

where C is an arbitrary constant,
$$R_+^2 = (x-\xi)^2 + (y+\eta)^2, \qquad R_-^2 = (x-\xi)^2 + (y-\eta)^2.$$

▶ **Domain: $0 \leq x \leq l_1$, $0 \leq y \leq l_2$. The sides of the plate are hinged.**

A rectangle is considered. Boundary conditions are prescribed:

$$u = u_{xx} = 0 \quad \text{at} \quad x = 0, \qquad u = u_{xx} = 0 \quad \text{at} \quad x = l_1,$$
$$u = u_{yy} = 0 \quad \text{at} \quad y = 0, \qquad u = u_{yy} = 0 \quad \text{at} \quad y = l_2.$$

Solution:

$$u(x,y) = \int_0^{l_1} \int_0^{l_2} \Phi(\xi,\eta) G(x,y,\xi,\eta)\, d\eta\, d\xi,$$

where

$$G(x,y,\xi,\eta) = \frac{4}{l_1 l_2} \sum_{n=1}^{\infty} \sum_{k=1}^{\infty} \frac{\sin(p_n x)\sin(q_k y)\sin(p_n \xi)\sin(q_k \eta)}{(p_n^2 + q_k^2)^2}, \quad p_n = \frac{\pi n}{l_1}, \quad q_k = \frac{\pi k}{l_2}.$$

▶ More solutions to linear equations and problems for linear partial differential equations of the second and higher orders can be found in the specialized reference books by Polyanin (2002) and Polyanin & Nazaikinskii (2016); see also Butkovskiy (1982) and Carslaw & Jaeger (1984). The main methods for finding analytical solutions (in the form of infinite series and definite integrals) to linear equations of mathematical physics are described, for example, in the books by Dezin (1987), Tikhonov & Samarskii (1990), Duffy (2004), Polyanin & Manzhirov (2007), Polyanin & Nazaikinskii (2016), Constanda (2022), and Zwillinger (2022).

References

Bandelli, R., Rajagopal, K.R., and Galdi, G.P., On some unsteady motions of fluids of second grade, *Arch. Mech.*, Vol. 47, pp. 661–667, 1995.

Barenblatt, G.I., On certain boundary-value problems for the equations of seepage of a liquid in fissured rocks, *J. Appl. Math. & Mech. (PMM)*, Vol. 27, No. 2, pp. 348–350, 1963.

Barenblatt, G.I., Zheltov, Yu.P., and Kochina, I.N., Basic concepts in the theory of seepage of homogeneous liquids in fissured rocks, *J. Appl. Math. & Mech. (PMM)*, Vol. 24, No. 5, pp. 1286–1303, 1960.

Butkovskiy, A.G., *Green's Functions and Transfer Functions Handbook*, Halstead Press–John Wiley & Sons, New York, 1982.

Carslaw, H.S. and Jaeger, J.C., *Conduction of Heat in Solids*, Clarendon Press, Oxford, 1984.

Cebeci, T. and Bradshaw, P., *Physical and Computational Aspects of Convective Heat Transfer*, Springer, Berlin–Heidelberg, 1984.

Christov, I.C., Stokes' first problem for some non-Newtonian fluids: Results and mistakes, *Mech. Research Comm.*, Vol. 37, No. 8, pp. 717–723, 2010.

Christov, I.C. and Christov, C.I., Comment on "On a class of exact solutions of the equations of motion of a second grade fluid" by C. Fetecâu and J. Zierep (*Acta Mech.* 150, 135–138, 2001), *Acta Mech.*, Vol. 215, pp. 25–28, 2010.

Christov, I.C. and Jordan, P.M., Comments on: "Starting solutions for some unsteady unidirectional flows of a second grade fluid," (*Int. J. Eng. Sci.* 43 (2005) 781), *Int. J. Eng. Sci.*, Vol. 51, pp. 326–332, 2012.

Constanda, C., *Solution Techniques for Elementary Partial Differential Equations*, 4th ed., Chapman & Hall/CRC Press, Boca Raton, 2022.

Davis, E.J., Exact solutions for a class of heat and mass transfer problems, *Can. J. Chem. Eng.,* Vol. 51, No. 5, pp. 562–572, 1973.

Deavours, C.A., An exact solution for the temperature distribution in parallel plate Poiseuille flow, *Trans. ASME, J. Heat Transfer,* Vol. 96, No. 4, 1974.

Dezin, A.A., *Partial Differential Equations. An Introduction to a General Theory of Linear Boundary Value Problems,* Springer-Verlag, Berlin, 1987.

Duffy, D.G., *Transform Methods for Solving Partial Differential Equations, 2nd Edition,* Chapman & Hall/CRC Press, Boca Raton, 2004.

Faminskii, A.V., Mixed problems for the Korteweg–de Vries equation, *Sbornik: Mathematics,* Vol. 190, No. 6, pp. 903–935, 1999.

Graetz, L., *Über die Warmeleitungsfähigkeit von Flüssigkeiten,* Annln. Phys., Bd. 18, S. 79–84, 1883.

Levich, V.G., *Physicochemical Hydrodynamics,* Prentice-Hall, Englewood Cliffs, New Jersey, 1962.

Miller, W., Jr., *Symmetry and Separation of Variables,* Addison-Wesley, London, 1977.

Nusselt, W., *Abhängigkeit der Wärmeübergangzahl con der Rohränge,* VDI Zeitschrift, Bd. 54, No. 28, S. 1154–1158, 1910.

Polyanin, A.D., *Handbook of Linear Partial Differential Equations for Engineers and Scientists,* Chapman & Hall/CRC Press, Boca Raton, 2002.

Polyanin, A.D., Kutepov, A.M., Vyazmin, A.V., and Kazenin, D.A., *Hydrodynamics, Mass and Heat Transfer in Chemical Engineering,* Taylor & Francis, London–New York, 2002.

Polyanin, A.D. and Manzhirov, A.V., *Handbook of Mathematics for Engineers and Scientists (Chapters 14 and T8),* Chapman & Hall/CRC Press, Boca Raton–London, 2007.

Polyanin, A.D. and Nazaikinskii, V.E., *Handbook of Linear Partial Differential Equations for Engineers and Scientists, 2nd ed.,* Chapman & Hall/CRC Press, Boca Raton–London, 2016.

Puri, P., Impulsive motion of a flat plate in a Rivlin–Ericksen fluid, *Rheol. Acta,* Vol. 23, pp. 451–453, 1984.

Rotem, Z., and Neilson, J.E., Exact solution for diffusion to flow down an incline, *Can. J. Chem. Eng.,* Vol. 47, pp. 341–346, 1966.

Sutton, W.G.L., On the equation of diffusion in a turbulent medium, *Proc. Roy. Soc., Ser. A,* Vol. 138, No. 988, pp. 48–75, 1943.

Tanner, R.I., Note on the Rayleigh problem for a visco-elastic fluid, *Z. Angew. Math. Phys. (ZAMP),* Vol. 13, pp. 573–580, 1962.

Tikhonov, A.N. and Samarskii, A.A., *Equations of Mathematical Physics,* Dover Publications, New York, 1990.

Zwillinger, D., *Handbook of Differential Equations, 4th ed.,* CRC Press, Boca Raton, 2022.

Chapter 6

Nonlinear Equations of Mathematical Physics

▶ **Preliminary remarks.** Nonlinear equations of mathematical physics are usually called nonlinear partial differential equations (PDEs) of the second or higher orders, which are used to describe various natural phenomena or processes.

Exact solutions of nonlinear equations of mathematical physics are understood as the following solutions:

- solutions that are expressed in terms of elementary functions, functions included in the equation (this is necessary when the equation depends on arbitrary functions), and indefinite integrals,
- solutions that are expressed in terms of solutions of ordinary differential equations (ODEs) or systems of such equations.

Note that exact solutions can be presented in explicit, implicit or parametric form. Sometimes exact solutions may additionally contain arbitrary functions (similar to the integration constants in ODE solutions) that are not included in the equation under consideration.

This chapter adopts a simple and straightforward classification of the most common solutions by their appearance, which is unrelated to the type and properties or appearance of the PDEs concerned (see Table 6.1).

TABLE 6.1. Most common types of exact solutions for nonlinear PDEs in two independent variables, x and t, and one unknown function, u.

No.	Type of solution	Solution structure (x and t can be swapped)
1	Additive separable solution	$u = \varphi(x) + \psi(t)$
2	Multiplicative separable solution	$u = \varphi(x)\psi(t)$
3	Traveling wave solution	$u = U(z)$, $z = \alpha x + \beta t$, $\alpha\beta \neq 0$
4	Self-similar solution	$u = t^\alpha F(z)$, $z = xt^\beta$
5	Generalized self-similar solution	$u = \varphi(t)F(z)$, $z = \psi(t)x$
6	Generalized traveling wave solution	$u = U(z)$, $z = \varphi(t)x + \psi(t)$
7	Generalized separable solution	$u = \varphi_1(x)\psi_1(t) + \cdots + \varphi_n(x)\psi_n(t)$
8	Functional separable solution (special case)	$u = U(z)$, $z = \varphi(x) + \psi(t)$
9	Functional separable solution	$u = U(z)$, $z = \varphi_1(x)\psi_1(t) + \cdots + \varphi_n(x)\psi_n(t)$

DOI: 10.1201/9781003051329-6

The chapter provides a brief description of exact solutions to various nonlinear equations of mathematical physics in two independent variables and some other nonlinear second- and higher-order PDEs.

When selecting suitable material, the author gave the preference to the following two essential types of PDEs:

- nonlinear equations that arise in various applications (theory of heat and mass transfer, wave theory, fluid dynamics, gas dynamics, nonlinear optics, biology, chemical engineering science, and more) and

- nonlinear equations of a reasonably general form that involve arbitrary functions (exact solutions to such equations are of significant interest for testing numerical and approximate analytical methods).

Degenerate solutions of nonlinear PDEs that depend on only one of the independent variables are not considered here.

6.1. Parabolic Equations

6.1.1. Quasilinear Heat Equations with a Source of the Form $u_t = au_{xx} + f(u)$

▶ Equations of this form admit traveling wave solutions $u = u(z)$, $z = \kappa x + \lambda t$, where κ and λ are arbitrary constants, and the function $u(z)$ satisfies the second-order autonomous ODE: $a\kappa^2 u''_{zz} - \lambda u'_z + f(u) = 0$.

1. $u_t = au_{xx} - bu^2$.

1°. Exact solutions (Barannyk, 2002):

$$u = \frac{a}{b} \frac{12(4-\sqrt{6})x^2 + 12(4-\sqrt{6})C_1 x + 120(12-5\sqrt{6})at + 12(2-\sqrt{6})C_2 + 6C_1^2}{\left[x^2 + C_1 x + 10(3-\sqrt{6})at + C_2\right]^2},$$

$$u = \frac{a}{b} \frac{12(4+\sqrt{6})x^2 + 12(4+\sqrt{6})C_1 x + 120(12+5\sqrt{6})at + 12(2+\sqrt{6})C_2 + 6C_1^2}{\left[x^2 + C_1 x + 10(3+\sqrt{6})at + C_2\right]^2},$$

where C_1 and C_2 are arbitrary constants.

2°. Self-similar solution:

$$u = t^{-1} U(\xi), \quad \xi = xt^{-1/2},$$

where the function $U = U(\xi)$ is described by the second-order ODE:

$$aU''_{\xi\xi} + \tfrac{1}{2}\xi U'_\xi + U - bU^2 = 0$$

2. $u_t = u_{xx} + au(1-u)$.

Fisher's equation. This equation arises in heat and mass transfer, biology, and ecology.

Traveling wave solutions (Ablowitz & Zeppetella, 1979):

$$u(x,t) = \left[1 + C\exp\left(-\tfrac{5}{6}at \pm \tfrac{1}{6}\sqrt{6a}\,x\right)\right]^{-2},$$

$$u(x,t) = \frac{1 + 2C\exp\left(-\tfrac{5}{6}at \pm \tfrac{1}{6}\sqrt{-6a}\,x\right)}{\left[1 + C\exp\left(-\tfrac{5}{6}at \pm \tfrac{1}{6}\sqrt{-6a}\,x\right)\right]^2},$$

where C is an arbitrary constant.

3. $u_t = au_{xx} - bu^3$.

1°. Exact solutions:

$$u(x,t) = \pm\sqrt{\frac{2a}{b}}\,\frac{2C_1 x + C_2}{C_1 x^2 + C_2 x + 6aC_1 t + C_3}.$$

2°. Self-similar solution:

$$u = t^{-1/2}\theta(\xi), \quad \xi = xt^{-1/2},$$

where the function $\theta(\xi)$ is described by the ODE:

$$a\theta''_{\xi\xi} + \tfrac{1}{2}\xi\theta'_\xi + \tfrac{1}{2}\theta - b\theta^3 = 0.$$

3°. Solution:

$$u = xU(\zeta), \quad \zeta = t + \frac{1}{6a}x^2,$$

where the function $U(\zeta)$ is described by the ODE: $U''_{\zeta\zeta} - 9abU^3 = 0$.

4. $u_t = u_{xx} + au - bu^3$.

1°. Exact solutions with $a > 0$ and $b > 0$:

$$u(x,t) = \pm\sqrt{\frac{a}{b}}\,\frac{C_1 \exp\left(\tfrac{1}{2}\sqrt{2a}\,x\right) - C_2 \exp\left(-\tfrac{1}{2}\sqrt{2a}\,x\right)}{C_1 \exp\left(\tfrac{1}{2}\sqrt{2a}\,x\right) + C_2 \exp\left(-\tfrac{1}{2}\sqrt{2a}\,x\right) + C_3 \exp\left(-\tfrac{3}{2}at\right)},$$

$$u(x,t) = \pm\sqrt{\frac{a}{b}}\left[\frac{2C_1 \exp\left(\sqrt{2a}\,x\right) + C_2 \exp\left(\tfrac{1}{2}\sqrt{2a}\,x - \tfrac{3}{2}at\right)}{C_1 \exp\left(\sqrt{2a}\,x\right) + C_2 \exp\left(\tfrac{1}{2}\sqrt{2a}\,x - \tfrac{3}{2}at\right) + C_3} - 1\right],$$

where C_1, C_2, and C_3 are arbitrary constants.

2°. Exact solutions with $a < 0$ and $b > 0$:

$$u(x,t) = \pm\sqrt{\frac{|a|}{b}}\,\frac{\sin\left(\tfrac{1}{2}\sqrt{2|a|}\,x + C_1\right)}{\cos\left(\tfrac{1}{2}\sqrt{2|a|}\,x + C_1\right) + C_2 \exp\left(-\tfrac{3}{2}at\right)}.$$

3°. Solution with $a > 0$ (generalizes the first solution of Item 1°):

$$u = \left[C_1 \exp\left(\tfrac{1}{2}\sqrt{2a}\,x + \tfrac{3}{2}at\right) - C_2 \exp\left(-\tfrac{1}{2}\sqrt{2a}\,x + \tfrac{3}{2}at\right)\right]U(z),$$

$$z = C_1 \exp\left(\tfrac{1}{2}\sqrt{2a}\,x + \tfrac{3}{2}at\right) + C_2 \exp\left(-\tfrac{1}{2}\sqrt{2a}\,x + \tfrac{3}{2}at\right) + C_3,$$

where C_1, C_2, and C_3 are arbitrary constants, and the function $U = U(z)$ is described by the autonomous ODE: $aU''_{zz} = 2bU^3$ (whose solution can be written in implicit form).

4°. Solution with $a < 0$ (generalizes the solution of Item 2°):
$$u = \exp(\tfrac{3}{2}at)\sin(\tfrac{1}{2}\sqrt{2|a|}\,x + C_1)V(\xi),$$
$$\xi = \exp(\tfrac{3}{2}at)\cos(\tfrac{1}{2}\sqrt{2|a|}\,x + C_1) + C_2,$$
where C_1 and C_2 are arbitrary constants, and the function $V = V(\xi)$ is described by the autonomous ODE: $aV''_{\xi\xi} = -2bV^3$ (whose solution can be written in implicit form).

5°. References for PDE 6.1.1.4: Cariello & Tabor (1989), Nucci & Clarkson (1992), and Polyanin & Zaitsev (2012).

5. $u_t = u_{xx} - u(1-u)(a-u)$.

FitzHugh–Nagumo equation. This equation arises in genetics, biology, and heat and mass transfer.

1°. Solutions:
$$u(x,t) = \frac{A\exp(z_1) + aB\exp(z_2)}{A\exp(z_1) + B\exp(z_2) + C},$$
$$z_1 = \pm\tfrac{\sqrt{2}}{2}x + (\tfrac{1}{2} - a)t, \quad z_2 = \pm\tfrac{\sqrt{2}}{2}ax + a(\tfrac{1}{2}a - 1)t,$$
where A, B, and C are arbitrary constants.

2°. References for PDE 6.1.1.5: Kawahara & Tanaka (1983) and Nucci & Clarkson (1992).

6. $u_t = u_{xx} + au + bu^k$.

Kolmogorov–Petrovskii–Piskunov equation (a special case). This equation arises in heat and mass transfer, combustion theory, biology, and ecology.

1°. Traveling wave solutions (the signs may be chosen arbitrarily):
$$u(x,t) = \left[\pm\beta + C\exp(\lambda t \pm \mu x)\right]^{\frac{2}{1-k}},$$
$$\beta = \sqrt{-\frac{b}{a}}, \quad \lambda = \frac{a(1-k)(k+3)}{2(k+1)}, \quad \mu = \sqrt{\frac{a(1-k)^2}{2(k+1)}}.$$
where C is an arbitrary constant.

2°. For $a = 0$, there is a self-similar solution of the form
$$u(x,t) = t^{1/(1-k)}U(z), \quad z = xt^{-1/2},$$
where the function $U = U(z)$ is described by the ODE
$$U''_{zz} + \tfrac{1}{2}zU'_z + \frac{1}{1-k}U + bU^k = 0.$$

7. $u_t = u_{xx} + au + bu^k + cu^{2k-1}$.

1°. Traveling wave solutions (Polyanin & Zaitsev, 2012):
$$u(x,t) = \left[\beta + C\exp(\lambda t + \mu x)\right]^{\frac{1}{1-k}},$$
$$\lambda = \frac{1-k}{k}\left[a(1+k) + \frac{b}{\beta}\right], \quad \mu = \pm(1-k)\sqrt{\tfrac{1}{k}\left(a + \frac{b}{\beta}\right)}, \tag{1}$$

where C is an arbitrary constant, and β is a root of the quadratic equation

$$a\beta^2 + b\beta + c = 0. \tag{2}$$

In the general case, formulas (1)–(2) gives four sets of the parameters, which generate four exact solutions of the original PDE.

2°. The substitution

$$v = u^{1-k}$$

leads to the PDE with quadratic nonlinearity

$$vv_t = vv_{xx} + \frac{k}{1-k}v_x^2 + a(1-k)v^2 + b(1-k)v + c(1-k). \tag{3}$$

Solutions (1) corresponds to particular solutions of PDE (3), which have the form $v = \beta + C\exp(\lambda t + \mu x)$.

For $a = 0$, equation (3) has also other traveling wave solutions:

$$v(x,t) = (1-k)\left(bt \pm \sqrt{-\frac{c}{k}}\,x\right) + C.$$

8. $u_t = u_{xx} + a + be^{\lambda u}$.

1°. Traveling wave solutions for $a \neq 0$ (the signs may be chosen arbitrarily):

$$u(x,t) = -\frac{2}{\lambda}\ln\bigl[\pm\beta + C\exp(\pm\mu x - \tfrac{1}{2}a\lambda t)\bigr], \quad \beta = \sqrt{-\frac{b}{a}},\ \mu = \sqrt{\frac{a\lambda}{2}},$$

where C is an arbitrary constant.

2°. For $a = 0$, there is an exact solution of the form

$$u = w(z) - \frac{1}{\lambda}\ln t, \quad z = \frac{x}{\sqrt{t}},$$

where the function $w = w(z)$ is determined by the ODE:

$$w''_{zz} + \frac{1}{2}zw'_z + \frac{1}{\lambda} + be^{\lambda w} = 0.$$

9. $u_t = u_{xx} + a + be^{\lambda u} + ce^{2\lambda u}$.

Equations of this form are encountered in problems of heat and mass transfer and combustion theory.

1°. Traveling wave solutions for $a \neq 0$ (Polyanin & Zaitsev, 2012):

$$u(x,t) = -\frac{1}{\lambda}\ln\bigl[\beta + C\exp(\pm\mu x - a\lambda t)\bigr], \quad \mu = \frac{1}{\beta}\sqrt{-c\lambda},$$

where C is an arbitrary constant, and the parameter β is determined by solving the quadratic equation

$$a\beta^2 + b\beta + c = 0.$$

2°. Traveling wave solutions for $a = 0$:

$$u(x,t) = -\frac{1}{\lambda}\ln\bigl(\pm\sqrt{-c\lambda}\,x - b\lambda t + C\bigr).$$

10. $u_t = u_{xx} + au \ln u$.

Functional separable solutions (Dorodnitsyn, 1982):
$$u(x,t) = \exp\left(Ae^{at}x + \frac{A^2}{a}e^{2at} + Be^{at}\right),$$
$$u(x,t) = \exp\left[\tfrac{1}{2} - \tfrac{1}{4}a(x+A)^2 + Be^{at}\right],$$
$$u(x,t) = \exp\left[-\frac{a(x+A)^2}{4(1+Be^{-at})} + \frac{1}{2B}e^{at}\ln(1+Be^{-at}) + Ce^{at}\right],$$

where A, B, and C are arbitrary constants.

11. $u_t = u_{xx} + au \ln^2 u$.

1°. The substitution $u = e^w$ leads to a PDE with quadratic nonlinearity:
$$w_t = w_{xx} + w_x^2 + aw^2. \tag{1}$$

2°. Exact solutions of PDE (1) for $a < 0$:
$$w(x,t) = C_1 \exp(-at \pm x\sqrt{-a}),$$
$$w(x,t) = \frac{1}{C_1 - at} + \frac{C_2}{(C_1 - at)^2}\exp(-at \pm x\sqrt{-a}), \tag{2}$$

where C_1 and C_2 are arbitrary constants. The first solution is a traveling wave solution and the second one is a generalized separable solution.

3°. PDE (1), in addition to solutions (2), has also generalized separable solutions of the following forms (Galaktionov & Posashkov, 1989):
$$w(x,t) = \varphi(t) + \psi(t)\left[C_1 \exp(-x\sqrt{-a}) + C_2 \exp(x\sqrt{-a})\right] \quad \text{if} \quad a < 0,$$
$$w(x,t) = \varphi(t) + \psi(t)\left[C_1 \sin(x\sqrt{a}) + C_2 \cos(x\sqrt{a})\right] \quad \text{if} \quad a > 0,$$

where the functions φ and ψ are described by some ODEs.

6.1.2. Burgers Type Equations and Related PDEs

1. $u_t + uu_x = u_{xx}$.

Burgers equation. It is used for describing wave processes in acoustics and hydrodynamics.

1°. Exact solutions:
$$u(x,t) = -\lambda - \frac{2}{x + \lambda t + A},$$
$$u(x,t) = -\frac{4x + 2A}{x^2 + Ax + 2t + B},$$
$$u(x,t) = -\frac{6(x^2 + 2t + A)}{x^3 + 6xt + 3Ax + B},$$
$$u(x,t) = -\frac{2\lambda}{1 + A\exp(-\lambda^2 t - \lambda x)},$$
$$u(x,t) = \lambda - A\frac{\exp[A(x - \lambda t)] - B}{\exp[A(x - \lambda t)] + B},$$
$$u(x,t) = -\frac{2\lambda \cos(\lambda x + A)}{B\exp(\lambda^2 t) + \sin(\lambda x + A)},$$

where A, B, and λ are arbitrary constants.

2°. Other solutions can be obtained using the following formula (*Hopf–Cole transformation*):
$$u(x,t) = -\frac{2}{Z} Z_x,$$
where $Z = Z(x,t)$ is a solution of the linear heat equation, $Z_t = Z_{xx}$ (see Section 5.1.1).

3°. The Cauchy problem for the Burgers equation with the initial condition
$$u = f(x) \quad \text{at} \quad t = 0 \quad (-\infty < x < \infty).$$
Solution:
$$u(x,t) = -2\frac{\partial}{\partial x}\ln Z(x,t), \quad Z(x,t) = \frac{1}{\sqrt{4\pi t}}\int_{-\infty}^{\infty}\exp\left[-\frac{(x-\xi)^2}{4t} - \frac{1}{2}\int_0^{\xi} f(\zeta)\,d\zeta\right]d\xi.$$

4°. References for PDE 6.1.2.1: Hopf (1950), Cole (1951), Benton & Platzman (1972), Ibragimov (1994), and Qin et al. (2007).

2. $u_t + buu_x = au_{xx}$.

Unnormalized Burgers equation. The scaling of the independent variables $x = \frac{a}{b}z$, $t = \frac{a}{b^2}\tau$ leads to an equation of the form 6.1.2.1:
$$u_\tau + uu_z = u_{zz}.$$

3. $u_t + buu_x = au_{xx} + c$.

The transformation
$$u = \xi(z,t) + ct, \quad z = x + \tfrac{1}{2}bct^2,$$
leads to the Burgers equation 6.1.2.2:
$$\xi_t + b\xi\xi_z = a\xi_{zz}.$$

4. $u_t + \sigma u u_x = au_{xx} + b_0 + b_1 u + b_2 u^2 + b_3 u^3$.

For $b_0 = 0$, this PDE is an unnormalized *Burgers–Huxley equation*, which describes wall motion of the fluid in liquid crystals as well as the dynamics of populations taking into account reproduction, mortality, nutrition and diffusion motion. For $\sigma = b_0 = 0$, it is the *Huxley equation*, which models the propagation of an electric pulse along a nerve fiber.

Exact solutions of the equation are given by
$$u(x,t) = \frac{\beta}{z}\frac{\partial z}{\partial x} + \lambda, \tag{*}$$

Here β and λ are any of the roots of the respective quadratic and cubic equations
$$b_3\beta^2 + \sigma\beta + 2a = 0,$$
$$b_3\lambda^3 + b_2\lambda^2 + b_1\lambda + b_0 = 0,$$
and the specific form of the function $z = z(x,t)$ depends on the equation coefficients.

1°. *Case* $b_3 \neq 0$. Introduce the notation:

$$p_1 = -\beta\sigma - 3a, \quad p_2 = \lambda\sigma + \beta b_2 + 3\beta\lambda b_3,$$
$$q_1 = -\frac{\beta b_2 + 3\beta\lambda b_3}{\beta\sigma + 2a}, \quad q_2 = -\frac{3b_3\lambda^2 + 2b_2\lambda + b_1}{\beta\sigma + 2a}.$$

Five main situations are possible, which are considered below in order.

1.1. For $q_2 \neq 0$ and $q_1^2 > 4q_2$, we have

$$z(x,t) = C_1 \exp(k_1 x + s_1 t) + C_2 \exp(k_2 x + s_2 t) + C_3,$$
$$k_n = -\tfrac{1}{2}q_1 \pm \tfrac{1}{2}\sqrt{q_1^2 - 4q_2}, \quad s_n = -k_n^2 p_1 - k_n p_2, \quad n = 1, 2,$$

where C_1, C_2, and C_3 are arbitrary constants.

1.2. For $q_2 \neq 0$ and $q_1^2 < 4q_2$, we find that

$$z(x,t) = [C_1 \sin(k_1 x + s_1 t) + C_2 \cos(k_1 x + s_1 t)] \exp(k_2 x + s_2 t) + C_3,$$
$$k_1 = \tfrac{1}{2}\sqrt{4q_2 - q_1^2}, \quad s_1 = \tfrac{1}{2}(p_1 q_1 - p_2)\sqrt{4q_2 - q_1^2},$$
$$k_2 = -\tfrac{1}{2}q_1, \quad s_2 = p_1(q_2 - \tfrac{1}{4}q_1^2) + \tfrac{1}{2}p_2 q_1,$$

where C_1, C_2, and C_3 are arbitrary constants.

1.3. For $q_2 \neq 0$ and $q_1^2 = 4q_2$, we get

$$z(x,t) = C_1 \exp(kx + s_1 t) + C_2(kx + s_2 t) \exp(kx + s_1 t) + C_3,$$
$$k = -\tfrac{1}{2}q_1, \quad s_1 = -\tfrac{1}{4}p_1 q_1^2 + \tfrac{1}{2}p_2 q_1, \quad s_2 = -\tfrac{1}{2}p_1 q_1^2 + \tfrac{1}{2}p_2 q_1.$$

1.4. For $q_2 = 0$ and $q_1 \neq 0$,

$$z(x,t) = C_1(x - p_2 t) + C_2 \exp[-q_1 x + q_1(p_2 - p_1 q_1)t] + C_3.$$

1.5. For $q_2 = q_1 = 0$,

$$z(x,t) = C_1(x - p_2 t)^2 + C_2(x - p_2 t) - 2C_1 p_1 t + C_3.$$

2°. *Case* $b_3 = 0$, $b_2 \neq 0$. The solutions are defined by formula $(*)$ where

$$z(x,t) = C_1 + C_2 \exp\left[Ax + A\left(\frac{b_1\sigma}{2b_2} + \frac{2ab_2}{\sigma}\right)t\right],$$
$$\beta = -\frac{2a}{\sigma}, \quad A = \frac{\sigma(b_1 + 2b_2\lambda)}{2ab_2},$$

and $\lambda = \lambda_{1,2}$ are roots of the quadratic equation

$$b_2\lambda^2 + b_1\lambda + b_0 = 0.$$

3°. References for PDE 6.1.2.4: Estévez & Gordoa (1990), Kudryashov (1993), and Estévez (1994).

5. $u_t + buu_x = \dfrac{a}{x^2}[x(xu_x)_x - u]$.

Cylindrical Burgers equation. The variable x plays the role of the radial coordinate.

Solution (Nerney et al., 1996):

$$u(x,t) = -\frac{2a}{b}\frac{Z_x}{Z},$$

where the function $Z = Z(x,t)$ satisfies the linear heat equation with axial symmetry

$$Z_t = \frac{a}{x}(xZ_x)_x.$$

6. $u_t = au_{xx} + bu_x^2.$

Potential Burgers equation.

1°. Exact solutions:

$$u(x,t) = A^2 bt \pm Ax + B,$$

$$u(x,t) = -\frac{(x+A)^2}{4bt} - \frac{a}{2b}\ln t + B,$$

$$u(x,t) = \frac{a}{b}\ln|x^2 + 2at + Ax + B| + C,$$

$$u(x,t) = \frac{a}{b}\ln|x^3 + 6axt + Ax + B| + C,$$

$$u(x,t) = \frac{a}{b}\ln|x^4 + 12ax^2 t + 12a^2 t^2 + A| + B,$$

$$u(x,t) = -\frac{a^2\lambda^2}{b}t + \frac{a}{b}\ln|\cos(\lambda x + A)| + B,$$

where A, B, C, and λ are arbitrary constants.

2°. The substitution

$$u(x,t) = \frac{a}{b}\ln|z(x,t)|$$

leads to the linear heat equation $z_t = az_{xx}.$

7. $u_t = au_{xx} + bu_x^2 + c.$

The substitution

$$u(x,t) = \frac{a}{b}\ln|z(x,t)| + ct$$

leads to the linear heat equation $z_t = az_{xx}.$

8. $u_t = u_{xx} + f(u)u_x^2.$

Generalized potential Burgers equation.

The substitution

$$z = \int F(u)\,du, \quad \text{where} \quad F(u) = \exp\left[\int f(u)\,du\right],$$

leads to the linear heat equation for the function $z = z(x,t)$:

$$z_t = az_{xx}.$$

6.1.3. Reaction-Diffusion Equations of the Form $u_t = [f(u)u_x]_x + g(u)$

▶ Equations of this form admit traveling wave solutions $u = u(z)$, $z = \kappa x + \lambda t$, where κ and λ are arbitrary constants, and the function $u(z)$ satisfies the second-order autonomous ODE: $\kappa^2[f(u)u'_z]'_z - \lambda u'_z + f(u) = 0$.

▶ **Equations with power nonlinearities.**

1. $u_t = a(u^k u_x)_x$.

This equation occurs in nonlinear problems of heat and mass transfer and flows in porous media. To $k > 0$ there corresponds *slow diffusion* and to $k < 0$, *fast diffusion*.

1°. Solutions:

$$u(x,t) = \left[\frac{k\lambda}{a}(\pm x + \lambda t) + A\right]^{\frac{1}{k}},$$

$$u(x,t) = \left[\frac{k(x-A)^2}{2a(k+2)(B-t)}\right]^{\frac{1}{k}},$$

$$u(x,t) = \left[A|t+B|^{-\frac{k}{k+2}} - \frac{k}{2a(k+2)}\frac{(x+C)^2}{t+B}\right]^{\frac{1}{k}},$$

$$u(x,t) = \left[\frac{k(x+A)^2}{\varphi(t)} + B|x+A|^{\frac{k}{k+1}}|\varphi(t)|^{-\frac{k(2k+3)}{2(k+1)^2}}\right]^{\frac{1}{k}}, \quad \varphi(t) = C - 2a(k+2)t,$$

where A, B, C, and λ are arbitrary constants. The second solution for $B > 0$ corresponds to *blow-up regime* (the solution increases without bound on a finite time interval).

2°. There are solutions of the following forms:

$$u(x,t) = (t+C)^{-1/k}F(x) \quad \text{(multiplicative separable solution)};$$

$$u(x,t) = t^\lambda G(\xi), \quad \xi = xt^{-\frac{k\lambda+1}{2}} \quad \text{(self-similar solution)};$$

$$u(x,t) = e^{-2\lambda t}H(\eta), \quad \eta = xe^{k\lambda t} \quad \text{(generalized self-similar solution)};$$

$$u(x,t) = t^{-1/k}U(\zeta), \quad \zeta = x + \lambda \ln t,$$

where C and λ are arbitrary constants.

3°. References for PDE 6.1.3.1: Zel'dovich & Kompaneets (1950), Barenblatt (1952), Barenblatt & Zel'dovich (1957), Ovsiannikov (1959, 1982), Ibragimov (1994), Rudykh & Semenov (1998), and Polyanin & Zaitsev (2012).

2. $u_t = a(u^k u_x)_x + bu$.

By the transformation $u(x,t) = e^{bt}v(x,\tau)$, $\tau = \frac{1}{bk}e^{bkt} + C$ the original equation can be reduced to a PDE of the form 6.1.3.1: $v_\tau = a(v^k v_x)_x$.

3. $u_t = a(u^k u_x)_x + bu^{k+1}$.

1°. Multiplicative separable solution ($a = b = 1$, $k > 0$):

$$u(x,t) = \begin{cases} \left[\dfrac{2(k+1)}{k(k+2)} \dfrac{\cos^2(\pi x/L)}{(t_0 - t)}\right]^{1/k} & \text{for } |x| \leq \dfrac{L}{2}, \\ 0 & \text{for } |x| > \dfrac{L}{2}, \end{cases}$$

where $L = 2\pi(k+1)^{1/2}/k$. This solution describes a blow-up regime that exists on a limited time interval $t \in [0, t_0)$. The solution is localized in the interval $|x| < L/2$.

2°. Multiplicative separable solution:

$$u(x,t) = \left(\frac{Ae^{\mu x} + Be^{-\mu x} + D}{k\lambda t + C}\right)^{1/k},$$

$$B = \frac{\lambda^2(k+1)^2}{4b^2 A(k+2)^2}, \quad D = -\frac{\lambda(k+1)}{b(k+2)}, \quad \mu = k\sqrt{-\frac{b}{a(k+1)}},$$

where A, C, and λ are arbitrary constants, $ab(k+1) < 0$.

3°. Functional separable solutions [it is assumed that $ab(k+1) < 0$]:

$$u(x,t) = \left[F(t) + C_2|F(t)|^{\frac{k+2}{k+1}} e^{\lambda x}\right]^{1/k},$$

$$F(t) = \frac{1}{C_1 - bkt}, \quad \lambda = \pm k\sqrt{\frac{-b}{a(k+1)}},$$

where C_1 and C_2 are arbitrary constants.

4°. There are functional separable solutions of the following forms:

$$u(x,t) = \left[f(t) + g(t)(Ae^{\lambda x} + Be^{-\lambda x})\right]^{1/k}, \quad \lambda = k\sqrt{\frac{-b}{a(k+1)}};$$

$$u(x,t) = \left[f(t) + g(t)\cos(\lambda x + C)\right]^{1/k}, \quad \lambda = k\sqrt{\frac{b}{a(k+1)}},$$

where A, B, and C are arbitrary constants.

3°. References for PDE 6.1.3.3: Bertsch et al. (1985), Galaktionov & Posashkov (1989), and Polyanin & Zaitsev (2012).

4. $u_t = a(u^k u_x)_x + bu^{k+1} + cu$.

1°. Multiplicative separable solutions:

$$u(x,t) = e^{ct}\left[A\cos(\beta x) + B\sin(\beta x)\right]^{\frac{1}{k+1}} \quad \text{if} \quad b(k+1)/a = \beta^2 > 0,$$

$$u(x,t) = e^{ct}\left[A\exp(\beta x) + B\exp(-\beta x)\right]^{\frac{1}{k+1}} \quad \text{if} \quad b(k+1)/a = -\beta^2 < 0,$$

where A and B are arbitrary constants.

2°. Multiplicative separable solution for $k = -1$:
$$u = A\exp\left(ct - \frac{b}{2a}x^2 + Bx\right),$$
where A and B are arbitrary constants.

3°. The transformation
$$u = e^{ct}w(x,\tau), \quad \tau = \frac{1}{ck}e^{ckt} + \text{const}$$
leads to a simpler PDE of the form 6.1.3.3:
$$w_\tau = a(w^k w_x)_x + bw^{k+1}.$$

5. $u_t = a(u^k u_x)_x + bu^{1-k}.$

Functional separable solution (Kersner, 1978):
$$u(x,t) = \left[\frac{(x+A)^2}{F(t)} + B|F(t)|^{-\frac{k}{k+2}} - \frac{bk^2}{4a(k+1)}F(t)\right]^{1/k}, \quad F(t) = C - \frac{2a(k+2)}{k}t,$$
where A, B, and C are arbitrary constants.

6. $u_t = a(u^k u_x)_x + b + cu^{-k}.$

Functional separable solution:
$$u = \left[c(k+1)t - \frac{b(k+1)}{2a}x^2 + C_1 x + C_2\right]^{\frac{1}{k+1}},$$
where C_1 and C_2 are arbitrary constants.

7. $u_t = a(u^k u_x)_x + bu^{k+1} + cu + du^{1-k}.$

Functional separable solutions (Galaktionov & Posashkov, 1989):
$$\begin{aligned} u &= \{\varphi(t)[C_1 \cos(\beta x) + C_2 \sin(\beta x)] + \psi(t)\}^{1/k} & \text{if } ab(k+1) > 0, \\ u &= \{\varphi(t)[C_1 \cosh(\beta x) + C_2 \sinh(\beta x)] + \psi(t)\}^{1/k} & \text{if } ab(k+1) < 0, \end{aligned} \quad (1)$$
where C_1 and C_2 are arbitrary constants,
$$\beta = \sqrt{\frac{|b|k^2}{|a(k+1)|}},$$
while $\varphi = \varphi(t)$ and $\psi = \psi(t)$ are functions that satisfy the system of ODEs
$$\begin{aligned} \varphi'_t &= \frac{bk(k+2)}{k+1}\varphi\psi + ck\varphi, \\ \psi'_t &= k(b\psi^2 + c\psi + d) + \frac{bk}{k+1}(C_1^2 \pm C_2^2)\varphi^2. \end{aligned} \quad (2)$$

The upper sign in the second equation corresponds to the first solution in (1), and the lower sign to the second solution in (1).

For $C_1 = C_2$, the last equation in (2) (lower sign) can be satisfied if we set $\psi = \text{const}$, where ψ is a root of the quadratic equation $b\psi^2 + c\psi + d = 0$. In this case, the general solution to the first equation of (2) is given by

$$\varphi = C_3 \exp\left[\left(\frac{bk(k+2)}{k+1}\psi + ck\right)t\right],$$

where C_3 is an arbitrary constant.

8. $u_t = a(u^{2k} u_x)_x + bu^{1-k}$.

Generalized traveling wave solution (Polyanin & Zaitsev, 2004):

$$u(x,t) = \left[\pm\frac{x+C_1}{\sqrt{C_2-st}} - \frac{bk^2}{3a(k+1)}(C_2-st)\right]^{1/k}, \quad s = \frac{2a(k+1)}{k},$$

where C_1 and C_2 are arbitrary constants.

9. $u_t = a(u^{2k} u_x)_x + bu + cu^{1-k}$.

Generalized traveling wave solutions for $b \neq 0$ (Polyanin & Zaitsev, 2004):

$$u(x,t) = \left[\varphi(t)(\pm x + C_1) + ck\varphi(t)\int\frac{dt}{\varphi(t)}\right]^{1/k}, \quad \varphi(t) = \left[C_2 e^{-2bkt} - \frac{a(k+1)}{bk^2}\right]^{-1/2},$$

where C_1 and C_2 are arbitrary constants.

10. $u_t = a(u^k u_x)_x + bu^m$.

There are solutions of the following forms (Dorodnitsyn, 1982):

$$u(x,t) = U(z), \quad z = \pm x + \lambda t \qquad \text{(traveling wave solutions)};$$

$$u(x,t) = t^{\frac{1}{1-m}} V(\xi), \quad \xi = xt^{\frac{m-k-1}{2(1-m)}} \quad \text{(self-similar solution)}.$$

11. $u_t = [(au^{2k} + bu^k)u_x]_x$.

This is a special case of PDE 6.1.3.24 with $f(u) = au^{2k} + bu^k$.

1°. Traveling wave solutions:

$$u(x,t) = \left(\pm\sqrt{2C_1 kx + 2aC_1^2 kt + C_2} - \frac{b}{a}\right)^{1/k},$$

where C_1 and C_2 are arbitrary constants.

2°. Self-similar solutions:

$$u(x,t) = \left[\pm\frac{x+C_1}{\sqrt{C_2-st}} - \frac{b}{a(k+1)}\right]^{1/k}, \quad s = \frac{2a(k+1)}{k}.$$

12. $u_t = [(au^{2k} + bu^k)u_x]_x + cu^{1-k}$.

Generalized traveling wave solutions (Polyanin & Zaitsev, 2004):

$$u(x,t) = \left[\pm\frac{x+C_1}{\sqrt{C_2-st}} - \frac{ck^2}{3a(k+1)}(C_2-st) - \frac{b}{a(k+1)}\right]^{1/k}, \quad s = \frac{2a(k+1)}{k},$$

where C_1 and C_2 are arbitrary constants.

13. $u_t = [(au^{2k} + bu^k)u_x]_x + cu + du^{1-k}$.

Generalized traveling wave solutions (Polyanin & Zaitsev, 2012):

$$u(x,t) = \left[\varphi(t)(\pm x + C_1) + \frac{b}{k}\varphi(t)\int \varphi(t)\,dt + dk\varphi(t)\int \frac{dt}{\varphi(t)}\right]^{1/k},$$

$$\varphi(t) = \left[C_2 e^{-2ckt} - \frac{a(k+1)}{ck^2}\right]^{-1/2},$$

where C_1 and C_2 are arbitrary constants.

▶ **Equations with exponential nonlinearities.**

14. $u_t = a(e^{\lambda u} u_x)_x$.

1°. Exact solutions:

$$u(x,t) = \frac{2}{\lambda}\ln\left(\frac{\pm x + A}{\sqrt{B - 2at}}\right),$$

$$u(x,t) = \frac{1}{\lambda}\ln\frac{A + Bx - Cx^2}{D + 2aCt},$$

where A, B, C, and D are arbitrary constants.

2°. There are solutions of the following forms:

$$u(x,t) = F(z), \quad z = kx + \beta t \quad \text{(traveling wave solution)};$$
$$u(x,t) = G(\xi), \quad \xi = xt^{-1/2} \quad \text{(self-similar solution)};$$
$$u(x,t) = H(\eta) + 2kt, \quad \eta = xe^{-k\lambda t};$$
$$u(x,t) = U(\zeta) - \lambda^{-1}\ln t, \quad \zeta = x + k\ln t,$$

where k and β are arbitrary constants.

3°. References for PDE 6.1.3.14: Ovsiannikov (1959, 1982), Ibragimov (1994), and Polyanin & Zaitsev (2012).

15. $u_t = a(e^{\lambda u} u_x)_x + b$.

1°. Additive separable solutions:

$$u(x,t) = \frac{1}{\lambda}\ln(C_1 x + C_2) + bt + C_3,$$

$$u(x,t) = \frac{2}{\lambda}\ln(x + C_1) - \ln\left(C_2 e^{-b\lambda t} - \frac{2a}{b\lambda}\right).$$

2°. The transformation

$$u(x,t) = w(x,\tau) + bt, \quad \tau = \frac{1}{b\lambda}e^{b\lambda t} + C$$

leads the original PDE to an equation of the form 6.1.3.14:

$$w_\tau = a(e^{\lambda w} w_x)_x.$$

16. $u_t = (ae^{\lambda u} u_x)_x + be^{\beta u}$.

1°. Traveling wave solution:
$$u = u(z), \quad z = k_2 x + k_1 t,$$

where k_1 and k_2 are arbitrary constants, and the function $u(z)$ is described by the autonomous ODE
$$ak_2^2 (e^{\lambda u} u'_z)'_z - k_1 u'_z + be^{\beta u} = 0.$$

2°. Solution (Dorodnitsyn, 1982):
$$u = U(\xi) - \frac{1}{\beta} \ln t, \quad \xi = x t^{\frac{\lambda - \beta}{2\beta}},$$

where the function $U(\xi)$ is described by the ODE
$$\frac{\lambda - \beta}{2\beta} \xi U'_\xi - \frac{1}{\beta} = (ae^{\lambda U} U'_\xi)'_\xi + be^{\beta U}.$$

17. $u_t = a(e^{\lambda u} u_x)_x + be^{\lambda u} + c + de^{-\lambda u}$.

Functional separable solutions (Galaktionov & Posashkov, 1989):
$$u = \frac{1}{\lambda} \ln\{e^{\sigma t}[C_1 \cos(x\sqrt{\beta}) + C_2 \sin(x\sqrt{\beta})] + \gamma\} \quad \text{if } ab\lambda > 0,$$
$$u = \frac{1}{\lambda} \ln\{e^{\sigma t}[C_1 \cosh(x\sqrt{-\beta}) + C_2 \sinh(x\sqrt{-\beta})] + \gamma\} \quad \text{if } ab\lambda < 0.$$

Here C_1 and C_2 are arbitrary constants, and
$$\sigma = \lambda(b\gamma + c), \quad \beta = b\lambda/a,$$

where $\gamma = \gamma_{1,2}$ are roots of the quadratic equation $b\gamma^2 + c\gamma + d = 0$.

18. $u_t = a(u e^{\lambda u} u_x)_x$.

This is a special case of PDE 6.1.3.24 with $f(u) = aue^{\lambda u}$.

Traveling wave solution:
$$u(x,t) = \frac{1}{\lambda} \ln\left(C_1 x + \frac{a}{\lambda} C_1^2 t + C_2\right),$$

where C_1 and C_2 are arbitrary constants.

19. $u_t = a(u e^{\lambda u} u_x)_x + b$.

Generalized traveling wave solution:
$$u(x,t) = \frac{1}{\lambda} \ln\left(C_1 e^{b\lambda t} x + \frac{aC_1^2}{b\lambda^2} e^{2b\lambda t} + C_2 e^{b\lambda t}\right),$$

where C_1 and C_2 are arbitrary constants.

20. $u_t = a(ue^{\lambda u}u_x)_x + be^{-\lambda u}$.

Traveling wave solution:
$$u(x,t) = \frac{1}{\lambda}\ln\left[C_1 x + \left(\frac{aC_1^2}{\lambda} + b\lambda\right)t + C_2\right],$$
where C_1 and C_2 are arbitrary constants.

21. $u_t = a(ue^{\lambda u}u_x)_x + b + ce^{-\lambda u}$.

Generalized traveling wave solution (Polyanin & Zaitsev, 2004):
$$u(x,t) = \frac{1}{\lambda}\ln\left(C_1 e^{b\lambda t}x + \frac{aC_1^2}{b\lambda^2}e^{2b\lambda t} + C_2 e^{b\lambda t} - \frac{c}{b}\right),$$
where C_1 and C_2 are arbitrary constants.

22. $u_t = [(ae^{2\lambda u} + bue^{\lambda u})u_x]_x$.

This is a special case of PDE 6.1.3.24 with $f(u) = ae^{2\lambda u} + bue^{\lambda u}$.

Self-similar solutions:
$$u(x,t) = \frac{1}{\lambda}\ln\left(\frac{\pm x + C_1}{\sqrt{C_2 - 2at}} - \frac{b}{a\lambda}\right),$$
where C_1 and C_2 are arbitrary constants.

23. $u_t = [(ae^{2\lambda u} + bue^{\lambda u})u_x]_x + c$.

Generalized traveling wave solutions (Polyanin & Zaitsev, 2012):
$$u(x,t) = \frac{1}{\lambda}\ln\left[\pm\varphi(t)x + C_1\varphi(t) + \frac{b}{\lambda}\varphi(t)\int\varphi(t)\,dt\right], \quad \varphi(t) = \left(C_2 e^{-2c\lambda t} - \frac{a}{c\lambda}\right)^{-1/2},$$
where C_1 and C_2 are arbitrary constants.

▶ **Equations containing arbitrary functions.**

24. $u_t = [f(u)u_x]_x$.

This equation occurs in nonlinear problems of heat and mass transfer and flows in porous media.

1°. Traveling wave solution in implicit form:
$$k^2\int\frac{f(u)\,du}{\lambda u + C_1} = kx + \lambda t + C_2,$$
where C_1, C_2, k, and λ are arbitrary constants. To $\lambda = 0$ there corresponds a stationary solution.

2°. Self-similar solution (Ovsiannikov, 1959):
$$u = u(z), \quad z = xt^{-1/2},$$
where the function $u(z)$ is described by the ODE: $[f(u)u'_z]'_z + \frac{1}{2}zu'_z = 0$.

25. $u_t = [f(u)u_x]_x + g(u)$.

Reaction-diffusion equation of the general form. This PDE occurs in nonlinear problems of heat and mass transfer with volume reaction.

$1°$. Traveling wave solutions:
$$u = u(z), \quad z = \pm x + \lambda t,$$
where the function $u(z)$ is described by the autonomous ODE $[f(u)u'_z]'_z - \lambda u'_z + g(u) = 0$.

$2°$. Let the function $f = f(u)$ be arbitrary and let $g = g(u)$ be defined by
$$g(u) = \frac{A}{f(u)} + B,$$
where A and B are some numbers. In this case, there is a functional separable solution that is defined implicitly by
$$\int f(u)\, du = At - \frac{1}{2}Bx^2 + C_1 x + C_2,$$
where C_1 and C_2 are arbitrary constants.

$3°$. Let the function $f = f(u)$ be arbitrary and let $g = g(u)$ be defined by
$$g(u) = \frac{aF(u) + b}{f(u)} + c[aF(u) + b], \quad F(u) = \int f(u)\, du,$$
where a, b, and c are some numbers. Then there are functional separable solutions defined implicitly by
$$\int f(u)\, du = e^{at}\left[C_1 \cos(x\sqrt{ac}) + C_2 \sin(x\sqrt{ac})\right] - \frac{b}{a} \quad \text{if } ac > 0,$$
$$\int f(u)\, du = e^{at}\left[C_1 \cosh(x\sqrt{-ac}) + C_2 \sinh(x\sqrt{-ac})\right] - \frac{b}{a} \quad \text{if } ac < 0.$$

$4°$. Now let $g = g(u)$ be arbitrary and let $f = f(u)$ be defined by
$$f(u) = \frac{A_1 A_2 u + B}{g(u)} + \frac{A_2 A_3}{g(u)} \int Z\, du, \tag{1}$$
$$Z = -A_2 \int \frac{du}{g(u)}, \tag{2}$$
where A_1, A_2, A_3, and B are some numbers. Then there are generalized traveling wave solutions of the form
$$u = u(Z), \quad Z = \frac{\pm x + C_2}{\sqrt{2A_3 t + C_1}} - \frac{A_1}{A_3} - \frac{A_2}{3A_3}(2A_3 t + C_1),$$
where the function $u(Z)$ is determined by the inversion of (2), and C_1 and C_2 are arbitrary constants.

$5°$. Let $g = g(u)$ be arbitrary and let $f = f(u)$ be defined by
$$f(u) = \frac{1}{g(u)}\left(A_1 u + A_3 \int Z\, du\right) \exp\left[-A_4 \int \frac{du}{g(u)}\right], \tag{3}$$
$$Z = \frac{1}{A_4} \exp\left[-A_4 \int \frac{du}{g(u)}\right] - \frac{A_2}{A_4}, \tag{4}$$

where A_1, A_2, A_3, and A_4 are some numbers ($A_4 \neq 0$). In this case, there are generalized traveling wave solutions of the form

$$u = u(Z), \quad Z = \varphi(t)x + \psi(t),$$

where the function $u(Z)$ is determined by the inversion of (4),

$$\varphi(t) = \pm\left(C_1 e^{2A_4 t} - \frac{A_3}{A_4}\right)^{-1/2}, \quad \psi(t) = -\varphi(t)\left[A_1 \int \varphi(t)\,dt + A_2 \int \frac{dt}{\varphi(t)} + C_2\right],$$

and C_1 and C_2 are arbitrary constants.

6°. Let $f(u)$ and $g(u)$ be as follows:

$$f(u) = u\varphi_u'(u), \quad g(u) = a\left[u + 2\frac{\varphi(u)}{\varphi_u'(u)}\right],$$

where $\varphi(u)$ is an arbitrary function and a is some number. Then there are functional separable solutions defined implicitly by

$$\varphi(u) = C_1 e^{2at} - \tfrac{1}{2}a(x + C_2)^2.$$

7°. Let $f(u)$ and $g(u)$ be as follows:

$$f(u) = A\frac{V(z)}{V_z'(z)}, \quad g(u) = B\left[2z^{-1/2}V_z'(z) + z^{-3/2}V(z)\right],$$

where $V(z)$ is an arbitrary function, A and B are some numbers ($AB \neq 0$), and the function $z = z(u)$ is defined implicitly by

$$u = \int z^{-1/2} V_z'(z)\,dz + C_1; \tag{5}$$

where C_1 is an arbitrary constant. Then there are functional separable solutions of the form (5), where

$$z = -\frac{(x + C_3)^2}{4At + C_2} + 2Bt + \frac{BC_2}{2A},$$

and C_2 and C_3 are arbitrary constants.

8°. References for PDE 6.1.3.25: Dorodnitsyn (1982), Galaktionov (1994), Ibragimov (1994), Polyanin & Vyazmina (2005), Polyanin & Zaitsev (2012), and Cherniha et al. (2018).

6.1.4. Other Reaction-Diffusion and Heat PDEs with Variable Transfer Coefficient

1. $u_t = [a(x)u_x]_x + b(x)u + cu \ln u.$

Multiplicative separable solution:

$$u = \exp(Ae^{ct})\varphi(x),$$

where A is an arbitrary constant, and the function $\varphi = \varphi(x)$ is described by the second-order ODE $[a(x)\varphi_x']_x' + b(x)\varphi + c\varphi \ln \varphi = 0$.

2. $u_t = [a(x)u_x]_x + \dfrac{x^2}{a(x)} f(u)$.

Exact solution (Polyanin, 2019a):
$$u = U(z), \quad z = t + \int \frac{x\,dx}{a(x)},$$
where the function $U = U(z)$ is described by a second-order autonomous ODE of the form 2.3.1.1: $U''_{zz} + f(U) = 0$.

3. $u_t = ax^{-n}(x^n u^k u_x)_x$.

This is a special case of PDE 6.1.4.4 with $f(u) = au^k$ and $g(u) = 0$. It occurs in nonlinear problems of heat and mass transfer and other applications. For $n = 0$, see equation 6.1.3.1. Two-dimensional problems with axial symmetry correspond to $n = 1$, and three-dimensional spherically symmetric problems correspond to $n = 2$. Equation with $n = 5$ are encountered in the theory of static turbulence. To $k > 0$ there corresponds *slow diffusion* and to $k < 0$, *fast diffusion*.

1°. Exact solutions:
$$u(x,t) = \left(\frac{kx^2}{A - st}\right)^{\frac{1}{k}}, \quad s = 2a(nk + k + 2),$$
$$u(x,t) = \left(A|st + B|^{-\frac{k(n+1)}{nk+k+2}} - \frac{kx^2}{st + B}\right)^{\frac{1}{k}}, \quad s = 2a(nk + k + 2),$$
$$u(x,t) = \left[A \exp\left(-\frac{4a\lambda}{k} t\right) + \lambda x^2\right]^{\frac{1}{k}}, \quad n = -\frac{k+2}{k},$$
where A, B, and λ are arbitrary constants.

2°. Exact solutions with $k = -\dfrac{2}{n+1}$:
$$u(x,t) = e^{-a\lambda(n+1)t}\left[\frac{\lambda}{n+1}(C + x^2 e^{-2a\lambda t})\right]^{-\frac{n+1}{2}},$$
$$u(x,t) = t^{\frac{1+n}{1-n}}\left(\frac{\xi^2}{n-1}\ln\frac{\xi}{C}\right)^{-\frac{n+1}{2}}, \quad \xi = xt^{\frac{1+n}{2(1-n)}},$$
where C and λ are arbitrary constants.

3°. References for PDE 6.1.4.3: Zel'dovich & Kompaneets (1950), Barenblatt (1951), Pattle (1959), King (1990), Ivanova (2008), and Polyanin & Zaitsev (2012).

4. $u_t = x^{-n}[x^n f(u) u_x]_x + g(u)$.

This is a nonlinear equation of heat and mass transfer in the radial symmetric case ($n = 1$ corresponds to a plane problem and $n = 2$ to a spatial one). For $n = 0$, see equation 6.1.3.25.

1°. Let the function $f = f(u)$ be arbitrary and let $g = g(u)$ be defined by

$$g(u) = \left(\frac{a}{f(u)} + b\right)\left(\int f(u)\,du + c\right),$$

where a, b, and c are some numbers. In this case, there is a functional separable solution that is defined implicitly by

$$\int f(u)\,du + c = e^{at} z(x),$$

where the function $z = z(x)$ is determined by the linear ODE:

$$z''_{xx} + \frac{n}{x} z'_x + bz = 0.$$

Its general solution can be expressed in terms of Bessel functions of modified Bessel functions.

2°. Let $f(u)$ and $g(u)$ be defined as

$$f(u) = u\varphi'_u(u), \quad g(u) = a(n+1)u + 2a\frac{\varphi(u)}{\varphi'_u(u)},$$

where $\varphi(u)$ is an arbitrary function, and a is some number; the prime denotes a derivative with respect to u. In this case, there is a functional separable solution defined implicitly by

$$\varphi(u) = Ce^{2at} - \tfrac{1}{2}ax^2,$$

where C is an arbitrary constant.

3°. Let $f(u)$ and $g(u)$ be defined as follows:

$$f(u) = a\varphi^{-\frac{n+1}{2}} \varphi' \int \varphi^{\frac{n+1}{2}} du, \quad g(u) = b\frac{\varphi}{\varphi'},$$

where $\varphi = \varphi(u)$ is an arbitrary function (the prime denotes a derivative with respect to u). In this case, there is a functional separable solution defined implicitly by

$$\varphi(u) = \frac{bx^2}{Ce^{-bt} - 4a}.$$

4°. Let $f(u)$ and $g(u)$ be defined as

$$f(u) = A\frac{V(z)}{V'_z(z)}, \quad g(u) = B\left[2z^{-\frac{n+1}{2}} V'_z(z) + (n+1)z^{-\frac{n+3}{2}} V(z)\right],$$

where $V(z)$ is an arbitrary function, A and B are some numbers ($AB \neq 0$), and the function $z = z(u)$ is defined implicitly by

$$u = \int z^{-\frac{n+1}{2}} V'_z(z)\,dz + C_1, \qquad (*)$$

where C_1 is an arbitrary constant. Then there are functional separable solutions of the form $(*)$ where

$$z = -\frac{x^2}{4At + C_2} + 2Bt + \frac{BC_2}{2A},$$

and C_2 is an arbitrary constant.

5°. Self-similar solution for $g(u) \equiv 0$:

$$u = u(z), \quad z = xt^{-1/2},$$

where the function $u(z)$ is described by the ODE: $z^{-n}[z^n f(u)u'_z]'_z + \frac{1}{2}zu'_z = 0$.

6°. References for PDE 6.1.4.4: Galaktionov (1994), Polyanin & Vyazmina (2005), and Polyanin & Zaitsev (2012).

5. $u_t = [a(x)u^k u_x]_x + b(x)u^{k+1}$.

Multiplicative separable solution:

$$u = (t+C)^{-1/k}\varphi(x),$$

where C is an arbitrary constant, and the function $\varphi = \varphi(x)$ is described by the ODE: $[a(x)\varphi^k \varphi'_x]'_x + b(x)\varphi^{k+1} + (1/k)\varphi = 0$.

6. $u_t = [x^k f(u)u_x]_x + x^{k-2}g(u), \quad k \neq 2$.

Self-similar solution (Polyanin, 2019b):

$$u = U(z), \quad z = x(t+C)^{\frac{1}{k-2}},$$

where C is an arbitrary constant, and the function $U = U(z)$ is described by the second-order ODE:

$$[z^k f(U)U'_z]'_z + \frac{1}{2-k}zU'_z + z^{k-2}g(U) = 0.$$

7. $u_t = [x^2 f(u)u_x]_x + g(u)$.

Exact solution:

$$u = U(z), \quad z = \ln x + \lambda t$$

where λ is an arbitrary constant, and the function $U = U(z)$ is described by the second-order nonlinear ODE:

$$[f(U)U'_z]'_z + [f(U) - \lambda]U'_z + g(U) = 0.$$

8. $u_t = [a(x)e^{\lambda u}u_x]_x + b(x)e^{\lambda u}$.

Additive separable solution:

$$u = \frac{1}{\lambda}\ln \varphi(x) - \frac{1}{\lambda}\ln(t+C),$$

where C is an arbitrary constant, and the function $\varphi = \varphi(x)$ is described by the second-order linear ODE:

$$[a(x)\varphi'_x]'_x + \lambda b(x)\varphi + 1 = 0.$$

9. $u_t = [e^{\lambda x} f(u)u_x]_x + e^{\lambda x}g(u), \quad \lambda \neq 0$.

Exact solution (Polyanin, 2019b):

$$u = U(z), \quad z = \lambda x + \ln t$$

where λ is an arbitrary constant, and the function $U = U(z)$ is described by the second-order nonlinear ODE:
$$\lambda^2 [e^z f(U) U'_z]'_z - U'_z + e^z g(U) = 0.$$

10. $u_t = [a(x) f(u) u_x]_x + b(x) + \dfrac{k}{f(u)}.$

Exact solution in implicit form (Polyanin, 2019b):
$$\int f(u)\, du = kt - \int \frac{1}{a(x)} \left(\int b(x)\, dx \right) dx + C_1 \int \frac{dx}{a(x)} + C_2,$$
where C_1 and C_2 are arbitrary constants.

11. $u_t = [a(x) f(u) u_x]_x + k \dfrac{a'_x(x)}{\sqrt{a(x)}} u.$

Exact solution in implicit form (Polyanin, 2019b):
$$\int \frac{f(u)}{u}\, du = 4k^2 t - 2k \int \frac{dx}{\sqrt{a(x)}} + C,$$
where C is an arbitrary constant.

12. $u_t = [a(x) f(u) u_x]_x + b(x) g(u).$

For exact solutions of such nonlinear PDEs, which are not considered in this section, with various functions $a(x)$, $b(x)$, $f(u)$, and $g(u)$, see Vaneeva et al. (2007, 2012), Polyanin & Zaitsev (2012), Polyanin (2019a, 2019b, 2020), Cherniha et al. (2018), and Polyanin & Zhurov (2022).

6.1.5. Convection-Diffusion Type PDEs

▶ **Equations with power nonlinearities.**

1. $u_t = a(u^k u_x)_x + b u_x.$

The transformation
$$u = \xi(z, t), \quad z = x + bt,$$
leads to a simpler equation of the form 6.1.3.1:
$$\xi_t = a(\xi^k \xi_z)_z.$$

2. $u_t = a(u^k u_x)_x + b u u_x.$

$1°$. Traveling wave solution in implicit form:
$$2a \int \frac{u^k\, du}{-bu^2 + 2\lambda u + C_1} = x + \lambda t + C_2,$$
where C_1, C_2, and λ are arbitrary constants.

2°. Self-similar solution for $k \neq 2$:
$$u(x,t) = U(z)t^{1/(k-2)}, \quad z = xt^{-(k-1)/(k-2)},$$
where the function $U = U(z)$ is described by the ODE
$$aU^k U''_{zz} + 2akU^{k-1}(U'_z)^2 + \left(bU + \frac{k-1}{k-2}z\right)U'_z - \frac{1}{k-2}U = 0.$$

3. $u_t = a(u^k u_x)_x + (bu^k + c)u_x$.

Generalized traveling wave solution:
$$u(x,t) = \left[\frac{C_2 - x}{b(t+C_1)} + \frac{a \ln|t+C_1|}{b^2 k(t+C_1)} - \frac{c}{b}\right]^{1/k},$$
where C_1 and C_2 are arbitrary constants.

4. $u_t = a(u^{2k} u_x)_x + bu^k u_x$.

   1°. Traveling wave solution in implicit form:
$$a\int \frac{u^{2k}\, du}{C_1 u + C_2 - \frac{b}{k+1}u^{k+1}} = x + C_1 t + C_3 \quad \text{if} \quad k \neq -1,$$
$$a\int \frac{du}{u^2(C_1 u + C_2 - b\ln|x|)} = x + C_1 t + C_3 \quad \text{if} \quad k = -1,$$
where C_1, C_2, and C_3 are arbitrary constants.

   2°. Multiplicative separable solution:
$$u(x,t) = (x+C_1)^{1/k}\varphi(t),$$
where C_1 is an arbitrary constant, and the function $\varphi = \varphi(t)$ is determined by the separable first-order ordinary differential equation
$$\varphi'_t = a\frac{k+1}{k^2}\varphi^{2k+1} + \frac{b}{k}\varphi^{k+1}.$$
Integrating yields the general solution in implicit form
$$\varphi^{-k} - \frac{a(k+1)}{bk}\ln\left[\varphi^{-k} + \frac{a(k+1)}{bk}\right] = -bt + C_2,$$
where C_2 is an arbitrary constant.

   3°. References for PDE 6.1.5.4: Yung et al. (1994), Polyanin & Zaitsev (2004), and Ivanova (2008).

5. $u_t = a(u^k u_x)_x + bu^n u_x$.

   1°. Traveling wave solution in implicit form:
$$a\int \frac{u^k\, du}{C_1 u + C_2 - \frac{b}{n+1}u^{n+1}} = x + C_1 t + C_3 \quad \text{if} \quad n \neq -1,$$
$$a\int \frac{u^k\, du}{C_1 u + C_2 - b\ln|x|} = x + C_1 t + C_3 \quad \text{if} \quad n = -1,$$
where C_1, C_2, and C_3 are arbitrary constants.

2°. Self-similar solution:
$$u(x,t) = t^{\frac{1}{k-2n}} U(z), \quad z = xt^{\frac{n-k}{k-2n}},$$
where the function $U = U(z)$ is described by the ODE
$$\frac{1}{k-2n}[U + (n-k)zU'_z] = a(U^k U'_z)'_z + bU^n U'_z.$$

3°. For $n = k$, there is the generalized traveling wave solution
$$u(x,t) = \left[\frac{C_2 - x}{b(t + C_1)} + \frac{a\ln|t + C_1|}{b^2 n(t + C_1)}\right]^{1/n},$$
where C_1 and C_2 are arbitrary constants (this solution is a special case of the one given in equation 6.1.5.3).

4°. References for PDE 6.1.5.5: Yung et al. (1994) and Ivanova (2008).

6. $u_t = [(au^{2k} + bu^k)u_x]_x + (cu^k + s)u_x.$

Generalized traveling wave solution (Polyanin & Zaitsev, 2004):
$$u(x,t) = \left[\varphi(t)x + (st + C_1)\varphi(t) + \frac{b}{k}\varphi(t)\int \varphi(t)\,dt\right]^{1/k},$$
where C_1 is an arbitrary constant, and the function $\varphi(t)$ is determined by the first-order separable ordinary differential equation
$$\varphi'_t = \frac{a(k+1)}{k}\varphi^3 + c\varphi^2.$$

▶ **Equations with exponential nonlinearities.**

7. $u_t = a(e^{\lambda u} u_x)_x + buu_x.$

1°. Traveling wave solution in implicit form:
$$2a\int \frac{e^{\lambda u}\,du}{-bu^2 + 2\beta u + C_1} = x + \beta t + C_2,$$
where C_1, C_2, and β are arbitrary constants.

2°. Solution:
$$u(x,t) = U(z) + \frac{1}{\lambda}\ln t, \quad z = \frac{x}{t} + \frac{b}{\lambda}\ln t,$$
where the function $U = U(z)$ is described by the ODE
$$\frac{1}{\lambda} = a(e^{\lambda}UU'_z)'_z + \left(bU + z - \frac{b}{\lambda}\right)U'_z.$$

8. $u_t = a(e^{\lambda u} u_x)_x + be^{\beta u} u_x.$

1°. Traveling wave solution in implicit form:
$$a\int \frac{e^{\lambda u}\,du}{C_1 u + C_2 - (b/\beta)e^{\beta u}} = x + C_1 t + C_3,$$
where C_1, C_2, and C_3 are arbitrary constants.

2°. Solution with $\lambda \neq 2\beta$:
$$u(x,t) = U(\zeta) + \frac{1}{\lambda - 2\beta} \ln t, \quad \zeta = xt^{\frac{\beta-\lambda}{\lambda-2\beta}},$$
where the function $U = U(\zeta)$ is described by the ODE
$$\frac{1}{\lambda - 2\beta}[(\beta - \lambda)\zeta U'_\zeta + 1] = a(e^{\lambda U} U'_\zeta)'_\zeta + be^{\beta U} U'_\zeta.$$

3°. Additive separable solution with $\lambda = 2\beta$:
$$u(x,t) = \varphi(t) + \frac{1}{\beta} \ln x,$$
where the function $\varphi = \varphi(t)$ is satisfies the separable first-order ordinary differential equation
$$\beta \varphi'_t = ae^{2\beta\varphi} + be^{\beta\varphi}.$$

4°. References for PDE 6.1.5.8: Edwards (1994) and Ivanova (2008).

9. $u_t = [(ae^{2u} + ce^u)u_x]_x + (ke^u + s)u_x.$
This is a special case of PDE 6.1.5.11 with $b = 0$ and $\lambda = 1$.

10. $u_t = [(ae^{2u} + bue^u)u_x]_x + (ke^u + s)u_x.$
This is a special case of PDE with $c = 0$ and $\lambda = 1$.

Generalized traveling wave solution (Polyanin & Zaitsev, 2004):
$$u(x,t) = \ln\left[\varphi(t)x + (st + C_1)\varphi(t) + b\varphi(t)\int \varphi(t)\,dt\right],$$
where C_1 is an arbitrary constant, and the function $\varphi(t)$ is determined by the first-order separable ODE
$$\varphi'_t = a\varphi^3 + k\varphi^2,$$
whose general solution can be written in implicit form:
$$-\frac{1}{k\varphi} + \frac{a}{k^2} \ln\left|\frac{a\varphi + k}{\varphi}\right| + C_2 = t.$$
Here C_2 is an arbitrary constant, and $k \neq 0$.

In special cases, we have
$$\varphi(t) = (C_2 - 2at)^{-1/2} \quad \text{if } k = 0,$$
$$\varphi(t) = (C_2 - kt)^{-1} \quad \text{if } a = 0.$$

11. $u_t = [(ae^{2\lambda u} + bue^{\lambda u} + ce^{\lambda u})u_x]_x + (ke^{\lambda u} + s)u_x.$
Generalized traveling wave solution:
$$u = \frac{1}{\lambda} \ln\left[\varphi(t)x + (st + C_1)\varphi(t) + \frac{b}{\lambda}\varphi(t)\int \varphi(t)\,dt\right],$$
where C_1 is an arbitrary constant, and the function $\varphi = \varphi(t)$ is determined by the first-order separable ODE
$$\varphi'_t = a\varphi^3 + k\varphi^2.$$
For solutions to this ODE, see the previous equation 6.1.5.10.

▶ **Equations containing arbitrary functions.**

12. $u_t = [a(x)u_x]_x - xf(u)u_x.$

Exact solution in implicit form (Polyanin, 2019c):

$$\int \left(\int f(u)\,du + C_1\right)^{-1} du = t + \int \frac{x\,dx}{a(x)} + C_2,$$

where C_1 and C_2 are arbitrary constants.

13. $u_t = [a(x)f(u)u_x]_x - \tfrac{1}{2}a'_x(x)f(u)u_x.$

Exact solution in implicit form (Polyanin, 2019d):

$$\int \frac{f(u)}{u}\,du = C_1^2 t + C_1 \int \frac{dx}{\sqrt{a(x)}} + C_2,$$

where C_1 and C_2 are arbitrary constants.

14. $u_t = [a(x)u^k u_x]_x + b(x)u^k u_x.$

Multiplicative separable solution:

$$u = (t + C)^{-1/k} \varphi(x),$$

where C is an arbitrary constant, and the function $\varphi = \varphi(x)$ is described by the ODE:
$[a(x)\varphi^k \varphi'_x]'_x + b(x)\varphi^k \varphi'_x + (1/k)\varphi = 0.$

15. $u_t = [f(u)u_x]_x + g(u)u_x.$

Traveling wave solution in implicit form:

$$k^2 \int \frac{f(u)\,du}{\lambda u - kG(u) + C_1} = kx + \lambda t + C_2, \quad \text{where} \quad G(u) = \int g(u)\,du,$$

and $C_1, C_2, k,$ and λ are arbitrary constants.

16. $u_t = [x^k f(u)u_x]_x + x^{k-1} g(u)u_x, \quad k \neq 2.$

Self-similar solution:

$$u = U(z), \quad z = x(t + C)^{\frac{1}{k-2}},$$

where C is an arbitrary constant, and the function $U = U(z)$ is described by the second-order ODE:

$$[z^k f(U)U'_z]'_z + \frac{1}{2-k} zU'_z + z^{k-1} g(U)U'_z = 0.$$

17. $u_t = [x^2 f(u)u_x]_x + xg(u)u_x.$

Exact solution:

$$u = U(z), \quad z = \ln x + \lambda t$$

where λ is an arbitrary constant, and the function $U = U(z)$ is described by the second-order ODE: $[f(U)U'_z]'_z + [f(U) + g(U) - \lambda]U'_z = 0.$ This equation, by substituting $U'_z = \theta(U)$, is reduced to a first-order ODE with separable variables.

18. $u_t = [a(x)e^{\lambda u}u_x]_x + b(x)e^{\lambda u}u_x.$

Additive separable solution (Polyanin, 2020):
$$u = \frac{1}{\lambda}\ln\varphi(x) - \frac{1}{\lambda}\ln(t+C),$$
where C is an arbitrary constant, and the function $\varphi = \varphi(x)$ is described by the second-order linear ODE: $[a(x)\varphi'_x]'_x + b(x)\varphi'_x + 1 = 0$. This equation is reduced, by substituting $\varphi'_x = \theta(x)$, to a first-order linear ODE.

19. $u_t = [e^{\lambda x}f(u)u_x]_x + e^{\lambda x}g(u)u_x, \quad \lambda \neq 0.$

Exact solution (Polyanin, 2020):
$$u = U(z), \quad z = \lambda x + \ln t$$
where λ is an arbitrary constant, and the function $U = U(z)$ is described by the second-order ODE: $\lambda^2[e^z f(U)U'_z]'_z + [\lambda e^z g(U) - 1]U'_z = 0$.

20. $u_t = [a(x)f(u)u_x]_x + b(x)g(u)u_x.$

For exact solutions of such nonlinear PDEs, which are not considered in this section, with various functions $a(x)$, $b(x)$, $f(u)$, and $g(u)$, see Ivanova & Sophocleous (2006), Ivanova (2008), Vaneeva et al. (2010), Polyanin & Zaitsev (2012), Cherniha et al. (2018), Polyanin (2019c, 2020), and Polyanin & Zhurov (2022).

21. $u_t = [a(x)f(u)u_x]_x + b(x)g(u)u_x + c(x)h(u).$

Reaction-convection-diffusion type equations.

For exact solutions of such nonlinear PDEs, see, for example, and Polyanin & Zaitsev (2012), Cherniha et al. (2018), Polyanin (2020), and Polyanin & Zhurov (2022).

6.1.6. Nonlinear Schrödinger Equations and Related PDEs

▶ *In equations 6.1.6.1–6.1.6.7, u is a complex function of real variables x and t; $i^2 = -1$.*

1. $iu_t + u_{xx} + a|u|^2 u = 0.$

Schrödinger equation with a cubic nonlinearity. Here, a is a real number. This equation occurs in various fields of physics, including nonlinear optics, superconductivity, and plasma physics.

1°. Exact solutions:
$$u(x,t) = C_1 \exp\{i[C_2 x + (aC_1^2 - C_2^2)t + C_3]\},$$
$$u(x,t) = \pm C_1\sqrt{\frac{2}{a}}\,\frac{\exp[i(C_1^2 t + C_2)]}{\cosh(C_1 x + C_3)},$$
$$u(x,t) = \pm A\sqrt{\frac{2}{a}}\,\frac{\exp[iBx + i(A^2 - B^2)t + iC_1]}{\cosh(Ax - 2ABt + C_2)},$$
$$u(x,t) = \frac{C_1}{\sqrt{t}}\exp\left[i\frac{(x+C_2)^2}{4t} + i(aC_1^2\ln t + C_3)\right],$$
where A, B, C_1, C_2, and C_3 are arbitrary real constants. The second and third solutions are valid for $a > 0$. The third solution describes the motion of a soliton in a rapidly decaying case.

2°. N-soliton solutions for $a > 0$:

$$u(x,t) = \sqrt{\frac{2}{a} \frac{\det \mathbb{R}(x,t)}{\det \mathbb{M}(x,t)}}.$$

Here $\mathbb{M}(x,t)$ is an $N \times N$ matrix with entries

$$M_{n,m}(x,t) = \frac{1 + \bar{g}_n(x,t) g_n(x,t)}{\bar{\lambda}_n - \lambda_m}, \quad g_n(x,t) = \gamma_n e^{i(\lambda_n x - \lambda_n^2 t)}, \quad n, m = 1, \ldots, N,$$

where λ_n and γ_n are arbitrary complex numbers that satisfy the constraints $\operatorname{Im} \lambda_n > 0$ ($\lambda_n \neq \lambda_m$ if $n \neq m$) and $\gamma_n \neq 0$; the bar over a symbol denotes the complex conjugate. The matrix $\mathbb{R}(x,t)$ is square of order $N+1$; it is obtained by augmenting $\mathbb{M}(x,t)$ with a column on the right and a row at the bottom. The entries of $\mathbb{R}$ are defined as

$$\begin{aligned}
&R_{n,m}(x,t) = M_{n,m}(x,t) &&\text{for } n,m = 1, \ldots, N &&\text{(bulk of the matrix)}, \\
&R_{n,N+1}(x,t) = g_n(x,t) &&\text{for } n = 1, \ldots, N &&\text{(rightmost column)}, \\
&R_{N+1,n}(x,t) = 1 &&\text{for } n = 1, \ldots, N &&\text{(bottom row)}, \\
&R_{N+1,N+1}(x,t) = 0 &&&&\text{(lower right diagonal entry)}.
\end{aligned}$$

The above solution can be represented, for $t \to \pm\infty$, as the sum of N single-soliton solutions.

3°. For other exact solutions, see equation 6.1.6.2 with $n = 1$ and equation 6.1.6.3 with $f(u) = au^2$.

4°. References for PDE 6.1.6.1: Ablowitz & Segur (1981), Dodd et al. (1982), Novikov et al. (1984), Faddeev & Takhtajan (1987), Korepin et al. (1993), Akhmediev & Ankiewicz (1997), Zhidkov (2001), Polyanin & Zaitsev (2012), and Khawaja & Sakkaf (2019).

2. $iu_t + u_{xx} + a|u|^{2k}u = 0.$

Schrödinger equation with a power-law nonlinearity. The numbers a and k are assumed real.

1°. Exact solutions:

$$u(x,t) = C_1 \exp\{i[C_2 x + (a|C_1|^{2k} - C_2^2)t + C_3]\},$$

$$u(x,t) = \pm \left[\frac{(k+1)C_1^2}{a \cosh^2(C_1 kx + C_2)}\right]^{\frac{1}{2k}} \exp[i(C_1^2 t + C_3)],$$

$$u(x,t) = \frac{C_1}{\sqrt{t}} \exp\left[i\frac{(x+C_2)^2}{4t} + i\left(\frac{aC_1^{2k}}{1-k} t^{1-k} + C_3\right)\right],$$

where C_1, C_2, and C_3 are arbitrary real constants.

2°. There is a self-similar solution of the form $u = t^{-1/(2k)} Z(\xi)$, where $\xi = xt^{-1/2}$.

3°. For other exact solutions, see equation 6.1.6.3 with $f(u) = au^{2k}$.

3. $iu_t + u_{xx} + f(|u|)u = 0.$

Schrödinger equation of general form; $f(u)$ is a real-valued function of a real variable.

1°. Suppose $u(x,t)$ is a solution of the Schrödinger equation in question. Then the function
$$u_1 = e^{-i(\lambda x + \lambda^2 t + C_1)} u(x + 2\lambda t + C_2, t + C_3),$$
where C_1, C_2, C_3, and λ are arbitrary real constants, is also a solution of the equation.

2°. Traveling wave solution:
$$u(x,t) = C_1 \exp[i\varphi(x,t)], \quad \varphi(x,t) = C_2 x - C_2^2 t + f(|C_1|)t + C_3.$$

3°. Multiplicative separable solution:
$$u(x,t) = Z(x) e^{i(C_1 t + C_2)},$$
where the function $Z = Z(x)$ is defined implicitly by
$$\int \frac{dZ}{\sqrt{C_1 Z^2 - 2F(Z) + C_3}} = C_4 \pm x, \quad F(Z) = \int Z f(|Z|)\, dZ.$$

Here $C_1, \ldots, C_4$ are arbitrary real constants.

4°. Solution:
$$u(x,t) = U(\xi) e^{i(Ax + Bt + C)}, \quad \xi = x - 2At, \tag{1}$$
where A, B, and C are arbitrary constants, and the function $U = U(\xi)$ is described by the autonomous ODE
$$U''_{\xi\xi} + f(|U|)U - (A^2 + B)U = 0.$$
Integrating yields the general solution in implicit form:
$$\int \frac{dU}{\sqrt{(A^2+B)U^2 - 2F(U) + C_1}} = C_2 \pm \xi, \quad F(U) = \int U f(|U|)\, dU. \tag{2}$$

Relations (1) and (2) involve arbitrary real constants A, B, C, C_1, and C_2.

5°. Solution (Polyanin & Zaitsev, 2004):
$$u(x,t) = \psi(z) \exp\left[i\left(Axt - \tfrac{2}{3}A^2 t^3 + Bt + C\right)\right], \quad z = x - At^2,$$
where A, B, and C are arbitrary constants, and the function $\psi = \psi(z)$ is described by the ODE
$$\psi''_{zz} + f(|\psi|)\psi - (Az + B)\psi = 0.$$

6°. Exact solutions (Polyanin & Zaitsev, 2004):
$$u(x,t) = \pm \frac{1}{\sqrt{C_1 t}} \exp[i\varphi(x,t)], \quad \varphi(x,t) = \frac{(x+C_2)^2}{4t} + \int f(|C_1 t|^{-1/2})\, dt + C_3,$$
where C_1, C_2, and C_3 are arbitrary real constants.

7°. Solution (Polyanin & Zaitsev, 2004):

$$u(x,t) = \theta(x)\exp[i\varphi(x,t)], \quad \varphi(x,t) = C_1 t + C_2 \int \frac{dx}{\theta^2(x)} + C_3,$$

where C_1, C_2, and C_3 are arbitrary real constants, and the function $\theta = \theta(x)$ is described by the autonomous ODE

$$\theta''_{xx} - C_1\theta - C_2^2\theta^{-3} + f(|\theta|)\theta = 0.$$

8°. There is an exact solution of the form

$$u(x,t) = \varphi(z)\exp[iAt + i\psi(z)], \quad z = \kappa x + \lambda t,$$

where A, κ, and λ are arbitrary real constants, and the functions $\varphi(z)$ and $\psi(z)$ are determined by a system of ODEs.

4. $u_t + x^{-n}(x^n u_x)_x + f(|u|)u = 0.$

Schrödinger equation of general form; $f(u)$ is a real-valued function of a real variable. To $n = 1$ there corresponds a two-dimensional Schrödinger equation with axial symmetry and to $n = 2$, a three-dimensional Schrödinger equation with central symmetry.

1°. Multiplicative separable solution:

$$u(x,t) = U(x)e^{i(C_1 t + C_2)},$$

where C_1 and C_2 are arbitrary real constants, and the function $U = U(x)$ is described by the ODE

$$x^{-n}(x^n U'_x)'_x - C_1 U + f(|U|)U = 0.$$

2°. Solution (Polyanin & Zaitsev, 2004):

$$u(x,t) = z(x)\exp[i\varphi(x,t)], \quad \varphi(x,t) = C_1 t + C_2 \int \frac{dx}{x^n z^2(x)} + C_3,$$

where C_1, C_2, and C_3 are arbitrary real constants, and the function $z = z(x)$ is described by the ODE

$$x^{-n}(x^n z'_x)'_x - C_1 z - C_2^2 x^{-2n} z^{-3} + f(|z|)z = 0.$$

3°. Solution (Polyanin & Zaitsev, 2004):

$$u(x,t) = C_1 t^{-\frac{n+1}{2}}\exp[i\varphi(x,t)], \quad \varphi(x,t) = \frac{x^2}{4t} + \int f\left(|C_1|t^{-\frac{n+1}{2}}\right)dt + C_2,$$

where C_1 and C_2 are arbitrary real constants.

5. $u_t = (a + ib)u_{xx} + [f(|u|) + ig(|u|)]u.$

Generalized Ginzburg–Landau equation, where $f(u)$ and $g(u)$ are real-valued functions of a real variable, and a and b are real numbers. Equations of this form are used for studying second-order phase transitions in superconductivity theory (see Ginzburg & Landau, 1950) and to describe two-component reaction-diffusion systems near a point of bifurcation (Kuramoto & Tsuzuki, 1975).

1°. Traveling wave solutions:
$$u(x,t) = C_1 \exp[i\varphi(x,t)], \quad \varphi(x,t) = \pm x\sqrt{\frac{f(|C_1|)}{a}} + t\left[g(|C_1|) - \frac{b}{a}f(|C_1|)\right] + C_2,$$
where C_1 and C_2 are arbitrary real constants.

2°. Solution (Berman & Danilov, 1981):
$$u(x,t) = U(t)\exp[i\varphi(x,t)], \quad \varphi(x,t) = C_1 x - C_1^2 bt + \int g(|U|)\,dt + C_2,$$
where the function $U = U(t)$ is described by the ODE: $U'_t = f(|U|)U - aC_1^2 U$, whose general solution can be represented in implicit form as
$$\int \frac{dU}{f(|U|)U - aC_1^2 U} = t + C_3.$$

3°. Solution (Polyanin & Zaitsev, 2004):
$$u(x,t) = V(z)\exp[iC_1 t + i\theta(z)], \quad z = x + \lambda t,$$
where C_1 and λ are arbitrary real constants, and the functions $V = V(z)$ and $\theta = \theta(z)$ are described by the ODE system
$$aV''_{zz} - aV(\theta'_z)^2 - bV\theta''_{zz} - 2bV'_z\theta'_z - \lambda V'_z + f(|V|)V = 0,$$
$$aV\theta''_{zz} - bV(\theta'_z)^2 + bV''_{zz} + 2aV'_z\theta'_z - \lambda V\theta'_z - C_1 V + g(|V|)V = 0.$$

6. $iu_t + u_{xx} + i[f(|u|)u]_x = 0.$

1°. Solution (Polyanin & Zaitsev, 2004):
$$u(x,t) = \xi(t)\exp[i\eta(x,t)], \quad \eta(x,t) = \varphi(t)x^2 + \psi(t)x + \chi(t),$$
where the functions $\xi = \xi(t)$, $\varphi = \varphi(t)$, $\psi = \psi(t)$, and $\chi = \chi(t)$ are described by the ODE system
$$\xi'_t + 2\varphi\xi = 0,$$
$$\varphi'_t + 4\varphi^2 = 0,$$
$$\psi'_t + 4\varphi\psi + 2\varphi f(\xi) = 0,$$
$$\chi'_t + \psi^2 + \psi f(\xi) = 0.$$

Integrating yields
$$\xi = \frac{C_2}{\sqrt{t+C_1}}, \quad \varphi = \frac{1}{4(t+C_1)}, \quad \psi = -2\varphi\int f(\xi)\,dt + C_3\varphi, \quad \chi = -\int[\psi^2 + \psi f(\xi)]\,dt + C_4,$$
where $C_1, \ldots, C_4$ are arbitrary real constants.

2°. Solution:
$$u(x,t) = U(z)\exp[i\beta t + iV(z)], \quad z = kx + \lambda t,$$
where k, β, and λ are arbitrary real constants, and the functions $U = U(z)$ and $V = V(z)$ are described by the ODE system
$$\lambda U'_z + k^2(UV'_z)'_z + k^2 U'_z V'_z + k[f(U)U]'_z = 0,$$
$$-U(\beta + \lambda V'_z) + k^2 U''_{zz} - k^2 U(V'_z)^2 - kf(U)UV'_z = 0.$$

7. $iu_t + [f(|u|)u]_{xx} = 0$.

1°. Solution (Polyanin & Zaitsev, 2004):
$$u(x,t) = \xi(t)\exp[i\eta(x,t)], \quad \eta(x,t) = \varphi(t)x^2 + \psi(t)x + \chi(t),$$

where the functions $\xi = \xi(t)$, $\varphi = \varphi(t)$, $\psi = \psi(t)$, and $\chi = \chi(t)$ are described by the ODE system
$$\begin{aligned}\xi'_t + 2\xi\varphi f(\xi) &= 0,\\ \varphi'_t + 4\varphi^2 f(\xi) &= 0,\\ \psi'_t + 4\varphi\psi f(\xi) &= 0,\\ \chi'_t + \psi^2 f(\xi) &= 0.\end{aligned}$$

Integrating yields
$$\varphi = C_1\xi^2, \quad \psi = C_2\xi^2, \quad \chi = -C_2^2\int \xi^4 f(\xi)\,dt + C_3,$$

where C_1, C_2, and C_3 are arbitrary real constants, and the function $\xi = \xi(t)$ is defined implicitly as (C_4 is an arbitrary constant):
$$\int \frac{d\xi}{\xi^3 f(\xi)} + 2C_1 t + C_4 = 0.$$

2°. There is a solution of the form
$$u(x,t) = U(z)\exp[i\beta t + iV(z)], \quad z = kx + \lambda t,$$

where k, β, and λ are arbitrary real constants, and the functions $U = U(z)$ and $V = V(z)$ are determined by an appropriate system of ordinary differential equations (which is not written out here).

4°. There is a self-similar solution of the form $u(x,t) = w(\xi)$, where $\xi = x^2/t$.

6.2. Hyperbolic Equations

6.2.1. Nonlinear Klein–Gordon Equations of the Form $u_{tt} = au_{xx} + f(u)$

▶ Equations of this form admit traveling wave solutions $u = u(z)$, $z = \kappa x + \lambda t$ (with $\lambda \neq \pm\kappa$), which can be represented in the implicit form:
$$\int \left[C_1 + \frac{2}{\lambda^2 - \kappa^2} F(u)\right]^{-1/2} du = C_2 \pm (\kappa x + \lambda t), \quad F(u) = \int f(u)\,du.$$

where C_1, C_2, κ, and λ are arbitrary constants.

1. $u_{tt} = u_{xx} + au^k$.

1°. Exact solutions:
$$u(x,t) = \left[\frac{a(1-k)^2}{2(1+k)(C_2^2 - C_1^2)}(C_1 x + C_2 t + C_3)^2\right]^{\frac{1}{1-k}};$$
$$u(x,t) = \{\tfrac{1}{4}a(1-k)^2[(t+C_1)^2 - (x+C_2)^2]\}^{\frac{1}{1-k}},$$

where C_1, C_2, and C_3 are arbitrary constants.

2°. The solutions of Item 1° are special cases of solutions of the following forms:
$$u(x,t) = F(z), \quad z = C_1 x + C_2 t;$$
$$u(x,t) = G(\rho), \quad \rho = (t+C_1)^2 - (x+C_2)^2.$$

3°. Self-similar solution:
$$u(x,t) = (t+C_1)^{\frac{2}{1-k}} \theta(\zeta), \quad \zeta = \frac{x+C_2}{t+C_1},$$

where the function $\theta(\zeta)$ is described by the ODE
$$(1-\zeta^2)\theta''_{\zeta\zeta} + \frac{2(1+k)}{1-k}\zeta\theta'_\zeta - \frac{2(1+k)}{(1-k)^2}\theta + a\theta^k = 0.$$

2. $u_{tt} = u_{xx} + au + bu^k$.

1°. Traveling wave solutions for $a > 0$:
$$u = \left[\frac{2b\sinh^2 z}{a(k+1)}\right]^{\frac{1}{1-k}}, \quad z = \frac{\sqrt{a}}{2}(1-k)(x\sinh C_1 \pm t\cosh C_1) + C_2 \quad \text{if } b(k+1) > 0,$$
$$u = \left[-\frac{2b\cosh^2 z}{a(k+1)}\right]^{\frac{1}{1-k}}, \quad z = \frac{\sqrt{a}}{2}(1-k)(x\sinh C_1 \pm t\cosh C_1) + C_2 \quad \text{if } b(k+1) < 0,$$

where C_1 and C_2 are arbitrary constants.

2°. Traveling wave solutions for $a < 0$ and $b(k+1) > 0$:
$$u = \left[-\frac{2b\cos^2 z}{a(k+1)}\right]^{\frac{1}{1-k}}, \quad z = \frac{\sqrt{|a|}}{2}(1-k)(x\sinh C_1 \pm t\cosh C_1) + C_2.$$

3°. For other exact solutions of this equation, see equation 6.2.1.8 with $f(u) = au + bu^k$.

3. $u_{tt} = u_{xx} + au^k + bu^{2k-1}$.

Exact solutions:
$$u(x,t) = \left[\frac{a(1-k)^2}{2(k+1)}(x\sinh C_1 \pm t\cosh C_1 + C_2)^2 - \frac{b(k+1)}{2ak}\right]^{\frac{1}{1-k}},$$
$$u(x,t) = \left\{\frac{1}{4}a(1-k)^2[(t+C_1)^2 - (x+C_2)^2] - \frac{b}{ak}\right\}^{\frac{1}{1-k}},$$

where C_1 and C_2 are arbitrary constants.

4. $u_{tt} = a^2 u_{xx} + be^{\beta u}$.

1°. Traveling wave solutions:
$$u(x,t) = \frac{1}{\beta}\ln\left[\frac{2(B^2 - a^2 A^2)}{b\beta(Ax + Bt + C)^2}\right],$$
$$u(x,t) = \frac{1}{\beta}\ln\left[\frac{2(a^2 A^2 - B^2)}{b\beta \cosh^2(Ax + Bt + C)}\right],$$

$$u(x,t) = \frac{1}{\beta} \ln\left[\frac{2(B^2 - a^2A^2)}{b\beta \sinh^2(Ax + Bt + C)}\right],$$

$$u(x,t) = \frac{1}{\beta} \ln\left[\frac{2(B^2 - a^2A^2)}{b\beta \cos^2(Ax + Bt + C)}\right],$$

where A, B, and C are arbitrary constants.

2°. Functional separable solutions:

$$u(x,t) = \frac{1}{\beta} \ln\left(\frac{8a^2C}{b\beta}\right) - \frac{2}{\beta} \ln|(x + A)^2 - a^2(t + B)^2 + C|,$$

$$u(x,t) = -\frac{2}{\beta} \ln\left[C_1 e^{\lambda x} \pm \frac{\sqrt{2b\beta}}{2a\lambda} \sinh(a\lambda t + C_2)\right],$$

$$u(x,t) = -\frac{2}{\beta} \ln\left[C_1 e^{\lambda x} \pm \frac{\sqrt{-2b\beta}}{2a\lambda} \cosh(a\lambda t + C_2)\right],$$

$$u(x,t) = -\frac{2}{\beta} \ln\left[C_1 e^{a\lambda t} \pm \frac{\sqrt{-2b\beta}}{2a\lambda} \sinh(\lambda x + C_2)\right],$$

$$u(x,t) = -\frac{2}{\beta} \ln\left[C_1 e^{a\lambda t} \pm \frac{\sqrt{2b\beta}}{2a\lambda} \cosh(\lambda x + C_2)\right],$$

where A, B, C, C_1, C_2, and λ are arbitrary constants.

3°. General solution:

$$u(x,t) = \frac{1}{\beta}[f(z) + g(y)] - \frac{2}{\beta} \ln\left|k \int \exp[f(z)]\,dz - \frac{b\beta}{8a^2k} \int \exp[g(y)]\,dy\right|,$$

$$z = x - at, \quad y = x + at,$$

where $f = f(z)$ and $g = g(y)$ are arbitrary functions and k is an arbitrary constant.

4°. References for PDE 6.2.1.4: Liouville (1853), Bullough & Caudrey (1980), and Polyanin & Zaitsev (2004).

5. $u_{tt} = u_{xx} + ae^{\beta u} + be^{2\beta u}$.

1°. Traveling wave solutions:

$$u(x,t) = -\frac{1}{\beta} \ln\left[\frac{a\beta}{C_1^2 - C_2^2} + C_3 E + \frac{a^2\beta^2 + b\beta(C_1^2 - C_2^2)}{4C_3(C_1^2 - C_2^2)^2 E}\right], \quad E = \exp(C_1 x + C_2 t);$$

$$u(x,t) = -\frac{1}{\beta} \ln\left[\frac{a\beta}{C_2^2 - C_1^2} + \frac{\sqrt{a^2\beta^2 + b\beta(C_2^2 - C_1^2)}}{C_2^2 - C_1^2} \sin(C_1 x + C_2 t + C_3)\right],$$

where C_1, C_2, and C_3 are arbitrary constants.

2°. For other exact solutions of this equation, see equation 6.2.1.8 with $f(u) = ae^{\beta u} + be^{2\beta u}$.

6. $u_{tt} = au_{xx} + b\sinh(\lambda u)$.

Sinh-Gordon equation. It appears in quantum field theory, kink dynamics, fluid dynamics, and other applications.

1°. Traveling wave solutions:
$$u(x,t) = \pm \frac{2}{\lambda} \ln\left[\tan\frac{b\lambda(kx+\mu t+\theta_0)}{2\sqrt{b\lambda(\mu^2-ak^2)}}\right],$$
$$u(x,t) = \pm \frac{4}{\lambda} \operatorname{artanh}\left[\exp\frac{b\lambda(kx+\mu t+\theta_0)}{\sqrt{b\lambda(\mu^2-ak^2)}}\right],$$

where k, μ, and θ_0 are arbitrary constants. It is assumed that $b\lambda(\mu^2-ak^2) > 0$ in both formulas.

2°. Functional separable solution:
$$u(x,t) = \frac{4}{\lambda} \operatorname{artanh}[f(t)g(x)], \qquad \operatorname{artanh} z = \frac{1}{2}\ln\frac{1+z}{1-z},$$

where the functions $f = f(t)$ and $g = g(x)$ are determined by the first-order autonomous ordinary differential equations
$$(f'_t)^2 = Af^4 + Bf^2 + C, \quad a(g'_x)^2 = Cg^4 + (B-b\lambda)g^2 + A,$$

where A, B, and C are arbitrary constants.

3°. For other exact solutions of this equation, see equation 6.2.1.8 with $f(u) = b\sinh(\lambda u)$.

4°. References for PDE 6.2.1.6: Grauel (1985), Grundland & Infeld (1992), Musette & Conte (1994), Zhdanov (1994), Andreev et al. (1999), and Hoenselaers (2007).

7. $u_{tt} = au_{xx} + b\sin(\lambda u)$.

Sine-Gordon equation. It arises in differential geometry and various areas of physics.

Remark 6.1. *This equation is sometimes called the Enneper equation; see historical aspects in Seeger & Wesolowski (1981).*

1°. Traveling wave solutions:
$$u(x,t) = \frac{4}{\lambda}\arctan\left\{\exp\left[\pm\frac{b\lambda(kx+\mu t+\theta_0)}{\sqrt{b\lambda(\mu^2-ak^2)}}\right]\right\} \qquad \text{if} \quad b\lambda(\mu^2-ak^2) > 0,$$
$$u(x,t) = -\frac{\pi}{\lambda} + \frac{4}{\lambda}\arctan\left\{\exp\left[\pm\frac{b\lambda(kx+\mu t+\theta_0)}{\sqrt{b\lambda(ak^2-\mu^2)}}\right]\right\} \qquad \text{if} \quad b\lambda(\mu^2-ak^2) < 0,$$

where k, μ, and θ_0 are arbitrary constants. The first expression corresponds to a single-soliton solution.

2°. Functional separable solutions:
$$u(x,t) = \frac{4}{\lambda}\arctan\left[\frac{\mu\sinh(kx+A)}{k\sqrt{a}\cosh(\mu t+B)}\right], \quad \mu^2 = ak^2 + b\lambda > 0;$$
$$u(x,t) = \frac{4}{\lambda}\arctan\left[\frac{\mu\sin(kx+A)}{k\sqrt{a}\cosh(\mu t+B)}\right], \quad \mu^2 = b\lambda - ak^2 > 0;$$
$$u(x,t) = \frac{4}{\lambda}\arctan\left[\frac{\gamma}{\mu}\frac{e^{\mu(t+A)}+ak^2 e^{-\mu(t+A)}}{e^{k\gamma(x+B)}+e^{-k\gamma(x+B)}}\right], \quad \mu^2 = ak^2\gamma^2 + b\lambda > 0,$$

where A, B, k, and γ are arbitrary constants.

3°. Two-soliton periodic solution, called a *breather*, is given by ($a = 1$, $b = -1$, and $\lambda = 1$):

$$w = 4 \arctan\left[\frac{\sqrt{1-\omega^2}}{\omega \cosh(\sqrt{1-\omega^2}\, x \sin \omega t)}\right],$$

where ω is an arbitrary constant ($0 < \omega < 1$).

4°. For $a = 1$, $b = -1$, and $\lambda = 1$, an *N-soliton solution* is given by formulas (Bullough & Caudrey, 1980):

$$u(x,t) = \arccos\left[1 - 2\left(\frac{\partial^2}{\partial x^2} - \frac{\partial^2}{\partial t^2}\right)(\ln F)\right], \quad F = \det[M_{ij}],$$

$$M_{ij} = \frac{2}{a_i + a_j} \cosh\left(\frac{z_i + z_j}{2}\right), \quad z_i = \pm \frac{x - \mu_i t + C_i}{\sqrt{1-\mu_i^2}}, \quad a_i = \pm\sqrt{\frac{1-\mu_i}{1+\mu_i}},$$

where μ_i and C_i are arbitrary constants.

5°. For other exact solutions of the original equation, see equation 6.2.1.8 with $f(u) = b \sin(\lambda u)$.

6°. The sine-Gordon equation is integrated by the inverse scattering method; see the book by Novikov et al. (1984).

7°. References for PDE 6.2.1.7: Steuerwald (1936), Bullough & Caudrey, 1980, Seeger & Wesolowski (1981), Novikov et al. (1984), Ablowitz & Clarkson (1991), Musette & Conte (1994), Zhdanov (1994), and Polyanin & Zaitsev (2012).

8. $u_{tt} = u_{xx} + f(u)$.

Nonlinear Klein–Gordon equation of the general form.

1°. Suppose $u = u(x,t)$ is a solution of the equation in question. Then the functions

$$u_1 = u(\pm x + C_1, \pm t + C_2),$$
$$u_2 = u(x \cosh \beta + t \sinh \beta, \, t \cosh \beta + x \sinh \beta),$$

where C_1, C_2, and β are arbitrary constants, are also solutions of the equation (the plus or minus signs in u_1 may be chosen arbitrarily).

2°. Traveling wave solution in implicit form:

$$\int \left[C_1 + \frac{2}{\lambda^2 - \kappa^2} \int f(u)\, du\right]^{-1/2} du = \kappa x + \lambda t + C_2,$$

where C_1, C_2, κ, and λ are arbitrary constants.

3°. Functional separable solution:

$$u = u(\xi), \quad \xi = \tfrac{1}{4}(t + C_1)^2 - \tfrac{1}{4}(x + C_2)^2,$$

where C_1 and C_2 are arbitrary constants, and the function $u = u(\xi)$ is described by the ODE: $\xi u''_{\xi\xi} + u'_\xi - f(u) = 0$.

4°. Grundland & Infeld (1992) found all functions $f(u)$ for which PDE 6.2.1.8 admits functional separable solutions of the form $u = U(\varphi(x) + \psi(t))$ (see also Zhdanov, 1994 and Polyanin & Zhurov, 2022). The results are summarized in Table 6.2.

TABLE 6.2. Nonlinear Klein–Gordon equations $u_{tt} - u_{xx} = f(u)$ that admit functional separable solutions of the form $u = u(z)$, where $z = \varphi(x) + \psi(t)$.

No.	Right-hand side of equation $f(u)$	Solution $u(z)$	Equations for $\psi(t)$ and $\varphi(x)$
1	$au\ln u + bu$	e^z	$(\psi'_t)^2 = C_1 e^{-2\psi} + a\psi - \tfrac{1}{2}a + b + A$, $(\varphi'_x)^2 = C_2 e^{-2\varphi} - a\varphi + \tfrac{1}{2}a + A$
2	$ae^u + be^{-2u}$	$\ln|z|$	$(\psi'_t)^2 = 2a\psi^3 + A\psi^2 + C_1\psi + C_2$, $(\varphi'_x)^2 = -2a\varphi^3 + A\varphi^2 - C_1\varphi + C_2 + b$
3	$a\sin u + b\left(\sin u \ln\tan\tfrac{u}{4} + 2\sin\tfrac{u}{4}\right)$	$4\arctan e^z$	$(\psi'_t)^2 = C_1 e^{2\psi} + C_2 e^{-2\psi} + b\psi + a + A$, $(\varphi'_x)^2 = -C_2 e^{2\varphi} - C_1 e^{-2\varphi} - b\varphi + A$
4	$a\sinh u + b\left(\sinh u \ln\tanh\tfrac{u}{4} + 2\sinh\tfrac{u}{2}\right)$	$2\ln\left|\coth\tfrac{z}{2}\right|$	$(\psi'_t)^2 = C_1 e^{2\psi} + C_2 e^{-2\psi} - \sigma b\psi + a + A$, $(\varphi'_x)^2 = C_2 e^{2\varphi} + C_1 e^{-2\varphi} + \sigma b\varphi + A$
5	$a\sinh u + 2b\left(\sinh u \arctan e^{u/2} + \cosh\tfrac{u}{2}\right)$	$2\ln\left|\tan\tfrac{z}{2}\right|$	$(\psi'_t)^2 = C_1\sin 2\psi + C_2\cos 2\psi + \sigma b\psi + a + A$, $(\varphi'_x)^2 = -C_1\sin 2\varphi + C_2\cos 2\varphi - \sigma b\varphi + A$

Notation: a, b are free parameters, A, C_1, C_2 are arbitrary constants; $\sigma = 1$ for $z > 0$ and $\sigma = -1$ for $z < 0$.

6.2.2. Other Nonlinear Wave Type Equations

1. $u_{tt} = a(uu_x)_x$.

This is a special case of PDE 6.2.2.11 with $f(u) = u$.

1°. Exact solutions:
$$u(x,t) = \tfrac{1}{2}aA^2 t^2 + Bt + Ax + C,$$
$$u(x,t) = \tfrac{1}{12}aA^{-2}(At+B)^4 + Ct + D + x(At+B),$$
$$u(x,t) = \frac{1}{a}\left(\frac{x+A}{t+B}\right)^2,$$
$$u(x,t) = (At+B)\sqrt{Cx+D},$$
$$u(x,t) = \pm\sqrt{A(x+a\lambda t)+B} + a\lambda^2,$$

where A, B, C, D, and λ are arbitrary constants.

2°. Generalized separable solution quadratic in x:
$$u(x,t) = \frac{1}{at^2}x^2 + \left(\frac{C_1}{t^2} + C_2 t^3\right)x + \frac{aC_1^2}{4t^2} + \frac{C_3}{t} + C_4 t^2 + \tfrac{1}{2}aC_1 C_2 t^3 + \frac{1}{54}aC_2^2 t^8,$$

where $C_1, \ldots, C_4$ are arbitrary constants.

3°. Solution:
$$u = U(z) + 4aC_1^2 t^2 + 4aC_1 C_2 t, \quad z = x + aC_1 t^2 + aC_2 t,$$

where C_1 and C_2 are arbitrary constants, and the function $U(z)$ is described by the first-order ODE:
$$(U - aC_2^2)U'_z - 2C_1 U = 8C_1^2 z + C_3.$$

4°. See also equation 6.2.2.11 with $f(u) = au$.

2. $u_{tt} = a(u^k u_x)_x + bu^m$.

There are solutions of the following forms:

$$u(x,t) = U(z), \quad z = C_1 x + C_2 t \quad \text{(traveling wave solution)};$$

$$u(x,t) = t^{\frac{2}{1-m}} V(\xi), \quad \xi = xt^{\frac{m-k-1}{1-m}} \quad \text{(self-similar solution)},$$

where C_1 and C_2 are arbitrary constants.

3. $u_{tt} = a(e^{\lambda u} u_x)_x, \quad a > 0$.

1°. Additive separable solutions:

$$u(x,t) = \frac{1}{\lambda} \ln|Ax + B| + Ct + D,$$

$$u(x,t) = \frac{2}{\lambda} \ln|Ax + B| - \frac{2}{\lambda} \ln|\pm A\sqrt{a}\, t + C|,$$

$$u(x,t) = \frac{1}{\lambda} \ln(aA^2 x^2 + Bx + C) - \frac{2}{\lambda} \ln(aAt + D),$$

$$u(x,t) = \frac{1}{\lambda} \ln(Ax^2 + Bx + C) + \frac{1}{\lambda} \ln\left[\frac{p^2}{aA\cos^2(pt+q)}\right],$$

$$u(x,t) = \frac{1}{\lambda} \ln(Ax^2 + Bx + C) + \frac{1}{\lambda} \ln\left[\frac{p^2}{aA\sinh^2(pt+q)}\right],$$

$$u(x,t) = \frac{1}{\lambda} \ln(Ax^2 + Bx + C) + \frac{1}{\lambda} \ln\left[\frac{-p^2}{aA\cosh^2(pt+q)}\right],$$

where A, B, C, D, p, and q are arbitrary constants.

2°. There are solutions of the following forms:

$$u(x,t) = F(z), \quad z = kx + \beta t \quad \text{(traveling wave solution)};$$
$$u(x,t) = G(\xi), \quad \xi = x/t \quad \text{(self-similar solution)};$$
$$u(x,t) = H(\eta) + 2(k-1)\lambda^{-1} \ln t, \quad \eta = xt^{-k};$$
$$u(x,t) = U(\zeta) - 2\lambda^{-1} \ln|t|, \quad \zeta = x + k\ln|t|;$$
$$u(x,t) = V(\zeta) - 2\lambda^{-1} t, \quad \eta = xe^t,$$

where k and β are arbitrary constants.

4. $u_{tt} = (ae^{\lambda u} u_x)_x + be^{\beta u}$.

1°. Traveling wave solution:

$$u = u(z), \quad z = k_2 x + k_1 t,$$

where k_1 and k_2 are arbitrary constants, and the function $u(z)$ is described by the second-order autonomous ODE:

$$k_1^2 u''_{zz} - ak_2^2 (e^{\lambda u} u'_z)'_z = be^{\beta u}.$$

The substitution $\Theta(u) = (u'_z)^2$ leads to the first-order linear ODE:
$$(k_1^2 - ak_2^2 e^{\lambda u})\Theta'_u - 2ak_2^2 \lambda e^{\lambda u}\Theta = 2be^{\beta u}.$$

2°. Solution:
$$u = U(\xi) - \frac{2}{\beta}\ln t, \quad \xi = xt^{\frac{\lambda-\beta}{\beta}},$$

where the function $U(\xi)$ is described by the second-order ODE:
$$\frac{2}{\beta} + \frac{(\lambda-\beta)(\lambda-2\beta)}{\beta^2}\xi U'_\xi + \frac{(\lambda-\beta)^2}{\beta^2}\xi^2 U''_{\xi\xi} = (ae^{\lambda U}U'_\xi)'_\xi + be^{\beta U}.$$

5. $u_{tt} = (ax^k u_x)_x + f(u).$

This equation describes the propagation of nonlinear waves in an inhomogeneous medium.

1°. Functional separable solution for $k \neq 2$ (Polyanin & Zaitsev, 2004):
$$u = U(\zeta), \quad \zeta^2 = s\left[\frac{1}{4}(t+C)^2 - \frac{x^{2-k}}{a(2-k)^2}\right],$$

where s and the expression in square brackets must have like signs, C is an arbitrary constant, and the function $U(\zeta)$ is described by the ODE:
$$U''_{\zeta\zeta} + \frac{2}{(2-k)\zeta}U'_\zeta = \frac{4}{s}f(U).$$

2°. Solution for $k = 2$:
$$u = u(z), \quad z = At + B\ln|x|,$$

where A and B are arbitrary constants, and the function $u = u(z)$ is described by the autonomous ODE:
$$(aB^2 - A^2)u''_{zz} + aBu'_z + f(u) = 0.$$

Solution of this equation with $A = \pm B\sqrt{a}$ in implicit form:
$$aB\int \frac{du}{f(u)} = -z + C,$$

where C is an arbitrary constant.

6. $u_{tt} = ax^{-n}(x^n u_x)_x + f(u), \quad a > 0.$

To $n = 1$ and $n = 2$ there correspond nonlinear waves with axial and central symmetry, respectively.

Functional separable solution:
$$u = u(\xi), \quad \xi = \sqrt{ak(t+C)^2 - kx^2},$$

where $u(\xi)$ is described by the ODE: $u''_{\xi\xi} + (1+n)\xi^{-1}u'_\xi = (ak)^{-1}f(u).$

7. $u_{tt} = (ae^{\lambda x} u_x)_x + f(u)$.

Functional separable solution (Polyanin & Zaitsev, 2004):

$$u = U(z), \quad z = \left[4ke^{-\lambda x} - ak\lambda^2(t+C)^2\right]^{1/2}, \quad k = \pm 1,$$

where C is an arbitrary constant, and the function $U(z)$ is described by a second-order autonomous ODE of the form 2.3.1.1:

$$U''_{zz} + \frac{1}{ak\lambda^2} f(U) = 0.$$

8. $u_{tt} = [a(x)u_x]_x + b(x)u + cu \ln u$.

Multiplicative separable solution:

$$u = \varphi(x)\psi(t),$$

where the functions $\varphi = \varphi(x)$ and $\psi = \psi(t)$ are described by the second-order ODEs:

$$[a(x)\varphi'_x]'_x + b(x)\varphi + c\varphi \ln \varphi = 0,$$
$$\psi''_{tt} - c\psi \ln \psi = 0.$$

9. $u_{tt} = [a(x)u_x]_x - \dfrac{a'_x(x)}{\sqrt{a(x)}} f(u)$.

Exact solutions in implicit form (Polyanin, 2019e):

$$\int \frac{du}{f(u)} = \pm 2t + 2 \int \frac{dx}{\sqrt{a(x)}} + C,$$

where C is an arbitrary constant.

10. $u_{tt} = [a(x)u^k u_x]_x + b(x)u^{k+1}$.

Multiplicative separable solution (Polyanin, 2019e):

$$u = \varphi(x)\psi(t).$$

The functions $\varphi = \varphi(x)$ and $\psi = \psi(t)$ are described by ODEs

$$[a(x)\varphi^k \varphi'_x]'_x + b(x)\varphi^{k+1} = C\varphi,$$
$$\psi''_{tt} = C\psi^{k+1},$$

where C is an arbitrary constant.

11. $u_{tt} = [f(u)u_x]_x$.

This equation is encountered in wave and gas dynamics.

1°. Traveling wave solution in implicit form:

$$\lambda^2 u - \int f(u)\, du = A(x + \lambda t) + B,$$

where A, B, and λ are arbitrary constants.

2°. Self-similar solution (Ames et al., 1981):

$$u = u(z), \quad z = \frac{x+A}{t+B},$$

where the function $u(z)$ is described by the ODE: $(z^2 u'_z)'_z = [f(u)u'_z]'_z$, which admits the first integral

$$[z^2 - f(u)]u'_z = C.$$

The special case $C = 0$ corresponds to a solution in implicit form: $z^2 = f(u)$.

3°. Exact solutions in implicit form:

$$x - t\sqrt{f(u)} = \varphi_1(u),$$
$$x + t\sqrt{f(u)} = \varphi_2(u),$$

where $\varphi_1(u)$ and $\varphi_2(u)$ are arbitrary functions.

4°. The original equation can be represented as the system of first-order PDEs

$$f(u)u_x = v_t, \quad u_t = v_x. \tag{1}$$

The *hodograph transformation*

$$x = x(u,v), \quad t = t(u,v),$$

where u and v are treated as the independent variables and x and t as the dependent ones, brings (1) to the linear system of first-order PDEs

$$f(u)t_v = x_u, \quad x_v = t_u. \tag{2}$$

Eliminating t, we obtain a linear second-order PDE for $x = x(u,v)$:

$$[x_u/f(u)]_u - x_{vv} = 0. \tag{3}$$

Likewise, from system (2) we obtain another linear PDE for $t = t(u,v)$:

$$t_{uu} - f(u)t_{vv} = 0.$$

5°. A number of exact solutions in the parametric form of the original nonlinear PDE, as well as linear PDE (3), are given in Zaitsev & Polyanin (2001) and Polyanin & Zaitsev (2012).

12. $u_{tt} = [x^k f(u)u_x]_x + x^{k-2}g(u), \quad k \neq 2.$

Self-similar solution (Polyanin, 2019e):

$$u = U(z), \quad z = x(t+C)^{\frac{2}{k-2}},$$

where C is an arbitrary constant, and the function $U = U(z)$ is described by the second-order ODE:

$$\frac{4}{(2-k)^2}z(zU'_z)'_z + \frac{2}{2-k}zU'_z = [z^k f(U)U'_z]'_z + z^{k-2}g(U).$$

13. $u_{tt} = [x^2 f(u) u_x]_x + g(u).$

Exact solution:
$$u = U(z), \quad z = \ln x + \lambda t$$
where λ is an arbitrary constant, and the function $U = U(z)$ is described by the second-order autonomous ODE:
$$[f(U)U'_z]'_z + [f(U) - \lambda^2]U'_z + g(U) = 0.$$

14. $u_{tt} = [a(x) e^{\lambda u} u_x]_x + b(x) e^{\lambda u}.$

Additive separable solution (Polyanin, 2019e):
$$u = \frac{1}{\lambda} \ln \varphi(x) - \frac{2}{\lambda} \ln(t + C),$$
where C is an arbitrary constant, and the function $\varphi = \varphi(x)$ is described by the second-order linear ODE: $[a(x)\varphi'_x]'_x + \lambda b(x)\varphi - 2 = 0.$

15. $u_{tt} = [e^{\lambda x} f(u) u_x]_x + e^{\lambda x} g(u), \quad \lambda \neq 0.$

Exact solution (Polyanin, 2019e):
$$u = U(z), \quad z = \lambda x + 2 \ln t$$
where λ is an arbitrary constant, and the function $U = U(z)$ is described by the second-order ODE: $4U''_{zz} - 2U'_z = \lambda^2 [e^z f(U) U'_z]'_z + e^z g(U).$

16. $u_{tt} = [a(x) f(u) u_x]_x + b(x) g(u).$

Exact solutions of such nonlinear PDEs, which are not considered in this section, for various functions $a(x)$, $b(x)$, $f(u)$, and $g(u)$ can be found in Hu & Qu (2007), Huang & Zhou (2010), Huang et al. (2017), Polyanin & Zaitsev (2012), Polyanin (2019e), and Polyanin & Zhurov (2022).

16. $u_{tt} = a u_x u_{xx}.$

Guderley equation. This equation is employed to describe transonic gas flows, where $\gamma = a - 1$ is the *adiabatic index* (Guderley, 1962).

1°. Additive separable solutions:
$$u = \frac{1}{2} a C_1 t^2 + C_2 t + C_3 \pm \frac{1}{3C_1}(2C_1 x + C_4)^{3/2}.$$

2°. Generalized separable solution (Titov, 1988):
$$u = \varphi_1(t) x^3 + \varphi_2(t) x^{3/2} + \varphi_3(t),$$
where the functions $\varphi_n = \varphi_n(t)$ are described by the system of second-order ODEs:
$$\varphi''_1 - 18a\varphi_1^2 = 0,$$
$$\varphi''_2 - \tfrac{45}{4} a \varphi_1 \varphi_2 = 0,$$
$$\varphi''_3 - \tfrac{9}{8} a \varphi_2^2 = 0.$$

It can be shown that this system admits the exact solution

$$\varphi_1 = \frac{1}{3a}(t+C_1)^{-2},$$
$$\varphi_2 = C_2(t+C_1)^{5/2} + C_3(t+C_1)^{-3/2},$$
$$\varphi_3 = \frac{3a}{112}C_2^2(t+C_1)^7 + \frac{3}{8}aC_2C_3(t+C_1)^3 + \frac{9}{16}aC_3^2(t+C_1)^{-1} + C_4 t + C_5,$$

where $C_1, \ldots, C_5$ are arbitrary constants.

3°. *Generalized separable solution in the form of a cubic polynomial in x* (Galaktionov & Svirshchevskii, 2007):

$$u = \psi_1(t) + \psi_2(t)x + \psi_3(t)x^2 + \psi_4(t)x^3.$$

where the functions $\psi_n = \psi_n(t)$ are described by the system of second-order ODEs:

$$\psi_1'' = 2a\psi_2\psi_3,$$
$$\psi_2'' = 2a(3\psi_2\psi_4 + 2\psi_3^2),$$
$$\psi_3'' = 18a\psi_3\psi_4,$$
$$\psi_4'' = 18a\psi_4^2.$$

This system is easy to integrate for $\psi_4 = 0$ to obtain two simple solutions quadratic in x,

$$u = C_1 x^2 + 2aC_1^2 t^2 x + \tfrac{1}{3}a^2 C_1^3 t^4,$$
$$u = C_1 t x^2 + (\tfrac{1}{3}aC_1^2 t^4 + C_2 t + C_3)x$$
$$\quad + \tfrac{1}{63}a^2 C_1^3 t^7 + \tfrac{1}{6}aC_1 C_2 t^4 + \tfrac{1}{3}aC_1 C_3 t^3 + C_4 t + C_5.$$

4°. *Multiplicative separable solution:*

$$u = \varphi(x)\psi(t).$$

The functions $\varphi(x)$ and $\psi(t)$ are described by the first-order autonomous ODEs:

$$(\varphi_x')^3 = \tfrac{3}{2}C_1\varphi^2 + C_3, \quad (\psi_t')^2 = \tfrac{2}{3}aC_1\psi^3 + C_2,$$

where C_1, C_2, and C_3 are arbitrary constants. Solving these equations with respect to derivatives, we obtain separable ODEs. The general solutions to these ODEs can be written in implicit form.

5°. *Self-similar solution:*

$$u = t^{-3\beta-2}U(z), \quad z = t^\beta x,$$

where β is an arbitrary constant, and the function $U(z)$ is described by the ODE:

$$3(\beta+1)(3\beta+2)U - 5\beta(\beta+1)zU_z' + \beta^2 z^2 U_{zz}'' = aU_z' U_{zz}''.$$

6°. *Solution (generalizes the solution from Item 1°):*

$$u = \tfrac{1}{2}aC_1 t^2 + C_2 t + \theta(z), \quad z = x + aC_3 t,$$

where C_1, C_2, and C_3 are arbitrary constants, and the function $\theta = \theta(z)$ is described by the first-order ODE:
$$C_1 z + a C_3^2 \theta'_z + C_4 = \tfrac{1}{2}(\theta'_z)^2,$$
the general solution of which has the form
$$\theta = a C_3^2 z + C_5 \pm \frac{1}{3 C_1} \left(2 C_1 z + a^2 C_4^4 + 2 C_4 \right)^{3/2}.$$

17. $u_{tt} = f(u_x) u_{xx}$.

Zabusky equation modeling the dynamics of nonlinear strings.

1°. Additive separable solution:
$$u(x,t) = A t^2 + B t + \varphi(x),$$
where A and B are arbitrary constants, and the function $\varphi = \varphi(x)$ is described by the ODE: $2A = f(\varphi'_x) \varphi''_{xx}$. Its general solution can be represented in parametric form:
$$x = \frac{1}{2A} \int f(\xi)\, d\xi + C_1, \quad \varphi = \frac{1}{2A} \int \xi f(\xi)\, d\xi + C_2,$$
where C_1 and C_2 are arbitrary constants.

2°. A more general solution:
$$u(x,t) = A t^2 + B t + \varphi(z), \quad z = x + \lambda t,$$
where A, B, and λ are arbitrary constants, and the function $\varphi = \varphi(z)$ is described by the ODE: $2A = [f(\varphi'_z) - \lambda^2] \varphi''_{zz}$. Its general solution can be represented in parametric form:
$$z = \frac{1}{2A} \int f(\xi)\, d\xi - \frac{\lambda^2}{2A} \xi + C_1, \quad \varphi = \frac{1}{2A} \int \xi f(\xi)\, d\xi - \frac{\lambda^2}{4A} \xi^2 + C_2,$$
where C_1 and C_2 are arbitrary constants.

3°. Self-similar solution:
$$u = x \psi(y), \quad y = x/t,$$
where the function $\psi = \psi(y)$ is described by the ODE:
$$[f(y\psi'_y + \psi) - y^2](y\psi''_{yy} + 2\psi'_y) = 0.$$
Equating the expression in square brackets with zero, we have the first-order ODE:
$$f(y\psi'_y + \psi) - y^2 = 0.$$
The general solution of this equation in parametric form:
$$y = \pm\sqrt{f(\tau)}, \quad \psi = \frac{1}{2\sqrt{f(\tau)}} \int \frac{\tau f'_\tau(\tau)}{\sqrt{f(\tau)}}\, d\tau + C.$$

4°. The *Legendre transformation*
$$w(z,\tau) = t z + x \tau - u(x,t), \quad z = u_t, \quad \tau = u_x,$$
where w is the new dependent variable, and z and τ are the new independent variables, leads to the linear PDE:
$$w_{\tau\tau} = f(\tau) w_{zz}.$$

5°. The substitution $v(x,t) = u_x$ leads to a PDE of the form 6.2.2.11:
$$v_{tt} = [f(v) v_x]_x.$$

6°. References for PDE 6.2.2.17: Ibragimov (1994) and Polyanin & Zaitsev (2012).

6.3. Elliptic Equations

6.3.1. Heat Equations with Nonlinear Source of the Form $u_{xx} + u_{yy} = f(u)$

1. $u_{xx} + u_{yy} = au + bu^k$.

1°. Traveling wave solutions for $a > 0$:

$$u(x,y) = \left[\frac{2b\sinh^2 z}{a(k+1)}\right]^{\frac{1}{1-k}}, \quad z = \tfrac{1}{2}\sqrt{a}\,(1-k)(x\sin A + y\cos A) + B \quad \text{if } b(k+1) > 0,$$

$$u(x,y) = \left[-\frac{2b\cosh^2 z}{a(k+1)}\right]^{\frac{1}{1-k}}, \quad z = \tfrac{1}{2}\sqrt{a}\,(1-k)(x\sin A + y\cos A) + B \quad \text{if } b(k+1) < 0,$$

where A and B are arbitrary constants.

2°. Traveling wave solutions for $a < 0$ and $b(k+1) > 0$:

$$u(x,y) = \left[-\frac{2b\cos^2 z}{a(k+1)}\right]^{\frac{1}{1-k}}, \quad z = \tfrac{1}{2}\sqrt{|a|}\,(1-k)(x\sin A + y\cos A) + B.$$

3°. For $a = 0$, there is a self-similar solution of the form $u = x^{\frac{2}{1-k}} F(z)$, where $z = y/x$.

4°. For other exact solutions of this equation, see equation 6.3.1.7 with $f(u) = au + bu^k$.

2. $u_{xx} + u_{yy} = au^k + bu^{2k-1}$.

Exact solutions:

$$u(x,y) = \left[\frac{a(1-k)^2}{2(k+1)}(x\sin C_1 + y\cos C_1 + C_2)^2 - \frac{b(k+1)}{2ak}\right]^{\frac{1}{1-k}},$$

$$u(x,y) = \left\{\frac{1}{4}a(1-k)^2[(x+C_1)^2 + (y+C_2)^2] - \frac{b}{ak}\right\}^{\frac{1}{1-k}},$$

where C_1 and C_2 are arbitrary constants.

3. $u_{xx} + u_{yy} = ae^{\beta u}$.

This equation occurs in combustion theory and is a special case of equation 6.3.1.7 with $f(u) = ae^{\beta u}$.

1°. Exact solutions:

$$u(x,y) = \frac{1}{\beta}\ln\left[\frac{2(A^2 + B^2)}{a\beta(Ax + By + C)^2}\right] \quad \text{if } a\beta > 0,$$

$$u(x,y) = \frac{1}{\beta}\ln\left[\frac{2(A^2 + B^2)}{a\beta\sinh^2(Ax + By + C)}\right] \quad \text{if } a\beta > 0,$$

$$u(x,y) = \frac{1}{\beta}\ln\left[\frac{-2(A^2 + B^2)}{a\beta\cosh^2(Ax + By + C)}\right] \quad \text{if } a\beta < 0,$$

$$u(x,y) = \frac{1}{\beta} \ln\left[\frac{2(A^2+B^2)}{a\beta \cos^2(Ax+By+C)}\right] \qquad \text{if } a\beta > 0,$$

$$u(x,y) = \frac{1}{\beta} \ln\left(\frac{8C}{a\beta}\right) - \frac{2}{\beta} \ln|(x+A)^2 + (y+B)^2 - C|,$$

where A, B, and C are arbitrary constants. The first four solutions are of traveling wave type and the last one is a radial symmetric solution with center at the point $(-A, -B)$.

2°. Functional separable solutions (Aristov, 1999):

$$u(x,y) = -\frac{2}{\beta} \ln\left[C_1 e^{ky} \pm \frac{\sqrt{2a\beta}}{2k} \cos(kx + C_2)\right],$$

$$u(x,y) = \frac{1}{\beta} \ln \frac{2k^2(B^2 - A^2)}{a\beta[A\cosh(kx+C_1) + B\sin(ky+C_2)]^2},$$

$$u(x,y) = \frac{1}{\beta} \ln \frac{2k^2(A^2 + B^2)}{a\beta[A\sinh(kx+C_1) + B\cos(ky+C_2)]^2},$$

where A, B, C_1, C_2, and k are arbitrary constants (x and y can be swapped to give another three solutions).

3°. General solution (Vekua, 1960 and Sabitov, 2001):

$$u(x,y) = -\frac{2}{\beta} \ln \frac{|1 - 2a\beta \Phi(z)\overline{\Phi(z)}|}{4|\Phi'_z(z)|},$$

where $\Phi = \Phi(z)$ is an arbitrary analytic (holomorphic) function of the complex variable $z = x + iy$ with nonzero derivative, and the bar over a symbol denotes the complex conjugate.

4. $u_{xx} + u_{yy} = ae^{\beta u} + be^{2\beta u}$.

1°. Traveling wave solutions:

$$u = -\frac{1}{\beta} \ln\left[-\frac{a\beta}{A^2+B^2} + C\exp(Ax+By) + \frac{a^2\beta^2 - b\beta(A^2+B^2)}{4C(A^2+B^2)^2} \exp(-Ax-By)\right],$$

$$u = -\frac{1}{\beta} \ln\left[\frac{a\beta}{A^2+B^2} + \frac{\sqrt{a^2\beta^2 + b\beta(A^2+B^2)}}{A^2+B^2} \sin(Ax+By+C)\right],$$

where A, B, and C are arbitrary constants.

2°. For other exact solutions of this equation, see equation 6.3.1.7 with $f(u) = ae^{\beta u} + be^{2\beta u}$.

5. $u_{xx} + u_{yy} = au\ln(\beta u)$.

1°. Exact solutions:

$$u = \frac{1}{\beta} \exp\left[\tfrac{1}{4}a(x+A)^2 + \tfrac{1}{4}a(y+B)^2 + 1\right],$$

$$u = \frac{1}{\beta} \exp\left[A(x+B)^2 \pm \sqrt{Aa - 4A^2}\,(x+B)(y+C) + (\tfrac{1}{4}a - A)(y+C)^2 + \tfrac{1}{2}\right],$$

where A, B, and C are arbitrary constants.

2°. There are exact solutions of the following forms:

$$u(x, y) = F(z), \quad z = C_1 x + C_2 y,$$
$$u(x, y) = G(r), \quad r = \sqrt{(x + C_1)^2 + (y + C_2)^2},$$
$$u(x, y) = f(x) g(y).$$

3°. References for PDE 6.3.1.5: Shercliff (1977) and Zaitsev & Polyanin (2004).

6. $u_{xx} + u_{yy} = a \sin(\beta u)$.

1°. Functional separable solution for $a = \beta = 1$:

$$u(x, y) = 4 \arctan\left(\cot A \, \frac{\cosh F}{\cosh G} \right),$$
$$F = \frac{\cos A}{\sqrt{1 + B^2}} (x - By), \quad G = \frac{\sin A}{\sqrt{1 + B^2}} (y + Bx),$$

where A and B are arbitrary constants.

2°. Functional separable solution:

$$u(x, y) = \frac{4}{\beta} \arctan\bigl[f(x) g(y)\bigr],$$

where the functions $f = f(x)$ and $g = g(y)$ are determined by the first-order autonomous ordinary differential equations

$$(f'_x)^2 = A f^4 + B f^2 + C, \quad (g'_y)^2 = C g^4 + (a\beta - B) g^2 + A,$$

and A, B, and C are arbitrary constants.

3°. For other exact solutions of this equation, see equation 6.3.1.7 with $f(u) = a \sin(\beta u)$.

4°. References for PDE 6.3.1.6: Bullough & Caudrey (1980) and Miller & Rubel (1993).

7. $u_{xx} + u_{yy} = f(u)$.

This is a *stationary heat equation with a nonlinear source*.

1°. Suppose $u = u(x, y)$ is a solution of the equation in question. Then the functions

$$u_1 = u(\pm x + C_1, \pm y + C_2),$$
$$u_2 = u(x \cos \beta - y \sin \beta, \, x \sin \beta + y \cos \beta),$$

where C_1, C_2, and β are arbitrary constants, are also solutions of the equation (the plus or minus signs in u_1 may be chosen arbitrarily).

2°. Traveling wave solution in implicit form:

$$\int \left[C + \frac{2}{A^2 + B^2} F(u) \right]^{-1/2} du = Ax + By + D, \quad F(u) = \int f(u) \, du,$$

where A, B, C, and D are arbitrary constants.

TABLE 6.3. Nonlinear heat-type equations $u_{xx} + u_{yy} = f(u)$ that admit functional separable solutions of the form $u = u(z)$, where $z = \varphi(x) + \psi(y)$.

No.	Right-hand side of equation $f(u)$	Solution $u(z)$	Equations for $\varphi(x)$ and $\psi(y)$
1	$a\ln u + bu$	e^z	$(\varphi'_x)^2 = C_1 e^{-2\varphi} + a\varphi - \tfrac{1}{2}a + b + A$, $(\psi'_y)^2 = C_2 e^{-2\psi} + a\psi - \tfrac{1}{2}a - A$
2	$ae^u + be^{-2u}$	$\ln\lvert z\rvert$	$(\varphi'_x)^2 = 2a\varphi^3 + A\varphi^2 + C_1\varphi + C_2$, $(\psi'_y)^2 = 2a\psi^3 - A\psi^2 + C_1\psi - C_2 - b$
3	$a\sin u + b\left(\sin u \ln\tan\dfrac{u}{4} + 2\sin\dfrac{u}{4}\right)$	$4\arctan e^z$	$(\varphi'_x)^2 = C_1 e^{2\varphi} + C_2 e^{-2\varphi} + b\varphi + a + A$, $(\psi'_y)^2 = C_2 e^{2\psi} + C_1 e^{-2\psi} + b\psi - A$
4	$a\sinh u + b\left(\sinh u \ln\tanh\dfrac{u}{4} + 2\sinh\dfrac{u}{2}\right)$	$2\ln\left\lvert\coth\dfrac{z}{2}\right\rvert$	$(\varphi'_x)^2 = C_1 e^{2\varphi} + C_2 e^{-2\varphi} - \sigma b\varphi + a + A$, $(\psi'_y)^2 = -C_2 e^{2\psi} - C_1 e^{-2\psi} - \sigma b\psi - A$
5	$a\sinh u + 2b\left(\sinh u \arctan e^{u/2} + \cosh\dfrac{u}{2}\right)$	$2\ln\left\lvert\tan\dfrac{z}{2}\right\rvert$	$(\varphi'_x)^2 = C_1\sin 2\varphi + C_2\cos 2\varphi + \sigma b\varphi + a + A$, $(\psi'_y)^2 = C_1\sin 2\psi - C_2\cos 2\psi + \sigma b\psi - A$

Notation: A, C_1, C_2 are arbitrary constants; $\sigma = 1$ for $z > 0$ and $\sigma = -1$ for $z < 0$

3°. Solution with central symmetry about the point $(-C_1, -C_2)$:
$$u = u(\zeta), \qquad \zeta = \sqrt{(x + C_1)^2 + (y + C_2)^2},$$
where C_1 and C_2 are arbitrary constants and the function $u = u(\zeta)$ is described by the ODE
$$u''_{\zeta\zeta} + \zeta^{-1} u'_\zeta = f(u).$$

4°. Miller & Rubel (1993) found all functions $f(u)$ for which PDE 6.3.1.3 admits solutions with functional separation of variables of the form $u = U(\varphi(x) + \psi(y))$ (see also Andreev et al., 1999 and Polyanin & Zhurov, 2022). The results are summarized in Table 6.3.

6.3.2. Stationary Anisotropic Heat/Diffusion Equations of the Form $[f(x)u_x]_x + [g(y)u_y]_y = h(u)$

1. $(ax^n u_x)_x + (by^m u_y)_y = f(u), \qquad ab > 0.$

Functional separable solution for $n \neq 2$ and $m \neq 2$ (Polyanin & Zaitsev, 2004):
$$u = u(r), \qquad r = \left[b(2 - m)^2 x^{2-n} + a(2 - n)^2 y^{2-m}\right]^{1/2}.$$

Here the function $u(r)$ is described by the second-order ODE
$$u''_{rr} + A r^{-1} u'_r = B f(u),$$

where $A = \dfrac{4 - nm}{(2 - n)(2 - m)}$, $B = \dfrac{4}{ab(2 - n)^2(2 - m)^2}$.

2. $au_{xx} + (be^{\mu y} u_y)_y = f(u)$, $\quad ab > 0$.

Functional separable solution for $\mu \neq 0$ (Polyanin & Zaitsev, 2004):
$$u = u(\xi), \quad \xi = \left[b\mu^2(x+C_1)^2 + 4ae^{-\mu y}\right]^{1/2},$$
where C_1 is an arbitrary constant, and the function $u(\xi)$ is defined implicitly by
$$\int \left[C_2 + \frac{2}{ab\mu^2} F(u)\right]^{-1/2} du = C_3 \pm \xi, \quad F(u) = \int f(u)\,du,$$
with C_2 and C_3 being arbitrary constants.

3. $(ae^{\beta x} u_x)_x + (be^{\mu y} u_y)_y = f(u)$, $\quad ab > 0$.

Functional separable solution for $\beta\mu \neq 0$ (Polyanin & Zaitsev, 2004):
$$u = u(\xi), \quad \xi = \left(b\mu^2 e^{-\beta x} + a\beta^2 e^{-\mu y}\right)^{1/2},$$
where the function $u(\xi)$ is described by the second-order ODE:
$$u''_{\xi\xi} - \xi^{-1} u'_\xi = A f(u), \quad A = 4/(ab\beta^2\mu^2).$$

4. $[f(x)u_x]_x + [g(y)u_y]_y = ku\ln u$.

Multiplicative separable solution (Polyanin & Zaitsev, 2004):
$$u(x,y) = \varphi(x)\psi(y),$$
where the functions $\varphi(x)$ and $\psi(y)$ are described by the second-order ODEs:
$$[f(x)\varphi'_x]'_x = k\varphi\ln\varphi + C\varphi, \quad [g(y)\psi'_y]'_y = k\psi\ln\psi - C\psi,$$
and C is an arbitrary constant.

6.3.3. Stationary Anisotropic Heat/Diffusion Equations of the Form $[f(u)u_x]_x + [g(u)u_y]_y = h(u)$

1. $u_{xx} + [(au+b)u_y]_y = 0$.

Stationary Khokhlov–Zabolotskaya equation. It arises in acoustics and nonlinear mechanics.

1°. Exact solutions:
$$u(x,y) = Ay - \tfrac{1}{2}A^2 ax^2 + C_1 x + C_2,$$
$$u(x,y) = (Ax+B)y - \frac{a}{12A^2}(Ax+B)^4 + C_1 x + C_2,$$
$$u(x,y) = -\frac{1}{a}\left(\frac{y+A}{x+B}\right)^2 + \frac{C_1}{x+B} + C_2(x+B)^2 - \frac{b}{a},$$
$$u(x,y) = -\frac{1}{a}\left[b + \lambda^2 \pm \sqrt{A(y+\lambda x) + B}\right],$$
$$u(x,y) = (Ax+B)\sqrt{C_1 y + C_2} - \frac{b}{a},$$
where A, B, C_1, C_2, and λ are arbitrary constants.

2°. Generalized separable solution quadratic in y (generalizes the third solution of Item 1°):

$$u(x,y) = -\frac{1}{a(x+A)^2}y^2 + \left[\frac{B_1}{(x+A)^2} + B_2(x+A)^3\right]y$$
$$+ \frac{C_1}{x+A} + C_2(x+A)^2 - \frac{b}{a} - \frac{aB_1^2}{4(x+A)^2} - \frac{1}{2}aB_1B_2(x+A)^3 - \frac{1}{54}aB_2^2(x+A)^8,$$

where A, B_1, B_2, C_1, and C_2 are arbitrary constants.

3°. See also equation 6.3.3.5 with $f(u) = 1$ and $g(u) = au + b$.

4°. Other exact solutions of PDE 6.3.3.1 can be found in Polyanin & Zaitsev (2012).

2. $a(u^k u_x)_x + b(u^m u_y)_y = 0$.

1°. Multiplicative separable solution:

$$u(x,y) = f(x)g(y).$$

The functions $f(x)$ and $g(y)$ are determined by the second-order autonomous ODEs:

$$(f^k f'_x)'_x = Abf^{m+1}, \qquad (g^m g'_y)'_y = -Aag^{k+1},$$

where A is an arbitrary constant.

2°. There are exact solutions of the following forms:

$$\begin{aligned}
u(x,y) &= F(z), & z &= C_1 x + C_2 y & &\text{traveling wave solution,} \\
u(x,y) &= x^{-2s} G(\xi), & \xi &= y x^{(m-k)s-1} & &\text{self-similar solution,} \\
u(x,y) &= x^{\frac{2}{k-m}} H(\eta), & \eta &= y + s \ln x, & & \\
u(x,y) &= e^{2x} Q(\zeta), & \zeta &= y e^{(k-m)x}, & &
\end{aligned}$$

where C_1, C_2, and s are arbitrary constants.

3°. See also PDE 6.3.3.5 with $f(u) = au^k$ and $g(u) = bu^m$.

3. $a(u^k u_x) + b(u^m u_y)_y = cu^n$.

There are exact solutions of the following forms:

$$\begin{aligned}
u(x,y) &= F(z), & z &= C_1 x + C_2 y & &\text{traveling wave solution;} \\
u(x,y) &= x^{\frac{2}{k-n+1}} U(z), & z &= y x^{\frac{n-m-1}{k-n+1}} & &\text{self-similar solution.}
\end{aligned}$$

4. $u_{xx} + (ae^{\beta u} u_y)_y = 0$, $\quad a > 0$.

1°. Additive separable solutions:

$$u(x,y) = \frac{1}{\beta} \ln(Ay + B) + Cx + D,$$
$$u(x,y) = \frac{1}{\beta} \ln(-aA^2 y^2 + By + C) - \frac{2}{\beta} \ln(-aAx + D),$$
$$u(x,y) = \frac{1}{\beta} \ln(Ay^2 + By + C) + \frac{1}{\beta} \ln\left[\frac{p^2}{aA \cosh^2(px+q)}\right],$$

$$u(x,y) = \frac{1}{\beta} \ln(Ay^2 + By + C) + \frac{1}{\beta} \ln\left[\frac{p^2}{-aA\cos^2(px+q)}\right],$$

$$u(x,y) = \frac{1}{\beta} \ln(Ay^2 + By + C) + \frac{1}{\beta} \ln\left[\frac{p^2}{-aA\sinh^2(px+q)}\right],$$

where A, B, C, D, p, and q are arbitrary constants.

2°. There are exact solutions of the following forms:

$$u(x,y) = F(r), \quad r = C_1 x + C_2 y;$$
$$u(x,y) = G(z), \quad z = y/x;$$
$$u(x,y) = H(\xi) - 2(k+1)\beta^{-1} \ln|x|, \quad \xi = y|x|^k;$$
$$u(x,y) = U(\eta) - 2\beta^{-1} \ln|x|, \quad \eta = y + k\ln|x|;$$
$$u(x,y) = V(\zeta) - 2\beta^{-1} x, \quad \zeta = ye^x,$$

where C_1, C_2, and k are arbitrary constants.

3°. For other solutions, see equation 6.3.3.5 with $f(u) = 1$ and $g(u) = ae^{\beta u}$.

5. $[f(u)u_x]_x + [g(u)u_y]_y = 0.$

This is a *stationary anisotropic heat (diffusion) equation*.

1°. Traveling wave solution in implicit form:

$$\int [A^2 f(u) + B^2 g(u)]\, du = C_1(Ax + By) + C_2,$$

where A, B, C_1, and C_2 are arbitrary constants.

2°. Self-similar solution:

$$u = u(\zeta), \quad \zeta = \frac{x+A}{y+B},$$

where the function $u(\zeta)$ is described by the ODE:

$$[f(u)u'_\zeta]'_\zeta + [\zeta^2 g(u)u'_\zeta]'_\zeta = 0. \tag{1}$$

Integrating (1) and taking u to be the independent variable, one obtains the Riccati equation $C\zeta'_u = g(u)\zeta^2 + f(u)$, where C is an arbitrary constant.

3°. The original equation can be represented as the system of first-order PDEs:

$$f(u)u_x = v_y, \quad -g(u)u_y = v_x. \tag{2}$$

The hodograph transformation

$$x = x(u,v), \quad y = y(u,v),$$

where u, v are treated as the independent variables and x, y as the dependent ones, brings (2) to the linear system

$$f(u)y_v = x_u, \quad -g(u)x_v = y_u. \tag{3}$$

Eliminating y yields the following second-order linear PDE for $x = x(u,v)$:

$$[x_u/f(u)]_u + g(u)x_{vv} = 0. \tag{4}$$

Likewise, we can obtain another second-order linear PDE for $y = y(u,v)$ from system (3).

4°. For the special case $g(u) = kf(u)$, the transformation

$$\bar{x} = x, \quad \bar{y} = k^{-1/2}y, \quad \bar{u} = \int f(u)\,du$$

leads to the Laplace equation $\Delta \bar{u} = 0$, where Δ is the Laplace operator in the variables $\bar{x}$ and $\bar{y}$.

5°. A number of exact solutions in the parametric form of the original nonlinear PDE, as well as linear PDE (4), are given in Zaitsev & Polyanin (2001) and Polyanin & Zaitsev (2012).

6.4. Other Second-Order Equations

6.4.1. Equations of Transonic Gas Flow

1. $au_x u_{xx} + u_{yy} = 0.$

This is an *equation of steady transonic gas flow*.

1°. Suppose $u(x, y)$ is a solution of the equation in question. Then the function

$$u_1 = C_1^{-3} C_2^2 u(C_1 x + C_3, C_2 y + C_4) + C_5 y + C_6,$$

where $C_1, \ldots, C_6$ are arbitrary constants, is also a solution of the equation.

2°. Exact solutions:

$$u = -\frac{(x+C_1)^3}{3a(y+C_2)^2} + C_3 y + C_4,$$

$$u = \frac{a^2 C_1^3}{39}(y+A)^{13} + \frac{2}{3}aC_1^2(y+A)^8(x+B) + 3C_1(y+A)^3(x+B)^2 - \frac{(x+B)^3}{3a(y+A)^2},$$

$$u = -aC_1 y^2 + C_2 y + C_3 \pm \frac{4}{3C_1}(C_1 x + C_4)^{3/2},$$

$$u = -aA^3 y^2 - \frac{B^2}{aA^2}x + C_1 y + C_2 \pm \frac{4}{3}(Ax + By + C_3)^{3/2},$$

$$u = \frac{1}{3}(Ay+B)(2C_1 x + C_2)^{3/2} - \frac{aC_1^3}{12A^2}(Ay+B)^4 + C_3 y + C_4,$$

$$u = -\frac{9aA^2}{y+C_1} + 4A\left(\frac{x+C_2}{y+C_1}\right)^{3/2} - \frac{(x+C_2)^3}{3a(y+C_1)^2} + C_3 y + C_4,$$

$$u = -\frac{3}{7}aA^2(y+C_1)^7 + 4A(x+C_2)^{3/2}(y+C_1)^{5/2} - \frac{(x+C_2)^3}{3a(y+C_1)^2} + C_3 y + C_4,$$

where $A, B, C_1, \ldots, C_4$ are arbitrary constants (the first solution is degenerate).

3°. There are solutions of the following forms:

$u = y^{-3k-2} U(z), \quad z = xy^k$ (self-similar solution);

$u = \varphi_1(y) + \varphi_2(y) x^{3/2} + \varphi_3(y) x^3$ (generalized separable solution);

$u = \psi_1(y) + \psi_2(y) x + \psi_3(y) x^2 + \psi_4(y) x^3$ (generalized separable solution);

$u = \psi_1(y)\varphi(x) + \psi_2(y)$ (generalized separable solution),

where k is an arbitrary constant.

4°. References for PDE 6.4.1.1: Guderley (1962), Titov (1988), Galaktionov & Svirshchevskii (2007), and Polyanin & Zaitsev (2012).

2. $u_{yy} + \dfrac{a}{y} u_y + b u_x u_{xx} = 0$.

1°. Suppose $u(x,y)$ is a solution of the equation in question. Then the function

$$u_1 = C_1^{-3} C_2^2 u(C_1 x + C_3, C_2 y) + C_4 y^{1-a} + C_5,$$

where $C_1, \ldots, C_5$ are arbitrary constants, is also a solution of the equation.

2°. Additive separable solutions:

$$u = -\dfrac{bC_1}{4(a+1)} y^2 + C_2 y^{1-a} + C_3 \pm \dfrac{2}{3C_1} (C_1 x + C_4)^{3/2},$$

where $C_1, \ldots, C_4$ are arbitrary constants.

3°. Generalized separable solutions:

$$u = -\dfrac{9A^2 b}{16(n+1)(2n+1+a)} y^{2n+2} + A y^n (x+C)^{3/2} + \dfrac{a-3}{9b} \dfrac{(x+C)^3}{y^2},$$

where A and C are arbitrary constants, and $n = n_{1,2}$ are roots of the quadratic equation $n^2 + (a-1)n + \tfrac{5}{4}(a-3) = 0$.

4°. Generalized separable solution:

$$u = (A y^{1-a} + B)(2C_1 x + C_2)^{3/2} + 9b C_1^3 \theta(y),$$

$$\theta(y) = -\dfrac{B^2}{2(a+1)} y^2 - \dfrac{AB}{3-a} y^{3-a} - \dfrac{A^2}{2(2-a)(3-a)} y^{4-2a} + C_3 y^{1-a} + C_4,$$

where A, B, C_1, C_2, C_3, and C_4 are arbitrary constants.

5°. There are solutions of the following forms:

$u = y^{-3k-2} U(z), \quad z = x y^k$ (self-similar solution);

$u = \varphi_1(y) + \varphi_2(y) x^{3/2} + \varphi_3(y) x^3$ (generalized separable solution);

$u = \psi_1(y) + \psi_2(y) x + \psi_3(y) x^2 + \psi_4(y) x^3$ (generalized separable solution);

$u = \psi_1(y) \varphi(x) + \psi_2(y)$ (generalized separable solution),

where k is an arbitrary constant.

6°. References for PDE 6.4.1.2: Guderley (1962), Titov (1988), Galaktionov & Svirshchevskii (2007), and Polyanin & Zaitsev (2012).

6.4.2. Monge–Ampère Type Equations

1. $u_{xy}^2 - u_{xx} u_{yy} = 0$.

Homogeneous Monge–Ampère equation.

1°. General solution in parametric form (Goursat, 1933):
$$u = tx + \varphi(t)y + \psi(t),$$
$$x + \varphi'(t)y + \psi'(t) = 0,$$
where t is the parameter, $\varphi = \varphi(t)$ and $\psi = \psi(t)$ are arbitrary functions, the prime denotes the derivative with respect to t.

2°. Exact solutions involving one arbitrary function (Khabirov, 1990 and Ibragimov, 1994):
$$u = \varphi(C_1 x + C_2 y) + C_3 x + C_4 y + C_5,$$
$$u = (C_1 x + C_2 y)\varphi\left(\frac{y}{x}\right) + C_3 x + C_4 y + C_5,$$
$$u = (C_1 x + C_2 y + C_3)\varphi\left(\frac{C_4 x + C_5 y + C_6}{C_1 x + C_2 y + C_3}\right) + C_7 x + C_8 y + C_9,$$
where $C_1, \ldots, C_9$ are arbitrary constants and $\varphi = \varphi(z)$ is an arbitrary function.

2. $u_{xy}^2 - u_{xx} u_{yy} = A.$

Nonhomogeneous Monge–Ampère equation.

1°. General solution in parametric form for $A = a^2 > 0$ (Goursat, 1933):
$$x = \frac{\beta - \lambda}{2a}, \quad y = \frac{\psi'(\lambda) - \varphi'(\beta)}{2a},$$
$$u = \frac{(\beta + \lambda)[\psi'(\lambda) - \varphi'(\beta)] + 2\varphi(\beta) - 2\psi(\lambda)}{4a},$$
where β and λ are the parameters, $\varphi = \varphi(\beta)$ and $\psi = \psi(\lambda)$ are arbitrary functions.

2°. Exact solutions:
$$u = \pm \frac{\sqrt{A}}{C_2} x(C_1 x + C_2 y) + \varphi(C_1 x + C_2 y) + C_3 x + C_4 y,$$
$$u = C_1 y^2 + C_2 xy + \frac{1}{4C_1}(C_2^2 - A)x^2 + C_3 y + C_4 x + C_5,$$
$$u = \frac{1}{x + C_1}\left(C_2 y^2 + C_3 y + \frac{C_3^2}{4C_2}\right) - \frac{A}{12C_2}(x^3 + 3C_1 x^2) + C_4 y + C_5 x + C_6,$$
$$u = \pm \frac{2\sqrt{A}}{3C_1 C_2}(C_1 x - C_2^2 y^2 + C_3)^{3/2} + C_4 x + C_5 y + C_6,$$
where $C_1, \ldots, C_6$ are arbitrary constants and $\varphi = \varphi(z)$ is an arbitrary function.

3. $u_{xy}^2 - u_{xx} u_{yy} = f(x).$

1°. Generalized separable solutions quadratic in y (Polyanin & Zaitsev, 2004):
$$u = C_1 y^2 + C_2 xy + \frac{C_2^2}{4C_1} x^2 - \frac{1}{2C_1} \int_0^x (x - t) f(t)\, dt + C_3 y + C_4 x + C_5,$$
$$u = \frac{1}{x + C_1}\left(C_2 y^2 + C_3 y + \frac{C_3^2}{4C_2}\right)$$
$$- \frac{1}{2C_2} \int_0^x (x - t)(t + C_1) f(t)\, dt + C_4 y + C_5 x + C_6,$$
where $C_1, \ldots, C_6$ are arbitrary constants.

2°. Generalized separable solutions for $f(x) > 0$:

$$u = \pm y \int \sqrt{f(x)}\, dx + \varphi(x) + C_1 y,$$

where $\varphi(x)$ is an arbitrary function.

4. $u_{xy}^2 - u_{xx} u_{yy} = f(x) y^k.$

1°. Generalized separable solutions (Polyanin & Zaitsev, 2004):

$$u = \frac{C_1 y^{k+2}}{(k+1)(k+2)} - \frac{1}{C_1} \int_a^x (x-t) f(t)\, dt + C_2 x + C_3 y + C_4,$$

$$u = \frac{y^{k+2}}{(C_1 x + C_2)^{k+1}} - \frac{1}{(k+1)(k+2)} \int_a^x (x-t)(C_1 t + C_2)^{k+1} f(t)\, dt + C_3 x + C_4 y + C_5,$$

where $C_1, \ldots, C_5$ are arbitrary constants.

2°. Multiplicative separable solution:

$$u = \varphi(x) y^{\frac{k+2}{2}},$$

where the function $\varphi = \varphi(x)$ is described by the ODE

$$k(k+2) \varphi \varphi''_{xx} - (k+2)^2 (\varphi'_x)^2 + 4 f(x) = 0.$$

5. $u_{xy}^2 - u_{xx} u_{yy} = f(x) y^{2k+2} + g(x) y^k.$

Generalized separable solution (Polyanin & Zaitsev, 2004):

$$u(x,y) = \varphi(x) y^{k+2} - \frac{1}{(k+1)(k+2)} \int_a^x (x-t) \frac{g(t)}{\varphi(t)}\, dt + C_1 x + C_2 y + C_3,$$

where the function $\varphi = \varphi(x)$ is described by the ODE

$$(k+1)(k+2) \varphi \varphi''_{xx} - (k+2)^2 (\varphi'_x)^2 + f(x) = 0.$$

6. $u_{xy}^2 - u_{xx} u_{yy} = f(x) e^{\lambda y}.$

1°. Generalized separable solutions (Polyanin & Zaitsev, 2004):

$$u = C_1 \int_a^x (x-t) f(t)\, dt + C_2 x - \frac{1}{C_1 \lambda^2} e^{\lambda y} + C_3 y + C_4,$$

$$u = C_1 e^{\beta x + \lambda y} - \frac{1}{C_1 \lambda^2} \int_a^x (x-t) e^{-\beta t} f(t)\, dt + C_2 x + C_3 y + C_4,$$

where $C_1, \ldots, C_4$ and β are arbitrary constants.

2°. Multiplicative separable solution:

$$u = \varphi(x) \exp\left(\tfrac{1}{2} \lambda y\right),$$

where the function $\varphi = \varphi(x)$ is described by the ODE

$$\varphi \varphi''_{xx} - (\varphi'_x)^2 + 4 \lambda^{-2} f(x) = 0.$$

7. $u_{xy}^2 - u_{xx}u_{yy} = f(x)e^{2\lambda y} + g(x)e^{\lambda y}.$

Generalized separable solution (Polyanin & Zaitsev, 2004):
$$u(x,y) = \varphi(x)e^{\lambda y} - \frac{1}{\lambda^2}\int_a^x (x-t)\frac{g(t)}{\varphi(t)}\,dt + C_1 x + C_2 y + C_3,$$

where the function $\varphi = \varphi(x)$ is described by the ODE
$$\varphi\varphi_{xx}'' - (\varphi_x')^2 + \lambda^{-2}f(x) = 0.$$

8. $u_{xy}^2 - u_{xx}u_{yy} = x^{2\alpha}f(x^\beta y).$

Self-similar solution (Khabirov, 1990):
$$u(x,y) = x^{\alpha-\beta+1}U(z) \quad z = x^\beta y,$$

where the function $U = U(z)$ is described by the ODE
$$[\beta(\beta+1)zU_z' + (\alpha-\beta)(\beta-\alpha-1)U]U_{zz}'' + (\alpha+1)^2(U_z')^2 - f(z) = 0.$$

9. $u_{xy}^2 - u_{xx}u_{yy} = e^{\alpha x}f(e^{\beta x}y).$

Solution (Khabirov, 1990):
$$u(x,y) = e^{\mu x}U(z), \quad z = e^{\beta x}y, \quad \mu = \tfrac{1}{2}\alpha - \beta,$$

where the function $U = U(z)$ is described by the ODE
$$\beta^2 z U_z' U_{zz}'' - \mu^2 U U_{zz}'' + (\beta+\mu)^2(U_z')^2 - f(z) = 0.$$

10. $u_{xy}^2 - u_{xx}u_{yy} = f(u).$

Functional separable solution (Polyanin & Zaitsev, 2004):
$$u = U(z), \quad z = ax^2 + bxy + cy^2 + kx + sy,$$

where a, b, c, k, and s are arbitrary constants and the function $U(z)$ is described by the ODE
$$2\bigl[(4ac - b^2)z + as^2 + ck^2 - bks\bigr]U_z'U_{zz}'' + (4ac - b^2)(U_z')^2 + f(U) = 0.$$

Remark 6.2. Monge–Ampère type equations of the form $u_{xy}^2 - u_{xx}u_{yy} = f(x,y,u)$ are encountered in differential geometry, gas dynamics, and meteorology. For these nonlinear PDEs, their transformations, and some solutions and applications, see Martin (1953), Khabirov (1990), Ibragimov (1994), and Polyanin & Zaitsev (2012).

6.5. Higher-Order Equations

6.5.1. Third-Order Equations

▶ **Korteweg–de Vries equation and related integrable PDEs.**

1. $u_t + u_{xxx} - 6uu_x = 0.$

Korteweg–de Vries equation (*KdV equation* for short). It is used in many sections of nonlinear mechanics and physics.

1°. Suppose $u(x,t)$ is a solution of the Korteweg–de Vries equation. Then the function
$$u_1 = C_1^2 u(C_1 x + 6C_1 C_2 t + C_3, C_1^3 t + C_4) + C_2,$$
where $C_1, \ldots, C_4$ are arbitrary constants, is also a solution of the equation.

2°. One-soliton solution (it is a traveling wave solution):
$$u(x,t) = -\frac{a}{2\cosh^2\left[\frac{1}{2}\sqrt{a}\,(x - at - b)\right]},$$
where a and b are arbitrary constants.

3°. Two-soliton solution (Hirota, 1971):
$$u(x,t) = -2\frac{\partial^2}{\partial x^2}\ln\!\left(1 + B_1 e^{\theta_1} + B_2 e^{\theta_2} + AB_1 B_2 e^{\theta_1 + \theta_2}\right),$$
$$\theta_1 = a_1 x - a_1^3 t, \quad \theta_2 = a_2 x - a_2^3 t, \quad A = \left(\frac{a_1 - a_2}{a_1 + a_2}\right)^2,$$
where B_1, B_2, a_1, and a_2 are arbitrary constants.

4°. N-soliton solution (Calogero & Degasperis, 1982):
$$u(x,t) = -2\frac{\partial^2}{\partial x^2}\Big\{\ln\det\big[\mathbb{I} + \mathbb{C}(x,t)\big]\Big\}.$$
Here $\mathbb{I}$ is the $N \times N$ identity matrix and $\mathbb{C}(x,t)$ is an $N \times N$ symmetric matrix with entries
$$C_{mn}(x,t) = \frac{\sqrt{\rho_m(t)\rho_n(t)}}{p_m + p_n}\exp\!\big[-(p_m + p_n)x\big],$$
where the normalizing factors $\rho_n(t)$ are given by
$$\rho_n(t) = \rho_n(0)\exp\!\big(8p_n^3 t\big), \quad n = 1, 2, \ldots, N.$$
The solution involves $2N$ arbitrary constants p_n and $\rho_n(0)$.

For $t \to \pm\infty$, the N-soliton solution can be represented as the sum of N single-soliton solutions.

5°. "One soliton + one pole" solution:
$$u(x,t) = -2p^2\big[\cosh^{-2}(pz) - (1 + px)^{-2}\tanh^2(pz)\big]\big[1 - (1 + px)^{-1}\tanh(pz)\big]^{-2},$$
$$z = x - 4p^2 t - c,$$
where p and c are arbitrary constants.

6°. "N solitons + one pole" solution (Calogero & Degasperis, 1982):
$$u(x,t) = -2\frac{\partial^2}{\partial x^2}\Big\{x\ln\det\big[\mathbb{I} + \mathbb{D}(x,t)\big]\Big\}.$$
Here $\mathbb{I}$ is the $N \times N$ identity matrix and $\mathbb{D}(x,t)$ is an $N \times N$ symmetric matrix with entries
$$D_{mn}(x,t) = c_m(t)c_n(t)\big[(p_m + p_n)^{-1} + (p_m p_n x)^{-1}\big]\exp\!\big[-(p_m + p_n)x\big],$$
where the normalizing factors $c_n(t)$ are given by
$$c_n(t) = c_n(0)\exp\!\big(4p_n^3 t\big), \quad n = 1, 2, \ldots, N.$$
The solution involves $2N$ arbitrary constants p_n and $c_n(0)$.

7°. Rational solutions (Ablowitz & Segur, 1981):
$$u(x,t) = \frac{6x(x^3 - 24t)}{(x^3 + 12t)^2},$$
$$u(x,t) = -2\frac{\partial^2}{\partial x^2}\ln(x^6 + 60x^3 t - 720t^2).$$

8°. There is a self-similar solution of the form $u = t^{-2/3}U(z)$, where $z = t^{-1/3}x$ (see Calogero & Degasperis, 1982).

9°. Solution:
$$u(x,t) = 2\varphi(z) + 2C_1 t, \quad z = x + 6C_1 t^2 + C_2 t,$$
where C_1 and C_2 are arbitrary constants, and the function $\varphi(z)$ is described by the second-order ODE: $\varphi''_{zz} = 6\varphi^2 - C_2\varphi - C_1 z + C_3$.

10°. General similarity solutions (Clarkson & Kruskal, 1989):
$$u = \varphi^2 U(z) + \frac{1}{6\varphi}(\varphi'_t x + \psi'_t), \quad z = \varphi(t)x + \psi(t).$$
The functions $\varphi = \varphi(t)$ and $\psi = \psi(t)$ are given by
$$\varphi(t) = (3At + C_1)^{-1/3}, \quad \psi(t) = C_2(3At + C_1)^{2/3} + C_3(3At + C_1)^{-1/3},$$
where A, C_1, C_2, and C_3 are arbitrary constants, and the function $U(z)$ is described by the ODE:
$$U'''_{zzz} - 6UU'_z - AU + \tfrac{2}{3}A^2 z = 0.$$

11°. The Korteweg–de Vries equation is solved by the inverse scattering method. Any rapidly decreasing function $F = F(x,y;t)$ as $x \to +\infty$ that simultaneously satisfies the two linear PDEs:
$$\frac{\partial^2 F}{\partial x^2} - \frac{\partial^2 F}{\partial y^2} = 0, \quad \frac{\partial F}{\partial t} + \left(\frac{\partial}{\partial x} + \frac{\partial}{\partial y}\right)^3 F = 0$$
generates a solution of the Korteweg–de Vries equation in the form
$$u = -2\frac{d}{dx}K(x,x;t),$$
where $K(x,y;t)$ is a solution of the linear Gelfand–Levitan–Marchenko integral equation
$$K(x,y;t) + F(x,y;t) + \int_x^\infty K(x,z;t)F(z,y;t)\,dz = 0.$$
Time t appears in this equation as a parameter.

12°. For the Korteweg–de Vries equation, see also Gardner et al. (1967), Miura (1968), Hirota (1971), Zakharov & Faddeev (1971), Bullough & Caudrey (1980), Ablowitz & Segur (1981), Calogero & Degasperis (1982), Dodd et al. (1982), Novikov et al. (1984), Zhidkov (2001), and Polyanin & Zaitsev (2012).

2. $u_t + u_{xxx} - 6uu_x + \dfrac{1}{2t}u = 0.$

Cylindrical Korteweg–de Vries equation.

1°. The transformation

$$u(x,t) = -\frac{x}{12t} - \frac{1}{2t}u(z,\tau), \quad x = \frac{z}{\tau}, \quad t = -\frac{1}{2\tau^2}$$

leads to the Korteweg–de Vries equation 6.5.1.1:

$$u_\tau + u_{zzz} - 6uu_z = 0.$$

2°. References for PDE 6.5.1.2: Johnson (1972), Calogero & Degasperis (1982), Zhu & Chen (1986), and Bluman & Kumei (1989).

3. $u_t + u_{xxx} - 6u^2 u_x = 0.$

Modified Korteweg–de Vries equation in canonical form (or *mKdV equation* for short).

1°. Suppose $u(x,t)$ is a solution of the equation in question. Then the function

$$u_1 = C_1 u(C_1 x + C_2, C_1^3 t + C_3),$$

where C_1, C_2, and C_3 are arbitrary constants, is also a solution of the equation.

2°. One-soliton solution:

$$u = \frac{2kC \exp(kx - k^3 t)}{1 - C^2 \exp[2(kx - k^3 t)]},$$

where C and k are arbitrary constants.

3°. Traveling wave solutions:

$$u = \pm \frac{\sqrt{c}}{\sinh\left[\sqrt{c}\,(x - ct + B)\right]} \qquad \text{if } c > 0,$$

$$u = \pm\sqrt{-c}\,\tanh\left[\sqrt{-c}\,(x - 2ct + B)\right] \qquad \text{if } c < 0,$$

$$u = \pm\sqrt{-c}\,\coth\left[\sqrt{-c}\,(x - 2ct + B)\right] \qquad \text{if } c < 0,$$

$$u = \pm \frac{\sqrt{-c}}{\cos\left[\sqrt{-c}\,(x - ct + B)\right]} \qquad \text{if } c < 0,$$

$$u = \pm\tfrac{1}{2}\sqrt{-c}\left\{2 + \tan^2\left[\tfrac{1}{2}\sqrt{-c}\,(x - ct + B)\right]\right.$$
$$\left. + \cot^2\left[\tfrac{1}{2}\sqrt{-c}\,(x - ct + B)\right]\right\}^{1/2} \qquad \text{if } c < 0,$$

where B and c are arbitrary constants.

4°. There is a self-similar solution of the form $u = t^{-1/3} f(y)$, where $y = t^{-1/3} x$.

5°. Suppose $u(x,t)$ is a solution of the equation in question. Then the function $u(x,t)$ obtained with the Miura transformation

$$z(x,t) = u_x + u^2 \tag{1}$$

satisfies the Korteweg–de Vries equation 6.5.1.1:

$$z_t + z_{xxx} - 6zz_x = 0. \tag{2}$$

6°. The modified Korteweg–de Vries equation is an integrable equation and can be solved by inverse scattering method.

Solutions of the modified Korteweg–de Vries equation

$$u_t + u_{xxx} - 6\sigma u^2 u_x = 0, \quad \sigma = \pm 1 \tag{3}$$

may be obtained from solutions of the linear Gelfand–Levitan–Marchenko integral equation. Any function $F = F(x, y; t)$ rapidly decaying as $x \to +\infty$ and satisfying simultaneously the two linear PDEs:

$$\frac{\partial F}{\partial x} - \frac{\partial F}{\partial y} = 0, \quad \frac{\partial F}{\partial t} + \left(\frac{\partial}{\partial x} + \frac{\partial}{\partial y}\right)^3 F = 0 \tag{4}$$

generates a solution of equation (3) in the form

$$u = K(x, x; t),$$

where $K(x, y; t)$ is a solution of the linear Gelfand–Levitan–Marchenko integral equation,

$$K(x, y; t) = F(x, y; t) + \frac{\sigma}{4} \int_x^\infty \int_x^\infty K(x, z; t) F(z, u; t) F(u, y; t)\, dz\, du. \tag{5}$$

Time t appears in (5) as a parameter. It follows from the first equation in (4) that $F(x, y; t) = F(x + y; t)$.

7°. References for PDE 6.5.1.3: Whitham (1965), Miura (1968), Wadati (1972), Ablowitz & Segur (1981), Calogero & Degasperis (1982), Hietarinta (1987), Drazin & Johnson (1996), and Wazwaz (2008).

4. $u_t + u_{xxx} + 6u^2 u_x = 0.$

Modified Korteweg–de Vries equation (another form).

1°. One-soliton solution (Ablowitz & Segur, 1981):

$$u(x, t) = a + \frac{k^2}{\sqrt{4a^2 + k^2}\,\cosh z + 2a}, \quad z = kx - (6a^2 k + k^3)t + b,$$

where a, b, and k are arbitrary constants.

2°. Two-soliton solution (Dodd et al., 1982):

$$u(x,t) = 2\frac{a_1 e^{\theta_1} + a_2 e^{\theta_2} + A a_2 e^{2\theta_1 + \theta_2} + A a_1 e^{\theta_1 + 2\theta_2}}{1 + e^{2\theta_1} + e^{2\theta_2} + 2(1-A)e^{\theta_1+\theta_2} + A^2 e^{2(\theta_1+\theta_2)}},$$

$$\theta_1 = a_1 x - a_1^3 t + b_1, \quad \theta_2 = a_2 x - a_2^3 t + b_2, \quad A = \left(\frac{a_1 - a_2}{a_1 + a_2}\right)^2,$$

where a_1, a_2, b_1, and b_2 are arbitrary constants.

3°. Rational solutions (algebraic solitons):

$$u(x, t) = a - \frac{4a}{4a^2 z^2 + 1}, \quad z = x - 6a^2 t,$$

$$u(x, t) = a - \frac{12a\left(z^4 + \frac{3}{2}a^{-2}z^2 - \frac{3}{16}a^{-4} - 24tz\right)}{4a^2\left(z^3 + 12t - \frac{3}{4}a^{-2}z\right)^2 + 3\left(z^2 + \frac{1}{4}a^{-2}\right)^2},$$

where a is an arbitrary constant.

4°. There is a self-similar solution of the form $u = t^{-1/3}U(z)$, where $z = t^{-1/3}x$.

5°. Traveling wave solutions involving hyperbolic functions:
$$u = \pm\tfrac{1}{2}\sqrt{c}\left\{2 - \tanh^2\left[\tfrac{1}{2}\sqrt{c}\,(x - ct + b)\right] - \coth^2\left[\tfrac{1}{2}\sqrt{c}\,(x - ct + b)\right]\right\}^{1/2},$$
where b and $c > 0$ are arbitrary constants.

6°. Traveling wave solutions involving trigonometric functions:
$$u = \pm\frac{2\sqrt{c/3}\,\cos^2\left[\tfrac{1}{2}\sqrt{c}\,(x - ct + b)\right]}{3 - 2\cos^2\left[\tfrac{1}{2}\sqrt{c}\,(x - ct + b)\right]}.$$

Setting $u = 0$ for $\tfrac{1}{2}\sqrt{c}\,|x - ct + b| \geq \tfrac{\pi}{2}$, one obtains compacton-like solutions, which are located in an interval of length $2\pi/\sqrt{c}$.

7°. The modified Korteweg–de Vries equation is an integrable equation and can be solved by the inverse scattering method; see Item 6° in equation 6.5.1.3 with $\sigma = -1$.

8°. References for PDE 6.5.1.4: Ono (1976), Ablowitz & Segur (1981), Dodd et al. (1982), Hietarinta (1987), Drazin & Johnson (1996), Rosenau (2000), and Polyanin & Zaitsev (2012).

5. $u_t + u_{xxx} + au_x^2 = 0.$

Potential Korteweg–de Vries equation (or *potential KdV equation* for short).

1°. One-soliton solution:
$$u = \frac{2kC\exp\left(\tfrac{1}{3}akx - \tfrac{1}{27}a^3k^3t\right)}{1 - C\exp\left(\tfrac{1}{3}akx - \tfrac{1}{27}a^3k^3t\right)},$$
where C and k are arbitrary constants.

2°. Traveling wave solutions (Wazwaz, 2008):

$u = A + \dfrac{3\sqrt{c}}{a}\tanh\left[\dfrac{\sqrt{c}}{2}(x - ct + B)\right]$ if $c > 0$,

$u = A + \dfrac{3\sqrt{c}}{a}\coth\left[\dfrac{\sqrt{c}}{2}(x - ct + B)\right]$ if $c > 0$,

$u = A + \left\{\sqrt{\dfrac{3cB^2 - aB\sqrt{c}}{3c}} + B\tanh\left[\dfrac{\sqrt{c}}{2}(x - ct + C)\right]\right\}^{-1}$ if $c > 0$,

$u = A + \left\{\sqrt{\dfrac{3cB^2 - aB\sqrt{c}}{3c}} + B\coth\left[\dfrac{\sqrt{c}}{2}(x - ct + C)\right]\right\}^{-1}$ if $c > 0$,

$u = A - \dfrac{3\sqrt{-c}}{a}\tan\left[\sqrt{-c}\,(x - ct + B)\right] - \dfrac{3\sqrt{-c}}{a}\sec\left[\sqrt{-c}\,(x - ct + B)\right]$ if $c < 0$,

$u = A + \dfrac{3\sqrt{-c}}{a}\cot\left[\sqrt{-c}\,(x - ct + B)\right] + \dfrac{3\sqrt{-c}}{a}\csc\left[\sqrt{-c}\,(x - ct + B)\right]$ if $c < 0$,

where A, B, C, and c are arbitrary constants.

3°. The Bäcklund transformation (Ibragimov, 1985):

$$u_x = -bz, \quad u_t = bz_{xx} - 3baz^2, \quad \text{where} \quad b = 3/a, \tag{1}$$

links the equation in question with the Korteweg–de Vries equation 6.5.1.1:

$$z_t + z_{xxx} - 6zz_x = 0. \tag{2}$$

Let $z = z(x,t)$ be a solution of equation (2). Then the linear system of first-order equations (1) enables us to find the corresponding solution $u = u(x,t)$ of the original equation.

6. $u_t + u_{xxx} - au_x^3 = 0.$

Potential modified Korteweg–de Vries equation (or *potential mKdV equation* for short).
The Bäcklund transformation (Ibragimov, 1985):

$$u_x = bz, \quad u_t = -bz_{xx} + 2bz^3, \quad \text{where} \quad b = \pm\sqrt{2/a}, \tag{1}$$

links the equation in question with the modified Korteweg–de Vries equation 6.5.1.3:

$$z_t + z_{xxx} - 6z^2 z_x = 0. \tag{2}$$

Let $z = z(x,t)$ be a solution of equation (2). Then the linear system of first-order equations (1) enables us to find the corresponding solution $u = u(x,t)$ of the original equation.

▶ **More complex Korteweg–de Vries type PDEs.**

7. $u_t + (2au - 3bu^2)u_x + u_{xxx} = 0.$

Gardner equation or the *combined KdV-mKdV equation*. It is used to model nonlinear phenomena in plasma and solid-state physics and in quantum field theory. The solutions below are obtained by Fu et al. (2004) and Wazwaz (2007, 2008).

1°. Traveling wave solutions:

$$u = \frac{a}{3b}\left\{1 \pm \tanh\left[\frac{a}{3\sqrt{2b}}\left(x - \frac{2a^2}{9b}t + B\right)\right]\right\} \quad \text{if } b > 0;$$

$$u = \frac{a}{3b}\left\{1 \pm \coth\left[\frac{a}{3\sqrt{2b}}\left(x - \frac{2a^2}{9b}t + B\right)\right]\right\} \quad \text{if } b > 0;$$

$$u = \frac{6c}{2a \pm \sqrt{4a^2 - 18bc}\,\cosh[\sqrt{c}\,(x - ct + B)]} \quad \text{if } c > 0 \text{ and } 2a^2 > 9bc;$$

$$u = \frac{6c}{2a \pm \sqrt{18bc - 4a^2}\,\sinh[\sqrt{c}\,(x - ct + B)]} \quad \text{if } c > 0 \text{ and } 2a^2 < 9bc,$$

where B and c are arbitrary constants.

2°. Traveling wave solution with $b > 0$ and $k = a^2 - 3bc > 0$:

$$u = \frac{a}{3b} + \frac{\sqrt{3k}}{3b}\operatorname{csch}\left[\sqrt{\frac{2k}{3b}}\,(x - ct + B)\right] + \frac{\sqrt{3k}}{3b}\coth\left[\sqrt{\frac{2k}{3b}}\,(x - ct + B)\right].$$

3°. Traveling wave solution with $k = 9b^2 - 16a^2 > 0$:

$$u = \operatorname{csch}\xi\left(1 + \frac{3b - \sqrt{k}}{4a}\operatorname{csch}\xi\right)^{-1}, \quad \xi = \frac{\sqrt{9b + 3\sqrt{k}}}{6}\left(x - \frac{3b + \sqrt{k}}{12}t + B\right),$$

where B and c are arbitrary constants.

4°. Traveling wave solutions:
$$u = \operatorname{sech}\xi \left(1 + \frac{3b \pm \sqrt{k}}{4a}\operatorname{sech}\xi\right)^{-1}, \quad \xi = \pm\frac{\sqrt{-9b \pm 3\sqrt{k}}}{6}\left(x - \frac{-3b \pm \sqrt{k}}{12}t + B\right),$$
where $k = 9b^2 + 16a^2$, and B and c are arbitrary constants.

5°. The transformation
$$u = U(z,t) + \frac{a}{3b}, \quad z = x - \frac{a^2}{3b}t$$
leads to the equation
$$U_t - 3bU^2U_z + U_{zzz} = 0,$$
that reduces, with the simple substitution $U = \sqrt{2/|b|}\, V$ and depending on the sign of the coefficient b, to either PDE 6.5.1.3 or PDE 6.5.1.4 (for the function V).

8. $u_t + u_{xxx} + au^k u_x = 0.$

Korteweg–de Vries type equation with a power-law nonlinearity. For $k = 1$ and $k = 2$ see equations 6.5.1.1, 6.5.1.3, and 6.5.1.4. For $k = 1/2$, the equation describes ion-acoustic waves in a cold-ion plasma with nonisothermal electrons.

1°. Traveling wave solutions:
$$u = \left\{\frac{c(k+1)(k+2)}{2a\cosh^2\left[\frac{1}{2}k\sqrt{c}\,(x - ct + B)\right]}\right\}^{1/k} \quad \text{if } c > 0,$$

$$u = \left\{-\frac{c(k+1)(k+2)}{2a\sinh^2\left[\frac{1}{2}k\sqrt{c}\,(x - ct + B)\right]}\right\}^{1/k} \quad \text{if } c > 0,$$

$$u = \left\{\frac{c(k+1)(k+2)}{2a\cos^2\left[\frac{1}{2}k\sqrt{-c}\,(x - ct + B)\right]}\right\}^{1/k} \quad \text{if } c < 0,$$

where B and c are arbitrary constants. The first solution is the soliton solution.

2°. Traveling wave solutions:
$$u = \left\{\frac{c(k+1)(k+2)}{8a}\left(2 - \tanh^2\left[\tfrac{1}{4}k\sqrt{c}\,(x-ct+B)\right] - \coth^2\left[\tfrac{1}{4}k\sqrt{c}\,(x-ct+B)\right]\right)\right\}^{1/k}$$
if $c > 0$,

$$u = \left\{\frac{c(k+1)(k+2)}{8a}\left(2 + \tan^2\left[\tfrac{1}{4}k\sqrt{-c}\,(x-ct+B)\right] + \cot^2\left[\tfrac{1}{4}k\sqrt{-c}\,(x-ct+B)\right]\right)\right\}^{1/k}$$
if $c < 0$,

where B and c are arbitrary constants.

3°. Self-similar solution:
$$u(x,t) = t^{-\frac{2}{3k}}U(z), \quad z = xt^{-\frac{1}{3}},$$
where the function $U = U(z)$ is described by the ODE:
$$-\frac{2}{3k}U - \frac{1}{3}zU'_z + U'''_{zzz} + aU^k U'_z = 0.$$

4°. References for PDE 6.5.1.8: Schamel (1973), Ablowitz & Segur (1981), Drazin & Johnson (1996), and Wazwaz (2008).

9. $u_t + (au^k - bu^{2k})u_x + u_{xxx} = 0.$

This equation describes the propagation of nonlinear long acoustic-type waves. The solutions below are obtained by Wazwaz (2006, 2008) and Polyanin & Zaitsev (2012).

1°. Traveling wave solutions:

$$u = \left\{ \frac{a(2k+1)}{2b(k+2)} \left[1 \pm \tanh(mkx - 4m^3kt + C)\right] \right\}^{1/k}, \quad m = \frac{a}{2(k+2)}\sqrt{\frac{2k+1}{b(k+1)}};$$

$$u = \left\{ \frac{a(2k+1)}{2b(k+2)} \left[1 \pm \coth(mkx - 4m^3kt + C)\right] \right\}^{1/k}, \quad m = \frac{a}{2(k+2)}\sqrt{\frac{2k+1}{b(k+1)}},$$

where C is an arbitrary constant; the first two solutions with tanh are kink solutions.

2°. Traveling wave solutions:

$$u = \left\{ \frac{(k+1)(k+2)c}{a \pm a\sqrt{m}\,\cosh[k\sqrt{c}\,(x-ct) + B]} \right\}^{1/k}, \quad m = 1 - \frac{k^3 + 5k^2 + 8k + 4}{2k+1}\frac{bc}{a^2}$$
if $c > 0$;

$$u = \left\{ \frac{(k+1)(k+2)c}{a \pm a\sqrt{s}\,\sinh[k\sqrt{c}\,(x-ct) + B]} \right\}^{1/k}, \quad s = \frac{k^3 + 5k^2 + 8k + 4}{2k+1}\frac{bc}{a^2} - 1$$
if $c > 0$;

$$u = \left\{ \frac{(k+1)(k+2)c}{a \pm a\sqrt{m}\,\cos[k\sqrt{-c}\,(x-ct) + B]} \right\}^{1/k}, \quad m = 1 - \frac{k^3 + 5k^2 + 8k + 4}{2k+1}\frac{bc}{a^2}$$
if $c < 0$,

where B and c are arbitrary constants.

10. $u_t + u_{xxx} + ae^u u_x = 0.$

Korteweg–de Vries type equation with an exponential nonlinearity.

1°. Traveling wave solution:

$$u = u(z), \quad z = x + \lambda t,$$

where the function $u(z)$ is determined by the second-order autonomous ordinary differential equation

$$u''_{zz} + \lambda u + ae^u = C,$$

and λ and C are arbitrary constants.

2°. Solution:

$$u(x,t) = U(\xi) - \tfrac{2}{3}\ln t, \quad \xi = xt^{-\frac{1}{3}},$$

where the function $U = U(\xi)$ is described by the ODE:

$$U'''_{\xi\xi\xi} + (ae^U - \tfrac{1}{3}\xi)U'_\xi - \tfrac{2}{3} = 0.$$

11. $u_t = au_{xxx} + (b\ln u + c)u_x.$

Korteweg–de Vries type equation with a logarithmic nonlinearity.

1°. Suppose $u(x,t)$ is a solution of the equation in question. Then the function
$$u_1 = e^{C_1} u(x + bC_1 t + C_2, t + C_3),$$
where C_1, C_2, and C_3 are arbitrary constants, is also a solution of the equation.

2°. Generalized traveling wave solution (Fushchich et al., 1991):
$$u(x,t) = \exp\left[\frac{C_2 - x}{bt + C_1} + \frac{a}{b}\frac{1}{(bt + C_1)^2} - \frac{c}{b}\right],$$
where C_1 and C_2 are arbitrary constants.

3°. Solution (Polyanin & Zaitsev, 2012):
$$u(x,t) = e^{\lambda t} U(z), \quad z = x + \tfrac{1}{2} b\lambda t^2 + kt,$$
where k and λ are arbitrary constants, and the function $U(z)$ is described by the autonomous ODE:
$$aU'''_{zzz} + (b\ln U + c - k) U'_z - \lambda U = 0.$$
To $\lambda = 0$ there corresponds a traveling wave solution.

12. $u_t = au_{xxx} + (b\operatorname{arsinh} u + c) u_x.$

Generalized traveling wave solution (Fushchich et al., 1991):
$$u(x,t) = \sinh\left[\frac{C_2 - x}{bt + C_1} + \frac{a}{b}\frac{1}{(bt + C_1)^2} - \frac{c}{b}\right],$$
where C_1 and C_2 are arbitrary constants.

13. $u_t = au_{xxx} + (b\operatorname{arcosh} u + c) u_x.$

Generalized traveling wave solution:
$$u(x,t) = \cosh\left[\frac{C_2 - x}{bt + C_1} + \frac{a}{b}\frac{1}{(bt + C_1)^2} - \frac{c}{b}\right],$$
where C_1 and C_2 are arbitrary constants.

14. $u_t = au_{xxx} + (b\arcsin u + c) u_x.$

Generalized traveling wave solution (Fushchich et al., 1991):
$$u(x,t) = \sin\left[\frac{C_2 - x}{bt + C_1} - \frac{a}{b}\frac{1}{(bt + C_1)^2} - \frac{c}{b}\right],$$
where C_1 and C_2 are arbitrary constants.

15. $u_t = au_{xxx} + (b\arccos u + c) u_x.$

Generalized traveling wave solution:
$$u(x,t) = \cos\left[\frac{C_2 - x}{bt + C_1} - \frac{a}{b}\frac{1}{(bt + C_1)^2} - \frac{c}{b}\right],$$
where C_1 and C_2 are arbitrary constants.

16. $u_t = [f(u)u_x]_{xx} + \dfrac{a}{f(u)} + b.$

Functional separable solution in implicit form (Polyanin & Zaitsev, 2004):
$$\int f(u)\,du = at - \tfrac{1}{6}bx^3 + C_1 x^2 + C_2 x + C_3,$$
where C_1, C_2, and C_3 are arbitrary constants.

17. $u_t = [f(u)u_x]_{xx} + \dfrac{aF(u)+b}{f(u)} + c[aF(u)+b], \quad F(u) = \int f(u)\,du.$

$1°$. Functional separable solution in implicit form with $a \neq 0$:
$$\int f(u)\,du = \tfrac{1}{a}[\rho(x)e^{at} - b],$$
where
$$\rho(x) = \begin{cases} C_1 + C_2 x + C_3 x^2 & \text{if } c = 0, \\ C_1 e^{-kx} + e^{kx/2}\!\left(C_2 \cos \dfrac{kx\sqrt{3}}{2} + C_3 \sin \dfrac{kx\sqrt{3}}{2}\right)\!, \quad k = (ac)^{1/3} & \text{if } c \neq 0, \end{cases}$$
and C_1, C_2, and C_3 are arbitrary constants.

$2°$. Functional separable solution in implicit form with $a = 0$:
$$\int f(u)\,du = bt + C_1 + C_2 x + C_3 x^2 - \tfrac{1}{6}bcx^3.$$

▶ **Hydrodynamics equations.**

18. $u_{xt} + u_x^2 - uu_{xx} = au_{xxx} + f(t).$

This equation arises in fluid dynamics and describes some classes of exact solutions to the Navier–Stokes equations. The function $f = f(t)$ can be defined arbitrarily.

$1°$. Suppose $u = u(x,t)$ is a solution of the equation in question. Then the function
$$u_1 = u(x + \psi(t), t) + \psi'_t(t),$$
where $\psi(t)$ is an arbitrary function, is also a solution of the equation.

$2°$. Exact solutions for $f(t) \equiv 0$:
$$u(x,t) = \dfrac{6a}{x + \psi(t)} + \psi'_t(t),$$
$$u(x,t) = C_1 \exp[-\lambda x + \lambda \psi(t)] - \psi'_t(t) + a\lambda,$$
where $\psi(t)$ is an arbitrary function and C_1, C_2, and λ are arbitrary constants.

$3°$. Generalized separable solutions for $f(t) = Ae^{-\beta t}$, $A > 0$, $\beta > 0$:
$$u(x,t) = Be^{-\frac{1}{2}\beta t}\sin[\lambda x + \lambda\psi(t)] + \psi'_t(t),$$
$$u(x,t) = Be^{-\frac{1}{2}\beta t}\cos[\lambda x + \lambda\psi(t)] + \psi'_t(t), \qquad B = \pm\sqrt{\dfrac{2Aa}{\beta}}, \quad \lambda = \sqrt{\dfrac{\beta}{2a}},$$
where $\psi(t)$ is an arbitrary function.

4°. Generalized separable solutions for $f(t) = Ae^{\beta t}$, $A > 0$, $\beta > 0$:

$$u(x,t) = Be^{\frac{1}{2}\beta t}\sinh[\lambda x + \lambda\psi(t)] + \psi'_t(t), \qquad B = \pm\sqrt{\frac{2Aa}{\beta}}, \quad \lambda = \sqrt{\frac{\beta}{2a}},$$

where $\psi(t)$ is an arbitrary function.

5°. Generalized separable solutions for $f(t) = Ae^{\beta t}$, $A < 0$, $\beta > 0$:

$$u(x,t) = Be^{\frac{1}{2}\beta t}\cosh[\lambda x + \lambda\psi(t)] + \psi'_t(t), \qquad B = \pm\sqrt{\frac{2|A|a}{\beta}}, \quad \lambda = \sqrt{\frac{\beta}{2a}},$$

where $\psi(t)$ is an arbitrary function.

6°. Generalized separable solutions for $f(t) = Ae^{\beta t}$, A is any, $\beta > 0$:

$$u(x,t) = \psi(t)e^{\lambda x} - \frac{Ae^{\beta t - \lambda x}}{4\lambda^2 \psi(t)} + \frac{\psi'_t(t)}{\lambda\psi(t)} - a\lambda, \quad \lambda = \pm\sqrt{\frac{\beta}{2a}},$$

where $\psi(t)$ is an arbitrary function.

7°. Self-similar solution for $f(t) = At^{-2}$:

$$u(x,t) = t^{-1/2}\left[\theta(z) - \tfrac{1}{2}z\right], \quad z = xt^{-1/2},$$

where the function $\theta = \theta(z)$ is described by the autonomous ODE

$$\tfrac{3}{4} - A - 2\theta'_z + (\theta'_z)^2 - \theta\theta''_{zz} = a\theta'''_{zzz},$$

whose order can be reduced by one.

8°. Traveling wave solution for $f(t) = A$:

$$u = U(\xi), \quad \xi = x + \lambda t,$$

where the function $U(\xi)$ is described by the autonomous ODE

$$-A + \lambda U''_{\xi\xi} + (U'_\xi)^2 - UU''_{\xi\xi} = aU'''_{\xi\xi\xi},$$

whose order can be reduced by one.

9°. The original equation admits order reduction. Denote

$$\eta = u_x, \quad \Phi = u_{xx}. \qquad (3)$$

Moving the term $-u_x^2$ to the right-hand side of the original equation, dividing the resulting equation by $u_{xx} = \Phi$, differentiating with respect to x, and taking into account (3), we obtain

$$\frac{\Phi_t}{\Phi} - \frac{u_{xt}\Phi_x}{\Phi^2} + \eta = \frac{\partial}{\partial x}\frac{a\Phi_x + \eta^2 + q(t)\eta + p(t)}{\Phi}. \qquad (4)$$

In (4), let us switch from the old variables t, x, and $u = u(x,t)$ to the new variables t, η, and $\Phi = \Phi(t,\eta)$ with η and Φ defined by formulas (3) (it is a *Crocco transformation*). The derivatives are transformed as follows:

$$\frac{\partial}{\partial x} = \frac{\partial \eta}{\partial x}\frac{\partial}{\partial \eta} = u_{xx}\frac{\partial}{\partial \eta} = \Phi\frac{\partial}{\partial \eta}, \quad \frac{\partial}{\partial t} = \frac{\partial}{\partial t} + \frac{\partial \eta}{\partial t}\frac{\partial}{\partial \eta} = \frac{\partial}{\partial t} + u_{xt}\frac{\partial}{\partial \eta}.$$

As a result, equation (4) reduces to the second-order PDE

$$\Phi_t + [f(t) - \eta^2]\Phi_\eta + \eta\Phi = a\Phi^2\Phi_{\eta\eta}. \qquad (5)$$

Note that in the degenerate case (inviscid fluid with $a = 0$), the original nonlinear second-order PDE reduces to the linear first-order PDE (5), which can be integrated using the method of characteristics.

$10°$. References for PDE 6.5.1.18: Polyanin (2001a), Galaktionov & Svirshchevskii (2007), and Polyanin & Zaitsev (2012).

19. $u_{tx} + uu_{xx} - u_x^2 = \nu u_{xxx} + q(t)u_x + p(t).$

This equation arises in fluid dynamics and describes some classes of exact solutions to the three-dimensional Navier–Stokes equations (ν is the kinematic viscosity of the fluid). The functions $q = q(t)$ and $p = p(t)$ can be defined arbitrarily. In the special case where $q \equiv 0$, setting $\nu = a$ and $p(t) = -f(t)$ and replacing $-u$ with u yields PDE 6.5.1.18.

$1°$. Suppose $u = u(x,t)$ is a solution of the equation in question. Then the function

$$u_1 = u(x + \psi(t), t) - \psi_t'(t),$$

where $\psi(t)$ is an arbitrary function, is also a solution of the equation.

$2°$. Functional separable solution:

$$u = -a_t'(t) + b(t)[x + a(t)] - \frac{6\nu}{x + a(t)}, \quad q = -4b, \quad p = b_t' + 3b^2,$$

where $a = a(t)$ and $b = b(t)$ are arbitrary functions.

$3°$. Generalized separable solution exponentially dependent on x:

$$u = a(t)e^{-\sigma x} + b(t), \quad p = 0, \quad q = \frac{a_t'}{a} - \sigma b - \sigma^2 \nu,$$

where $a = a(t)$ and $b = b(t)$ are arbitrary functions. On choosing periodic $a(t)$ and $b(t)$, one arrives at solutions periodic in time t.

$4°$. Multiplicative separable solution:

$$u = a(t)(C_1 e^{\sigma x} + C_2 e^{-\sigma x}), \quad p = 4C_1 C_2 \sigma^2 a^2(t), \quad q = \frac{a_t'}{a} - \sigma^2 \nu,$$

where $a = a(t)$ is an arbitrary function and C_1, C_2, and σ are arbitrary constants.

$5°$. Traveling wave solution:

$$u = u(z), \quad z = x - \lambda t, \quad p = \text{const}, \quad q = \text{const},$$

where p, q, and λ are arbitrary constants, and the function $u = u(z)$ is described by the ODE

$$(u - \lambda)u_{zz}'' - (u_z')^2 = \nu u_{zzz}''' + qu_z' + p.$$

Special case 5.1. Monotonic traveling wave solution:

$$u = -6\nu\sigma \tanh[\sigma(x - \lambda t) + B] + \lambda, \quad p = 0, \quad q = 8\nu\sigma^2,$$

where B, σ, and λ are arbitrary constants.

Special case 5.2. Unbounded periodic traveling wave solution:

$$u = 6\nu\sigma \tan[\sigma(x - \lambda t) + B] + \lambda, \quad p = 0, \quad q = -8\nu\sigma^2,$$

where B, σ, and λ are arbitrary constants.

6°. Multiplicative separable solution periodic in x:

$$u = a(t)\sin(\sigma x + B), \quad a(t) = C\exp\left[-\nu\sigma^2 t + \int q(t)\,dt\right],$$

$$p = -\sigma^2 a^2(t), \quad q = q(t) \text{ is an arbitrary function,}$$

where B, C, and σ are arbitrary constants. For $q(t) = \nu\sigma^2 + \varphi'_t(t)$, where $\varphi(t)$ is a periodic function, we have a periodic solution in both arguments x and t.

7°. Functional separable solution:

$$u = \frac{a(t)}{\lambda(t)}\exp[-\lambda(t)x] + b(t) + c(t)x,$$

where the functions $a = a(t)$, $b = b(t)$, $c = c(t)$, and $\lambda = \lambda(t)$ satisfy the system of ordinary differential equations

$$\lambda'_t = -c\lambda, \quad a'_t = (\nu\lambda^2 + q + 2c + b\lambda)a, \quad c'_t = c^2 + qc + p.$$

Three out of the six functions $a(t)$, $b(t)$, $c(t)$, $\lambda(t)$, $p(t)$, and $q(t)$ can be defined arbitrarily.

8°. Functional separable solution:

$$u = \omega(t)x + \frac{\xi(t)}{\theta(t)}\sin[\theta(t)x + a], \qquad (1)$$

where a is an arbitrary constant, and the functions $\omega = \omega(t)$, $\xi = \xi(t)$, and $\theta = \theta(t)$ are determined by the system of ordinary differential equations

$$\theta'_t = -\omega\theta, \quad \omega'_t = \omega^2 + q(t)\omega + p(t) + \xi^2, \quad \xi'_t = [2\omega - \nu\theta^2 + q(t)]\xi. \qquad (2)$$

The functions $\theta(t)$ and $\xi(t)$ in system (2) can be regarded as prescribed (in an arbitrary way). Then the functions $\omega(t)$, $p(t)$, and $q(t)$ are easily determinable (without integrals). To periodic $\theta(t)$ and $\xi(t)$, there corresponds a periodic solution (1).

9°. Functional separable solution:

$$u = \omega(t)x + \frac{\xi(t)}{\theta(t)}\left[C_1 e^{\theta(t)x} + C_2 e^{-\theta(t)x}\right],$$

where C_1 and C_2 are arbitrary constants, and the functions $\omega = \omega(t)$, $\xi = \xi(t)$, and $\theta = \theta(t)$ are determined by the system of ordinary differential equations

$$\theta'_t = -\omega\theta, \quad \omega'_t = \omega^2 + q(t)\omega + p(t) - 4C_1 C_2 \xi^2, \quad \xi'_t = [2\omega + \nu\theta^2 + q(t)]\xi.$$

10°. Self-similar solution:

$$u = t^{-1/2} U(\zeta), \quad \zeta = xt^{-1/2}; \quad p = At^{-2}, \quad q = Bt^{-1},$$

where A and B are arbitrary constants, and the function $U = U(\zeta)$ is described by the ODE

$$(U - \tfrac{1}{2}\zeta)U''_{\zeta\zeta} - (U'_\zeta)^2 = \nu U'''_{\zeta\zeta\zeta} + (B+1)U'_\zeta + A.$$

11°. The Crocco transformation $\eta = u_x$, $\Phi = u_{xx}$ (for details, see Item 9° of equation 6.5.1.19) reduces the considered third-order PDE to the second-order PDE

$$\Phi_t + (\eta^2 + q\eta + p)\Phi_\eta = (\eta + q)\Phi + \nu\Phi^2\Phi_{\eta\eta}. \qquad (5)$$

Whenever a solution to the original equation is known, formulas (3) give a solution to equation (5) in parametric form.

12°. References for PDE 6.5.1.19: Aristov & Polyanin (2009) and Aristov et al. (2009).

20. $u_y u_{xy} - u_x u_{yy} = a u_{yyy}$.

This is an *equation of a steady laminar boundary layer* on a flat plate (u is the stream function, a is the kinematic viscosity of the fluid).

1°. Suppose $u(x, y)$ is a solution of the equation in question. Then the function
$$u_1 = C_1 u(C_2 x + C_3, C_1 C_2 y + \varphi(x)) + C_4,$$
where $\varphi(x)$ is an arbitrary function and $C_1, \ldots, C_5$ are arbitrary constants, is also a solution of the equation (Pavlovskii, 1961 and Ovsiannikov, 1982).

2°. Exact solutions involving arbitrary functions (Ignatovich, 1993 and Polyanin, 2001b):
$$u(x, y) = C_1 y + \varphi(x),$$
$$u(x, y) = C_1 y^2 + \varphi(x) y + \frac{1}{4C_1} \varphi^2(x) + C_2,$$
$$u(x, y) = \frac{6ax + C_1}{y + \varphi(x)} + \frac{C_2}{[y + \varphi(x)]^2} + C_3,$$
$$u(x, y) = \varphi(x) \exp(-C_1 y) + aC_1 x + C_2,$$
$$u(x, y) = C_1 \exp[-C_2 y - C_2 \varphi(x)] + C_3 y + C_3 \varphi(x) + aC_2 x + C_4,$$
$$u(x, y) = 6aC_1 x^{1/3} \tanh \xi + C_2, \quad \xi = C_1 \frac{y}{x^{2/3}} + \varphi(x),$$
$$u(x, y) = -6aC_1 x^{1/3} \tan \xi + C_2, \quad \xi = C_1 \frac{y}{x^{2/3}} + \varphi(x),$$
where $C_1, \ldots, C_4$ are arbitrary constants and $\varphi(x)$ is an arbitrary function. The first two solutions are degenerate–they are independent of a and correspond to inviscid fluid flows.

3°. Generalized separable solution linear in x:
$$u(x, y) = xF(y) + G(y), \tag{1}$$
where the functions $F = F(y)$ and $G = G(y)$ are described by the system of autonomous ODEs
$$(F'_y)^2 - FF''_{yy} = aF'''_{yyy}, \tag{2}$$
$$F'_y G'_y - FG''_{yy} = aG'''_{yyy}. \tag{3}$$

Equation (2) has the following particular solutions:
$$F = 6a(y + C)^{-1},$$
$$F = Ce^{\lambda y} - a\lambda,$$
where C and λ are arbitrary constants.

Let $F = F(y)$ be a solution of equation (2) ($F \not\equiv \text{const}$). Then the corresponding general solution of equation (3) can be written in the form
$$G(y) = C_1 + C_2 F + C_3 \left(F \int \psi \, dy - \int F\psi \, dy \right), \quad \text{where } \psi = \frac{1}{(F'_y)^2} \exp\left(-\frac{1}{a} \int F \, dy \right).$$

References for Item 3°: Hiemenz (1911), Wang (1991), and Polyanin (2001b).

4°. Table 6.4 lists some other exact solutions to the hydrodynamic boundary layer equation (Pavlovskii, 1961 and Schlichting, 1981). Solution 1 is self-similar, and solution 2 is generalized self-similar. In the special case $\beta = 0$, solution 3 becomes a self-similar solution (see solution 1 with $\lambda = -1$). All ODEs for U in Table 6.4 are autonomous and generalized homogeneous; hence, their order can be reduced by two.

TABLE 6.4. Exact solutions to the hydrodynamic boundary layer equation (C_1, C_2, C_3, β, and λ are arbitrary constants).

No.	Form of solution $u = u(x,y)$	ODE for the function $U = U(z)$
1	$u = x^{\lambda+1}U(z)$, $z = x^\lambda y$	$(2\lambda+1)(U'_z)^2 - (\lambda+1)UU''_{zz} = aU'''_{zzz}$
2	$u = e^{\lambda x}U(z)$, $z = e^{\lambda x}y$	$2\lambda(U'_z)^2 - \lambda UU''_{zz} = aU'''_{zzz}$
3	$u = U(z) + \beta \ln|x|$, $z = y/x$	$-(U'_z)^2 - \beta U''_{zz} = aU'''_{zzz}$

21. $u_y u_{xy} - u_x u_{yy} = a u_{yyy} + f(x)$.

This is a *hydrodynamic boundary layer equation with pressure gradient*. For $f(x) \equiv 0$, see equation 6.5.1.20.

1°. Suppose $u(x,y)$ is a solution of the equation in question. Then the functions
$$u_1 = \pm u(x, \pm y + \varphi(x)) + C,$$
where either the upper or lower signs are taken, $\varphi(x)$ is an arbitrary function, and C is an arbitrary constant, are also solutions of the equation (Pavlovskii, 1961 and Ovsiannikov, 1982).

2°. Degenerate solutions (linear and quadratic in y) for arbitrary $f(x)$:
$$u(x,y) = \pm y \left[2 \int f(x)\,dx + C_1 \right]^{1/2} + \varphi(x),$$
$$u(x,y) = C_1 y^2 + \varphi(x) y + \frac{1}{4C_1}\left[\varphi^2(x) - 2\int f(x)\,dx \right] + C_2,$$
where $\varphi(x)$ is an arbitrary function and C_1 and C_2 are arbitrary constants. These solutions are independent of a and correspond to inviscid fluid flows.

3°. Additive separable solution for $f(x) = b$:
$$u = kx + U(y), \quad U(y) = \begin{cases} C_1 \exp\left(-\frac{k}{a}y\right) - \frac{b}{2k}y^2 + C_2 y + C_3 & \text{if } k \neq 0, \\ -\frac{b}{6a}y^3 + C_1 y^2 + C_2 y + C_3 & \text{if } k = 0, \end{cases}$$
where C_1, C_2, C_3, and k are arbitrary constants.

4°. Generalized separable solution for $f(x) = bx + c$:
$$u(x,y) = xF(y) + G(y),$$
where the functions $F = F(y)$ and $G = G(y)$ are described by the system of autonomous ODEs
$$(F'_y)^2 - FF''_{yy} = aF'''_{yyy} + b, \quad F'_y G'_y - FG''_{yy} = aG'''_{yyy} + c.$$

5°. Exact solutions for $f(x) = -bx^{-5/3}$ (Burde, 1996):

$$u(x,y) = \frac{6ax}{y+\varphi(x)} \pm \frac{\sqrt{3b}}{x^{1/3}}[y+\varphi(x)],$$

where $\varphi(x)$ is an arbitrary function.

6°. Exact solutions for $f(x) = bx^{-1/3} - cx^{-5/3}$ (Burde, 1996):

$$u(x,y) = \pm\sqrt{3c}\,z + x^{2/3}\theta(z), \quad z = yx^{-1/3},$$

where the function $\theta = \theta(z)$ is described by the ODE $\frac{1}{3}(\theta'_z)^2 - \frac{2}{3}\theta\theta''_{zz} = a\theta'''_{zzz} + b$.

7°. Generalized separable solution for $f(x) = be^{\beta x}$ (Polyanin, 2001b):

$$u(x,y) = \varphi(x)e^{\lambda y} - \frac{b}{2\beta\lambda^2\varphi(x)}e^{\beta x - \lambda y} - a\lambda x + \frac{2a\lambda^2}{\beta}y + \frac{2a\lambda}{\beta}\ln|\varphi(x)|,$$

where $\varphi(x)$ is an arbitrary function and λ is an arbitrary constant.

8°. Table 6.5 lists some other exact solutions to the boundary layer equation with pressure gradient (Pavlovskii, 1961 and Schlichting, 1981).

TABLE 6.5. Exact solutions to the hydrodynamic boundary layer equation with pressure gradient (b, k, and β are arbitrary constants).

No.	Function $f(x)$	Form of solution $u = u(x,y)$	ODE for the function $U = U(z)$
1	$f(x) = bx^k$	$u = x^{\frac{k+3}{4}}U(z)$, $z = x^{\frac{k-1}{4}}y$	$\frac{k+1}{2}(U'_z)^2 - \frac{k+3}{4}UU''_{zz} = aU'''_{zzz} + b$
2	$f(x) = be^{\beta x}$	$u = e^{\frac{1}{4}\beta x}U(z)$, $z = e^{\frac{1}{4}\beta x}y$	$\frac{1}{2}\beta(U'_z)^2 - \frac{1}{4}\beta UU''_{zz} = aU'''_{zzz} + b$
3	$f(x) = bx^{-3}$	$u = U(z) + \beta\ln|x|$, $z = y/x$	$-(U'_z)^2 - \beta U''_{zz} = aU'''_{zzz} + b$

9°. Below are two transformations that reduce the order of the boundary layer equation.

The *von Mises transformation*

$$\xi = x, \quad \eta = u, \quad \Phi(\xi,\eta) = u_y, \quad \text{where} \quad u = u(x,y),$$

leads to a second-order nonlinear PDE:

$$\Phi\Phi_\xi = a\Phi(\Phi\Phi_\eta)_\eta + f(\xi).$$

The Crocco transformation

$$\xi = x, \quad \zeta = u_y, \quad \Psi(\xi,\zeta) = u_{yy}, \quad \text{where} \quad u = u(x,y),$$

leads to a second-order nonlinear PDE (Loitsyanskiy, 1996):

$$\zeta\Psi_\xi = f(\xi)\Psi_\zeta + a\Psi^2\Psi_{\zeta\zeta}.$$

22. $u_y(\Delta u)_x - u_x(\Delta u)_y = 0, \qquad \Delta u = u_{xx} + u_{yy}.$

This is an *equation of motion of an ideal fluid* (u is the stream function); it derives from the two-dimensional stationary Euler equations.

1°. Suppose $u(x,y)$ is a solution of the equation in question. Then the functions
$$u_1 = C_1 u(C_2 x + C_3, C_2 y + C_4) + C_5,$$
$$u_2 = u(x\cos\alpha + y\sin\alpha, -x\sin\alpha + y\cos\alpha),$$
where $C_1, \ldots, C_5$ and α are arbitrary constants, are also solutions of the equation.

2°. Exact solutions of general form:
$$u(x,y) = \varphi_1(\xi), \quad \xi = a_1 x + b_1 y;$$
$$u(x,y) = \varphi_2(r), \quad r = \sqrt{(x-a_2)^2 + (y-b_2)^2};$$
where $\varphi_1(\xi)$ and $\varphi_2(r)$ are arbitrary functions; a_1, b_1, a_2, and b_2 are arbitrary constants.

3°. Any solutions of the linear PDEs

$\Delta u = 0$ (Laplace equation),

$\Delta u = C$ (Poisson equation),

$\Delta u = \lambda u$ (Helmholtz equation),

$\Delta u = \lambda u + C$ (nonhomogeneous Helmholtz equation),

where C and λ are arbitrary constants, are also solutions of the original equation.

Solutions of the Laplace equation $\Delta u = 0$ correspond to irrotational (potential) solutions of the original equation.

4°. The Jacobian of the functions u and $v = \Delta u$ appears on the left-hand side of the equation in question. The fact that the Jacobian of two functions is zero means that the two functions are functionally dependent. Hence, v must be a function of u, so that
$$\Delta u = f(u), \tag{1}$$
where $f(u)$ is an arbitrary function. Any solution of the second-order nonlinear heat-type PDE (1) for arbitrary $f(u)$ is a solution of the original equation.

The results of Item 6° correspond to special cases of the linear function $f(u) = \lambda u + C$.

5°. Additive separable solutions:
$$u(x,y) = A_1 x^2 + A_2 x + B_1 y^2 + B_2 y + C,$$
$$u(x,y) = A_1 \exp(\lambda x) + A_2 \exp(-\lambda x) + B_1 \exp(\lambda y) + B_2 \exp(-\lambda y) + C,$$
$$u(x,y) = A_1 \sin(\lambda x) + A_2 \cos(\lambda x) + B_1 \sin(\lambda y) + B_2 \cos(\lambda y) + C,$$
where A_1, A_2, B_1, B_2, C, and λ are arbitrary constants. These solutions are special cases of solutions from Item 3°.

6°. Generalized separable solutions:
$$u(x,y) = (Ax+B)e^{-\lambda y} + C,$$
$$u(x,y) = [A_1 \sin(\beta x) + A_2 \cos(\beta x)][B_1 \sin(\lambda y) + B_2 \cos(\lambda y)] + C,$$
$$u(x,y) = [A_1 \sin(\beta x) + A_2 \cos(\beta x)][B_1 \sinh(\lambda y) + B_2 \cosh(\lambda y)] + C,$$
$$u(x,y) = [A_1 \sinh(\beta x) + A_2 \cosh(\beta x)][B_1 \sin(\lambda y) + B_2 \cos(\lambda y)] + C,$$
$$u(x,y) = [A_1 \sinh(\beta x) + A_2 \cosh(\beta x)][B_1 \sinh(\lambda y) + B_2 \cosh(\lambda y)] + C,$$
$$u(x,y) = A e^{\alpha x + \beta y} + B e^{\gamma x + \lambda y} + C, \quad \alpha^2 + \beta^2 = \gamma^2 + \lambda^2,$$
where A, B, C, D, k, β, and λ are arbitrary constants. These solutions are special cases of solutions from Item 3°.

7°. Solution:
$$u(x,y) = F(z)x + G(z), \quad z = y + kx,$$
where k is an arbitrary constant, and the functions $F = F(z)$ and $G = G(z)$ are determined by the autonomous system of third-order ODEs:
$$F'_z F''_{zz} - F F'''_{zzz} = 0, \tag{2}$$
$$G'_z F''_{zz} - F G'''_{zzz} = \frac{2k}{(k^2+1)} F F''_{zz}. \tag{3}$$

On integrating the system once, we arrive at the following second-order ODEs:
$$(F'_z)^2 - F F''_{zz} = A_1, \tag{4}$$
$$G'_z F'_z - F G''_{zz} = \frac{2k}{k^2+1} \int F F''_{zz}\, dz + A_2, \tag{5}$$
where A_1 and A_2 are arbitrary constants.

The autonomous equation (4) can be reduced, with the change of variable $Z(F) = (F'_z)^2$, to a first-order linear ODE.

The general solution of ODE (2), or (4), is given by
$$F(z) = B_1 z + B_2, \qquad A_1 = B_1^2;$$
$$F(z) = B_1 \exp(\lambda z) + B_2 \exp(-\lambda z), \quad A_1 = -4\lambda^2 B_1 B_2;$$
$$F(z) = B_1 \sin(\lambda z) + B_2 \cos(\lambda z), \qquad A_1 = \lambda^2(B_1^2 + B_2^2),$$
where B_1, B_2, and λ are arbitrary constants.

The general solution of ODE (3), or (5), is expressed as
$$G = C_1 \int F\, dz - \int F \left(\int \frac{\psi\, dz}{F^2} \right) dz + C_2,$$
$$F = F(z), \quad \psi = \frac{2k}{k^2+1} \int F F''_{zz}\, dz + A_2,$$
where C_1 and C_2 are arbitrary constants.

8°. There are exact solutions of the following forms:
$$u(x,y) = x^a U(\zeta), \qquad \zeta = y/x;$$
$$u(x,y) = e^{ax} V(\rho), \qquad \rho = bx + cy;$$
$$u(x,y) = W(\zeta) + a \ln|x|, \quad \zeta = y/x,$$
where a, b, and c are arbitrary constants.

9°. References for PDE 6.5.1.22: Buchnev (1971), Andreev et al. (1999), and Polyanin & Zaitsev (2004).

6.5.2. Fourth-Order Equations

1. $u_{tt} + (uu_x)_x + u_{xxxx} = 0.$

Boussinesq equation. This equation arises in hydrodynamics and some physical applications.

1°. Suppose $u(x,t)$ is a solution of the equation in question. Then the functions

$$u_1 = C_1^2 u(C_1 x + C_2, \pm C_1^2 t + C_3),$$

where C_1, C_2, and C_3 are arbitrary constants, are also solutions of the equation.

2°. Exact solutions:

$$u(x,t) = 2C_1 x - 2C_1^2 t^2 + C_2 t + C_3,$$

$$u(x,t) = (C_1 t + C_2)x - \frac{1}{12C_1^2}(C_1 t + C_2)^4 + C_3 t + C_4,$$

$$u(x,t) = -\frac{(x+C_1)^2}{(t+C_2)^2} + \frac{C_3}{t+C_2} + C_4(t+C_2)^2,$$

$$u(x,t) = -\frac{x^2}{t^2} + C_1 t^3 x - \frac{C_1^2}{54} t^8 + C_2 t^2 + \frac{C_4}{t},$$

$$u(x,t) = -\frac{(x+C_1)^2}{(t+C_2)^2} - \frac{12}{(x+C_1)^2},$$

$$u(x,t) = -3\lambda^2 \cos^{-2}\left[\tfrac{1}{2}\lambda(x \pm \lambda t) + C_1\right],$$

where $C_1, \ldots, C_4$ and λ are arbitrary constants.

3°. Traveling wave solution that generalizes the last solution from Item 2° (Nishitani & Tajiri, 1982):

$$u(x,t) = U(\zeta), \quad \zeta = x + \lambda t,$$

where the function $U(\zeta)$ is described by the ODE $U''_{\zeta\zeta} + U^2 + 2\lambda^2 U + C_1 \zeta + C_2 = 0$.

4°. Self-similar solution (Nishitani & Tajiri, 1982):

$$u(x,t) = t^{-1}\theta(z), \quad z = xt^{-1/2},$$

where the function $\theta = \theta(z)$ is described by the ODE:

$$\theta''''_{zzzz} + (\theta \theta'_z)'_z + \tfrac{1}{4}z^2 \theta''_{zz} + \tfrac{7}{4} z \theta'_z + 2\theta = 0.$$

5°. There are exact solutions of the following forms (Clarkson & Kruskal, 1989):

$$u(x,t) = (x+C)^2 F(t) - 12(x+C)^{-2};$$

$$u(x,t) = G(\xi) - 4C_1^2 t^2 - 4C_1 C_2 t, \quad \xi = x - C_1 t^2 - C_2 t;$$

$$u(x,t) = \frac{1}{t}H(\eta) - \frac{1}{4}\left(\frac{x}{t} + Ct\right)^2, \quad \eta = \frac{x}{\sqrt{t}} - \frac{1}{3}Ct^{3/2};$$

$$u(x,t) = (a_1 t + a_0)^2 U(\zeta) - \left(\frac{a_1 x + b_1}{a_1 t + a_0}\right)^2, \quad \zeta = x(a_1 t + a_0) + b_1 t + b_0,$$

where C, C_1, C_2, a_1, a_0, b_1, and b_0 are arbitrary constants.

6°. The Boussinesq equation is solved by the inverse scattering method. Any rapidly decaying function $F = F(x,y;t)$ as $x \to +\infty$ and satisfying simultaneously the two linear PDEs:

$$\frac{1}{\sqrt{3}}\frac{\partial F}{\partial t} + \frac{\partial^2 F}{\partial x^2} - \frac{\partial^2 F}{\partial y^2} = 0, \quad \frac{\partial^3 F}{\partial x^3} + \frac{\partial^3 F}{\partial y^3} = 0$$

generates a solution of the Boussinesq equation in the form

$$u = 12\frac{d}{dx}K(x,x;t),$$

where $K(x,y;t)$ is a solution of the linear Gelfand–Levitan–Marchenko integral equation

$$K(x,y;t) + F(x,y;t) + \int_x^\infty K(x,s;t)F(s,y;t)\,ds = 0.$$

Time t appears here as a parameter.

7°. References for PDE 6.5.2.1: Zakharov & Shabat (1974), Ablowitz & Segur (1981), Clarkson & Kruskal (1989), and Polyanin & Zaitsev (2012).

2. $u_y(\Delta u)_x - u_x(\Delta u)_y = a\Delta\Delta u, \qquad \Delta u = u_{xx} + u_{yy}.$

There is a two-dimensional stationary equation of motion of a viscous incompressible fluid; it is obtained from the *Navier–Stokes equations* by the introduction of the stream function u.

1°. Suppose $u(x,y)$ is a solution of the equation in question. Then the functions

$$u_1 = -u(y,x),$$
$$u_2 = u(C_1 x + C_2, C_1 y + C_3) + C_4,$$
$$u_3 = u(x\cos\alpha + y\sin\alpha, -x\sin\alpha + y\cos\alpha),$$

where $C_1, \ldots, C_4$ and α are arbitrary constants, are also solutions of the equation (Pukhnachov, 1960).

2°. Any solution of the Poisson equation

$$\Delta u = C,$$

where C is an arbitrary constant, is also a solution of the original equation (these are "inviscid" solutions). For the use of these solutions in the hydrodynamics of ideal fluids, see, for example, Lamb (1945) and Loitsyanskiy (1996).

3°. Additive separable solutions (Pukhnachov, 1960 and Polyanin, 2001c):

$$u(y) = C_1 y^3 + C_2 y^2 + C_3 y + C_4,$$
$$u(x,y) = C_1 x^2 + C_2 x + C_3 y^2 + C_4 y + C_5,$$
$$u(x,y) = C_1 \exp(-\lambda y) + C_2 y^2 + C_3 y + C_4 + a\lambda x,$$
$$u(x,y) = C_1 \exp(\lambda x) - a\lambda x + C_2 \exp(\lambda y) + a\lambda y + C_3,$$
$$u(x,y) = C_1 \exp(\lambda x) + a\lambda x + C_2 \exp(-\lambda y) + a\lambda y + C_3,$$

where $C_1, \ldots, C_5$ and λ are arbitrary constants.

4°. Generalized separable solutions (Pukhnachov, 1960 and Polyanin, 2001c):

$$u(x,y) = A(kx + \lambda y)^3 + B(kx + \lambda y)^2 + C(kx + \lambda y) + D,$$
$$u(x,y) = Ae^{-\lambda(y+kx)} + B(y + kx)^2 + C(y + kx) + a\lambda(k^2 + 1)x + D,$$
$$u(x,y) = 6ax(y + \lambda)^{-1} + A(y + \lambda)^3 + B(y + \lambda)^{-1} + C(y + \lambda)^{-2} + D,$$

$$u(x,y) = (Ax+B)e^{-\lambda y} + a\lambda x + C,$$

$$u(x,y) = \left[A\sinh(\beta x) + B\cosh(\beta x)\right]e^{-\lambda y} + \frac{a}{\lambda}(\beta^2+\lambda^2)x + C,$$

$$u(x,y) = \left[A\sin(\beta x) + B\cos(\beta x)\right]e^{-\lambda y} + \frac{a}{\lambda}(\lambda^2-\beta^2)x + C,$$

$$u(x,y) = Ae^{\lambda y + \beta x} + Be^{\gamma x} + a\gamma y + \frac{a}{\lambda}\gamma(\beta-\gamma)x + C, \quad \gamma = \pm\sqrt{\lambda^2+\beta^2},$$

where A, B, C, D, k, β, and λ are arbitrary constants.

5°. Generalized separable solution linear in x:

$$u(x,y) = F(y)x + G(y),$$

where the functions $F = F(y)$ and $G = G(y)$ are determined by the autonomous system of fourth-order ordinary differential equations

$$F'_y F''_{yy} - F F'''_{yyy} = a F''''_{yyyy}, \tag{1}$$

$$G'_y F''_{yy} - F G'''_{yyy} = a G''''_{yyyy}. \tag{2}$$

Equation (1) has the following particular solutions:

$$F = by + c,$$
$$F = 6a(y+b)^{-1},$$
$$F = be^{-\lambda y} + a\lambda,$$

where b, c, and λ are arbitrary constants.

Let $F = F(y)$ be a solution of equation (1) ($F \not\equiv \text{const}$). Then the corresponding general solution of equation (2) can be written in the form

$$G = \int U\, dy + C_4, \quad U = C_1 U_1 + C_2 U_2 + C_3\left(U_2 \int \frac{U_1}{\Phi}\, dy - U_1 \int \frac{U_2}{\Phi}\, dy\right),$$

where C_1, C_2, C_3, and C_4 are arbitrary constants, and

$$U_1 = \begin{cases} F''_{yy} & \text{if } F''_{yy} \not\equiv 0, \\ F & \text{if } F''_{yy} \equiv 0, \end{cases} \quad U_2 = U_1 \int \frac{\Phi\, dy}{U_1^2}, \quad \Phi = \exp\left(-\frac{1}{a}\int F\, dy\right).$$

References for Item 5°: Hiemenz (1911), Rott (1956), Stuart (1959), Berker (1963), Dorrepaal (1986), Polyanin (2001c), and Drazin & Riley (2006).

6°. There is an exact solution of the form (generalizes the solution of Item 5°):

$$u(x,y) = F(z)x + G(z), \quad z = y + kx, \quad k \text{ is any number}.$$

7°. Self-similar solution (Loitsyanskiy, 1996):

$$u = \int F(z)\, dz + C_1, \quad z = \arctan\left(\frac{x}{y}\right).$$

Here, the function F is determined by the first-order autonomous ordinary differential equation

$$3a(F'_z)^2 - 2F^3 + 12aF^2 + C_2 F + C_3 = 0$$

where C_1, C_2, and C_3 are arbitrary constants.

8°. There is an exact solution of the form (generalizes the solution of Item 7°):
$$u = C_1 \ln |x| + \int V(z)\,dz + C_2, \quad z = \arctan\left(\frac{x}{y}\right).$$

3. $u_y(\Delta u)_x - u_x(\Delta u)_y = \nu \Delta \Delta u + f(y), \qquad \Delta u = u_{xx} + u_{yy}.$

This PDE describes the plane flow of a viscous incompressible fluid under the action of a transverse force, u is the stream function.

1°. Solution in the form of a one-argument function:
$$u(y) = -\frac{1}{6\nu}\int_0^y (y-z)^3 f(z)\,dz + C_1 y^3 + C_2 y^2 + C_3 y + C_4,$$
where $C_1, \dots, C_4$ are arbitrary constants.

2°. Additive separable solution for arbitrary $f(y)$ (Polyanin & Zaitsev, 2012):
$$u(x,y) = -\frac{1}{2\nu}\int_0^y (y-z)^2 \Phi(z)\,dz + C_1 e^{-\lambda y} + C_2 y^2 + C_3 y + C_4 + \nu \lambda x,$$
$$\Phi(z) = e^{-\lambda z}\int e^{\lambda z} f(z)\,dz,$$
where $C_1, \dots, C_4$ and λ are arbitrary constants.

For $f(y) = a\beta \cos(\beta y)$, it follows from the preceding formula with $C_1 = C_2 = C_4 = 0$ and $B = -\nu\lambda$ that
$$u(x,y) = -\frac{a}{\beta^2(B^2 + \nu^2\beta^2)}\bigl[B\sin(\beta y) + \nu\beta\cos(\beta y)\bigr] + Cy - Bx,$$
where B and C are arbitrary constants. This solution describes a flow with a periodic structure.

3°. Additive separable solution for $f(y) = Ae^{\lambda y} + Be^{-\lambda y}$ (Polyanin & Zaitsev, 2012):
$$u(x,y) = C_1 e^{-\lambda x} + C_2 x - \frac{A}{\lambda^3(C_2 + \nu\lambda)}e^{\lambda y} + \frac{B}{\lambda^3(C_2 - \nu\lambda)}e^{-\lambda y} - \nu\lambda y,$$
where C_1 and C_2 are arbitrary constants.

4°. Generalized separable solution linear in x:
$$u(x,y) = \varphi(y)x + \psi(y),$$
where the functions $\varphi = \varphi(y)$ and $\psi = \psi(y)$ are determined by the system of fourth-order ordinary differential equations
$$\varphi'_y \varphi''_{yy} - \varphi\varphi'''_{yyy} = \nu \varphi''''_{yyyy}, \tag{1}$$
$$\psi'_y \varphi''_{yy} - \varphi\psi'''_{yyy} = \nu \psi''''_{yyyy} + f(y). \tag{2}$$

On integrating once, we obtain the system of third-order ODEs:
$$(\varphi'_y)^2 - \varphi\varphi''_{yy} = \nu\varphi'''_{yyy} + A, \tag{3}$$
$$\psi'_y \varphi'_y - \varphi\psi''_{yy} = \nu\psi'''_{yyy} + \int f(y)\,dy + B, \tag{4}$$

where A and B are arbitrary constants. The order of the autonomous equation (3) can be reduced by one.

Equation (1) has the following particular solutions:

$$\varphi(y) = ay + b, \quad \varphi(y) = 6\nu(y+a)^{-1}, \quad \varphi(y) = ae^{-\lambda y} + \lambda\nu,$$

where a, b, and λ are arbitrary constants.

In the general case, equation (4) is reduced, with the substitution $U = \psi'_y$, to a second-order linear nonhomogeneous ODE, the general solution of which is given in the book by Polyanin & Zaitsev (2012).

▶ Many exact solutions of other nonlinear mathematical physics equations and other nonlinear PDEs of the second and higher orders can be found in the specialized reference books by Polyanin & Zaitsev (2004, 2012). The main methods for finding exact solutions to nonlinear mathematical physics equations are described, for example, in the books by Calogero & Degasperis (1982), Ovsiannikov (1982), Novikov et al. (1984), Bluman & Kumei (1989), Ablowitz & Clarkson (1991), Ibragimov (1994), Galaktionov & Svirshchevskii (2007), Polyanin & Zaitsev (2012), and Polyanin & Zhurov (2022).

References

Ablowitz, M.J. and Clarkson, P.A., *Solitons, Nonlinear Evolution Equations and Inverse Scattering*, Cambridge Univ. Press, Cambridge, 1991.

Ablowitz, M.J. and Segur, H., *Solitons and the Inverse Scattering Transform*, Society for Industrial and Applied Mathematics, Philadelphia, 1981.

Ablowitz, M.J. and Zeppetella, A., Explicit solutions of Fisher's equation for a special wave speed, *Bull. Math. Biol.*, Vol. 41, pp. 835–840, 1979.

Akhmediev, N. N. and Ankiewicz, A., *Solitons. Nonlinear Pulses and Beams*, Chapman & Hall, London, 1997.

Ames, W.F., Lohner, J.R., and Adams E., Group properties of $u_{tt} = [F(u)u_x]_x$, *Int. J. Nonlinear Mech.*, Vol. 16, No. 5–6, pp. 439–447, 1981.

Andreev, V.K., Kaptsov, O.V., Pukhnachov, V.V., and Rodionov, A.A., *Applications of Group-Theoretical Methods in Hydrodynamics*, Kluwer Academic, Dordrecht, 1999.

Aristov, S.N., Exact periodic and localized solutions of the equation $h_t = \Delta \ln h$, *J. Appl. Mech. & Tech. Phys.*, Vol. 40, No. 1, pp. 16–19, 1999.

Aristov, S N., Knyazev, D.V., and Polyanin, A.D., Exact solutions of the Navier–Stokes equations with the linear dependence of velocity components of two space variables, *Theor. Found. Chem. Eng.*, Vol. 43, No. 5, pp. 642–662, 2009.

Aristov, S.N. and Polyanin, A.D., Exact solutions of unsteady three-dimensional Navier–Stokes equations, *Dokl. Phys.*, Vol. 54, No. 7, pp. 316–321, 2009.

Barannyk, T., Symmetry and exact solutions for systems of nonlinear reaction-diffusion equations, *Proc. of Inst. of Mathematics of NAS of Ukraine*, Vol. 43, Part 1, pp. 80–85, 2002.

Barenblatt, G.I., On nonsteady motions of gas and fluid in porous medium, *Appl. Math. Mech. (PMM)*, Vol. 16, No. 1, pp. 67–78, 1952.

Barenblatt, G.I. and Zel'dovich, Ya.B., On dipole-type solutions in problems of nonstationary filtration of gas under polytropic regime [in Russian], *Prikl. Math. Mech. (PMM)*, Vol. 21, pp. 718–720, 1957.

Benton, E.R. and Platzman, G.W., A table of solutions of the one-dimensional Burgers equation, *Quart. Appl. Math.*, Vol. 30, pp. 195–212, 1972.

Berker, R., Intégration des équations du mouvement d'un fluide visqueux incompressible, In: *Encyclopedia of Physics, Vol. VIII/2* (Ed. S. Flügge), pp. 1–384. Springer, Berlin, 1963.

Berman, V.S. and Danilov, Yu.A., Group properties of the generalized Landau–Ginzburg equation [in Russian], *Dokl. Akad. Nauk SSSR,* Vol. 258, No. 1, pp. 67–70, 1981.

Bertsch, M., Kersner, R., and Peletier, L.A., Positivity versus localization in degenerate diffusion equations, *Nonlinear Analys., Theory, Meth. Appl.,* Vol. 9, No. 9, pp. 987–1008, 1985.

Bluman, G.W. and Kumei, S., *Symmetries and Differential Equations,* Springer-Verlag, New York, 1989.

Buchnev, A.A., A Lie group admitted by the ideal incompressible fluid equations [in Russian], In: *Dinamika Sploshnoi Sredy,* No. 7, Inst. gidrodinamiki AN USSR, Novosibirsk, pp. 212–214, 1971.

Bullough, R.K. and Caudrey, P.J. (eds.), *Solitons,* Springer-Verlag, Berlin, 1980.

Burde, G.I., New similarity reductions of the steady-state boundary-layer equations, *J. Phys. A: Math. Gen.,* Vol. 29, No. 8, pp. 1665–1683, 1996.

Calogero, F. and Degasperis, A., *Spectral Transform and Solitons: Tolls to Solve and Investigate Nonlinear Evolution Equations,* North-Holland, Amsterdam, 1982.

Cariello, F. and Tabor, M., Painlevé expansions for nonintegrable evolution equations, *Physica D,* Vol. 39, No. 1, pp. 77–94, 1989.

Cherniha R., Serov M., and Pliukhin O., *Nonlinear Reaction-Diffusion-Convection Equations: Lie and Conditional Symmetry, Exact Solutions and Their Applications.* Chapman & Hall/CRC Press, Boca Raton, 2018.

Clarkson, P.A. and Kruskal, M.D., New similarity reductions of the Boussinesq equation, *J. Math. Phys.,* Vol. 30, No. 10, pp. 2201–2213, 1989.

Cole, J.D., On a quasi-linear parabolic equation occurring in aerodynamics, *Quart. Appl. Math.,* Vol. 9, No. 3, pp. 225–236, 1951.

Dodd, R.K., Eilbeck, J.C., Gibbon, J.D., and Morris, H.C., *Solitons and Nonlinear Wave Equations,* Academic Press, London, 1982.

Dorrepaal, J. M., An exact solution of the Navier–Stokes equation which describes non-orthogonal stagnation point flow in two dimensions, *J. Fluid Mech.,* Vol. 163, pp. 141–147, 1986.

Dorodnitsyn, V.A., On invariant solutions of the equation of non-linear heat conduction with a source, *USSR Comput. Math. Math. Phys.,* Vol. 22, No. 6, pp. 115–122, 1982.

Drazin, P.G. and Johnson, R.S., *Solitons: An Introduction,* Cambridge Univ. Press, Cambridge, 1996.

Drazin, P.G. and Riley, N., *The Navier–Stokes Equations: A Classification of Flows and Exact Solutions,* Cambridge Univ. Press, Cambridge, 2006.

Edwards, M. P., Classical symmetry reductions of nonlinear diffusion-convection equations, *Phys. Lett. A,* Vol. 190, pp. 149–154, 1994.

Estévez, P.G., Non-classical symmetry and the singular manifold: the Burgers and the Burgers–Huxley equation, *J. Phys. A: Math. Gen.,* Vol. 27, pp. 2113–2127, 1994.

Estévez, P.G. and Gordoa, P.R., Painleve analysis of the generalized Burgers–Huxley equation, *J. Phys. A: Math. Gen.,* Vol. 23, 4831, 1990.

Faddeev, L.D. and Takhtajan, L.A., *Hamiltonian Methods in the Theory of Solitons,* Springer-Verlag, Berlin, 1987.

Fu, Z., Liu, S., and Liu, Sh., New kinds of solutions to Gardner equation, *Chaos, Solitons, and Fractals,* Vol. 20, pp. 301–309, 2004.

Fushchich, W.I., Serov, N.I., and Ahmerov, T.K., On the conditional symmetry of the generalized KdV equation, *Dokl. Akad. Nauk Ukr. SSR, Ser. A,* No. 12, pp. 28–30, 1991.

Galaktionov, V.A., Quasilinear heat equations with first-order sign-invariants and new explicit solutions, *Nonlinear Analys., Theory, Meth. Applications,* Vol. 23, pp. 1595–1621, 1994.

Galaktionov, V.A. and Posashkov, S.A., New exact solutions of parabolic equations with quadratic nonlinearities, *USSR Comput. Math. Math. Phys.,* Vol. 29, No. 2, pp. 112–119, 1989.

Galaktionov, V.A. and Svirshchevskii, S.R., *Exact Solutions and Invariant Subspaces of Nonlinear Partial Differential Equations in Mechanics and Physics*, Chapman & Hall/CRC Press, Boca Raton, 2007.

Gardner, C.S., Greene, J.M., Kruskal, M.D., and Miura, R.M., Method for solving the Korteweg–de Vries equation, *Phys. Rev. Lett.*, Vol. 19, No. 19, pp. 1095–1097, 1967.

Ginzburg, V.L. and Landau, L.D., On the theory of superconductivity, *Sov. Phys. JETP*, Vol. 20, 1064-1082, 1950.

Goursat, E., *A Course of Mathematical Analysis, Vol. 3, Part 1* [Russian translation from French], Gostekhizdat, Moscow, 1933.

Grauel, A., Sinh-Gordon equation, Painlevé property and Bäcklund transformation, *Physica A*, Vol. 132, pp. 557–568, 1985.

Grundland, A.M. and Infeld, E., A family of non-linear Klein–Gordon equations and their solutions, *J. Math. Phys.*, Vol. 33, pp. 2498–2503, 1992.

Guderley, K.G., *The Theory of Transonic Flow*, Pergamon, Oxford, 1962.

Hiemenz, K., Die Grenzschicht an einem in den gleichförmigen Flüssigkeitsstrom eingetauchten geraden Kreiszylinder, *Dinglers Polytech. J.*, Vol. 326, pp. 321–324, 344–348, 357–362, 372–374, 407–410, 1911.

Hietarinta, J., A search for bilinear equations passing Hirota's three-soliton condition. II. mKdV-type bilinear equations, *J. Math. Phys.*, Vol. 28, No. 9, pp. 2094–2101, 1987.

Hirota, R., Exact solution of the Korteweg–de Vries equation for multiple collisions of solutions, *Phys. Rev. Lett.*, Vol. 27, pp. 1192–1194, 1971.

Hopf, E., The partial differential equation $u_t + u u_x = \mu u_{xx}$, *Comm. Pure and Appl. Math.*, Vol. 3, pp. 201–230, 1950.

Hu, J. and Qu, C., Functionally separable solutions to nonlinear wave equations by group foliation method, *J. Math. Anal. Appl.*, Vol. 330, pp. 298–311, 2007.

Huang, D.J. and Zhou, S., Group properties of generalized quasi-linear wave equations, *J. Math. Anal. Appl.*, Vol. 366, pp. 460–472, 2010.

Huang, D.J., Zhu, Y., and Yang, Q., Reduction operators and exact solutions of variable coefficient nonlinear wave equations with power nonlinearities, *Symmetry*, Vol. 9, No. 1, 3, 2017.

Ibragimov, N.H., *Transformation Groups Applied in Mathematical Physics*, D. Reidel Publ., Dordrecht, 1985.

Ibragimov, N.H. (ed.), *CRC Handbook of Lie Group Analysis of Differential Equations, Vol. 1, Symmetries, Exact Solutions and Conservation Laws*, CRC Press, Boca Raton, 1994.

Ignatovich, N.V., Solutions that are not reducible to invariant ones, and partially invariant solutions of equations of a stationary boundary layer, *Math. Notes*, Vol. 53, No. 1, pp. 98–100, 1993.

Ivanova, N.M., Exact solutions of diffusion-convection equations, *Dyn. PDEs*, Vol. 5, No. 2, pp. 139–171, 2008.

Ivanova, N.M. and Sophocleous, C., On the group classification of variable-coefficient nonlinear diffusion-convection equations, *J. Comput. Appl. Math.*, Vol. 197, No. 2, pp. 322–344, 2006.

Johnson, R S., On the inverse scattering transform, the cylindrical Korteweg–de Vries equation and similarity solutions, *Phys. Lett.*, Ser. A, Vol. 72, No. 2, p. 197, 1979.

Kawahara, T. and Tanaka, M., Interactions of traveling fronts: an exact solution of a nonlinear diffusion equation, *Phys. Lett.*, Vol. 97, p. 311, 1983.

Kersner, R., On some properties of weak solutions of quasilinear degenerate parabolic equations, *Acta Math. Academy of Sciences, Hung.*, Vol. 32, No. 3–4, pp. 301–330, 1978.

Khabirov, S.V., Nonisentropic one-dimensional gas motions obtained with the help of the contact group of the nonhomogeneous Monge–Ampère equation [in Russian], *Mat. Sbornik*, Vol. 181, No. 12, pp. 1607–1622, 1990.

Khawaja, U. and Sakkaf, L., *Handbook of Exact Solutions to the Nonlinear Schrödinger Equations*, Inst. Physics Publ. (IOP), Bristol, 2019.

King, J.R., Exact similarity solutions to some nonlinear diffusion equations, *J. Phys. A: Math. Gen.,* Vol. 23, pp. 3681–3697, 1990.

Korepin, V.E., Bogoliubov, N.N., and Izergin A.G., *Quantum Inverse Scattering Method and Correlation Functions,* Cambridge Univ. Press, Cambridge, 1993.

Kudryashov, N.A., On exact solutions of families of Fisher equations, *Theor. Math. Phys.,* Vol. 94, No. 2, pp. 211–218, 1993.

Kuramoto, Y. and Tsuzuki, T., On the formation of dissipative structures in reaction-diffusion systems, *Progr. Theor. Phys.,* Vol. 54, No. 3, pp. 687–699, 1975.

Lamb, H., *Hydrodynamics,* Dover Publ., New York, 1945.

Liouville, J., Sur l'équation aux différences partielles: $\partial^2 \log \lambda / \partial u \partial v \pm \lambda / 2a^2 = 0$, *J. Math.,* Vol. 18, pp. 71–72, 1853.

Loitsyanskiy, L.G., *Mechanics of Liquids and Gases,* Begell House, New York, 1996.

Martin, M. N., The propagation of a plane shock into a quiet atmosphere, *Canad. J. Math.,* Vol. 3, pp. 165–187, 1953.

Miller, W. (Jr.) and Rubel, L. A., Functional separation of variables for Laplace equations in two dimensions, *J. Phys. A,* Vol. 26, pp. 1901–1913, 1993.

Miura, R.M., Korteweg–de Vries equation and generalizations. I. A remarkable explicit nonlinear transformation, *J. Math. Phys.,* Vol. 9, No. 8, pp. 1202–1204, 1968.

Musette, M. and Conte, R., The two-singular-manifold method: Modified Korteweg–de Vries and the sine-Gordon equations, *Phys. A, Math. Gen.,* Vol. 27, No. 11, pp. 3895–3913, 1994.

Nerney, S., Schmahl, E. J., and Musielak Z. E., Analytic solutions of the vector Burgers' equation, *Quart. Appl. Math.,* Vol. LIV, No. 1, pp. 63–71, 1996.

Nishitani, T. and Tajiri, M., On similarity solutions of the Boussinesq equation, *Phys. Lett. A,* Vol. 89, pp. 379–380, 1982.

Hoenselaers, C., Solutions of the hyperbolic sine–Gordon equations, *Int. J. Theor. Phys.,* Vol. 46, 1096–1099, 2007.

Novikov, S.P., Manakov, S.V., Pitaevskii, L.B., and Zakharov, V.E., *Theory of Solitons. The Inverse Scattering Method,* Plenum Press, New York, 1984.

Nucci, M.C. and Clarkson, P.A., The nonclassical method is more general than the direct method for symmetry reductions. An example of the Fitzhugh–Nagumo equation, *Phys. Lett. A,* Vol. 164, pp. 49–56, 1992.

Ono, H., Algebraic soliton of the modified Korteweg–de Vries' equation, *J. Soc. Japan,* Vol. 41, pp. 1817–1818, 1976.

Ovsiannikov, L.V., Group properties of nonlinear heat equations [in Russian], *Doklady Acad. Nauk USSR,* Vol. 125, No. 3, pp. 492–495, 1959.

Ovsiannikov, L.V., *Group Analysis of Differential Equations,* Academic Press, New York, 1982.

Pattle, R.E., Diffusion from an instantaneous point source with a concentration-dependent coefficient, *J. Mech. Appl. Math.,* Vol. 12, pp. 407–409, 1959.

Pavlovsky, Yu.N., Research into some invariant solutions of boundary layer equations, *U.S.S.R. Comput. Math. Math. Phys.,* Vol. 1, No. 2, pp. 321–339, 1962.

Polyanin, A.D., Transformations and exact solutions containing arbitrary functions for boundary-layer equations, *Dokl. Phys.,* Vol. 46, No. 7, pp. 526–531, 2001a.

Polyanin, A.D., Exact solutions and transformations of the equations of a stationary laminar boundary layer, *Theor. Found. Chem. Eng.,* Vol. 35, No. 4, pp. 319–328, 2001b.

Polyanin, A.D., Exact solutions to the Navier–Stokes equations with generalized separation of variables, *Dokl. Phys.,* Vol. 46, No. 10, pp. 726–731, 2001c.

Polyanin, A.D., Functional separable solutions of nonlinear reaction–diffusion equations with variable coefficients, *Appl. Math. Comput.,* Vol. 347, pp. 282–292, 2019a.

Polyanin, A.D., Construction of exact solutions in implicit form for PDEs: New functional separable solutions of non-linear reaction–diffusion equations with variable coefficients, *Int. J. Non-Linear Mech.,* Vol. 111, pp. 95–105, 2019b.

Polyanin, A.D., Functional separable solutions of nonlinear convection–diffusion equations with variable coefficients, *Commun. Nonlinear Sci. Numer. Simulat.,* Vol. 73, pp. 379–390, 2019c.

Polyanin, A.D., Comparison of the effectiveness of different methods for constructing exact solutions to nonlinear PDEs. Generalizations and new solutions, *Mathematics,* Vol. 7, No. 5, 386, 2019d.

Polyanin, A.D., Construction of functional separable solutions in implicit form for non-linear Klein–Gordon type equations with variable coefficients, *Int. J. Non-Linear Mech.,* Vol. 114, pp. 29–40, 2019e.

Polyanin, A.D., Functional separation of variables in nonlinear PDEs: General approach, new solutions of diffusion-type equations, *Mathematics,* Vol. 8, No. 1, 90, 2020.

Polyanin, A.D. and Vyazmina, E.A., New classes of exact solutions to general nonlinear heat (diffusion) equations, *Dokl. Math.,* Vol. 72, No. 2, pp. 798–801, 2005.

Polyanin, A.D. and Zaitsev, V.F., *Handbook of Nonlinear Partial Differential Equations,* Chapman & Hall/CRC Press, Boca Raton–London, 2004 (1st ed.) and 2012 (2nd ed.).

Polyanin, A.D. and Zhurov, A.I., *Separation of Variables and Exact Solutions to Nonlinear PDEs,* CRC Press, Boca Raton–London, 2022.

Pukhnachov, V. V., Group properties of the Navier–Stokes equations in the plane case, *J. Appl. Math. Tech. Phys.,* No. 1, pp. 83–90, 1960.

Qin, M., Mei, F., and Fan, G., New explicit solutions of the Burgers equation, *Nonlinear Dyn.,* Vol. 48, pp. 91–96, 2007.

Rosenau, P., Compact and noncompact dispersive patterns, *Phys. Lett. A.,* Vol. 275, No. 3, pp. 193–203, 2000.

Rott, N., Unsteady viscous flow in the vicinity of a stagnation point, *Quart. Appl. Math.,* Vol. 13, No. 4, pp. 444–451, 1956.

Rudykh, G.A. and Semenov, E.I., On new exact solutions of a one-dimensional nonlinear diffusion equation with a source (sink), *Comput. Math. Math. Phys.,* Vol. 38, No. 6, pp. 930–936, 1998.

Sabitov, I.Kh., Solutions of the equation $\Delta u = f(x,y)e^{cu}$ in some special cases, *Sbornik: Math.,* Vol. 192, No. 6, pp. 879–894, 2001.

Schamel, H., A modified Korteweg–de Vries equation for ion acoustic waves due to resonant electrons, *J. Plasma Phys.,* Vol. 9, pp. 377–387, 1973.

Schlichting, H., *Boundary Layer Theory,* McGraw-Hill, New York, 1981.

Seeger, A. and Wesolowski, Z., Standing-wave solutions of the Enneper (sine-Gordon) equation, *Int. J. Eng. Sci.,* Vol. 19, pp. 1535–1549, 1981.

Shercliff, J. A., Simple rotational flows, *J. Fluid Mech.,* Vol. 82, No. 4, pp. 687–703, 1977.

Steuerwald, R., Über enneper'sche Flächen und Bäcklund'sche Transformation, *Abh. Bayer. Akad. Wiss.* (Muench.), Vol. 40, pp. 1–105, 1936.

Stuart, J.T., The viscous flow near a stagnation point when the external flow has uniform vorticity, *J. Aerosp. Sci.,* Vol. 26, pp. 124–125, 1959.

Titov, S.S., A method of finite-dimensional rings for solving nonlinear equations of mathematical physics [in Russian], In: *Aerodynamics* (ed. T.P. Ivanova), Saratov Univ., Saratov, pp. 104–110, 1988.

Vaneeva, O.O., Johnpillai, A.G., Popovych, R.O., and Sophocleous, C., Extended group analysis of variable coefficient reaction-diffusion equations with power nonlinearities, *J. Math. Anal. Appl.,* Vol. 330, pp. 1363–1386, 2007.

Vaneeva, O.O., Popovych, R.O., and Sophocleous, C., Group analysis of variable coefficient diffusion-convection equations. I Enhanced group classification, *Lobachevskii J. Math.,* Vol. 31, No. 2, pp. 100–122, 2010.

Vaneeva, O.O., Popovych, R.O., and Sophocleous, C., Extended group analysis of variable coefficient reaction-diffusion equations with exponential nonlinearities, *J. Math. Anal. Appl.,* Vol. 396, pp. 225–242, 2012.

Vekua, I. N., Remarks on the properties of solutions to equation $\Delta u = -Ke^{2u}$ [in Russian], *Sib. Matem. Zhurn.*, Vol. 1, No. 3, pp. 331–342, 1960.

Wadati, M., The exact solution of the modified Korteweg–de Vries equation, *J. Phys. Soc. Japan*, Vol. 32, pp. 1681–1687, 1972.

Wang, C.Y., Exact solutions for the steady-state Navier–Stokes equations, *Ann. Rev. Fluid Mech.*, Vol. 23, pp. 159–177, 1991.

Wazwaz, A.M., Kinks and solitons solutions for the generalized KdV equation with two power nonlinearities, *Appl. Math. Comput.*, Vol. 183, No. 2, 1181–1189, 2006.

Wazwaz, A.M., New solitons and kink solutions for the Gardner equation, *Commun. Nonlinear Sci. Numer. Simul.*, Vol. 12, No. 8, 1395–1404, 2007.

Wazwaz, A.M., The KdV equation, In: *Handbook of Differential Equations, Evolutionary Equations, Vol. 4* (eds. C. M. Dafermos and M. Pokorný), Elsevier, Amsterdam, 2008.

Whitham, G.B., Non-linear dispersive waves, *Proc. Roy. Soc. London*, Ser. A, Vol. 283, pp. 238–261, 1965.

Yung, C.M., Verburg, K., and Baveye, P., Group classifications and symmetry reductions of the nonlinear diffusion-convection equation $u_t = (D(u)u_x)_x - K'(u)u_x$, *Int. J. Non-Linear Mech.*, Vol. 29, pp. 273–278, 1994.

Zaitsev, V.F. and Polyanin, A.D., Exact solutions and transformations of nonlinear heat and wave equations, *Dokl. Math.*, Vol. 64, No. 3, pp. 416–420, 2001.

Zakharov, V. E. and Faddeev, L. D., The Korteweg–de Vries equation: a completely integrable Hamiltonian system, *Funct. Anal. Appl.*, Vol. 5, pp. 280–287, 1971.

Zakharov, V.E. and Shabat, A.B., A scheme for integrating the nonlinear equations of mathematical physics by the method of the inverse scattering problem, *Funct. Anal. Its Appl.*, Vol. 8, pp. 226–235, 1974.

Zel'dovich, Ya.B. and Kompaneets, A.S., On the theory of propagation of heat with the heat conductivity depending upon the temperature, pp. 61–71, In: *Collection in Honor of the Seventieth Birthday of Academician A. F. Ioffe* [in Russian], Izdat. Akad. Nauk USSR, Moscow, 1950.

Zhdanov, R.Z., Separation of variables in the non-linear wave equation, *J. Phys. A*, Vol. 27, pp. L291–L297, 1994.

Zhidkov, P.E., *Korteweg–de Vries and Nonlinear Schrödinger Equations: Qualitative Theory*, Springer, Berlin, 2001.

Zhu, G.C. & Chen, H.H., Symmetries and integrability of the cylindrical Korteweg–de Vries equation, *J. Math. Phys.*, Vol. 27, No. 1, pp. 100–103, 1986.

Chapter 7

Systems of Partial Differential Equations

▶ **Preliminary remarks.** This chapter describes exact solutions to various linear and nonlinear systems consisting of two coupled partial differential equations of the first or second order in two independent variables, as well as some nonlinear PDE systems of general form. We also consider reductions leading to one ODE, systems of ODEs, one PDE, or two independent PDEs.

The main attention is paid to nonlinear PDE systems that occur in the theory of heat and mass transfer, wave theory, and gas dynamics, as well as nonlinear systems of a fairly general form that involve arbitrary functions (exact solutions of such equations are of considerable interest for testing numerical methods).

7.1. Systems of Two First-Order PDEs

7.1.1. Linear Systems of Two First-Order PDEs

1. $u_t = au_x + f_1(t)u + g_1(t)w$, $\quad w_t = aw_x + f_2(t)u + g_2(t)w$.

First-order linear homogeneous system of PDEs with variable coefficients.

General solution (Polyanin & Nazaikinskii, 2016):

$$u = \varphi_1(t)U(x+at) + \varphi_2(t)W(x+at),$$
$$w = \psi_1(t)U(x+at) + \psi_2(t)W(x+at),$$

where $U = U(z)$ and $W = W(z)$ are arbitrary functions and the two pairs of functions $\varphi_1 = \varphi_1(t)$, $\psi_1 = \psi_1(t)$ and $\varphi_2 = \varphi_2(t)$, $\psi_2 = \psi_2(t)$ are linearly independent (fundamental) solutions of the linear homogeneous system of ODEs:

$$\varphi'_t = f_1(t)\varphi + g_1(t)\psi, \quad \psi'_t = f_2(t)\varphi + g_2(t)\psi.$$

2. $u_t = a(t)u_x + f_1(t)u + g_1(t)w + h_1(t)$,
   $w_t = a(t)w_x + f_2(t)u + g_2(t)w + h_2(t)$.

First-order linear nonhomogeneous system of PDEs with variable coefficients.

General solution (Polyanin & Nazaikinskii, 2016):

$$u = \varphi_1(t)U(z) + \varphi_2(t)W(z) + u_0(t),$$
$$w = \psi_1(t)U(z) + \psi_2(t)W(z) + w_0(t), \quad z = x + \int a(t)\,dt,$$

where $U = U(z)$ and $W = W(z)$ are arbitrary functions. The pairs of functions $\varphi_1 = \varphi_1(t)$,

$\psi_1 = \psi_1(t)$ and $\varphi_2 = \varphi_2(t)$, $\psi_2 = \psi_2(t)$ are linearly independent (fundamental) solutions of the linear homogeneous system of ODEs

$$\varphi'_t = f_1(t)\varphi + g_1(t)\psi, \quad \psi'_t = f_2(t)\varphi + g_2(t)\psi.$$

The functions $u_0 = u_0(t)$, $w_0 = w_0(t)$ are a solution of the linear nonhomogeneous system of ODEs

$$u'_0 = f_1(t)u_0 + g_1(t)w_0 + h_1(t), \quad w'_0 = f_2(t)u_0 + g_2(t)w_0 + h_2(t).$$

7.1.2. Nonlinear Systems of the Form $u_x = F(u, w)$, $w_t = G(u, w)$

▶ **Preliminary remarks.** Such systems of equations arise in the theory of chemical reactors, in studying fluid flows through porous media, and in chromatography.

Note that the more general first-order PDE systems

$$u_\tau + a_1 u_\xi = F(u, w), \quad w_\tau + a_2 w_\xi = G(u, w),$$

which describe convective mass transfer in a two-component medium with a volume chemical reaction where the diffusion of both components can be neglected, is reduced to the system in question by changing from ξ and τ to the characteristic variables x and t defined by

$$x = \frac{\xi - a_2\tau}{a_1 - a_2}, \quad t = \frac{\xi - a_1\tau}{a_2 - a_1} \quad (a_1 \neq a_2).$$

If the first (resp., second) component is quiescent, then $a_1 = 0$ (resp., $a_2 = 0$).

The systems in question are invariant under translations in the independent variables and, hence, admit traveling wave solutions: $u = u(kx - \lambda t)$, $w = w(kx - \lambda t)$. Neither these solutions nor degenerate solutions, where one of the desired functions is identically zero or constant, are considered in what follows.

Below $f(z)$, $g(z)$, $h(z)$, and $r(z)$ are arbitrary functions of their argument, and $z = z(u, w)$. The PDE systems are listed in order of complexity of the argument.

1. $u_x = auw, \quad w_t = buw.$

General solution (Berman, 1979):

$$u = -\frac{\psi'_t(t)}{a\varphi(x) + b\psi(t)}, \quad w = -\frac{\varphi'_x(x)}{a\varphi(x) + b\psi(t)},$$

where $\varphi(x)$ and $\psi(t)$ are arbitrary functions.

2. $u_x = auw, \quad w_t = bu^k.$

General solution (Polyanin & Vyazmina, 2006):

$$u = \left[\frac{1}{b}\psi'_t(t)E(x)\right]^{1/k} \left[\psi(t) + \frac{1}{2}ak\int E(x)\,dx\right]^{-2/k},$$

$$w = \varphi(x) - E(x)\left[\psi(t) + \frac{1}{2}ak\int E(x)\,dx\right]^{-1}, \quad E(x) = \exp\left[ak\int \varphi(x)\,dx\right],$$

where $\varphi(x)$ and $\psi(t)$ are arbitrary functions.

3. $u_x = auw^n$, $w_t = bu^k w$.

General solution (Polyanin & Vyazmina, 2006):
$$u = \left(\frac{-\psi'_t(t)}{bn\psi(t) - ak\varphi(x)}\right)^{1/k}, \quad w = \left(\frac{\varphi'_x(x)}{bn\psi(t) - ak\varphi(x)}\right)^{1/n},$$
where $\varphi(x)$ and $\psi(t)$ are arbitrary functions.

4. $u_x = au^{1-k}w^n$, $w_t = bu^k w^{1-n}$.

The transformation $U = u^k$, $W = w^n$ leads to a linear system of PDEs with constant coefficients:
$$U_x = akW, \quad W_t = bnU.$$
Excluding the function W, we obtain a linear hyperbolic equation, $U_{xt} = abknU$.

5. $u_x = aw$, $w_t = be^{\lambda u}$.

Eliminating w gives the *Liouville equation* (Berman, 1981):
$$u_{xt} = abe^{\lambda u},$$
the general solution of which has the form
$$u = \frac{1}{\lambda}[\varphi(x) + \psi(y)] - \frac{2}{\lambda}\ln\left|k\int \exp[\varphi(x)]\,dx + \frac{ab\lambda}{2k}\int \exp[\psi(y)]\,dy\right|,$$
where $\varphi(x)$ and $\psi(y)$ are arbitrary functions and k is an arbitrary constant.

6. $u_x = uf(w)$, $w_t = u^k g(w)$.

$1°$. The transformation of the dependent variables
$$U = u^k, \quad W = \int \frac{dw}{g(w)} \tag{1}$$
leads to the simpler system of PDEs
$$U_x = \Phi(W)U, \quad W_t = U, \tag{2}$$
where the function $\Phi(W)$ is defined parametrically:
$$\Phi = kf(w), \quad W = \int \frac{dw}{g(w)}, \tag{3}$$
with w playing the role of the parameter. By replacing U in the first equation of system (2) with the left-hand side of the second equation, one arrives at a second-order PDE for W:
$$W_{xt} = \Phi(W)W_t.$$
Integrating with respect to t gives
$$W_x = \int \Phi(W)\,dW + \theta(x), \tag{4}$$
where $\theta(x)$ is an arbitrary function.

In (4), returning to the original variable w using formulas (1) and (3), one obtains (Polyanin & Vyazmina, 2006):
$$w_x = kg(w)\int \frac{f(w)}{g(w)}\,dw + \theta(x)g(w). \tag{5}$$

The first integral (5) may be treated as a first-order ODE in x. On finding its general solution, one should replace the constant of integration C with an arbitrary function of time $\psi(t)$, since w is dependent on x and t.

2°. To the special case $\theta(x) = \text{const}$ in (5) there correspond special solutions of the form (Polyanin & Vyazmina, 2006):

$$w = w(z), \quad u = [\psi'_t(t)]^{1/k} v(z), \quad z = x + \psi(t)$$

involving one arbitrary function $\psi(t)$, with the prime denoting a derivative. The functions $w(z)$ and $v(z)$ are described by the autonomous ODE system

$$v'_z = f(w)v, \quad w'_z = g(w)v^k,$$

the general solution of which can be written in implicit form as

$$\int \frac{dw}{g(w)[kF(w) + C_1]} = z + C_2, \quad v = [kF(w) + C_1]^{1/k}, \quad F(w) = \int \frac{f(w)}{g(w)} dw.$$

7. $u_x = f(a_1 u + b_1 w), \quad w_t = g(a_2 u + b_2 w).$

Let $\Delta = a_1 b_2 - a_2 b_1 \neq 0$.

Additive separable solution (Polyanin & Vyazmina, 2006):

$$u = \frac{1}{\Delta}[b_2 \varphi(x) - b_1 \psi(t)], \quad w = \frac{1}{\Delta}[a_1 \psi(t) - a_2 \varphi(x)],$$

where the functions $\varphi(x)$ and $\psi(t)$ are described by the autonomous ODEs

$$\frac{b_2}{\Delta} \varphi'_x = f(\varphi), \quad \frac{a_1}{\Delta} \psi'_t = g(\psi).$$

Integrating yields

$$\frac{b_2}{\Delta} \int \frac{d\varphi}{f(\varphi)} = x + C_1, \quad \frac{a_1}{\Delta} \int \frac{d\psi}{g(\psi)} = t + C_2.$$

8. $u_x = f(au + bw), \quad w_t = g(au + bw).$

Exact solution (Polyanin & Vyazmina, 2006):

$$u = b(k_1 x - \lambda_1 t) + y(\xi), \quad w = -a(k_1 x - \lambda_1 t) + z(\xi), \quad \xi = k_2 x - \lambda_2 t,$$

where k_1, k_2, λ_1, and λ_2 are arbitrary constants, and the functions $y(\xi)$ and $z(\xi)$ are described by the autonomous ODE system

$$k_2 y'_\xi + bk_1 = f(ay + bz), \quad -\lambda_2 z'_\xi + a\lambda_1 = g(ay + bz).$$

This system has a simple particular solution $y = y_0$, $z = z_0$, where y_0 and z_0 are some constants.

9. $u_x = f(au - bw), \quad w_t = ug(au - bw) + wh(au - bw) + r(au - bw).$

Here, $f(z), g(z), h(z)$, and $r(z)$ are arbitrary functions.

Generalized separable solution (Polyanin & Manzhirov, 2007):

$$u = \varphi(t) + b\theta(t)x, \quad w = \psi(t) + a\theta(t)x.$$

The functions $\varphi = \varphi(t)$, $\psi = \psi(t)$, and $\theta = \theta(t)$ are determined by a system involving one algebraic (transcendental) and two ODEs:

$$b\theta = f(a\varphi - b\psi),$$
$$a\theta'_t = b\theta g(a\varphi - b\psi) + a\theta h(a\varphi - b\psi),$$
$$\psi'_t = \varphi g(a\varphi - b\psi) + \psi h(a\varphi - b\psi) + r(a\varphi - b\psi).$$

10. $u_x = f(au - bw) + cw$, $\quad w_t = ug(au - bw) + wh(au - bw) + r(au - bw)$.

Here, $f(z)$, $g(z)$, $h(z)$, and $r(z)$ are arbitrary functions.

Generalized separable solution (Polyanin & Manzhirov, 2007):

$$u = \varphi(t) + b\theta(t)e^{\lambda x}, \quad w = \psi(t) + a\theta(t)e^{\lambda x}, \quad \lambda = ac/b.$$

The functions $\varphi = \varphi(t)$, $\psi = \psi(t)$, and $\theta = \theta(t)$ are determined by a system involving one algebraic (transcendental) and two ODEs:

$$f(a\varphi - b\psi) + c\psi = 0,$$
$$\psi'_t = \varphi g(a\varphi - b\psi) + \psi h(a\varphi - b\psi) + r(a\varphi - b\psi),$$
$$a\theta'_t = b\theta g(a\varphi - b\psi) + a\theta h(a\varphi - b\psi).$$

11. $u_x = e^{\lambda u} f(\lambda u - \sigma w)$, $\quad w_t = e^{\sigma w} g(\lambda u - \sigma w)$.

Exact solution (Polyanin & Manzhirov, 2007):

$$u = y(\xi) - \frac{1}{\lambda}\ln(C_1 t + C_2), \quad w = z(\xi) - \frac{1}{\sigma}\ln(C_1 t + C_2), \quad \xi = \frac{x + C_3}{C_1 t + C_2},$$

where the functions $y(\xi)$ and $z(\xi)$ are described by the ODE system

$$y'_\xi = e^{\lambda y} f(\lambda y - \sigma z), \quad -C_1 \xi z'_\xi - (C_1/\sigma) = e^{\sigma z} g(\lambda y - \sigma z).$$

12. $u_x = u^k f(u^n w^m)$, $\quad w_t = w^s g(u^n w^m)$.

Self-similar solution with $s \neq 1$ and $n \neq 0$ (Polyanin & Manzhirov, 2007):

$$u = t^{\frac{m}{n(s-1)}} y(\xi), \quad w = t^{-\frac{1}{s-1}} z(\xi), \quad \xi = xt^{\frac{m(k-1)}{n(s-1)}},$$

where the functions $y(\xi)$ and $z(\xi)$ are described by the ODE system

$$y'_\xi = y^k f(y^n z^m), \quad m(k-1)\xi z'_\xi - nz = n(s-1) z^s g(y^n z^m).$$

13. $u_x = u^k f(u^n w^m)$, $\quad w_t = wg(u^n w^m)$.

1°. Exact solution (Polyanin & Manzhirov, 2007):

$$u = e^{mt} y(\xi), \quad w = e^{-nt} z(\xi), \quad \xi = e^{m(k-1)t} x,$$

where the functions $y(\xi)$ and $z(\xi)$ are described by the ODE system

$$y'_\xi = y^k f(y^n z^m), \quad m(k-1)\xi z'_\xi - nz = zg(y^n z^m).$$

2°. Exact solution for $k \neq 1$ (Polyanin & Manzhirov, 2007):
$$u = x^{-\frac{1}{k-1}}\varphi(\zeta), \quad w = x^{\frac{n}{m(k-1)}}\psi(\zeta), \quad \zeta = t + a\ln|x|,$$
where a is an arbitrary constant, and the functions $\varphi(\zeta)$ and $\psi(\zeta)$ are described by the ODE system
$$a\varphi'_\zeta + \frac{1}{1-k}\varphi = \varphi^k f(\varphi^n \psi^m), \quad \psi'_\zeta = \psi g(\varphi^n \psi^m).$$

14. $u_x = uf(u^n w^m), \quad w_t = wg(u^n w^m).$

Exact solution (Polyanin & Manzhirov, 2007):
$$u = e^{m(kx-\lambda t)} y(\xi), \quad w = e^{-n(kx-\lambda t)} z(\xi), \quad \xi = \alpha x - \beta t,$$
where k, α, β, and λ are arbitrary constants, and the functions $y(\xi)$ and $z(\xi)$ are described by the autonomous ODE system
$$\alpha y'_\xi + kmy = yf(y^n z^m), \quad -\beta z'_\xi + n\lambda z = zg(y^n z^m).$$

15. $u_x = uf(u^n w^m), \quad w_t = wg(u^k w^s).$

Let $\Delta = sn - km \neq 0$.

Multiplicative separable solutions (Polyanin & Manzhirov, 2007):
$$u = [\varphi(x)]^{s/\Delta} [\psi(t)]^{-m/\Delta}, \quad w = [\varphi(x)]^{-k/\Delta} [\psi(t)]^{n/\Delta},$$
where the functions $\varphi(x)$ and $\psi(t)$ are described by the autonomous ODEs
$$\frac{s}{\Delta}\varphi'_x = \varphi f(\varphi), \quad \frac{n}{\Delta}\psi'_t = \psi g(\psi).$$

Integrating yields
$$\frac{s}{\Delta}\int \frac{d\varphi}{\varphi f(\varphi)} = x + C_1, \quad \frac{n}{\Delta}\int \frac{d\psi}{\psi g(\psi)} = t + C_2.$$

16. $u_x = au\ln u + uf(u^n w^m), \quad w_t = wg(u^n w^m).$

Exact solution (Polyanin & Manzhirov, 2007):
$$u = \exp(Cme^{ax}) y(\xi), \quad w = \exp(-Cne^{ax}) z(\xi), \quad \xi = kx - \lambda t,$$
where C, k, and λ are arbitrary constants, and the functions $y(\xi)$ and $z(\xi)$ are described by the autonomous ODE system
$$ky'_\xi = ay\ln y + yf(y^n z^m), \quad -\lambda z'_\xi = zg(y^n z^m).$$

17. $u_x = uf(au^n + bw), \quad w_t = u^k g(au^n + bw).$

Exact solution (Polyanin & Manzhirov, 2007):
$$u = (C_1 t + C_2)^{\frac{1}{n-k}} \theta(x), \quad w = \varphi(x) - \frac{a}{b}(C_1 t + C_2)^{\frac{n}{n-k}} [\theta(x)]^n,$$
where C_1 and C_2 are arbitrary constants, and the functions $\theta = \theta(x)$ and $\varphi = \varphi(x)$ are determined by the system of differential-algebraic equations
$$\theta'_x = \theta f(b\varphi), \quad \theta^{n-k} = \frac{b(k-n)}{aC_1 n} g(b\varphi).$$

7.1.3. Gas Dynamic Type Systems Linearizable with the Hodograph Transformation

1. $u_t = w_x, \quad w_t = -uu_x.$

Equations of a stationary transonic plane-parallel gas flow. This is a special case of the PDE system 7.1.3.6, where the independent variables play the role of spatial variables x and $t = y$.

1°. Exact solutions:
$$u = -\frac{(x+C_1)^2}{(t+C_2)^2}, \quad w = \frac{2}{3}\frac{(x+C_1)^3}{(t+C_2)^3} + C_3,$$
$$u = -\frac{1}{C_1^2}x^2\wp(C_1 t + C_2), \quad w = -\frac{1}{3C_1}x^3\wp'(C_1 t + C_2) + C_3,$$

where C_1, C_2, and C_3 are arbitrary constants, $\wp(z) = \wp(z, 0, 4)$ is the Weierstrass elliptic function; the prime designates the derivative with respect to the argument.

3°. Solution:
$$u = u(z), \quad w = -C\ln t - \int zu_z'(z)\,dz, \quad z = \frac{x}{t},$$

where C is an arbitrary constant, and the function u is described by the first-order ODE: $(z^2 + u)u_z' = C$.

4°. The system is linearized using the hodograph transformation
$$t_u - x_w = 0, \quad x_u + ut_w = 0, \tag{1}$$

where u and w are treated as the independent variables and x and t as the dependent variables.

Eliminating x from (1) gives the second-order linear PDE:
$$t_{uu} + ut_{ww} = 0. \tag{2}$$

This equation admits polynomial solutions in w: $t = \sum_{k=0}^{n}\varphi_k(u)w^k$. Some polynomial solutions of equation (2) and the corresponding solutions of system (1) are listed below:

(a) $t = C_1 uw + C_2 u + C_3 w + C_4,$
$x = C_1(\frac{1}{2}w^2 - \frac{1}{3}u^3) + C_2(w - \frac{1}{2}u^2) + C_5;$

(b) $t = C_1(w^2 - \frac{1}{3}u^3) + C_2 uw + C_3 u + C_4 w + C_5,$
$x = -C_1 u^2 w + C_2(\frac{1}{2}w^2 - \frac{1}{3}u^3) + C_3 w - \frac{1}{2}C_4 u^2 + C_6;$

(c) $t = C_1(w^3 - u^3 w) + C_2 uw + C_3 u + C_4 w + C_5,$
$x = C_1(\frac{1}{5}u^5 - \frac{3}{2}u^2 w^2) + C_2(\frac{1}{2}w^2 - \frac{1}{3}u^3) + C_3 w - \frac{1}{2}C_4 u^2 + C_6,$

where $C_1, \ldots, C_6$ are arbitrary constants. These are exact implicit-form solutions of the original system.

5°. References for the system of PDEs 7.1.3.1: Ovsiannikov (1962), Polyanin & Zaitsev (2012).

2. $u_t - w_x = 0$, $\quad w_t - [f(u)]_x = 0$.

This is a special case of the PDE system 7.1.3.6. The system describes nonlinear one-dimensional longitudinal vibrations of an elastic rod, with u denoting the deformation gradient (strain), w the rate of deformation (strain rate), and $f(u)$ the stress. The condition $f'(u) > 0$ expresses the hyperbolicity of the system, with the prime denoting the derivative with respect to u.

1°. Trivial solutions:
$$u = C_1, \quad w = C_2,$$
where C_1 and C_2 are arbitrary constants.

2°. Self-similar solutions dependent on the ratio of the independent variables x/t:
$$w - \int \sqrt{f'(u)}\, du = C_1, \quad \sqrt{f'(u)} = -\frac{x}{t};$$
$$w + \int \sqrt{f'(u)}\, du = C_2, \quad \sqrt{f'(u)} = \frac{x}{t},$$
where C_1 and C_2 are arbitrary constants, and the prime denotes a derivative.

3°. Exact solutions in implicit form:
$$w - \int \sqrt{f'(u)}\, du = C_1, \quad x + t\sqrt{f'(u)} = \Phi_1(u);$$
$$w + \int \sqrt{f'(u)}\, du = C_2, \quad x - t\sqrt{f'(u)} = \Phi_2(u),$$
where $\Phi_m(u)$ are arbitrary functions and C_m are arbitrary constants ($m = 1, 2$). These solutions describe Riemann simple waves and are characterized by a functional relationship between the unknowns, $u = u(w)$. In the special cases $\Phi_m(w) \equiv 0$, these formulas become self-similar solutions from Item 2°.

4°. The system can be linearized with the hodograph transformation (for details see Item 5° of system 7.1.3.6):
$$t_u - x_w = 0, \quad x_u - f'(u) t_w = 0, \tag{1}$$
where u and w are treated as the independent variables and x and t as the dependent variables.

Eliminating x from (1) gives
$$t_{uu} = f'(u) t_{ww}. \tag{2}$$

This second-order linear PDE admits polynomial solutions in w: $t = \sum_{k=0}^{n} \varphi_k(u) w^k$. Some polynomial solutions of equation (2) and the corresponding solutions of system (1) are listed below (Polyanin & Zaitsev, 2012):

(a) $t = C_1 uw + C_2 u + C_3 w + C_4$,
$$x = \tfrac{1}{2}C_1 w^2 + C_2 w + (C_1 u + C_3) f(u) - C_1 \int f(u)\, du + C_5;$$

(b) $t = C_1 w^2 + C_2 uw + C_3 w + C_4 u + 2C_1 \int f(u)\, du + C_5$,
$$x = 2C_1 f(u) w + \tfrac{1}{2} C_2 w^2 + C_4 w + (C_2 u + C_3) f(u) - C_2 \int f(u)\, du + C_6;$$

(c) $t = C_1 w^3 + C_2 uw + C_3 u + C_4 w + 6 C_1 w \int f(u)\, du + C_5$,
$$x = 3 C_1 w^2 f(u) + \tfrac{1}{2} C_2 w^2 + C_3 w + \int \left[C_2 u + C_4 + 6 C_1 \int f(u)\, du \right] f'_u(u)\, du + C_6,$$

where $C_1, \ldots, C_6$ are arbitrary constants.

5°. Eliminating w from the original system leads to a nonlinear wave equation of the form 6.2.2.11:
$$u_{tt} = [f'(u) u_x]_x.$$

3. $u_t + u u_x + b w_x = 0$, $\quad w_t + u w_x + w u_x = 0$.

Shallow water equations. This is a special case of the PDE system 7.1.3.4 with $n = 2$ and $a = \tfrac{1}{2} b$, where u is the horizontal velocity averaged over the height w of the water level and b is the acceleration due to gravity.

4. $u_t + u u_x + a n w^{n-2} w_x = 0$, $\quad w_t + w u_x + u w_x = 0$.

This is a special case of the PDE system 7.1.3.5 with $p(w) = a w^n + b$. It describes *one-dimensional polytropic ideal gas flow*, where u is the gas velocity and w is the gas density.

For $n \neq 1$, the original PDE system is often written as
$$u_t + u u_x + \frac{2}{n-1} c c_x = 0, \quad c_t + u c_x + \frac{n-1}{2} c u_x = 0, \tag{1}$$
where $c = \sqrt{p'(w)} = \sqrt{a n w^{n-1}}$ is the sound speed.

1°. Let $u = u(x, t)$, $w = w(x, t)$ be a solution of the system in question. Then the pair of functions
$$u_1 = B_1^{n-1} u(B_1^{1-n} B_2 x + B_1^{1-n} B_2 B_3 t + B_4,\ B_2 t + B_5) - B_3,$$
$$w_1 = B_1^2 w(B_1^{1-n} B_2 x + B_1^{1-n} B_2 B_3 t + B_4,\ B_2 t + B_5),$$
where $B_1, \ldots, B_5$ are arbitrary constants, is also a solution of the system.

2°. Trivial solutions:
$$u = B_1, \quad w = B_2,$$
where B_1 and B_2 are arbitrary constants.

3°. Self-similar solutions dependent on the ratio of the dependent variables x/t:
$$u = \frac{2}{n+1} \frac{x}{t} + B_1, \quad c = \frac{n-1}{n+1} \frac{x}{t} - B_1, \quad c = \sqrt{a n w^{n-1}};$$
$$u = \frac{2}{n+1} \frac{x}{t} + B_2, \quad c = B_2 - \frac{n-1}{n+1} \frac{x}{t}, \quad c = \sqrt{a n w^{n-1}},$$
where B_1 and B_2 are arbitrary constants.

Remark 7.1. *The solutions from Items 2° and 3°, which are appropriately "glued" together along the straight lines $x/t = $ const, allow one to construct solutions to many gas dynamics problems.*

4°. More general self-similar solutions:
$$u = t^{k(1-n)}U(z), \quad w = t^{-2k}W(z), \quad z = t^{nk-k-1}x,$$
where k is an arbitrary constant, and the functions $U(z)$ and $W(z)$ are described by the ODE system
$$k(1-n)U + (nk-k-1)zU'_z + UU'_z + anW^{n-2}W'_z = 0,$$
$$-2kW + (nk-k-1)zW'_z + WU'_z + UW'_z = 0.$$

5°. Exact solutions in implicit form:
$$u = \frac{2}{n-1}c + B_1, \quad x - t\left(\frac{n+1}{n-1}c + B_1\right) = \Phi_1(w), \quad c = \sqrt{anw^{n-1}};$$
$$u = -\frac{2}{n-1}c - B_2, \quad x + t\left(\frac{n+1}{n-1}c + B_2\right) = \Phi_2(w), \quad c = \sqrt{anw^{n-1}},$$
where $\Phi_m(w)$ are arbitrary functions and B_m are arbitrary constants ($m = 1, 2$). These solutions describe *Riemann simple waves* and are characterized by a functional relationship between the unknowns, $u = u(w)$. In the special cases $\Phi_m(w) \equiv 0$, these formulas become self-similar solutions from Item 3°.

6°. For $n = 3$, the general solution of system (1) is expressed implicitly by the formulas
$$x = (u+c)t + F_1(u+c),$$
$$x = (u-c)t + F_2(u-c),$$
where $F_1(z_1)$ and $F_2(z_2)$ are arbitrary functions. This solution, as well as the general solutions of the PDE system in question with $n = 5/3$ and $n = 7/5$, were obtained by Aksenov (2012).

7°. The original system can be linearized with the hodograph transformation (for details, see Item 5° of system 7.1.3.6):
$$ut_w - x_w - anw^{n-2}t_u = 0, \quad wt_w - ut_u + x_u = 0. \tag{2}$$
Here u and w are taken to be the independent variables and x and t to be the dependent variables. For arbitrary n, the general solution of system (2) can be expressed in terms of Gauss's hypergeometric function.

8°. References for the system of PDEs 7.1.3.4: Courant (1964), Ovsiannikov (1981), Stanyukovich (1971), Rozhdestvenskii & Yanenko (1983), Courant & Friedrichs (1985), Chernyi (1988), Akhatov et al. (1991), Aksenov (2012), and Polyanin & Zaitsev (2012).

5. $u_t + uu_x + \dfrac{1}{w}[f(w)]_x = 0, \quad w_t + wu_x + uw_x = 0.$

This is a special case of the PDE system 7.1.3.6. The system describes *one-dimensional barotropic flows of an ideal compressible gas*, with u denoting the gas velocity, w the gas density, and $f(w)$ the pressure. The speed of sound is given by $c = \sqrt{f'(w)}$, where the prime stands for a derivative; the $c > 0$ expresses the hyperbolicity of the system.

$1°$. Let $u = u(x,t)$, $w = w(x,t)$ be a solution of the system in question. Then the pair of functions

$$u_1 = u(C_1 x + C_1 C_2 t + C_3,\ C_1 t + C_4) - C_2, \quad w_1 = w(C_1 x + C_1 C_2 t + C_3,\ C_1 t + C_4),$$

where $C_1, \ldots, C_4$ are arbitrary constants, is also a solution of the system.

$2°$. Trivial solutions:
$$u = C_1, \quad w = C_2,$$
where C_1 and C_2 are arbitrary constants.

$3°$. Solution:
$$u = \frac{x + C_1}{t + C_2}, \quad w = \frac{C_3}{t + C_2},$$
where C_1, C_2, and C_3 are arbitrary constants.

$4°$. Self-similar solutions, dependent on the ratio of the independent variables x/t, in implicit form:

$$u = \int \sqrt{f'(w)}\,\frac{dw}{w} + A_1, \quad \int \sqrt{f'(w)}\,\frac{dw}{w} + \sqrt{f'(w)} + A_1 = \frac{x}{t};$$

$$u = -\int \sqrt{f'(w)}\,\frac{dw}{w} - A_2, \quad \int \sqrt{f'(w)}\,\frac{dw}{w} + \sqrt{f'(w)} + A_2 = -\frac{x}{t},$$

where A_1 and A_2 are arbitrary constants, and the prime denotes a derivative.

Remark 7.2. *The solutions from Items $2°$ and $4°$, which are appropriately "glued" together along the straight lines $x/t = \text{const}$, allow one to construct solutions to many gas dynamics problems and, in particular, make it possible to obtain solutions to the discontinuity decay problem.*

$5°$. Solution:
$$u = C_1 t + C_2 + \theta(z), \quad w = \frac{C_3}{\theta(z)}, \quad z = x - \frac{1}{2}C_1 t^2 - C_2 t,$$

where the function $\theta = \theta(z)$ is defined implicitly by

$$C_1 z + \frac{1}{2}\theta^2 - \int g\left(\frac{C_3}{\theta}\right)\frac{d\theta}{\theta} = C_4, \quad g(w) = f'(w),$$

and C_1, C_2, C_3, and C_4 are arbitrary constants.

$6°$. Exact solutions in implicit form:

$$u = \int \sqrt{f'(w)}\,\frac{dw}{w} + A_1, \quad x - t\left[\int \sqrt{f'(w)}\,\frac{dw}{w} + \sqrt{f'(w)} + A_1\right] = \Phi_1(w);$$

$$u = -\int \sqrt{f'(w)}\,\frac{dw}{w} - A_2, \quad x + t\left[\int \sqrt{f'(w)}\,\frac{dw}{w} + \sqrt{f'(w)} + A_2\right] = \Phi_2(w),$$

where $\Phi_m(w)$ are arbitrary functions and A_m are arbitrary constants ($m = 1, 2$). These solutions describe *Riemann simple waves* and are characterized by a functional relationship between the unknowns, $u = u(w)$. In the special cases $\Phi_m(w) \equiv 0$, these formulas determine the self-similar solutions from Item $4°$.

7°. The original system can be linearized with the hodograph transformation (for details, see Item 5° of system 7.1.3.6).

8°. References for the system of PDEs 7.1.3.5: Courant (1964), Stanyukovich (1971), Ovsiannikov (1981), Rozhdestvenskii & Yanenko (1983), Courant & Friedrichs (1985), Chernyi (1988), Akhatov et al. (1991), and Polyanin & Zaitsev (2012).

6. $f_1(u,w)u_t + g_1(u,w)w_t + h_1(u,w)u_x + k_1(u,w)w_x = 0,$
    $f_2(u,w)u_t + g_2(u,w)w_t + h_2(u,w)u_x + k_2(u,w)w_x = 0.$

1°. Let $u = u(x,t)$, $w = w(x,t)$ be a solution of the system. Then the pair of functions

$$u_1 = u(C_1 x + C_2, C_1 t + C_3), \quad w_1 = w(C_1 x + C_2, C_1 t + C_3),$$

where C_1, C_2, and C_3 are arbitrary constants, is also a solution of the system.

2°. For any functions $f_j(u,w)$, $g_j(u,w)$, $h_j(u,w)$, and $k_j(u,w)$ ($j = 1, 2$), the system in question admits the trivial solution

$$u = C_1, \quad w = C_2,$$

where C_1 and C_2 are arbitrary constants.

3°. The system admits self-similar solutions of the form

$$u = u(\xi), \quad w = w(\xi), \quad \xi = x/t.$$

4°. We look for exact solutions that would generalize the solutions from Item 3° and be characterized by a functional relationship between the unknowns:

$$w = w(u). \tag{1}$$

Substituting (1) into the original system, one obtains two equations for the function $u = u(x,t)$:

$$\begin{aligned} (f_1 + g_1 w'_u)u_t + (h_1 + k_1 w'_u)u_x &= 0, \\ (f_2 + g_2 w'_u)u_t + (h_2 + k_2 w'_u)u_x &= 0. \end{aligned} \tag{2}$$

The relationship (1) must be chosen so that the equations of (2) are consistent. This condition results in the following nonlinear first-order ordinary differential equation for $w(u)$:

$$(g_1 k_2 - g_2 k_1)(w'_u)^2 + (f_1 k_2 + g_1 h_2 - f_2 k_1 - g_2 h_1)w'_u + f_1 h_2 - f_2 h_1 = 0. \tag{3}$$

Treating (3) as a quadratic equation for the derivative w'_u, let us require that its discriminant is positive (which corresponds to the hyperbolicity condition of the system):

$$(f_1 k_2 + g_1 h_2 - f_2 k_1 - g_2 h_1)^2 - 4(f_1 h_2 - f_2 h_1)(g_1 k_2 - g_2 k_1) > 0. \tag{4}$$

In this case, equation (3) has two distinct real roots and is equivalent to two different first-order ordinary differential equations solved for the derivative:

$$w'_u = \Lambda_m(u,w), \quad m = 1, 2. \tag{5}$$

Having determined a solution of this equation (for each $m = 1, 2$, we have a different solution), $w = w(u)$, we substitute it into either equation of (2) to obtain a quasilinear first-order partial differential equation for $u = u(x, t)$:

$$(f_1 + g_1 \Lambda_m)u_t + (h_1 + k_1 \Lambda_m)u_x = 0, \qquad w = w(u). \tag{6}$$

The general solution of this equation can be obtained by the method of characteristics (see Section 4.2.1).

Solutions of equations (5) and (6), with $m = 1, 2$, that depend on an arbitrary function and an arbitrary constant are called *Riemann simple waves*.

5°. Let us perform the *hodograph transformation*

$$x = x(u, w), \quad t = t(u, w), \tag{7}$$

so that u and w are treated as the independent variables and x and t as the dependent ones. Differentiating the relations of (7) with respect to x and t (as composite functions) and eliminating the partial derivatives u_t, w_t, u_x, and w_x from the resulting equations, we obtain

$$u_t = -Jx_w, \quad w_t = Jx_u, \quad u_x = Jt_w, \quad w_x = -Jt_u, \tag{8}$$

where $J = u_x w_t - u_t w_x$ is the Jacobian of the functions $u = u(x, t)$ and $w = w(x, t)$. Replacing in the original system the derivatives by using (8) and dividing by J, we arrive at the linear system of first-order PDEs:

$$g_1(u, w)x_u - k_1(u, w)t_u - f_1(u, w)x_w + h_1(u, w)t_w = 0,$$
$$g_2(u, w)x_u - k_2(u, w)t_u - f_2(u, w)x_w + h_2(u, w)t_w = 0.$$

Remark 7.3. *The hodograph transformation (7) is inapplicable if $J \equiv 0$. In this degenerate case, u and w will be functionally dependent and, hence, cannot be used as independent variables. In this case, relation (1), which determines Riemann simple waves, holds. Consequently, the use of the hodograph transformation (7) results in the loss of solutions (1), describing Riemann simple waves.*

8°. References for the system of PDEs 7.1.3.6: Courant (1964), Stanyukovich (1971), Ovsiannikov (1981), Rozhdestvenskii & Yanenko (1983), Courant & Friedrichs (1985), Chernyi (1988), Polyanin & Manzhirov (2007), and Polyanin & Zaitsev (2012).

7.2. Systems of Two Second-Order PDEs

7.2.1. Linear Systems of Two Second-Order PDEs

1. $u_t = au_{xx} + b_1 u + c_1 w, \quad w_t = aw_{xx} + b_2 u + c_2 w.$

Constant-coefficient second-order linear system of parabolic type PDEs.

General solution (Polyanin & Manzhirov, 2007):

$$u = \frac{b_1 - \lambda_2}{b_2(\lambda_1 - \lambda_2)} e^{\lambda_1 t} U - \frac{b_1 - \lambda_1}{b_2(\lambda_1 - \lambda_2)} e^{\lambda_2 t} W, \quad w = \frac{1}{\lambda_1 - \lambda_2}\left(e^{\lambda_1 t} U - e^{\lambda_2 t} W\right),$$

where λ_1 and λ_2 are roots of the quadratic equation

$$\lambda^2 - (b_1 + c_2)\lambda + b_1 c_2 - b_2 c_1 = 0,$$

and the functions $U = U(x,t)$ and $W = W(x,t)$ satisfy the independent linear heat equations
$$U_t = aU_{xx}, \quad W_t = aW_{xx}.$$

2. $u_t = au_{xx} + f_1(t)u + g_1(t)w, \quad w_t = aw_{xx} + f_2(t)u + g_2(t)w.$

Variable-coefficient second-order linear system of parabolic type PDEs.

General solution (Polyanin & Manzhirov, 2007):
$$u = \varphi_1(t)U(x,t) + \varphi_2(t)W(x,t), \quad w = \psi_1(t)U(x,t) + \psi_2(t)W(x,t),$$

where the pairs of functions $\varphi_1 = \varphi_1(t), \psi_1 = \psi_1(t)$ and $\varphi_2 = \varphi_2(t), \psi_2 = \psi_2(t)$ are linearly independent (fundamental) solutions to the system of linear ordinary differential equations
$$\varphi'_t = f_1(t)\varphi + g_1(t)\psi, \quad \psi'_t = f_2(t)\varphi + g_2(t)\psi,$$

and the functions $U = U(x,t)$ and $W = W(x,t)$ satisfy the independent linear heat equations
$$U_t = aU_{xx}, \quad W_t = aW_{xx}.$$

3. $u_{tt} = ku_{xx} + a_1 u + b_1 w, \quad w_{tt} = kw_{xx} + a_2 u + b_2 w.$

Constant-coefficient second-order linear system of hyperbolic type PDEs.

General solution (Polyanin & Manzhirov, 2007):
$$u = \frac{a_1 - \lambda_2}{a_2(\lambda_1 - \lambda_2)}U - \frac{a_1 - \lambda_1}{a_2(\lambda_1 - \lambda_2)}W, \quad w = \frac{1}{\lambda_1 - \lambda_2}(U - W),$$

where λ_1 and λ_2 are roots of the quadratic equation
$$\lambda^2 - (a_1 + b_2)\lambda + a_1 b_2 - a_2 b_1 = 0,$$

and the functions $\theta_n = \theta_n(x,t)$ satisfy the independent linear Klein–Gordon equations
$$U_{tt} = kU_{xx} + \lambda_1 U, \quad W_{tt} = kW_{xx} + \lambda_2 W.$$

4. $u_{xx} + u_{yy} = a_1 u + b_1 w, \quad w_{xx} + w_{yy} = a_2 u + b_2 w.$

Constant-coefficient second-order linear system of elliptic type.

General solution (Polyanin & Manzhirov, 2007):
$$u = \frac{a_1 - \lambda_2}{a_2(\lambda_1 - \lambda_2)}U - \frac{a_1 - \lambda_1}{a_2(\lambda_1 - \lambda_2)}W, \quad w = \frac{1}{\lambda_1 - \lambda_2}(U - W),$$

where λ_1 and λ_2 are roots of the quadratic equation
$$\lambda^2 - (a_1 + b_2)\lambda + a_1 b_2 - a_2 b_1 = 0,$$

and the functions $\theta_n = \theta_n(x,y)$ satisfy the independent linear Helmholtz equations
$$U_{xx} + U_{yy} = \lambda_1 U, \quad W_{xx} + W_{yy} = \lambda_2 W.$$

Remark 7.4. Many analytical solutions to more complex systems of linear PDEs encountered in various fields of physics and continuum mechanics are given in Polyanin & Nazaikinskii (2016).

7.2.2. Nonlinear Parabolic Systems of the Form
$$u_t = au_{xx} + F(u,w), \quad w_t = bw_{xx} + G(u,w)$$

▶ **Preliminary remarks.** Reaction-diffusion systems of this form often arise in the theory of heat and mass transfer in chemically reactive media, theory of chemical reactors, combustion theory, mathematical biology, and biophysics.

Such systems are invariant under translations in the independent variables (and under reflection, the change of x to $-x$) and admit traveling wave solutions $u = u(kx - \lambda t)$, $w = w(kx - \lambda t)$. These solutions as well as those where one of the unknowns is identically zero are not considered in this section.

The functions $f(\varphi)$, $g(\varphi)$, and $h(\varphi)$ appearing below are arbitrary functions of their argument, $\varphi = \varphi(u,w)$. The equations are arranged in order of complexity of this argument.

▶ **Arbitrary functions depend on a linear combination of the unknowns.**

1. $u_t = au_{xx} + ue^{kw/u} f(u), \quad w_t = aw_{xx} + e^{kw/u}[wf(u) + g(u)].$

Exact solution (Barannyk, 2002):
$$u = y(\xi), \quad w = -\frac{2}{k} \ln|bx| y(\xi) + z(\xi), \quad \xi = \frac{x + C_3}{\sqrt{C_1 t + C_2}},$$

where C_1, C_2, C_3, and b are arbitrary constants, and the functions $y = y(\xi)$ and $z = z(\xi)$ are described by the ODE system

$$ay''_{\xi\xi} + \frac{1}{2} C_1 \xi y'_\xi + \frac{1}{b^2\xi^2} y \exp\left(k\frac{z}{y}\right) f(y) = 0,$$

$$az''_{\xi\xi} + \frac{1}{2} C_1 \xi z'_\xi - \frac{4a}{k\xi} y'_\xi + \frac{2a}{k\xi^2} y + \frac{1}{b^2\xi^2} \exp\left(k\frac{z}{y}\right)[zf(y) + g(y)] = 0.$$

2. $u_t = au_{xx} + f(bu + cw), \quad w_t = aw_{xx} + g(bu + cw).$

This is a special case of the PDE system 7.2.2.3 with $a_1 = a_2 = a$.

Exact solution (Polyanin & Vyazmina, 2006):
$$u = c\theta(x,t) + y(\xi), \quad w = -b\theta(x,t) + z(\xi), \quad \xi = kx - \lambda t,$$

where the functions $y(\xi)$ and $z(\xi)$ are described by the autonomous ODE system
$$ak^2 y''_{\xi\xi} + \lambda y'_\xi + f(by + cz) = 0,$$
$$ak^2 z''_{\xi\xi} + \lambda z'_\xi + g(by + cz) = 0,$$

and the function $\theta = \theta(x,t)$ satisfies the linear heat equation
$$\theta_t = a\theta_{xx}.$$

3. $u_t = a_1 u_{xx} + f(bu + cw), \quad w_t = a_2 w_{xx} + g(bu + cw).$

Exact solution (Polyanin & Vyazmina, 2006):
$$u = c(\alpha x^2 + \beta x + \gamma t) + y(\xi), \quad w = -b(\alpha x^2 + \beta x + \gamma t) + z(\xi), \quad \xi = kx - \lambda t,$$

where k, α, β, γ, and λ are arbitrary constants, and the functions $y(\xi)$ and $z(\xi)$ are described by the autonomous ODE system

$$a_1 k^2 y''_{\xi\xi} + \lambda y'_\xi + 2a_1 c\alpha - c\gamma + f(by + cz) = 0,$$
$$a_2 k^2 z''_{\xi\xi} + \lambda z'_\xi - 2a_2 b\alpha + b\gamma + g(by + cz) = 0.$$

4. $u_t = a u_{xx} + c_2 f(b_1 u + c_1 w) + c_1 g(b_2 u + c_2 w),$
   $w_t = a w_{xx} - b_2 f(b_1 u + c_1 w) - b_1 g(b_2 u + c_2 w).$

It is assumed that $b_1 c_2 - b_2 c_1 \neq 0$.

Multiplying the equations by suitable constants and adding together, one arrives at two independent PDEs (Polyanin & Zaitsev, 2012):

$$U_t = a U_{xx} + (b_1 c_2 - b_2 c_1) f(U), \quad U = b_1 u + c_1 w;$$
$$W_t = a W_{xx} - (b_1 c_2 - b_2 c_1) g(W), \quad W = b_2 u + c_2 w.$$

In the general case, these PDEs admit traveling wave solutions propagating at different speeds

$$U = U(k_1 x - \lambda_1 t), \quad W = W(k_2 x - \lambda_2 t),$$

where k_m and λ_m are arbitrary constants. The corresponding solution of the original system is the superposition (linear combination) of two nonlinear traveling waves.

For exact solutions of the independent PDEs for U and W for some functions $f(U)$ and $g(W)$, see Section 6.1.1.

5. $u_t = a u_{xx} + u f(bu - cw) + g(bu - cw),$
   $w_t = a w_{xx} + w f(bu - cw) + h(bu - cw).$

This is a special case of the PDE system 7.2.2.6.

1°. Solution:

$$u = \varphi(t) + c \exp\left[\int f(b\varphi - c\psi)\, dt\right] \theta(x,t), \quad w = \psi(t) + b \exp\left[\int f(b\varphi - c\psi)\, dt\right] \theta(x,t),$$

where $\varphi = \varphi(t)$ and $\psi = \psi(t)$ are described by the autonomous ODE system

$$\varphi'_t = \varphi f(b\varphi - c\psi) + g(b\varphi - c\psi), \quad \psi'_t = \psi f(b\varphi - c\psi) + h(b\varphi - c\psi),$$

and the function $\theta = \theta(x,t)$ satisfies the linear heat equation

$$\theta_t = a \theta_{xx}.$$

2°. Let us multiply the first equation by b and the second one by $-c$ and add the results together to obtain

$$\zeta_t = a \zeta_{xx} + \zeta f(\zeta) + bg(\zeta) - ch(\zeta), \quad \zeta = bu - cw. \tag{1}$$

This equation will be considered in conjunction with the first equation of the original system

$$u_t = a u_{xx} + u f(\zeta) + g(\zeta). \tag{2}$$

Equation (1) can be treated separately. An extensive list of exact solutions to equations of this form for various kinetic functions $F(\zeta) = \zeta f(\zeta) + bg(\zeta) - ch(\zeta)$ can be found in the book by Polyanin and Zaitsev (2012). Given a solution $\zeta = \zeta(x,t)$ to equation (1), the function $u = u(x,t)$ can be determined by solving the linear equation (2) and the function $w = w(x,t)$ is found as $w = (bu - \zeta)/c$.

Note two important solutions to equation (1):

(i) In the general case, equation (1) admits traveling wave solutions $\zeta = \zeta(z)$, where $z = kx - \lambda t$. Then the corresponding exact solutions to equation (2) are expressed as $u = u_0(z) + \sum e^{\beta_n t} u_n(z)$.

(ii) If the condition $\zeta f(\zeta) + bg(\zeta) - ch(\zeta) = k_1 \zeta + k_0$ holds, equation (1) is linear,

$$\zeta_t = a\zeta_{xx} + k_1\zeta + k_0,$$

and, hence, can be reduced to the linear heat equation.

3°. References for the system of PDEs 7.2.2.5: Polyanin (2004, 2005), Polyanin & Manzhirov (2007), and Polyanin & Zaitsev (2012).

6. $u_t = a_1 u_{xx} + k_1 u + u f_1(bu - cw) + g_1(bu - cw),$
$w_t = a_2 w_{xx} + k_2 w + w f_2(bu - cw) + g_2(bu - cw).$

Here, $f_1(z)$, $f_2(z)$, $g_1(z)$, and $g_2(z)$ are arbitrary functions.

1°. Let
$$u = u° = \text{const}, \quad w = w° = \text{const}$$

is a stationary point (simplest solution) of the PDE system which is determined from a nonlinear algebraic system:

$$u°[k_1 + f_1(bu° - cw°)] + g_1(bu° - cw°) = 0,$$
$$w°[k_2 + f_2(bu° - cw°)] + g_2(bu° - cw°) = 0.$$

This system usually has multiple roots.

More complex exact solutions of the original PDE system have the form

$$u = u° + ce^{\lambda t}\theta(x), \quad w = w° + be^{\lambda t}\theta(x),$$

where the constant λ and the function $\theta(x)$ are defined by the formulas

$$\lambda = \frac{a_2 k_1 - a_1 k_2 + a_2 f_1° - a_1 f_2°}{a_2 - a_1}, \quad f_1° = f_1(bu° - cw°), \quad f_2° = f_2(bu° - cw°);$$

$$\theta(x) = \begin{cases} C_1 \cos(\sqrt{|\mu|}\, x) + C_2 \sin(\sqrt{|\mu|}\, x) & \text{if } \mu < 0, \\ C_1 \exp(-\sqrt{\mu}\, x) + C_2 \exp(\sqrt{\mu}\, x) & \text{if } \mu > 0, \end{cases} \quad \mu = \frac{k_1 - k_2 + f_1° - f_2°}{a_2 - a_1},$$

and C_1 and C_2 are arbitrary constants.

2°. Let us put
$$f_1(z) = a_1 f(z), \quad f_2(z) = a_2 f(z),$$

where $f(z)$ is an arbitrary function, and the functions $\varphi = \varphi(x)$ and $\psi = \psi(x)$ form a stationary solution of the original PDE system, i.e., they satisfy the system of second-order ODEs

$$a_1 \varphi''_{xx} + (k_1 + a_1 \hat{f})\varphi + \hat{g}_1 = 0,$$
$$a_2 \psi''_{xx} + (k_2 + a_2 \hat{f})\psi + \hat{g}_2 = 0.$$
(1)

Here we use the notation $\hat{f} = f(b\varphi - c\psi)$, $\hat{g}_1 = g_1(b\varphi - c\psi)$, and $\hat{g}_2 = g_2(b\varphi - c\psi)$.

Non-stationary exact solutions of original PDE system have the form

$$u = ce^{\lambda t}\theta(x) + \varphi(x), \quad w = be^{\lambda t}\theta(x) + \psi(x), \quad \lambda = \frac{a_2 k_1 - a_1 k_2}{a_2 - a_1},$$

where the functions $\varphi = \varphi(x)$ and $\psi = \psi(x)$ satisfy the system of ODEs (1), and the function $\theta = \theta(x)$ satisfies the second-order linear ODE with variable coefficients

$$\theta''_{xx} + \left(\frac{k_2 - k_1}{a_2 - a_1} + \hat{f}\right)\theta = 0, \quad \hat{f} = f(b\varphi - c\psi).$$

3°. Let us put

$$f_1(z) = f_2(z) = f(z),$$

where $f(z)$ is an arbitrary function, and the functions $\varphi = \varphi(t)$ and $\psi = \psi(t)$ describe a spatially homogeneous non-stationary solution of the original PDE system and satisfy the first-order ODE system

$$\begin{aligned}\varphi'_t &= k_1\varphi + \varphi f(b\varphi - c\psi) + g_1(b\varphi - c\psi),\\ \psi'_t &= k_2\psi + \psi f(b\varphi - c\psi) + g_2(b\varphi - c\psi).\end{aligned} \quad (2)$$

The original PDE system has more complex non-stationary exact solutions of the form

$$u = c\rho(t)\theta(x) + \varphi(t), \quad w = b\rho(t)\theta(x) + \psi(t),$$

where the functions $\varphi = \varphi(t)$ and $\psi = \psi(t)$ satisfy the ODE system (2), the functions $\theta = \theta(x)$ and $\rho = \rho(t)$ are defined by the formulas

$$\rho(t) = A \exp\left[\frac{a_2 k_1 - a_1 k_2}{a_2 - a_1} t + \int f(b\varphi - c\psi)\, dt\right],$$

$$\theta(x) = \begin{cases} C_1 \cos(\sqrt{|\mu|}\, x) + C_2 \sin(\sqrt{|\mu|}\, x) & \text{if } \mu < 0, \\ C_1 \exp(-\sqrt{\mu}\, x) + C_2 \exp(\sqrt{\mu}\, x) & \text{if } \mu > 0, \end{cases} \quad \mu = \frac{k_1 - k_2}{a_2 - a_1},$$

and C_1 and C_2 are arbitrary constants.

4°. References for PDE 7.2.2.6: Polyanin & Sorokin (2022) and Polyanin et al. (2023) (in these works, a more general system was considered, additionally containing several delays). In the special case of $f_1(z) = s_1 z$, $f_2(z) = s_2 z$, and $g(z) = h(z) = 0$, the system under consideration becomes a Lotka–Volterra type system; for exact solutions of such PDE systems with quadratic nonlinearity, see Cherniha & Davydovych (2017 and 2021), Polyanin & Zhurov (2022), and Polyanin et al. (2023).

7. $u_t = au_{xx} + e^{\lambda u}f(\lambda u - \sigma w), \quad w_t = bw_{xx} + e^{\sigma w}g(\lambda u - \sigma w).$

1°. Exact solution (Polyanin, 2004):

$$u = y(\xi) - \frac{1}{\lambda}\ln(C_1 t + C_2), \quad w = z(\xi) - \frac{1}{\sigma}\ln(C_1 t + C_2), \quad \xi = \frac{x + C_3}{\sqrt{C_1 t + C_2}},$$

where C_1, C_2, and C_3 are arbitrary constants, and the functions $y = y(\xi)$ and $z = z(\xi)$ are described by the ODE system

$$\begin{aligned}ay''_{\xi\xi} + \tfrac{1}{2}C_1\xi y'_\xi + \frac{C_1}{\lambda} + e^{\lambda y}f(\lambda y - \sigma z) &= 0,\\ bz''_{\xi\xi} + \tfrac{1}{2}C_1\xi z'_\xi + \frac{C_1}{\sigma} + e^{\sigma z}g(\lambda y - \sigma z) &= 0.\end{aligned}$$

2°. Exact solution for $b = a$ (Polyanin & Manzhirov, 2007):
$$u = U(x,t), \quad w = \frac{\lambda}{\sigma}U(x,t) - \frac{k}{\sigma},$$
where k is a root of the algebraic (transcendental) equation
$$\lambda f(k) = \sigma e^{-k} g(k),$$
and the function $U(x,t)$ is determined by the PDE:
$$U_t = aU_{xx} + f(k)e^{\lambda U}.$$
This equation has an exact solution of the form
$$U = \theta(\xi) - \frac{1}{\lambda}\ln(C_1 t + C_2), \quad \xi = \frac{x + C_3}{\sqrt{C_1 t + C_2}},$$
where the function $\theta(\xi)$ is determined by the ODE:
$$a\theta''_{\xi\xi} + \frac{1}{2}C_1 \xi \theta'_\xi + \frac{C_1}{\lambda} + f(k)e^{\lambda\theta} = 0.$$

▶ **Arbitrary functions depend on the ratio of the unknowns.**

8. $u_t = au_{xx} + uf\left(\dfrac{u}{w}\right), \quad w_t = bw_{xx} + wg\left(\dfrac{u}{w}\right).$

1°. Multiplicative separable solution:
$$u = [C_1 \sin(kx) + C_2 \cos(kx)]\varphi(t), \quad w = [C_1 \sin(kx) + C_2 \cos(kx)]\psi(t),$$
where C_1, C_2, and k are arbitrary constants, and the functions $\varphi = \varphi(t)$ and $\psi = \psi(t)$ are described by the autonomous ODE system
$$\varphi'_t = -ak^2\varphi + \varphi f(\varphi/\psi), \quad \psi'_t = -bk^2\psi + \psi g(\varphi/\psi).$$

2°. Multiplicative separable solution:
$$u = [C_1 \exp(kx) + C_2 \exp(-kx)]U(t), \quad w = [C_1 \exp(kx) + C_2 \exp(-kx)]W(t),$$
where C_1, C_2, and k are arbitrary constants, and the functions $U = U(t)$ and $W = W(t)$ are described by the autonomous ODE system
$$U'_t = ak^2 U + U f(U/W), \quad W'_t = bk^2 W + W g(U/W).$$

3°. Degenerate solution:
$$u = (C_1 x + C_2)U(t), \quad w = (C_1 x + C_2)W(t),$$
where C_1 and C_2 are arbitrary constants, and the functions $U = U(t)$ and $W = W(t)$ are described by the autonomous ODE system
$$U'_t = U f(U/W), \quad W'_t = W g(U/W).$$

This autonomous system can be integrated since it is reduced, after eliminating t, to a homogeneous first-order equation. The systems presented in Items 1° and 2° can be integrated likewise.

4°. Multiplicative separable solution:
$$u = e^{-\lambda t} y(x), \quad w = e^{-\lambda t} z(x),$$
where λ is an arbitrary constant, and the functions $y = y(x)$ and $z = z(x)$ are described by the autonomous ODE system
$$ay''_{xx} + \lambda y + y f(y/z) = 0, \quad b z''_{xx} + \lambda z + z g(y/z) = 0.$$

5°. Solution (generalizes the solution of Item 4°):
$$u = e^{kx - \lambda t} y(\xi), \quad w = e^{kx - \lambda t} z(\xi), \quad \xi = \beta x - \gamma t,$$
where k, λ, β, and γ are arbitrary constants, and the functions $y = y(\xi)$ and $z = z(\xi)$ are described by the autonomous ODE system
$$a\beta^2 y''_{\xi\xi} + (2ak\beta + \gamma) y'_\xi + (ak^2 + \lambda) y + y f(y/z) = 0,$$
$$b\beta^2 z''_{\xi\xi} + (2bk\beta + \gamma) z'_\xi + (bk^2 + \lambda) z + z g(y/z) = 0.$$

To the special case $k = \lambda = 0$ there corresponds a traveling wave solution. If $k = \gamma = 0$ and $\beta = 1$, we have the solution of Item 4°.

6°. References for the system of PDEs 7.2.2.8: Polyanin (2004, 2005) and Polyanin & Zaitsev (2012).

9. $u_t = a u_{xx} + u f\left(\dfrac{u}{w}\right), \quad w_t = a w_{xx} + w g\left(\dfrac{u}{w}\right).$

This PDE system is a special case of the previous system 7.2.2.8 with $b = a$, and hence, it admits all the above solutions. In addition, it has some interesting properties and other solutions, which are given below.

1°. Suppose the pair of functions $u = u(x, t)$, $w = w(x, t)$ is a solution of the system. Then the functions
$$u_1 = A u(\pm x + C_1, t + C_2), \quad w_1 = A w(\pm x + C_1, t + C_2);$$
$$u_2 = \exp(\lambda x + a\lambda^2 t) u(x + 2a\lambda t, t), \quad w_2 = \exp(\lambda x + a\lambda^2 t) w(x + 2a\lambda t, t),$$
where A, C_1, C_2, and λ are arbitrary constants, are also solutions of this system.

2°. Point-source solution:
$$u = \exp\left(-\frac{x^2}{4at}\right) \varphi(t), \quad w = \exp\left(-\frac{x^2}{4at}\right) \psi(t),$$
where the functions $\varphi = \varphi(t)$ and $\psi = \psi(t)$ are described by the autonomous ODE system
$$\varphi'_t = -\frac{1}{2t} \varphi + \varphi f\left(\frac{\varphi}{\psi}\right), \quad \psi'_t = -\frac{1}{2t} \psi + \psi g\left(\frac{\varphi}{\psi}\right).$$

3°. Functional separable solution:
$$u = \exp\left(kxt + \tfrac{2}{3} ak^2 t^3 - \lambda t\right) y(\xi),$$
$$w = \exp\left(kxt + \tfrac{2}{3} ak^2 t^3 - \lambda t\right) z(\xi),$$
$\xi = x + akt^2,$

where k and λ are arbitrary constants, and the functions $y = y(\xi)$ and $z = z(\xi)$ are described by the autonomous ODE system

$$ay''_{\xi\xi} + (\lambda - k\xi)y + yf(y/z) = 0, \quad az''_{\xi\xi} + (\lambda - k\xi)z + zg(y/z) = 0.$$

4°. Let k be a root of the algebraic (transcendental) equation

$$f(k) = g(k). \tag{1}$$

Solution:
$$u = ke^{\lambda t}\theta, \quad w = e^{\lambda t}\theta, \quad \lambda = f(k),$$

where the function $\theta = \theta(x,t)$ satisfies the linear heat equation

$$\theta_t = a\theta_{xx}.$$

5°. Periodic solution:
$$u = Ak\exp(-\mu x)\sin(\beta x - 2a\beta\mu t + B),$$
$$w = A\exp(-\mu x)\sin(\beta x - 2a\beta\mu t + B), \quad \beta = \sqrt{\mu^2 + \frac{1}{a}f(k)},$$

where A, B, and μ are arbitrary constants, and k is a root of the algebraic (transcendental) equation (1).

6°. Solution:
$$u = \varphi(t)\exp\left[\int g(\varphi(t))\,dt\right]\theta(x,t), \quad w = \exp\left[\int g(\varphi(t))\,dt\right]\theta(x,t),$$

where the function $\varphi = \varphi(t)$ is described by the separable first-order ordinary differential equation

$$\varphi'_t = [f(\varphi) - g(\varphi)]\varphi, \tag{2}$$

and the function $\theta = \theta(x,t)$ satisfies the linear heat equation

$$\theta_t = a\theta_{xx}.$$

To the particular solution $\varphi = k = \text{const}$ of equation (2) there corresponds the solution given in Item 4°. The general solution of equation (2) is written out in implicit form as

$$\int \frac{d\varphi}{[f(\varphi) - g(\varphi)]\varphi} = t + C.$$

7°. The transformation
$$u = a_1 U + b_1 W, \quad w = a_2 U + b_2 W,$$

where a_n and b_n are arbitrary constants ($n = 1, 2$), leads to a system of a similar form for U and W.

8°. References for the system of PDEs 7.2.2.9: Polyanin (2004, 2005) and Polyanin & Zaitsev (2012).

10. $u_t = au_{xx} + uf\left(\dfrac{u}{w}\right) + g\left(\dfrac{u}{w}\right), \quad w_t = aw_{xx} + wf\left(\dfrac{u}{w}\right) + h\left(\dfrac{u}{w}\right).$

Let k be a root of the algebraic (transcendental) equation

$$g(k) = kh(k).$$

1°. Solution with $f(k) \neq 0$ (Polyanin, 2004):

$$u(x,t) = k\left(\exp[f(k)t]\theta(x,t) - \dfrac{h(k)}{f(k)}\right), \quad w(x,t) = \exp[f(k)t]\theta(x,t) - \dfrac{h(k)}{f(k)},$$

where the function $\theta = \theta(x,t)$ satisfies the linear heat equation

$$\theta_t = a\theta_{xx}. \tag{1}$$

2°. Solution with $f(k) = 0$:

$$u(x,t) = k[\theta(x,t) + h(k)t], \quad w(x,t) = \theta(x,t) + h(k)t,$$

where the function $\theta = \theta(x,t)$ satisfies the linear heat equation (1).

11. $u_t = au_{xx} + uf\left(\dfrac{u}{w}\right) + \dfrac{u}{w}h\left(\dfrac{u}{w}\right), \quad w_t = aw_{xx} + wg\left(\dfrac{u}{w}\right) + h\left(\dfrac{u}{w}\right).$

Solution (Polyanin, 2004):

$$u = \varphi(t)G(t)\left[\theta(x,t) + \int \dfrac{h(\varphi)}{G(t)}\,dt\right], \quad G(t) = \exp\left[\int g(\varphi)\,dt\right],$$

$$w = G(t)\left[\theta(x,t) + \int \dfrac{h(\varphi)}{G(t)}\,dt\right],$$

where the function $\varphi = \varphi(t)$ is described by the separable first-order ordinary differential equation

$$\varphi'_t = [f(\varphi) - g(\varphi)]\varphi,$$

and the function $\theta = \theta(x,t)$ satisfies the linear heat equation

$$\theta_t = a\theta_{xx}.$$

12. $u_t = au_{xx} + uf_1\left(\dfrac{w}{u}\right) + wg_1\left(\dfrac{w}{u}\right), \quad w_t = aw_{xx} + uf_2\left(\dfrac{w}{u}\right) + wg_2\left(\dfrac{w}{u}\right).$

Solution:

$$u = \exp\left\{\int [f_1(\varphi) + \varphi g_1(\varphi)]\,dt\right\}\theta(x,t),$$

$$w = \varphi(t)\exp\left\{\int [f_1(\varphi) + \varphi g_1(\varphi)]\,dt\right\}\theta(x,t),$$

where the function $\varphi = \varphi(t)$ is described by the separable first-order ordinary differential equation

$$\varphi'_t = f_2(\varphi) + \varphi g_2(\varphi) - \varphi[f_1(\varphi) + \varphi g_1(\varphi)],$$

and the function $\theta = \theta(x,t)$ satisfies the linear heat equation

$$\theta_t = a\theta_{xx}.$$

13. $u_t = au_{xx} + u^3 f\left(\dfrac{u}{w}\right)$, $\quad w_t = aw_{xx} + u^3 g\left(\dfrac{u}{w}\right)$.

Solution (Barannyk, 2002 and Barannyk & Nikitin, 2004):

$$u = (x+C_1)\varphi(z), \quad w = (x+C_1)\psi(z), \quad z = t + \frac{1}{6a}(x+C_1)^2 + C_2,$$

where C_1 and C_2 are arbitrary constants, and the functions $\varphi = \varphi(z)$ and $\psi = \psi(z)$ are described by the autonomous ODE system

$$\varphi''_{zz} + 9a\varphi^3 f(\varphi/\psi) = 0, \quad \psi''_{zz} + 9a\varphi^3 g(\varphi/\psi) = 0.$$

14. $u_t = u_{xx} + au - u^3 f\left(\dfrac{u}{w}\right)$, $\quad w_t = w_{xx} + aw - u^3 g\left(\dfrac{u}{w}\right)$.

$1°$. Solution with $a > 0$:

$$u = \left[C_1 \exp\!\left(\tfrac{1}{2}\sqrt{2a}\,x + \tfrac{3}{2}at\right) - C_2 \exp\!\left(-\tfrac{1}{2}\sqrt{2a}\,x + \tfrac{3}{2}at\right)\right]\varphi(z),$$
$$w = \left[C_1 \exp\!\left(\tfrac{1}{2}\sqrt{2a}\,x + \tfrac{3}{2}at\right) - C_2 \exp\!\left(-\tfrac{1}{2}\sqrt{2a}\,x + \tfrac{3}{2}at\right)\right]\psi(z),$$
$$z = C_1 \exp\!\left(\tfrac{1}{2}\sqrt{2a}\,x + \tfrac{3}{2}at\right) + C_2 \exp\!\left(-\tfrac{1}{2}\sqrt{2a}\,x + \tfrac{3}{2}at\right) + C_3,$$

where C_1, C_2, and C_3 are arbitrary constants, and the functions $\varphi = \varphi(z)$ and $\psi = \psi(z)$ are described by the autonomous ODE system

$$a\varphi''_{zz} = 2\varphi^3 f(\varphi/\psi), \quad a\psi''_{zz} = 2\varphi^3 g(\varphi/\psi).$$

$2°$. Solution with $a < 0$:

$$u = \exp\!\left(\tfrac{3}{2}at\right)\sin\!\left(\tfrac{1}{2}\sqrt{2|a|}\,x + C_1\right)U(\xi),$$
$$w = \exp\!\left(\tfrac{3}{2}at\right)\sin\!\left(\tfrac{1}{2}\sqrt{2|a|}\,x + C_1\right)W(\xi),$$
$$\xi = \exp\!\left(\tfrac{3}{2}at\right)\cos\!\left(\tfrac{1}{2}\sqrt{2|a|}\,x + C_1\right) + C_2,$$

where C_1 and C_2 are arbitrary constants, and the functions $U = U(\xi)$ and $W = W(\xi)$ are described by the autonomous ODE system

$$aU''_{\xi\xi} = -2U^3 f(U/W), \quad aW''_{\xi\xi} = -2U^3 g(U/W).$$

$3°$. References for the system of PDEs 7.2.2.14: Barannyk (2002), Polyanin (2004), and Polyanin & Manzhirov (2007).

15. $u_t = au_{xx} + u^n f\left(\dfrac{u}{w}\right)$, $\quad w_t = bw_{xx} + w^n g\left(\dfrac{u}{w}\right)$.

If $f(z) = kz^{-m}$ and $g(z) = -kz^{n-m}$, the system describes an nth-order chemical reaction (of order $n-m$ in the component u and of order m in the component w).

$1°$. Self-similar solution with $n \neq 1$ (Polyanin, 2004):

$$u = (C_1 t + C_2)^{\frac{1}{1-n}} y(\xi), \quad w = (C_1 t + C_2)^{\frac{1}{1-n}} z(\xi), \quad \xi = \frac{x + C_3}{\sqrt{C_1 t + C_2}},$$

where C_1, C_2, and C_3 are arbitrary constants, and the functions $y = y(\xi)$ and $z = z(\xi)$ are described by the ODE system

$$ay''_{\xi\xi} + \tfrac{1}{2}C_1 \xi y'_\xi + \frac{C_1}{n-1} y + y^n f\!\left(\tfrac{y}{z}\right) = 0, \quad bz''_{\xi\xi} + \tfrac{1}{2}C_1 \xi z'_\xi + \frac{C_1}{n-1} z + z^n g\!\left(\tfrac{y}{z}\right) = 0.$$

2°. Solution with $b = a$:
$$u(x,t) = k\theta(x,t), \quad w(x,t) = \theta(x,t),$$
where k is a root of the algebraic (transcendental) equation
$$k^{n-1} f(k) = g(k),$$
and the function $\theta = \theta(x,t)$ satisfies the heat equation with a power-law nonlinearity
$$\theta_t = a\theta_{xx} + g(k)\theta^n.$$

16. $u_t = au_{xx} + uf\left(\dfrac{u}{w}\right)\ln u + ug\left(\dfrac{u}{w}\right),$
$w_t = aw_{xx} + wf\left(\dfrac{u}{w}\right)\ln w + wh\left(\dfrac{u}{w}\right).$

Solution (Polyanin & Manzhirov, 2007):
$$u(x,t) = \varphi(t)\psi(t)\theta(x,t), \quad w(x,t) = \psi(t)\theta(x,t),$$
where the functions $\varphi = \varphi(t)$ and $\psi = \psi(t)$ are determined by solving the first-order autonomous ordinary differential equations
$$\varphi'_t = \varphi[g(\varphi) - h(\varphi) + f(\varphi)\ln\varphi], \tag{1}$$
$$\psi'_t = \psi[h(\varphi) + f(\varphi)\ln\psi], \tag{2}$$
and the function $\theta = \theta(x,t)$ is determined by the PDE:
$$\theta_t = a\theta_{xx} + f(\varphi)\theta\ln\theta. \tag{3}$$

The separable ODE (1) can be solved to obtain a solution in implicit form. Equation (2) is easy to integrate — with the change of variable $\psi = e^\zeta$, it is reduced to a linear ODE. PDE (3) admits exact solutions of the form
$$\theta = \exp\bigl[\sigma_2(t)x^2 + \sigma_1(t)x + \sigma_0(t)\bigr],$$
where the functions $\sigma_n(t)$ are described by the ODE system
$$\sigma'_2 = f(\varphi)\sigma_2 + 4a\sigma_2^2,$$
$$\sigma'_1 = f(\varphi)\sigma_1 + 4a\sigma_1\sigma_2,$$
$$\sigma'_0 = f(\varphi)\sigma_0 + a\sigma_1^2 + 2a\sigma_2.$$
This system can be integrated directly, since the first equation is a Bernoulli equation and the second and third ones are linear in the unknown. Note that the first equation has a particular solution $\sigma_2 = 0$.

Remark 7.5. Equation (1) has a special solution $\varphi = k = \text{const}$, where k is a root of the algebraic (transcendental) equation $g(k) - h(k) + f(k)\ln k = 0$.

17. $u_t = au_{xx} + uf\left(\dfrac{w}{u}\right) - wg\left(\dfrac{w}{u}\right) + \dfrac{u}{\sqrt{u^2+w^2}} h\left(\dfrac{w}{u}\right),$
$w_t = aw_{xx} + wf\left(\dfrac{w}{u}\right) + ug\left(\dfrac{w}{u}\right) + \dfrac{w}{\sqrt{u^2+w^2}} h\left(\dfrac{w}{u}\right).$

Solution (Polyanin & Manzhirov, 2007):
$$u = r(x,t)\cos\varphi(t), \quad w = r(x,t)\sin\varphi(t),$$

where the function $\varphi = \varphi(t)$ is determined from the separable first-order ODE

$$\varphi'_t = g(\tan \varphi),$$

and the function $r = r(x,t)$ satisfies the linear PDE:

$$r_t = a r_{xx} + r f(\tan \varphi) + h(\tan \varphi).$$

The change of variable

$$r = F(t)\left[Z(x,t) + \int \frac{h(\tan \varphi)\, dt}{F(t)}\right], \quad F(t) = \exp\left[\int f(\tan \varphi)\, dt\right],$$

brings the resulting PDE to the linear heat equation $Z_t = a Z_{xx}$.

18. $\quad u_t = a u_{xx} + u f\left(\dfrac{w}{u}\right) + w g\left(\dfrac{w}{u}\right) + \dfrac{u}{\sqrt{u^2 - w^2}} h\left(\dfrac{w}{u}\right),$

$\quad\; w_t = a w_{xx} + w f\left(\dfrac{w}{u}\right) + u g\left(\dfrac{w}{u}\right) + \dfrac{w}{\sqrt{u^2 - w^2}} h\left(\dfrac{w}{u}\right).$

Solution (Polyanin & Manzhirov, 2007):

$$u = r(x,t)\cosh \varphi(t), \quad w = r(x,t)\sinh \varphi(t),$$

where the function $\varphi = \varphi(t)$ is determined from the separable first-order ODE

$$\varphi'_t = g(\tanh \varphi),$$

and the function $r = r(x,t)$ satisfies the linear PDE:

$$r_t = a r_{xx} + r f(\tanh \varphi) + h(\tanh \varphi).$$

The change of variable

$$r = F(t)\left[Z(x,t) + \int \frac{h(\tanh \varphi)\, dt}{F(t)}\right], \quad F(t) = \exp\left[\int f(\tanh \varphi)\, dt\right],$$

brings the resulting PDE to the linear heat equation $Z_t = a Z_{xx}$.

▶ **Arbitrary functions depend on the product of powers of the unknowns.**

19. $\quad u_t = a u_{xx} + u f(u^n w^m), \quad w_t = b w_{xx} + w g(u^n w^m).$

Solution (Polyanin & Manzhirov, 2007):

$$u = e^{m(kx - \lambda t)} y(\xi), \quad w = e^{-n(kx - \lambda t)} z(\xi), \quad \xi = \beta x - \gamma t,$$

where k, λ, β, and γ are arbitrary constants, and the functions $y = y(\xi)$ and $z = z(\xi)$ are described by the autonomous ODE system

$$a\beta^2 y''_{\xi\xi} + (2akm\beta + \gamma) y'_\xi + m(ak^2 m + \lambda) y + y f(y^n z^m) = 0,$$
$$b\beta^2 z''_{\xi\xi} + (-2bkn\beta + \gamma) z'_\xi + n(bk^2 n - \lambda) z + z g(y^n z^m) = 0.$$

To the special case $k = \lambda = 0$ there corresponds a traveling wave solution.

20. $u_t = au_{xx} + u^{1+kn}f(u^n w^m)$, $w_t = bw_{xx} + w^{1-km}g(u^n w^m)$.

Self-similar solution (Polyanin & Manzhirov, 2007):

$$u = (C_1 t + C_2)^{-\frac{1}{kn}} y(\xi), \quad w = (C_1 t + C_2)^{\frac{1}{km}} z(\xi), \quad \xi = \frac{x + C_3}{\sqrt{C_1 t + C_2}},$$

where C_1, C_2, and C_3 are arbitrary constants, and the functions $y = y(\xi)$ and $z = z(\xi)$ are described by the ODE system

$$ay''_{\xi\xi} + \frac{1}{2}C_1\xi y'_\xi + \frac{C_1}{kn}y + y^{1+kn}f(y^n z^m) = 0,$$

$$bz''_{\xi\xi} + \frac{1}{2}C_1\xi z'_\xi - \frac{C_1}{km}z + z^{1-km}g(y^n z^m) = 0.$$

21. $u_t = au_{xx} + cu \ln u + uf(u^n w^m)$, $w_t = bw_{xx} + cw \ln w + wg(u^n w^m)$.

Solution (Polyanin & Manzhirov, 2007):

$$u = \exp(Ame^{ct})y(\xi), \quad w = \exp(-Ane^{ct})z(\xi), \quad \xi = kx - \lambda t,$$

where A, k, and λ are arbitrary constants, and the functions $y = y(\xi)$ and $z = z(\xi)$ are described by the autonomous ODE system

$$ak^2 y''_{\xi\xi} + \lambda y'_\xi + cy \ln y + yf(y^n z^m) = 0,$$
$$bk^2 z''_{\xi\xi} + \lambda z'_\xi + cz \ln z + zg(y^n z^m) = 0.$$

To the special case $A = 0$ there corresponds a traveling wave solution. For $\lambda = 0$, we have a solution in the form of the product of two functions dependent on time t and the coordinate x.

▶ **Arbitrary functions depend on the sum or difference of squares of the unknowns.**

22. $u_t = au_{xx} + uf(u^2 + w^2) - wg(u^2 + w^2)$,
    $w_t = aw_{xx} + ug(u^2 + w^2) + wf(u^2 + w^2)$.

1°. Periodic solution in the spatial coordinate (Polyanin & Manzhirov, 2007):

$$u = \psi(t)\cos\varphi(x,t), \quad w = \psi(t)\sin\varphi(x,t), \quad \varphi(x,t) = C_1 x + \int g(\psi^2)\,dt + C_2,$$

where C_1 and C_2 are arbitrary constants, and the function $\psi = \psi(t)$ is described by the separable first-order ordinary differential equation

$$\psi'_t = \psi f(\psi^2) - aC_1^2 \psi.$$

Its general solution can be represented in implicit form as

$$\int \frac{d\psi}{\psi f(\psi^2) - aC_1^2 \psi} = t + C_3.$$

2°. Periodic solution in time (Polyanin & Manzhirov, 2007):

$$u = r(x)\cos[\theta(x) + C_1 t + C_2], \quad w = r(x)\sin[\theta(x) + C_1 t + C_2],$$

where C_1 and C_2 are arbitrary constants, and the functions $r = r(x)$ and $\theta = \theta(x)$ are described by the autonomous ODE system

$$ar''_{xx} - ar(\theta'_x)^2 + rf(r^2) = 0,$$
$$ar\theta''_{xx} + 2ar'_x\theta'_x - C_1 r + rg(r^2) = 0.$$

3°. Solution (generalizes the solution of Item 2°):

$$u = r(z)\cos[\theta(z) + C_1 t + C_2], \quad w = r(z)\sin[\theta(z) + C_1 t + C_2], \quad z = x + \lambda t,$$

where C_1, C_2, and λ are arbitrary constants, and the functions $r = r(z)$ and $\theta = \theta(z)$ are described by the ODE system

$$ar''_{zz} - ar(\theta'_z)^2 - \lambda r'_z + rf(r^2) = 0,$$
$$ar\theta''_{zz} + 2ar'_z\theta'_z - \lambda r\theta'_z - C_1 r + rg(r^2) = 0.$$

23. $u_t = au_{xx} + uf(u^2 - w^2) + wg(u^2 - w^2),$
    $w_t = aw_{xx} + ug(u^2 - w^2) + wf(u^2 - w^2).$

1°. Solution (Polyanin & Manzhirov, 2007):

$$u = \psi(t)\cosh\varphi(x,t), \quad w = \psi(t)\sinh\varphi(x,t), \quad \varphi(x,t) = C_1 x + \int g(\psi^2)\,dt + C_2,$$

where C_1 and C_2 are arbitrary constants, and the function $\psi = \psi(t)$ is described by the separable first-order ODE:

$$\psi'_t = \psi f(\psi^2) + aC_1^2 \psi,$$

whose general solution can be represented in implicit form as

$$\int \frac{d\psi}{\psi f(\psi^2) + aC_1^2 \psi} = t + C_3.$$

2°. Solution (Polyanin & Manzhirov, 2007):

$$u = r(x)\cosh[\theta(x) + C_1 t + C_2], \quad w = r(x)\sinh[\theta(x) + C_1 t + C_2],$$

where C_1 and C_2 are arbitrary constants, and the functions $r = r(x)$ and $\theta = \theta(x)$ are described by the autonomous ODE system

$$ar''_{xx} + ar(\theta'_x)^2 + rf(r^2) = 0,$$
$$ar\theta''_{xx} + 2ar'_x\theta'_x + rg(r^2) - C_1 r = 0.$$

3°. Solution (generalizes the solution of Item 2°):

$$u = r(z)\cosh[\theta(z) + C_1 t + C_2], \quad w = r(z)\sinh[\theta(z) + C_1 t + C_2], \quad z = x + \lambda t,$$

where C_1, C_2, and λ are arbitrary constants, and the functions $r = r(z)$ and $\theta = \theta(z)$ are described by the autonomous ODE system

$$ar''_{zz} + ar(\theta'_z)^2 - \lambda r'_z + rf(r^2) = 0,$$
$$ar\theta''_{zz} + 2ar'_z\theta'_z - \lambda r\theta'_z - C_1 r + rg(r^2) = 0.$$

▶ **Arbitrary functions depend on the unknowns in a complex way.**

24. $u_t = au_{xx} + uf(u^2+w^2) - wg\left(\frac{w}{u}\right)$, $\quad w_t = aw_{xx} + ug\left(\frac{w}{u}\right) + wf(u^2+w^2)$.

Solution (Polyanin & Manzhirov, 2007):
$$u = r(x,t)\cos\varphi(t), \quad w = r(x,t)\sin\varphi(t),$$

where the function $\varphi = \varphi(t)$ is described by the autonomous ODE
$$\varphi'_t = g(\tan\varphi), \tag{1}$$

and the function $r = r(x,t)$ satisfies the PDE
$$r_t = ar_{xx} + rf(r^2). \tag{2}$$

The general solution of equation (1) is expressed in implicit form as
$$\int \frac{d\varphi}{g(\tan\varphi)} = t + C.$$

PDE (2) has a time-independent exact solution $r = r(x)$. In addition, this equation also admits a more complex traveling wave solution $r = r(z)$, where $z = kx - \lambda t$ with arbitrary constants k and λ, and the function $r(z)$ is described by the autonomous ODE
$$ak^2 r''_{zz} + \lambda r'_z + rf(r^2) = 0.$$

25. $u_t = au_{xx} + uf(u^2 - w^2) + wg\left(\frac{w}{u}\right)$,
$\quad w_t = aw_{xx} + ug\left(\frac{w}{u}\right) + wf(u^2 - w^2)$.

Solution (Polyanin & Manzhirov, 2007):
$$u = r(x,t)\cosh\varphi(t), \quad w = r(x,t)\sinh\varphi(t),$$

where the function $\varphi = \varphi(t)$ is described by the autonomous ODE
$$\varphi'_t = g(\tanh\varphi), \tag{1}$$

and the function $r = r(x,t)$ satisfies the PDE
$$r_t = ar_{xx} + rf(r^2). \tag{2}$$

The general solution of equation (1) is expressed in implicit form as
$$\int \frac{d\varphi}{g(\tanh\varphi)} = t + C.$$

Equation (2) admits an exact, traveling wave solution $r = r(z)$, where $z = kx - \lambda t$ with arbitrary constants k and λ, and the function $r(z)$ is described by the autonomous ODE
$$ak^2 r''_{zz} + \lambda r'_z + rf(r^2) = 0.$$

For other exact solutions to equation (2) for various functions f, see Polyanin and Zaitsev (2012).

26. $u_t = au_{xx} + uf(u^2+w^2) - wg(u^2+w^2) - w\arctan\left(\dfrac{w}{u}\right)h(u^2+w^2)$,

$w_t = aw_{xx} + wf(u^2+w^2) + ug(u^2+w^2) + u\arctan\left(\dfrac{w}{u}\right)h(u^2+w^2)$.

Functional separable solution (Polyanin & Manzhirov, 2007):
$$u = r(t)\cos[\varphi(t)x + \psi(t)], \quad w = r(t)\sin[\varphi(t)x + \psi(t)],$$
where the functions $r = r(t)$, $\varphi = \varphi(t)$, and $\psi = \psi(t)$ are described by the autonomous ODE system
$$r'_t = -ar\varphi^2 + rf(r^2), \quad \varphi'_t = h(r^2)\varphi, \quad \psi'_t = h(r^2)\psi + g(r^2).$$
Note that for a fixed t, this solution defines a structure periodic in x.

27. $u_t = au_{xx} + uf(u^2-w^2) + wg(u^2-w^2) + w\operatorname{artanh}\left(\dfrac{w}{u}\right)h(u^2-w^2)$,

$w_t = aw_{xx} + wf(u^2-w^2) + ug(u^2-w^2) + u\operatorname{artanh}\left(\dfrac{w}{u}\right)h(u^2-w^2)$.

Functional separable solution (Polyanin & Manzhirov, 2007):
$$u = r(t)\cosh[\varphi(t)x + \psi(t)], \quad w = r(t)\sinh[\varphi(t)x + \psi(t)],$$
where the functions $r = r(t)$, $\varphi = \varphi(t)$, and $\psi = \psi(t)$ are described by the autonomous ODE system
$$r'_t = ar\varphi^2 + rf(r^2), \quad \varphi'_t = h(r^2)\varphi, \quad \psi'_t = h(r^2)\psi + g(r^2).$$

28. $u_t = au_{xx} + u^{k+1}f(\varphi), \quad \varphi = u\exp\left(-\dfrac{w}{u}\right)$,

$w_t = aw_{xx} + u^{k+1}[f(\varphi)\ln u + g(\varphi)]$.

Solution (Barannyk, 2002):
$$u = (C_1 t + C_2)^{-\frac{1}{k}} y(\xi), \quad \xi = \frac{x + C_3}{\sqrt{C_1 t + C_2}},$$
$$w = (C_1 t + C_2)^{-\frac{1}{k}}\left[z(\xi) - \frac{1}{k}\ln(C_1 t + C_2) y(\xi)\right],$$
where C_1, C_2, and C_3 are arbitrary constants, and the functions $y = y(\xi)$ and $z = z(\xi)$ are described by the ODE system
$$ay''_{\xi\xi} + \frac{1}{2}C_1\xi y'_\xi + \frac{C_1}{k} y + y^{k+1} f(\varphi) = 0, \quad \varphi = y\exp\left(-\frac{z}{y}\right),$$
$$az''_{\xi\xi} + \frac{1}{2}C_1\xi z'_\xi + \frac{C_1}{k} z + \frac{C_1}{k} y + y^{k+1}[f(\varphi)\ln y + g(\varphi)] = 0.$$

Remark 7.6. *Exact solutions of many reaction-diffusion PDE systems of this type containing free parameters can be found in the book by Cherniha & Davydovych (2017); see also Polyanin & Zaitsev (2012).*

7.2.3. Nonlinear Parabolic Systems of the Form
$$u_t = ax^{-n}(x^n u_x)_x + F(u,w), \quad w_t = bx^{-n}(x^n w_x)_x + G(u,w)$$

▶ **Preliminary remarks.** These are nonlinear systems of reaction-diffusion PDEs in the radial symmetric case ($n = 1$ corresponds to a plane problem and $n = 2$ to a spatial one).

▶ **Arbitrary functions depend on a linear combination of the unknowns.**

1. $u_t = ax^{-n}(x^n u_x)_x + uf(bu - cw) + g(bu - cw),$
   $w_t = ax^{-n}(x^n w_x)_x + wf(bu - cw) + h(bu - cw).$

   1°. Solution (Polyanin & Manzhirov, 2007):

   $$u = \varphi(t) + c \exp\left[\int f(b\varphi - c\psi)\, dt\right] \theta(x,t), \quad w = \psi(t) + b \exp\left[\int f(b\varphi - c\psi)\, dt\right] \theta(x,t),$$

   where $\varphi = \varphi(t)$ and $\psi = \psi(t)$ are described by the autonomous ODE system

   $$\varphi'_t = \varphi f(b\varphi - c\psi) + g(b\varphi - c\psi),$$
   $$\psi'_t = \psi f(b\varphi - c\psi) + h(b\varphi - c\psi),$$

   and the function $\theta = \theta(x,t)$ satisfies the linear heat equation

   $$\theta_t = ax^{-n}(x^n \theta_x)_x. \tag{1}$$

   2°. Let us add together the first equation multiplied by b and the second one multiplied by $-c$ to obtain

   $$\zeta_t = ax^{-n}(x^n \zeta_x)_x + \zeta f(\zeta) + bg(\zeta) - ch(\zeta), \quad \zeta = bu - cw. \tag{2}$$

   This equation will be considered in conjunction with the first equation of the original system

   $$u_t = ax^{-n}(x^n u_x)_x + uf(\zeta) + g(\zeta). \tag{3}$$

   Equation (2) can be treated separately. Given a solution $\zeta = \zeta(x,t)$ to equation (2), the function $u = u(x,t)$ can be determined by solving the linear equation (3) and the function $w = w(x,t)$ is found as $w = (bu - \zeta)/c$.

   Note two important solutions to equation (2):

   (i) In the general case, equation (2) admits steady-state solutions $\zeta = \zeta(x)$. The corresponding exact solutions to equation (3) are expressed as $u = u_0(x) + \sum e^{\beta_n t} u_n(x)$.

   (ii) If the condition $\zeta f(\zeta) + bg(\zeta) - ch(\zeta) = k_1 \zeta + k_0$ holds, equation (2) is linear,

   $$\zeta_t = ax^{-n}(x^n \zeta_x)_x + k_1 \zeta + k_0,$$

   and is hence reducible to the linear heat equation (1) with the substitution $\zeta = e^{k_1 t} \bar{\zeta} - k_0 k_1^{-1}$.

2. $u_t = ax^{-n}(x^n u_x)_x + e^{\lambda u} f(\lambda u - \sigma w),$
   $w_t = bx^{-n}(x^n w_x)_x + e^{\sigma w} g(\lambda u - \sigma w).$

   Solution (Polyanin & Manzhirov, 2007):

   $$u = y(\xi) - \frac{1}{\lambda} \ln(C_1 t + C_2), \quad w = z(\xi) - \frac{1}{\sigma} \ln(C_1 t + C_2), \quad \xi = \frac{x}{\sqrt{C_1 t + C_2}},$$

   where C_1 and C_2 are arbitrary constants, and the functions $y = y(\xi)$ and $z = z(\xi)$ are described by the ODE system

   $$a\xi^{-n}(\xi^n y'_\xi)'_\xi + \frac{1}{2} C_1 \xi y'_\xi + \frac{C_1}{\lambda} + e^{\lambda y} f(\lambda y - \sigma z) = 0,$$
   $$b\xi^{-n}(\xi^n z'_\xi)'_\xi + \frac{1}{2} C_1 \xi z'_\xi + \frac{C_1}{\sigma} + e^{\sigma z} g(\lambda y - \sigma z) = 0.$$

▶ **Arbitrary functions depend on the ratio of the unknowns.**

3. $u_t = ax^{-n}(x^n u_x)_x + uf\left(\dfrac{u}{w}\right)$, $\quad w_t = bx^{-n}(x^n w_x)_x + wg\left(\dfrac{u}{w}\right)$.

1°. Multiplicative separable solution:
$$u = x^{\frac{1-n}{2}}[C_1 J_\nu(kx) + C_2 Y_\nu(kx)]\varphi(t), \quad \nu = \tfrac{1}{2}|n-1|,$$
$$w = x^{\frac{1-n}{2}}[C_1 J_\nu(kx) + C_2 Y_\nu(kx)]\psi(t),$$

where C_1, C_2, and k are arbitrary constants, $J_\nu(z)$ and $Y_\nu(z)$ are Bessel functions, and the functions $\varphi = \varphi(t)$ and $\psi = \psi(t)$ are described by the autonomous ODE system
$$\varphi'_t = -ak^2\varphi + \varphi f(\varphi/\psi), \quad \psi'_t = -bk^2\psi + \psi g(\varphi/\psi).$$

2°. Multiplicative separable solution:
$$u = x^{\frac{1-n}{2}}[C_1 I_\nu(kx) + C_2 K_\nu(kx)]\varphi(t), \quad \nu = \tfrac{1}{2}|n-1|,$$
$$w = x^{\frac{1-n}{2}}[C_1 I_\nu(kx) + C_2 K_\nu(kx)]\psi(t),$$

where C_1, C_2, and k are arbitrary constants, $I_\nu(z)$ and $K_\nu(z)$ are modified Bessel functions, and the functions $\varphi = \varphi(t)$ and $\psi = \psi(t)$ are described by the autonomous ODE system
$$\varphi'_t = ak^2\varphi + \varphi f(\varphi/\psi), \quad \psi'_t = bk^2\psi + \psi g(\varphi/\psi).$$

3°. Multiplicative separable solution:
$$u = e^{-\lambda t}y(x), \quad w = e^{-\lambda t}z(x),$$

where λ is an arbitrary constant, and the functions $y = y(x)$ and $z = z(x)$ are described by the ODE system
$$ax^{-n}(x^n y'_x)'_x + \lambda y + yf(y/z) = 0, \quad bx^{-n}(x^n z'_x)'_x + \lambda z + zg(y/z) = 0.$$

4°. Consider the special case $b = a$. Let k be a root of the algebraic (transcendental) equation
$$f(k) = g(k).$$

Solution:
$$u = ke^{\lambda t}\theta, \quad w = e^{\lambda t}\theta, \quad \lambda = f(k),$$

where the function $\theta = \theta(x, t)$ satisfies the linear heat equation
$$\theta_t = ax^{-n}(x^n \theta_x)_x. \tag{1}$$

5°. Solution for the special case $b = a$:
$$u = \varphi(t)\exp\left[\int g(\varphi(t))\, dt\right]\theta(x,t), \quad w = \exp\left[\int g(\varphi(t))\, dt\right]\theta(x,t),$$

where the function $\varphi = \varphi(t)$ is described by the separable first-order ODE:
$$\varphi'_t = [f(\varphi) - g(\varphi)]\varphi, \tag{2}$$

and the function $\theta = \theta(x, t)$ satisfies the linear heat equation (1).

To the particular solution $\varphi = k = \text{const}$ of equation (2), there corresponds the solution presented in Item 4°. The general solution of equation (2) is written out in implicit form as

$$\int \frac{d\varphi}{[f(\varphi) - g(\varphi)]\varphi} = t + C.$$

6°. References for the system of PDEs 7.2.3.3: Polyanin & Vyazmina (2004) and Polyanin & Manzhirov (2007).

4. $u_t = ax^{-n}(x^n u_x)_x + uf\left(\dfrac{u}{w}\right) + \dfrac{u}{w}h\left(\dfrac{u}{w}\right),$

   $w_t = ax^{-n}(x^n w_x)_x + wg\left(\dfrac{u}{w}\right) + h\left(\dfrac{u}{w}\right).$

Solution (Polyanin & Manzhirov, 2007):

$$u = \varphi(t)G(t)\left[\theta(x,t) + \int \frac{h(\varphi)}{G(t)}\,dt\right], \quad G(t) = \exp\left[\int g(\varphi)\,dt\right],$$

$$w = G(t)\left[\theta(x,t) + \int \frac{h(\varphi)}{G(t)}\,dt\right],$$

where the function $\varphi = \varphi(t)$ is described by the separable first-order ODE:

$$\varphi'_t = [f(\varphi) - g(\varphi)]\varphi,$$

and the function $\theta = \theta(x,t)$ satisfies the linear heat equation

$$\theta_t = ax^{-n}(x^n \theta_x)_x.$$

5. $u_t = ax^{-n}(x^n u_x)_x + uf_1\left(\dfrac{w}{u}\right) + wg_1\left(\dfrac{w}{u}\right),$

   $w_t = ax^{-n}(x^n w_x)_x + uf_2\left(\dfrac{w}{u}\right) + wg_2\left(\dfrac{w}{u}\right).$

Solution (Polyanin & Manzhirov, 2007):

$$u = \exp\left\{\int [f_1(\varphi) + \varphi g_1(\varphi)]\,dt\right\}\theta(x,t), \quad w = \varphi(t)\exp\left\{\int [f_1(\varphi) + \varphi g_1(\varphi)]\,dt\right\}\theta(x,t),$$

where the function $\varphi = \varphi(t)$ is described by the separable first-order ODE:

$$\varphi'_t = f_2(\varphi) + \varphi g_2(\varphi) - \varphi[f_1(\varphi) + \varphi g_1(\varphi)],$$

and the function $\theta = \theta(x,t)$ satisfies the linear heat equation

$$\theta_t = ax^{-n}(x^n \theta_x)_x.$$

6. $u_t = ax^{-n}(x^n u_x)_x + u^k f\left(\dfrac{u}{w}\right), \quad w_t = bx^{-n}(x^n w_x)_x + w^k g\left(\dfrac{u}{w}\right).$

Self-similar solution (Polyanin & Manzhirov, 2007):

$$u = (C_1 t + C_2)^{\frac{1}{1-k}} y(\xi), \quad w = (C_1 t + C_2)^{\frac{1}{1-k}} z(\xi), \quad \xi = \frac{x}{\sqrt{C_1 t + C_2}},$$

where C_1 and C_2 are arbitrary constants, and the functions $y = y(\xi)$ and $z = z(\xi)$ are described by the ODE system

$$a\xi^{-n}(\xi^n y'_\xi)'_\xi + \frac{1}{2}C_1 \xi y'_\xi + \frac{C_1}{k-1}y + y^k f(y/z) = 0,$$

$$b\xi^{-n}(\xi^n z'_\xi)'_\xi + \frac{1}{2}C_1 \xi z'_\xi + \frac{C_1}{k-1}z + z^k g(y/z) = 0.$$

7. $u_t = ax^{-n}(x^n u_x)_x + uf\left(\dfrac{u}{w}\right)\ln u + ug\left(\dfrac{u}{w}\right),$
   $w_t = ax^{-n}(x^n w_x)_x + wf\left(\dfrac{u}{w}\right)\ln w + wh\left(\dfrac{u}{w}\right).$

Solution (Polyanin & Manzhirov, 2007):

$$u = \varphi(t)\psi(t)\theta(x,t), \quad w = \psi(t)\theta(x,t),$$

where the functions $\varphi = \varphi(t)$ and $\psi = \psi(t)$ are determined by solving the autonomous ODE:

$$\begin{aligned}\varphi'_t &= \varphi[g(\varphi) - h(\varphi) + f(\varphi)\ln\varphi],\\ \psi'_t &= \psi[h(\varphi) + f(\varphi)\ln\psi],\end{aligned} \quad (1)$$

and the function $\theta = \theta(x,t)$ is determined by the PDE:

$$\theta_t = ax^{-n}(x^n \theta_x)_x + f(\varphi)\theta\ln\theta. \quad (2)$$

The first equation in (1) is a separable equation; its solution can be written out in implicit form. The second equation in (1) can be solved using the change of variable $\psi = e^\zeta$ (it reduces to a linear equation for ζ).

Equation (2) admits exact solutions of the form

$$\theta = \exp[\sigma_1(t)x^2 + \sigma_2(t)],$$

where the functions $\sigma_n(t)$ are described by the ODE system

$$\begin{aligned}\sigma'_1 &= f(\varphi)\sigma_1 + 4a\sigma_1^2,\\ \sigma'_2 &= f(\varphi)\sigma_2 + 2a(n+1)\sigma_1.\end{aligned}$$

This system can be successively integrated, since the first ODE is a Bernoulli equation and the second one is linear in the unknown.

If $f = \text{const}$, equation (2) also has a traveling wave solution $\theta = \theta(kx - \lambda t)$.

▶ **Arbitrary functions depend on the product of powers of the unknowns.**

8. $u_t = ax^{-n}(x^n u_x)_x + uf(x, u^k w^m),$
   $w_t = bx^{-n}(x^n w_x)_x + wg(x, u^k w^m).$

Multiplicative separable solution (Polyanin & Manzhirov, 2007):

$$u = e^{-m\lambda t}y(x), \quad w = e^{k\lambda t}z(x),$$

where λ is an arbitrary constant, and the functions $y = y(x)$ and $z = z(x)$ are described by the ODE system

$$\begin{aligned}ax^{-n}(x^n y'_x)'_x + m\lambda y + yf(x, y^k z^m) &= 0,\\ bx^{-n}(x^n z'_x)'_x - k\lambda z + zg(x, y^k z^m) &= 0.\end{aligned}$$

9. $u_t = ax^{-n}(x^n u_x)_x + u^{1+kn} f(u^n w^m),$
   $w_t = bx^{-n}(x^n w_x)_x + w^{1-km} g(u^n w^m).$

Self-similar solution (Polyanin & Manzhirov, 2007):
$$u = (C_1 t + C_2)^{-\frac{1}{kn}} y(\xi), \quad w = (C_1 t + C_2)^{\frac{1}{km}} z(\xi), \quad \xi = \frac{x}{\sqrt{C_1 t + C_2}},$$

where C_1 and C_2 are arbitrary constants, and the functions $y = y(\xi)$ and $z = z(\xi)$ are described by the ODE system
$$a\xi^{-n}(\xi^n y'_\xi)'_\xi + \frac{1}{2}C_1 \xi y'_\xi + \frac{C_1}{kn} y + y^{1+kn} f(y^n z^m) = 0,$$
$$b\xi^{-n}(\xi^n z'_\xi)'_\xi + \frac{1}{2}C_1 \xi z'_\xi - \frac{C_1}{km} z + z^{1-km} g(y^n z^m) = 0.$$

10. $u_t = ax^{-n}(x^n u_x)_x + cu \ln u + u f(x, u^k w^m),$
    $w_t = bx^{-n}(x^n w_x)_x + cw \ln w + w g(x, u^k w^m).$

Multiplicative separable solution (Polyanin & Manzhirov, 2007):
$$u = \exp(Ame^{ct}) y(x), \quad w = \exp(-Ake^{ct}) z(x),$$

where A is an arbitrary constant, and the functions $y = y(x)$ and $z = z(x)$ are described by the ODE system
$$ax^{-n}(x^n y'_x)'_x + cy \ln y + y f(x, y^k z^m) = 0,$$
$$bx^{-n}(x^n z'_x)'_x + cz \ln z + z g(x, y^k z^m) = 0.$$

▶ **Arbitrary functions depend on the sum or difference of squares of the unknowns.**

11. $u_t = ax^{-n}(x^n u_x)_x + u f(u^2 + w^2) - w g(u^2 + w^2),$
    $w_t = ax^{-n}(x^n w_x)_x + w f(u^2 + w^2) + u g(u^2 + w^2).$

Time-periodic solution (Polyanin & Manzhirov, 2007):
$$u = r(x) \cos[\theta(x) + C_1 t + C_2], \quad w = r(x) \sin[\theta(x) + C_1 t + C_2],$$

where C_1 and C_2 are arbitrary constants, and the functions $r = r(x)$ and $\theta = \theta(x)$ are described by the ODE system
$$ar''_{xx} - ar(\theta'_x)^2 + \frac{an}{x} r'_x + r f(r^2) = 0,$$
$$ar\theta''_{xx} + 2ar'_x \theta'_x + \frac{an}{x} r\theta'_x + r g(r^2) - C_1 r = 0.$$

12. $u_t = ax^{-n}(x^n u_x)_x + u f(u^2 - w^2) + w g(u^2 - w^2),$
    $w_t = ax^{-n}(x^n w_x)_x + w f(u^2 - w^2) + u g(u^2 - w^2).$

Solution (Polyanin & Manzhirov, 2007):
$$u = r(x) \cosh[\theta(x) + C_1 t + C_2], \quad w = r(x) \sinh[\theta(x) + C_1 t + C_2],$$

where C_1 and C_2 are arbitrary constants, and the functions $r = r(x)$ and $\theta = \theta(x)$ are described by the ODE system
$$ar''_{xx} + ar(\theta'_x)^2 + \frac{an}{x} r'_x + r f(r^2) = 0,$$
$$ar\theta''_{xx} + 2ar'_x \theta'_x + \frac{an}{x} r\theta'_x + r g(r^2) - C_1 r = 0.$$

▶ **Arbitrary functions have different arguments.**

13. $u_t = ax^{-n}(x^n u_x)_x + uf(u^2 + w^2) - wg\left(\dfrac{w}{u}\right),$
    $w_t = ax^{-n}(x^n w_x)_x + wf(u^2 + w^2) + ug\left(\dfrac{w}{u}\right).$

Solution (Polyanin & Manzhirov, 2007):
$$u = r(x,t)\cos\varphi(t), \quad w = r(x,t)\sin\varphi(t),$$
where the function $\varphi = \varphi(t)$ is described by the autonomous ODE:
$$\varphi'_t = g(\tan\varphi), \tag{1}$$
and the function $r = r(x,t)$ is determined by the PDE:
$$r_t = ax^{-n}(x^n r_x)_x + rf(r^2). \tag{2}$$
The general solution of equation (1) is expressed in implicit form as
$$\int \frac{d\varphi}{g(\tan\varphi)} = t + C.$$
The nonlinear PDE (2) admits a time-independent exact solution $r = r(x)$.

14. $u_t = ax^{-n}(x^n u_x)_x + uf(u^2 - w^2) + wg\left(\dfrac{w}{u}\right),$
    $w_t = ax^{-n}(x^n w_x)_x + wf(u^2 - w^2) + ug\left(\dfrac{w}{u}\right).$

Solution (Polyanin & Manzhirov, 2007):
$$u = r(x,t)\cosh\varphi(t), \quad w = r(x,t)\sinh\varphi(t),$$
where the function $\varphi = \varphi(t)$ is described by the autonomous ODE:
$$\varphi'_t = g(\tanh\varphi), \tag{1}$$
and the function $r = r(x,t)$ is determined by the PDE:
$$r_t = ax^{-n}(x^n r_x)_x + rf(r^2). \tag{2}$$
The general solution of equation (1) is expressed in implicit form as
$$\int \frac{d\varphi}{g(\tanh\varphi)} = t + C.$$
The nonlinear PDE (2) admits a time-independent exact solution $r = r(x)$.

7.2.4. Nonlinear Hyperbolic Systems of the Form $u_{tt} = au_{xx} + F(u,w),\ w_{tt} = bw_{xx} + G(u,w)$

▶ **Preliminary remarks.** Such systems are invariant under translations in the independent variables (and under reflection, the change of x to $-x$) and admit traveling wave solutions $u = u(kx - \lambda t),\ w = w(kx - \lambda t)$. These solutions as well as those where one of the unknowns is identically zero are not considered in this section.

▶ **Arbitrary functions depend on a linear combination of the unknowns.**

1. $u_{tt} = au_{xx} + c_2 f(b_1 u + c_1 w) + c_1 g(b_2 u + c_2 w),$
   $w_{tt} = aw_{xx} - b_2 f(b_1 u + c_1 w) - b_1 g(b_2 u + c_2 w).$

The condition $b_1 c_2 - b_2 c_1 \neq 0$ is assumed to hold.

Multiplying the equations by appropriate constants and then adding together, one arrives at two independent equations (Polyanin & Zaitsev, 2012):

$$U_{tt} = aU_{xx} + (b_1c_2 - b_2c_1)f(U), \quad U = b_1u + c_1w;$$
$$W_{tt} = aW_{xx} - (b_1c_2 - b_2c_1)g(W), \quad W = b_2u + c_2w.$$

For solutions of these equations, see equation 6.2.1.8.

2. $u_{tt} = au_{xx} + uf(bu - cw) + g(bu - cw),$
$w_{tt} = aw_{xx} + wf(bu - cw) + h(bu - cw).$

This is a special case of the PDE system 7.2.4.3.

1°. Solution (Polyanin & Zaitsev, 2012):

$$u = \varphi(t) + c\theta(x,t), \quad w = \psi(t) + b\theta(x,t),$$

where $\varphi = \varphi(t)$ and $\psi = \psi(t)$ are described by the autonomous ODE system

$$\varphi''_{tt} = \varphi f(b\varphi - c\psi) + g(b\varphi - c\psi),$$
$$\psi''_{tt} = \psi f(b\varphi - c\psi) + h(b\varphi - c\psi),$$

and the function $\theta = \theta(x,t)$ satisfies the linear equation

$$\theta_{tt} = a\theta_{xx} + f(b\varphi - c\psi)\theta.$$

For $f = \text{const}$, this equation can be solved by separation of variables.

2°. Solution (Polyanin & Zaitsev, 2012):

$$u = \tilde{\varphi}(x) + c\tilde{\theta}(x,t), \quad w = \tilde{\psi}(x) + b\tilde{\theta}(x,t),$$

where the functions $\tilde{\varphi} = \tilde{\varphi}(x)$ and $\tilde{\psi} = \tilde{\psi}(x)$ are described by the ODE system

$$a\tilde{\varphi}''_{xx} + \tilde{\varphi}f(b\tilde{\varphi} - c\tilde{\psi}) + g(b\tilde{\varphi} - c\tilde{\psi}) = 0,$$
$$a\tilde{\psi}''_{xx} + \tilde{\psi}f(b\tilde{\varphi} - c\tilde{\psi}) + h(b\tilde{\varphi} - c\tilde{\psi}) = 0,$$

and the function $\tilde{\theta} = \tilde{\theta}(x,t)$ satisfies the linear equation

$$\tilde{\theta}_{tt} = a\tilde{\theta}_{xx} + f(b\tilde{\varphi} - c\tilde{\psi})\tilde{\theta}.$$

For $f = \text{const}$, this equation can be solved by separation of variables.

3°. Let us add together the first equation multiplied by b and the second one multiplied by $-c$ to obtain

$$\zeta_{tt} = a\zeta_{xx} + \zeta f(\zeta) + bg(\zeta) - ch(\zeta), \quad \zeta = bu - cw. \tag{1}$$

This equation will be considered in conjunction with the first equation of the original system

$$u_{tt} = au_{xx} + uf(\zeta) + g(\zeta). \tag{2}$$

Equation (1) can be treated separately. Given a solution $\zeta = \zeta(x,t)$ to equation (1), the function $u = u(x,t)$ can be determined by solving equation (2) and the function $w = w(x,t)$ is found as $w = (bu - \zeta)/c$.

Note three important solutions to equation (1):

(i) In the general case, equation (1) admits a spatially homogeneous solution $\zeta = \zeta(t)$. The corresponding solution to the original system is given in Item 1° in another form.

(ii) In the general case, equation (1) admits a steady-state solution $\zeta = \zeta(x)$. The corresponding exact solutions to equation (2) are expressed as $u = u_0(x) + \sum e^{-\beta_n t} u_n(x)$ and $u = u_0(x) + \sum \cos(\beta_n t) u_n^{(1)}(x) + \sum \sin(\beta_n t) u_n^{(2)}(x)$.

(iii) If the condition $\zeta f(\zeta) + bg(\zeta) - ch(\zeta) = k_1 \zeta + k_0$ holds, equation (1) is linear,

$$\zeta_{tt} = a\zeta_{xx} + k_1\zeta + k_0,$$

and, hence, can be solved by separation of variables.

For other solutions to equation (1), see equation 6.2.1.8.

3. $u_{tt} = a_1 u_{xx} + k_1 u + u f_1(bu - cw) + g_1(bu - cw),$
   $w_{tt} = a_2 w_{xx} + k_2 w + w f_2(bu - cw) + g_2(bu - cw).$

Here, $f_1(z)$, $f_2(z)$, $g_1(z)$, and $g_2(z)$ are arbitrary functions.

1°. Let

$$u = u^\circ = \text{const}, \quad w = w^\circ = \text{const}$$

be a stationary point (simplest solution) of the PDE system which is determined from a nonlinear algebraic system,

$$u^\circ[k_1 + f_1(bu^\circ - cw^\circ)] + g_1(bu^\circ - cw^\circ) = 0,$$
$$w^\circ[k_2 + f_2(bu^\circ - cw^\circ)] + g_2(bu^\circ - cw^\circ) = 0.$$

This system usually has multiple roots.

We introduce the notation

$$\beta = \frac{a_2 k_1 - a_1 k_2 + a_2 f_1^\circ - a_1 f_2^\circ}{a_2 - a_1}, \quad \mu = \frac{k_1 - k_2 + f_1^\circ - f_2^\circ}{a_2 - a_1},$$
$$f_1^\circ = f_1(bu^\circ - cw^\circ), \quad f_2^\circ = f_2(bu^\circ - cw^\circ).$$

For $\beta > 0$, the original PDE system has an exact solution of the form

$$u = u^\circ + c[C_1 \exp(-\sqrt{\beta}\, t) + C_2 \exp(\sqrt{\beta}\, t)]\theta(x),$$
$$w = w^\circ + b[C_1 \exp(-\sqrt{\beta}\, t) + C_2 \exp(\sqrt{\beta}\, t)]\theta(x),$$

where C_1 and C_2 are arbitrary constants, and the function $\theta(x)$ is defined by the formulas

$$\theta(x) = \begin{cases} C_3 \exp(-\sqrt{\mu}\, x) + C_4 \exp(\sqrt{\mu}\, x) & \text{if } \mu > 0, \\ C_3 \cos(\sqrt{|\mu|}\, x) + C_4 \sin(\sqrt{|\mu|}\, x) & \text{if } \mu < 0, \end{cases} \quad (1)$$

where C_3 and C_4 are arbitrary constants.

For $\beta < 0$, the original PDE system has an exact solution of the form

$$u = u^\circ + c[C_1 \cos(\sqrt{|\beta|}\, t) + C_2 \sin(\sqrt{|\beta|}\, t)]\theta(x),$$
$$w = w^\circ + b[C_1 \cos(\sqrt{|\beta|}\, t) + C_2 \sin(\sqrt{|\beta|}\, t)]\theta(x),$$

where C_1 and C_2 are arbitrary constants, and the function $\theta(x)$ is defined by formulas (1).

2°. Let us consider the special case
$$f_1(z) = a_1 f(z), \quad f_2(z) = a_2 f(z),$$
where $f(z)$ is an arbitrary function, and the functions $\varphi = \varphi(x)$ and $\psi = \psi(x)$ are a stationary solution of the original PDE system, i.e., they satisfy the system of second-order ODEs
$$\begin{aligned} a_1 \varphi''_{xx} + k_1 \varphi + a_1 \varphi f(b\varphi - c\psi) + g_1(b\varphi - c\psi) &= 0, \\ a_2 \psi''_{xx} + k_2 \psi + a_2 \psi f(b\varphi - c\psi) + g_2(b\varphi - c\psi) &= 0. \end{aligned} \quad (2)$$

We introduce the notation
$$\beta = \frac{a_2 k_1 - a_1 k_2}{a_2 - a_1}.$$

For $\beta > 0$, the original PDE system has an exact solution of the form
$$\begin{aligned} u &= \varphi(x) + c[C_1 \exp(-\sqrt{\beta}\, t) + C_2 \exp(\sqrt{\beta}\, t)]\theta(x), \\ w &= \psi(x) + b[C_1 \exp(-\sqrt{\beta}\, t) + C_2 \exp(\sqrt{\beta}\, t)]\theta(x), \end{aligned}$$
where C_1 and C_2 are arbitrary constants, the functions $\varphi = \varphi(x)$ and $\psi = \psi(x)$ satisfy the system of ODEs (2), and the function $\theta = \theta(x)$ is described by a second-order linear ODE with variable coefficients,
$$\theta''_{xx} + \left(\frac{k_2 - k_1}{a_2 - a_1} + \hat{f}\right)\theta = 0, \quad \hat{f} = f(b\varphi - c\psi). \quad (3)$$

For $\beta < 0$, the original PDE system has an exact solution of the form
$$\begin{aligned} u &= \varphi(x) + c[C_1 \cos(\sqrt{|\beta|}\, t) + C_2 \sin(\sqrt{|\beta|}\, t)]\theta(x), \\ w &= \psi(x) + b[C_1 \cos(\sqrt{|\beta|}\, t) + C_2 \sin(\sqrt{|\beta|}\, t)]\theta(x), \end{aligned}$$
where C_1 and C_2 are arbitrary constants, and the function $\theta(x)$ satisfies the linear ODE (3).

3°. Consider another special case:
$$f_1(z) = f_2(z) = f(z),$$
where $f(z)$ is an arbitrary function, and the functions $\varphi = \varphi(t)$ and $\psi = \psi(t)$ describe a spatially homogeneous non-stationary solution of the original PDE system and satisfy the second-order ODE system
$$\begin{aligned} \varphi''_{tt} &= k_1 \varphi + \varphi f(b\varphi - c\psi) + g_1(b\varphi - c\psi), \\ \psi''_{tt} &= k_2 \psi + \psi f(b\varphi - c\psi) + g_2(b\varphi - c\psi). \end{aligned} \quad (4)$$

There are more complex spatially nonhomogeneous non-stationary exact solutions of the original PDE system:
$$u = c\rho(t)\theta(x) + \varphi(t), \quad w = b\rho(t)\theta(x) + \psi(t),$$
$$\theta(x) = \begin{cases} C_1 \exp(-\sqrt{\mu}\, x) + C_2 \exp(\sqrt{\mu}\, x) & \text{if } \mu > 0, \\ C_1 \cos(\sqrt{|\mu|}\, x) + C_2 \sin(\sqrt{|\mu|}\, x) & \text{if } \mu < 0, \end{cases} \quad \mu = \frac{k_1 - k_2}{a_2 - a_1},$$

where the functions $\varphi = \varphi(t)$ and $\psi = \psi(t)$ satisfy the ODE system (4), and the function $\rho = \rho(t)$ is described by the second-order ODE:

$$\rho''_{tt} = \left[\frac{a_2 k_1 - a_1 k_2}{a_2 - a_1} + f(b\varphi - c\psi)\right]\rho.$$

4. $u_{tt} = au_{xx} + e^{\lambda u} f(\lambda u - \sigma w)$, $\quad w_{tt} = bw_{xx} + e^{\sigma w} g(\lambda u - \sigma w)$.

1°. Solution (Polyanin & Zaitsev, 2012):

$$u = y(\xi) - \frac{2}{\lambda}\ln(C_1 t + C_2), \quad w = z(\xi) - \frac{2}{\sigma}\ln(C_1 t + C_2), \quad \xi = \frac{x}{C_1 t + C_2},$$

where C_1 and C_2 are arbitrary constants, and the functions $y = y(\xi)$ and $z = z(\xi)$ are described by the ODE system

$$C_1^2 (\xi^2 y'_\xi)'_\xi + 2C_1^2 \lambda^{-1} = a\xi^{-n}(\xi^n y'_\xi)'_\xi + e^{\lambda y} f(\lambda y - \sigma z),$$
$$C_1^2 (\xi^2 z'_\xi)'_\xi + 2C_1^2 \sigma^{-1} = b\xi^{-n}(\xi^n z'_\xi)'_\xi + e^{\sigma z} g(\lambda y - \sigma z).$$

2°. Solution with $b = a$:

$$u = \theta(x, t), \quad w = \frac{\lambda}{\sigma}\theta(x, t) - \frac{k}{\sigma},$$

where k is a root of the algebraic (transcendental) equation

$$\lambda f(k) = \sigma e^{-k} g(k),$$

and the function $\theta = \theta(x, t)$ is described by the equation

$$\theta_{tt} = a\theta_{xx} + f(k)e^{\lambda \theta}.$$

It is a solvable equation; see equation 6.2.1.4.

▶ **Arbitrary functions depend on the ratio of the unknowns.**

5. $u_{tt} = au_{xx} + uf\left(\dfrac{u}{w}\right)$, $\quad w_{tt} = bw_{xx} + wg\left(\dfrac{u}{w}\right)$.

1°. Periodic multiplicative separable solution (Polyanin & Zaitsev, 2012):

$$u = [C_1 \cos(kt) + C_2 \sin(kt)]y(x), \quad w = [C_1 \cos(kt) + C_2 \sin(kt)]z(x), \quad (1)$$

where C_1, C_2, and k are arbitrary constants, and the functions $y = y(x)$ and $z = z(x)$ are described by the autonomous ODE system

$$ay''_{xx} + k^2 y + yf(y/z) = 0,$$
$$bz''_{xx} + k^2 z + zg(y/z) = 0.$$

2°. Multiplicative separable solution (Polyanin & Zaitsev, 2012):

$$u = [C_1 \exp(kt) + C_2 \exp(-kt)]y(x), \quad w = [C_1 \exp(kt) + C_2 \exp(-kt)]z(x), \quad (2)$$

where C_1, C_2, and k are arbitrary constants, and the functions $y = y(x)$ and $z = z(x)$ are described by the autonomous ODE system

$$ay''_{xx} - k^2 y + yf(y/z) = 0,$$
$$bz''_{xx} - k^2 z + zg(y/z) = 0.$$

$3°$. Degenerate multiplicative separable solution:

$$u = (C_1 t + C_2) y(x), \quad w = (C_1 t + C_2) z(x), \tag{3}$$

where the functions $y = y(x)$ and $z = z(x)$ are described by the autonomous ODE system

$$ay''_{xx} + yf(y/z) = 0,$$
$$bz''_{xx} + zg(y/z) = 0.$$

Remark 7.7. *In solutions (1)–(3), the variables t and x can be swapped (this will result in slight changes in the determining systems of the ordinary differential equations for y and z).*

$4°$. Solution (Polyanin & Zaitsev, 2012):

$$u = e^{kx - \lambda t} y(\xi), \quad w = e^{kx - \lambda t} z(\xi), \quad \xi = \beta x - \gamma t,$$

where k, λ, β, and γ are arbitrary constants, and the functions $y = y(\xi)$ and $z = z(\xi)$ are described by the autonomous ODE system

$$(a\beta^2 - \gamma^2) y''_{\xi\xi} + 2(ak\beta - \gamma\lambda) y'_\xi + (ak^2 - \lambda^2) y + yf(y/z) = 0,$$
$$(b\beta^2 - \gamma^2) z''_{\xi\xi} + 2(bk\beta - \gamma\lambda) z'_\xi + (bk^2 - \lambda^2) z + zg(y/z) = 0.$$

To the special case $k = \lambda = 0$ there corresponds a traveling wave solution.

$5°$. Solution with $b = a$ (Polyanin & Zaitsev, 2012):

$$u = k\theta(x, t), \quad w = \theta(x, t),$$

where k is a root of the algebraic (transcendental) equation $f(k) = g(k)$, and the function $\theta = \theta(x, t)$ is described by the linear Klein–Gordon equation

$$\theta_{tt} = a\theta_{xx} + f(k)\theta.$$

For details about this equation, see Section 5.2.3.

6. $u_{tt} = a u_{xx} + u f\!\left(\dfrac{u}{w}\right) + \dfrac{u}{w} h\!\left(\dfrac{u}{w}\right),$

$w_{tt} = \dfrac{a}{x^n} \dfrac{\partial}{\partial x}\!\left(x^n \dfrac{\partial u}{\partial w}\right) + w g\!\left(\dfrac{u}{w}\right) + h\!\left(\dfrac{u}{w}\right).$

Solution (Polyanin & Zaitsev, 2012):

$$u = k\theta(x, t), \quad w = \theta(x, t),$$

where k is a root of the algebraic (transcendental) equation $f(k) = g(k)$, and the function $\theta = \theta(x, t)$ is described by the linear equation

$$\theta_{tt} = a\theta_{xx} + f(k)\theta + h(k).$$

7. $u_{tt} = a u_{xx} + u^k f\!\left(\dfrac{u}{w}\right), \quad w_{tt} = b w_{xx} + w^k g\!\left(\dfrac{u}{w}\right).$

Self-similar solution (Polyanin & Zaitsev, 2012):

$$u = (C_1 t + C_2)^{\frac{2}{1-k}} y(\xi), \quad w = (C_1 t + C_2)^{\frac{2}{1-k}} z(\xi), \quad \xi = \frac{x}{C_1 t + C_2},$$

where C_1 and C_2 are arbitrary constants, and the functions $y = y(\xi)$ and $z = z(\xi)$ are described by the ODE system

$$C_1^2\xi^2 y''_{\xi\xi} + \frac{2C_1^2(k+1)}{k-1}\xi y'_\xi + \frac{C_1^2(k+1)}{(k-1)^2}y = \frac{a}{\xi^n}(\xi^n y'_\xi)'_\xi + y^k f\left(\frac{y}{z}\right),$$

$$C_1^2\xi^2 z''_{\xi\xi} + \frac{2C_1^2(k+1)}{k-1}\xi z'_\xi + \frac{C_1^2(k+1)}{(k-1)^2}z = \frac{b}{\xi^n}(\xi^n z'_\xi)'_\xi + z^k g\left(\frac{y}{z}\right).$$

▶ **Other hyperbolic systems.**

8. $u_{tt} = au_{xx} + uf(x, u^k w^m), \qquad w_{tt} = bw_{xx} + wg(x, u^k w^m).$

Multiplicative separable solution (Polyanin & Zaitsev, 2012):

$$u = e^{-m\lambda t} y(x), \quad w = e^{k\lambda t} z(x),$$

where λ is an arbitrary constant, and the functions $y = y(x)$ and $z = z(x)$ are described by the ODE system

$$ay''_{xx} - m^2\lambda^2 y + yf(x, y^k z^m) = 0,$$
$$bz''_{xx} - k^2\lambda^2 z + zg(x, y^k z^m) = 0.$$

9. $u_{tt} = au_{xx} + uf(u^2 + w^2) - wg(u^2 + w^2),$
   $w_{tt} = aw_{xx} + wf(u^2 + w^2) + ug(u^2 + w^2).$

1°. Periodic solution in t (Polyanin & Zaitsev, 2012):

$$u = r(x)\cos[\theta(x) + C_1 t + C_2], \quad w = r(x)\sin[\theta(x) + C_1 t + C_2],$$

where C_1 and C_2 are arbitrary constants, and the functions $r = r(x)$ and $\theta(x)$ are described by the ODE system

$$ar''_{xx} - ar(\theta'_x)^2 + C_1^2 r + rf(r^2) = 0,$$
$$ar\theta''_{xx} + 2ar'_x\theta'_x + rg(r^2) = 0.$$

2°. There is an exact solution of the form (generalizes the solution of Item 1°):

$$u = r(z)\cos[\theta(z) + C_1 t + C_2], \quad z = kx - \lambda t,$$
$$w = r(z)\sin[\theta(z) + C_1 t + C_2].$$

10. $u_{tt} = au_{xx} + uf(u^2 - w^2) + wg(u^2 - w^2),$
    $w_{tt} = aw_{xx} + wf(u^2 - w^2) + ug(u^2 - w^2).$

1°. Solution (Polyanin & Zaitsev, 2012):

$$u = r(x)\cosh[\theta(x) + C_1 t + C_2], \quad w = r(x)\sinh[\theta(x) + C_1 t + C_2],$$

where C_1 and C_2 are arbitrary constants, and the functions $r = r(x)$ and $\theta(x)$ are described by the ODE system

$$ar''_{xx} + ar(\theta'_x)^2 - C_1^2 r + rf(r^2) = 0,$$
$$ar\theta''_{xx} + 2ar'_x\theta'_x + rg(r^2) = 0.$$

2°. There is an exact solution of the form (generalizes the solution of Item 1°):

$$u = r(z)\cosh[\theta(z) + C_1 t + C_2], \quad w = r(z)\sinh[\theta(z) + C_1 t + C_2], \quad z = kx - \lambda t.$$

7.2.5. Nonlinear Hyperbolic Systems of the Form
$$u_{tt} = ax^{-n}(x^n u_x)_x + F(u,w), \quad w_{tt} = bx^{-n}(x^n w_x)_x + G(u,w)$$

Preliminary remarks. These are nonlinear systems of wave-type PDEs in the radial symmetric case ($n = 1$ corresponds to a plane problem and $n = 2$ to a spatial one).

▶ **Arbitrary functions depend on a linear combination of the unknowns.**

1. $u_{tt} = ax^{-n}(x^n u_x)_x + uf(bu - cw) + g(bu - cw),$
   $w_{tt} = ax^{-n}(x^n w_x)_x + wf(bu - cw) + h(bu - cw).$

1°. Solution:
$$u = \varphi(t) + c\theta(x,t), \quad w = \psi(t) + b\theta(x,t),$$

where $\varphi = \varphi(t)$ and $\psi = \psi(t)$ are described by the autonomous ODE system
$$\varphi''_{tt} = \varphi f(b\varphi - c\psi) + g(b\varphi - c\psi), \quad \psi''_{tt} = \psi f(b\varphi - c\psi) + h(b\varphi - c\psi),$$

and the function $\theta = \theta(x,t)$ satisfies the linear PDE
$$\theta_{tt} = ax^{-n}(x^n \theta_x)_x + f(b\varphi - c\psi)\theta.$$

For $f = \text{const}$, this equation can be solved by separation of variables.

2°. Let us add together the first equation multiplied by b and the second one multiplied by $-c$ to obtain
$$\zeta_{tt} = ax^{-n}(x^n \zeta_x)_x + \zeta f(\zeta) + bg(\zeta) - ch(\zeta), \quad \zeta = bu - cw. \qquad (1)$$

This equation will be considered in conjunction with the first equation of the original system
$$u_{tt} = ax^{-n}(x^n u_x)_x + uf(\zeta) + g(\zeta). \qquad (2)$$

Equation (1) can be treated separately. Given a solution $\zeta = \zeta(x,t)$ to equation (1), the function $u = u(x,t)$ can be determined by solving equation (2), and the function $w = w(x,t)$ can be found as $w = (bu - \zeta)/c$.

Note three important solutions to equation (1):

(i) In the general case, equation (1) admits a spatially homogeneous solution $\zeta = \zeta(t)$. The corresponding solution to the original system is given in Item 1° in another form.

(ii) In the general case, equation (1) admits a steady-state solution $\zeta = \zeta(x)$. The corresponding exact solutions to equation (2) are expressed as $u = u_0(x) + \sum e^{-\beta_n t} u_n(x)$ and $u = u_0(x) + \sum \cos(\beta_n t) u_n^{(1)}(x) + \sum \sin(\beta_n t) u_n^{(2)}(x)$.

(iii) If the condition $\zeta f(\zeta) + bg(\zeta) - ch(\zeta) = k_1 \zeta + k_0$ holds, equation (1) is linear,
$$\zeta_{tt} = ax^{-n}(x^n \zeta_x)_x + k_1 \zeta + k_0,$$

and, hence, can be solved by separation of variables.

3°. References for the system of PDEs 7.2.5.1: Polyanin & Manzhirov (2007) and Polyanin & Zaitsev (2012).

2. $u_{tt} = ax^{-n}(x^n u_x)_x + e^{\lambda u} f(\lambda u - \sigma w)$,
   $w_{tt} = bx^{-n}(x^n w_x)_x + e^{\sigma w} g(\lambda u - \sigma w)$.

   1°. Solution (Polyanin & Manzhirov, 2007):
   $$u = y(\xi) - \frac{2}{\lambda} \ln(C_1 t + C_2), \quad w = z(\xi) - \frac{2}{\sigma} \ln(C_1 t + C_2), \quad \xi = \frac{x}{C_1 t + C_2},$$
   where C_1 and C_2 are arbitrary constants, and the functions $y = y(\xi)$ and $z = z(\xi)$ are described by the ODE system
   $$C_1^2 (\xi^2 y'_\xi)'_\xi + 2C_1^2 \lambda^{-1} = a\xi^{-n} (\xi^n y'_\xi)'_\xi + e^{\lambda y} f(\lambda y - \sigma z),$$
   $$C_1^2 (\xi^2 z'_\xi)'_\xi + 2C_1^2 \sigma^{-1} = b\xi^{-n} (\xi^n z'_\xi)'_\xi + e^{\sigma z} g(\lambda y - \sigma z).$$

   2°. Solution with $b = a$:
   $$u = \theta(x,t), \quad w = \frac{\lambda}{\sigma} \theta(x,t) - \frac{k}{\sigma},$$
   where k is a root of the algebraic (transcendental) equation
   $$\lambda f(k) = \sigma e^{-k} g(k),$$
   and the function $\theta = \theta(x,t)$ is described by the PDE:
   $$\theta_{tt} = ax^{-n}(x^n \theta_x)_x + f(k) e^{\lambda \theta}.$$
   This equation is solvable for $n = 0$; for its exact solutions, see PDE 6.2.1.4.

▶ **Arbitrary functions depend on the ratio of the unknowns.**

3. $u_{tt} = ax^{-n}(x^n u_x)_x + u f\left(\frac{u}{w}\right), \quad w_{tt} = bx^{-n}(x^n w_x)_x + w g\left(\frac{u}{w}\right).$

   1°. Periodic multiplicative separable solution:
   $$u = [C_1 \cos(kt) + C_2 \sin(kt)] y(x), \quad w = [C_1 \cos(kt) + C_2 \sin(kt)] z(x),$$
   where C_1, C_2, and k are arbitrary constants, and the functions $y = y(x)$ and $z = z(x)$ are described by the ODE system
   $$ax^{-n}(x^n y'_x)'_x + k^2 y + y f(y/z) = 0, \quad bx^{-n}(x^n z'_x)'_x + k^2 z + z g(y/z) = 0.$$

   2°. Multiplicative separable solution:
   $$u = [C_1 \exp(kt) + C_2 \exp(-kt)] y(x), \quad w = [C_1 \exp(kt) + C_2 \exp(-kt)] z(x),$$
   where C_1, C_2, and k are arbitrary constants, and the functions $y = y(x)$ and $z = z(x)$ are described by the ODE system
   $$ax^{-n}(x^n y'_x)'_x - k^2 y + y f(y/z) = 0, \quad bx^{-n}(x^n z'_x)'_x - k^2 z + z g(y/z) = 0.$$

   3°. Degenerate multiplicative separable solution:
   $$u = (C_1 t + C_2) y(x), \quad w = (C_1 t + C_2) z(x),$$

where the functions $y = y(x)$ and $z = z(x)$ are described by the ODE system

$$ax^{-n}(x^n y'_x)'_x + yf(y/z) = 0, \quad bx^{-n}(x^n z'_x)'_x + zg(y/z) = 0.$$

4°. Multiplicative separable solution:

$$u = x^{\frac{1-n}{2}}[C_1 J_\nu(kx) + C_2 Y_\nu(kx)]\varphi(t), \quad \nu = \tfrac{1}{2}|n-1|,$$
$$w = x^{\frac{1-n}{2}}[C_1 J_\nu(kx) + C_2 Y_\nu(kx)]\psi(t),$$

where C_1, C_2, and k are arbitrary constants, $J_\nu(z)$ and $Y_\nu(z)$ are Bessel functions, and the functions $\varphi = \varphi(t)$ and $\psi = \psi(t)$ are described by the autonomous ODE system

$$\varphi''_{tt} = -ak^2\varphi + \varphi f(\varphi/\psi), \quad \psi''_{tt} = -bk^2\psi + \psi g(\varphi/\psi).$$

5°. Multiplicative separable solution:

$$u = x^{\frac{1-n}{2}}[C_1 I_\nu(kx) + C_2 K_\nu(kx)]\varphi(t), \quad \nu = \tfrac{1}{2}|n-1|,$$
$$w = x^{\frac{1-n}{2}}[C_1 I_\nu(kx) + C_2 K_\nu(kx)]\psi(t),$$

where C_1, C_2, and k are arbitrary constants, $I_\nu(z)$ and $K_\nu(z)$ are modified Bessel functions, and the functions $\varphi = \varphi(t)$ and $\psi = \psi(t)$ are described by the autonomous ODE system

$$\varphi''_{tt} = ak^2\varphi + \varphi f(\varphi/\psi),$$
$$\psi''_{tt} = bk^2\psi + \psi g(\varphi/\psi).$$

6°. Solution with $b = a$:

$$u = k\theta(x,t), \quad w = \theta(x,t),$$

where k is a root of the algebraic (transcendental) equation $f(k) = g(k)$, and the function $\theta = \theta(x,t)$ is described by the linear Klein–Gordon equation

$$\theta_{tt} = ax^{-n}(x^n \theta_x)_x + f(k)\theta.$$

7°. References for the system of PDEs 7.2.5.3: Polyanin & Manzhirov (2007) and Polyanin & Zaitsev (2012).

4. $u_{tt} = ax^{-n}(x^n u_x)_x + uf\left(\dfrac{u}{w}\right) + \dfrac{u}{w}h\left(\dfrac{u}{w}\right),$
$w_{tt} = ax^{-n}(x^n w_x)_x + wg\left(\dfrac{u}{w}\right) + h\left(\dfrac{u}{w}\right).$

Solution (Polyanin & Manzhirov, 2007):

$$u = k\theta(x,t), \quad w = \theta(x,t),$$

where k is a root of the algebraic (transcendental) equation $f(k) = g(k)$, and the function $\theta = \theta(x,t)$ is described by the linear PDE

$$\theta_{tt} = ax^{-n}(x^n \theta_x)_x + f(k)\theta + h(k).$$

5. $u_{tt} = ax^{-n}(x^n u_x)_x + u^k f\left(\dfrac{u}{w}\right), \quad w_{tt} = bx^{-n}(x^n w_x)_x + w^k g\left(\dfrac{u}{w}\right).$

Self-similar solution (Polyanin & Manzhirov, 2007):
$$u = (C_1 t + C_2)^{\frac{2}{1-k}} y(\xi), \quad w = (C_1 t + C_2)^{\frac{2}{1-k}} z(\xi), \quad \xi = \dfrac{x}{C_1 t + C_2},$$

where C_1 and C_2 are arbitrary constants, and the functions $y = y(\xi)$ and $z = z(\xi)$ are described by the ODE system

$$C_1^2 \xi^2 y_{\xi\xi}'' + \dfrac{2C_1^2(k+1)}{k-1}\xi y_\xi' + \dfrac{C_1^2(k+1)}{(k-1)^2} y = \dfrac{a}{\xi^n}(\xi^n y_\xi')_\xi' + y^k f\left(\dfrac{y}{z}\right),$$
$$C_1^2 \xi^2 z_{\xi\xi}'' + \dfrac{2C_1^2(k+1)}{k-1}\xi z_\xi' + \dfrac{C_1^2(k+1)}{(k-1)^2} z = \dfrac{b}{\xi^n}(\xi^n z_\xi')_\xi' + z^k g\left(\dfrac{y}{z}\right).$$

▶ **Other hyperbolic systems.**

6. $u_{tt} = ax^{-n}(x^n u_x)_x + u f(x, u^k w^m),$
$w_{tt} = bx^{-n}(x^n w_x)_x + w g(x, u^k w^m).$

Multiplicative separable solution (Polyanin & Manzhirov, 2007):
$$u = e^{-m\lambda t} y(x), \quad w = e^{k\lambda t} z(x),$$

where λ is an arbitrary constant, and the functions $y = y(x)$ and $z = z(x)$ are described by the ODE system

$$ax^{-n}(x^n y_x')_x' - m^2\lambda^2 y + y f(x, y^k z^m) = 0,$$
$$bx^{-n}(x^n z_x')_x' - k^2\lambda^2 z + z g(x, y^k z^m) = 0.$$

7. $u_{tt} = ax^{-n}(x^n u_x)_x + u f(u^2 + w^2) - w g(u^2 + w^2),$
$w_{tt} = ax^{-n}(x^n w_x)_x + w f(u^2 + w^2) + u g(u^2 + w^2).$

1°. Periodic solution in t (Polyanin & Manzhirov, 2007):
$$u = r(x)\cos[\theta(x) + C_1 t + C_2], \quad w = r(x)\sin[\theta(x) + C_1 t + C_2],$$

where C_1 and C_2 are arbitrary constants, and the functions $r = r(x)$ and $\theta(x)$ are described by the ODE system

$$a r_{xx}'' - a r(\theta_x')^2 + \dfrac{an}{x} r_x' + C_1^2 r + r f(r^2) = 0,$$
$$a r \theta_{xx}'' + 2 a r_x' \theta_x' + \dfrac{an}{x} r \theta_x' + r g(r^2) = 0.$$

2°. For $n = 0$, there is an exact solution of the form

$$u = r(z)\cos[\theta(z) + C_1 t + C_2], \quad w = r(z)\sin[\theta(z) + C_1 t + C_2], \quad z = kx - \lambda t.$$

8. $u_{tt} = ax^{-n}(x^n u_x)_x + u f(u^2 - w^2) + w g(u^2 - w^2),$
$w_{tt} = ax^{-n}(x^n w_x)_x + w f(u^2 - w^2) + u g(u^2 - w^2).$

1°. Solution (Polyanin & Manzhirov, 2007):
$$u = r(x)\cosh[\theta(x) + C_1 t + C_2],$$

$$w = r(x)\sinh[\theta(x) + C_1 t + C_2],$$

where C_1 and C_2 are arbitrary constants, and the functions $r = r(x)$ and $\theta(x)$ are described by the ODE system

$$ar''_{xx} + ar(\theta'_x)^2 + \frac{an}{x}r'_x - C_1^2 r + rf(r^2) = 0,$$
$$ar\theta''_{xx} + 2ar'_x\theta'_x + \frac{an}{x}r\theta'_x + rg(r^2) = 0.$$

$2°$. For $n = 0$, there is an exact solution of the form

$$u = r(z)\cosh[\theta(z) + C_1 t + C_2], \quad w = r(z)\sinh[\theta(z) + C_1 t + C_2], \quad z = kx - \lambda t.$$

7.2.6. Nonlinear Elliptic Systems of the Form $\Delta u = F(u, w)$, $\Delta w = G(u, w)$

▶ **Arbitrary functions depend on a linear combination of the unknowns.**

1. $u_{xx} + u_{yy} = uf(au - bw) + g(au - bw),$
   $w_{xx} + w_{yy} = wf(au - bw) + h(au - bw).$

$1°$. Solution (Polyanin & Manzhirov, 2007):

$$u = \varphi(x) + b\theta(x, y), \quad w = \psi(x) + a\theta(x, y),$$

where $\varphi = \varphi(x)$ and $\psi = \psi(x)$ are described by the autonomous ODE system

$$\varphi''_{xx} = \varphi f(a\varphi - b\psi) + g(a\varphi - b\psi),$$
$$\psi''_{xx} = \psi f(a\varphi - b\psi) + h(a\varphi - b\psi),$$

and the function $\theta = \theta(x, y)$ satisfies the linear Schrödinger equation of the special form

$$\theta_{xx} + \theta_{yy} = F(x)\theta, \quad F(x) = f(a\varphi - b\psi).$$

Its solutions are determined by separation of variables.

$2°$. Let us add together the first equation multiplied by a and the second one multiplied by $-b$ to obtain

$$\zeta_{xx} + \zeta_{yy} = \zeta f(\zeta) + ag(\zeta) - bh(\zeta), \quad \zeta = au - bw. \tag{1}$$

This equation will be considered in conjunction with the first equation of the original system

$$u_{xx} + u_{yy} = uf(\zeta) + g(\zeta). \tag{2}$$

Equation (1) can be treated separately.

Note two important solutions to equation (1):

(i) In the general case, equation (1) admits an exact, traveling wave solution $\zeta = \zeta(z)$, where $z = k_1 x + k_2 y$ with arbitrary constants k_1 and k_2.

(ii) If the condition $\zeta f(\zeta) + ag(\zeta) - bh(\zeta) = c_1\zeta + c_0$ holds, equation (1) is a linear Helmholtz equation.

Given a solution $\zeta = \zeta(x, y)$ to equation (1), the function $u = u(x, y)$ can be determined by solving the linear equation (2), and the function $w = w(x, y)$ is found as $w = (bu - \zeta)/c$.

2. $u_{xx} + u_{yy} = e^{\lambda u} f(\lambda u - \sigma w)$, $w_{xx} + w_{yy} = e^{\sigma w} g(\lambda u - \sigma w)$.

1°. Solution (Polyanin & Manzhirov, 2007):
$$u = U(\xi) - \frac{2}{\lambda} \ln|x + C_1|, \quad w = W(\xi) - \frac{2}{\sigma} \ln|x + C_1|, \quad \xi = \frac{y + C_2}{x + C_1},$$

where C_1 and C_2 are arbitrary constants, and the functions $U = U(\xi)$ and $W = W(\xi)$ are described by the ODE system

$$(1 + \xi^2) U''_{\xi\xi} + 2\xi U'_\xi + \frac{2}{\lambda} = e^{\lambda U} f(\lambda U - \sigma W),$$
$$(1 + \xi^2) W''_{\xi\xi} + 2\xi W'_\xi + \frac{2}{\sigma} = e^{\sigma W} g(\lambda U - \sigma W).$$

2°. Solution (Polyanin & Manzhirov, 2007):
$$u = \theta(x, y), \quad w = \frac{\lambda}{\sigma} \theta(x, y) - \frac{k}{\sigma},$$

where k is a root of the algebraic (transcendental) equation
$$\lambda f(k) = \sigma e^{-k} g(k),$$

and the function $\theta = \theta(x, y)$ is described by the solvable equation
$$\theta_{xx} + \theta_{yy} = f(k) e^{\lambda \theta}.$$

This equation is encountered in combustion theory; for its exact solutions, see equation 6.3.1.3 and Polyanin & Zaitsev (2012).

▶ **Arbitrary functions depend on the ratio of the unknowns.**

3. $u_{xx} + u_{yy} = u f\left(\dfrac{u}{w}\right)$, $w_{xx} + w_{yy} = w g\left(\dfrac{u}{w}\right)$.

1°. Space-periodic multiplicative separable solution (another solution is obtained by swapping x and y):
$$u = [C_1 \sin(kx) + C_2 \cos(kx)] \varphi(y), \quad w = [C_1 \sin(kx) + C_2 \cos(kx)] \psi(y),$$

where C_1, C_2, and k are arbitrary constants, and the functions $\varphi = \varphi(y)$ and $\psi = \psi(y)$ are described by the autonomous ODE system
$$\varphi''_{yy} = k^2 \varphi + \varphi f(\varphi/\psi), \quad \psi''_{yy} = k^2 \psi + \psi g(\varphi/\psi).$$

2°. Multiplicative separable solution:
$$u = [C_1 \exp(kx) + C_2 \exp(-kx)] U(y), \quad w = [C_1 \exp(kx) + C_2 \exp(-kx)] W(y),$$

where C_1, C_2, and k are arbitrary constants, and the functions $U = U(y)$ and $W = W(y)$ are described by the autonomous ODE system
$$U''_{yy} = -k^2 U + U f(U/W), \quad W''_{yy} = -k^2 W + W g(U/W).$$

3°. Degenerate multiplicative separable solution:
$$u = (C_1 x + C_2) U(y), \quad w = (C_1 x + C_2) W(y),$$
where C_1 and C_2 are arbitrary constants, and the functions $U = U(y)$ and $W = W(y)$ are described by the autonomous ODE system
$$U''_{yy} = U f(U/W), \quad W''_{yy} = W g(U/W).$$

4°. Solution in multiplicative form:
$$u = e^{a_1 x + b_1 y} \xi(z), \quad w = e^{a_1 x + b_1 y} \eta(z), \quad z = a_2 x + b_2 y,$$
where a_1, a_2, b_1, and b_2 are arbitrary constants, and the functions $\xi = \xi(z)$ and $\eta = \eta(z)$ are described by the autonomous ODE system
$$(a_2^2 + b_2^2)\xi''_{zz} + 2(a_1 a_2 + b_1 b_2)\xi'_z + (a_1^2 + b_1^2)\xi = \xi f(\xi/\eta),$$
$$(a_2^2 + b_2^2)\eta''_{zz} + 2(a_1 a_2 + b_1 b_2)\eta'_z + (a_1^2 + b_1^2)\eta = \eta g(\xi/\eta).$$

5°. Solution:
$$u = k\theta(x, y), \quad w = \theta(x, y),$$
where k is a root of the algebraic (transcendental) equation $f(k) = g(k)$, and the function $\theta = \theta(x, y)$ is described by the linear Helmholtz equation
$$\theta_{xx} + \theta_{yy} = f(k)\theta.$$

For its exact solutions, see Section 5.3.3.

6°. References for the system of PDEs 7.2.5.3: Polyanin & Manzhirov (2007) and Polyanin & Zaitsev (2012).

4. $u_{xx} + u_{yy} = u f\left(\dfrac{u}{w}\right) + \dfrac{u}{w} h\left(\dfrac{u}{w}\right), \quad w_{xx} + w_{yy} = w g\left(\dfrac{u}{w}\right) + h\left(\dfrac{u}{w}\right).$

Solution (Polyanin & Manzhirov, 2007):
$$u = kw, \quad w = \theta(x, y) - \frac{h(k)}{f(k)},$$
where k is a root of the algebraic (transcendental) equation
$$f(k) = g(k),$$
and the function $\theta = \theta(x, y)$ satisfies the linear Helmholtz equation
$$\theta_{xx} + \theta_{yy} = f(k)\theta.$$

5. $u_{xx} + u_{yy} = u^n f\left(\dfrac{u}{w}\right), \quad w_{xx} + w_{yy} = w^n g\left(\dfrac{u}{w}\right).$

For $f(z) = kz^{-m}$ and $g(z) = -kz^{n-m}$, the system describes an nth-order chemical reaction (of order $n - m$ in the component u and of order m in the component w); to $n = 2$ and $m = 1$ there corresponds a second-order reaction, which often occurs in applications.

1°. Solution (Polyanin & Manzhirov, 2007):

$$u = r^{\frac{2}{1-n}} U(\theta), \quad w = r^{\frac{2}{1-n}} W(\theta), \quad r = \sqrt{(x+C_1)^2 + (y+C_2)^2}, \quad \theta = \frac{y+C_2}{x+C_1},$$

where C_1 and C_2 are arbitrary constants, and the functions $y = y(\xi)$ and $z = z(\xi)$ are determined by the autonomous ODE system

$$U''_{\theta\theta} + \frac{4}{(1-n)^2} U = U^n f\left(\frac{U}{W}\right), \quad W''_{\theta\theta} + \frac{4}{(1-n)^2} W = W^n g\left(\frac{U}{W}\right).$$

2°. Solution:

$$u = k\zeta(x,y), \quad w = \zeta(x,y),$$

where k is a root of the algebraic (transcendental) equation

$$k^{n-1} f(k) = g(k),$$

and the function $\zeta = \zeta(x,y)$ satisfies the PDE with a power-law nonlinearity

$$\zeta_{xx} + \zeta_{yy} = g(k)\zeta^n.$$

▶ Other elliptic systems.

6. $u_{xx} + u_{yy} = u f(u^n w^m), \quad w_{xx} + w_{yy} = w g(u^n w^m).$

Solution in multiplicative form (Polyanin & Manzhirov, 2007):

$$u = e^{m(a_1 x + b_1 y)} \xi(z), \quad w = e^{-n(a_1 x + b_1 y)} \eta(z), \quad z = a_2 x + b_2 y,$$

where a_1, a_2, b_1, and b_2 are arbitrary constants, and the functions $\xi = \xi(z)$ and $\eta = \eta(z)$ are described by the autonomous ODE system

$$(a_2^2 + b_2^2)\xi''_{zz} + 2m(a_1 a_2 + b_1 b_2)\xi'_z + m^2(a_1^2 + b_1^2)\xi = \xi f(\xi^n \eta^m),$$
$$(a_2^2 + b_2^2)\eta''_{zz} - 2n(a_1 a_2 + b_1 b_2)\eta'_z + n^2(a_1^2 + b_1^2)\eta = \eta g(\xi^n \eta^m).$$

7. $u_{xx} + u_{yy} = u f(u^2 + w^2) - w g(u^2 + w^2),$
   $w_{xx} + w_{yy} = w f(u^2 + w^2) + u g(u^2 + w^2).$

1°. Periodic solution in y (Polyanin & Manzhirov, 2007):

$$u = r(x) \cos[\theta(x) + C_1 y + C_2],$$
$$w = r(x) \sin[\theta(x) + C_1 y + C_2],$$

where C_1 and C_2 are arbitrary constants, and the functions $r = r(x)$ and $\theta = \theta(x)$ are described by the autonomous ODE system

$$r''_{xx} = r(\theta'_x)^2 + C_1^2 r + r f(r^2),$$
$$r\theta''_{xx} = -2r'_x \theta'_x + r g(r^2).$$

2°. Solution that generalizes the solution from Item 1° (Polyanin & Manzhirov, 2007):
$$u = r(z)\cos[\theta(z) + C_1 y + C_2], \quad w = r(z)\sin[\theta(z) + C_1 y + C_2], \quad z = k_1 x + k_2 y,$$
where C_1, C_2, k_1, and k_2 are arbitrary constants, and the functions $r = r(z)$ and $\theta = \theta(z)$ are described by the autonomous ODE system
$$(k_1^2 + k_2^2)r''_{zz} = k_1^2 r(\theta'_z)^2 + r(k_2 \theta'_z + C_1)^2 + rf(r^2),$$
$$(k_1^2 + k_2^2)r\theta''_{zz} = -2[(k_1^2 + k_2^2)\theta'_z + C_1 k_2]r'_z + rg(r^2).$$

8. $u_{xx} + u_{yy} = uf(u^2 - w^2) + wg(u^2 - w^2),$
   $w_{xx} + w_{yy} = wf(u^2 - w^2) + ug(u^2 - w^2).$

Solution (Polyanin & Manzhirov, 2007):
$$u = r(z)\cosh[\theta(z) + C_1 y + C_2], \quad w = r(z)\sinh[\theta(z) + C_1 y + C_2], \quad z = k_1 x + k_2 y,$$
where C_1, C_2, k_1, and k_2 are arbitrary constants, and the functions $r = r(z)$ and $\theta = \theta(z)$ are described by the autonomous ODE system
$$(k_1^2 + k_2^2)r''_{zz} + k_1^2 r(\theta'_z)^2 + r(k_2\theta'_z + C_1)^2 = rf(r^2),$$
$$(k_1^2 + k_2^2)r\theta''_{zz} + 2[(k_1^2 + k_2^2)\theta'_z + C_1 k_2]r'_z = rg(r^2).$$

7.3. PDE Systems of General Form

7.3.1. Linear Systems

1. $u_t = L[u] + f_1(t)u + g_1(t)w, \quad w_t = L[w] + f_2(t)u + g_2(t)w.$

Here, L is an arbitrary linear differential operator with respect to the coordinates $x_1, \ldots, x_n$ (of any order in the derivatives), whose coefficients can be dependent on $x_1, \ldots, x_n, t$. It is assumed that $L[\text{const}] = 0$.

Solution (Polyanin & Manzhirov, 2007):
$$u = \varphi_1(t)U(x_1, \ldots, x_n, t) + \varphi_2(t)W(x_1, \ldots, x_n, t),$$
$$w = \psi_1(t)U(x_1, \ldots, x_n, t) + \psi_2(t)W(x_1, \ldots, x_n, t),$$
where the two pairs of functions $\varphi_1 = \varphi_1(t)$, $\psi_1 = \psi_1(t)$ and $\varphi_2 = \varphi_2(t)$, $\psi_2 = \psi_2(t)$ are linearly independent (fundamental) solutions to the system of first-order linear ODEs
$$\varphi'_t = f_1(t)\varphi + g_1(t)\psi, \quad \psi'_t = f_2(t)\varphi + g_2(t)\psi,$$
and the functions $U = U(x_1, \ldots, x_n, t)$ and $W = W(x_1, \ldots, x_n, t)$ satisfy the independent linear PDEs
$$U_t = L[U], \quad W_t = L[W].$$

2. $u_{tt} = L[u] + a_1 u + b_1 w, \quad w_{tt} = L[w] + a_2 u + b_2 w.$

Here, L is an arbitrary linear differential operator with respect to the coordinates $x_1, \ldots, x_n$ (of any order in the derivatives).

Solution (Polyanin & Manzhirov, 2007):
$$u = \frac{a_1 - \lambda_2}{a_2(\lambda_1 - \lambda_2)}U - \frac{a_1 - \lambda_1}{a_2(\lambda_1 - \lambda_2)}W, \quad w = \frac{1}{\lambda_1 - \lambda_2}(U - W),$$

where λ_1 and λ_2 are roots of the quadratic equation
$$\lambda^2 - (a_1 + b_2)\lambda + a_1 b_2 - a_2 b_1 = 0,$$

and the functions $U = U(x_1, \ldots, x_n, t)$ and $W = W(x_1, \ldots, x_n, t)$ satisfy the independent PDEs
$$U_{tt} = L[U] + \lambda_1 U, \quad W_{tt} = L[W] + \lambda_2 W.$$

7.3.2. Nonlinear Systems of Two Equations Involving the First Derivatives with Respect to t

1. $u_t = L[u] + u f(t, bu - cw) + g(t, bu - cw),$
    $w_t = L[w] + w f(t, bu - cw) + h(t, bu - cw).$

Here, L is an arbitrary linear differential operator (of any order) with respect to the spatial variables $x_1, \ldots, x_n$; its coefficients can be dependent on $x_1, \ldots, x_n, t$.

1°. Solution (Polyanin & Manzhirov, 2007):
$$u = \varphi(t) + c \exp\left[\int f(t, b\varphi - c\psi)\, dt\right] \theta(x, t), \quad w = \psi(t) + b \exp\left[\int f(t, b\varphi - c\psi)\, dt\right] \theta(x, t),$$

where $\varphi = \varphi(t)$ and $\psi = \psi(t)$ are described by the ODE system
$$\varphi'_t = \varphi f(t, b\varphi - c\psi) + g(t, b\varphi - c\psi),$$
$$\psi'_t = \psi f(t, b\varphi - c\psi) + h(t, b\varphi - c\psi),$$

and the function $\theta = \theta(x_1, \ldots, x_n, t)$ satisfies the linear PDE:
$$\theta_t = L[\theta].$$

2°. Let us add together the first equation multiplied by b and the second one multiplied by $-c$ to obtain
$$\zeta_t = L[\zeta] + \zeta f(t, \zeta) + b g(t, \zeta) - c h(t, \zeta), \quad \zeta = bu - cw. \tag{1}$$

This PDE will be considered in conjunction with the first equation of the original system
$$u_t = L[u] + u f(t, \zeta) + g(t, \zeta). \tag{2}$$

PDE (1) can be treated separately. Given a solution of this equation, $\zeta = \zeta(x_1, \ldots, x_n, t)$, the function $u = u(x_1, \ldots, x_n, t)$ can be determined by solving the linear PDE (2), and the function $w = w(x_1, \ldots, x_n, t)$ is found as $w = (bu - \zeta)/c$.

2. $u_t = L_1[u] + u f\left(\dfrac{u}{w}\right), \quad w_t = L_2[w] + w g\left(\dfrac{u}{w}\right).$

Here, L_1 and L_2 are arbitrary constant-coefficient linear differential operators (of any order) with respect to x.

1°. Solution (Polyanin & Manzhirov, 2007):
$$u = e^{kx-\lambda t} y(\xi), \quad w = e^{kx-\lambda t} z(\xi), \qquad \xi = \beta x - \gamma t,$$
where k, λ, β, and γ are arbitrary constants, and the functions $y = y(\xi)$ and $z = z(\xi)$ are described by the ODE system
$$M_1[y] + \lambda y + y f(y/z) = 0, \quad M_2[z] + \lambda z + z g(y/z) = 0,$$
$$M_1[y] = e^{-kx} L_1[e^{kx} y(\xi)], \quad M_2[z] = e^{-kx} L_2[e^{kx} z(\xi)].$$
To the special case $k = \lambda = 0$ there corresponds a traveling wave solution.

2°. If the operators L_1 and L_2 contain only even derivatives, there are solutions of the form
$$u = [C_1 \sin(kx) + C_2 \cos(kx)] \varphi(t), \quad w = [C_1 \sin(kx) + C_2 \cos(kx)] \psi(t);$$
$$u = [C_1 \exp(kx) + C_2 \exp(-kx)] \varphi(t), \quad w = [C_1 \exp(kx) + C_2 \exp(-kx)] \psi(t);$$
$$u = (C_1 x + C_2) \varphi(t), \quad w = (C_1 x + C_2) \psi(t),$$
where C_1, C_2, and k are arbitrary constants. Note that the third solution is degenerate.

3. $u_t = L[u] + u f\left(t, \dfrac{u}{w}\right), \quad w_t = L[w] + w g\left(t, \dfrac{u}{w}\right).$

Here, L is an arbitrary linear differential operator with respect to the variables $x_1, \ldots, x_n$ (of any order in derivatives), whose coefficients can be dependent on $x_1, \ldots, x_n, t$:
$$L[u] = \sum A_{k_1 \ldots k_n}(x_1, \ldots, x_n, t) \frac{\partial^{k_1 + \cdots + k_n} u}{\partial x_1^{k_1} \ldots \partial x_n^{k_n}}.$$

1°. Solution (Polyanin & Manzhirov, 2007):
$$u = \varphi(t) \exp\left[\int g(t, \varphi(t))\, dt\right] \theta(x_1, \ldots, x_n, t),$$
$$w = \exp\left[\int g(t, \varphi(t))\, dt\right] \theta(x_1, \ldots, x_n, t),$$
where the function $\varphi = \varphi(t)$ is described by the nonlinear first-order ODE:
$$\varphi'_t = [f(t, \varphi) - g(t, \varphi)] \varphi,$$
and the function $\theta = \theta(x_1, \ldots, x_n, t)$ satisfies the linear PDE $\theta_t = L[\theta]$.

2°. The transformation
$$u = a_1(t) U + b_1(t) W, \quad w = a_2(t) U + b_2(t) W,$$
where $a_n(t)$ and $b_n(t)$ are arbitrary functions ($n = 1, 2$), leads to a similar system for U and W.

4. $u_t = L[u] + u f\left(\dfrac{u}{w}\right) + g\left(\dfrac{u}{w}\right), \quad w_t = L[w] + w f\left(\dfrac{u}{w}\right) + h\left(\dfrac{u}{w}\right).$

Here, L is an arbitrary linear differential operator with respect to $x_1, \ldots, x_n$ (of any order in derivatives), whose coefficients can be dependent on $x_1, \ldots, x_n, t$:
$$L[u] = \sum A_{k_1 \ldots k_n}(x_1, \ldots, x_n, t) \frac{\partial^{k_1 + \cdots + k_n} u}{\partial x_1^{k_1} \ldots \partial x_n^{k_n}},$$
where $k_1 + \cdots + k_n \geq 1$.

Let λ be a root of the algebraic (transcendental) equation
$$g(\lambda) = \lambda h(\lambda). \tag{1}$$

1°. Solution if $f(\lambda) \neq 0$ (Polyanin & Manzhirov, 2007):

$$u(x,t) = \lambda\left(\exp[f(\lambda)t]\theta(x,t) - \frac{h(\lambda)}{f(\lambda)}\right), \quad w(x,t) = \exp[f(\lambda)t]\theta(x,t) - \frac{h(\lambda)}{f(\lambda)},$$

where the function $\theta = \theta(x_1, \ldots, x_n, t)$ satisfies the linear PDE:

$$\theta_t = L[\theta]. \tag{2}$$

2°. Solution if $f(\lambda) = 0$:

$$u(x,t) = \lambda[\theta(x,t) + h(\lambda)t], \quad w(x,t) = \theta(x,t) + h(\lambda)t,$$

where the function $\theta = \theta(x_1, \ldots, x_n, t)$ satisfies the linear equation (2).

5. $u_t = L[u] + uf\left(t, \frac{u}{w}\right) + \frac{u}{w}h\left(t, \frac{u}{w}\right), \quad w_t = L[w] + wg\left(t, \frac{u}{w}\right) + h\left(t, \frac{u}{w}\right).$

Solution (Polyanin & Manzhirov, 2007):

$$u = \varphi(t)G(t)\left[\theta(x_1,\ldots,x_n,t) + \int \frac{h(t,\varphi)}{G(t)}\,dt\right], \quad G(t) = \exp\left[\int g(t,\varphi)\,dt\right],$$

$$w = G(t)\left[\theta(x_1,\ldots,x_n,t) + \int \frac{h(t,\varphi)}{G(t)}\,dt\right],$$

where the function $\varphi = \varphi(t)$ is described by the nonlinear first-order ODE:

$$\varphi'_t = [f(t,\varphi) - g(t,\varphi)]\varphi,$$

and the function $\theta = \theta(x_1, \ldots, x_n, t)$ satisfies the linear PDE:

$$\theta_t = L[\theta].$$

6. $u_t = L[u] + uf\left(t, \frac{u}{w}\right)\ln u + ug\left(t, \frac{u}{w}\right),$

   $w_t = L[w] + wf\left(t, \frac{u}{w}\right)\ln w + wh\left(t, \frac{u}{w}\right).$

Solution (Polyanin & Manzhirov, 2007):

$$u(x,t) = \varphi(t)\psi(t)\theta(x_1,\ldots,x_n,t),$$
$$w(x,t) = \psi(t)\theta(x_1,\ldots,x_n,t),$$

where the functions $\varphi = \varphi(t)$ and $\psi = \psi(t)$ are determined by solving the ODE system

$$\begin{aligned}\varphi'_t &= \varphi[g(t,\varphi) - h(t,\varphi) + f(t,\varphi)\ln\varphi],\\ \psi'_t &= \psi[h(t,\varphi) + f(t,\varphi)\ln\psi],\end{aligned} \tag{1}$$

and the function $\theta = \theta(x_1, \ldots, x_n, t)$ is determined by the PDE:

$$\theta_t = L[\theta] + f(t,\varphi)\theta\ln\theta. \tag{2}$$

Given a solution to the first ODE in (1), the second ODE can be solved with the change of variable $\psi = e^\zeta$ by reducing it to a linear ODE for ζ. If L is a constant-coefficient one-dimensional operator ($n = 1$) and $f = $ const, then equation (2) has a traveling-wave solution $\theta = \theta(kx - \lambda t)$.

7. $u_t = L[u] + uf(au + bw) - bwg\left(\dfrac{w}{u}\right),$
   $w_t = L[w] + wf(au + bw) + awg\left(\dfrac{w}{u}\right).$

Here, L is an arbitrary linear differential operator in x (of any order in the derivatives) whose coefficients can depend on x.

1°. Solution (Polyanin & Zaitsev, 2012):
$$u = br(x,t)\cos^2\varphi(t), \quad w = ar(x,t)\sin^2\varphi(t),$$
where the function $\varphi = \varphi(t)$ is determined from the separable first-order ODE:
$$\varphi'_t = \tfrac{1}{2}a\tan\varphi\, g\!\left(\tfrac{a}{b}\tan^2\varphi\right), \tag{1}$$
and the function $r = r(x,t)$ satisfies the PDE:
$$r_t = L[r] + rf(abr). \tag{2}$$

The general solution to equation (1) can be written out in implicit form:
$$2\int \dfrac{d\varphi}{a\tan\varphi\, g(ab^{-1}\tan^2\varphi)} = t + C.$$

The nonlinear PDE (2) admits a time-independent solution of the form $r = r(x)$. If L is a linear differential operator with constant coefficients, then equation (2) admits a traveling wave solution $r = r(kx - \lambda t)$, where k and λ are arbitrary constants.

2°. Solution:
$$u = br(x,t)\cosh^2\psi(t), \quad w = -ar(x,t)\sinh^2\psi(t),$$
where the function $\psi = \psi(t)$ is determined by the separable first-order ODE:
$$\psi'_t = \tfrac{1}{2}a\tanh\psi\, g\!\left(-\tfrac{a}{b}\tanh^2\psi\right),$$
and the function $r = r(x,t)$ satisfies PDE (2).

3°. Multiplying the first equation by a and the second one by b and adding them together, one obtains the PDE:
$$z_t = L[z] + zf(z), \quad z = au + bw.$$

Remark 7.8. The solutions presented in Items 1° and 2° are easy to generalize to the case where the linear differential operator L depends on n spatial coordinates $x_1, \ldots, x_n$.

8. $u_t = L[u] + uf(u^2 + w^2) - wg\left(\dfrac{w}{u}\right), \quad w_t = L[w] + wf(u^2 + w^2) + ug\left(\dfrac{w}{u}\right).$

Here, L is an arbitrary linear differential operator in x (of any order in the derivatives) whose coefficients can depend on x.

Solution (Polyanin & Zaitsev, 2012):
$$u = r(x,t)\cos\varphi(t), \quad w = r(x,t)\sin\varphi(t),$$
where the function $\varphi = \varphi(t)$ is determined by the separable first-order ODE:
$$\varphi'_t = g(\tan\varphi), \tag{1}$$

and $r = r(x,t)$ satisfies the PDE:
$$r_t = L[r] + rf(r^2). \qquad (2)$$

The general solution to equation (1) can be written out in implicit form:
$$\int \frac{d\varphi}{g(\tan \varphi)} = t + C.$$

It is noteworthy that the nonlinear PDE (2) admits a time-independent solution of the form $r = r(x)$. If the coefficients of the linear differential operator L are constant, then equation (2) admits a traveling wave solution $r = r(kx - \lambda t)$, where k and λ are arbitrary constants.

Remark 7.9. *The above solution is easy to generalize to the case where the linear differential operator L depends on n spatial coordinates $x_1, \ldots, x_n$.*

9. $u_t = L[u] + uf(u^2 - w^2) + wg\left(\dfrac{w}{u}\right), \quad w_t = L[w] + wf(u^2 - w^2) + ug\left(\dfrac{w}{u}\right).$

Here, L is an arbitrary linear differential operator in x (of any order in the derivatives) whose coefficients can depend on x.

Solution (Polyanin & Zaitsev, 2012):
$$u = r(x,t)\cosh\varphi(t), \quad w = r(x,t)\sinh\varphi(t),$$
where the function $\varphi = \varphi(t)$ is described by the separable first-order ODE
$$\varphi'_t = g(\tanh\varphi), \qquad (1)$$
and $r = r(x,t)$ satisfies the PDE:
$$r_t = L[r] + rf(r^2). \qquad (2)$$

The general solution to equation (1) can be represented in implicit form:
$$\int \frac{d\varphi}{g(\tanh\varphi)} = t + C.$$

Note that the nonlinear PDE (2) admits a time-independent solution of the form $r = r(x)$. If the coefficients of the linear differential operator L are constant, then equation (2) admits a traveling wave solution $r = r(kx - \lambda t)$, where k and λ are arbitrary constants.

Remark 7.10. *The above solution is easy to generalize to the case where the linear differential operator L depends on n spatial coordinates $x_1, \ldots, x_n$.*

7.3.3. Nonlinear Systems of Two Equations Involving the Second Derivatives with Respect to t

1. $u_{tt} = L[u] + uf(t, au - bw) + g(t, au - bw),$
    $w_{tt} = L[w] + wf(t, au - bw) + h(t, au - bw).$

Here, L is an arbitrary linear differential operator (of any order) with respect to the spatial variables $x_1, \ldots, x_n$, whose coefficients can be dependent on $x_1, \ldots, x_n, t$. It is assumed that $L[\text{const}] = 0$.

1°. Solution (Polyanin & Manzhirov, 2007):
$$u = \varphi(t) + a\theta(x_1, \ldots, x_n, t), \quad w = \psi(t) + b\theta(x_1, \ldots, x_n, t),$$

where $\varphi = \varphi(t)$ and $\psi = \psi(t)$ are described by the ODE system

$$\varphi''_{tt} = \varphi f(t, a\varphi - b\psi) + g(t, a\varphi - b\psi), \qquad \psi''_{tt} = \psi f(t, a\varphi - b\psi) + h(t, a\varphi - b\psi),$$

and the function $\theta = \theta(x_1, \ldots, x_n, t)$ satisfies the linear PDE:

$$\theta_{tt} = L[\theta] + f(t, a\varphi - b\psi)\theta.$$

2°. Let us add together the first equation multiplied by a and the second one multiplied by $-b$ to obtain

$$\zeta_{tt} = L[\zeta] + \zeta f(t, \zeta) + ag(t, \zeta) - bh(t, \zeta), \qquad \zeta = au - bw. \qquad (1)$$

This PDE will be considered in conjunction with the first equation of the original system

$$u_{tt} = L[u] + uf(t, \zeta) + g(t, \zeta). \qquad (2)$$

Equation (1) can be treated separately. Given a solution $\zeta = \zeta(x_1, \ldots, x_n, t)$ to equation (1), the function $u = u(x_1, \ldots, x_n, t)$ can be determined by solving the linear equation (2), and the function $w = w(x_1, \ldots, x_n, t)$ is found as $w = (au - \zeta)/b$.

Note three important cases where equation (1) admits exact solutions:

(i) Equation (1) admits a spatially homogeneous solution $\zeta = \zeta(t)$.

(ii) Suppose the coefficients of L and the functions f, g, h are not explicitly dependent on t. Then equation (1) admits a steady-state solution $\zeta = \zeta(x_1, \ldots, x_n)$.

(iii) If the condition $\zeta f(t, \zeta) + bg(t, \zeta) - ch(t, \zeta) = k_1\zeta + k_0$ holds, equation (1) is linear. If the operator L is constant-coefficient, the method of separation of variables can be used to obtain solutions.

2. $u_t = L[u] + b_2 f(a_1 u + b_1 w) + b_1 g(a_2 u + b_2 w),$
   $w_t = L[w] - a_2 f(a_1 u + b_1 w) - a_1 g(a_2 u + b_2 w).$

Here, L is an arbitrary linear differential operator in x (of any order in the derivatives) whose coefficients can be dependent on x and t. It is assumed that $a_1 b_2 - a_2 b_1 \neq 0$.

1°. Multiplying the equations by suitable constants and adding together, one obtains two independent equations

$$\begin{aligned} U_t &= L[U] + (a_1 b_2 - a_2 b_1) f(U), & U &= a_1 u + b_1 w; \\ W_t &= L[W] - (a_1 b_2 - a_2 b_1) g(W), & W &= a_2 u + b_2 w. \end{aligned} \qquad (1)$$

2°. If L is an arbitrary linear differential operator with constant coefficients, then equations (1) admits traveling wave solutions

$$U = U(k_1 x - \lambda_1 t), \quad W = W(k_2 x - \lambda_2 t),$$

where k_m and λ_m are arbitrary constants. The corresponding solution of the original system will be a superposition (linear combination) of two nonlinear traveling waves.

3°. If the coefficients of the linear operator L depend on x alone, then equations (1) have simple solutions of the forms

$$U = U(t), \quad W = W(x); \qquad U = U(x), \quad W = W(t).$$

Remark 7.11. *The case where the linear differential operator L depends on several spatial coordinates $x_1, \ldots, x_n$ is treated likewise.*

3. $u_{tt} = L_1[u] + uf\left(\dfrac{u}{w}\right)$, $w_{tt} = L_2[w] + wg\left(\dfrac{u}{w}\right)$.

Here, L_1 and L_2 are arbitrary constant-coefficient linear differential operators with respect to x. It is assumed that $L_1[\text{const}] = 0$ and $L_2[\text{const}] = 0$.

1°. Solution in the form of the product of two waves traveling at different speeds (Polyanin & Manzhirov, 2007):

$$u = e^{kx - \lambda t} y(\xi), \quad w = e^{kx - \lambda t} z(\xi), \quad \xi = \beta x - \gamma t,$$

where k, λ, β, and γ are arbitrary constants, and the functions $y = y(\xi)$ and $z = z(\xi)$ are described by the ODE system

$$\gamma^2 y''_{\xi\xi} + 2\lambda \gamma y'_\xi + \lambda^2 y = M_1[y] + yf(y/z), \quad \gamma^2 z''_{\xi\xi} + 2\lambda\gamma z'_\xi + \lambda^2 z = M_2[z] + zg(y/z),$$
$$M_1[y] = e^{-kx} L_1[e^{kx} y(\xi)], \qquad\qquad M_2[z] = e^{-kx} L_2[e^{kx} z(\xi)].$$

To the special case $k = \lambda = 0$ there corresponds a traveling wave solution.

2°. Periodic multiplicative separable solution (Polyanin & Manzhirov, 2007):

$$u = [C_1 \sin(kt) + C_2 \cos(kt)] \varphi(x), \quad w = [C_1 \sin(kt) + C_2 \cos(kt)] \psi(x),$$

where C_1, C_2, and k are arbitrary constants, and the functions $\varphi = \varphi(x)$ and $\psi = \psi(x)$ are described by the ODE system

$$L_1[\varphi] + k^2 \varphi + \varphi f(\varphi/\psi) = 0, \quad L_2[\psi] + k^2 \psi + \psi g(\varphi/\psi) = 0.$$

3°. Multiplicative separable solution (Polyanin & Manzhirov, 2007):

$$u = [C_1 \sinh(kt) + C_2 \cosh(kt)] \varphi(x), \quad w = [C_1 \sinh(kt) + C_2 \cosh(kt)] \psi(x),$$

where C_1, C_2, and k are arbitrary constants, and the functions $\varphi = \varphi(x)$ and $\psi = \psi(x)$ are described by the ODE system

$$L_1[\varphi] - k^2 \varphi + \varphi f(\varphi/\psi) = 0, \quad L_2[\psi] - k^2 \psi + \psi g(\varphi/\psi) = 0.$$

4°. Degenerate multiplicative separable solution:

$$u = (C_1 t + C_2) \varphi(x), \quad w = (C_1 t + C_2) \psi(x),$$

where C_1 and C_2 are arbitrary constants, and the functions $\varphi = \varphi(x)$ and $\psi = \psi(x)$ are described by the ODE system

$$L_1[\varphi] + \varphi f(\varphi/\psi) = 0, \quad L_2[\psi] + \psi g(\varphi/\psi) = 0.$$

Remark 7.12. The coefficients of L_1, L_2 and the functions f and g in Items 2°–4° can be dependent on x.

Remark 7.13. If L_1 and L_2 contain only even derivatives, there are solutions of the form

$$u = [C_1 \sin(kx) + C_2 \cos(kx)] U(t), \quad w = [C_1 \sin(kx) + C_2 \cos(kx)] W(t);$$
$$u = [C_1 \exp(kx) + C_2 \exp(-kx)] U(t), \quad w = [C_1 \exp(kx) + C_2 \exp(-kx)] W(t);$$
$$u = (C_1 x + C_2) U(t), \quad w = (C_1 x + C_2) W(t),$$

where C_1, C_2, and k are arbitrary constants. Note that the third solution is degenerate.

4. $u_{tt} = L[u] + uf\left(t, \dfrac{u}{w}\right), \quad w_{tt} = L[w] + wg\left(t, \dfrac{u}{w}\right).$

Here, L is an arbitrary linear differential operator with respect to the coordinates $x_1, \ldots, x_n$ (of any order in the derivatives), whose coefficients can be dependent on the coordinates.

Solution (Polyanin & Manzhirov, 2007):

$$u = \varphi(t)\theta(x_1, \ldots, x_n), \quad w = \psi(t)\theta(x_1, \ldots, x_n),$$

where the functions $\varphi = \varphi(t)$ and $\psi = \psi(t)$ are described by the nonlinear system of second-order ODEs:

$$\varphi''_{tt} = a\varphi + \varphi f(t, \varphi/\psi), \quad \psi''_{tt} = a\psi + \psi g(t, \varphi/\psi),$$

a is an arbitrary constant, and the function $\theta = \theta(x_1, \ldots, x_n)$ satisfies the linear steady-state PDE:

$$L[\theta] = a\theta.$$

5. $u_{tt} = L[u] + uf\left(\dfrac{u}{w}\right) + g\left(\dfrac{u}{w}\right), \quad w_{tt} = L[w] + wf\left(\dfrac{u}{w}\right) + h\left(\dfrac{u}{w}\right).$

Here, L is an arbitrary linear differential operator with respect to the coordinates $x_1, \ldots, x_n$ (of any order in the derivatives), whose coefficients can be dependent on $x_1, \ldots, x_n, t$.

Solution (Polyanin & Manzhirov, 2007):

$$u = k\theta(x_1, \ldots, x_n, t), \quad w = \theta(x_1, \ldots, x_n, t),$$

where k is a root of the algebraic (transcendental) equation $g(k) = kh(k)$, and the function $\theta = \theta(x, t)$ satisfies the linear PDE:

$$\theta_{tt} = L[\theta] + f(k)\theta + h(k).$$

6. $u_{tt} = L[u] + au \ln u + uf\left(t, \dfrac{u}{w}\right), \quad w_{tt} = L[w] + aw \ln w + wg\left(t, \dfrac{u}{w}\right).$

Here, L is an arbitrary linear differential operator with respect to the coordinates $x_1, \ldots, x_n$ (of any order in the derivatives), whose coefficients can be dependent on the coordinates.

Solution (Polyanin & Manzhirov, 2007):

$$u = \varphi(t)\theta(x_1, \ldots, x_n), \quad w = \psi(t)\theta(x_1, \ldots, x_n),$$

where the functions $\varphi = \varphi(t)$ and $\psi = \psi(t)$ are described by the nonlinear system of second-order ODEs:

$$\varphi''_{tt} = a\varphi \ln \varphi + b\varphi + \varphi f(t, \varphi/\psi), \quad \psi''_{tt} = a\psi \ln \psi + b\psi + \psi g(t, \varphi/\psi),$$

b is an arbitrary constant, and the function $\theta = \theta(x_1, \ldots, x_n)$ satisfies the steady-state PDE:

$$L[\theta] + a\theta \ln \theta - b\theta = 0.$$

▶ For exact solutions to other nonlinear systems of equations of mathematical physics with two or more PDEs (including multidimensional PDEs), see, for example, the books by Drazin & Riley (2006), Stephani et al. (2009), Polyanin & Zaitsev (2012), Cherniha & Davydovych (2017), and Khawaja & Sakkaf (2019).

References

Akhatov, I.Sh., Gazizov, R.K., and Ibragimov, N.H., Nonlocal symmetries. Heuristic approach, *J. Soviet Math.*, Vol. 55, No. 1, pp. 1401–1450, 1991.

Aksenov, A.V., Nonlinear periodic waves in a gas, *Fluid Dyn.*, Vol. 47, No. 5, pp. 636–646, 2012.

Barannyk, T., Symmetry and exact solutions for systems of nonlinear reaction-diffusion equations, *Proc. Inst. Math. NAS Ukraine,* Vol. 43, Part 1, pp. 80–85, 2002.

Barannyk, T.A. and Nikitin, A.G., *Proc. Inst. Math. NAS Ukraine,* Vol. 50, Part 1, pp. 34–39, 2004.

Berman, V.S., Group properties of a hyperbolic system of two differential equations encountered in the theory of mass transfer, *Fluid Dyn.*, Vol. 16, No. 3, pp. 367–373, 1981.

Berman, V.S., Galin, L.A., and Churmaev, O.M., Analysis of a simple model of a bubble-liquid reactor, *Fluid Dyn.*, Vol. 14, No. 5, pp. 740–747, 1979.

Cherniha, R. and Davydovych, V., *Nonlinear Reaction-Diffusion Systems: Conditional Symmetry, Exact Solutions and Their Applications in Biology*, Springer, Cham, 2017.

Cherniha, R. and Davydovych, V., New conditional symmetries and exact solutions of the diffusive two-component Lotka–Volterra system. *Mathematics*, Vol. 9, No. 16, 1984, 2021.

Chernyi, G. G., *Gas Dynamics* [in Russian], Nauka, Moscow, 1988.

Courant, R., *Partial Differential Equations*, InterScience, New York, 1962.

Courant, R. and Friedrichs, R., *Supersonic Flow and Shock Waves*, Springer-Verlag, New York, 1985.

Drazin, P. G. and Riley, N., *The Navier–Stokes Equations: A Classification of Flows and Exact Solutions*, Cambridge Univ. Press, Cambridge, 2006.

Khawaja, U. and Sakkaf, L., *Handbook of Exact Solutions to the Nonlinear Schrödinger Equations*, Inst. Physics Publ. (IOP), Bristol, 2019.

Nikitin, A.G. and Wiltshire, R.J., Systems of reaction-diffusion equations and their symmetry properties, *J. Math. Phys.*, Vol. 42, No. 4, pp. 1667–1688, 2001.

Ovsiannikov, L.V., *Group Properties of Differential Equations* [in Russian], Izd-vo SO AN USSR, Novosibirsk, 1962 (English translation by G. Bluman, 1967).

Ovsiannikov, L.V., *Lectures on Fundamentals of Gas Dynamics* [in Russian], Nauka, Moscow, 1981.

Polyanin, A. D., Exact solutions of nonlinear sets of equations of the theory of heat and mass transfer in reactive media and mathematical biology, *Theor. Found. Chem. Eng.*, Vol. 38, No. 6, pp. 622–635, 2004.

Polyanin, A. D., Exact solutions of nonlinear systems of diffusion equations for reacting media and mathematical biology, *Dokl. Math.*, Vol. 71, No. 1, pp. 148–154, 2005.

Polyanin, A.D. and Manzhirov, A.V., *Handbook of Mathematics for Engineers and Scientists*, Chapman & Hall/CRC Press, Boca Raton–London, 2007.

Polyanin, A.D. and Nazaikinskii, V.E., *Handbook of Linear Partial Differential Equations for Engineers and Scientists, 2nd ed.*, Chapman & Hall/CRC Press, Boca Raton–London, 2016.

Polyanin, A.D. and Sorokin, V.G., Reductions and exact solutions of Lotka–Volterra and more complex reaction-diffusion systems with delays, *Appl. Math. Lett.*, Vol. 125, 107731, 2022.

Polyanin, A.D., Sorokin, V.G., and Zhurov, A.I., *Delay Ordinary and Partial Differential Equations*, CRC Press, Boca Raton, 2023.

Polyanin, A.D. and Vyazmina, E.A., New classes of exact solutions to nonlinear systems of reaction-diffusion equations, *Dokl. Math.*, Vol. 74, No. 1, pp. 597–602, 2006.

Polyanin, A.D. and Zaitsev, V.F., *Handbook of Nonlinear Partial Differential Equations, 2nd ed.*, Chapman & Hall/CRC Press, Boca Raton–London, 2012.

Polyanin, A.D. and Zhurov, A.I., Multi-parameter reaction-diffusion systems with quadratic nonlinearity and delays: New exact solutions in elementary functions. *Mathematics*, Vol. 10, No. 9, 529, 2022.

Rozhdestvenskii, B.L. and Yanenko, N.N., *Systems of Quasilinear Equations and Their Applications to Gas Dynamics*, American Mathematical Society, Providence, 1983.

Stanyukovich, K.P., *Unsteady Motions of a Continuous Medium, 2nd ed.* [in Russian], Nauka, Moscow, 1971.

Stephani, H., Kramer, D., MacCallum, M., Hoenselaers, C., and Herlt, E., *Exact Solutions of Einstein's Field Equations, 2nd ed.*, Cambridge Univ. Press, Cambridge, 2009.

Chapter 8

Integral Equations

▶ **Preliminary remarks.** Integral equations are mathematical equations that contain an unknown function under the integral sign.

This chapter describes exact solutions of various linear and nonlinear integral equations of the first and second kind (in the first case, the unknown function is only included in the integrand, while in the second, it is contained both under the sign of the integral and outside it). Both Volterra-type equations with a variable integration limit and Fredholm-type equations with constant integration limits are considered. Reductions that lead some integral equations to ordinary differential equations, which are usually easier to solve than the original equations, are also described. Particular attention is paid to integral equations of a fairly general form that involve arbitrary functions, and their solutions.

8.1. Integral Equations of the First Kind with Variable Limit of Integration

8.1.1. Linear Volterra Integral Equations of the First Kind

▶ **Equations with kernels containing power law functions.**

1. $\int_a^x y(t)\, dt = f(x), \qquad f(a) = 0.$

Solution: $y(x) = f'_x(x)$.

2. $\int_a^x (x-t) y(t)\, dt = f(x), \qquad f(a) = f'_x(a) = 0.$

Solution: $y(x) = f''_{xx}(x)$.

3. $\int_a^x (Ax + Bt + C) y(t)\, dt = f(x), \qquad f(a) = 0.$

   1°. Solution for $B \neq -A$:
   $$y(x) = \frac{d}{dx}\left\{ [(A+B)x + C]^{-\frac{A}{A+B}} \int_a^x [(A+B)t + C]^{-\frac{B}{A+B}} f'_t(t)\, dt \right\}.$$

   2°. Solution for $B = -A$:
   $$y(x) = \frac{1}{C}\frac{d}{dx}\left[\exp\left(-\frac{A}{C}x\right) \int_a^x \exp\left(\frac{A}{C}t\right) f'_t(t)\, dt \right].$$

DOI: 10.1201/9781003051329-8

4. $\int_a^x (x-t)^n y(t)\,dt = f(x), \quad n = 1, 2, \ldots$

It is assumed that the right-hand side of the equation satisfies the conditions $f(a) = f'_x(a) = \cdots = f_x^{(n)}(a) = 0$.

Solution:
$$y(x) = \frac{1}{n!} f_x^{(n+1)}(x).$$

5. $\int_a^x (x^n - t^n) y(t)\,dt = f(x), \quad f(a) = f'_x(a) = 0, \quad n = 1, 2, \ldots$

Solution:
$$y(x) = \frac{1}{n} \frac{d}{dx}\left[\frac{f'_x(x)}{x^{n-1}}\right].$$

6. $\int_a^x \sqrt{x-t}\, y(t)\,dt = f(x), \quad f(a) = 0.$

Solution:
$$y(x) = \frac{2}{\pi} \frac{d^2}{dx^2} \int_a^x \frac{f(t)\,dt}{\sqrt{x-t}}.$$

7. $\int_a^x \frac{y(t)\,dt}{\sqrt{x-t}} = f(x).$

Abel's equation.

Solution (Whittaker & Watson, 1958):
$$y(x) = \frac{1}{\pi} \frac{d}{dx} \int_a^x \frac{f(t)\,dt}{\sqrt{x-t}} = \frac{f(a)}{\pi\sqrt{x-a}} + \frac{1}{\pi} \int_a^x \frac{f'_t(t)\,dt}{\sqrt{x-t}}.$$

8. $\int_a^x (x-t)^\lambda y(t)\,dt = f(x), \quad f(a) = 0, \quad 0 < \lambda < 1.$

Solution (Gakhov, 1990):
$$y(x) = \frac{\sin(\pi\lambda)}{\pi\lambda} \frac{d^2}{dx^2} \int_a^x \frac{f(t)\,dt}{(x-t)^\lambda}.$$

9. $\int_a^x \frac{y(t)\,dt}{(x-t)^\lambda} = f(x), \quad 0 < \lambda < 1.$

Generalized Abel equation.

Solution (Whittaker & Watson, 1958):
$$y(x) = \frac{\sin(\pi\lambda)}{\pi} \frac{d}{dx} \int_a^x \frac{f(t)\,dt}{(x-t)^{1-\lambda}} = \frac{\sin(\pi\lambda)}{\pi}\left[\frac{f(a)}{(x-a)^{1-\lambda}} + \int_a^x \frac{f'_t(t)\,dt}{(x-t)^{1-\lambda}}\right].$$

▶ **Equations with kernels containing exponential functions.**

10. $\int_a^x e^{\lambda(x-t)} y(t)\,dt = f(x), \quad f(a) = 0.$

Solution:
$$y(x) = f'_x(x) - \lambda f(x).$$

11. $\int_a^x e^{\lambda x + \beta t} y(t)\, dt = f(x)$, $f(a) = 0$.

Solution:
$$y(x) = e^{-(\lambda+\beta)x}\left[f'_x(x) - \lambda f(x)\right].$$

12. $\int_a^x \left[e^{\lambda(x-t)} - 1\right] y(t)\, dt = f(x)$, $f(a) = f'_x(a) = 0$.

Solution:
$$y(x) = \frac{1}{\lambda} f''_{xx}(x) - f'_x(x).$$

13. $\int_a^x \left[e^{\lambda(x-t)} + b\right] y(t)\, dt = f(x)$, $f(a) = 0$.

For $b = -1$, see equation 8.1.1.12.
  Solution for $b \neq -1$:

$$y(x) = \frac{f'_x(x)}{b+1} - \frac{\lambda}{(b+1)^2} \int_a^x \exp\left[\frac{\lambda b}{b+1}(x-t)\right] f'_t(t)\, dt.$$

14. $\int_a^x \left[e^{\lambda(x-t)} - e^{\mu(x-t)}\right] y(t)\, dt = f(x)$, $f(a) = f'_x(a) = 0$.

Solution:
$$y(x) = \frac{1}{\lambda - \mu}\left[f''_{xx} - (\lambda + \mu) f'_x + \lambda\mu f\right], \qquad f = f(x).$$

15. $\int_a^x \frac{y(t)\, dt}{\sqrt{e^{\lambda x} - e^{\lambda t}}} = f(x)$, $\lambda > 0$.

Solution:
$$y(x) = \frac{\lambda}{\pi}\frac{d}{dx}\int_a^x \frac{e^{\lambda t} f(t)\, dt}{\sqrt{e^{\lambda x} - e^{\lambda t}}}.$$

16. $\int_a^x (e^{\lambda x} - e^{\lambda t})^\mu y(t)\, dt = f(x)$, $f(a) = 0$, $\lambda > 0$, $0 < \mu < 1$.

Solution:
$$y(x) = k e^{\lambda x}\left(e^{-\lambda x}\frac{d}{dx}\right)^2 \int_a^x \frac{e^{\lambda t} f(t)\, dt}{(e^{\lambda x} - e^{\lambda t})^\mu}, \qquad k = \frac{\sin(\pi\mu)}{\pi\mu}.$$

17. $\int_a^x \frac{y(t)\, dt}{(e^{\lambda x} - e^{\lambda t})^\mu} = f(x)$, $\lambda > 0$, $0 < \mu < 1$.

Solution:
$$y(x) = \frac{\lambda \sin(\pi\mu)}{\pi}\frac{d}{dx}\int_a^x \frac{e^{\lambda t} f(t)\, dt}{(e^{\lambda x} - e^{\lambda t})^{1-\mu}}.$$

18. $\int_a^x (x-t)^\lambda e^{\mu(x-t)} y(t)\, dt = f(x)$, $f(a) = 0$, $0 < \lambda < 1$.

Solution:
$$y(x) = k e^{\mu x}\frac{d^2}{dx^2}\int_a^x \frac{e^{-\mu t} f(t)\, dt}{(x-t)^\lambda}, \qquad k = \frac{\sin(\pi\lambda)}{\pi\lambda}.$$

▶ **Equations with kernels containing hyperbolic functions.**

19. $\int_a^x \cosh[\lambda(x-t)]y(t)\,dt = f(x), \qquad f(a) = 0.$

Solution:
$$y(x) = f'_x(x) - \lambda^2 \int_a^x f(x)\,dx.$$

20. $\int_a^x \{\cosh[\lambda(x-t)] - 1\}y(t)\,dt = f(x), \qquad f(a) = f'_x(a) = f''_{xx}(x) = 0.$

Solution:
$$y(x) = \frac{1}{\lambda^2} f'''_{xxx}(x) - f'_x(x).$$

21. $\int_a^x \{\cosh[\lambda(x-t)] + b\}y(t)\,dt = f(x), \qquad f(a) = 0.$

For $b = 0$, see equation 8.1.1.19. For $b = -1$, see equation 8.1.1.20.

1°. Solution for $b(b+1) < 0$:
$$y(x) = \frac{f'_x(x)}{b+1} - \frac{\lambda^2}{k(b+1)^2}\int_a^x \sin[k(x-t)]f'_t(t)\,dt, \qquad \text{where}\quad k = \lambda\sqrt{\frac{-b}{b+1}}.$$

2°. Solution for $b(b+1) > 0$:
$$y(x) = \frac{f'_x(x)}{b+1} - \frac{\lambda^2}{k(b+1)^2}\int_a^x \sinh[k(x-t)]f'_t(t)\,dt, \qquad \text{where}\quad k = \lambda\sqrt{\frac{b}{b+1}}.$$

22. $\int_a^x \cosh^2[\lambda(x-t)]y(t)\,dt = f(x), \qquad f(a) = 0.$

Solution:
$$y(x) = f'_x(x) - \frac{2\lambda^2}{k}\int_a^x \sinh[k(x-t)]f'_t(t)\,dt, \qquad \text{where}\quad k = \lambda\sqrt{2}.$$

23. $\int_a^x \sinh[\lambda(x-t)]y(t)\,dt = f(x), \qquad f(a) = f'_x(a) = 0.$

Solution:
$$y(x) = \frac{1}{\lambda} f''_{xx}(x) - \lambda f(x).$$

24. $\int_a^x \{\sinh[\lambda(x-t)] + b\}y(t)\,dt = f(x), \qquad f(a) = 0.$

For $b = 0$, see equation 8.1.1.23.

Solution for $b \neq 0$:
$$y(x) = \frac{1}{b} f'_x(x) + \int_a^x R(x-t)f'_t(t)\,dt,$$

$$R(x) = \frac{\lambda}{b^2}\exp\left(-\frac{\lambda x}{2b}\right)\left[\frac{\lambda}{2bk}\sinh(kx) - \cosh(kx)\right], \qquad k = \frac{\lambda\sqrt{1+4b^2}}{2b}.$$

25. $\int_a^x \sinh(\lambda\sqrt{x-t})y(t)\,dt = f(x), \qquad f(a) = 0.$

Solution: $y(x) = \dfrac{2}{\pi\lambda}\dfrac{d^2}{dx^2}\displaystyle\int_a^x \dfrac{\cos(\lambda\sqrt{x-t})}{\sqrt{x-t}} f(t)\,dt.$

▶ **Equations with kernels containing logarithmic functions.**

26. $\int_a^x (\ln x - \ln t) y(t)\, dt = f(x)$, $\quad f(a) = f'_x(a) = 0$.

Solution:
$$y(x) = x f''_{xx}(x) + f'_x(x).$$

27. $\int_0^x \ln(x - t) y(t)\, dt = f(x)$.

Solution (Krasnov et al., 1971):
$$y(x) = -\int_0^x f''_{tt}(t)\, dt \int_0^\infty \frac{(x-t)^z e^{-Cz}}{\Gamma(z+1)}\, dz - f'_x(0) \int_0^\infty \frac{x^z e^{-Cz}}{\Gamma(z+1)}\, dz,$$

where $C = \lim\limits_{k\to\infty} \left(1 + \dfrac{1}{2} + \cdots + \dfrac{1}{k+1} - \ln k\right) = 0.5772\ldots$ is the Euler constant and $\Gamma(z)$ is the gamma function.

28. $\int_a^x [\ln(x-t) + A] y(t)\, dt = f(x)$, $\quad f(a) = 0$.

For $A = 0$, see equation 8.1.1.27.

Solution for $A \neq 0$ (Samko et al. (1993)):
$$y(x) = -\frac{d}{dx} \int_a^x \nu_A(x-t) f(t)\, dt, \quad \nu_A(x) = \frac{d}{dx} \int_0^\infty \frac{x^z e^{(A-C)z}}{\Gamma(z+1)}\, dz,$$

where $C = 0.5772\ldots$ is the Euler constant and $\Gamma(z)$ is the gamma function.

For $a = 0$, the solution can be written in the form
$$y(x) = -\int_0^x f''_{tt}(t)\, dt \int_0^\infty \frac{(x-t)^z e^{(A-C)z}}{\Gamma(z+1)}\, dz - f'_x(0) \int_0^\infty \frac{x^z e^{(A-C)z}}{\Gamma(z+1)}\, dz.$$

29. $\int_a^x (x-t) \big[\ln(x-t) + A\big] y(t)\, dt = f(x)$, $\quad f(a) = 0$.

Solution:
$$y(x) = -\frac{d^2}{dx^2} \int_a^x \nu_A(x-t) f(t)\, dt, \quad \nu_A(x) = \frac{d}{dx} \int_0^\infty \frac{x^z e^{(A-C)z}}{\Gamma(z+1)}\, dz,$$

where $C = 0.5772\ldots$ is the Euler constant and $\Gamma(z)$ is the gamma function.

▶ **Equations with kernels containing trigonometric functions.**

30. $\int_a^x \cos[\lambda(x-t)] y(t)\, dt = f(x)$, $\quad f(a) = 0$.

Solution:
$$y(x) = f'_x(x) + \lambda^2 \int_a^x f(x)\, dx.$$

31. $\int_a^x \{\cos[\lambda(x-t)] - 1\} y(t)\, dt = f(x)$, $\quad f(a) = f'_x(a) = f''_{xx}(a) = 0$.

Solution:
$$y(x) = -\frac{1}{\lambda^2} f'''_{xxx}(x) - f'_x(x).$$

32. $\int_a^x \{\cos[\lambda(x-t)] + b\} y(t)\, dt = f(x), \qquad f(a) = 0.$

For $b = 0$, see equation 8.1.1.30. For $b = -1$, see equation 8.1.1.31.

1°. Solution for $b(b+1) > 0$:

$$y(x) = \frac{f'_x(x)}{b+1} + \frac{\lambda^2}{k(b+1)^2} \int_a^x \sin[k(x-t)] f'_t(t)\, dt, \qquad \text{where} \quad k = \lambda\sqrt{\frac{b}{b+1}}.$$

2°. Solution for $b(b+1) < 0$:

$$y(x) = \frac{f'_x(x)}{b+1} + \frac{\lambda^2}{k(b+1)^2} \int_a^x \sinh[k(x-t)] f'_t(t)\, dt, \qquad \text{where} \quad k = \lambda\sqrt{\frac{-b}{b+1}}.$$

33. $\int_a^x \sin[\lambda(x-t)] y(t)\, dt = f(x), \qquad f(a) = f'_x(a) = 0.$

Solution:

$$y(x) = \frac{1}{\lambda} f''_{xx}(x) + \lambda f(x).$$

34. $\int_a^x \sin\left(\lambda\sqrt{x-t}\right) y(t)\, dt = f(x), \qquad f(a) = 0.$

Solution:

$$y(x) = \frac{2}{\pi\lambda} \frac{d^2}{dx^2} \int_a^x \frac{\cosh\left(\lambda\sqrt{x-t}\right)}{\sqrt{x-t}} f(t)\, dt.$$

▶ **Equations with kernels containing special functions.**

35. $\int_a^x J_0(\lambda(x-t)) y(t)\, dt = f(x).$

Here, $J_0(z)$ is the Bessel function of the first kind.

If $f(a) = f'_x(a) = 0$ then the solution is (Srivastava & Buschman, 1977 and Prudnikov et al., 1992):

$$y(x) = \int_a^x J_0(\lambda(x-t)) \left(\frac{d^2}{dt^2} + \lambda^2\right) f(t)\, dt.$$

36. $\int_a^x J_0(\lambda\sqrt{x-t}) y(t)\, dt = f(x).$

Here, $J_0(z)$ is the Bessel function of the first kind.

If $f(a) = f'_x(a) = 0$ then the solution is (Prudnikov et al., 1992)

$$y(x) = \frac{d^2}{dx^2} \int_a^x I_0(\lambda\sqrt{x-t}) f(t)\, dt.$$

37. $\int_a^x I_0(\lambda(x-t)) y(t)\, dt = f(x).$

Here, $I_0(z)$ is the modified Bessel function of the first kind.

If $f(a) = f'_x(a) = 0$, then the solution is (Prudnikov et al., 1992):

$$y(x) = \int_a^x I_0(\lambda(x-t)) \left(\frac{d^2}{dt^2} - \lambda^2\right) f(t)\, dt.$$

38. $\int_a^x I_0(\lambda\sqrt{x-t})y(t)\,dt = f(x).$

Here, $I_0(z)$ is the modified Bessel function of the first kind. If $f(a) = f'_x(a) = 0$ then the solution is

$$y(x) = \frac{d^2}{dx^2}\int_a^x J_0(\lambda\sqrt{x-t})f(t)\,dt.$$

▶ **Equations with kernels containing arbitrary functions.**

39. $\int_a^x [g(x) - g(t)]y(t)\,dt = f(x).$

It is assumed that $f(a) = f'_x(a) = 0$ and $f'_x/g'_x \neq \text{const}$.

Solution: $y(x) = \dfrac{d}{dx}\left[\dfrac{f'_x(x)}{g'_x(x)}\right].$

40. $\int_a^x [g(x) - g(t) + b]y(t)\,dt = f(x), \qquad f(a) = 0.$

For $b = 0$, see equation 8.1.1.39.

Solution for $b \neq 0$:

$$y(x) = \frac{1}{b}f'_x(x) - \frac{1}{b^2}g'_x(x)\int_a^x \exp\left[\frac{g(t) - g(x)}{b}\right]f'_t(t)\,dt.$$

41. $\int_a^x [g(x) + h(t)]y(t)\,dt = f(x), \qquad f(a) = 0.$

For $h(t) = -g(t)$, see equation 8.1.1.40.

Solution:

$$y(x) = \frac{d}{dx}\left[\frac{\Phi(x)}{g(x) + h(x)}\int_a^x \frac{f'_t(t)\,dt}{\Phi(t)}\right], \qquad \Phi(x) = \exp\left[\int_a^x \frac{h'_t(t)\,dt}{g(t) + h(t)}\right].$$

42. $\int_a^x [g(x) - g(t)]^n y(t)\,dt = f(x), \qquad n = 1, 2, \ldots$

The right-hand side of the equation is assumed to satisfy the conditions $f(a) = f'_x(a) = \cdots = f_x^{(n)}(a) = 0$.

Solution: $y(x) = \dfrac{1}{n!}g'_x(x)\left(\dfrac{1}{g'_x(x)}\dfrac{d}{dx}\right)^{n+1} f(x).$

43. $\int_a^x \sqrt{g(x) - g(t)}\, y(t)\,dt = f(x), \qquad f(a) = 0, \quad g'_x(x) > 0.$

Solution:

$$y(x) = \frac{2}{\pi}g'_x(x)\left(\frac{1}{g'_x(x)}\frac{d}{dx}\right)^2 \int_a^x \frac{f(t)g'_t(t)\,dt}{\sqrt{g(x) - g(t)}}.$$

44. $\displaystyle\int_a^x \frac{y(t)\,dt}{\sqrt{g(x)-g(t)}} = f(x), \qquad g_x'(x) > 0.$

Solution:
$$y(x) = \frac{1}{\pi}\frac{d}{dx}\int_a^x \frac{f(t)g_t'(t)\,dt}{\sqrt{g(x)-g(t)}}.$$

45. $\displaystyle\int_a^x [g(x)-g(t)]^\lambda y(t)\,dt = f(x), \qquad f(a) = 0, \quad 0 < \lambda < 1.$

Solution:
$$y(x) = kg_x'(x)\left(\frac{1}{g_x'(x)}\frac{d}{dx}\right)^2 \int_a^x \frac{g_t'(t)f(t)\,dt}{[g(x)-g(t)]^\lambda}, \qquad k = \frac{\sin(\pi\lambda)}{\pi\lambda}.$$

46. $\displaystyle\int_a^x \frac{h(t)y(t)\,dt}{[g(x)-g(t)]^\lambda} = f(x), \qquad g_x' > 0, \quad 0 < \lambda < 1.$

Solution:
$$y(x) = \frac{\sin(\pi\lambda)}{\pi h(x)}\frac{d}{dx}\int_a^x \frac{f(t)g_t'(t)\,dt}{[g(x)-g(t)]^{1-\lambda}}.$$

47. $\displaystyle\int_a^x K(x-t)y(t)\,dt = f(x).$

$1°$. Let $K(0) = 1$ and $f(a) = 0$. Differentiating the equation with respect to x yields a Volterra equation of the second kind:
$$y(x) + \int_a^x K_x'(x-t)y(t)\,dt = f_x'(x).$$

The solution of this equation can be represented in the form
$$y(x) = f_x'(x) + \int_a^x R(x-t)f_t'(t)\,dt.$$

Here the resolvent $R(x)$ is related to the kernel $K(x)$ of the original equation by
$$R(x) = \mathfrak{L}^{-1}\!\left[\frac{1}{p\tilde{K}(p)} - 1\right], \qquad \tilde{K}(p) = \mathfrak{L}[K(x)],$$

where $\mathfrak{L}$ and $\mathfrak{L}^{-1}$ are the operators of the direct and inverse Laplace transforms, respectively,
$$\tilde{K}(p) = \mathfrak{L}[K(x)] = \int_0^\infty e^{-px}K(x)\,dx, \qquad R(x) = \mathfrak{L}^{-1}[\tilde{R}(p)] = \frac{1}{2\pi i}\int_{c-i\infty}^{c+i\infty} e^{px}\tilde{R}(p)\,dp.$$

$2°$. Let $K(x)$ have an integrable power-law singularity at $x = 0$. Denote by $w = w(x)$ the solution of the simpler auxiliary equation (compared with the original equation) with $a = 0$ and constant right-hand side $f \equiv 1$:
$$\int_0^x K(x-t)w(t)\,dt = 1.$$

Then the solution of the original integral equation with arbitrary right-hand side is expressed in terms of w as follows:
$$y(x) = \frac{d}{dx}\int_a^x w(x-t)f(t)\,dt = f(a)w(x-a) + \int_a^x w(x-t)f_t'(t)\,dt.$$

$3°$. For this integral equation, see also Titchmarsh (1986) and Samko et al. (1993).

8.1.2. Nonlinear Volterra Integral Equations of the First Kind

▶ **Equations with quadratic nonlinearity.**

1. $\int_0^x y(t)y(x-t)\,dt = Ax + B, \quad A, B > 0.$

Solutions:
$$y(x) = \pm\sqrt{B}\left[\frac{1}{\sqrt{\pi x}}\exp\left(-\frac{A}{B}x\right) + \sqrt{\frac{A}{B}}\,\mathrm{erf}\left(\sqrt{\frac{A}{B}}\,x\right)\right],$$

where $\mathrm{erf}\,z = \dfrac{2}{\sqrt{\pi}}\displaystyle\int_0^z \exp(-t^2)\,dt$ is the error function.

2. $\int_0^x y(t)y(x-t)\,dt = A^2 x^\lambda.$

Solutions (Polyanin & Manzhirov, 2008):
$$y(x) = \pm A\frac{\sqrt{\Gamma(\lambda+1)}}{\Gamma\left(\frac{\lambda+1}{2}\right)}\,x^{\frac{\lambda-1}{2}},$$

where $\Gamma(z)$ is the gamma function.

3. $\int_0^x y(t)y(x-t)\,dt = A^2 e^{\lambda x}.$

Solutions (Polyanin & Manzhirov, 2008): $y(x) = \pm\dfrac{A}{\sqrt{\pi x}}e^{\lambda x}.$

4. $\int_0^x y(t)y(x-t)\,dt = A^2 \cosh(\lambda x).$

Solutions:
$$y(x) = \pm\frac{A}{\sqrt{\pi}}\frac{d}{dx}\int_0^x \frac{I_0(\lambda t)\,dt}{\sqrt{x-t}},$$

where $I_0(z)$ is the modified Bessel function.

5. $\int_0^x y(t)y(x-t)\,dt = A\sinh(\lambda x).$

Solutions: $y = \pm\sqrt{A\lambda}\,I_0(\lambda x)$, where $I_0(z)$ is the modified Bessel function.

6. $\int_0^x y(t)y(x-t)\,dt = A\sinh(\lambda\sqrt{x}\,).$

Solutions:
$$y = \pm\sqrt{A}\,\pi^{1/4}2^{-7/8}\lambda^{3/4}x^{-1/8}I_{-1/4}\left(\lambda\sqrt{\tfrac{1}{2}x}\right),$$

where $I_{-1/4}(z)$ is the modified Bessel function.

7. $\int_0^x y(t)y(x-t)\,dt = A^2\cos(\lambda x).$

Solutions:
$$y(x) = \pm\frac{A}{\sqrt{\pi}}\frac{d}{dx}\int_0^x \frac{J_0(\lambda t)\,dt}{\sqrt{x-t}},$$

where $J_0(z)$ is the Bessel function.

8. $\int_0^x y(t)y(x-t)\,dt = A\sin(\lambda x).$

Solutions: $y = \pm\sqrt{A\lambda}\,J_0(\lambda x)$, where $J_0(z)$ is the Bessel function.

9. $\int_0^x y(t)y(x-t)\,dt = A\sin(\lambda\sqrt{x}).$

Solutions:
$$y = \pm\sqrt{A}\,\pi^{1/4}2^{-7/8}\lambda^{3/4}x^{-1/8}J_{-1/4}\!\left(\lambda\sqrt{\tfrac{1}{2}x}\right),$$

where $J_{-1/4}(z)$ is the Bessel function.

10. $\int_0^x t^k y(t)y(x-t)\,dt = Ax^\mu e^{\lambda x}.$

Solutions:
$$y(x) = \pm\left[\frac{A\Gamma(\mu+1)}{\Gamma\!\left(\frac{\mu+k+1}{2}\right)\Gamma\!\left(\frac{\mu-k+1}{2}\right)}\right]^{1/2} x^{\frac{\mu-k-1}{2}} e^{\lambda x},$$

where $\Gamma(z)$ is the gamma function.

11. $\int_0^x \dfrac{y(t)y(x-t)}{ax+bt}\,dt = Ax^\mu e^{\lambda x}.$

Solutions (Polyanin & Manzhirov, 2008):
$$y(x) = \pm\sqrt{\frac{A}{I}}\,x^{\mu/2}e^{\lambda x}, \qquad I = \int_0^1 z^{\mu/2}(1-z)^{\mu/2}\,\frac{dz}{a+bz}.$$

12. $\int_0^x y(t)y(xt)\,dt = Ax^\mu.$

Solutions (Polyanin & Manzhirov, 2008):
$$y(x) = \pm\sqrt{\tfrac{1}{3}A(2\mu+1)}\,x^{\frac{\mu-1}{3}} \qquad (A>0,\ \mu\geq 0).$$

13. $\int_a^x K(t)y(x)y(t)\,dt = f(x).$

Solutions:
$$y(x) = \pm f(x)\left[2\int_a^x K(t)f(t)\,dt\right]^{-1/2}.$$

14. $\int_0^x f\!\left(\dfrac{t}{x}\right)y(t)y(x-t)\,dt = Ax^\mu e^{\lambda x}.$

Solutions:
$$y(x) = \pm\sqrt{\frac{A}{I}}\,x^{\frac{\mu-1}{2}}e^{\lambda x}, \qquad I = \int_0^1 f(z) z^{\frac{\mu-1}{2}}(1-z)^{\frac{\mu-1}{2}}\,dz.$$

▶ **Other integral equations.**

15. $\int_a^x f(t, y(t))\, dt = g(x), \qquad g(a) = 0.$

Solution in implicit form:
$$f(x, y) - g'_x(x) = 0.$$

16. $\int_a^x (x - t) f(t, y(t))\, dt = g(x), \qquad g(a) = g'(a) = 0.$

Solution in implicit form:
$$f(x, y) - g''_{xx}(x) = 0.$$

17. $\int_a^x e^{\lambda(x-t)} f(t, y(t))\, dt = g(x), \qquad g(a) = 0.$

Solution in implicit form:
$$f(x, y) + \lambda g(x) - g'_x(x) = 0.$$

18. $\int_a^x \sinh[\lambda(x - t)] f(t, y(t))\, dt = g(x), \qquad g(a) = g'(a) = 0.$

Solution in implicit form:
$$\lambda f(x, y) + \lambda^2 g(x) - g''_{xx}(x) = 0.$$

19. $\int_a^x \cosh[\lambda(x - t)] f(t, y(t))\, dt = g(x), \qquad g(a) = 0.$

Solution in implicit form:
$$f(x, y) + \lambda^2 \int_a^x g(t)\, dt - g'_x(x) = 0.$$

20. $\int_a^x \sin[\lambda(x - t)] f(t, y(t))\, dt = g(x), \qquad g(a) = g'(a) = 0.$

Solution in implicit form:
$$\lambda f(x, y) - \lambda^2 g(x) - g''_{xx}(x) = 0.$$

21. $\int_a^x \cos[\lambda(x - t)] f(t, y(t))\, dt = g(x), \qquad g(a) = 0.$

Solution in implicit form:
$$f(x, y) - \lambda^2 \int_a^x g(t)\, dt - g'_x(x) = 0.$$

22. $\int_a^x [h(x) - h(t)] f(t, y(t))\, dt = g(x).$

It is assumed that $g(a) = g'_x(a) = 0$ and $g'_x/h'_x \neq \text{const}$.

Solution in implicit form:
$$f(x, y) = \frac{d}{dx}\left[\frac{g'_x(x)}{h'_x(x)}\right].$$

8.2. Integral Equations of the Second Kind with Variable Limit of Integration

8.2.1. Linear Volterra Integral Equations of the Second Kind

▶ **Equations with kernels containing power law functions.**

1. $y(x) - \lambda \int_a^x y(t)\, dt = f(x).$

Solution: $y(x) = f(x) + \lambda \int_a^x e^{\lambda(x-t)} f(t)\, dt.$

2. $y(x) + \lambda \int_a^x (x-t) y(t)\, dt = f(x).$

   1°. Solution for $\lambda > 0$:
   $$y(x) = f(x) - k \int_a^x \sin[k(x-t)] f(t)\, dt, \qquad k = \sqrt{\lambda}.$$

   2°. Solution for $\lambda < 0$:
   $$y(x) = f(x) + k \int_a^x \sinh[k(x-t)] f(t)\, dt, \qquad k = \sqrt{-\lambda}.$$

3. $y(x) + \int_a^x [A + B(x-t)]\, y(t)\, dt = f(x).$

   1°. Solution with $A^2 > 4B$:
   $$y(x) = f(x) - \int_a^x R(x-t) f(t)\, dt,$$
   $$R(x) = \exp\!\left(-\tfrac{1}{2}Ax\right)\!\left[A\cosh(\beta x) + \frac{2B - A^2}{2\beta}\sinh(\beta x)\right], \qquad \beta = \sqrt{\tfrac{1}{4}A^2 - B}.$$

   2°. Solution with $A^2 < 4B$:
   $$y(x) = f(x) - \int_a^x R(x-t) f(t)\, dt,$$
   $$R(x) = \exp\!\left(-\tfrac{1}{2}Ax\right)\!\left[A\cos(\beta x) + \frac{2B - A^2}{2\beta}\sin(\beta x)\right], \qquad \beta = \sqrt{B - \tfrac{1}{4}A^2}.$$

   3°. Solution with $A^2 = 4B$:
   $$y(x) = f(x) - \int_a^x R(x-t) f(t)\, dt, \qquad R(x) = \exp\!\left(-\tfrac{1}{2}Ax\right)\!\left(A - \tfrac{1}{4}A^2 x\right).$$

4. $y(x) + \lambda \int_a^x (x-t)^2 y(t)\, dt = f(x).$

Solution:
$$y(x) = f(x) - \int_a^x R(x-t) f(t)\, dt,$$
$$R(x) = \tfrac{2}{3} k e^{-2kx} - \tfrac{2}{3} k e^{kx}\!\left[\cos(\sqrt{3}\, kx) - \sqrt{3}\sin(\sqrt{3}\, kx)\right], \qquad k = \left(\tfrac{1}{4}\lambda\right)^{1/3}.$$

5. $y(x) + \lambda \int_a^x (x-t)^3 y(t)\, dt = f(x).$

Solution:
$$y(x) = f(x) - \int_a^x R(x-t) f(t)\, dt,$$

where

$$R(x) = \begin{cases} k\big[\cosh(kx)\sin(kx) - \sinh(kx)\cos(kx)\big], & k = \big(\tfrac{3}{2}\lambda\big)^{1/4} & \text{for } \lambda > 0, \\ \tfrac{1}{2} s[\sin(sx) - \sinh(sx)], & s = (-6\lambda)^{1/4} & \text{for } \lambda < 0. \end{cases}$$

6. $y(x) + A \int_a^x (x-t)^n y(t)\, dt = f(x), \qquad n = 1, 2, \dots$

1°. Differentiating the equation $n+1$ times with respect to x yields an $(n+1)$st-order linear ordinary differential equation with constant coefficients for $y = y(x)$:

$$y_x^{(n+1)} + An!\, y = f_x^{(n+1)}(x).$$

This equation under the initial conditions $y(a) = f(a)$, $y'_x(a) = f'_x(a)$, ..., $y_x^{(n)}(a) = f_x^{(n)}(a)$ determines the solution of the original integral equation.

2°. Solution:
$$y(x) = f(x) + \int_a^x R(x-t) f(t)\, dt,$$

$$R(x) = \frac{1}{n+1} \sum_{k=0}^{n} \exp(\sigma_k x) \big[\sigma_k \cos(\beta_k x) - \beta_k \sin(\beta_k x)\big],$$

where the coefficients σ_k and β_k are given by

$$\sigma_k = |An!|^{\frac{1}{n+1}} \cos\!\left(\frac{2\pi k}{n+1}\right), \quad \beta_k = |An!|^{\frac{1}{n+1}} \sin\!\left(\frac{2\pi k}{n+1}\right) \quad \text{for } A < 0;$$

$$\sigma_k = |An!|^{\frac{1}{n+1}} \cos\!\left(\frac{2\pi k + \pi}{n+1}\right), \quad \beta_k = |An!|^{\frac{1}{n+1}} \sin\!\left(\frac{2\pi k + \pi}{n+1}\right) \quad \text{for } A > 0.$$

7. $y(x) + A \int_x^\infty (t-x)^n y(t)\, dt = f(x), \qquad n = 1, 2, \dots$

The Picard–Goursat equation.

1°. A solution of the homogeneous equation ($f \equiv 0$) is

$$y(x) = C e^{-\lambda x}, \qquad \lambda = (-An!)^{\frac{1}{n+1}},$$

where C is an arbitrary constant and $A < 0$. This is a unique solution for $n = 0, 1, 2, 3$.

The general solution of the homogeneous equation for any sign of A has the form

$$y(x) = \sum_{k=1}^{s} C_k \exp(-\lambda_k x). \tag{1}$$

Here C_k are arbitrary constants and λ_k are the roots of the algebraic equation $\lambda^{n+1} + An! = 0$ that satisfy the condition $\operatorname{Re} \lambda_k > 0$. The number of terms in (1) is determined by the inequality $s \leq 2\big[\tfrac{n}{4}\big] + 1$, where $[a]$ stands for the integral part of a number a.

2°. For $f(x) = \sum_{k=1}^{m} a_k \exp(-\beta_k x)$, where $\beta_k > 0$, a solution of the original equation has the form
$$y(x) = \sum_{k=1}^{m} \frac{a_k \beta_k^{n+1}}{\beta_k^{n+1} + An!} \exp(-\beta_k x), \qquad (2)$$
where $\beta_k^{n+1} + An! \neq 0$. For $A > 0$, this formula can also be used for arbitrary $f(x)$ expandable into a convergent exponential series (which corresponds to $m = \infty$).

3°. For $f(x) = e^{-\beta x} \sum_{k=1}^{m} a_k x^k$, where $\beta > 0$, a solution of the original equation has the form
$$y(x) = e^{-\beta x} \sum_{k=0}^{m} B_k x^k, \qquad (3)$$
where the constants B_k are found by the method of undetermined coefficients.

4°. For $f(x) = \cos(\beta x) \sum_{k=1}^{m} a_k \exp(-\mu_k x)$, a solution of the original equation has the form
$$y(x) = \cos(\beta x) \sum_{k=1}^{m} B_k \exp(-\mu_k x) + \sin(\beta x) \sum_{k=1}^{m} C_k \exp(-\mu_k x), \qquad (4)$$
where the constants B_k and C_k are found by the method of undetermined coefficients.

5°. For $f(x) = \sin(\beta x) \sum_{k=1}^{m} a_k \exp(-\mu_k x)$, a solution of the original equation has the form
$$y(x) = \cos(\beta x) \sum_{k=1}^{m} B_k \exp(-\mu_k x) + \sin(\beta x) \sum_{k=1}^{m} C_k \exp(-\mu_k x), \qquad (5)$$
where the constants B_k and C_k are found by the method of undetermined coefficients.

6°. To obtain the general solution in Items 2°–5°, the solution (1) of the homogeneous equation must be added to each right-hand side of (2)–(5).

8. $y(x) - \lambda \int_0^x \dfrac{y(t)\,dt}{x+t} = f(x)$.

Dixon's equation.

1°. The solution of the homogeneous equation ($f \equiv 0$) is
$$y(x) = Cx^\beta \qquad (\beta > -1,\ \lambda > 0). \qquad (1)$$

Here C is an arbitrary constant, and $\beta = \beta(\lambda)$ is determined by the transcendental equation
$$\lambda I(\beta) = 1, \qquad \text{where} \quad I(\beta) = \int_0^1 \frac{z^\beta\,dz}{1+z}. \qquad (2)$$

2°. For a polynomial right-hand side,
$$f(x) = \sum_{n=0}^{N} A_n x^n$$

the solution bounded at zero is given by

$$y(x) = \begin{cases} \displaystyle\sum_{n=0}^{N} \frac{A_n}{1 - (\lambda/\lambda_n)} x^n & \text{if } \lambda < \lambda_0, \\ \displaystyle\sum_{n=0}^{N} \frac{A_n}{1 - (\lambda/\lambda_n)} x^n + Cx^\beta & \text{if } \lambda > \lambda_0 \text{ and } \lambda \neq \lambda_n, \end{cases}$$

$$\lambda_n = \frac{1}{I(n)}, \qquad I(n) = (-1)^n \left[\ln 2 + \sum_{m=1}^{n} \frac{(-1)^m}{m} \right],$$

where C is an arbitrary constant, and $\beta = \beta(\lambda)$ is determined by the transcendental equation (2).

For special $\lambda = \lambda_n$ ($n = 1, 2, \ldots$), the solution differs in one term and has the form

$$y(x) = \sum_{m=0}^{n-1} \frac{A_m}{1 - (\lambda_n/\lambda_m)} x^m + \sum_{m=n+1}^{N} \frac{A_m}{1 - (\lambda_n/\lambda_m)} x^m - A_n \frac{\bar{\lambda}_n}{\lambda_n} x^n \ln x + Cx^n,$$

where $\bar{\lambda}_n = (-1)^{n+1} \left[\dfrac{\pi^2}{12} + \displaystyle\sum_{k=1}^{n} \dfrac{(-1)^k}{k^2} \right]^{-1}$.

Remark 8.1. For arbitrary $f(x)$, expandable into power series, the formulas of Item 2° can be used, in which one should set $N = \infty$. In this case, the radius of convergence of the solution $y(x)$ is equal to the radius of convergence of $f(x)$.

3°. For logarithmic-polynomial right-hand side,

$$f(x) = \ln x \left(\sum_{n=0}^{N} A_n x^n \right),$$

the solution with logarithmic singularity at zero is given by

$$y(x) = \begin{cases} \ln x \displaystyle\sum_{n=0}^{N} \frac{A_n}{1 - (\lambda/\lambda_n)} x^n + \sum_{n=0}^{N} \frac{A_n D_n \lambda}{[1 - (\lambda/\lambda_n)]^2} x^n & \text{if } \lambda < \lambda_0, \\ \ln x \displaystyle\sum_{n=0}^{N} \frac{A_n}{1 - (\lambda/\lambda_n)} x^n + \sum_{n=0}^{N} \frac{A_n D_n \lambda}{[1 - (\lambda/\lambda_n)]^2} x^n + Cx^\beta & \text{if } \lambda > \lambda_0 \text{ and } \lambda \neq \lambda_n, \end{cases}$$

$$\lambda_n = \frac{1}{I(n)}, \quad I(n) = (-1)^n \left[\ln 2 + \sum_{k=1}^{n} \frac{(-1)^k}{k} \right], \quad D_n = (-1)^{n+1} \left[\frac{\pi^2}{12} + \sum_{k=1}^{n} \frac{(-1)^k}{k^2} \right].$$

9. $y(x) + \lambda \displaystyle\int_a^x \dfrac{y(t)\,dt}{\sqrt{x-t}} = f(x).$

Abel equation of the second kind. This equation is encountered in problems of heat and mass transfer.

Solution (Brakhage et al., 1965):

$$y(x) = F(x) + \pi\lambda^2 \int_a^x \exp[\pi\lambda^2 (x-t)] F(t)\,dt,$$

where

$$F(x) = f(x) - \lambda \int_a^x \frac{f(t)\,dt}{\sqrt{x-t}}.$$

10. $y(x) - \lambda \int_0^x \dfrac{y(t)\,dt}{(x-t)^\alpha} = f(x), \qquad 0 < \alpha < 1.$

Generalized Abel equation of the second kind.

1°. Assume that the number α can be represented in the form

$$\alpha = 1 - \frac{m}{n}, \qquad \text{where} \quad m = 1, 2, \ldots, \quad n = 2, 3, \ldots \quad (m < n).$$

In this case, the solution of the generalized Abel equation of the second kind can be written in closed form (Brakhage et al., 1965):

$$y(x) = f(x) + \int_0^x R(x-t) f(t)\, dt,$$

where

$$R(x) = \sum_{\nu=1}^{n-1} \frac{\lambda^\nu \Gamma^\nu(m/n)}{\Gamma(\nu m/n)} x^{(\nu m/n)-1} + \frac{b}{m} \sum_{\mu=0}^{m-1} \varepsilon_\mu \exp(\varepsilon_\mu b x)$$

$$+ \frac{b}{m} \sum_{\nu=1}^{n-1} \frac{\lambda^\nu \Gamma^\nu(m/n)}{\Gamma(\nu m/n)} \left[\sum_{\mu=0}^{m-1} \varepsilon_\mu \exp(\varepsilon_\mu b x) \int_0^x t^{(\nu m/n)-1} \exp(-\varepsilon_\mu b t)\, dt \right],$$

$$b = \lambda^{n/m} \Gamma^{n/m}(m/n), \quad \varepsilon_\mu = \exp\left(\frac{2\pi \mu i}{m}\right), \quad i^2 = -1, \quad \mu = 0, 1, \ldots, m-1.$$

2°. Solution for any α from $0 < \alpha < 1$:

$$y(x) = f(x) + \int_0^x R(x-t) f(t)\, dt, \qquad \text{where} \quad R(x) = \sum_{n=1}^{\infty} \frac{[\lambda \Gamma(1-\alpha) x^{1-\alpha}]^n}{x\, \Gamma[n(1-\alpha)]}.$$

▶ **Equations with kernels containing exponential functions.**

11. $y(x) + A \int_a^x e^{\lambda(x-t)} y(t)\, dt = f(x).$

Solution: $y(x) = f(x) - A \int_a^x e^{(\lambda - A)(x-t)} f(t)\, dt.$

12. $y(x) + A \int_a^x \left[e^{\lambda(x-t)} - 1 \right] y(t)\, dt = f(x).$

1°. Solution for $D \equiv \lambda(\lambda - 4A) > 0$:

$$y(x) = f(x) - \frac{2A\lambda}{\sqrt{D}} \int_a^x R(x-t) f(t)\, dt, \qquad R(x) = \exp\left(\tfrac{1}{2}\lambda x\right) \sinh\left(\tfrac{1}{2}\sqrt{D}\, x\right).$$

2°. Solution for $D \equiv \lambda(\lambda - 4A) < 0$:

$$y(x) = f(x) - \frac{2A\lambda}{\sqrt{|D|}} \int_a^x R(x-t) f(t)\, dt, \qquad R(x) = \exp\left(\tfrac{1}{2}\lambda x\right) \sin\left(\tfrac{1}{2}\sqrt{|D|}\, x\right).$$

3°. Solution for $\lambda = 4A$:
$$y(x) = f(x) - 4A^2 \int_a^x (x-t) \exp[2A(x-t)] f(t)\, dt.$$

13. $y(x) + \int_a^x [Ae^{\lambda(x-t)} + B]\, y(t)\, dt = f(x).$

1°. The structure of the solution depends on the sign of the discriminant
$$D \equiv (A - B - \lambda)^2 + 4AB \tag{1}$$
of the quadratic equation
$$\mu^2 + (A + B - \lambda)\mu - B\lambda = 0. \tag{2}$$

2°. If $D > 0$, then equation (2) has the different real roots
$$\mu_1 = \tfrac{1}{2}(\lambda - A - B) + \tfrac{1}{2}\sqrt{D}, \quad \mu_2 = \tfrac{1}{2}(\lambda - A - B) - \tfrac{1}{2}\sqrt{D}.$$
In this case, the original integral equation has the solution
$$y(x) = f(x) + \int_a^x \left[E_1 e^{\mu_1(x-t)} + E_2 e^{\mu_2(x-t)}\right] f(t)\, dt,$$
where
$$E_1 = A\frac{\mu_1}{\mu_2 - \mu_1} + B\frac{\mu_1 - \lambda}{\mu_2 - \mu_1}, \quad E_2 = A\frac{\mu_2}{\mu_1 - \mu_2} + B\frac{\mu_2 - \lambda}{\mu_1 - \mu_2}.$$

3°. If $D < 0$, then equation (2) has the complex conjugate roots
$$\mu_1 = \sigma + i\beta, \quad \mu_2 = \sigma - i\beta, \quad \sigma = \tfrac{1}{2}(\lambda - A - B), \quad \beta = \tfrac{1}{2}\sqrt{-D}.$$
In this case, the original integral equation has the solution
$$y(x) = f(x) + \int_a^x \left\{E_1 e^{\sigma(x-t)} \cos[\beta(x-t)] + E_2 e^{\sigma(x-t)} \sin[\beta(x-t)]\right\} f(t)\, dt,$$
where
$$E_1 = -A - B, \quad E_2 = \frac{1}{\beta}(-A\sigma - B\sigma + B\lambda).$$

14. $y(x) + \int_a^x \left[A_1 e^{\lambda_1(x-t)} + A_2 e^{\lambda_2(x-t)}\right] y(t)\, dt = f(x).$

1°. Introduce the notation
$$I_1 = \int_a^x e^{\lambda_1(x-t)} y(t)\, dt, \quad I_2 = \int_a^x e^{\lambda_2(x-t)} y(t)\, dt.$$
Differentiating the integral equation twice yields (the first line is the original equation)
$$y + A_1 I_1 + A_2 I_2 = f, \quad f = f(x), \tag{1}$$
$$y'_x + (A_1 + A_2)y + A_1\lambda_1 I_1 + A_2\lambda_2 I_2 = f'_x, \tag{2}$$
$$y''_{xx} + (A_1 + A_2)y'_x + (A_1\lambda_1 + A_2\lambda_2)y + A_1\lambda_1^2 I_1 + A_2\lambda_2^2 I_2 = f''_{xx}. \tag{3}$$
Eliminating I_1 and I_2 from (1)–(3), we arrive at the second-order linear ordinary differential equation with constant coefficients
$$y''_{xx} + (A_1 + A_2 - \lambda_1 - \lambda_2)y'_x + (\lambda_1\lambda_2 - A_1\lambda_2 - A_2\lambda_1)y = f''_{xx} - (\lambda_1 + \lambda_2)f'_x + \lambda_1\lambda_2 f. \tag{4}$$
Substituting $x = a$ into (1) and (2) yields the initial conditions
$$y(a) = f(a), \quad y'_x(a) = f'_x(a) - (A_1 + A_2)f(a). \tag{5}$$
Solving the differential equation (4) under conditions (5), we can find the solution of the integral equation.

2°. Consider the characteristic equation

$$\mu^2 + (A_1 + A_2 - \lambda_1 - \lambda_2)\mu + \lambda_1\lambda_2 - A_1\lambda_2 - A_2\lambda_1 = 0 \tag{6}$$

which corresponds to the homogeneous linear ODE (4) (with $f(x) \equiv 0$). The structure of the solution of the integral equation depends on the sign of the discriminant

$$D \equiv (A_1 - A_2 - \lambda_1 + \lambda_2)^2 + 4A_1A_2$$

of the quadratic equation (6).

If $D > 0$, the quadratic equation (6) has the different real roots

$$\mu_1 = \tfrac{1}{2}(\lambda_1 + \lambda_2 - A_1 - A_2) + \tfrac{1}{2}\sqrt{D}, \quad \mu_2 = \tfrac{1}{2}(\lambda_1 + \lambda_2 - A_1 - A_2) - \tfrac{1}{2}\sqrt{D}.$$

In this case, the solution of the original integral equation has the form

$$y(x) = f(x) + \int_a^x \left[B_1 e^{\mu_1(x-t)} + B_2 e^{\mu_2(x-t)} \right] f(t)\,dt,$$

where

$$B_1 = A_1 \frac{\mu_1 - \lambda_2}{\mu_2 - \mu_1} + A_2 \frac{\mu_1 - \lambda_1}{\mu_2 - \mu_1}, \quad B_2 = A_1 \frac{\mu_2 - \lambda_2}{\mu_1 - \mu_2} + A_2 \frac{\mu_2 - \lambda_1}{\mu_1 - \mu_2}.$$

If $D < 0$, the quadratic equation (6) has the complex conjugate roots

$$\mu_1 = \sigma + i\beta, \quad \mu_2 = \sigma - i\beta, \qquad \sigma = \tfrac{1}{2}(\lambda_1 + \lambda_2 - A_1 - A_2), \quad \beta = \tfrac{1}{2}\sqrt{-D}.$$

In this case, the solution of the original integral equation has the form

$$y(x) = f(x) + \int_a^x \left\{ B_1 e^{\sigma(x-t)} \cos[\beta(x-t)] + B_2 e^{\sigma(x-t)} \sin[\beta(x-t)] \right\} f(t)\,dt,$$

where

$$B_1 = -A_1 - A_2, \quad B_2 = \frac{1}{\beta}\left[A_1(\lambda_2 - \sigma) + A_2(\lambda_1 - \sigma)\right].$$

15. $y(x) + \displaystyle\int_a^x \left[\sum_{k=1}^n A_k e^{\lambda_k(x-t)} \right] y(t)\,dt = f(x).$

1°. This integral equation can be reduced to an nth-order linear nonhomogeneous ordinary differential equation with constant coefficients.

The solution of the equation can be represented in the form

$$y(x) = f(x) + \int_a^x \left[\sum_{k=1}^n B_k e^{\mu_k(x-t)} \right] f(t)\,dt. \tag{1}$$

The unknown constants μ_k are roots of the algebraic equation

$$\sum_{k=1}^n \frac{A_k}{z - \lambda_k} + 1 = 0, \tag{2}$$

which is reduced (by separating the numerator) to the problem of finding the roots of an nth-order characteristic polynomial.

After the μ_k have been calculated, the coefficients B_k can be found from the following linear system of algebraic equations:

$$\sum_{k=1}^{n} \frac{B_k}{\lambda_m - \mu_k} + 1 = 0, \qquad m = 1, \ldots, n. \tag{3}$$

Another way of determining the B_k is presented in Item 2° below.

If all the roots μ_k of equation (2) are real and different, then the solution of the original integral equation can be calculated by formula (1).

To a pair of complex conjugate roots $\mu_{k,k+1} = \alpha \pm i\beta$ of the characteristic polynomial (2) there corresponds a pair of complex conjugate coefficients $B_{k,k+1}$ in equations (3). In this case, the corresponding terms $B_k e^{\mu_k(x-t)} + B_{k+1} e^{\mu_{k+1}(x-t)}$ in solution (1) can be written in the form $\overline{B}_k e^{\alpha(x-t)} \cos[\beta(x-t)] + \overline{B}_{k+1} e^{\alpha(x-t)} \sin[\beta(x-t)]$, where $\overline{B}_k$ and $\overline{B}_{k+1}$ are real coefficients.

2°. For $a = 0$, the solution of the original integral equation is given by

$$y(x) = f(x) - \int_0^x R(x-t) f(t)\, dt, \qquad R(x) = \mathcal{L}^{-1}[\overline{R}(p)],$$

where $\mathcal{L}^{-1}[\overline{R}(p)]$ is the inverse Laplace transform of the function

$$\overline{R}(p) = \frac{\overline{K}(p)}{1 + \overline{K}(p)}, \qquad \overline{K}(p) = \sum_{k=1}^{n} \frac{A_k}{p - \lambda_k}.$$

The transform $\overline{R}(p)$ of the resolvent $R(x)$ can be represented as a regular fractional function:

$$\overline{R}(p) = \frac{Q(p)}{P(p)}, \qquad P(p) = (p - \mu_1)(p - \mu_2) \ldots (p - \mu_n),$$

where $Q(p)$ is a polynomial in p of degree $< n$. The roots μ_k of the polynomial $P(p)$ coincide with the roots of equation (2). If all μ_k are real and different, then the resolvent can be determined by the formula

$$R(x) = \sum_{k=1}^{n} B_k e^{\mu_k x}, \qquad B_k = \frac{Q(\mu_k)}{P'(\mu_k)},$$

where the prime stands for differentiation.

16. $y(x) + A \displaystyle\int_a^x (x-t) e^{\lambda(x-t)} y(t)\, dt = f(x).$

1°. Solution for $A > 0$:

$$y(x) = f(x) - k \int_a^x e^{\lambda(x-t)} \sin[k(x-t)] f(t)\, dt, \qquad k = \sqrt{A}.$$

2°. Solution for $A < 0$:

$$y(x) = f(x) + k \int_a^x e^{\lambda(x-t)} \sinh[k(x-t)] f(t)\, dt, \qquad k = \sqrt{-A}.$$

▶ **Equations with kernels containing hyperbolic functions.**

17. $y(x) - A \int_a^x \cosh(\lambda x) y(t)\, dt = f(x)$.

Solution:
$$y(x) = f(x) + A \int_a^x \cosh(\lambda x) \exp\left\{\frac{A}{\lambda}[\sinh(\lambda x) - \sinh(\lambda t)]\right\} f(t)\, dt.$$

18. $y(x) - A \int_a^x \cosh(\lambda t) y(t)\, dt = f(x)$.

Solution:
$$y(x) = f(x) + A \int_a^x \cosh(\lambda t) \exp\left\{\frac{A}{\lambda}[\sinh(\lambda x) - \sinh(\lambda t)]\right\} f(t)\, dt.$$

19. $y(x) + A \int_a^x \cosh[\lambda(x - t)] y(t)\, dt = f(x)$.

Solution:
$$y(x) = f(x) + \int_a^x R(x - t) f(t)\, dt,$$
$$R(x) = \exp\left(-\tfrac{1}{2}Ax\right)\left[\frac{A^2}{2k}\sinh(kx) - A\cosh(kx)\right], \quad k = \sqrt{\lambda^2 + \tfrac{1}{4}A^2}.$$

20. $y(x) + \int_a^x \left\{\sum_{k=1}^{n} A_k \cosh[\lambda_k(x - t)]\right\} y(t)\, dt = f(x)$.

This equation can be reduced to an equation of the form 8.2.1.15 by using the identity $\cosh z \equiv \tfrac{1}{2}(e^z + e^{-z})$.

21. $y(x) - A \int_a^x \sinh(\lambda x) y(t)\, dt = f(x)$.

Solution:
$$y(x) = f(x) + A \int_a^x \sinh(\lambda x) \exp\left\{\frac{A}{\lambda}[\cosh(\lambda x) - \cosh(\lambda t)]\right\} f(t)\, dt.$$

22. $y(x) - A \int_a^x \sinh(\lambda t) y(t)\, dt = f(x)$.

Solution:
$$y(x) = f(x) + A \int_a^x \sinh(\lambda t) \exp\left\{\frac{A}{\lambda}[\cosh(\lambda x) - \cosh(\lambda t)]\right\} f(t)\, dt.$$

23. $y(x) + A \int_a^x \sinh[\lambda(x - t)] y(t)\, dt = f(x)$.

1°. Solution for $\lambda(A - \lambda) > 0$:
$$y(x) = f(x) - \frac{A\lambda}{k} \int_a^x \sin[k(x - t)] f(t)\, dt, \quad \text{where} \quad k = \sqrt{\lambda(A - \lambda)}.$$

2°. Solution for $\lambda(A - \lambda) < 0$:

$$y(x) = f(x) - \frac{A\lambda}{k} \int_a^x \sinh[k(x - t)] f(t)\, dt, \qquad \text{where} \quad k = \sqrt{\lambda(\lambda - A)}.$$

3°. Solution for $A = \lambda$:

$$y(x) = f(x) - \lambda^2 \int_a^x (x - t) f(t)\, dt.$$

24. $y(x) + \displaystyle\int_a^x \left\{ \sum_{k=1}^n A_k \sinh[\lambda_k(x - t)] \right\} y(t)\, dt = f(x).$

1°. This equation can be reduced to an equation of the form 8.2.1.15 by using the identity $\sinh z \equiv \frac{1}{2}(e^z - e^{-z})$.

2°. Let us find the roots z_k of the algebraic equation

$$\sum_{k=1}^n \frac{\lambda_k A_k}{z - \lambda_k^2} + 1 = 0. \tag{1}$$

By reducing it to a common denominator, we arrive at the problem of determining the roots of an nth-degree characteristic polynomial.

Assume that all z_k are real, different, and nonzero. Let us divide the roots into two groups

$$z_1 > 0, \quad z_2 > 0, \quad \ldots, \quad z_s > 0 \qquad \text{(positive roots)};$$
$$z_{s+1} < 0, \quad z_{s+2} < 0, \quad \ldots, \quad z_n < 0 \qquad \text{(negative roots)}.$$

Then the solution of the integral equation can be written in the form

$$y(x) = f(x) + \int_a^x \left\{ \sum_{k=1}^s B_k \sinh[\mu_k(x - t)] + \sum_{k=s+1}^n C_k \sin[\mu_k(x - t)] \right\} f(t)\, dt, \tag{2}$$

where $\mu_k = \sqrt{|z_k|}$. The coefficients B_k and C_k in (2) are determined from the following system of linear algebraic equations:

$$\sum_{k=0}^s \frac{B_k \mu_k}{\lambda_m^2 - \mu_k^2} + \sum_{k=s+1}^n \frac{C_k \mu_k}{\lambda_m^2 + \mu_k^2} + 1 = 0, \qquad \mu_k = \sqrt{|z_k|}, \quad m = 1, \ldots, n. \tag{3}$$

In the case of a zero root $z_s = 0$, we can introduce the new constant $D = B_s \mu_s$ and proceed to the limit $\mu_s \to 0$. As a result, the term $D(x - t)$ appears in solution (2) instead of $B_s \sinh[\mu_s(x - t)]$ and the corresponding terms $D\lambda_m^{-2}$ appear in system (3).

▶ **Equations with kernels containing trigonometric functions.**

25. $y(x) - A \displaystyle\int_a^x \cos(\lambda x) y(t)\, dt = f(x).$

Solution:

$$y(x) = f(x) + A \int_a^x \cos(\lambda x) \exp\left\{ \frac{A}{\lambda}[\sin(\lambda x) - \sin(\lambda t)] \right\} f(t)\, dt.$$

26. $y(x) - A \int_a^x \cos(\lambda t) y(t)\, dt = f(x)$.

Solution:
$$y(x) = f(x) + A \int_a^x \cos(\lambda t) \exp\left\{\frac{A}{\lambda}[\sin(\lambda x) - \sin(\lambda t)]\right\} f(t)\, dt.$$

27. $y(x) + A \int_a^x \cos[\lambda(x-t)]\, y(t)\, dt = f(x)$.

The solution of this integral equation is reduced to solving the following second-order nonhomogeneous linear ODE with constant coefficients:
$$y''_{xx} + A y'_x + \lambda^2 y = f''_{xx} + \lambda^2 f, \qquad f = f(x),$$

with the initial conditions
$$y(a) = f(a), \qquad y'_x(a) = f'_x(a) - A f(a).$$

1°. Solution with $|A| > 2|\lambda|$:
$$y(x) = f(x) + \int_a^x R(x-t) f(t)\, dt,$$
$$R(x) = \exp\left(-\tfrac{1}{2}Ax\right)\left[\frac{A^2}{2k}\sinh(kx) - A\cosh(kx)\right], \qquad k = \sqrt{\tfrac{1}{4}A^2 - \lambda^2}.$$

2°. Solution with $|A| < 2|\lambda|$:
$$y(x) = f(x) + \int_a^x R(x-t) f(t)\, dt,$$
$$R(x) = \exp\left(-\tfrac{1}{2}Ax\right)\left[\frac{A^2}{2k}\sin(kx) - A\cos(kx)\right], \qquad k = \sqrt{\lambda^2 - \tfrac{1}{4}A^2}.$$

3°. Solution with $\lambda = \pm\tfrac{1}{2}A$:
$$y(x) = f(x) + \int_a^x R(x-t) f(t)\, dt, \qquad R(x) = \exp\left(-\tfrac{1}{2}Ax\right)\left(\tfrac{1}{2}A^2 x - A\right).$$

28. $y(x) + \int_a^x \left\{\sum_{k=1}^n A_k \cos[\lambda_k(x-t)]\right\} y(t)\, dt = f(x)$.

This integral equation is reduced to a linear nonhomogeneous ordinary differential equation of order $2n$ with constant coefficients. We set
$$I_k(x) = \int_a^x \cos[\lambda_k(x-t)]\, y(t)\, dt. \tag{1}$$

Differentiating (1) with respect to x twice yields
$$I'_k = y(x) - \lambda_k \int_a^x \sin[\lambda_k(x-t)]\, y(t)\, dt,$$
$$I''_k = y'_x(x) - \lambda_k^2 \int_a^x \cos[\lambda_k(x-t)]\, y(t)\, dt, \tag{2}$$

where the primes stand for differentiation with respect to x. Comparing (1) and (2), we see that

$$I_k'' = y_x'(x) - \lambda_k^2 I_k, \qquad I_k = I_k(x). \tag{3}$$

With the aid of (1), the integral equation can be rewritten in the form

$$y(x) + \sum_{k=1}^{n} A_k I_k = f(x). \tag{4}$$

Differentiating (4) with respect to x twice taking into account (3) yields

$$y_{xx}''(x) + \sigma_n y_x'(x) - \sum_{k=1}^{n} A_k \lambda_k^2 I_k = f_{xx}''(x), \qquad \sigma_n = \sum_{k=1}^{n} A_k. \tag{5}$$

Eliminating the integral I_n from (4) and (5), we obtain

$$y_{xx}''(x) + \sigma_n y_x'(x) + \lambda_n^2 y(x) + \sum_{k=1}^{n-1} A_k (\lambda_n^2 - \lambda_k^2) I_k = f_{xx}''(x) + \lambda_n^2 f(x). \tag{6}$$

Differentiating (6) with respect to x twice followed by eliminating I_{n-1} from the resulting expression with the aid of (6) yields a similar equation whose left-hand side is a fourth-order differential operator (acting on y) with constant coefficients plus the sum $\sum_{k=1}^{n-2} B_k I_k$. Successively eliminating the terms I_{n-2}, I_{n-3}, ... using double differentiation and formula (3), we finally arrive at a linear nonhomogeneous ordinary differential equation of order $2n$ with constant coefficients.

The initial conditions for $y(x)$ can be obtained by setting $x = a$ in the integral equation and all its derivative equations.

29. $y(x) - A \int_a^x \sin(\lambda x) y(t)\, dt = f(x).$

Solution:

$$y(x) = f(x) + A \int_a^x \sin(\lambda x) \exp\left\{\frac{A}{\lambda}[\cos(\lambda t) - \cos(\lambda x)]\right\} f(t)\, dt.$$

30. $y(x) - A \int_a^x \sin(\lambda t) y(t)\, dt = f(x).$

Solution:

$$y(x) = f(x) + A \int_a^x \sin(\lambda t) \exp\left\{\frac{A}{\lambda}[\cos(\lambda t) - \cos(\lambda x)]\right\} f(t)\, dt.$$

31. $y(x) + A \int_a^x \sin[\lambda(x-t)]\, y(t)\, dt = f(x).$

1°. Solution with $\lambda(A + \lambda) > 0$:

$$y(x) = f(x) - \frac{A\lambda}{k} \int_a^x \sin[k(x-t)] f(t)\, dt, \qquad \text{where} \quad k = \sqrt{\lambda(A + \lambda)}.$$

2°. Solution with $\lambda(A+\lambda) < 0$:

$$y(x) = f(x) - \frac{A\lambda}{k}\int_a^x \sinh[k(x-t)]f(t)\,dt, \qquad \text{where} \quad k = \sqrt{-\lambda(\lambda+A)}.$$

3°. Solution with $A = -\lambda$:

$$y(x) = f(x) + \lambda^2 \int_a^x (x-t)f(t)\,dt.$$

32. $y(x) + \displaystyle\int_a^x \left\{\sum_{k=1}^n A_k \sin[\lambda_k(x-t)]\right\} y(t)\,dt = f(x).$

1°. This integral equation can be reduced to a linear nonhomogeneous ordinary differential equation of order $2n$ with constant coefficients.

2°. Let us find the roots z_k of the algebraic equation

$$\sum_{k=1}^n \frac{\lambda_k A_k}{z + \lambda_k^2} + 1 = 0. \tag{1}$$

By reducing it to a common denominator, we arrive at the problem of determining the roots of an algebraic equation of nth-degree.

Assume that all z_k are real, different, and nonzero. Let us divide the roots into two groups

$$z_1 > 0, \quad z_2 > 0, \quad \ldots, \quad z_s > 0 \quad \text{(positive roots)};$$
$$z_{s+1} < 0, \quad z_{s+2} < 0, \quad \ldots, \quad z_n < 0 \quad \text{(negative roots)}.$$

Then the solution of the integral equation can be written in the form

$$y(x) = f(x) + \int_a^x \left\{\sum_{k=1}^s B_k \sinh[\mu_k(x-t)] + \sum_{k=s+1}^n C_k \sin[\mu_k(x-t)]\right\} f(t)\,dt, \tag{2}$$

where $\mu_k = \sqrt{|z_k|}$. The coefficients B_k and C_k in (2) are determined from the following system of linear algebraic equations:

$$\sum_{k=0}^s \frac{B_k \mu_k}{\lambda_m^2 + \mu_k^2} + \sum_{k=s+1}^n \frac{C_k \mu_k}{\lambda_m^2 - \mu_k^2} - 1 = 0, \qquad \mu_k = \sqrt{|z_k|}, \quad m = 1,\ldots,n. \tag{3}$$

In the case of a zero root $z_s = 0$, we can introduce the new constant $D = B_s \mu_s$ and proceed to the limit $\mu_s \to 0$. As a result, the term $D(x-t)$ appears in solution (2) instead of $B_s \sinh[\mu_s(x-t)]$ and the corresponding terms $D\lambda_m^{-2}$ appear in system (3).

▶ **Integral equations with other kernels.**

33. $y(x) - \lambda \displaystyle\int_0^x J_0(x-t)y(t)\,dt = f(x).$

Here, $J_0(z)$ is the Bessel function of the first kind.

Solution:

$$y(x) = f(x) + \int_0^x R(x-t)f(t)\,dt,$$

where

$$R(x) = \lambda \cos(\sqrt{1-\lambda^2}\, x) + \frac{\lambda^2}{\sqrt{1-\lambda^2}} \sin(\sqrt{1-\lambda^2}\, x)$$
$$+ \frac{\lambda}{\sqrt{1-\lambda^2}} \int_0^x \sin[\sqrt{1-\lambda^2}\,(x-t)] \frac{J_1(t)}{t}\, dt.$$

34. $y(x) - \int_a^x g(x)h(t)y(t)\, dt = f(x).$

Solution:

$$y(x) = f(x) + \int_a^x R(x,t) f(t)\, dt, \qquad \text{where} \quad R(x,t) = g(x)h(t) \exp\left[\int_t^x g(s)h(s)\, ds\right].$$

35. $y(x) + \int_a^x (x-t) g(x) y(t)\, dt = f(x).$

1°. Solution:

$$y(x) = f(x) + \frac{1}{W} \int_a^x \bigl[Y_1(x) Y_2(t) - Y_2(x) Y_1(t)\bigr] g(x) f(t)\, dt, \tag{1}$$

where $Y_1 = Y_1(x)$ and $Y_2 = Y_2(x)$ are two linearly independent solutions ($Y_1/Y_2 \not\equiv$ const) of the second-order linear homogeneous ODE $Y''_{xx} + g(x) Y = 0$. In this case, the Wronskian is a constant: $W = Y_1(Y_2)'_x - Y_2(Y_1)'_x \equiv$ const.

2°. Given only one nontrivial solution $Y_1 = Y_1(x)$ of the linear homogeneous differential equation $Y''_{xx} + g(x) Y = 0$, one obtains the solution of the integral equation by formula (1) with

$$W = 1, \qquad Y_2(x) = Y_1(x) \int_b^x \frac{d\xi}{Y_1^2(\xi)},$$

where b is an arbitrary number.

36. $y(x) + \int_a^x (x-t) g(t) y(t)\, dt = f(x).$

1°. Solution:

$$y(x) = f(x) + \frac{1}{W} \int_a^x \bigl[Y_1(x) Y_2(t) - Y_2(x) Y_1(t)\bigr] g(t) f(t)\, dt, \tag{1}$$

where $Y_1 = Y_1(x)$ and $Y_2 = Y_2(x)$ are two linearly independent solutions ($Y_1/Y_2 \not\equiv$ const) of the second-order linear homogeneous differential equation $Y''_{xx} + g(x) Y = 0$. In this case, the Wronskian is a constant: $W = Y_1(Y_2)'_x - Y_2(Y_1)'_x \equiv$ const.

2°. Given only one nontrivial solution $Y_1 = Y_1(x)$ of the linear homogeneous differential equation $Y''_{xx} + g(x) Y = 0$, one obtains the solution of the integral equation by formula (1) with

$$W = 1, \qquad Y_2(x) = Y_1(x) \int_b^x \frac{d\xi}{Y_1^2(\xi)},$$

where b is an arbitrary number.

37. $y(x) + \int_a^x [g(x) - g(t)] y(t)\, dt = f(x).$

1°. Differentiating the equation with respect to x yields

$$y'_x(x) + g'_x(x) \int_a^x y(t)\, dt = f'_x(x). \qquad (1)$$

Introducing the new variable $Y(x) = \int_a^x y(t)\, dt$, we obtain the second-order linear ordinary differential equation

$$Y''_{xx} + g'_x(x) Y = f'_x(x), \qquad (2)$$

which must be supplemented by the initial conditions

$$Y(a) = 0, \quad Y'_x(a) = f(a). \qquad (3)$$

Conditions (3) follow from the original equation and the definition of $Y(x)$.

For exact solutions of second-order linear ordinary differential equations (2) with various $f(x)$, see Kamke (1977), Murphy (1960), and Polyanin & Zaitsev (2003, 2018).

2°. Let $Y_1 = Y_1(x)$ and $Y_2 = Y_2(x)$ be two linearly independent solutions ($Y_1/Y_2 \not\equiv$ const) of the second-order linear homogeneous differential equation $Y''_{xx} + g'_x(x) Y = 0$, which follows from (2) for $f(x) \equiv 0$. In this case, the Wronskian is a constant:

$$W = Y_1(Y_2)'_x - Y_2(Y_1)'_x \equiv \text{const}.$$

Solving the nonhomogeneous equation (2) under the initial conditions (3) with arbitrary $f = f(x)$ and taking into account $y(x) = Y'_x(x)$, we obtain the solution of the original integral equation in the form

$$y(x) = f(x) + \frac{1}{W} \int_a^x [Y'_1(x) Y'_2(t) - Y'_2(x) Y'_1(t)] f(t)\, dt, \qquad (4)$$

where the primes stand for the differentiation with respect to the argument specified in the parentheses.

38. $y(x) + \int_a^x [g(x) + h(t)] y(t)\, dt = f(x).$

1°. Differentiating the equation with respect to x yields

$$y'_x(x) + [g(x) + h(x)] y(x) + g'_x(x) \int_a^x y(t)\, dt = f'_x(x).$$

Introducing the new variable $Y(x) = \int_a^x y(t)\, dt$, we obtain the second-order linear ordinary differential equation

$$Y''_{xx} + [g(x) + h(x)] Y'_x + g'_x(x) Y = f'_x(x), \qquad (1)$$

which must be supplemented by the initial conditions

$$Y(a) = 0, \quad Y'_x(a) = f(a). \qquad (2)$$

Conditions (2) follow from the original equation and the definition of $Y(x)$.

For exact solutions of second-order linear ordinary differential equations (1) with various $f(x)$, see Kamke (1977), Murphy (1960), and Polyanin & Zaitsev (2003, 2018).

2°. Let $Y_1 = Y_1(x)$ and $Y_2 = Y_2(x)$ be two linearly independent solutions ($Y_1/Y_2 \neq$ const) of the second-order linear homogeneous differential equation $Y''_{xx} + [g(x) + h(x)]Y'_x + g'_x(x)Y = 0$, which follows from (1) for $f(x) \equiv 0$.

Solving the nonhomogeneous equation (1) under the initial conditions (2) with arbitrary $f = f(x)$ and taking into account $y(x) = Y'_x(x)$, we obtain the solution of the original integral equation in the form

$$y(x) = f(x) + \int_a^x R(x,t)f(t)\,dt,$$

$$R(x,t) = \frac{\partial^2}{\partial x \partial t}\left[\frac{Y_1(x)Y_2(t) - Y_2(x)Y_1(t)}{W(t)}\right], \quad W(x) = Y_1(x)Y'_2(x) - Y_2(x)Y'_1(x),$$

where $W(x)$ is the Wronskian and the primes stand for the differentiation with respect to the argument specified.

39. $y(x) + \int_a^x \cosh[\lambda(x-t)]g(t)y(t)\,dt = f(x).$

Differentiating the equation with respect to x twice yields

$$y'_x(x) + g(x)y(x) + \lambda \int_a^x \sinh[\lambda(x-t)]g(t)y(t)\,dt = f'_x(x), \tag{1}$$

$$y''_{xx}(x) + [g(x)y(x)]'_x + \lambda^2 \int_a^x \cosh[\lambda(x-t)]g(t)y(t)\,dt = f''_{xx}(x). \tag{2}$$

Eliminating the integral term from (2) with the aid of the original equation, we arrive at the second-order linear ordinary differential equation

$$y''_{xx} + [g(x)y]'_x - \lambda^2 y = f''_{xx}(x) - \lambda^2 f(x). \tag{3}$$

By setting $x = a$ in the original equation and (1), we obtain the initial conditions for $y = y(x)$:

$$y(a) = f(a), \quad y'_x(a) = f'_x(a) - f(a)g(a). \tag{4}$$

Equation (3) under conditions (4) determines the solution of the original integral equation.

40. $y(x) + \int_a^x \sinh[\lambda(x-t)]g(t)y(t)\,dt = f(x).$

1°. Differentiating the equation with respect to x twice yields

$$y'_x(x) + \lambda \int_a^x \cosh[\lambda(x-t)]g(t)y(t)\,dt = f'_x(x), \tag{1}$$

$$y''_{xx}(x) + \lambda g(x)y(x) + \lambda^2 \int_a^x \sinh[\lambda(x-t)]g(t)y(t)\,dt = f''_{xx}(x). \tag{2}$$

Eliminating the integral term from (2) with the aid of the original equation, we arrive at the second-order linear ordinary differential equation

$$y''_{xx} + \lambda[g(x) - \lambda]y = f''_{xx}(x) - \lambda^2 f(x). \tag{3}$$

By setting $x = a$ in the original equation and (1), we obtain the initial conditions for $y = y(x)$:
$$y(a) = f(a), \quad y'_x(a) = f'_x(a). \tag{4}$$

For exact solutions of second-order linear ordinary differential equations (3) with various $g(x)$, see Kamke (1977), Murphy (1960), and Polyanin & Zaitsev (2003, 2018).

2°. Let $y_1 = y_1(x)$ and $y_2 = y_2(x)$ be two linearly independent solutions ($y_1/y_2 \not\equiv \text{const}$) of the homogeneous linear ODE $y''_{xx} + \lambda[g(x) - \lambda]y = 0$, which follows from (3) for $f(x) \equiv 0$. In this case, the Wronskian is a constant:
$$W = y_1(y_2)'_x - y_2(y_1)'_x \equiv \text{const}.$$

The solution of the nonhomogeneous equation (3) under conditions (4) with arbitrary $f = f(x)$ has the form
$$y(x) = f(x) + \frac{\lambda}{W} \int_a^x [y_1(x)y_2(t) - y_2(x)y_1(t)]g(t)f(t)\,dt \tag{5}$$
and determines the solution of the original integral equation.

3°. Given only one nontrivial solution $y_1 = y_1(x)$ of the linear homogeneous differential equation $y''_{xx} + \lambda[g(x) - \lambda]y = 0$, one can obtain the solution of the nonhomogeneous linear ODE (3) under the initial conditions (4) by formula (5) with
$$W = 1, \quad y_2(x) = y_1(x) \int_b^x \frac{d\xi}{y_1^2(\xi)},$$
where b is an arbitrary number.

41. $y(x) + \displaystyle\int_a^x \cos[\lambda(x-t)]g(t)y(t)\,dt = f(x).$

Differentiating the equation with respect to x twice yields
$$y'_x(x) + g(x)y(x) - \lambda \int_a^x \sin[\lambda(x-t)]g(t)y(t)\,dt = f'_x(x), \tag{1}$$
$$y''_{xx}(x) + [g(x)y(x)]'_x - \lambda^2 \int_a^x \cos[\lambda(x-t)]g(t)y(t)\,dt = f''_{xx}(x). \tag{2}$$

Eliminating the integral term from (2) with the aid of the original equation, we arrive at the second-order linear ordinary differential equation
$$y''_{xx} + [g(x)y]'_x + \lambda^2 y = f''_{xx}(x) + \lambda^2 f(x).$$

By setting $x = a$ in the original equation and (1), we obtain the initial conditions for $y = y(x)$:
$$y(a) = f(a), \quad y'_x(a) = f'_x(a) - f(a)g(a).$$

42. $y(x) + \displaystyle\int_a^x \sin[\lambda(x-t)]g(t)y(t)\,dt = f(x).$

1°. Differentiating the equation with respect to x twice yields
$$y'_x(x) + \lambda \int_a^x \cos[\lambda(x-t)]g(t)y(t)\,dt = f'_x(x), \tag{1}$$
$$y''_{xx}(x) + \lambda g(x)y(x) - \lambda^2 \int_a^x \sin[\lambda(x-t)]g(t)y(t)\,dt = f''_{xx}(x). \tag{2}$$

Eliminating the integral term from (2) with the aid of the original equation, we arrive at the second-order linear ordinary differential equation

$$y''_{xx} + \lambda[g(x) + \lambda]y = f''_{xx}(x) + \lambda^2 f(x). \tag{3}$$

By setting $x = a$ in the original equation and (1), we obtain the initial conditions for $y = y(x)$:

$$y(a) = f(a), \quad y'_x(a) = f'_x(a). \tag{4}$$

For exact solutions of second-order linear ordinary differential equations (3) with various $f(x)$, see Kamke (1977) and Polyanin & Zaitsev (2003, 2018).

2°. Let $y_1 = y_1(x)$ and $y_2 = y_2(x)$ be two linearly independent solutions ($y_1/y_2 \not\equiv$ const) of the homogeneous linear ODE $y''_{xx} + \lambda[g(x) - \lambda]y = 0$, which follows from (3) for $f(x) \equiv 0$. In this case, the Wronskian is a constant:

$$W = y_1(y_2)'_x - y_2(y_1)'_x \equiv \text{const}.$$

The solution of the nonhomogeneous equation (3) under conditions (4) with arbitrary $f = f(x)$ has the form

$$y(x) = f(x) + \frac{\lambda}{W} \int_a^x [y_1(x)y_2(t) - y_2(x)y_1(t)]g(t)f(t)\,dt \tag{5}$$

and determines the solution of the original integral equation.

3°. Given only one nontrivial solution $y_1 = y_1(x)$ of the linear homogeneous differential equation $y''_{xx} + \lambda[g(x) + \lambda]y = 0$, one obtains the solution of the nonhomogeneous equation (3) under the initial conditions (4) by formula (5) with

$$W = 1, \quad y_2(x) = y_1(x) \int_b^x \frac{d\xi}{y_1^2(\xi)},$$

where b is an arbitrary number.

43. $y(x) + \int_a^x K(x-t)y(t)\,dt = f(x).$

Renewal equation.

1°. To solve this integral equation, direct and inverse Laplace transforms are used. The solution can be represented in the form

$$y(x) = f(x) - \int_a^x R(x-t)f(t)\,dt. \tag{1}$$

Here the resolvent $R(x)$ is expressed via the kernel $K(x)$ of the original equation as follows:

$$R(x) = \frac{1}{2\pi i} \int_{c-i\infty}^{c+i\infty} \tilde{R}(p)e^{px}\,dp, \quad \tilde{R}(p) = \frac{\tilde{K}(p)}{1+\tilde{K}(p)}, \quad \tilde{K}(p) = \int_0^\infty K(x)e^{-px}\,dx.$$

2°. Let $w = w(x)$ be the solution of the simpler auxiliary equation with $a = 0$ and $f \equiv 1$:

$$w(x) + \int_0^x K(x-t)w(t)\,dt = 1. \tag{2}$$

Then the solution of the original integral equation with arbitrary $f = f(x)$ is expressed via the solution of the auxiliary equation (2) as (Bellman & Cooke, 1963):

$$y(x) = \frac{d}{dx} \int_a^x w(x-t)f(t)\,dt = f(a)w(x-a) + \int_a^x w(x-t)f'_t(t)\,dt.$$

8.2.2. Nonlinear Volterra Integral Equations of the Second Kind

▶ **Equations with power-law or exponential nonlinearity.**

1. $y(x) + \int_a^x f(t)y^2(t)\,dt = A.$

Solution:
$$y(x) = A\left[1 + A\int_a^x f(t)\,dt\right]^{-1}.$$

2. $y(x) + \int_a^x g(x)h(t)y^2(t)\,dt = f(x).$

Differentiating the equation with respect to x yields

$$y'_x + g(x)h(x)y^2 + g'_x(x)\int_a^x h(t)y^2(t)\,dt = f'_x(x). \qquad (1)$$

Eliminating the integral term from (1) with the aid of the original equation, we arrive at a Riccati ordinary differential equation,

$$y'_x + g(x)h(x)y^2 - \frac{g'_x(x)}{g(x)}y = f'_x(x) - \frac{g'_x(x)}{g(x)}f(x), \qquad (2)$$

under the initial condition $y(a) = f(a)$. Equation (2) can be reduced to a second-order linear ordinary differential equation. For the exact solutions of equation (2) with various specific functions f, g, and h, see, for example, Kamke (1977) and Polyanin & Zaitsev (2003, 2018).

3. $y(x) + \int_a^x f(t)y^k(t)\,dt = A.$

Solution:
$$y(x) = \left[A^{1-k} + (k-1)\int_a^x f(t)\,dt\right]^{\frac{1}{1-k}}.$$

4. $y(x) - \int_a^x f(x)g(t)y^k(t)\,dt = 0.$

1°. Differentiating the equation with respect to x and eliminating the integral term (using the original equation), we obtain the Bernoulli ordinary differential equation

$$y'_x - f(x)g(x)y^k - \frac{f'_x(x)}{f(x)}y = 0, \qquad y(a) = 0.$$

2°. Solution with $k < 1$:

$$y(x) = f(x)\left[(1-k)\int_a^x f^k(t)g(t)\,dt\right]^{\frac{1}{1-k}}.$$

Additionally, for $k > 0$, there is the trivial solution $y(x) \equiv 0$.

5. $y(x) + \int_a^x f(t)\exp[\lambda y(t)]\,dt = A.$

Solution:
$$y(x) = -\frac{1}{\lambda}\ln\left[\lambda \int_a^x f(t)\,dt + e^{-A\lambda}\right].$$

6. $y(x) + \int_a^x g(t)\exp[\lambda y(t)]\,dt = f(x).$

1°. By differentiation, this integral equation can be reduced to the first-order ordinary differential equation
$$y'_x + g(x)e^{\lambda y} = f'_x(x) \qquad (1)$$

under the initial condition $y(a) = f(a)$. The substitution $w = e^{-\lambda y}$ reduces (1) to the linear ODE
$$w'_x + \lambda f'_x(x)w - \lambda g(x) = 0, \qquad w(a) = \exp[-\lambda f(a)].$$

2°. Solution:
$$y(x) = f(x) - \frac{1}{\lambda}\ln\left\{1 + \lambda \int_a^x g(t)\exp[\lambda f(t)]\,dt\right\}.$$

▶ **Other integral equations.**

7. $y(x) + \int_a^x g(t)f(y(t))\,dt = A.$

Solution in an implicit form:
$$\int_A^y \frac{du}{f(u)} + \int_a^x g(t)\,dt = 0.$$

8. $y(x) + \int_a^x f(t, y(t))\,dt = g(x).$

The solution of this integral equation is determined by the solution of the first-order ordinary differential equation
$$y'_x + f(x, y) - g'_x(x) = 0$$

under the initial condition $y(a) = g(a)$. For the exact solutions of the first-order differential equations with various $f(x, y)$ and $g(x)$, see Kamke (1977) and Polyanin & Zaitsev (2003, 2018).

9. $y(x) + \int_a^x (x-t)f(t, y(t))\,dt = g(x).$

Differentiating the equation with respect to x yields
$$y'_x + \int_a^x f(t, y(t))\,dt = g'_x(x). \qquad (1)$$

In turn, differentiating this equation with respect to x yields the second-order nonlinear ordinary differential equation

$$y''_{xx} + f(x, y) - g''_{xx}(x) = 0. \qquad (2)$$

By setting $x = a$ in the original equation and equation (1), we obtain the initial conditions for $y = y(x)$:

$$y(a) = g(a), \quad y'_x(a) = g'_x(a). \qquad (3)$$

Equation (2) under conditions (3) defines the solution of the original integral equation. For the exact solutions of the second-order differential equation (2) with various $f(x, y)$ and $g(x)$, see Polyanin & Zaitsev (2003, 2018).

10. $y(x) + \int_a^x e^{\lambda(x-t)} f(t, y(t))\, dt = g(x).$

Differentiating the equation with respect to x yields

$$y'_x + f(x, y(x)) + \lambda \int_a^x e^{\lambda(x-t)} f(t, y(t))\, dt = g'_x(x).$$

Eliminating the integral term with the aid of the original equation, we obtain the first-order nonlinear ordinary differential equation

$$y'_x + f(x, y) - \lambda y + \lambda g(x) - g'_x(x) = 0.$$

The unknown function $y = y(x)$ must satisfy the initial condition $y(a) = g(a)$. For the exact solutions of the first-order differential equations with various $f(x, y)$ and $g(x)$, see Kamke (1977) and Polyanin & Zaitsev (2003, 2018).

11. $y(x) + \int_a^x \cosh[\lambda(x - t)] f(t, y(t))\, dt = g(x).$

Differentiating the equation with respect to x twice yields

$$y'_x(x) + f(x, y(x)) + \lambda \int_a^x \sinh[\lambda(x - t)] f(t, y(t))\, dt = g'_x(x), \qquad (1)$$

$$y''_{xx}(x) + [f(x, y(x))]'_x + \lambda^2 \int_a^x \cosh[\lambda(x - t)] f(t, y(t))\, dt = g''_{xx}(x). \qquad (2)$$

Eliminating the integral term from (2) with the aid of the original equation, we arrive at the second-order nonlinear ordinary differential equation

$$y''_{xx} + [f(x, y)]'_x - \lambda^2 y + \lambda^2 g(x) - g''_{xx}(x) = 0. \qquad (3)$$

By setting $x = a$ in the original equation and in (1), we obtain the initial conditions for $y = y(x)$:

$$y(a) = g(a), \quad y'_x(a) = g'_x(a) - f(a, g(a)). \qquad (4)$$

ODE (3) under conditions (4) defines the solution of the original integral equation.

12. $y(x) + \int_a^x \sinh[\lambda(x-t)] f(t, y(t))\, dt = g(x).$

Differentiating the equation with respect to x twice yields

$$y'_x(x) + \lambda \int_a^x \cosh[\lambda(x-t)] f(t, y(t))\, dt = g'_x(x), \tag{1}$$

$$y''_{xx}(x) + \lambda f(x, y(x)) + \lambda^2 \int_a^x \sinh[\lambda(x-t)] f(t, y(t))\, dt = g''_{xx}(x). \tag{2}$$

Eliminating the integral term from (2) with the aid of the original equation, we arrive at the second-order nonlinear ordinary differential equation

$$y''_{xx} + \lambda f(x, y) - \lambda^2 y + \lambda^2 g(x) - g''_{xx}(x) = 0. \tag{3}$$

By setting $x = a$ in the original equation and in (1), we obtain the initial conditions for $y = y(x)$:

$$y(a) = g(a), \quad y'_x(a) = g'_x(a). \tag{4}$$

ODE (3) under conditions (4) defines the solution of the original integral equation. For the exact solutions of the second-order differential equation (3) with various $f(x, y)$ and $g(x)$, see Polyanin & Zaitsev (2003, 2018).

13. $y(x) + \int_a^x \cos[\lambda(x-t)] f(t, y(t))\, dt = g(x).$

Differentiating the equation with respect to x twice yields

$$y'_x(x) + f(x, y(x)) - \lambda \int_a^x \sin[\lambda(x-t)] f(t, y(t))\, dt = g'_x(x), \tag{1}$$

$$y''_{xx}(x) + [f(x, y(x))]'_x - \lambda^2 \int_a^x \cos[\lambda(x-t)] f(t, y(t))\, dt = g''_{xx}(x). \tag{2}$$

Eliminating the integral term from (2) with the aid of the original equation, we arrive at the second-order nonlinear ordinary differential equation

$$y''_{xx} + [f(x, y)]'_x + \lambda^2 y - \lambda^2 g(x) - g''_{xx}(x) = 0. \tag{3}$$

By setting $x = a$ in the original equation and in (1), we obtain the initial conditions for $y = y(x)$:

$$y(a) = g(a), \quad y'_x(a) = g'_x(a) - f(a, g(a)). \tag{4}$$

ODE (3) under conditions (4) defines the solution of the original integral equation.

14. $y(x) + \int_a^x \sin[\lambda(x-t)] f(t, y(t))\, dt = g(x).$

Differentiating the equation with respect to x twice yields

$$y'_x(x) + \lambda \int_a^x \cos[\lambda(x-t)] f(t, y(t))\, dt = g'_x(x), \tag{1}$$

$$y''_{xx}(x) + \lambda f(x, y(x)) - \lambda^2 \int_a^x \sin[\lambda(x-t)] f(t, y(t))\, dt = g''_{xx}(x). \tag{2}$$

Eliminating the integral term from (2) with the aid of the original equation, we arrive at the second-order nonlinear ordinary differential equation

$$y''_{xx} + \lambda f(x,y) + \lambda^2 y - \lambda^2 g(x) - g''_{xx}(x) = 0. \tag{3}$$

By setting $x = a$ in the original equation and in (1), we obtain the initial conditions for $y = y(x)$:

$$y(a) = g(a), \quad y'_x(a) = g'_x(a). \tag{4}$$

ODE (3) under conditions (4) defines the solution of the original integral equation. For the exact solutions of the second-order differential equation (3) with various $f(x,y)$ and $g(x)$, see Polyanin & Zaitsev (2003, 2018).

8.3. Equations of the First Kind with Constant Limits of Integration

8.3.1. Linear Fredholm Integral Equations of the First Kind

▶ **Equations with kernels containing power law functions.**

1. $\displaystyle\int_a^b |x-t|\, y(t)\, dt = f(x), \qquad 0 \le a < b < \infty.$

Solution:
$$y(x) = \tfrac{1}{2} f''_{xx}(x).$$

The right-hand side $f(x)$ of the integral equation must satisfy certain relations. The general form of $f(x)$ is as follows:

$$f(x) = F(x) + Ax + B,$$
$$A = -\tfrac{1}{2}\big[F'_x(a) + F'_x(b)\big], \quad B = \tfrac{1}{2}\big[aF'_x(a) + bF'_x(b) - F(a) - F(b)\big],$$

where $F(x)$ is an arbitrary bounded twice differentiable function (with bounded first derivative).

2. $\displaystyle\int_0^a \frac{y(t)}{\sqrt{|x-t|}}\, dt = f(x), \qquad 0 < a \le \infty.$

Solution:
$$y(x) = -\frac{A}{x^{1/4}} \frac{d}{dx}\left[\int_x^a \frac{dt}{(t-x)^{1/4}} \int_0^t \frac{f(s)\, ds}{s^{1/4}(t-s)^{1/4}}\right], \quad A = \frac{1}{\sqrt{8\pi}\,\Gamma^2(3/4)}.$$

3. $\displaystyle\int_a^b \frac{y(t)}{|x-t|^k}\, dt = f(x), \qquad |a| + |b| < \infty, \quad 0 < k < 1.$

Solution (Gakhov, 1990):
$$y(x) = \frac{1}{2\pi}\cot(\tfrac{1}{2}\pi k)\frac{d}{dx}\int_a^x \frac{f(t)\,dt}{(x-t)^{1-k}} - \frac{1}{\pi^2}\cos^2(\tfrac{1}{2}\pi k)\int_a^x \frac{Z(t)F(t)}{(x-t)^{1-k}}\, dt,$$

where

$$Z(t) = (t-a)^{\frac{1+k}{2}}(b-t)^{\frac{1-k}{2}}, \quad F(t) = \frac{d}{dt}\left[\int_a^t \frac{d\tau}{(t-\tau)^k}\int_\tau^b \frac{f(s)\,ds}{Z(s)(s-\tau)^{1-k}}\right].$$

4. $\int_0^b \dfrac{y(t)}{|x^\lambda - t^\lambda|^k}\, dt = f(x), \qquad 0 < k < 1, \quad \lambda > 0.$

Solution:

$$y(x) = -A x^{\frac{\lambda(k-1)}{2}} \frac{d}{dx}\left[\int_x^b \frac{t^{\frac{\lambda(3-2k)-2}{2}}}{(t^\lambda - x^\lambda)^{\frac{1-k}{2}}}\, dt \int_0^t \frac{s^{\frac{\lambda(k+1)-2}{2}} f(s)\, ds}{(t^\lambda - s^\lambda)^{\frac{1-k}{2}}}\right],$$

$$A = \frac{\lambda^2}{2\pi}\cos\left(\frac{\pi k}{2}\right)\Gamma(k)\left[\Gamma\left(\frac{1+k}{2}\right)\right]^{-2},$$

where $\Gamma(k)$ is the gamma function.

5. $\int_{-\infty}^{\infty} \dfrac{y(t)}{|x-t|^{1-\lambda}}\, dt = f(x), \qquad 0 < \lambda < 1.$

Solution (Prudnikov et al., 1992 and Samko et al., 1993):

$$y(x) = \frac{\lambda}{2\pi}\tan\left(\frac{\pi\lambda}{2}\right)\int_{-\infty}^{\infty} \frac{f(x) - f(t)}{|x-t|^{1+\lambda}}\, dt$$

$$= \frac{\lambda}{2\pi}\tan\left(\frac{\pi\lambda}{2}\right)\int_0^\infty \frac{2f(x) - f(x+t) - f(x-t)}{t^{1+\lambda}}\, dt.$$

It assumed that the condition $\int_{-\infty}^{\infty}|f(x)|^p\, dx < \infty$ is satisfied for some p, $1 < p < 1/\lambda$.

6. $\int_{-\infty}^{\infty} \dfrac{\text{sign}(x-t)}{|x-t|^{1-\lambda}}\, y(t)\, dt = f(x), \qquad 0 < \lambda < 1.$

Solution (Prudnikov et al., 1992 and Samko et al., 1993):

$$y(x) = \frac{\lambda}{2\pi}\cot\left(\frac{\pi\lambda}{2}\right)\int_{-\infty}^{\infty} \frac{f(x) - f(t)}{|x-t|^{1+\lambda}}\,\text{sign}(x-t)\, dt$$

$$= \frac{\lambda}{2\pi}\cot\left(\frac{\pi\lambda}{2}\right)\int_0^\infty \frac{f(x+t) - f(x-t)}{t^{1+\lambda}}\, dt.$$

7. $\int_{-\infty}^{\infty} \dfrac{a + b\,\text{sign}(x-t)}{|x-t|^{1-\lambda}}\, y(t)\, dt = f(x), \qquad 0 < \lambda < 1.$

Solution (Prudnikov et al., 1992 and Samko et al., 1993):

$$y(x) = c\int_{-\infty}^{\infty} \frac{a + b\,\text{sign}(x-t)}{|x-t|^{1+\lambda}}[f(x) - f(t)]\, dt$$

$$= c\int_0^\infty \frac{2af(x) - (a+b)f(x-t) - (a-b)f(x+t)}{t^{1+\lambda}}\, dt,$$

where

$$c = \frac{\lambda \sin(\pi\lambda)}{4\pi\left[a^2\cos^2\left(\frac{1}{2}\pi\lambda\right) + b^2\sin^2\left(\frac{1}{2}\pi\lambda\right)\right]}.$$

8. $\int_0^\infty \dfrac{y(x+t) - y(x-t)}{t}\, dt = f(x).$

Solution (Prudnikov et al., 1992):

$$y(x) = -\frac{1}{\pi^2}\int_0^\infty \frac{f(x+t) - f(x-t)}{t}\, dt.$$

9. $\int_0^1 y(xt)\,dt = f(x).$

Solution:
$$y(x) = xf'_x(x) + f(x).$$

The function $f(x)$ is assumed to satisfy the condition $[xf(x)]_{x=0} = 0$.

▶ In equations 8.3.1.10 and 8.3.1.11 and their solutions, all singular integrals are understood in the sense of the Cauchy principal value:

$$\int_a^b \frac{f(x)}{x-c}\,dx = \lim_{\varepsilon \to 0}\left[\int_a^{c-\varepsilon} \frac{f(x)}{x-c}\,dx + \int_{c+\varepsilon}^b \frac{f(x)}{x-c}\,dx\right], \quad a < c < b.$$

10. $\int_{-\infty}^{\infty} \frac{y(t)\,dt}{t-x} = f(x).$

Solution (Ditkin & Prudnikov, 1965):
$$y(x) = -\frac{1}{\pi^2}\int_{-\infty}^{\infty} \frac{f(t)\,dt}{t-x}.$$

The integral equation and its solution form a *Hilbert transform pair* (in the asymmetric form).

11. $\int_a^b \frac{y(t)\,dt}{t-x} = f(x).$

This equation is encountered in hydrodynamics in solving the problem on the flow of an ideal inviscid fluid around a thin profile ($a \le x \le b$). Depending on the conditions set at the ends of the segment $[a,b]$, where $|a|+|b|<\infty$, there are the following solutions (Gakhov, 1990):

1°. The solution bounded at the endpoints is

$$y(x) = -\frac{1}{\pi^2}\sqrt{(x-a)(b-x)}\int_a^b \frac{f(t)}{\sqrt{(t-a)(b-t)}}\frac{dt}{t-x},$$

provided that $\int_a^b \frac{f(t)\,dt}{\sqrt{(t-a)(b-t)}} = 0.$

2°. The solution bounded at the endpoint $x=a$ and unbounded at the endpoint $x=b$ is

$$y(x) = -\frac{1}{\pi^2}\sqrt{\frac{x-a}{b-x}}\int_a^b \sqrt{\frac{b-t}{t-a}}\frac{f(t)}{t-x}\,dt.$$

3°. The solution unbounded at the endpoints is

$$y(x) = -\frac{1}{\pi^2\sqrt{(x-a)(b-x)}}\left[\int_a^b \frac{\sqrt{(t-a)(b-t)}}{t-x}f(t)\,dt + C\right],$$

where C is an arbitrary constant. The formula $\int_a^b y(t)\,dt = \dfrac{C}{\pi}$ holds.

► **Equations with kernels containing exponential functions.**

12. $\int_{-\infty}^{\infty} e^{-\lambda|x-t|} y(t)\, dt = f(x), \qquad f(\pm\infty) = 0.$

Solution:
$$y(x) = \frac{1}{2\lambda}\left[\lambda^2 f(x) - f''_{xx}(x)\right].$$

13. $\int_{0}^{\infty} e^{-\lambda|x-t|} y(t)\, dt = f(x), \qquad f(\infty) = 0.$

Solution:
$$y(x) = \frac{1}{2\lambda} e^{-\lambda x}\frac{d}{dx} e^{2\lambda x}\frac{d}{dx} e^{-\lambda x} f(x).$$

14. $\int_{a}^{b} e^{\lambda|x-t|} y(t)\, dt = f(x), \qquad -\infty < a < b < \infty.$

Solution:
$$y(x) = \frac{1}{2\lambda}\left[f''_{xx}(x) - \lambda^2 f(x)\right].$$

The right-hand side $f(x)$ of the integral equation must satisfy following relations:
$$f'_x(a) + \lambda f(a) = 0, \quad f'_x(b) - \lambda f(b) = 0.$$

The general form of the right-hand side of the integral equation the is given by the formula
$$f(x) = F(x) + Ax + B,$$

where $F(x)$ is an arbitrary bounded, twice differentiable function and
$$A = \frac{1}{b\lambda - a\lambda - 2}\left[F'_x(a) + F'_x(b) + \lambda F(a) - \lambda F(b)\right], \quad B = -\frac{1}{\lambda}\left[F'_x(a) + \lambda F(a) + Aa\lambda + A\right].$$

15. $\int_{a}^{b} |e^{\lambda x} - e^{\lambda t}| y(t)\, dt = f(x), \qquad \lambda > 0.$

This is a special case of equation 8.3.1.34 with $g(x) = e^{\lambda x}$.

Solution:
$$y(x) = \frac{1}{2\lambda}\frac{d}{dx}\left[e^{-\lambda x} f'_x(x)\right].$$

The right-hand side $f(x)$ of the integral equation must satisfy certain relations (see Item 2° of equation 8.3.1.34).

16. $\int_{-\infty}^{\infty} e^{-(x-t)^2} y(t)\, dt = f(x).$

1°. Solution (Hirschman & Widder, 1955):
$$y(t) = \frac{1}{\pi^{3/2}} \int_0^{\infty} e^{s^2/4}\, ds \int_{-\infty}^{\infty} \cos(s(t-x)) f(x)\, dx$$
$$= \exp\left[-\frac{1}{4\sqrt{\pi}}\frac{d^2}{dt^2} f(t)\right] \equiv \sum_{k=0}^{\infty} \frac{1}{k!}\left(-\frac{1}{4\sqrt{\pi}}\right)^k \frac{d^{2k} f(t)}{dt^{2k}}.$$

2°. Solution (alternative representation):

$$y(x) = \frac{1}{\sqrt{\pi}} \sum_{n=0}^{\infty} \frac{f_x^{(n)}(0)}{2^n n!} H_n(x),$$

where $H_n(x)$ are the Hermite polynomials

$$H_m(x) = (-1)^m \exp(x^2) \frac{d^m}{dx^m} \exp(-x^2).$$

17. $\dfrac{1}{\sqrt{\pi\lambda}} \displaystyle\int_{-\infty}^{\infty} \exp\left[-\dfrac{(x-t)^2}{\lambda}\right] y(t)\, dt = f(x).$

This is the *Gauss transform* (the *Weierstrass transform* for $\lambda = 4$).

Solution (Hirschman & Widder, 1955):

$$y(t) = \frac{1}{\pi} \int_0^{\infty} e^{\lambda s^2/4}\, ds \int_{-\infty}^{\infty} \cos\big(s(t-x)\big) f(x)\, dx$$

$$= \exp\left[-\frac{\lambda}{4} \frac{d^2}{dt^2} f(t)\right] \equiv \sum_{k=0}^{\infty} \frac{1}{k!} \left(-\frac{\lambda}{4}\right)^k \frac{d^{2k} f(t)}{dt^{2k}}.$$

▶ **Equations with kernels containing logarithmic functions.**

18. $\displaystyle\int_a^b \ln|x-t|\, y(t)\, dt = f(x).$

Carleman's equation.

1°. Solution for $b - a \neq 4$ (Gakhov, 1977)):

$$y(x) = \frac{1}{\pi^2 \sqrt{(x-a)(b-x)}} \left[\int_a^b \frac{\sqrt{(t-a)(b-t)}\, f_t'(t)\, dt}{t - x} \right.$$
$$\left. + \frac{1}{\ln[\frac{1}{4}(b-a)]} \int_a^b \frac{f(t)\, dt}{\sqrt{(t-a)(b-t)}} \right].$$

2°. If $b - a = 4$, then for the equation to be solvable, the condition

$$\int_a^b f(t)(t-a)^{-1/2}(b-t)^{-1/2}\, dt = 0$$

must be satisfied. In this case, the solution has the form

$$y(x) = \frac{1}{\pi^2 \sqrt{(x-a)(b-x)}} \left[\int_a^b \frac{\sqrt{(t-a)(b-t)}\, f_t'(t)\, dt}{t-x} + C \right],$$

where C is an arbitrary constant.

19. $\displaystyle\int_a^b \big(\ln|x-t| + \beta\big) y(t)\, dt = f(x).$

By setting

$$x = e^{-\beta} z, \quad t = e^{-\beta} \tau, \quad y(t) = Y(\tau), \quad f(x) = e^{-\beta} g(z),$$

we arrive at an equation of the form 8.3.1.18:

$$\int_A^B \ln|z - \tau| Y(\tau)\, d\tau = g(z), \qquad A = ae^\beta,\ B = be^\beta.$$

20. $\displaystyle\int_{-a}^{a} \left(\ln \frac{A}{|x-t|}\right) y(t)\, dt = f(x), \qquad -a \le x \le a.$

Solution for $0 < a < 2A$ (Gohberg & Krein, 1970):

$$y(x) = \frac{1}{2M'(a)} \left[\frac{d}{da}\int_{-a}^{a} w(t,a) f(t)\, dt\right] w(x,a)$$
$$- \frac{1}{2}\int_{|x|}^{a} w(x,\xi) \frac{d}{d\xi}\left[\frac{1}{M'(\xi)} \frac{d}{d\xi}\int_{-\xi}^{\xi} w(t,\xi) f(t)\, dt\right] d\xi$$
$$- \frac{1}{2}\frac{d}{dx}\int_{|x|}^{a} \frac{w(x,\xi)}{M'(\xi)}\left[\int_{-\xi}^{\xi} w(t,\xi)\, df(t)\right] d\xi,$$

where the prime stands for a derivative and

$$M(\xi) = \left(\ln\frac{2A}{\xi}\right)^{-1}, \qquad w(x,\xi) = \frac{M(\xi)}{\pi\sqrt{\xi^2 - x^2}}.$$

21. $\displaystyle\int_0^a \ln\left|\frac{x+t}{x-t}\right| y(t)\, dt = f(x).$

Solution:

$$y(x) = -\frac{2}{\pi^2}\frac{d}{dx}\int_x^a \frac{F(t)\, dt}{\sqrt{t^2 - x^2}}, \qquad F(t) = \frac{d}{dt}\int_0^t \frac{s f(s)\, ds}{\sqrt{t^2 - s^2}}.$$

▶ **Equations with kernels containing trigonometric functions.**

22. $\displaystyle\int_0^\infty \cos(xt) y(t)\, dt = f(x).$

Solution: $y(x) = \dfrac{2}{\pi}\displaystyle\int_0^\infty \cos(xt) f(t)\, dt.$

Up to constant factors, the function $f(x)$ and solution $y(t)$ are the *Fourier cosine transform pair* (Ditkin & Prudnikov, 1965).

23. $\displaystyle\int_a^b \cos(xt) y(t)\, dt = f(x), \qquad 0 \le x < \infty.$

Solution:

$$y(t) = \begin{cases} \dfrac{2}{\pi}\displaystyle\int_0^\infty \cos(xt) f(x)\, dx & \text{if } a < t < b, \\ 0 & \text{if } 0 < t < a \text{ or } t > b, \end{cases}$$

where $0 \le a \le b \le \infty$.

24. $\displaystyle\int_a^b |\cos(\lambda x) - \cos(\lambda t)|\, y(t)\, dt = f(x).$

This is a special case of equation 8.3.1.34 with $g(x) = \cos(\lambda x)$.

Solution:
$$y(x) = -\frac{1}{2\lambda}\frac{d}{dx}\left[\frac{f'_x(x)}{\sin(\lambda x)}\right].$$

The right-hand side $f(x)$ of the integral equation must satisfy certain relations (see Item 2° of equation 8.3.1.34).

25. $\int_0^\infty \sin(xt)y(t)\,dt = f(x).$

Solution: $y(x) = \dfrac{2}{\pi}\int_0^\infty \sin(xt)f(t)\,dt.$

Up to constant factors, the function $f(x)$ and solution $y(t)$ are the *Fourier sine transform pair* (Ditkin & Prudnikov, 1965).

26. $\int_a^b \sin(xt)y(t)\,dt = f(x), \qquad 0 \le x < \infty.$

Solution:
$$y(t) = \begin{cases} \dfrac{2}{\pi}\int_0^\infty \sin(xt)f(x)\,dx & \text{if } a < t < b, \\ 0 & \text{if } 0 < t < a \text{ or } t > b, \end{cases}$$

where $0 \le a \le b \le \infty$.

27. $\int_{-\infty}^\infty \sin(\lambda|x-t|)y(t)\,dt = f(x), \qquad f(\pm\infty) = 0.$

Solution:
$$y(x) = \frac{1}{2\lambda}\left[f''_{xx}(x) + \lambda^2 f(x)\right].$$

28. $\int_{-\infty}^\infty [\cos(xt) + \sin(xt)]y(t)\,dt = f(x).$

Solution:
$$y(x) = \frac{1}{2\pi}\int_{-\infty}^\infty [\cos(xt) + \sin(xt)]f(t)\,dt.$$

Up to constant factors, the function $f(x)$ and solution $y(t)$ are the *Hartley transform pair* (Zwillinger, 1989).

29. $\int_0^{2\pi} \cot\left(\dfrac{t-x}{2}\right) y(t)\,dt = f(x), \qquad 0 \le x \le 2\pi.$

Here the integral is understood in the sense of the Cauchy principal value and the right-hand side is assumed to satisfy the condition $\int_0^{2\pi} f(t)\,dt = 0$.

Solution:
$$y(x) = -\frac{1}{4\pi^2}\int_0^{2\pi} \cot\left(\frac{t-x}{2}\right) f(t)\,dt + C,$$

where C is an arbitrary constant.

It follows from the solution that $\int_0^{2\pi} y(t)\,dt = 2\pi C$.

The equation and its solution form a Hilbert transform pair in the asymmetric form (Gakhov, 1990).

▶ **Other integral equations.**

30. $\int_0^{\pi/2} y(\xi)\, dt = f(x)$, $\quad \xi = x \sin t.$

Schlömilch equation.

Solution (Whittaker & Watson, 1958):

$$y(x) = \frac{2}{\pi}\left[f(0) + x \int_0^{\pi/2} f'_\xi(\xi)\, dt\right], \quad \xi = x \sin t.$$

31. $\int_0^\infty t J_\nu(xt) y(t)\, dt = f(x)$, $\quad \nu > -1.$

Here, $J_\nu(z)$ is the Bessel function of the first kind.

Solution:

$$y(x) = \int_0^\infty t J_\nu(xt) f(t)\, dt.$$

The function $f(x)$ and solution $y(t)$ are the *Hankel transform pair* (Ditkin & Prudnikov, 1965).

32. $\int_{-\infty}^\infty K_0(|x - t|) y(t)\, dt = f(x).$

Here, $K_0(z)$ is the modified Bessel function of the second kind.

Solution (Zwillinger, 1989):

$$y(x) = -\frac{1}{\pi^2}\left(\frac{d^2}{dx^2} - 1\right) \int_{-\infty}^\infty K_0(|x - t|) f(t)\, dt.$$

33. $\int_a^b [g_1(x) h_1(t) + g_2(x) h_2(t)] y(t)\, dt = f(x).$

This integral equation has solutions only if its right-hand side is representable in the form

$$f(x) = A_1 g_1(x) + A_2 g_2(x), \quad A_1 = \text{const},\ A_2 = \text{const}.$$

In this case, any function $y = y(x)$ satisfying the normalization type conditions

$$\int_a^b h_1(t) y(t)\, dt = A_1, \quad \int_a^b h_2(t) y(t)\, dt = A_2$$

is a solution of the integral equation. Otherwise, the equation has no solutions.

34. $\int_a^b |g(x) - g(t)|\, y(t)\, dt = f(x).$

Let $a \leq x \leq b$ and $a \leq t \leq b$; it is assumed in Items 1° and 2° that $0 < g'_x(x) < \infty$.

1°. Let us remove the modulus in the integrand:

$$\int_a^x [g(x) - g(t)] y(t)\, dt + \int_x^b [g(t) - g(x)] y(t)\, dt = f(x). \tag{1}$$

Differentiating (1) with respect to x yields

$$g'_x(x)\int_a^x y(t)\,dt - g'_x(x)\int_x^b y(t)\,dt = f'_x(x). \tag{2}$$

Divide both sides of (2) by $g'_x(x)$ and differentiate the resulting equation to obtain the solution

$$y(x) = \frac{1}{2}\frac{d}{dx}\left[\frac{f'_x(x)}{g'_x(x)}\right]. \tag{3}$$

2°. Let us demonstrate that the right-hand side $f(x)$ of the integral equation must satisfy certain relations. By setting $x = a$ and $x = b$, in (1), we obtain two corollaries

$$\int_a^b [g(t) - g(a)]y(t)\,dt = f(a), \quad \int_a^b [g(b) - g(t)]y(t)\,dt = f(b). \tag{4}$$

Substitute $y(x)$ of (3) into (4). Integrating by parts yields the desired constraints for $f(x)$:

$$\begin{aligned}[g(b) - g(a)]\frac{f'_x(b)}{g'_x(b)} &= f(a) + f(b), \\ [g(a) - g(b)]\frac{f'_x(a)}{g'_x(a)} &= f(a) + f(b).\end{aligned} \tag{5}$$

Let us point out a useful property of these constraints: $f'_x(b)g'_x(a) + f'_x(a)g'_x(b) = 0$.

Conditions (5) make it possible to find the admissible general form of the right-hand side of the integral equation:

$$f(x) = F(x) + Ax + B, \tag{6}$$

where $F(x)$ is an arbitrary bounded twice differentiable function (with bounded first derivative), and the coefficients A and B are given by

$$A = -\frac{g'_x(a)F'_x(b) + g'_x(b)F'_x(a)}{g'_x(a) + g'_x(b)},$$

$$B = -\tfrac{1}{2}A(a+b) - \tfrac{1}{2}[F(a) + F(b)] - \frac{g(b) - g(a)}{2g'_x(a)}[A + F'_x(a)].$$

3°. If $g(x)$ is representable in the form $g(x) = O(x-a)^k$ with $0 < k < 1$ in the vicinity of the point $x = a$ (in particular, the derivative g'_x is unbounded as $x \to a$), then the solution of the integral equation is given by formula (3) as well. In this case, the right-hand side of the integral equation must satisfy the conditions

$$f(a) + f(b) = 0, \quad f'_x(b) = 0. \tag{7}$$

As before, the right-hand side of the integral equation is given by (6), with

$$A = -F'_x(b), \quad B = \tfrac{1}{2}\big[(a+b)F'_x(b) - F(a) - F(b)\big].$$

4°. For $g'_x(a) = 0$, the right-hand side of the integral equation must satisfy the conditions

$$f'_x(a) = 0, \quad [g(b) - g(a)]f'_x(b) = [f(a) + f(b)]g'_x(b).$$

As before, the right-hand side of the integral equation is given by (6), with

$$A = -F'_x(a), \quad B = \tfrac{1}{2}[(a+b)F'_x(a) - F(a) - F(b)] + \frac{g(b)-g(a)}{2g'_x(b)}[F'_x(b) - F'_x(a)].$$

35. $\displaystyle\int_{-\infty}^{\infty} K(x-t)y(t)\,dt = f(x).$

Here $K(z)$ is a given function. The Fourier transform is used to solve this equation.

1°. Solution:

$$y(x) = \frac{1}{2\pi}\int_{-\infty}^{\infty} \frac{\tilde f(u)}{\tilde K(u)} e^{iux}\,du,$$

$$\tilde f(u) = \frac{1}{\sqrt{2\pi}}\int_{-\infty}^{\infty} f(x)e^{-iux}\,dx, \quad \tilde K(u) = \frac{1}{\sqrt{2\pi}}\int_{-\infty}^{\infty} K(x)e^{-iux}\,dx.$$

The following statement is valid. Let $f(x) \in L_2(-\infty,\infty)$ and $K(x) \in L_1(-\infty,\infty)$. Then for a solution $y(x) \in L_2(-\infty,\infty)$ of the integral equation to exist, it is necessary and sufficient that $\tilde f(u)/\tilde K(u) \in L_2(-\infty,\infty)$.

2°. Let the function $P(s)$ defined by the formula

$$\frac{1}{P(s)} = \int_{-\infty}^{\infty} e^{-st} K(t)\,dt$$

be a polynomial of degree n with real roots of the form

$$P(s) = \left(1 - \frac{s}{a_1}\right)\left(1 - \frac{s}{a_2}\right)\cdots\left(1 - \frac{s}{a_n}\right).$$

Then the solution of the integral equation is given by

$$y(x) = P(D)f(x), \quad D = \frac{d}{dx}.$$

36. $\displaystyle\int_0^{\infty} K(x-t)y(t)\,dt = f(x).$

Wiener–Hopf equation of the first kind. This equation is discussed in the books by Gakhov & Cherskii (1978), Mikhlin & Prössdorf (1986), Muskhelishvili (1992), and Polyanin & Manzhirov (2008) in detail.

8.3.2. Nonlinear Fredholm Integral Equations of the First Kind

1. $\displaystyle\int_a^b y(x)f(t,y(t))\,dt = g(x).$

A solution: $y(x) = \lambda g(x)$, where λ is determined by the algebraic (or transcendental) equation $\lambda \int_a^b f(t,\lambda g(t))\,dt = 1.$

2. $\displaystyle\int_a^b y^k(x)f(t,y(t))\,dt = g(x).$

A solution: $y(x) = \lambda[g(x)]^{1/k}$, where λ is determined from the algebraic (or transcendental) equation $\lambda^k \int_a^b f(t,\lambda g^{1/k}(t))\,dt = 1.$

3. $\int_a^b \varphi(y(x)) f(t, y(t)) \, dt = g(x).$

A solution in an implicit form:
$$\lambda \varphi(y(x)) - g(x) = 0, \qquad (1)$$
where λ is determined by the algebraic (or transcendental) equation
$$\lambda - F(\lambda) = 0, \qquad F(\lambda) = \int_a^b f(t, y(t)) \, dt. \qquad (2)$$
Here the function $y(x) = y(x, \lambda)$ obtained by solving (1) must be substituted into (2).

The number of solutions of the integral equation is determined by the number of the solutions obtained from (1) and (2).

4. $\int_0^\infty [\sin(xt) y(t) + \varphi(x) \Psi(t, y(t))] \, dt = f(x).$

Solutions (Polyanin & Manzhirov, 2008):
$$y_m(t) = Y_f(t) + A_m Y_\varphi(t),$$
where
$$Y_f(t) = \frac{2}{\pi} \int_0^\infty \sin(xt) f(x) \, dx, \qquad Y_\varphi(t) = \frac{2}{\pi} \int_0^\infty \sin(xt) \varphi(x) \, dx,$$
and A_m are roots of the algebraic (transcendental) equation
$$A + \int_a^b \Psi(t, Y_f(t) + A Y_\varphi(t)) \, dt = 0.$$

5. $\int_0^\infty [\cos(xt) y(t) + \varphi(x) \Psi(t, y(t))] \, dt = f(x).$

Solutions (Polyanin & Manzhirov, 2008):
$$y_m(t) = Y_f(t) + A_m Y_\varphi(t),$$
where
$$Y_f(t) = \frac{2}{\pi} \int_0^\infty \cos(xt) f(x) \, dx, \qquad Y_\varphi(t) = \frac{2}{\pi} \int_0^\infty \cos(xt) \varphi(x) \, dx,$$
and A_m are roots of the algebraic (transcendental) equation
$$A + \int_a^b \Psi(t, Y_f(t) + A Y_\varphi(t)) \, dt = 0.$$

6. $\int_0^\infty [t J_\nu(xt) y(t) + \varphi(x) \Psi(t, y(t))] \, dt = f(x), \qquad \nu > -1.$

Here $J_\nu(z)$ is the Bessel function of the first kind.

Solutions (Polyanin & Manzhirov, 2008):
$$y_m(t) = Y_f(t) + A_m Y_\varphi(t),$$
where
$$Y_f(t) = \int_0^\infty x J_\nu(xt) f(x) \, dx, \qquad Y_\varphi(t) = \int_0^\infty x J_\nu(xt) \varphi(x) \, dx,$$
and A_m are roots of the algebraic (transcendental) equation
$$A + \int_a^b \Psi(t, Y_f(t) + A Y_\varphi(t)) \, dt = 0.$$

8.4. Equations of the Second Kind with Constant Limits of Integration

8.4.1. Linear Fredholm Integral Equations of the Second Kind

▶ **Equations with kernels containing power law functions.**

1. $y(x) - \lambda \int_a^b (x-t)y(t)\,dt = f(x).$

Solution:
$$y(x) = f(x) + \lambda(A_1 x + A_2),$$
where
$$A_1 = \frac{12f_1 + 6\lambda(f_1\Delta_2 - 2f_2\Delta_1)}{\lambda^2\Delta_1^4 + 12}, \quad A_2 = \frac{-12f_2 + 2\lambda(3f_2\Delta_2 - 2f_1\Delta_3)}{\lambda^2\Delta_1^4 + 12},$$
$$f_1 = \int_a^b f(x)\,dx, \quad f_2 = \int_a^b xf(x)\,dx, \quad \Delta_n = b^n - a^n.$$

2. $y(x) + A \int_a^b |x-t|\,y(t)\,dt = f(x).$

1°. For $A < 0$, the solution is given by
$$y(x) = C_1\cosh(kx) + C_2\sinh(kx) + f(x) + k\int_a^x \sinh[k(x-t)]f(t)\,dt, \quad k = \sqrt{-2A}, \tag{1}$$
where the constants C_1 and C_2 are determined by the conditions
$$\begin{aligned} y'_x(a) + y'_x(b) &= f'_x(a) + f'_x(b), \\ y(a) + y(b) + (b-a)y'_x(a) &= f(a) + f(b) + (b-a)f'_x(a). \end{aligned} \tag{2}$$

2°. For $A > 0$, the solution is given by
$$y(x) = C_1\cos(kx) + C_2\sin(kx) + f(x) - k\int_a^x \sin[k(x-t)]f(t)\,dt, \quad k = \sqrt{2A}, \tag{3}$$
where the constants C_1 and C_2 are determined by conditions (2).

3°. In the special case $a = 0$ and $A > 0$, the solution of the integral equation is given by formula (3) with
$$C_1 = k\frac{I_s(1+\cos\lambda) - I_c(\lambda + \sin\lambda)}{2 + 2\cos\lambda + \lambda\sin\lambda}, \quad C_2 = k\frac{I_s\sin\lambda + I_c(1+\cos\lambda)}{2 + 2\cos\lambda + \lambda\sin\lambda},$$
$$k = \sqrt{2A}, \quad \lambda = bk, \quad I_s = \int_0^b \sin[k(b-t)]f(t)\,dt, \quad I_c = \int_0^b \cos[k(b-t)]f(t)\,dt.$$

3. $\int_{-1}^{1} \frac{y(x) - y(t)}{|x-t|}\,dt = \lambda y(x).$

The characteristic values of the equation:
$$\lambda_n = 2\left(1 + \frac{1}{2} + \cdots + \frac{1}{n}\right), \quad \text{where} \quad n = 1, 2, \ldots$$

The eigenfunctions of the equation:
$$y_n(x) = P_n(x), \quad \text{where} \quad n = 1, 2, \ldots$$
Here $P_n(x) = \dfrac{1}{n!\, 2^n} \dfrac{d^n}{dx^n}(x^2 - 1)^n$ are the Legendre polynomials.

▶ *In equations 8.4.1.4 and 8.4.1.5 and their solutions, all singular integrals are understood in the sense of the Cauchy principal value.*

4. $y(x) - \lambda \displaystyle\int_{-\infty}^{\infty} \dfrac{y(t)\, dt}{t - x} = f(x).$

Solution:
$$y(x) = \dfrac{1}{1 + \pi^2 \lambda^2}\left[f(x) + \lambda \int_{-\infty}^{\infty} \dfrac{f(t)\, dt}{t - x} \right].$$

5. $Ay(x) + \dfrac{B}{\pi} \displaystyle\int_{-1}^{1} \dfrac{y(t)\, dt}{t - x} = f(x), \quad -1 < x < 1.$

Without loss of generality, we may assume that $A^2 + B^2 = 1$.

1°. The solution bounded at the endpoints (Lifanov et al., 2004):
$$y(x) = Af(x) - \dfrac{B}{\pi} \int_{-1}^{1} \dfrac{g(x)}{g(t)} \dfrac{f(t)\, dt}{t - x}, \quad g(x) = (1 + x)^\alpha (1 - x)^{1-\alpha}, \tag{1}$$
where α is the solution of the trigonometric equation
$$A + B\cot(\pi\alpha) = 0 \tag{2}$$
on the interval $0 < \alpha < 1$. This solution $y(x)$ exists if and only if $\displaystyle\int_{-1}^{1} \dfrac{f(t)}{g(t)}\, dt = 0$.

2°. The solution bounded at the endpoint $x = 1$ and unbounded at the endpoint $x = -1$:
$$y(x) = Af(x) - \dfrac{B}{\pi} \int_{-1}^{1} \dfrac{g(x)}{g(t)} \dfrac{f(t)\, dt}{t - x}, \quad g(x) = (1 + x)^\alpha (1 - x)^{-\alpha}, \tag{3}$$
where α is the solution of the trigonometric equation (2) on the interval $-1 < \alpha < 0$.

3°. The solution unbounded at the endpoints:
$$y(x) = Af(x) - \dfrac{B}{\pi} \int_{-1}^{1} \dfrac{g(x)}{g(t)} \dfrac{f(t)\, dt}{t - x} + Cg(x), \quad g(x) = (1 + x)^\alpha (1 - x)^{-1-\alpha},$$
where C is an arbitrary constant and α is the solution of the trigonometric equation (2) on the interval $-1 < \alpha < 0$.

6. $y(x) - \lambda \displaystyle\int_0^1 \left(\dfrac{1}{t - x} - \dfrac{1}{x + t - 2xt} \right) y(t)\, dt = f(x), \quad 0 < x < 1.$

Tricomi's equation.

Solution (Zabreyko et al., 1975):
$$y(x) = \dfrac{1}{1 + \lambda^2 \pi^2}\left[f(x) + \int_0^1 \dfrac{t^\alpha (1 - x)^\alpha}{x^\alpha (1 - t)^\alpha}\left(\dfrac{1}{t - x} - \dfrac{1}{x + t - 2xt} \right) f(t)\, dt \right] + \dfrac{C(1 - x)^\beta}{x^{1+\beta}},$$
$$\alpha = \dfrac{2}{\pi} \arctan(\lambda \pi) \quad (-1 < \alpha < 1), \quad \tan\dfrac{\beta\pi}{2} = \lambda\pi \quad (-2 < \beta < 0),$$
where C is an arbitrary constant.

▶ **Equations with kernels containing exponential functions.**

7. $y(x) + \lambda \int_0^\infty e^{-|x-t|} y(t)\, dt = f(x)$.

Solution for $\lambda > -\tfrac{1}{2}$ (Gakhov & Cherskii, 1978):

$$y(x) = f(x) - \frac{\lambda}{\sqrt{1+2\lambda}} \int_0^\infty \exp\bigl(-\sqrt{1+2\lambda}\,|x-t|\bigr) f(t)\, dt$$
$$+ \left(1 - \frac{\lambda+1}{\sqrt{1+2\lambda}}\right) \int_0^\infty \exp\bigl[-\sqrt{1+2\lambda}\,(x+t)\bigr] f(t)\, dt.$$

8. $y(x) - \lambda \int_{-\infty}^\infty e^{-|x-t|} y(t)\, dt = 0, \qquad \lambda > 0.$

Lalesco–Picard equation.

Solution (Krasnov et al., 1971):

$$y(x) = \begin{cases} C_1 \exp\bigl(x\sqrt{1-2\lambda}\bigr) + C_2 \exp\bigl(-x\sqrt{1-2\lambda}\bigr) & \text{for } 0 < \lambda < \tfrac{1}{2}, \\ C_1 + C_2 x & \text{for } \lambda = \tfrac{1}{2}, \\ C_1 \cos\bigl(x\sqrt{2\lambda-1}\bigr) + C_2 \sin\bigl(x\sqrt{2\lambda-1}\bigr) & \text{for } \lambda > \tfrac{1}{2}, \end{cases}$$

where C_1 and C_2 are arbitrary constants.

9. $y(x) + \lambda \int_{-\infty}^\infty e^{-|x-t|} y(t)\, dt = f(x)$.

1°. Solution for $\lambda > -\tfrac{1}{2}$ (Gakhov & Cherskii, 1978):

$$y(x) = f(x) - \frac{\lambda}{\sqrt{1+2\lambda}} \int_{-\infty}^\infty \exp\bigl(-\sqrt{1+2\lambda}\,|x-t|\bigr) f(t)\, dt.$$

2°. If $\lambda \leq -\tfrac{1}{2}$, for the equation to be solvable, the conditions

$$\int_{-\infty}^\infty f(x) \cos(ax)\, dx = 0, \qquad \int_{-\infty}^\infty f(x) \sin(ax)\, dx = 0,$$

where $a = \sqrt{-1-2\lambda}$, must be satisfied. In this case, the solution has the form

$$y(x) = f(x) - \frac{a^2+1}{2a} \int_0^\infty \sin(at) f(x+t)\, dt \qquad (-\infty < x < \infty).$$

In the class of solutions not belonging to $L_2(-\infty, \infty)$, the homogeneous equation (with $f(x) \equiv 0$) has a nontrivial solution. In this case, the general solution of the corresponding nonhomogeneous equation with $\lambda \leq -\tfrac{1}{2}$ has the form

$$y(x) = C_1 \sin(ax) + C_2 \cos(ax) + f(x) - \frac{a^2+1}{4a} \int_{-\infty}^\infty \sin(a|x-t|) f(t)\, dt.$$

10. $y(x) + A \int_a^b e^{\lambda|x-t|} y(t)\, dt = f(x)$.

1°. The function $y = y(x)$ obeys the following second-order linear nonhomogeneous ordinary differential equation with constant coefficients:

$$y''_{xx} + \lambda(2A - \lambda) y = f''_{xx}(x) - \lambda^2 f(x). \tag{1}$$

The boundary conditions for (1) have the form
$$y'_x(a) + \lambda y(a) = f'_x(a) + \lambda f(a),$$
$$y'_x(b) - \lambda y(b) = f'_x(b) - \lambda f(b). \tag{2}$$

Equation (1) under the boundary conditions (2) determines the solution of the original integral equation.

2°. For $\lambda(2A - \lambda) < 0$, the general solution of equation (1) is given by
$$y(x) = C_1 \cosh(kx) + C_2 \sinh(kx) + f(x) - \frac{2A\lambda}{k} \int_a^x \sinh[k(x-t)]\, f(t)\, dt,$$
$$k = \sqrt{\lambda(\lambda - 2A)}, \tag{3}$$

where C_1 and C_2 are arbitrary constants.

For $\lambda(2A - \lambda) > 0$, the general solution of equation (1) is given by
$$y(x) = C_1 \cos(kx) + C_2 \sin(kx) + f(x) - \frac{2A\lambda}{k} \int_a^x \sin[k(x-t)]\, f(t)\, dt,$$
$$k = \sqrt{\lambda(2A - \lambda)}. \tag{4}$$

For $\lambda = 2A$, the general solution of equation (1) is given by
$$y(x) = C_1 + C_2 x + f(x) - 4A^2 \int_a^x (x-t) f(t)\, dt. \tag{5}$$

The constants C_1 and C_2 in solutions (3)–(5) are determined by conditions (2).

3°. In the special case of $a = 0$ and $\lambda(2A - \lambda) > 0$, the solution of the integral equation is given by formula (4) with
$$C_1 = \frac{A(kI_c - \lambda I_s)}{(\lambda - A)\sin\mu - k\cos\mu}, \quad C_2 = -\frac{\lambda}{k} \frac{A(kI_c - \lambda I_s)}{(\lambda - A)\sin\mu - k\cos\mu},$$
$$k = \sqrt{\lambda(2A - \lambda)}, \quad \mu = bk, \quad I_s = \int_0^b \sin[k(b-t)] f(t)\, dt, \quad I_c = \int_0^b \cos[k(b-t)] f(t)\, dt.$$

11. $y(x) + \lambda \int_{-\infty}^{\infty} \dfrac{y(t)\, dt}{\cosh[b(x-t)]} = f(x).$

Solution for $b > \pi|\lambda|$ (Gakhov & Cherskii, 1978):
$$y(x) = f(x) - \frac{2\lambda b}{\sqrt{b^2 - \pi^2 \lambda^2}} \int_{-\infty}^{\infty} \frac{\sinh[2k(x-t)]}{\sinh[2b(x-t)]} f(t)\, dt, \quad k = \frac{b}{\pi} \arccos\left(\frac{\pi\lambda}{b}\right).$$

12. $y(x) + A \int_a^b \sinh(\lambda|x - t|) y(t)\, dt = f(x).$

1°. The function $y = y(x)$ obeys the following second-order linear nonhomogeneous ordinary differential equation with constant coefficients:
$$y''_{xx} + \lambda(2A - \lambda) y = f''_{xx}(x) - \lambda^2 f(x). \tag{1}$$

The boundary conditions for (1) have the form
$$\sinh[\lambda(b-a)]\varphi'_x(b) - \lambda\cosh[\lambda(b-a)]\varphi(b) = \lambda\varphi(a),$$
$$\sinh[\lambda(b-a)]\varphi'_x(a) + \lambda\cosh[\lambda(b-a)]\varphi(a) = -\lambda\varphi(b), \qquad \varphi(x) = y(x) - f(x). \tag{2}$$

ODE (1) under the boundary conditions (2) determines the solution of the original integral equation.

$2°$. For $\lambda(2A - \lambda) = -k^2 < 0$, the general solution of equation (1) is given by

$$y(x) = C_1 \cosh(kx) + C_2 \sinh(kx) + f(x) - \frac{2A\lambda}{k} \int_a^x \sinh[k(x-t)]f(t)\,dt, \qquad (3)$$

where C_1 and C_2 are arbitrary constants.

For $\lambda(2A - \lambda) = k^2 > 0$, the general solution of equation (1) is given by

$$y(x) = C_1 \cos(kx) + C_2 \sin(kx) + f(x) - \frac{2A\lambda}{k} \int_a^x \sin[k(x-t)]f(t)\,dt. \qquad (4)$$

For $\lambda = 2A$, the general solution of equation (1) is

$$y(x) = C_1 + C_2 x + f(x) - 4A^2 \int_a^x (x-t)f(t)\,dt. \qquad (5)$$

The constants C_1 and C_2 in solutions (3)–(5) are determined by conditions (2).

▶ **Equations with kernels containing trigonometric functions.**

13. $y(x) - \lambda \int_0^\infty \cos(xt) y(t)\,dt = 0.$

Characteristic values: $\lambda = \pm\sqrt{2/\pi}$. For the characteristic values, the integral equation has infinitely many linearly independent eigenfunctions (Krasnov et al., 1971).

Eigenfunctions for $\lambda = +\sqrt{2/\pi}$ have the form

$$y_+(x) = f(x) + \sqrt{\frac{2}{\pi}} \int_0^\infty f(t) \cos(xt)\,dt, \qquad (1)$$

where $f = f(x)$ is any continuous function absolutely integrable on the interval $[0, \infty)$.

Eigenfunctions for $\lambda = -\sqrt{2/\pi}$ have the form

$$y_-(x) = f(x) - \sqrt{\frac{2}{\pi}} \int_0^\infty f(t) \cos(xt)\,dt, \qquad (2)$$

where $f = f(x)$ is any continuous function absolutely integrable on the interval $[0, \infty)$.

In particular, from (1) and (2) with $f(x) = e^{-ax}$ we obtain

$$y_+(x) = e^{-ax} + \sqrt{\frac{2}{\pi}} \frac{a}{a^2 + x^2} \qquad \text{for} \quad \lambda = +\sqrt{\frac{2}{\pi}},$$

$$y_-(x) = e^{-ax} - \sqrt{\frac{2}{\pi}} \frac{a}{a^2 + x^2} \qquad \text{for} \quad \lambda = -\sqrt{\frac{2}{\pi}},$$

where a is any positive number.

14. $y(x) - \lambda \int_0^\infty \cos(xt) y(t)\,dt = f(x).$

Solution (Krasnov et al., 1971):

$$y(x) = \frac{f(x)}{1 - \frac{\pi}{2}\lambda^2} + \frac{\lambda}{1 - \frac{\pi}{2}\lambda^2} \int_0^\infty \cos(xt) f(t)\,dt, \qquad \lambda \neq \pm\sqrt{2/\pi}.$$

15. $y(x) - \lambda \int_0^\infty \sin(xt) y(t)\, dt = 0.$

Characteristic values: $\lambda = \pm\sqrt{2/\pi}$. For the characteristic values, the integral equation has infinitely many linearly independent eigenfunctions (Krasnov et al., 1971).

Eigenfunctions for $\lambda = +\sqrt{2/\pi}$ have the form

$$y_+(x) = f(x) + \sqrt{\frac{2}{\pi}} \int_0^\infty f(t) \sin(xt)\, dt,$$

where $f = f(x)$ is any continuous function absolutely integrable on the interval $[0, \infty)$.

Eigenfunctions for $\lambda = -\sqrt{2/\pi}$ have the form

$$y_-(x) = f(x) - \sqrt{\frac{2}{\pi}} \int_0^\infty f(t) \sin(xt)\, dt,$$

where $f = f(x)$ is any continuous function absolutely integrable on the interval $[0, \infty)$.

16. $y(x) - \lambda \int_0^\infty \sin(xt) y(t)\, dt = f(x).$

Solution (Krasnov et al., 1971):

$$y(x) = \frac{f(x)}{1 - \frac{\pi}{2}\lambda^2} + \frac{\lambda}{1 - \frac{\pi}{2}\lambda^2} \int_0^\infty \sin(xt) f(t)\, dt, \qquad \lambda \neq \pm\sqrt{2/\pi}.$$

17. $y(x) + A \int_a^b \sin(\lambda |x - t|) y(t)\, dt = f(x).$

1°. The function $y = y(x)$ obeys the following second-order linear nonhomogeneous ordinary differential equation with constant coefficients:

$$y''_{xx} + \lambda(2A + \lambda) y = f''_{xx}(x) + \lambda^2 f(x). \tag{1}$$

The boundary conditions for (1) have the form

$$\sin[\lambda(b-a)]\varphi'_x(b) - \lambda \cos[\lambda(b-a)]\varphi(b) = \lambda \varphi(a),$$
$$\sin[\lambda(b-a)]\varphi'_x(a) + \lambda \cos[\lambda(b-a)]\varphi(a) = -\lambda \varphi(b), \qquad \varphi(x) = y(x) - f(x). \tag{2}$$

ODE (1) under the boundary conditions (2) determines the solution of the original integral equation.

2°. For $\lambda(2A + \lambda) = -k^2 < 0$, the general solution of ODE (1) is given by

$$y(x) = C_1 \cosh(kx) + C_2 \sinh(kx) + f(x) - \frac{2A\lambda}{k} \int_a^x \sinh[k(x-t)] f(t)\, dt, \tag{3}$$

where C_1 and C_2 are arbitrary constants.

For $\lambda(2A + \lambda) = k^2 > 0$, the general solution of ODE (1) is given by

$$y(x) = C_1 \cos(kx) + C_2 \sin(kx) + f(x) - \frac{2A\lambda}{k} \int_a^x \sin[k(x-t)] f(t)\, dt. \tag{4}$$

For $\lambda = 2A$, the general solution of ODE (1) is

$$y(x) = C_1 + C_2 x + f(x) + 4A^2 \int_a^x (x-t) f(t)\, dt. \tag{5}$$

The constants C_1 and C_2 in solutions (3)–(5) are determined by conditions (2).

18. $y(x) - \lambda \int_{-\infty}^{\infty} \frac{\sin(x-t)}{x-t} y(t)\, dt = f(x).$

Solution (Gakhov & Cherskii, 1978):

$$y(x) = f(x) + \frac{\lambda}{\sqrt{2\pi} - \pi\lambda} \int_{-\infty}^{\infty} \frac{\sin(x-t)}{x-t} f(t)\, dt, \qquad \lambda \neq \sqrt{2/\pi}.$$

19. $Ay(x) - \dfrac{B}{2\pi} \int_{0}^{2\pi} \cot\left(\dfrac{t-x}{2}\right) y(t)\, dt = f(x), \qquad 0 \leq x \leq 2\pi.$

Here the integral is understood in the sense of the Cauchy principal value. Without loss of generality we may assume that $A^2 + B^2 = 1$.

Solution (Lifanov et al., 2004):

$$y(x) = Af(x) + \frac{B}{2\pi} \int_{0}^{2\pi} \cot\left(\frac{t-x}{2}\right) f(t)\, dt + \frac{B^2}{2\pi A} \int_{0}^{2\pi} f(t)\, dt.$$

20. $y(x) - \lambda \int_{0}^{\infty} e^{\mu(x-t)} \cos(xt) y(t)\, dt = f(x).$

Solution:

$$y(x) = \frac{f(x)}{1 - \frac{\pi}{2}\lambda^2} + \frac{\lambda}{1 - \frac{\pi}{2}\lambda^2} \int_{0}^{\infty} e^{\mu(x-t)} \cos(xt) f(t)\, dt, \qquad \lambda \neq \pm\sqrt{2/\pi}.$$

21. $y(x) - \lambda \int_{0}^{\infty} e^{\mu(x-t)} \sin(xt) y(t)\, dt = f(x).$

Solution:

$$y(x) = \frac{f(x)}{1 - \frac{\pi}{2}\lambda^2} + \frac{\lambda}{1 - \frac{\pi}{2}\lambda^2} \int_{0}^{\infty} e^{\mu(x-t)} \sin(xt) f(t)\, dt, \qquad \lambda \neq \pm\sqrt{2/\pi}.$$

▶ **Equations with kernels containing arbitrary functions.**

22. $y(x) - \lambda \int_{a}^{b} g(x) h(t) y(t)\, dt = f(x).$

1°. Assume that $\lambda \neq \left(\int_{a}^{b} g(t) h(t)\, dt\right)^{-1}$.

Solution:

$$y(x) = f(x) + \lambda k g(x), \qquad \text{where} \quad k = \left(1 - \lambda \int_{a}^{b} g(t) h(t)\, dt\right)^{-1} \int_{a}^{b} h(t) f(t)\, dt.$$

2°. Assume that $\lambda = \left(\int_{a}^{b} ag(t) h(t)\, dt\right)^{-1}$.

For $\int_{a}^{b} h(t) f(t)\, dt = 0$, the solution has the form

$$y = f(x) + C g(x),$$

where C is an arbitrary constant.

For $\int_{a}^{b} h(t) f(t)\, dt \neq 0$, there is no solution.

Remark 8.2. The limits of integration may take the values $a = -\infty$ and/or $b = \infty$, provided that the corresponding improper integral converges.

23. $y(x) - \lambda \int_a^b [g(x) + h(t)] y(t)\, dt = f(x).$

The characteristic values of the equation:
$$\lambda_{1,2} = \frac{s_1 + s_3 \pm \sqrt{(s_1 - s_3)^2 + 4(b-a)s_2}}{2[s_1 s_3 - (b-a)s_2]},$$

where
$$s_1 = \int_a^b g(x)\, dx, \quad s_2 = \int_a^b g(x)h(x)\, dx, \quad s_3 = \int_a^b h(x)\, dx.$$

1°. Solution with $\lambda \neq \lambda_{1,2}$:
$$y(x) = f(x) + \lambda[A_1 g(x) + A_2],$$

where the constants A_1 and A_2 are given by
$$A_1 = \frac{f_1 - \lambda[f_1 s_3 - (b-a)f_2]}{[s_1 s_3 - (b-a)s_2]\lambda^2 - (s_1 + s_3)\lambda + 1},$$
$$A_2 = \frac{f_2 - \lambda(f_2 s_1 - f_1 s_2)}{[s_1 s_3 - (b-a)s_2]\lambda^2 - (s_1 + s_3)\lambda + 1},$$
$$f_1 = \int_a^b f(x)\, dx, \quad f_2 = \int_a^b f(x)h(x)\, dx.$$

2°. Solution with $\lambda = \lambda_1 \neq \lambda_2$ and $f_1 = f_2 = 0$:
$$y(x) = f(x) + Cy_1(x), \quad y_1(x) = g(x) + \frac{1 - \lambda_1 s_1}{\lambda_1(b-a)},$$

where C is an arbitrary constant and $y_1(x)$ is an eigenfunction of the equation corresponding to the characteristic value λ_1.

3°. The solution with $\lambda = \lambda_2 \neq \lambda_1$ and $f_1 = f_2 = 0$ is given by the formulas of Item 2° in which one must replace λ_1 and $y_1(x)$ by λ_2 and $y_2(x)$, respectively.

4°. Solution with $\lambda = \lambda_{1,2} = \lambda_*$ and $f_1 = f_2 = 0$ provided that $s_1 \neq \pm s_3$, where the characteristic value $\lambda_* = \dfrac{2}{s_1 + s_3}$ is double:
$$y(x) = f(x) + Cy_*(x), \quad y_*(x) = g(x) - \frac{s_1 - s_3}{2(b-a)}.$$

Here C is an arbitrary constant and $y_*(x)$ is an eigenfunction of the equation corresponding to λ_*.

The equation has no multiple characteristic values if $s_1 = \pm s_3$.

24. $y(x) = \lambda \int_{-\pi}^{\pi} K(x-t) y(t)\, dt, \quad K(x) = K(-x).$

Characteristic values:
$$\lambda_n = \frac{1}{\pi a_n}, \quad a_n = \frac{1}{\pi} \int_{-\pi}^{\pi} K(x) \cos(nx)\, dx \quad (n = 0, 1, 2, \ldots).$$

The corresponding eigenfunctions are

$$y_0(x) = 1, \quad y_n^{(1)}(x) = \cos(nx), \quad y_n^{(2)}(x) = \sin(nx) \quad (n = 1, 2, \dots).$$

For each value λ_n with $n \neq 0$, there are two corresponding linearly independent eigenfunctions $y_n^{(1)}(x)$ and $y_n^{(2)}(x)$.

25. $y(x) - \displaystyle\int_{-\infty}^{\infty} K(x-t) y(t)\, dt = f(x).$

Here $-\infty < x < \infty$, $f(x) \in L_1(-\infty, \infty)$, and $K(x) \in L_1(-\infty, \infty)$.

For the integral equation to be solvable (in L_1), it is necessary and sufficient that

$$1 - \sqrt{2\pi}\, \tilde{K}(u) \neq 0, \qquad -\infty < u < \infty,$$

where $\tilde{K}(u) = \dfrac{1}{\sqrt{2\pi}} \displaystyle\int_{-\infty}^{\infty} K(x) e^{-iux}\, dx$ is the Fourier transform of $K(x)$. In this case, the equation has a unique solution, which is given by the formulas (Ditkin & Prudnikov, 1965):

$$y(x) = f(x) + \int_{-\infty}^{\infty} R(x-t) f(t)\, dt,$$

$$R(x) = \frac{1}{\sqrt{2\pi}} \int_{-\infty}^{\infty} \tilde{R}(u) e^{iux}\, du, \qquad \tilde{R}(u) = \frac{\tilde{K}(u)}{1 - \sqrt{2\pi}\, \tilde{K}(u)}.$$

26. $y(x) - \displaystyle\int_0^{\infty} K(x-t) y(t)\, dt = f(x).$

Wiener–Hopf equation of the second kind. This equation is discussed in the books by Noble (1958), Gakhov & Cherskii (1978), and Polyanin & Manzhirov (2008) in detail.

27. $y(x) + \displaystyle\int_a^b |x-t| g(t) y(t)\, dt = f(x), \qquad a \leq x \leq b.$

1°. Let us remove the modulus in the integrand,

$$y(x) + \int_a^x (x-t) g(t) y(t)\, dt + \int_x^b (t-x) g(t) y(t)\, dt = f(x). \tag{1}$$

Differentiating (1) with respect to x yields

$$y'_x(x) + \int_a^x g(t) y(t)\, dt - \int_x^b g(t) y(t)\, dt = f'_x(x). \tag{2}$$

Differentiating (2), we arrive at a second-order ODE for $y = y(x)$:

$$y''_{xx} + 2g(x) y = f''_{xx}(x). \tag{3}$$

2°. Let us derive the boundary conditions for ODE (3). We assume that the limits of integration satisfy the conditions $-\infty < a < b < \infty$. By setting $x = a$ and $x = b$ in (1), we obtain two consequences

$$\begin{aligned} y(a) + \int_a^b (t-a) g(t) y(t)\, dt &= f(a), \\ y(b) + \int_a^b (b-t) g(t) y(t)\, dt &= f(b). \end{aligned} \tag{4}$$

Let us express $g(x)y$ from (3) via y''_{xx} and f''_{xx} and substitute the result into (4). Integrating by parts yields the desired boundary conditions for $y(x)$:

$$y(a) + y(b) + (b-a)[f'_x(b) - y'_x(b)] = f(a) + f(b),$$
$$y(a) + y(b) + (a-b)[f'_x(a) - y'_x(a)] = f(a) + f(b). \quad (5)$$

Note a useful consequence of (5),

$$y'_x(a) + y'_x(b) = f'_x(a) + f'_x(b), \quad (6)$$

which can be used together with one of conditions (5).

ODE (3) under the boundary conditions (5) determines the solution of the original integral equation. Conditions (5) make it possible to calculate the constants of integration that occur in the solution of the differential equation (3).

28. $\quad y(x) + \displaystyle\int_a^b e^{\lambda|x-t|} g(t) y(t)\, dt = f(x), \qquad a \le x \le b.$

$1°$. Let us remove the modulus in the integrand:

$$y(x) + \int_a^x e^{\lambda(x-t)} g(t) y(t)\, dt + \int_x^b e^{\lambda(t-x)} g(t) y(t)\, dt = f(x). \quad (1)$$

Differentiating (1) with respect to x twice yields

$$y''_{xx}(x) + 2\lambda g(x) y(x) + \lambda^2 \int_a^x e^{\lambda(x-t)} g(t) y(t)\, dt + \lambda^2 \int_x^b e^{\lambda(t-x)} g(t) y(t)\, dt = f''_{xx}(x). \quad (2)$$

Eliminating the integral terms from (1) and (2), we arrive at a second-order ordinary differential equation for $y = y(x)$:

$$y''_{xx} + 2\lambda g(x) y - \lambda^2 y = f''_{xx}(x) - \lambda^2 f(x). \quad (3)$$

$2°$. Let us derive the boundary conditions for ODE (3). We assume that the limits of integration satisfy the conditions $-\infty < a < b < \infty$. By setting $x = a$ and $x = b$ in (1), we obtain two consequences

$$y(a) + e^{-\lambda a} \int_a^b e^{\lambda t} g(t) y(t)\, dt = f(a),$$
$$y(b) + e^{\lambda b} \int_a^b e^{-\lambda t} g(t) y(t)\, dt = f(b). \quad (4)$$

Let us express $g(x)y$ from (3) via y''_{xx} and f''_{xx} and substitute the result into (4). Integrating by parts yields the conditions

$$e^{\lambda b} \varphi'_x(b) - e^{\lambda a} \varphi'_x(a) = \lambda e^{\lambda a} \varphi(a) + \lambda e^{\lambda b} \varphi(b),$$
$$e^{-\lambda b} \varphi'_x(b) - e^{-\lambda a} \varphi'_x(a) = \lambda e^{-\lambda a} \varphi(a) + \lambda e^{-\lambda b} \varphi(b), \qquad \varphi(x) = y(x) - f(x).$$

Finally, after some manipulations, we arrive at the desired boundary conditions for $y(x)$:

$$\varphi'_x(a) + \lambda \varphi(a) = 0, \quad \varphi'_x(b) - \lambda \varphi(b) = 0; \qquad \varphi(x) = y(x) - f(x). \quad (5)$$

ODE (3) under the boundary conditions (5) determines the solution of the original integral equation. Conditions (5) make it possible to calculate the constants of integration that occur in solving the differential equation (3).

29. $y(x) + \int_a^b \sinh(\lambda|x-t|)g(t)y(t)\,dt = f(x), \qquad a \le x \le b.$

$1°$. Let us remove the modulus in the integrand:

$$y(x) + \int_a^x \sinh[\lambda(x-t)]g(t)y(t)\,dt + \int_x^b \sinh[\lambda(t-x)]g(t)y(t)\,dt = f(x). \qquad (1)$$

Differentiating (1) with respect to x twice yields

$$y''_{xx}(x) + 2\lambda g(x)y(x) + \lambda^2 \int_a^x \sinh[\lambda(x-t)]g(t)y(t)\,dt$$
$$+ \lambda^2 \int_x^b \sinh[\lambda(t-x)]g(t)y(t)\,dt = f''_{xx}(x). \qquad (2)$$

Eliminating the integral terms from (1) and (2), we arrive at a second-order ordinary differential equation for $y = y(x)$:

$$y''_{xx} + 2\lambda g(x)y - \lambda^2 y = f''_{xx}(x) - \lambda^2 f(x). \qquad (3)$$

$2°$. Let us derive the boundary conditions for ODE (3). We assume that the limits of integration satisfy the conditions $-\infty < a < b < \infty$. By setting $x = a$ and $x = b$ in (1), we obtain two corollaries

$$y(a) + \int_a^b \sinh[\lambda(t-a)]g(t)y(t)\,dt = f(a),$$
$$y(b) + \int_a^b \sinh[\lambda(b-t)]g(t)y(t)\,dt = f(b). \qquad (4)$$

Let us express $g(x)y$ from (3) via y''_{xx} and f''_{xx} and substitute the result into (4). Integrating by parts yields the desired boundary conditions for $y(x)$:

$$\sinh[\lambda(b-a)]\varphi'_x(b) - \lambda\cosh[\lambda(b-a)]\varphi(b) = \lambda\varphi(a),$$
$$\sinh[\lambda(b-a)]\varphi'_x(a) + \lambda\cosh[\lambda(b-a)]\varphi(a) = -\lambda\varphi(b); \quad \varphi(x) = y(x) - f(x). \qquad (5)$$

ODE (3) under the boundary conditions (5) determines the solution of the original integral equation. Conditions (5) make it possible to calculate the constants of integration that occur in solving the differential equation (3).

30. $y(x) + \int_a^b \sin(\lambda|x-t|)g(t)y(t)\,dt = f(x), \qquad a \le x \le b.$

$1°$. Let us remove the modulus in the integrand:

$$y(x) + \int_a^x \sin[\lambda(x-t)]g(t)y(t)\,dt + \int_x^b \sin[\lambda(t-x)]g(t)y(t)\,dt = f(x). \qquad (1)$$

Differentiating (1) with respect to x twice yields

$$y''_{xx}(x) + 2\lambda g(x)y(x) - \lambda^2 \int_a^x \sin[\lambda(x-t)]g(t)y(t)\,dt$$
$$- \lambda^2 \int_x^b \sin[\lambda(t-x)]g(t)y(t)\,dt = f''_{xx}(x). \qquad (2)$$

Eliminating the integral terms from (1) and (2), we arrive at a second-order ordinary differential equation for $y = y(x)$:

$$y''_{xx} + 2\lambda g(x)y + \lambda^2 y = f''_{xx}(x) + \lambda^2 f(x). \tag{3}$$

2°. Let us derive the boundary conditions for ODE (3). We assume that the limits of integration satisfy the conditions $-\infty < a < b < \infty$. By setting $x = a$ and $x = b$ in (1), we obtain two consequences

$$\begin{aligned} y(a) + \int_a^b \sin[\lambda(t-a)]g(t)y(t)\,dt &= f(a), \\ y(b) + \int_a^b \sin[\lambda(b-t)]g(t)y(t)\,dt &= f(b). \end{aligned} \tag{4}$$

Let us express $g(x)y$ from (3) via y''_{xx} and f''_{xx} and substitute the result into (4). Integrating by parts yields the desired boundary conditions for $y(x)$:

$$\begin{aligned} \sin[\lambda(b-a)]\varphi'_x(b) - \lambda \cos[\lambda(b-a)]\varphi(b) &= \lambda\varphi(a), \\ \sin[\lambda(b-a)]\varphi'_x(a) + \lambda \cos[\lambda(b-a)]\varphi(a) &= -\lambda\varphi(b); \quad \varphi(x) = y(x) - f(x). \end{aligned} \tag{5}$$

ODE (3) under the boundary conditions (5) determines the solution of the original integral equation. Conditions (5) make it possible to calculate the constants of integration that occur in solving the differential equation (3).

8.4.2. Nonlinear Fredholm Integral Equations of the Second Kind

1. $y(x) + \int_a^b g(t)y(x)y(t)\,dt = f(x).$

Solutions:
$$y_1(x) = \lambda_1 f(x), \qquad y_2(x) = \lambda_2 f(x),$$

where λ_1 and λ_2 are the roots of the quadratic equation

$$I\lambda^2 + \lambda - 1 = 0, \qquad I = \int_a^b f(t)g(t)\,dt.$$

2. $y(x) + \int_a^b g(x)h(t)y(x)y(t)\,dt = f(x).$

A solution:
$$y(x) = \frac{f(x)}{1 + \lambda g(x)},$$

where λ is a root of the algebraic (or transcendental) equation

$$\lambda - \int_a^b \frac{f(t)h(t)\,dt}{1 + \lambda g(t)} = 0.$$

Different roots generate different solutions of the integral equation.

3. $y(x) + \int_a^b f(t, y(t))\, dt = g(x).$

A solution: $y(x) = g(x) + \lambda$, where λ is determined by the algebraic (or transcendental) equation

$$\lambda + F(\lambda) = 0, \qquad F(\lambda) = \int_a^b f(t, g(t) + \lambda)\, dt.$$

4. $y(x) + \int_a^b e^{\lambda(x-t)} f(t, y(t))\, dt = g(x).$

A solution: $y(x) = \beta e^{\lambda x} + g(x)$, where β is determined by the algebraic (or transcendental) equation

$$\beta + F(\beta) = 0, \qquad F(\beta) = \int_a^b e^{-\lambda t} f(t, \beta e^{\lambda t} + g(t))\, dt.$$

5. $y(x) + \int_a^b g(x) f(t, y(t))\, dt = h(x).$

A solution: $y(x) = \lambda g(x) + h(x)$, where λ is determined by the algebraic (or transcendental) equation

$$\lambda + F(\lambda) = 0, \qquad F(\lambda) = \int_a^b f(t, \lambda g(t) + h(t))\, dt.$$

6. $y(x) + \int_a^b |x - t| f(t, y(t))\, dt = g(x), \qquad a \leq x \leq b.$

1°. Let us remove the modulus in the integrand:

$$y(x) + \int_a^x (x - t) f(t, y(t))\, dt + \int_x^b (t - x) f(t, y(t))\, dt = g(x). \tag{1}$$

Differentiating (1) with respect to x yields

$$y'_x(x) + \int_a^x f(t, y(t))\, dt - \int_x^b f(t, y(t))\, dt = g'_x(x). \tag{2}$$

Differentiating (2), we arrive at a second-order ordinary differential equation for $y = y(x)$ (Polyanin & Manzhirov, 2008):

$$y''_{xx} + 2f(x, y) = g''_{xx}(x). \tag{3}$$

2°. Let us derive the boundary conditions for ODE (3). We assume that $-\infty < a < b < \infty$. By setting $x = a$ and $x = b$ in (1), we obtain the relations

$$y(a) + \int_a^b (t - a) f(t, y(t))\, dt = g(a),$$
$$y(b) + \int_a^b (b - t) f(t, y(t))\, dt = g(b). \tag{4}$$

Let us solve equation (3) for $f(x, y)$ and substitute the result into (4). Integrating by parts yields the desired boundary conditions for $y(x)$:

$$y(a) + y(b) + (b-a)\big[g'_x(b) - y'_x(b)\big] = g(a) + g(b),$$
$$y(a) + y(b) + (a-b)\big[g'_x(a) - y'_x(a)\big] = g(a) + g(b). \qquad (5)$$

Let us point out a useful consequence of (5):

$$y'_x(a) + y'_x(b) = g'_x(a) + g'_x(b). \qquad (6)$$

It can be used together with one of conditions (5).

ODE (3) under the boundary conditions (5) determines the solution of the original integral equation (it may have several solutions). Conditions (5) make it possible to calculate the constants of integration that occur in solving the differential equation (3).

7. $\displaystyle y(x) + \int_a^b e^{\lambda|x-t|} f(t, y(t))\, dt = g(x), \qquad a \le x \le b.$

1°. Let us remove the modulus in the integrand:

$$y(x) + \int_a^x e^{\lambda(x-t)} f(t, y(t))\, dt + \int_x^b e^{\lambda(t-x)} f(t, y(t))\, dt = g(x). \qquad (1)$$

Differentiating (1) with respect to x twice yields

$$y''_{xx}(x) + 2\lambda f(x, y(x)) + \lambda^2 \int_a^x e^{\lambda(x-t)} f(t, y(t))\, dt + \lambda^2 \int_x^b e^{\lambda(t-x)} f(t, y(t))\, dt = g''_{xx}(x). \qquad (2)$$

Eliminating the integral terms from (1) and (2), we arrive at a second-order ordinary differential equation for $y = y(x)$ (Polyanin & Manzhirov, 2008):

$$y''_{xx} + 2\lambda f(x, y) - \lambda^2 y = g''_{xx}(x) - \lambda^2 g(x). \qquad (3)$$

2°. Let us derive the boundary conditions for ODE (3). We assume that $-\infty < a < b < \infty$. By setting $x = a$ and $x = b$ in (1), we obtain the relations

$$y(a) + e^{-\lambda a} \int_a^b e^{\lambda t} f(t, y(t))\, dt = g(a),$$
$$y(b) + e^{\lambda b} \int_a^b e^{-\lambda t} f(t, y(t))\, dt = g(b). \qquad (4)$$

Let us solve equation (3) for $f(x, y)$ and substitute the result into (4). Integrating by parts yields

$$e^{\lambda b} \varphi'_x(b) - e^{\lambda a} \varphi'_x(a) = \lambda e^{\lambda a} \varphi(a) + \lambda e^{\lambda b} \varphi(b), \quad \varphi(x) = y(x) - g(x);$$
$$e^{-\lambda b} \varphi'_x(b) - e^{-\lambda a} \varphi'_x(a) = \lambda e^{-\lambda a} \varphi(a) + \lambda e^{-\lambda b} \varphi(b).$$

Hence, we obtain the boundary conditions for $y(x)$:

$$\varphi'_x(a) + \lambda \varphi(a) = 0, \quad \varphi'_x(b) - \lambda \varphi(b) = 0; \qquad \varphi(x) = y(x) - g(x). \qquad (5)$$

ODE (3) under the boundary conditions (5) determines the solution of the original integral equation (it may have several solutions). Conditions (5) make it possible to calculate the constants of integration that occur in solving the differential equation (3).

8. $y(x) + \int_a^b \sinh\left(\lambda|x-t|\right) f(t, y(t))\, dt = g(x), \qquad a \le x \le b.$

1°. Let us remove the modulus in the integrand:

$$y(x) + \int_a^x \sinh[\lambda(x-t)] f(t,y(t))\, dt + \int_x^b \sinh[\lambda(t-x)] f(t,y(t))\, dt = g(x). \qquad (1)$$

Differentiating (1) with respect to x twice yields

$$y''_{xx}(x) + 2\lambda f(x, y(x)) + \lambda^2 \int_a^x \sinh[\lambda(x-t)] f(t,y(t))\, dt$$
$$+ \lambda^2 \int_x^b \sinh[\lambda(t-x)] f(t,y(t))\, dt = g''_{xx}(x). \qquad (2)$$

Eliminating the integral terms from (1) and (2), we arrive at a second-order ordinary differential equation for $y = y(x)$ (Polyanin & Manzhirov, 2008):

$$y''_{xx} + 2\lambda f(x, y) - \lambda^2 y = g''_{xx}(x) - \lambda^2 g(x). \qquad (3)$$

2°. Let us derive the boundary conditions for ODE (3). We assume that $-\infty < a < b < \infty$. By setting $x = a$ and $x = b$ in (1), we obtain the relations

$$y(a) + \int_a^b \sinh[\lambda(t-a)] f(t,y(t))\, dt = g(a),$$
$$y(b) + \int_a^b \sinh[\lambda(b-t)] f(t,y(t))\, dt = g(b). \qquad (4)$$

Let us solve equation (3) for $f(x, y)$ and substitute the result into (4). Integrating by parts yields

$$\sinh[\lambda(b-a)]\varphi'_x(b) - \lambda\cosh[\lambda(b-a)]\varphi(b) = \lambda\varphi(a), \quad \varphi(x) = y(x) - g(x);$$
$$\sinh[\lambda(b-a)]\varphi'_x(a) + \lambda\cosh[\lambda(b-a)]\varphi(a) = -\lambda\varphi(b). \qquad (5)$$

ODE (3) under the boundary conditions (5) determines the solution of the original integral equation (there may be several solutions). Conditions (5) make it possible to calculate the constants of integration that occur in solving the differential equation (3).

9. $y(x) + \int_a^b \sin(\lambda|x-t|)\, f(t,y(t))\, dt = g(x), \qquad a \le x \le b.$

1°. Let us remove the modulus in the integrand:

$$y(x) + \int_a^x \sin[\lambda(x-t)] f(t,y(t))\, dt + \int_x^b \sin[\lambda(t-x)] f(t,y(t))\, dt = g(x). \qquad (1)$$

Differentiating (1) with respect to x twice yields

$$y''_{xx}(x) + 2\lambda f(x, y(x)) - \lambda^2 \int_a^x \sin[\lambda(x-t)] f(t,y(t))\, dt$$
$$- \lambda^2 \int_x^b \sin[\lambda(t-x)] f(t,y(t))\, dt = g''_{xx}(x). \qquad (2)$$

Eliminating the integral terms from (1) and (2), we arrive at a second-order ordinary differential equation for $y = y(x)$ (Polyanin & Manzhirov, 2008):

$$y''_{xx} + 2\lambda f(x, y) + \lambda^2 y = g''_{xx}(x) + \lambda^2 g(x). \tag{3}$$

$2°$. Let us derive the boundary conditions for ODE (3). We assume that $-\infty < a < b < \infty$. By setting $x = a$ and $x = b$ in (1), we obtain the relations

$$\begin{aligned} y(a) + \int_a^b \sin[\lambda(t-a)]\, f(t, y(t))\, dt = g(a), \\ y(b) + \int_a^b \sin[\lambda(b-t)]\, f(t, y(t))\, dt = g(b). \end{aligned} \tag{4}$$

Let us solve equation (3) for $f(x, y)$ and substitute the result into (4). Integrating by parts yields

$$\begin{aligned} \sin[\lambda(b-a)]\, \varphi'_x(b) - \lambda \cos[\lambda(b-a)]\, \varphi(b) = \lambda \varphi(a), \quad \varphi(x) = y(x) - g(x); \\ \sin[\lambda(b-a)]\, \varphi'_x(a) + \lambda \cos[\lambda(b-a)]\, \varphi(a) = -\lambda \varphi(b). \end{aligned} \tag{5}$$

ODE (3) under the boundary conditions (5) determines the solution of the original integral equation (there may be several solutions). Conditions (5) make it possible to calculate the constants of integration that occur in solving the differential equation (3).

10. $y(x) + \int_a^b [g_1(x) f_1(t, y(t)) + g_2(x) f_2(t, y(t))]\, dt = h(x).$

A solution:

$$y(x) = h(x) + \lambda_1 g_1(x) + \lambda_2 g_2(x),$$

where the constants λ_1 and λ_2 are determined from the algebraic (or transcendental) system of equations

$$\begin{aligned} \lambda_1 + \int_a^b f_1(t, h(t) + \lambda_1 g_1(t) + \lambda_2 g_2(t))\, dt = 0, \\ \lambda_2 + \int_a^b f_2(t, h(t) + \lambda_1 g_1(t) + \lambda_2 g_2(t))\, dt = 0. \end{aligned}$$

11. $y(x) + \int_a^b \left[\sum_{k=1}^n g_k(x) f_k(t, y(t)) \right] dt = h(x).$

Solutions:

$$y(x) = h(x) + \sum_{k=1}^n \lambda_k g_k(x),$$

where the coefficients λ_k are determined from the algebraic (or transcendental) system

$$\lambda_m + \int_a^b f_m\!\left(t, h(t) + \sum_{k=1}^n \lambda_k g_k(t)\right) dt = 0; \quad m = 1, \ldots, n.$$

Different roots of this system generate different solutions of the integral equation.

12. $y(x) + \int_a^b g(x,y(x)) f(t,y(t))\,dt = h(x).$

Solution in implicit form:

$$y(x) + \lambda g(x,y(x)) - h(x) = 0, \tag{1}$$

where λ is determined from the algebraic (or transcendental) equation

$$\lambda - F(\lambda) = 0, \qquad F(\lambda) = \int_a^b f(t,y(t))\,dt. \tag{2}$$

Here the function $y(x) = y(x,\lambda)$ obtained by solving (1) must be substituted into (2).

The number of solutions of the integral equation is determined by the number of the solutions obtained from (1) and (2).

▶ More exact solutions of linear and nonlinear integral equations can be found in the specialized reference book by Polyanin & Manzhirov (2008). The main methods for finding exact solutions to various types of integral equations are described, for example, in the books by Gakhov & Cherskii (1990), Gorenflo & Vessella (1991), Muskhelishvili (1992), Bitsadze (1995), Sakhnovich (1996), and Polyanin & Manzhirov (2007, 2008).

References

Bellman, R. and Cooke, K.L., *Differential–Difference Equations*, Academic Press, New York, 1963.

Bitsadze, A.V., *Integral Equation of the First Kind*, World Scientific Publishing Co., Singapore, 1995.

Brakhage, H., Nickel, K., and Rieder, P., Auflösung der Abelschen Integralgleichung 2. Art, *ZAMP*, Vol. 16, S. 295–298, 1965.

Ditkin, V.A. and Prudnikov, A.P., *Integral Transforms and Operational Calculus*, Pergamon Press, New York, 1965.

Gakhov, F.D., *Boundary Value Problems*, Dover Publications, New York, 1990.

Gakhov, F.D. and Cherskii, Yu.I., *Equations of Convolution Type* [in Russian], Nauka Publishers, Moscow, 1978.

Gohberg, I.T. and Krein M.G., *Theory and Applications of Volterra Operators in Hilbert Space*, American Mathematical Society, 1970.

Gorenflo, R. and Vessella, S., *Abel Integral Equations: Analysis and Applications*, Springer-Verlag, Berlin, 1991.

Hirschman, I.I. and Widder, D.V., *The Convolution Transform*, Princeton Univ. Press, Princeton, New Jersey, 1955.

Krasnov, M.L., Kiselev, A.I., and Makarenko, G.I., *Problems and Exercises in Integral Equations*, Mir Publishers, Moscow, 1971.

Lifanov, I.K., Poltavskii, L.N., and Vainikko, G., *Hypersingular Integral Equations and Their Applications*, Chapman & Hall/CRC Press, Boca Raton, 2004.

Mikhlin, S.G. and Prössdorf, S., *Singular Integral Operators*, Springer-Verlag, Berlin, 1986.

Muskhelishvili, N.I., *Singular Integral Equations: Boundary Problems of Function Theory and Their Applications to Mathematical Physics*, Dover Publications, New York, 1992.

Noble, B., *Methods Based on Wiener–Hopf Technique for the Solution of Partial Differential Equations*, Pergamon Press, London, 1958.

Polyanin, A.D. and Manzhirov, A.V., *Handbook of Integral Equations, 2nd ed.,* Chapman & Hall/CRC Press, Boca Raton–London, 2008.

Polyanin, A.D. and Manzhirov, A.V., *Handbook of Mathematics for Engineers and Scientists,* Chapman & Hall/CRC Press, Boca Raton–London, 2007.

Polyanin, A.D. and Zaitsev, V.F., *Handbook of Exact Solutions for Ordinary Differential Equations, 2nd ed.,* Chapman & Hall/CRC Press, Boca Raton–London, 2003.

Polyanin, A.D. and Zaitsev, V.F., *Handbook of Ordinary Differential Equations: Exact Solutions, Methods, and Problems,* CRC Press, Boca Raton–London, 2018.

Prudnikov, A.P., Brychkov, Yu.A., and Marichev, O.I., *Integrals and Series, Vol. 5, Inverse Laplace Transforms,* Gordon & Breach, New York, 1992.

Sakhnovich, L.A., *Integral Equations with Difference Kernels on Finite Intervals,* Birkhäuser Verlag, Basel, 1996.

Samko, S.G., Kilbas, A.A., and Marichev, O.I., *Fractional Integrals and Derivatives. Theory and Applications,* Gordon & Breach, New York, 1993.

Srivastava, H.M. and Buschman, R.G., *Convolution Integral Equations with Special Function Kernels,* Wiley Eastern, New Delhi, 1977.

Titchmarsh, E.C., *Introduction to the Theory of Fourier Integrals, 3rd ed.,* Chelsea Publishing, New York, 1986.

Whittaker, E.T. and Watson, G.N., *A Course of Modern Analysis,* Cambridge Univ. Press, Cambridge, 1958.

Zabreyko, P.P., Koshelev, A.I., et al., *Integral Equations: A Reference Text,* Noordhoff International Publishing, Leyden, 1975.

Zwillinger, D., *Handbook of Differential Equations.* San Diego: Academic Press, 1989.

Chapter 9

Difference and Functional Equations

▶ **Preliminary remarks.** Functional equations are mathematical equations that contain an unknown function with different arguments. Difference equations are the simplest functional equations containing the quantities $y(x + a_k)$, where a_k are given constants ($k = 1, \ldots, m$), and $y(z)$ is the unknown function.

This chapter describes exact solutions of difference and more complex functional equations of various types. The exact solutions are either expressed in terms of elementary functions or as an explicit or implicit expressions involving an arbitrary function of one or two arguments and the functions included in the equation. Some solutions are presented in parametric form. Useful reductions are described that convert the functional equations in question to simpler equations. In addition to equations with a single unknown function, some equations involving several unknown functions are also considered.

The chapter often uses the term "solution" to mean "general solution."

9.1. Difference Equations

9.1.1. Difference Equations with Discrete Argument

▶ **Preliminary remarks.** Difference equations with discrete argument defined on a discrete set of integer points $x = n$, where $n = 0, 1, 2, \ldots$ The difference equation includes a function of the integer argument $y_n = y(x)|_{x=n}$, which is to be found.

▶ **First-order linear difference equations with discrete argument.**

1. $y_{n+1} + a_n y_n = 0.$

A first-order homogeneous linear difference equation. Here, $n = 0, 1, 2, \ldots$
Solution:
$$y_n = C u_n, \quad u_n = (-1)^n a_0 a_1 \ldots a_{n-1}, \quad n = 1, 2, \ldots,$$
where $C = y_0$ is an arbitrary constant and u_n is a particular solution.

2. $y_{n+1} - a y_n = f_n.$

A first-order nonhomogeneous linear difference equation.
Solution:
$$y_n = C a^n + \sum_{j=0}^{n-1} a^{n-j-1} f_j.$$

3. $y_{n+1} + a_n y_n = f_n$.

A *first-order nonhomogeneous linear difference equation* (generalization of the previous equation).

Solution:
$$y_n = C u_n + \tilde{y}_n, \quad n = 1, 2, \ldots,$$

where C is an arbitrary constant, and

$$u_n = (-1)^n a_0 a_1 \ldots a_{n-1},$$

$$\tilde{y}_n = \sum_{j=0}^{n-1} \frac{u_n}{u_{j+1}} f_j = f_{n-1} - a_{n-1} f_{n-2} + a_{n-2} a_{n-1} f_{n-3} - \cdots + (-1)^{n-1} a_1 a_2 \ldots a_{n-1} f_0.$$

▶ **First-order nonlinear difference equations with discrete argument.**

4. $y_{n+1} = a y_n \left(1 - \dfrac{y_n}{b}\right)$.

Logistic difference equation ($a > 0$, $b > 0$, and $0 \le y_0 \le b$).

1°. Let
$$a = b = 4, \quad y_0 = 4 \sin^2 \theta \quad (0 \le \theta \le \tfrac{\pi}{2}).$$

Closed-form solution:
$$y_n = 4 \sin^2(2^n \theta), \quad n = 0, 1, \ldots$$

2°. Let
$$a = 4, \quad b = 1, \quad y_0 = \sin^2 \theta \quad (0 \le \theta \le \tfrac{\pi}{2}).$$

Closed-form solution:
$$y_n = \sin^2(2^n \theta), \quad n = 0, 1, \ldots$$

3°. Let $0 \le a \le 4$ and $b = 1$. In this case, the solutions of the logistic equation have the following properties:

(a) There are equilibrium solutions $y_n = 0$ and $y_n = (a-1)/a$.
(b) If $0 \le y_0 \le 1$, then $0 \le y_n \le 1$.
(c) If $a = 0$, then $y_n = 0$.
(d) If $0 < a \le 1$, then $y_n \to 0$ as $n \to \infty$.
(e) If $1 < a \le 3$, then $y_n \to (a-1)/a$ as $n \to \infty$.
(f) If $3 < a < 3.449\ldots$, then y_n oscillates between the two points:

$$y_\pm = \frac{1}{2a}\left(a + 1 \pm \sqrt{a^2 - 2a - 3}\right).$$

5. $y_n y_{n+1} = a_n y_{n+1} + b_n y_n + c_n$.

Riccati difference equation. Here, $n = 0, 1, \ldots$ and the constants a_n, b_n, and c_n satisfy the condition $a_n b_n + c_n \ne 0$.

$1°$. The substitution
$$y_n = \frac{u_{n+1}}{u_n} + a_n$$
leads to a second-order linear difference equation of the form 9.1.1.13:
$$u_{n+2} + (a_{n+1} - b_n)u_{n+1} - (a_n b_n + c_n)u_n = 0.$$

$2°$. Let y_n^* be a particular solution of the Riccati difference equation. Then the substitution
$$z_n = \frac{1}{y_n - y_n^*}, \qquad n = 0, 1, \ldots$$
reduces this equation to a first-order nonhomogeneous linear difference equation of the form 9.1.1.3:
$$z_{n+1} + \frac{(y_n^* - a_n)^2}{a_n b_n + c_n} z_n + \frac{y_n^* - a_n}{a_n b_n + c_n} = 0.$$

$3°$. Let $y_n^{(1)}$ and $y_n^{(2)}$ be two particular solutions of the Riccati difference equation with $y_n^{(1)} \neq y_n^{(2)}$. Then the substitution
$$w_n = \frac{1}{y_n - y_n^{(1)}} + \frac{1}{y_n^{(1)} - y_n^{(2)}}, \qquad n = 0, 1, \ldots,$$
reduces this equation to a first-order homogeneous linear difference equation of the form 9.1.1.1:
$$w_{n+1} + \frac{(y_n^{(1)} - a_n)^2}{a_n b_n + c_n} w_n = 0, \qquad n = 0, 1, \ldots$$

6. $y_{n+1} = a y_n^\beta.$

Solution:
$$y_n = a^{\frac{\beta^n - 1}{\beta - 1}} C^{\beta^n},$$
where $C = y_0$ is an arbitrary constant.

7. $y_{n+1} = a_n y_n^{\beta_n}, \quad a_n > 0.$

We take the logarithm of both sides of the equation, and then we make the substitution $u_n = \ln y_n$. As a result, we arrive at a first-order linear difference equation of the form 9.1.1.3:
$$u_{n+1} - \beta_n u_n = \ln a_n.$$

8. $y_{n+1} = (a_n y_n^\beta + b_n)^{1/\beta}.$

The substitution $u_n = y_n^\beta$ leads to a first-order linear difference equation of the form 9.1.1.3:
$$u_{n+1} - a_n u_n = b_n.$$

9. $y_{n+1} = f(y_n).$

Solution:
$$y_n = f^{[n]}(y_0).$$
Notation used: $f^{[1]}(y) = f(y)$, $f^{[2]}(y) = f(f(y))$, $\ldots$, $f^{[n]}(y) = f(f^{[n-1]}(y))$.

Any constant solution y_* of the original equation such that y_* is independent of n is called an *equilibrium solution*. Equilibrium solutions are determined from the algebraic (transcendental) equation: $y_* = f(y_*)$.

10. $y_{n+1} = F(n, y_n)$.

The general first-order difference equation resolved with respect to the leading term. Here, $n = 0, 1, 2, \ldots$

The *Cauchy problem* suggests finding a solution of this equation with a given initial value of y_0.

The next value y_1 is calculated by substituting the initial value into the right-hand side of the original equation for $n = 0$:

$$y_1 = F(0, y_0). \tag{1}$$

Then, taking $n = 1$ in the original equation, we get

$$y_2 = F(1, y_1). \tag{2}$$

Substituting the previous value (1) into relation (2), we find $y_2 = F\big(1, F(0, y_0)\big)$. Taking $n = 2$ in the original equation and using the calculated value y_2, we find y_3, etc. In a similar way, one finds subsequent values $y_4, y_5, \ldots$

Remark 9.1. *As a rule, solutions of nonlinear difference equations cannot be found in closed form (i.e., in terms of a single, non-recursive formula).*

▶ **Second-order linear difference equations with discrete argument.**

11. $y_{n+2} + a y_{n+1} + b y_n = 0$.

Second-order homogeneous linear difference equation with constant coefficients. Here, $n = 0, 1, 2, \ldots$

The general solution of this equation is determined by the roots of the quadratic equation

$$\lambda^2 + a\lambda + b = 0. \tag{1}$$

1°. Let $a^2 - 4b > 0$. Then equation (1) has two different real roots

$$\lambda_1 = \frac{-a + \sqrt{a^2 - 4b}}{2}, \quad \lambda_2 = \frac{-a - \sqrt{a^2 - 4b}}{2},$$

and the general solution of the original difference equation is given by the formula (Doetsch, 1974):

$$y_n = -C_1 \lambda_1 \lambda_2 \frac{\lambda_1^{n-1} - \lambda_2^{n-1}}{\lambda_1 - \lambda_2} + C_2 \frac{\lambda_1^n - \lambda_2^n}{\lambda_1 - \lambda_2}, \tag{2}$$

where C_1 and C_2 are arbitrary constants. Solution (2) satisfies the initial conditions $y_0 = C_1$ and $y_1 = C_2$.

2°. Let $a^2 - 4b = 0$. Then the quadratic equation (1) has one double real root

$$\lambda_1 = \lambda_2 = \lambda = -\frac{a}{2},$$

and the general solution of the original difference equation has the form

$$y_n = -C_1(n-1)\lambda^n + C_2 n \lambda^{n-1}.$$

This formula can be obtained from (2) by taking $\lambda_1 = \lambda$ and $\lambda_2 = \lambda(1-\varepsilon)$ and passing to the limit as $\varepsilon \to 0$.

3°. Let $a^2 - 4b < 0$. Then the quadratic equation (2) has two complex conjugate roots

$$\lambda_1 = \rho(\cos\varphi + i\sin\varphi), \quad \lambda_2 = \rho(\cos\varphi - i\sin\varphi),$$
$$\rho = \sqrt{b}, \quad \tan\varphi = -\frac{1}{a}\sqrt{4b - a^2},$$

and the general solution of the original difference equation is

$$y_n = C_1 \rho^n \frac{\sin[(n-1)\varphi]}{\sin\varphi} + C_2 \rho^{n-1} \frac{\sin(n\varphi)}{\sin\varphi}.$$

This formula can also be obtained from (2) by expressing λ_1 and λ_2 in terms of ρ and φ.

4°. The substitution $u_n = y_{n+1}/y_n$ leads the original equation to a first-order difference equation of the form 9.1.1.9:

$$u_{n+1} = -a - \frac{b}{u_n}.$$

12. $y_{n+2} + a y_{n+1} + b y_n = f_n.$

Second-order nonhomogeneous linear difference equation with constant coefficients.

Let λ_1 and λ_2 be roots of the characteristic equation

$$\lambda^2 + a\lambda + b = 0.$$

1°. If $\lambda_1 \neq \lambda_2$, the general solution of the difference equation has the form (Doetsch, 1974):

$$y_n = y_1 \frac{\lambda_1^n - \lambda_2^n}{\lambda_1 - \lambda_2} - y_0 b \frac{\lambda_1^{n-1} - \lambda_2^{n-1}}{\lambda_1 - \lambda_2} + \sum_{k=2}^{n} f_{n-k} \frac{\lambda_1^{k-1} - \lambda_2^{k-1}}{\lambda_1 - \lambda_2},$$

where y_1 and y_0 are arbitrary constants, the values of y at the first two points.

In the case of complex conjugate roots, one should separate the real and imaginary parts in the above solution.

2°. If $\lambda_1 = \lambda_2$, the general solution of the difference equation is given by

$$y_n = y_1 n \lambda_1^{n-1} - y_0 b(n-1) \lambda_1^{n-2} + \sum_{k=2}^{n} f_{n-k}(k-1) \lambda_1^{k-2}.$$

3°. In boundary value problems, the initial and final values of the unknown function, y_0 and y_N, are prescribed. It is required to find $y_1, \ldots, y_{N-1}$.

If $\lambda_1 \neq \lambda_2$, the solution is given by (Doetsch, 1974):
$$y_n = y_0 \frac{\lambda_1^N \lambda_2^n - \lambda_1^n \lambda_2^N}{\lambda_1^N - \lambda_2^N} + y_N \frac{\lambda_1^n - \lambda_2^n}{\lambda_1^N - \lambda_2^N}$$
$$+ \sum_{k=2}^{n} f_{n-k} \frac{\lambda_1^{k-1} - \lambda_2^{k-1}}{\lambda_1 - \lambda_2} - \frac{\lambda_1^n - \lambda_2^n}{\lambda_1^N - \lambda_2^N} \sum_{k=2}^{N} f_{N-k} \frac{\lambda_1^{k-1} - \lambda_2^{k-1}}{\lambda_1 - \lambda_2}.$$
For $n = 1$, the first sum is zero.

13. $a_n y_{n+2} + b_n y_{n+1} + c_n y_n = 0.$

Second-order homogeneous linear difference equation with variable coefficients.

$1°$. Let $y_n^{(1)}$ and $y_n^{(2)}$ be particular solutions of the original equation satisfying the condition
$$y_0^{(1)} y_1^{(2)} - y_0^{(2)} y_1^{(1)} \neq 0.$$
Then the general solution of this equation is given by
$$y_n = C_1 y_n^{(1)} + C_2 y_n^{(2)},$$
where C_1 and C_2 are arbitrary constants.

$2°$. Let y_n^* be a nontrivial particular solution of the original equation. Then the substitution
$$y_n = y_n^* u_n \tag{1}$$
gives the equation
$$a_n y_{n+2}^* u_{n+2} + b_n y_{n+1}^* u_{n+1} + c_n y_n^* u_n = 0. \tag{2}$$
Taking into account that the original equation holds for y_n^*, substituting
$$b_n y_{n+1}^* = -a_n y_{n+2}^* - c_n y_n^*$$
into (2), and rearranging, we obtain
$$a_n y_{n+2}^* (u_{n+2} - u_{n+1}) - c_n y_n^* (u_{n+1} - u_n) = 0.$$
Introducing the new variable
$$w_n = u_{n+1} - u_n, \tag{3}$$
we arrive at a first-order homogeneous linear difference equation of the form 9.1.1.1:
$$a_n y_{n+2}^* w_{n+1} - c_n y_n^* w_n = 0.$$
Solving this equation yields a solution of the first-order nonhomogeneous linear equation with constant coefficients (3) (see equation 9.1.1.2). Then, using (1), one finds a solution of the original equation.

$3°$. The substitution $u_n = y_{n+1}/y_n$ leads the original equation to a first-order difference equation:
$$u_{n+1} = -a_n - \frac{b_n}{u_n}.$$

Remark 9.2. *The trivial solution $y_n = 0$ is a particular solution of the original homogeneous difference equation.*

14. $a_n y_{n+2} + b_n y_{n+1} + c_n y_n = f_n.$

Second-order nonhomogeneous linear difference equation with variable coefficients.

1°. The general solution of the nonhomogeneous linear equation can be represented as the sum of the general solution of the homogeneous equation 9.1.1.13 and a particular solution $\widetilde{y}_n$ of the original nonhomogeneous equation:

$$y_n = C_1 y_n^{(1)} + C_2 y_n^{(2)} + \widetilde{y}_n,$$

where

$$\widetilde{y}_0 = \widetilde{y}_1 = 0, \quad \widetilde{y}_n = \sum_{j=0}^{n-2} \frac{y_{j+1}^{(1)} y_n^{(2)} - y_n^{(1)} y_{j+1}^{(2)}}{y_{j+1}^{(1)} y_{j+2}^{(2)} - y_{j+2}^{(1)} y_{j+1}^{(2)}} \frac{f_j}{a_j}, \quad n = 2, 3, \ldots$$

2°. Let y_n^* be a nontrivial particular solution of the homogeneous linear difference equation 9.1.1.13. Then the substitutions

$$y_n = y_n^* u_n, \quad w_n = u_{n+1} - u_n$$

yield a first-order nonhomogeneous linear difference equation of the form 9.1.1.3:

$$a_n y_{n+2}^* w_{n+1} - c_n y_n^* w_n = f_n.$$

▶ **Second-order nonlinear difference equations with discrete argument.**

15. $y_n y_{n+2} = a_n y_{n+1}^2 + b_n y_n y_{n+1} + c_n y_n^2.$

A special case of equation 9.1.1.18. The substitution $u_n = y_{n+1}/y_n$ leads to a first-order difference equation:

$$u_n u_{n+1} = a_n u_n^2 + b_n u_n + c_n.$$

16. $y_{n+1} y_{n+2} = a_n y_n y_{n+2} + b_n y_{n+1}^2 + c_n y_n y_{n+1}.$

A special case of equation 9.1.1.18. The substitution $u_n = y_{n+1}/y_n$ leads to the Riccati difference equation 9.1.1.5:

$$u_n u_{n+1} = a_n u_{n+1} + b_n u_n + c_n.$$

17. $y_{n+2} = a^2 y_n + F(n, y_{n+1} + a y_n).$

The substitution $u_n = y_{n+1} + a y_n$ leads to a first-order difference equation:

$$u_{n+1} = a u_n + F(n, u_n).$$

For $F(n, u_n) = f(u_n)$, it is a difference equation of the form 9.1.1.9.

18. $y_{n+2} = y_n F(n, y_{n+1}/y_n).$

Homogeneous second-order difference equation.

The substitution $u_n = y_{n+1}/y_n$ leads to a first-order difference equation:

$$u_{n+1} = F(n, u_n)/u_n.$$

For $F(n, u_n) = f(u_n)$, it is a difference equation of the form 9.1.1.9.

19. $y_{n+2} = y_n^{k^2} F(n, y_n^k y_{n+1}).$

This equation is a generalization of the previous difference equation. The substitution $u_n = y_n^k y_{n+1}$ leads to a first-order difference equation:

$$u_{n+1} = u_n^k F(n, u_n).$$

For $F(n, u_n) = f(u_n)$, it is a difference equation of the form 9.1.1.9.

▶ **Higher-order linear difference equations with discrete argument.**

20. $y_{n+m} + a_{m-1}y_{n+m-1} + \cdots + a_1 y_{n+1} + a_0 y_n = 0.$

Homogeneous mth-order linear difference equation.

The general solution of this equation is determined by the roots of the characteristic equation

$$P(\lambda) \equiv \lambda^m + a_{m-1}\lambda^{m-1} + \cdots + a_1\lambda + a_0 = 0. \qquad (1)$$

The following cases are possible:

$1°$. All roots $\lambda_1, \lambda_2, \ldots, \lambda_m$ of the characteristic equation (1) are real and distinct. Then the general solution of the original homogeneous linear differential equation has the form

$$y_n = C_1 \lambda_1^n + C_2 \lambda_2^n + \cdots + C_m \lambda_m^n, \qquad (2)$$

where $C_1, C_2, \ldots, C_m$ are arbitrary constants.

If all the roots of the characteristic equation (1) are real and distinct, then the solution of the Cauchy problem for the original equation with initial data $y_0, y_1, \ldots, y_m$ has the form (Doetsch, 1971):

$$y_n = \sum_{i=0}^{m-1} y_i \sum_{j=0}^{m-i-1} a_{i+j+1} \sum_{k=1}^{m} \frac{\lambda_k^{n+1}}{P'(\lambda_k)},$$

where the prime denotes a derivative.

$2°$. There are k equal real roots $\lambda_1 = \lambda_2 = \cdots = \lambda_k$ $(k \leq m)$, and the other roots are real and distinct. Then the general solution is given by

$$y_n = (C_1 + C_2 n + \cdots + C_k n^{k-1})\lambda_1^n + C_{k+1}\lambda_{k+1}^n + \cdots + C_m \lambda_m^n.$$

$3°$. There are k pairs of distinct complex conjugate roots $\lambda_j = \rho_j(\cos\varphi_j \pm i\sin\varphi_j)$ $(j = 1, \ldots, k;\ 2k \leq m)$, and the other roots are real and distinct. In this case, the general solution is

$$y_n = \rho_1^n[A_1\cos(n\varphi_1) + B_1\sin(n\varphi_1)] + \cdots + \rho_k^n[A_k\cos(n\varphi_k) + B_k\sin(n\varphi_k)] + C_{2k+1}\lambda_{2k+1}^n + \cdots + C_m \lambda_m^n,$$

where $A_1, \ldots, A_k, B_1, \ldots, B_k, C_{2k+1}, \ldots, C_m$ are arbitrary constants.

$4°$. In the general case, if there are r different roots $\lambda_1, \lambda_2, \ldots, \lambda_r$ of multiplicities $k_1, k_2, \ldots, k_r$, respectively, the left-hand side of the characteristic equation (1) can be represented as the product

$$P(\lambda) = (\lambda - \lambda_1)^{k_1}(\lambda - \lambda_2)^{k_2} \ldots (\lambda - \lambda_r)^{k_r},$$

where $k_1 + k_2 + \cdots + k_r = m$. The general solution of the original equation is given by the formula

$$y_n = \sum_{s=1}^{r}(C_{s,1} + C_{s,2}n + \cdots + C_{s,k_s}n^{k_s-1})\lambda_s^n,$$

where $C_{s,k}$ are arbitrary constants.

If the characteristic equation (1) has complex conjugate roots of the form $\lambda = \rho e^{\pm i\varphi}$, then in the above solution, one should extract the real part using the formula $\lambda^n = \rho^n e^{\pm in\varphi} = \rho^n[\cos(n\varphi) \pm i\sin(n\varphi)]$.

5°. The substitution $u_n = y_{n+1}/y_n$ takes the original equation to an $(m-1)$st-order difference equation.

Remark 9.3. *The trivial solution* $y_n = 0$ *is a particular solution of the original homogeneous difference equation.*

21. $y_{n+m} + a_{m-1} y_{n+m-1} + \cdots + a_1 y_{n+1} + a_0 y_n = f_n.$

Nonhomogeneous mth-order linear difference equation.

The general solution of the difference equation has the form $y(x) = Y(x) + \bar{y}(x)$, where $Y(x)$ is the general solution of the homogeneous equation (with $f_n \equiv 0$) and $\bar{y}(x)$ is any particular solution of the nonhomogeneous equation.

Let $\lambda_1, \lambda_2, \ldots, \lambda_m$ be roots of the characteristic equation

$$P(\lambda) \equiv \lambda^m + a_{m-1} \lambda^{m-1} + \cdots + a_1 \lambda + a_0 = 0. \tag{1}$$

If all the roots of equation (1) are distinct, then the general solution of the original difference equation has the form (Doetsch, 1971):

$$y_n = \sum_{i=0}^{m-1} y_i \sum_{j=0}^{m-i-1} a_{i+j+1} \sum_{k=1}^{m} \frac{\lambda_k^{n+1}}{P'(\lambda_k)} + \sum_{\nu=m}^{n} f_{n-\nu} \sum_{k=1}^{m} \frac{\lambda_k^{\nu-1}}{P'(\lambda_k)}, \tag{2}$$

where the prime denotes a derivative.

Formula (2) involves the initial values $y_0, y_1, \ldots, y_m$. They can be set arbitrarily.

In the case of complex conjugate roots, one should separate the real and imaginary parts in solution (2).

9.1.2. Difference Equations with Continuous Argument

▶ **First-order linear difference equations.**

1. $y(x+1) - y(x) = 0.$

This functional equation may be treated as a definition of periodic functions with unit period.

1°. Solution:
$$y(x) = \Theta(x),$$
where $\Theta(x) = \Theta(x+1)$ is an arbitrary periodic function with unit period.

2°. A periodic function $\Theta(x)$ with period 1 that satisfies the Dirichlet conditions can be expanded into a Fourier series:

$$\Theta(x) = \frac{a_0}{2} + \sum_{n=1}^{\infty} \left[a_n \cos(2\pi n x) + b_n \sin(2\pi n x) \right],$$

where

$$a_n = 2 \int_0^1 \Theta(x) \cos(2\pi n x) \, dx, \qquad b_n = 2 \int_0^1 \Theta(x) \sin(2\pi n x) \, dx.$$

2. $y(x+1) - y(x) = f(x)$.

$1°$. Solution:
$$y(x) = \Theta(x) + \bar{y}(x),$$
where $\Theta(x) = \Theta(x+1)$ is an arbitrary periodic function with period 1, and $\bar{y}(x)$ is any particular solution of the nonhomogeneous equation. Table 9.1 presents particular solutions of the linear nonhomogeneous difference equation for some specific $f(x)$.

$2°$. If $g(x) = g(x+1)$ is a given 1-periodic function, then the solution is
$$y(x) = \Theta(x) + xg(x).$$

$3°$. Let $x \in (a, \infty)$ with an arbitrary a. Suppose that the function $f(x)$ is monotone, strictly convex (or strictly concave), and satisfies the condition
$$\lim_{x \to \infty}[f(x+1) - f(x)] = 0,$$
and let $x_0 \in (a, \infty)$ be an arbitrary fixed point. Then for every y_0, there exists exactly one function $y(x)$ (monotone and strictly convex/concave) satisfying the original equation, together with the initial condition
$$y(x_0) = y_0.$$
This solution is given by the formula
$$y(x) = y_0 + (x-x_0)f(x_0) - \sum_{n=0}^{\infty}\left\{f(x+n) - f(x_0+n) - (x-x_0)[f(x_0+n+1) - f(x_0+n)]\right\}.$$

3. $y(x+1) + y(x) = f(x)$.

The transformation
$$y(x) = u\left(\frac{x+1}{2}\right) - u\left(\frac{x}{2}\right), \quad \xi = \frac{x}{2},$$
leads to an equation of the form 9.1.2.2:
$$u(\xi+1) - u(\xi) = f(2\xi).$$

4. $y(x+1) - ay(x) = 0$.

Homogeneous first-order constant-coefficient linear difference equation.

$1°$. Solution for $a > 0$:
$$y(x) = \Theta(x)a^x,$$
where $\Theta(x) = \Theta(x+1)$ is an arbitrary periodic function with period 1.
For $\Theta(x) \equiv \text{const}$, we get a particular solution $y(x) = Ca^x$, where C is an arbitrary constant.

$2°$. Solution for $a < 0$:
$$y(x) = \Theta_1(x)|a|^x \sin(\pi x) + \Theta_2(x)|a|^x \cos(\pi x),$$
where $\Theta_k(x) = \Theta_k(x+1)$, $k = 1, 2$, are arbitrary periodic functions with period 1.

5. $y(x+1) - ay(x) = f(x)$.

Nonhomogeneous first-order constant-coefficient linear difference equation.

TABLE 9.1. Particular solutions of the nonhomogeneous difference equation $y(x+1)-y(x)=f(x)$.

No.	Right-hand side of equation, $f(x)$	Particular solution, $\bar{y}(x)$
1	1	x
2	x	$\frac{1}{2}x(x-1)$
3	x^2	$\frac{1}{6}x(x-1)(2x-1)$
4	x^n, $n=0,1,2,\ldots$	$\frac{1}{n+1}B_n(x)$, where the $B_n(x)$ are Bernoulli polynomials. The generating function: $\dfrac{te^{xt}}{e^t-1} = \sum_{n=0}^{\infty} B_n(x)\dfrac{t^n}{n!}$
5	$\dfrac{1}{x}$	$\psi(x) = -\mathcal{C} + \displaystyle\int_0^1 \dfrac{1-t^{x-1}}{1-t}\,dt$ is the logarithmic derivative of the gamma function, $\mathcal{C} = 0.5772\ldots$ is the Euler constant
6	$\dfrac{1}{x(x+1)}$	$-\dfrac{1}{x}$
7	$a^{\lambda x}$, $a\neq 1$, $\lambda\neq 0$	$\dfrac{1}{a^\lambda-1}a^{\lambda x}$
8	$\sinh(a+2bx)$, $b>0$	$\dfrac{\cosh(a-b+2bx)}{2\sinh b}$
9	$\cosh(a+2bx)$, $b>0$	$\dfrac{\sinh(a-b+2bx)}{2\sinh b}$
10	xa^x, $a\neq 1$	$\dfrac{1}{a-1}a^x\left(x-\dfrac{a}{a-1}\right)$
11	$\ln x$, $x>0$	$\ln\Gamma(x)$, where $\Gamma(x) = \displaystyle\int_0^\infty t^{x-1}e^{-t}\,dt$ is the gamma function
12	$\sin(2ax)$, $a\neq \pi n$	$-\dfrac{\cos[a(2x-1)]}{2\sin a}$
13	$\sin(2\pi nx)$	$x\sin(2\pi nx)$
14	$\cos(2ax)$, $a\neq \pi n$	$\dfrac{\sin[a(2x-1)]}{2\sin a}$
15	$\cos(2\pi nx)$	$x\cos(2\pi nx)$
16	$\sin^2(ax)$, $a\neq \pi n$	$\dfrac{x}{2} - \dfrac{\sin[a(2x-1)]}{4\sin a}$
17	$\sin^2(\pi nx)$	$x\sin^2(\pi nx)$
18	$\cos^2(ax)$, $a\neq \pi n$	$\dfrac{x}{2} + \dfrac{\sin[a(2x-1)]}{4\sin a}$
19	$\cos^2(\pi nx)$	$x\cos^2(\pi nx)$
20	$x\sin(2ax)$, $a\neq \pi n$	$\dfrac{\sin(2ax)}{4\sin^2 a} - x\dfrac{\cos[a(2x-1)]}{2\sin a}$
21	$x\sin(2\pi nx)$	$\frac{1}{2}x(x-1)\sin(2\pi nx)$
22	$x\cos(2ax)$, $a\neq \pi n$	$\dfrac{\cos(2ax)}{4\sin^2 a} + x\dfrac{\sin[a(2x-1)]}{2\sin a}$
23	$x\cos(2\pi nx)$	$\frac{1}{2}x(x-1)\cos(2\pi nx)$
24	$a^x\sin(bx)$, $a>0$, $a\neq 1$	$a^x\dfrac{a\sin[b(x-1)] - \sin(bx)}{a^2 - 2a\cos b + 1}$
25	$a^x\cos(bx)$, $a>0$, $a\neq 1$	$a^x\dfrac{a\cos[b(x-1)] - \cos(bx)}{a^2 - 2a\cos b + 1}$

1°. Solution:
$$y(x) = \begin{cases} \Theta(x)a^x + \bar{y}(x) & \text{if } a > 0, \\ \Theta_1(x)|a|^x \sin(\pi x) + \Theta_2(x)|a|^x \cos(\pi x) + \bar{y}(x) & \text{if } a < 0, \end{cases}$$
where $\Theta(x)$, $\Theta_1(x)$, and $\Theta_2(x)$ are arbitrary periodic functions with period 1, and $\bar{y}(x)$ is any particular solution of the nonhomogeneous equation.

2°. For $f(x) = \sum_{k=0}^{n} b_k x^n$ and $a \neq 1$, the nonhomogeneous equation has a particular solution of the form $\bar{y}(x) = \sum_{k=0}^{n} A_k x^n$; the constants A_k are found by the method of undetermined coefficients.

3°. For $f(x) = \sum_{k=1}^{n} b_k e^{\lambda_k x}$, the nonhomogeneous equation has a particular solution of the form
$$\bar{y}(x) = \begin{cases} \displaystyle\sum_{k=1}^{n} \frac{b_k}{e^{\lambda_k} - a} e^{\lambda_k x} & \text{if } a \neq e^{\lambda_m}, \\ b_m x e^{\lambda_m(x-1)} + \displaystyle\sum_{k=1, k\neq m}^{n} \frac{b_k}{e^{\lambda_k} - a} e^{\lambda_k x} & \text{if } a = e^{\lambda_m}, \end{cases}$$
where $m = 1, \ldots, n$.

4°. For $f(x) = \sum_{k=1}^{n} b_k \cos(\beta_k x)$, the nonhomogeneous equation has a particular solution of the form
$$\bar{y}(x) = \sum_{k=1}^{n} \frac{b_k}{a^2 + 1 - 2a\cos\beta_k} \left[(\cos\beta_k - a)\cos(\beta_k x) + \sin\beta_k \sin(\beta_k x)\right].$$

5°. For $f(x) = \sum_{k=1}^{n} b_k \sin(\beta_k x)$, the nonhomogeneous equation has a particular solution of the form
$$\bar{y}(x) = \sum_{k=1}^{n} \frac{b_k}{a^2 + 1 - 2a\cos\beta_k} \left[(\cos\beta_k - a)\sin(\beta_k x) - \sin\beta_k \cos(\beta_k x)\right].$$

6. $y(x+1) - xy(x) = 0$.

Solution (Mirolyubov & Soldatov, 1981):
$$y(x) = \Theta(x)\Gamma(x), \qquad \Gamma(x) = \int_0^\infty t^{x-1} e^{-t}\, dt,$$
where $\Gamma(x)$ is the gamma function and $\Theta(x) = \Theta(x+1)$ is an arbitrary periodic function with period 1.

The simplest particular solution corresponds to $\Theta(x) \equiv 1$.

7. $y(x+1) - a(x-b)(x-c)y(x) = 0$.

Solution (Mirolyubov & Soldatov, 1981):
$$y(x) = \Theta(x)a^x \Gamma(x-b)\Gamma(x-c),$$
where $\Gamma(x)$ is the gamma function and $\Theta(x)$ is an arbitrary periodic function with period 1.

8. $y(x+1) - R(x)y(x) = 0$, $\quad R(x) = a\dfrac{(x-\lambda_1)(x-\lambda_2)\ldots(x-\lambda_n)}{(x-\mu_1)(x-\mu_2)\ldots(x-\mu_m)}$.

Solution (Mirolyubov & Soldatov, 1981):

$$y(x) = \Theta(x)a^x \frac{\Gamma(x-\lambda_1)\Gamma(x-\lambda_2)\ldots\Gamma(x-\lambda_n)}{\Gamma(x-\mu_1)\Gamma(x-\mu_2)\ldots\Gamma(x-\mu_m)},$$

where $\Gamma(x)$ is the gamma function, $\Theta(x)$ is an arbitrary periodic function with period 1.

9. $y(x+1) - ae^{\lambda x}y(x) = 0$.

Solution:

$$y(x) = \Theta(x)a^x \exp\left(\tfrac{1}{2}\lambda x^2 - \tfrac{1}{2}\lambda x\right),$$

where $\Theta(x)$ is an arbitrary periodic function with period 1.

10. $y(x+1) - ae^{\mu x^2 + \lambda x}y(x) = 0$.

Solution:

$$y(x) = \Theta(x)a^x \exp\left[\tfrac{1}{3}\mu x^3 + \tfrac{1}{2}(\lambda-\mu)x^2 + \tfrac{1}{6}(\mu - 3\lambda)x\right],$$

where $\Theta(x)$ is an arbitrary periodic function with period 1.

11. $y(x+1) - f(x)y(x) = 0$.

First-order homogeneous linear difference equation.

1°. Let $f(x)$ is a given continuous function defined on $[0, \infty)$ and let $y_1 = y_1(x)$ be a nontrivial particular solution of the original equation. Then the general solution of the equation is given by

$$y(x) = \Theta(x)y_1(x),$$

where $\Theta(x) = \Theta(x+1)$ is an arbitrary 1-periodic function.

2°. Let $f(x) = f(x+1)$ is a given continuous positive periodic function with period 1. Then the solution is

$$y(x) = \Theta(x)[f(x)]^x,$$

where $\Theta(x) = \Theta(x+1)$ is an arbitrary periodic function with period 1.

For $\Theta(x) \equiv \text{const}$, we have a particular solution $y(x) = C[f(x)]^x$, where C is an arbitrary constant.

3°. *Cauchy's problem*: Find a solution of the original equation with the initial condition

$$y(x) = \varphi(x) \quad \text{for} \quad 0 \leq x < 1,$$

where $\varphi(x)$ is a given continuous function defined on the interval $0 \leq x \leq 1$.

The solution of the Cauchy problem can be written as

$$y(x) = \varphi(\{x\}) \prod_{k=1}^{[x]} f(x - k),$$

where $[x]$ and $\{x\}$ denote, respectively, the integer and fractional parts of the argument x ($x = [x] + \{x\}$), and the product for $[x] = 0$ is assumed equal to unity. This solution is continuous if it is continuous at the integer points $x = 1, 2, \ldots$, and this brings us to the condition

$$\varphi(1) = f(0)\varphi(0).$$

12. $y(x+1) - f(x)y(x) = g(x)$.

First-order nonhomogeneous linear difference equation. Here $f(x)$ and $g(x)$ are given continuous functions defined on the interval $0 \leq x < \infty$.

1°. The general solution of the equation can be represented as the sum
$$y(x) = u(x) + \widetilde{y}(x),$$
where the first term $u(x)$ is the general solution of the homogeneous equation (with $g \equiv 0$), and the second term $\widetilde{y}(x)$ is a particular solution of the original equation.

2°. Let $g(x) = g(x+1)$ be a 1-periodic function and let $y_1(x)$ be a solution of the original equation in the special case of $g(x) \equiv 1$. Then the function
$$y(x) = g(x)y_1(x)$$
is a solution of the original equation.

3°. *Cauchy's problem*: Find a solution of the original equation subject to the initial condition
$$y(x) = \varphi(x) \quad \text{for} \quad 0 \leq x < 1,$$
where $\varphi(x)$ is a given continuous function defined on the interval $0 \leq x \leq 1$.

The solution of the Cauchy problem can be written as
$$y(x) = \varphi(\{x\}) \prod_{j=1}^{[x]} f(x-j) + \sum_{i=1}^{[x]} g(x-i) \prod_{j=1}^{i-1} f(x-j),$$
where $[x]$ and $\{x\}$ denote, respectively, the integer and the fractional parts of the argument x ($x = [x] + \{x\}$); the product and the sum over the empty index set (for $[x] = 0$) are assumed equal to 1 and 0, respectively.

If a nontrivial particular solution $y_1(x)$ of the homogeneous equation with $g(x) \equiv 0$ is known, then the solution of the Cauchy problem for the original nonhomogeneous equation is given by
$$y(x) = y_1(x) \left[\frac{\varphi(\{x\})}{y_1(0)} + \sum_{i=1}^{[x]} \frac{g(x-i)}{y_1(x-i+1)} \right].$$

13. $y(x+a) - by(x) = 0$.

1°. Solution for $b > 0$:
$$y(x) = \Theta(x) b^{x/a},$$
where $\Theta(x) = \Theta(x+a)$ is an arbitrary periodic function with period a.

For $\Theta(x) \equiv \text{const}$, we have a particular solution $y(x) = C b^{x/a}$, where C is an arbitrary constant.

2°. Solution for $b < 0$:
$$y(x) = \Theta_1(x) |b|^{x/a} \sin\left(\frac{\pi x}{a}\right) + \Theta_2(x) |b|^{x/a} \cos\left(\frac{\pi x}{a}\right),$$
where $\Theta_k(x) = \Theta_k(x+a)$ ($k = 1, 2$) are arbitrary periodic functions with period a.

14. $y(x+a) - by(x) = f(x)$.

1°. General solution:

$$y(x) = \begin{cases} \Theta(x)b^{x/a} + \bar{y}(x) & \text{if } b > 0, \\ \Theta_1(x)|b|^{x/a}\sin\left(\dfrac{\pi x}{a}\right) + \Theta_2(x)|b|^{x/a}\cos\left(\dfrac{\pi x}{a}\right) + \bar{y}(x) & \text{if } b < 0, \end{cases}$$

where $\Theta(x) = \Theta(x+a)$ and $\Theta_k(x) = \Theta_k(x+a)$ ($k=1, 2$) are arbitrary periodic functions with period a, and $\bar{y}(x)$ is any particular solution of the nonhomogeneous equation.

2°. For $f(x) = \sum_{k=0}^{n} A_k x^n$ and $b \neq 1$, the nonhomogeneous equation has a particular solution of the form $\bar{y}(x) = \sum_{k=0}^{n} B_k x^n$; the constants B_k are found by the method of undetermined coefficients.

3°. For $f(x) = \sum_{k=1}^{n} A_k \exp(\lambda_k x)$, the nonhomogeneous equation has a particular solution of the form $\bar{y}(x) = \sum_{k=1}^{n} B_k \exp(\lambda_k x)$; the constants B_k are found by the method of undetermined coefficients.

4°. For $f(x) = \sum_{k=1}^{n} A_k \cos(\lambda_k x)$, the nonhomogeneous equation has a particular solution of the form $\bar{y}(x) = \sum_{k=1}^{n} B_k \cos(\lambda_k x) + \sum_{k=1}^{n} D_k \sin(\lambda_k x)$; the constants B_k and D_k are found by the method of undetermined coefficients.

5°. For $f(x) = \sum_{k=1}^{n} A_k \sin(\lambda_k x)$, the nonhomogeneous equation has a particular solution of the form $\bar{y}(x) = \sum_{k=1}^{n} B_k \cos(\lambda_k x) + \sum_{k=1}^{n} D_k \sin(\lambda_k x)$; the constants B_k and D_k are found by the method of undetermined coefficients.

15. $y(x+a) - bxy(x) = 0$, $\quad a, b > 0$.

Solution:
$$y(x) = \Theta(x) \int_0^\infty t^{(x/a)-1} e^{-t/(ab)}\, dt,$$

where $\Theta(x) = \Theta(x+a)$ is an arbitrary periodic function with period a.

16. $y(x+a) - f(x)y(x) = 0$.

Here, $f(x) = f(x+a)$ is a given positive periodic function with period a.

Solution:
$$y(x) = \Theta(x)[f(x)]^{x/a},$$

where $\Theta(x) = \Theta(x+a)$ is an arbitrary periodic function with period a.

For $\Theta(x) \equiv \text{const}$, we have a particular solution $y(x) = C[f(x)]^{x/a}$, where C is an arbitrary constant.

▶ **First-order nonlinear difference equations.**

17. $y(x+a) = 2y(x) + by^2(x)$.

Solution (Pelyukh & Sharkovsky, 1974):

$$y(x) = \frac{1}{b}\{\exp[2^{\frac{x}{a}}\Theta(x)] - 1\},$$

where $\Theta(x) = \Theta(x+a)$ is an arbitrary periodic function with period a.

18. $y(x+1) - ay^2(x) = f(x)$.

A special case of equation 9.1.2.23.

19. $y(x)y(x+1) + a[y(x+1) - y(x)] = 0$.

Solution:

$$y(x) = \frac{a}{x + \Theta(x)},$$

where $\Theta(x) = \Theta(x+1)$ is an arbitrary periodic function with period 1.

20. $y(x)y(x+1) = f(x)y(x+1) + g(x)y(x) + h(x)$.

Riccati difference equation. Here, the functions $f(x)$, $g(x)$, $h(x)$ satisfy the condition $f(x)g(x) + h(x) \not\equiv 0$.

$1°$. The substitution

$$y(x) = \frac{u(x+1)}{u(x)} + f(x)$$

leads to a linear second-order difference equation:

$$u(x+2) + [a(x+1) - g(x)]u(x+1) - [f(x)g(x) + h(x)]u(x) = 0.$$

$2°$. Let $y_0(x)$ be a particular solution of Riccati difference equation. Then the substitution

$$z(x) = \frac{1}{y(x) - y_0(x)}$$

reduces this equation to the nonhomogeneous first-order linear difference equation

$$z(x+1) + \frac{[y_0(x) - f(x)]^2}{f(x)g(x) + h(x)}z(x) + \frac{y_0(x) - f(x)}{f(x)g(x) + h(x)} = 0.$$

21. $y(x+a) - by^\lambda(x) = f(x)$.

A special case of equation 9.1.2.23.

22. $y(x+1) = f(y(x))$.

A special case of equation 9.1.2.23, where $f(y)$ is a continuous function and $0 \leq x < \infty$.

Solution:

$$y(x) = f^{[n]}(\varphi(\{x\})), \quad n = [x],$$

where $[x]$ and $\{x\}$ are, respectively, the integer and the fractional parts of x, and $\varphi(x)$ is any continuous function on $[0, 1)$ such that $\varphi(0) = a$ and $\varphi(1-0) = f(a)$, and a is an arbitrary constant.

23. $F(x, y(x), y(x+a)) = 0.$

We assume that $a > 0$. Let us solve the equation for $y(x+a)$ to obtain

$$y(x+a) = f(x, y(x)). \tag{1}$$

1°. First, let us assume that the equation is defined on a discrete set of points $x = x_0 + ak$ with integer k. Given an initial value $y(x_0)$, one can make use of (1) to find sequentially $y(x_0 + a)$, $y(x_0 + 2a)$, etc.

Solving the original equation for $y(x)$ yields

$$y(x) = g(x, y(x+a)). \tag{2}$$

On setting $x = x_0 - a$ here, one can find $y(x_0 - a)$ and then likewise $y(x_0 - 2a)$ etc.

Thus, given initial data, one can use the equation to find $y(x)$ at all points $x_0 + ak$, where $k = 0, \pm 1, \pm 2, \ldots$

2°. Now we assume that x in the equation can vary continuously. Also assume that $y(x)$ is a continuous function defined arbitrarily on the semi-open interval $[0, a)$. Setting $x = 0$ in (1), one finds $y(a)$. Then, given $y(x)$ on $[0, a]$, one can use (1) to find $y(x)$ on $x \in [a, 2a]$, then on $x \in [2a, 3a]$, and so on.

Remark 9.4. *The case of $a < 0$ can be reduced, using the change of variable $z = x + a$, to an equation of the form $F(z+b, y(z+b), y(z)) = 0$ with $b = -a > 0$, which was already considered above.*

▶ **Second-order linear difference equations.**

24. $ay(x+2) + by(x+1) + cy(x) = 0, \quad ac \neq 0.$

Second-order homogeneous constant-coefficient linear difference equation.

This equation has the trivial solution $y(x) \equiv 0$.

The general solution of the original difference equation is determined by the roots of the characteristic equation (Math. Encyclopedia, 1979):

$$a\lambda^2 + b\lambda + c = 0. \tag{1}$$

1°. For $b^2 - 4ac > 0$, the quadratic equation (1) has two real distinct roots:

$$\lambda_1 = \frac{-b + \sqrt{b^2 - 4ac}}{2a}, \quad \lambda_2 = \frac{-b - \sqrt{b^2 - 4ac}}{2a}.$$

The general solution of the original difference equation is given by

$$\begin{aligned} y(x) &= \Theta_1(x)\lambda_1^x + \Theta_2(x)\lambda_2^x & \text{if} \quad ab < 0, \; ac > 0; \\ y(x) &= \Theta_1(x)\lambda_1^x + \Theta_2(x)|\lambda_2|^x \cos(\pi x) & \text{if} \quad ac < 0; \\ y(x) &= \Theta_1(x)|\lambda_1|^x \cos(\pi x) + \Theta_2(x)|\lambda_2|^x \cos(\pi x) & \text{if} \quad ab > 0, \; ac > 0, \end{aligned} \tag{2}$$

where $\Theta_1(x)$ and $\Theta_2(x)$ are arbitrary 1-periodic functions, $\Theta_k(x) = \Theta_k(x+1)$, $k = 1, 2$.

2°. For $b^2 - 4ac = 0$, the quadratic equation (1) has one double real root

$$\lambda = -\frac{b}{2a},$$

and the general solution of the original difference equation is given by

$$\begin{aligned} y &= \big[\Theta_1(x) + x\Theta_2(x)\big]\lambda^x & \text{if} \quad ab < 0, \\ y &= \big[\Theta_1(x) + x\Theta_2(x)\big]|\lambda|^x \cos(\pi x) & \text{if} \quad ab > 0. \end{aligned} \quad (3)$$

3°. For $b^2 - 4ac < 0$, the quadratic equation (1) has two complex conjugate roots

$$\lambda_{1,2} = \rho(\cos\beta \pm i\sin\beta), \qquad \rho = \frac{c}{a}, \qquad \beta = \arccos\left(-\frac{b}{2\sqrt{ac}}\right), \quad (4)$$

and the general solution of the original difference equation has the form

$$y = \Theta_1(x)\rho^x \cos(\beta x) + \Theta_2(x)\rho^x \sin(\beta x),$$

where $\Theta_1(x)$ and $\Theta_2(x)$ are arbitrary 1-periodic functions.

25. $y(x+2) + ay(x+1) + by(x) = f(x).$

Nonhomogeneous second-order constant-coefficient linear difference equation.

1°. General solution:

$$y(x) = Y(x) + \bar{y}(x),$$

where $Y(x)$ is the general solution of the homogeneous difference equation $Y(x+2) + aY(x+1) + bY(x) = 0$ (see the previous equation), and $\bar{y}(x)$ is any particular solution of the nonhomogeneous equation.

2°. For $f(x) = \sum_{k=0}^{n} A_k x^k$ and $a + b + 1 \neq 1$, the nonhomogeneous equation has a particular solution of the form $\bar{y}(x) = \sum_{k=0}^{n} B_k x^k$; the constants B_k are found by the method of undetermined coefficients.

3°. For $f(x) = \sum_{k=1}^{n} A_k \exp(\lambda_k x)$, the nonhomogeneous equation has a particular solution of the form $\bar{y}(x) = \sum_{k=1}^{n} B_k \exp(\lambda_k x)$; the constants B_k are found by the method of undetermined coefficients.

4°. For $f(x) = \sum_{k=1}^{n} A_k \cos(\lambda_k x)$, the nonhomogeneous equation has a particular solution of the form $\bar{y}(x) = \sum_{k=1}^{n} B_k \cos(\lambda_k x) + \sum_{k=1}^{n} D_k \sin(\lambda_k x)$; the constants B_k and D_k are found by the method of undetermined coefficients.

5°. For $f(x) = \sum_{k=1}^{n} A_k \sin(\lambda_k x)$, the nonhomogeneous equation has a particular solution of the form $\bar{y}(x) = \sum_{k=1}^{n} B_k \cos(\lambda_k x) + \sum_{k=1}^{n} D_k \sin(\lambda_k x)$; the constants B_k and D_k are found by the method of undetermined coefficients.

26. $y(x+2) + a(x+1)y(x+1) + bx(x+1)y(x) = 0.$

This difference equation has particular solutions of the form

$$y(x; \lambda) = \int_0^\infty t^{x-1} e^{-t/\lambda} \, dt, \tag{1}$$

where λ is a root of the quadratic equation

$$\lambda^2 + a\lambda + b = 0. \tag{2}$$

For the integral on the right-hand side of (1) to converge, the roots of equation (2) that satisfy the condition $\operatorname{Re} \lambda > 0$ should be selected. If both roots, λ_1 and λ_2, meet this condition, the general solution of the original functional equation is expressed as

$$y(x) = \Theta_1(x) y(x, \lambda_1) + \Theta_2(x) y(x, \lambda_2),$$

where $\Theta_1(x)$ and $\Theta_2(x)$ are arbitrary periodic functions with period 1.

27. $y(x+2) = f(x)y(x+1) + a[f(x) + a]y(x) + g(x).$

The substitution

$$u(x) = y(x+1) + ay(x)$$

leads to a first-order difference equation

$$u(x+1) = [f(x) + a]u(x) + g(x).$$

28. $y(x+2) + f(x)y(x+1) + g(x)y(x) = 0, \qquad g(x) \not\equiv 0.$

General second-order homogeneous linear difference equation with variable coefficients. The trivial solution, $y(x) = 0$, is a particular solution of the equation.

1°. Let $y_1(x)$ and $y_2(x)$ be two particular solutions of this equation with the condition*

$$D(x) \equiv y_1(x) y_2(x+1) - y_2(x) y_1(x+1) \neq 0. \tag{1}$$

Then the general solution of the original equation is given by

$$y(x) = \Theta_1(x) y_1(x) + \Theta_2(x) y_2(x),$$

where $\Theta_1(x)$ and $\Theta_2(x)$ are arbitrary 1-periodic functions, $\Theta_{1,2}(x) = \Theta_{1,2}(x+1)$.

2°. Let $y_0(x)$ be a nontrivial particular solution of the original equation. Then the substitution

$$y(x) = y_0(x) u(x) \tag{2}$$

results in the equation

$$y_0(x+2) u(x+2) + f(x) y_0(x+1) u(x+1) + g(x) y_0(x) u(x) = 0. \tag{3}$$

Taking into account that $y_0(x)$ satisfies the original equation, let us substitute the expression

$$f(x) y_0(x+1) = -y_0(x+2) - g(x) y_0(x)$$

---
*Condition (1) may be violated at singular points of the original equation.

into (3). After rearranging, we obtain

$$y_0(x+2)[u(x+2) - u(x+1)] - g(x)y_0(x)[u(x+1) - u(x)] = 0.$$

Introducing the new variable

$$w(x) = u(x+1) - u(x) \tag{4}$$

we arrive at the first-order difference equation

$$y_0(x+2)w(x+1) - g(x)y_0(x)w(x) = 0.$$

After solving this equation, one solves the nonhomogeneous first-order equation with constant coefficients (4), and then, using (2), finds a solution of the original equation.

▶ **Second-order nonlinear difference equations.**

29. $y(x)y(x+2) = f(x)y^2(x+1) + g(x)y(x)y(x+1) + h(x)y^2(x).$
A special case of equation 9.1.2.37. The substitution $u(x) = y(x+1)/y(x)$ leads to a first-order difference equation:

$$u(x)u(x+1) = f(x)u^2(x) + g(x)u(x) + h(x).$$

30. $y(x+1)y(x+2) = f(x)y(x)y(x+2) + g(x)y^2(x+1) + h(x)y(x)y(x+1).$
A special case of equation 9.1.2.37. The substitution $u(x) = y(x+1)/y(x)$ leads to the Riccati difference equation 9.1.2.20:

$$u(x)u(x+1) = f(x)u(x+1) + g(x)u(x) + h(x).$$

31. $y^{k-1}(x)y(x+2) = f(x)y^k(x+1) + g(x)y^{k-n}(x)y^n(x+1) + h(x)y^k(x).$
A special case of equation 9.1.2.37. The substitution $u(x) = y(x+1)/y(x)$ leads to a first-order difference equation:

$$u(x)u(x+1) = f(x)u^k(x) + g(x)u^n(x) + h(x).$$

32. $y(x+2) = a^2 y(x) + F(x, y(x+1) - ay(x)).$
1°. The substitution $u(x) = y(x+1) - ay(x)$ leads to a first-order difference equation:

$$u(x+1) = -au(x) + F(x, u(x)).$$

2°. For $F(x, u) = f(u)$ and $a > 0$, the equation admits a particular solution of the form:

$$y = \Theta(x)a^x + k,$$

where $\Theta(x) = \Theta(x+1)$ is an arbitrary periodic function with period 1 and k is a root of the algebraic (transcendental) equation $(a^2 - 1)k + f((1-a)k) = 0$.

33. $y(x+2) = x(x+1)y(x) + F(x, y(x+1) + xy(x)).$
The substitution $u(x) = y(x+1) + xy(x)$ leads to a first-order difference equation:

$$u(x+1) = (x+1)u(x) + F(x, u(x)).$$

34. $y(x+2) = x(x+1)y(x) + F(x, y(x+1) - xy(x))$.

The substitution $u(x) = y(x+1) - xy(x)$ leads to a first-order difference equation:

$$u(x+1) = -(x+1)u(x) + F(x, u(x)).$$

35. $y(x+2) = g(x)g(x+1)y(x) + F(x, y(x+1) + g(x)y(x))$.

The substitution $u(x) = y(x+1) + g(x)y(x)$ leads to a first-order difference equation:

$$u(x+1) = g(x+1)u(x) + F(x, u(x)).$$

36. $y(x+2) = y(x)F(y(x+1)/y(x))$.

A special case of equation 9.1.2.37.

Particular solution:

$$y(x) = \Theta(x)e^{\lambda x},$$

where $\Theta(x) = \Theta(x+1)$ is an arbitrary periodic function with period 1 and λ is a root of the algebraic (transcendental) equation $e^{2\lambda} = f(e^\lambda)$.

37. $y(x+2) = y(x)F(x, y(x+1)/y(x))$.

Homogeneous second-order difference equation.

The substitution $u(x) = y(x+1)/y(x)$ leads to a first-order difference equation:

$$u(x)u(x+1) = F(x, u(x)).$$

38. $F(x, y(x), y(x+1), y(x+2)) = 0$.

Second-order nonlinear difference equation of general form. A special case of equation 9.1.2.45.

▶ **Higher-order linear difference equations.**

39. $y(x+n) + a_{n-1}y(x+n-1) + \cdots + a_1 y(x+1) + a_0 y(x) = 0$.

Homogeneous nth-order constant-coefficient linear difference equation.

Let us write out the characteristic equation:

$$\lambda^n + a_{n-1}\lambda^{n-1} + \cdots + a_1 \lambda + a_0 = 0. \tag{1}$$

Consider the following cases.

1°. All roots $\lambda_1, \lambda_2, \ldots, \lambda_n$ of equation (1) are real and distinct. Then the general solution of the original difference equation has the form (Math. Encyclopedia, 1979):

$$y(x) = \Theta_1(x)\lambda_1^x + \Theta_2(x)\lambda_2^x + \cdots + \Theta_n(x)\lambda_n^x, \tag{2}$$

where $\Theta_1(x), \Theta_2(x), \ldots, \Theta_n(x)$ are arbitrary periodic functions with period 1.

For $\Theta_k(x) \equiv C_k$, formula (2) gives a particular solution

$$y(x) = C_1 \lambda_1^x + C_2 \lambda_2^x + \cdots + C_n \lambda_n^x,$$

where $C_1, C_2, \ldots, C_n$ are arbitrary constants.

2°. There are m equal real roots, $\lambda_1 = \lambda_2 = \cdots = \lambda_m$ ($m \leq n$), the other roots real and distinct. In this case, the general solution of the difference equation is expressed as

$$y = \left[\Theta_1(x) + x\Theta_2(x) + \cdots + x^{m-1}\Theta_m(x)\right]\lambda_1^x$$
$$+ \Theta_{m+1}(x)\lambda_{m+1}^x + \Theta_{m+2}(x)\lambda_{m+2}^x + \cdots + \Theta_n(x)\lambda_n^x.$$

3°. There are m equal complex conjugate roots, $\lambda = \rho(\cos\beta \pm i\sin\beta)$, $2m \leq n$, with the other roots all real and distinct. Then the original difference equation has a solution corresponding to $\Theta_n(x) \equiv \text{const}_n$:

$$y = \rho^x \cos(\beta x)(A_1 + A_2 x + \cdots + A_m x^{m-1})$$
$$+ \rho^x \sin(\beta x)(B_1 + B_2 x + \cdots + B_m x^{m-1})$$
$$+ C_{m+1}\lambda_{m+1}^x + C_{m+2}\lambda_{m+2}^x + \cdots + C_n\lambda_n^x,$$

where $A_1, \ldots, A_m, B_1, \ldots, B_m, C_{2m+1}, \ldots, C_n$ are arbitrary constants.

40. $y(x+n) + a_{n-1}y(x+n-1) + \cdots + a_1 y(x+1) + a_0 y(x) = f(x)$.

Nonhomogeneous nth-order constant-coefficient linear difference equation.

1°. General solution:

$$y(x) = Y(x) + \bar{y}(x),$$

where $Y(x)$ is the general solution of the corresponding homogeneous equation

$$Y(x+n) + a_{n-1}Y(x+n-1) + \cdots + a_1 Y(x+1) + a_0 Y(x) = 0$$

(see the previous equation), and $\bar{y}(x)$ is any particular solution of the nonhomogeneous equation.

2°. For $f(x) = \sum_{k=0}^{n} A_k x^n$, the nonhomogeneous equation has a particular solution of the form $\bar{y}(x) = \sum_{k=0}^{n} B_k x^n$; the constants B_k are found by the method of undetermined coefficients.

3°. For $f(x) = \sum_{k=1}^{n} A_k \exp(\lambda_k x)$, the nonhomogeneous equation has a particular solution of the form $\bar{y}(x) = \sum_{k=1}^{n} B_k \exp(\lambda_k x)$; the constants B_k are found by the method of undetermined coefficients.

4°. For $f(x) = \sum_{k=1}^{n} A_k \cos(\lambda_k x)$, the nonhomogeneous equation has a particular solution of the form $\bar{y}(x) = \sum_{k=1}^{n} B_k \cos(\lambda_k x) + \sum_{k=1}^{n} D_k \sin(\lambda_k x)$; the constants B_k and D_k are found by the method of undetermined coefficients.

5°. For $f(x) = \sum_{k=1}^{n} A_k \sin(\lambda_k x)$, the nonhomogeneous equation has a particular solution of the form $\bar{y}(x) = \sum_{k=1}^{n} B_k \cos(\lambda_k x) + \sum_{k=1}^{n} D_k \sin(\lambda_k x)$; the constants B_k and D_k are found by the method of undetermined coefficients.

41. $y(x+b_n) + a_{n-1}y(x+b_{n-1}) + \cdots + a_1 y(x+b_1) + a_0 y(x) = 0.$

There are particular solutions of the form $y(x) = \lambda_k^x$, where λ_k are roots of the transcendental (or algebraic) equation

$$\lambda^{b_n} + a_{n-1}\lambda^{b_{n-1}} + \cdots + a_1 \lambda^{b_1} + a_0 = 0.$$

▶ **Higher-order nonlinear difference equations.**

42. $y(x+n) = a^n y(x) + F(x, y(x+1) - ay(x)).$

1°. The substitution $u(x) = y(x+1) - ay(x)$ leads to an $(n-1)$st-order difference equation.

In particular, for $n = 3$, we get

$$u(x+2) + au(x+1) + a^2 u(x) = F(x, u(x)).$$

2°. For $F(x,u) = f(u)$ and $a > 0$, the equation admits a particular solution of the form:

$$y = \Theta(x) a^x + k,$$

where $\Theta(x) = \Theta(x+1)$ is an arbitrary periodic function with period 1 and k is a root of the algebraic (transcendental) equation $(a^n - 1)k + f((1-a)k) = 0$.

43. $y(x+n) = y(x) F\bigl(y(x+1)/y(x), y(x+2)/y(x), \ldots, y(x+n-1)/y(x)\bigr).$

A special case of equation 9.1.2.44.

Particular solution:

$$y(x) = \Theta(x) e^{\lambda x},$$

where $\Theta(x) = \Theta(x+1)$ is an arbitrary periodic function with period 1 and λ is a root of the algebraic (transcendental) equation $e^{n\lambda} = F(e^\lambda, e^{2\lambda}, \ldots, e^{(n-1)\lambda})$.

44. $y(x+n) = y(x) F\bigl(x, y(x+1)/y(x), y(x+2)/y(x), \ldots, y(x+n-1)/y(x)\bigr).$

Homogeneous nth-order difference equation.

The substitution $u(x) = y(x+1)/y(x)$ leads to an $(n-1)$st-order difference equation.

45. $F\bigl(x, y(x), y(x+1), \ldots, y(x+n)\bigr) = 0.$

An nth-order nonlinear difference equation of general form.

Let us solve the equation for $y(x+n)$ to obtain

$$y(x+n) = f\bigl(x, y(x), y(x+1), \ldots, y(x+n-1)\bigr). \tag{1}$$

1°. Let us assume that the equation is defined on a discrete set of points $x = x_0 + k$ with integer k. Given initial values $y(x_0), y(x_0 + 1), \ldots, y(x_0 + n - 1)$, one can make use of (1) to find sequentially $y(x_0 + n), y(x_0 + n + 1)$, etc.

Solving the original equation for $y(x)$ gives

$$y(x) = g\bigl(x, y(x+1), y(x+2), \ldots, y(x+n)\bigr). \tag{2}$$

On setting $x = x_0 - 1$ here, one can find $y(x_0 - 1)$, then likewise $y(x_0 - 2)$ etc.

Thus, given initial data, one can use the equation to find $y(x)$ at all points $x_0 + k$, where $k = 0, \pm 1, \pm 2, \ldots$

2°. Now we assume that x in the original equation can vary continuously. Also assume that $y(x)$ is a continuous function defined arbitrarily on the semi-open interval $[0, n)$. Setting $x = 0$ in (1), one finds $y(n)$. Then, given $y(x)$ on $[0, n]$, one can use (1) to find $y(x)$ on $x \in [n, n+1]$, then on $x \in [n+1, n+2]$, and so on.

9.2. Linear Functional Equations in One Independent Variable

9.2.1. Linear Functional Equations Involving Unknown Function with Two Different Arguments

▶ **Linear functional equations involving $y(x)$ and $y(ax)$.**

1. $y(ax) - by(x) = 0$, $\quad a, b > 0$.

Solution for $x > 0$:
$$y(x) = x^\lambda \Theta\left(\frac{\ln x}{\ln a}\right), \qquad \lambda = \frac{\ln b}{\ln a},$$
where $\Theta(z) = \Theta(z+1)$ is an arbitrary periodic function with period 1, $a \neq 1$.

For $\Theta(z) \equiv \text{const}$, we have a particular solution $y(x) = Cx^\lambda$, where C is an arbitrary constant.

2. $y(ax) - by(x) = f(x)$.

1°. General solution:
$$y(x) = Y(x) + \bar{y}(x),$$
where $Y(x)$ is the general solution of the homogeneous equation $Y(ax) - bY(x) = 0$ (see the previous equation), and $\bar{y}(x)$ is any particular solution of the nonhomogeneous equation.

2°. For $f(x) = \sum_{k=0}^{n} A_k x^n$, the nonhomogeneous equation has a particular solution of the form
$$\bar{y}(x) = \sum_{k=0}^{n} \frac{A_k}{a^k - b} x^k, \qquad a^k - b \neq 0.$$

3°. For $f(x) = \ln x \sum_{k=0}^{n} A_k x^k$, the nonhomogeneous equation has a particular solution of the form
$$\bar{y}(x) = \sum_{k=1}^{n} x^k \left(B_k \ln x + C_k\right), \qquad B_k = \frac{A_k}{a^k - b}, \qquad C_k = -\frac{A_k a^k \ln a}{(a^k - b)^2}.$$

3. $y(2x) - a \cos x \, y(x) = 0$.

Solution for $a > 0$ and $x > 0$ (Pelyukh & Sharkovsky, 1974):
$$y(x) = x^{\frac{\ln a}{\ln 2} - 1} \sin x \, \Theta\left(\frac{\ln x}{\ln 2}\right),$$
where $\Theta(x) = \Theta(x+1)$ is an arbitrary periodic function with period 1.

▶ **Linear reciprocal functional equations involving $y(x)$ and $y(a-x)$.**

4. $y(x) = y(-x)$.

This functional equation may be treated as a definition of even functions.

Solution:
$$y(x) = \frac{\varphi(x) + \varphi(-x)}{2},$$
where $\varphi(x)$ is an arbitrary function.

5. $y(x) = -y(-x).$

This functional equation may be treated as a definition of odd functions.

Solution:
$$y(x) = \frac{\varphi(x) - \varphi(-x)}{2},$$
where $\varphi(x)$ is an arbitrary function.

6. $y(x) - y(a - x) = 0.$

1°. Solution:
$$\begin{aligned} y(x) &= \Phi(x, a - x) \\ &= \Psi(x, a - x) + \Psi(a - x, x), \end{aligned}$$
where $\Phi(x, z) = \Phi(z, x)$ is any symmetric function with two arguments and $\Psi(x, z)$ is any function with two arguments.

2°. Specific particular solutions may be obtained using the formula
$$y(x) = \Omega\bigl(\varphi(x) + \varphi(a - x)\bigr)$$
by specifying the functions $\Omega(z)$ and $\varphi(x)$.

7. $y(x) + y(a - x) = 0.$

1°. Solution:
$$\begin{aligned} y(x) &= \Phi(x, a - x) \\ &= \Psi(x, a - x) - \Psi(a - x, x), \end{aligned}$$
where $\Phi(x, z) = -\Phi(z, x)$ is any antisymmetric function with two arguments and $\Psi(x, z)$ is any function with two arguments.

2°. Specific particular solutions may be obtained using the formula
$$y(x) = (2x - a)\Omega\bigl(\varphi(x) + \varphi(a - x)\bigr)$$
by specifying $\Omega(z)$ and $\varphi(x)$.

8. $y(x) + y(a - x) = b.$

Solution:
$$y(x) = \tfrac{1}{2}b + \Phi(x, a - x),$$
where $\Phi(x, z) = -\Phi(z, x)$ is any antisymmetric function with two arguments.

Particular solutions: $y(x) = b\sin^2\left(\dfrac{\pi x}{2a}\right)$ and $y(x) = b\cos^2\left(\dfrac{\pi x}{2a}\right)$.

9. $y(x) + y(a - x) = f(x)$.

Here, the function $f(x)$ must satisfy the condition $f(x) = f(a - x)$.
   Solution (Polyanin & Manzhirov, 1998):
$$y(x) = \tfrac{1}{2}f(x) + \Phi(x, a - x),$$
where $\Phi(x, z) = -\Phi(z, x)$ is any antisymmetric function with two arguments.

10. $y(x) - y(a - x) = f(x)$.

Here, the function $f(x)$ must satisfy the condition $f(x) = -f(a - x)$.
   Solution (Polyanin & Manzhirov, 1998):
$$y(x) = \tfrac{1}{2}f(x) + \Phi(x, a - x),$$
where $\Phi(x, z) = \Phi(z, x)$ is any symmetric function with two arguments.

11. $y(x) + g(x)y(a - x) = f(x)$.

Solution:
$$y(x) = \frac{f(x) - g(x)f(a - x)}{1 - g(x)g(a - x)} \qquad [\text{if } g(x)g(a - x) \not\equiv 1].$$

▶ **Linear reciprocal functional equations involving $y(x)$ and $y(a/x)$.**

12. $y(x) - y(a/x) = 0$.

   Solution:
$$y(x) = \Phi(x, a/x),$$
where $\Phi(x, z) = \Phi(z, x)$ is any symmetric function with two arguments.

13. $y(x) + y(a/x) = 0$.

Solution:
$$y(x) = \Phi(x, a/x),$$
where $\Phi(x, z) = -\Phi(z, x)$ is any antisymmetric function with two arguments.

14. $y(x) + y(a/x) = b$.

Solution:
$$y(x) = \tfrac{1}{2}b + \Phi(x, a/x),$$
where $\Phi(x, z) = -\Phi(z, x)$ is any antisymmetric function with two arguments.

15. $y(x) + y(a/x) = f(x)$.

The right-hand side must satisfy the condition $f(x) = f(a/x)$.
   Solution (Polyanin & Manzhirov, 1998):
$$y(x) = \tfrac{1}{2}f(x) + \Phi(x, a/x),$$
where $\Phi(x, z) = -\Phi(z, x)$ is any antisymmetric function with two arguments.

16. $y(x) - y(a/x) = f(x)$.

Here, the function $f(x)$ must satisfy the condition $f(x) = -f(a/x)$.

Solution (Polyanin & Manzhirov, 1998):
$$y(x) = \tfrac{1}{2}f(x) + \Phi(x, a/x),$$
where $\Phi(x, z) = \Phi(z, x)$ is any symmetric function with two arguments.

17. $y(x) + x^a y(1/x) = 0$.

Solution (Polyanin & Manzhirov, 2007):
$$y(x) = (1 - x^a)\Phi(x, 1/x),$$
where $\Phi(x, z) = \Phi(z, x)$ is any symmetric function with two arguments.

18. $y(x) - x^a y(1/x) = 0$.

Solution (Polyanin & Manzhirov, 2007):
$$y(x) = (1 + x^a)\Phi(x, 1/x),$$
where $\Phi(x, z) = \Phi(z, x)$ is any symmetric function with two arguments.

19. $y(x) + g(x)y(a/x) = f(x)$.

Solution:
$$y(x) = \frac{f(x) - g(x)f(a/x)}{1 - g(x)g(a/x)} \qquad [\text{if } g(x)g(a/x) \not\equiv 1].$$

▶ **Linear equations involving unknown function with rational argument.**

20. $y(x) - y\left(\dfrac{a-x}{1+bx}\right) = 0$.

Solution:
$$y(x) = \Phi\left(x, \frac{a-x}{1+bx}\right),$$
where $\Phi(x, z) = \Phi(z, x)$ is any symmetric function with two arguments.

21. $y(x) + y\left(\dfrac{a-x}{1+bx}\right) = 0$.

Solution:
$$y(x) = \Phi\left(x, \frac{a-x}{1+bx}\right),$$
where $\Phi(x, z) = -\Phi(z, x)$ is any antisymmetric function with two arguments.

22. $y(x) + y\left(\dfrac{a-x}{1+bx}\right) = f(x)$.

The function $f(x)$ must satisfy the condition $f(x) = f\left(\dfrac{a-x}{1+bx}\right)$.

Solution:
$$y(x) = \tfrac{1}{2}f(x) + \Phi\left(x, \frac{a-x}{1+bx}\right),$$
where $\Phi(x, z) = -\Phi(z, x)$ is any antisymmetric function with two arguments.

23. $y(x) - y\left(\dfrac{a-x}{1+bx}\right) = f(x)$.

Here, the function $f(x)$ must satisfy the condition $f(x) = -f\left(\dfrac{a-x}{1+bx}\right)$.

Solution:
$$y(x) = \tfrac{1}{2}f(x) + \Phi\left(x, \dfrac{a-x}{1+bx}\right),$$

where $\Phi(x, z) = \Phi(z, x)$ is any symmetric function with two arguments.

24. $y(x) - cy\left(\dfrac{a-x}{1+bx}\right) = f(x)$, $c \neq \pm 1$.

Solution:
$$y(x) = \dfrac{1}{1-c^2}f(x) + \dfrac{c}{1-c^2}f\left(\dfrac{a-x}{1+bx}\right).$$

25. $y(x) + g(x)y\left(\dfrac{a-x}{1+bx}\right) = f(x)$.

Solution:
$$y(x) = \dfrac{f(x) - g(x)f(z)}{1 - g(x)g(z)}, \qquad z = \dfrac{a-x}{1+bx}.$$

26. $y(x) + cy\left(\dfrac{ax-\beta}{x+b}\right) = f(x)$, $\beta = a^2 + ab + b^2$.

A special case of equation 9.2.2.9.

27. $y(x) + g(x)y\left(\dfrac{ax-\beta}{x+b}\right) = f(x)$, $\beta = a^2 + ab + b^2$.

A special case of equation 9.2.2.10.

▶ Linear functional equations involving $y(x)$ and $y(\sqrt{a^2 - x^2})$.

28. $y(x) - y(\sqrt{a^2 - x^2}) = 0$, $0 \leq x \leq a$.

Solution:
$$y(x) = \Phi(x, \sqrt{a^2 - x^2}),$$

where $\Phi(x, z) = \Phi(z, x)$ is any symmetric function with two arguments.

29. $y(x) + y(\sqrt{a^2 - x^2}) = 0$, $0 \leq x \leq a$.

Solution:
$$y(x) = \Phi(x, \sqrt{a^2 - x^2}),$$

where $\Phi(x, z) = -\Phi(z, x)$ is any antisymmetric function with two arguments.

30. $y(x) + y(\sqrt{a^2 - x^2}) = b$, $0 \leq x \leq a$.

Solution:
$$y(x) = \tfrac{1}{2}b + \Phi(x, \sqrt{a^2 - x^2}),$$

where $\Phi(x, z) = -\Phi(z, x)$ is any antisymmetric function with two arguments.

31. $y(x) + y(\sqrt{a^2 - x^2}) = f(x)$, $\quad 0 \le x \le a$.

Here, the function $f(x)$ must satisfy the condition $f(x) = f(\sqrt{a^2 - x^2})$.
   Solution (Polyanin & Manzhirov, 1998):
$$y(x) = \tfrac{1}{2} f(x) + \Phi(x, \sqrt{a^2 - x^2}),$$
where $\Phi(x, z) = -\Phi(z, x)$ is any antisymmetric function with two arguments.

32. $y(x) - y(\sqrt{a^2 - x^2}) = f(x)$, $\quad 0 \le x \le a$.

Here, the function $f(x)$ must satisfy the condition $f(x) = -f(\sqrt{a^2 - x^2})$.
   Solution (Polyanin & Manzhirov, 1998):
$$y(x) = \tfrac{1}{2} f(x) + \Phi(x, \sqrt{a^2 - x^2}),$$
where $\Phi(x, z) = \Phi(z, x)$ is any symmetric function with two arguments.

33. $y(x) + g(x) y(\sqrt{a^2 - x^2}) = f(x)$, $\quad 0 \le x \le a$.

Solution:
$$y(x) = \frac{f(x) - g(x) f(\sqrt{a^2 - x^2})}{1 - g(x) g(\sqrt{a^2 - x^2})}.$$

▶ **Linear functional equations involving $y(\sin x)$ and $y(\cos x)$.**

34. $y(\sin x) - y(\cos x) = 0$.

Solution in implicit form:
$$y(\sin x) = \Phi(\sin x, \cos x),$$
where $\Phi(x, z) = \Phi(z, x)$ is any symmetric function with two arguments.

35. $y(\sin x) + y(\cos x) = 0$.

Solution in implicit form:
$$y(\sin x) = \Phi(\sin x, \cos x),$$
where $\Phi(x, z) = -\Phi(z, x)$ is any antisymmetric function with two arguments.

36. $y(\sin x) + y(\cos x) = a$.

Solution in implicit form:
$$y(\sin x) = \tfrac{1}{2} a + \Phi(\sin x, \cos x),$$
where $\Phi(x, z) = -\Phi(z, x)$ is any antisymmetric function with two arguments.

37. $y(\sin x) + y(\cos x) = f(x)$.

Here, the function $f(x)$ must satisfy the condition $f(x) = f(\tfrac{\pi}{2} - x)$.
   Solution in implicit form (Polyanin & Manzhirov, 1998):
$$y(\sin x) = \tfrac{1}{2} f(x) + \Phi(\sin x, \cos x),$$
where $\Phi(x, z) = -\Phi(z, x)$ is any antisymmetric function with two arguments.

38. $y(\sin x) - y(\cos x) = f(x)$.

Here, the function $f(x)$ must satisfy the condition $f(x) = -f\left(\frac{\pi}{2} - x\right)$.

Solution in implicit form (Polyanin & Manzhirov, 1998):

$$y(\sin x) = \tfrac{1}{2} f(x) + \Phi(\sin x, \cos x),$$

where $\Phi(x, z) = \Phi(z, x)$ is any symmetric function with two arguments.

39. $y(\sin x) + g(x) y(\cos x) = f(x)$.

Solution in implicit form:

$$y(\sin x) = \frac{f(x) - g(x) f\left(\frac{\pi}{2} - x\right)}{1 - g(x) g\left(\frac{\pi}{2} - x\right)}.$$

▶ **Other equations involving unknown function with two different arguments.**

40. $y(x^a) - b y(x) = 0$, $a, b > 0$.

Solution:

$$y(x) = |\ln x|^p \Theta\left(\frac{\ln|\ln x|}{\ln a}\right), \qquad p = \frac{\ln b}{\ln a},$$

where $\Theta(z) = \Theta(z+1)$ is an arbitrary periodic function with period 1.

For $\Theta(z) \equiv \text{const}$, we have a particular solution $y(x) = C|\ln x|^p$, where C is an arbitrary constant.

41. $y(x) - y(\omega(x)) = 0$, where $\omega(\omega(x)) = x$.

Solution:

$$y(x) = \Phi(x, \omega(x)),$$

where $\Phi(x, z) = \Phi(z, x)$ is any symmetric function with two arguments.

42. $y(x) + y(\omega(x)) = 0$, where $\omega(\omega(x)) = x$.

Solution:

$$y(x) = \Phi(x, \omega(x)),$$

where $\Phi(x, z) = -\Phi(z, x)$ is any antisymmetric function with two arguments.

43. $y(x) + y(\omega(x)) = b$, where $\omega(\omega(x)) = x$.

Solution:

$$y(x) = \tfrac{1}{2} b + \Phi(x, \omega(x)),$$

where $\Phi(x, z) = -\Phi(z, x)$ is any antisymmetric function with two arguments.

44. $y(x) + y(\omega(x)) = f(x)$, where $\omega(\omega(x)) = x$.

Here, the function $f(x)$ must satisfy the condition $f(x) = f(\omega(x))$.

Solution (Polyanin & Manzhirov, 1998):

$$y(x) = \tfrac{1}{2} f(x) + \Phi(x, \omega(x)),$$

where $\Phi(x, z) = -\Phi(z, x)$ is any antisymmetric function with two arguments.

45. $y(x) - y(\omega(x)) = f(x)$,   where $\omega(\omega(x)) = x$.

Here, the function $f(x)$ must satisfy the condition $f(x) = -f(\omega(x))$.

Solution (Polyanin & Manzhirov, 1998):
$$y(x) = \tfrac{1}{2} f(x) + \Phi(x, \omega(x)),$$
where $\Phi(x, z) = \Phi(z, x)$ is any symmetric function with two arguments.

46. $y(x) + g(x) y(\omega(x)) = f(x)$,   where $\omega(\omega(x)) = x$.

Solution:
$$y(x) = \frac{f(x) - g(x) f(\omega(x))}{1 - g(x) g(\omega(x))}.$$

47. $y(f(x)) = y(x)$.

Solutions of this equation are called *automorphic functions*. If $f(x)$ is invertible on a set I, then the general solution of the original equation may be written in the form
$$y(x) = \sum_{n=-\infty}^{\infty} \varphi(f^{[n]}(x)),$$
where $\varphi(x)$ is an arbitrary function on I such that the series is convergent.

48. $y(f(x)) = y(x) + c$, $c \neq 0$.

Abel functional equation.

Note that without loss of generality, we can take $c = \pm 1$ (this can be achieved by switching to the normalized unknown function $\bar{y} = \pm y/c$).

1°. Let $\bar{y}(x)$ be a solution of the original equation. Then the function
$$y(x) = \bar{y}(x) + \Theta\!\left(\frac{\bar{y}(x)}{c}\right),$$
where $\Theta(z)$ is an arbitrary 1-periodic function, is also a solution of this equation.

2°. Suppose that the following conditions hold:

(i) $f(x)$ is strictly increasing and continuous for $0 \leq x \leq a$;

(ii) $f(0) = 0$ and $f(x) < x$ for $0 < x < a$;

(iii) the derivative $f'(x)$ exists, has bounded variation on the interval $0 < x < a$, and $\lim_{x \to +0} f'(x) = 1$.

Then, for any $x, x_0 \in (0, a]$, there exists the limit
$$y(x) = \lim_{n \to \infty} \frac{f^{[n]}(x) - f^{[n]}(x_0)}{f^{[n-1]}(x_0) - f^{[n]}(x_0)},$$
which defines a monotonically increasing function satisfying the Abel equation with $c = -1$ (this is the *Lévy solution*).

3°. The Abel functional equation of the special form
$$y(x^k) = y(x) + c, \qquad k > 0,$$
where $0 < x < 1$ or $1 < x < \infty$, admits a particular solution
$$y(x) = \frac{c \ln |\ln x|}{\ln k}.$$

4°. More information about the Abel equation can be found in the books by Aczél (2002, 2006), Laitochová (2007), and Polyanin & Manzhirov (2007).

49. $y(f(x)) = sy(x)$, $\quad s > 0$.

Schröder–Koenigs equation.

1°. Let $s \neq 1$ and $\bar{y}(x)$ be a solution of the original equation satisfying the condition $\bar{y}(x) \neq 0$. Then the function

$$y(x) = \bar{y}(x)\Theta\left(\frac{\ln|\bar{y}(x)|}{\ln s}\right),$$

where $\Theta(z)$ is an arbitrary 1-periodic function, is also a solution of this equation.

2°. The substitution $u(x) = \ln|y(x)|$ leads to the Abel functional equation 9.2.1.48:

$$u(f(x)) = u(x) + c, \quad c = \ln s.$$

3°. The Schröder–Koenigs equation of the special form

$$y(x^k) = sy(x),$$

admits a particular solution

$$y(x) = A \ln^m x, \quad m = \frac{\ln s}{\ln k},$$

where A is an arbitrary constant and $x > 1$.

4°. More information about the Schröder–Koenigs equation can be found in the books by Kuczma (1968), Aczél (2002, 2006), Laitochová (2007), and Polyanin & Manzhirov (2007).

9.2.2. Other Linear Functional Equations

▶ **Linear equations involving composite functions $y(y(x))$ or $y(y(y(x)))$.**

1. $y(y(x)) = 0$.

Solution:

$$y(x) = \begin{cases} \varphi_1(x) & \text{for } x \leq a, \\ 0 & \text{for } a \leq x \leq b, \\ \varphi_2(x) & \text{for } b \leq x, \end{cases}$$

where $a \leq 0$ and $b \geq 0$ are arbitrary numbers; $\varphi_1(x)$ and $\varphi_2(x)$ are arbitrary continuous functions satisfying the conditions

$$\varphi_1(a) = 0, \quad a \leq \varphi_1(x) \leq b \quad \text{if} \quad x \leq a;$$
$$\varphi_2(b) = 0, \quad a \leq \varphi_2(x) \leq b \quad \text{if} \quad b \leq x.$$

2. $y(y(x)) - x = 0$.

Babbage equation or the *equation of involutory functions.*

1°. Particular solutions:

$$y_1(x) = x, \quad y_2(x) = C - x, \quad y_3(x) = \frac{C}{x}, \quad y_4(x) = \frac{C_1 - x}{1 + C_2 x},$$

where C, C_1, and C_2 are arbitrary constants.

2°. On the interval $x \in (a, b)$, there exists a decreasing solution of the original equation involving an arbitrary function:
$$y(x) = \begin{cases} \varphi(x) & \text{for } x \in (a, c], \\ \varphi^{-1}(x) & \text{for } x \in (c, b), \end{cases}$$
where c is an arbitrary point belonging to the interval (a, b), and $\varphi(x)$ is an arbitrary continuous decreasing function on $(a, c]$ such that
$$\lim_{x \to a+0} \varphi(x) = b, \quad \varphi(c) = c.$$

3°. Solution in parametric form:
$$x = \Theta\left(\frac{t}{2}\right), \quad y = \Theta\left(\frac{t+1}{2}\right),$$
where $\Theta(t) = \Theta(t+1)$ is an arbitrary periodic function with period 1.

4°. Solution in parametric form:
$$x = \Theta_1(t) + \Theta_2(t) \sin(\pi t),$$
$$y = \Theta_1(t) - \Theta_2(t) \sin(\pi t),$$
where $\Theta_1(t)$ and $\Theta_2(t)$ are arbitrary periodic functions with period 1.

5°. The original functional equation has a single increasing solution: $y(x) = x$.

6°. Particular solutions of the equation may be represented in implicit form using the algebraic (or transcendental) equation
$$\Phi(x, y) = 0,$$
where $\Phi(x, y) = \Phi(y, x)$ is some symmetric function with two arguments.

3. $y(y(x)) + ay(x) + bx = 0.$

1°. General solution in parametric form (Math. Encyclopedia, 1985):
$$x = \Theta_1(t)\lambda_1^t + \Theta_2(t)\lambda_2^t,$$
$$y = \Theta_1(t)\lambda_1^{t+1} + \Theta_2(t)\lambda_2^{t+1},$$
where λ_1 and λ_2 are roots of the quadratic equation
$$\lambda^2 + a\lambda + b = 0$$
and $\Theta_1 = \Theta_1(t)$ and $\Theta_2 = \Theta_2(t)$ are arbitrary periodic functions with period 1.

2°. For $\Theta_1(t) = C_1 = \text{const}$ and $\Theta_2(t) = C_2 = \text{const}$, we have a particular solution in implicit form
$$\frac{\lambda_2 x - y(x)}{\lambda_2 - \lambda_1} = C_1 \left[\frac{\lambda_1 x - y(x)}{C_2(\lambda_1 - \lambda_2)}\right]^k, \quad k = \frac{\ln \lambda_1}{\ln \lambda_2}.$$

4. $y(y(y(x))) - x = 0.$

1°. Particular solutions:
$$y_1(x) = -\frac{C^2}{C + x}, \quad y_2(x) = C - \frac{C^2}{x}, \quad y_3(x) = C_1 - \frac{(C_1 + C_2)^2}{C_2 + x},$$
where C, C_1, C_2 are arbitrary constants.

2°. Solution in parametric form:
$$x = \Theta\left(\frac{t}{3}\right), \quad y = \Theta\left(\frac{t+1}{3}\right),$$
where $\Theta(t) = \Theta(t+1)$ is an arbitrary periodic function with period 1.

▶ **Equations involving unknown function with three different arguments.**

5. $y(a^2 x) = by(ax) + cy(x)$.

Here $a > 0$, $a \neq 1$, $x > 0$.
   Solution:
$$y(x) = x^{\frac{\ln \lambda_1}{\ln a}} \Theta_1\left(\frac{\ln x}{\ln a}\right) + x^{\frac{\ln \lambda_2}{\ln a}} \Theta_2\left(\frac{\ln x}{\ln a}\right), \qquad \lambda_{1,2} = \frac{b \pm \sqrt{b^2 + 4c}}{2},$$
where $\Theta_1(z)$ and $\Theta_2(z)$ are arbitrary periodic functions with period 1.

6. $Ay(ax) + By(bx) + y(x) = 0$.

1°. This equation has particular solutions of the form $y(x) = Cx^\beta$, where C is an arbitrary constant and β is a root of the transcendental equation
$$Aa^\beta + Bb^\beta + 1 = 0.$$

2°. The transformation
$$z = \ln|x|, \quad y(x) = u(z)$$
leads to a second-order linear difference equation:
$$Au(z + \ln|a|) + Bu(z + \ln|b|) + u(z) = 0.$$

7. $Ay(x^a) + By(x^b) + y(x) = 0$, $\quad x > 0$.

1°. This equation has particular solutions of the form $y(x) = C|\ln x|^p$, where C is an arbitrary constant and p is a root of the transcendental equation
$$A|a|^p + B|b|^p + 1 = 0.$$

2°. The transformation
$$z = \ln x, \quad y(x) = u(z)$$
leads to a functional equation of the form 9.2.2.6:
$$Au(az) + Bu(bz) + u(z) = 0.$$

8. $y(x + 1) = ay(x) + by(-x)$.

Solution (Pelyukh & Sharkovsky, 1974):
$$y(x) = \Theta(x)\lambda^x + \frac{\lambda - a}{b}\Theta(-x)\lambda^{-x}, \tag{1}$$
where
$$\lambda = \frac{a^2 - b^2 + 1}{2a} + \sqrt{\left(\frac{a^2 - b^2 + 1}{2a}\right)^2 - 1}, \tag{2}$$
and $\Theta(x) = \Theta(x+1)$ is an arbitrary periodic function with period 1. In solution (1), it is assumed that the constant λ, which is determined by formula (2), is a real positive number.
   If λ is a negative or complex number, then the real part of (1) must be taken.

9. $Ay(x) + By\left(\dfrac{ax-\beta}{x+b}\right) + Cy\left(\dfrac{bx+\beta}{a-x}\right) = f(x)$, $\beta = a^2 + ab + b^2$.

Let us substitute x first with $\dfrac{ax-\beta}{x+b}$ and then with $\dfrac{bx+\beta}{a-x}$ to obtain two more equations.
So we get the system (the original equation comes first)

$$\begin{aligned} Ay(x) + By(u) + Cy(w) &= f(x), \\ Ay(u) + By(w) + Cy(x) &= f(u), \\ Ay(w) + By(x) + Cy(u) &= f(w), \end{aligned} \qquad (1)$$

where $u = \dfrac{ax-\beta}{x+b}$ and $w = \dfrac{bx+\beta}{a-x}$.

Eliminating $y(u)$ and $y(w)$ from the system of linear algebraic equations (1) yields the solution of the original functional equation.

10. $f_1(x)y(x) + f_2(x)y\left(\dfrac{ax-\beta}{x+b}\right) + f_3(x)y\left(\dfrac{bx+\beta}{a-x}\right) = g(x)$, $\beta = a^2 + ab + b^2$.

Let us substitute x first with $\dfrac{ax-\beta}{x+b}$ and then with $\dfrac{bx+\beta}{a-x}$ to obtain two more equations.
So we get the system (the original equation comes first)

$$\begin{aligned} f_1(x)y(x) + f_2(x)y(u) + f_3(x)y(w) &= g(x), \\ f_1(u)y(u) + f_2(u)y(w) + f_3(u)y(x) &= g(u), \\ f_1(w)y(w) + f_2(w)y(x) + f_3(w)y(u) &= g(w), \end{aligned} \qquad (1)$$

where

$$u = \dfrac{ax-\beta}{x+b}, \qquad w = \dfrac{bx+\beta}{a-x}.$$

Eliminating $y(u)$ and $y(w)$ from the system of linear algebraic equations (1) yields the solution $y = y(x)$ of the original functional equation.

▶ **Equations involving unknown function with many different arguments.**

11. $y(a_n x) + b_{n-1} y(a_{n-1} x) + \cdots + b_1 y(a_1 x) + b_0 y(x) = 0$.

1°. This functional equation has particular solutions of the form $y = Cx^\beta$, where C is an arbitrary constant and β is a root of the transcendental equation

$$a_n^\beta + b_{n-1} a_{n-1}^\beta + \cdots + b_1 a_1^\beta + b_0 = 0.$$

2°. The transformation

$$z = \ln|x|, \quad y(x) = u(z)$$

leads to a linear difference equation:

$$u(z + \alpha_n) + b_{n-1} u(z + \alpha_{n-1}) + \cdots + b_1 y(z + \alpha_1) + b_0 y(x) = 0, \quad \alpha_k = \ln|a_k|.$$

12. $y(x^{a_n}) + b_{n-1} y(x^{a_{n-1}}) + \cdots + b_1 y(x^{a_1}) + b_0 y(x) = 0$.

1°. This functional equation has particular solutions of the form $y(x) = C|\ln x|^p$, where C is an arbitrary constant and p is a root of the transcendental equation

$$|a_n|^p + b_{n-1}|a_{n-1}|^p + \cdots + b_1|a_1|^p + b_0 = 0.$$

2°. The transformation
$$z = \ln x, \quad y(x) = u(z)$$
leads to a functional equation of the form 9.2.2.11:
$$u(a_n z) + b_{n-1} u(a_{n-1} z) + \cdots + b_1 u(a_1 z) + b_0 u(z) = 0.$$

13. $y^{[n]}(x) + a_{n-1} y^{[n-1]}(x) + \cdots + a_1 y(x) + a_0 x = 0.$

Notation used: $y^{[2]}(x) = y(y(x)), \ldots, y^{[n]}(x) = y(y^{[n-1]}(x))$.

1°. Solutions are sought in the parametric form
$$x = w(t), \quad y = w(t+1).$$
As a result, the original equation is reduced to an nth-order linear difference equation of the form 9.1.2.39:
$$w(t+n) + a_{n-1} w(t+n-1) + \cdots + a_1 w(t+1) + a_0 w(t) = 0.$$

2°. In the special case of $a_{n-1} = \cdots = a_1 = 0$ and $a_0 = -1$, we get the following solution in parametric form:
$$x = \Theta\!\left(\frac{t}{n}\right), \quad y = \Theta\!\left(\frac{t+1}{n}\right),$$
where $\Theta(t) = \Theta(t+1)$ is an arbitrary periodic function with period 1.

9.3. Nonlinear Functional Equations in One Independent Variable

9.3.1. Functional Equations with Quadratic Nonlinearity

▶ Functional equations involving $y(x)$ and $y(a-x)$.

1. $y(x) y(a-x) = b^2.$

Solutions:
$$y(x) = \pm b \exp[\Phi(x, a-x)],$$
where $\Phi(x, z) = -\Phi(z, x)$ is any antisymmetric function with two arguments.

On setting $\Phi(x, z) = C(x-z)$, we arrive at particular solutions of the form
$$y(x) = \pm b e^{C(2x-a)},$$
where C is an arbitrary constant.

2. $y(x) y(a-x) = -b^2.$

Substituting $x = a/2$ into the equation, we get the relation $y^2(a/2) = -b^2$, which cannot be satisfied for $b \neq 0$. Therefore, this equation has no continuous solutions.

Discontinuous solutions:
$$\pm y(x) = \begin{cases} b \varphi(x) & \text{if } x \geq a/2, \\ -\dfrac{b}{\varphi(a-x)} & \text{if } x < a/2, \end{cases}$$
where $\varphi(x)$ is an arbitrary function.

3. $y(x)y(a-x) = f^2(x)$.

The function $f(x)$ must satisfy the condition $f(x) = \pm f(a-x)$.

1°. The change of variable $y(x) = f(x)u(x)$ leads to one of the equations of the form 9.3.1.1 or 9.3.1.2.

2°. For $f(x) = f(a-x)$, there are solutions of the form
$$y(x) = \pm f(x) \exp[\Phi(x, a-x)],$$
where $\Phi(x, z) = -\Phi(z, x)$ is any antisymmetric function with two arguments.

On setting $\Phi(x, z) = C(x-z)$, we arrive at particular solutions
$$y(x) = \pm e^{C(2x-a)} f(x),$$
where C is an arbitrary constant.

4. $y^2(x) + y^2(a-x) = b^2$.

1°. Solutions:
$$y(x) = \pm\sqrt{\tfrac{1}{2}b^2 + \Phi(x, a-x)},$$
where $\Phi(x, z) = -\Phi(z, x)$ is any antisymmetric function of two arguments.

2°. Particular solutions:
$$y_{1,2}(x) = \pm\frac{b}{\sqrt{2}}, \quad y_{3,4}(x) = \pm b \sin\left(\frac{\pi x}{2a}\right), \quad y_{5,6}(x) = \pm b \cos\left(\frac{\pi x}{2a}\right).$$

5. $y^2(x) + Ay(x)y(a-x) + By^2(a-x) + Cy(x) + Dy(a-x) = f(x)$.

A special case of equation 9.3.3.2.

Solution in parametric form (t is the parameter):
$$y^2 + Ayt + Bt^2 + Cy + Dt = f(x),$$
$$t^2 + Ayt + By^2 + Ct + Dy = f(a-x).$$

Eliminating t gives the solutions in implicit form.

▶ **Functional equations involving $y(x)$ and $y(ax)$.**

6. $y(2x) - ay^2(x) = 0$.

A special case of equation 9.3.3.1.

Particular solutions:
$$y(x) = 0, \quad y(x) = \frac{1}{a}e^{Cx},$$
where C is an arbitrary constant.

7. $y(2x) - 2y^2(x) + a = 0$.

A special case of equation 9.3.3.1.

Particular solutions with $a = 1$:
$$y(x) = -\tfrac{1}{2}, \quad y(x) = 1, \quad y(x) = \cos(Cx), \quad y(x) = \cosh(Cx),$$
where C is an arbitrary constant.

8. $y(ax) = 2y(x) + by^2(x)$.

Solution:
$$y(x) = \frac{1}{b}\left\{\exp\left[x^{\frac{\ln 2}{\ln a}}\Theta\left(\frac{\ln x}{\ln a}\right)\right] - 1\right\},$$

where $\Theta(x) = \Theta(x+1)$ is an arbitrary periodic function with unit period.

9. $y(x)y(ax) = f(x)$.

A special case of equation 9.3.3.1.

▶ **Functional equations involving $y(x)$ and $y(a/x)$.**

10. $y(x)y(a/x) = b^2$.

Solutions:
$$y(x) = \pm b \exp[\Phi(x, a/x)],$$

where $\Phi(x, z) = -\Phi(z, x)$ is any antisymmetric function with two arguments.

On setting $\Phi(x, z) = C(\ln x - \ln z)$, one arrives at particular solutions of the form
$$y = \pm b a^{-C} x^{2C},$$

where C is an arbitrary constant.

11. $y(x)y(a/x) = f^2(x)$.

The function $f(x)$ must satisfy the condition $f(x) = \pm f(a/x)$. For definiteness, we take $f(x) = f(a/x)$.

Solutions:
$$y(x) = \pm f(x) \exp[\Phi(x, a/x)],$$

where $\Phi(x, z) = -\Phi(z, x)$ is any antisymmetric function with two arguments.

On setting $\Phi(x, z) = C(\ln x - \ln z)$, one arrives at particular solutions of the form
$$y = \pm a^{-C} x^{2C} f(x),$$

where C is an arbitrary constant.

12. $y^2(x) + Ay(x)y(a/x) + By^2(a/x) + Cy(x) + Dy(a/x) = f(x)$.

A special case of equation 9.3.3.3.

Solution in parametric form (t is the parameter):
$$y^2 + Ayt + Bt^2 + Cy + Dt = f(x),$$
$$t^2 + Ayt + By^2 + Ct + Dy = f(a/x).$$

Eliminating t gives the solutions in implicit form.

▶ **Other functional equations with quadratic nonlinearity.**

13. $y(x^2) - ay^2(x) = 0$.

Solution:
$$y(x) = \frac{1}{a}x^C,$$

where C is an arbitrary constant. In addition, $y(x) \equiv 0$ is also a continuous solution.

14. $y(x)y(x^a) = f(x)$, $\quad a > 0$.

A special case of equation 9.3.3.6.

15. $y(x)y\left(\dfrac{a-x}{1+bx}\right) = A^2$.

Solutions:
$$y(x) = \pm A \exp\left[\Phi\left(x, \dfrac{a-x}{1+bx}\right)\right],$$

where $\Phi(x, z) = -\Phi(z, x)$ is any antisymmetric function with two arguments.

16. $y(x)y\left(\dfrac{a-x}{1+bx}\right) = f^2(x)$.

The right-hand side function must satisfy the condition $f(x) = \pm f\left(\dfrac{a-x}{1+bx}\right)$. To be specific, we choose $f(x) = f\left(\dfrac{a-x}{1+bx}\right)$.

Solutions:
$$y(x) = \pm f(x) \exp\left[\Phi\left(x, \dfrac{a-x}{1+bx}\right)\right],$$

where $\Phi(x, z) = -\Phi(z, x)$ is any antisymmetric function with two arguments.

17. $y^2(x) + Ay(x)y\left(\dfrac{a-x}{1+bx}\right) + By(x) = f(x)$.

A special case of equation 9.3.3.4.

18. $y(x)y\left(\sqrt{a^2 - x^2}\right) = b^2$, $\quad 0 \le x \le a$.

Solutions:
$$y(x) = \pm b \exp\left[\Phi\left(x, \sqrt{a^2 - x^2}\right)\right],$$

where $\Phi(x, z) = -\Phi(z, x)$ is any antisymmetric function with two arguments.

19. $y(x)y\left(\sqrt{a^2 - x^2}\right) = f^2(x)$, $\quad 0 \le x \le a$.

The right-hand side function must satisfy the condition $f(x) = \pm f\left(\sqrt{a^2 - x^2}\right)$. For definiteness, we take $f(x) = f\left(\sqrt{a^2 - x^2}\right)$.

Solutions:
$$y(x) = \pm f(x) \exp\left[\Phi\left(x, \sqrt{a^2 - x^2}\right)\right],$$

where $\Phi(x, z) = -\Phi(z, x)$ is any antisymmetric function with two arguments.

20. $y(\sin x)y(\cos x) = a^2$.

Solutions in implicit form:
$$y(\sin x) = \pm a \exp\left[\Phi(\sin x, \cos x)\right],$$

where $\Phi(x, z) = -\Phi(z, x)$ is any antisymmetric function with two arguments.

21. $y(\sin x)y(\cos x) = f^2(x)$.

The right-hand side function must satisfy the condition $f(x) = \pm f\left(\dfrac{\pi}{2} - x\right)$. For definiteness, we take $f(x) = f\left(\dfrac{\pi}{2} - x\right)$.

Solutions in implicit form:

$$y(\sin x) = \pm f(x) \exp[\Phi(\sin x, \cos x)],$$

where $\Phi(x, z) = -\Phi(z, x)$ is any antisymmetric function with two arguments.

22. $y(x)y(\omega(x)) = b^2,$ where $\omega(\omega(x)) = x.$

Solutions:
$$y(x) = \pm b \exp[\Phi(x, \omega(x))],$$

where $\Phi(x, z) = -\Phi(z, x)$ is any antisymmetric function with two arguments.

23. $y(x)y(\omega(x)) = f^2(x),$ where $\omega(\omega(x)) = x.$

The right-hand side function must satisfy the condition $f(x) = \pm f(\omega(x))$. To be specific, we take $f(x) = f(\omega(x))$.

Solutions:
$$y(x) = \pm f(x) \exp[\Phi(x, \omega(x))],$$

where $\Phi(x, z) = -\Phi(z, x)$ is any antisymmetric function with two arguments.

9.3.2. Functional Equations with Power Nonlinearity

1. $y^k(x)y(a - x) = f(x).$

A special case of equation 9.3.3.2.

Solution:
$$y(x) = [f(x)]^{-\frac{k}{1-k^2}} [f(a-x)]^{\frac{1}{1-k^2}}.$$

2. $y^{2n+1}(x) + y^{2n+1}(a - x) = b,$ $n = 1, 2, \ldots$

The change of variable $w(x) = y^{2n+1}(x)$ leads to a linear equation of the form 9.2.1.8: $w(x) + w(a - x) = b.$

3. $y(ax) = by^k(x).$

Solution for $x > 0$, $a > 0$, $b > 0$, and $k > 0$ ($a \neq 1$ and $k \neq 1$):

$$y(x) = b^{\frac{1}{1-k}} \exp\left[x^{\frac{\ln k}{\ln a}} \Theta\left(\frac{\ln x}{\ln a}\right)\right],$$

where $\Theta(x) = \Theta(x + 1)$ is an arbitrary periodic function with period 1.

4. $y^k(x)y(a/x) = f(x).$

A special case of equation 9.3.3.3.

Solution for $k \neq \pm 1$:
$$y(x) = [f(x)]^{-\frac{k}{1-k^2}} [f(a/x)]^{\frac{1}{1-k^2}}.$$

5. $y^k(x)y\left(\dfrac{a-x}{1+bx}\right) = f(x).$

A special case of equation 9.3.3.4.

6. $y^k(x)y\left(\dfrac{ax-\beta}{x+b}\right) = f(x)$, $\beta = a^2 + ab + b^2$.

A special case of equation 9.3.3.10.

7. $y^k(x)y(x^a) = f(x)$.

A special case of equation 9.3.3.6.

8. $y^k(x)y(\sqrt{a^2-x^2}) = f(x)$.

A special case of equation 9.3.3.7.

9. $y^k(\sin x)y(\cos x) = f(x)$.

A special case of equation 9.3.3.8.

9.3.3. Nonlinear Functional Equation of General Form

1. $F(x, y(x), y(ax)) = 0$, $a > 0$.

The transformation $z = \ln x$, $w(z) = y(x)$ leads to an equation of the form 9.1.2.23:
$$F(e^z, w(z), w(z+b)) = 0, \quad b = \ln a.$$

2. $F(x, y(x), y(a-x)) = 0$.

A nonlinear reciprocal (cyclic) functional equation.

Substituting x with $a-x$, one obtains $F(a-x, y(a-x), y(x)) = 0$. Now, eliminating $y(a-x)$ from this equation and the original one, one arrives at an algebraic (or transcendental) equation of the form $\Psi(x, y(x)) = 0$.

To put it differently, the solution $y = y(x)$ of the original functional equation is defined parametrically by the system of two algebraic (or transcendental) equations
$$F(x, y, t) = 0, \quad F(a-x, t, y) = 0,$$
where t is the parameter.

3. $F(x, y(x), y(a/x)) = 0$.

A nonlinear reciprocal (cyclic) functional equation.

Substituting x with a/x yields $F(a/x, y(a/x), y(x)) = 0$. Eliminating $y(a/x)$ from this equation and the original one, we arrive at an algebraic (or transcendental) equation of the form $\Psi(x, y(x)) = 0$.

To put it differently, the solution $y = y(x)$ of the original functional equation is defined parametrically by the system of two algebraic (or transcendental) equations
$$F(x, y, t) = 0, \quad F(a/x, t, y) = 0,$$
where t is the parameter.

4. $F\left(x, y(x), y\left(\dfrac{a-x}{1+bx}\right)\right) = 0$.

Substituting x with $\dfrac{a-x}{1+bx}$ yields
$$F\left(\dfrac{a-x}{1+bx}, y\left(\dfrac{a-x}{1+bx}\right), y(x)\right) = 0.$$

Eliminating $y\left(\dfrac{a-x}{1+bx}\right)$ from this equation and the original one, we arrive at an algebraic (or transcendental) equation of the form $\Psi(x, y(x)) = 0$.

In other words, the solution $y = y(x)$ of the original functional equation is defined parametrically by the system of two algebraic (or transcendental) equations

$$F(x, y, t) = 0, \quad F\left(\dfrac{a-x}{1+bx}, t, y\right) = 0,$$

where t is the parameter.

5. $F\left(x,\, y(x),\, y\left(\dfrac{ax-\beta}{x+b}\right)\right) = 0, \qquad \beta = a^2 + ab + b^2.$

A special case of equation 9.3.3.10.

6. $F(x,\, y(x),\, y(x^a)) = 0.$

The transformation $\xi = \ln x$, $u(\xi) = y(x)$ leads to an equation of the form 9.3.3.1:

$$F(e^\xi,\, u(\xi),\, u(a\xi)) = 0.$$

7. $F(x,\, y(x),\, y(\sqrt{a^2 - x^2})) = 0, \qquad 0 \le x \le a.$

Substituting x with $\sqrt{a^2 - x^2}$ yields

$$F(\sqrt{a^2 - x^2},\, y(\sqrt{a^2 - x^2}),\, y(x)) = 0.$$

Eliminating $y(\sqrt{a^2 - x^2})$ from this equation and the original one, we arrive at an algebraic (or transcendental) equation of the form $\Psi(x, y(x)) = 0$.

In other words, the solution $y = y(x)$ of the original functional equation is defined parametrically by the system of two algebraic (or transcendental) equations

$$F(x, y, t) = 0, \quad F(\sqrt{a^2 - x^2}, t, y) = 0,$$

where t is the parameter.

8. $F(x,\, y(\sin x),\, y(\cos x)) = 0.$

Substituting x with $\tfrac{\pi}{2} - x$ yields $F(\tfrac{\pi}{2} - x,\, y(\cos x),\, y(\sin x)) = 0$. Eliminating $y(\cos x)$ from this equation and the original one, we arrive at an algebraic (or transcendental) equation of the form $\Psi(x, y(\sin x)) = 0$ for $y(\sin x)$.

9. $F(x,\, y(x),\, y(\omega(x))) = 0, \qquad$ where $\omega(\omega(x)) = x.$

Substituting x with $\omega(x)$ yields $F(\omega(x),\, y(\omega(x)),\, y(x)) = 0$. Eliminating $y(\omega(x))$ from this equation and the original one, we arrive at an algebraic (or transcendental) equation of the form $\Psi(x, y(x)) = 0$.

Thus, the solution $y = y(x)$ of the original functional equation is defined parametrically by the system of two algebraic (or transcendental) equations

$$F(x, y, t) = 0, \quad F(\omega(x), t, y) = 0,$$

where t is the parameter.

10. $F\left(x, y(x), y\left(\dfrac{ax-\beta}{x+b}\right), y\left(\dfrac{bx+\beta}{a-x}\right)\right) = 0$, $\qquad \beta = a^2 + ab + b^2$.

Let us substitute x first with $\dfrac{ax-\beta}{x+b}$ and then with $\dfrac{bx+\beta}{a-x}$ to obtain two more equations. As a result, we get the following system (the original equation comes first):

$$\begin{aligned} F(x, y(x), y(u), y(w)) &= 0, \\ F(u, y(u), y(w), y(x)) &= 0, \\ F(w, y(w), y(x), y(u)) &= 0. \end{aligned} \qquad (1)$$

The arguments u and w are expressed in terms of x as

$$u = \frac{ax-\beta}{x+b}, \qquad w = \frac{bx+\beta}{a-x}.$$

Eliminating $y(u)$ and $y(w)$ from the system of algebraic (transcendental) equations (1), we arrive at the solutions $y = y(x)$ of the original functional equation.

11. $F(x, y(a_1 x), y(a_2 x), \ldots, y(a_n x)) = 0$.

Here, $x > 0$ and $a_k > 0$, where $k = 1, \ldots, n$. The transformation

$$y(x) = w(z), \qquad z = \ln x$$

leads to the difference equation

$$F(e^z, w(z+b_1), w(z+b_2), \ldots, w(z+b_n)) = 0, \qquad b_k = \ln a_k.$$

12. $F(x, y(x^{a_1}), y(x^{a_2}), \ldots, y(x^{a_m})) = 0$, $\qquad x > 0$.

The transformation

$$y(x) = w(z), \qquad z = \ln x$$

leads to a difference equation of the form 9.3.3.11:

$$F(e^z, w(a_1 z), w(a_2 z), \ldots, w(a_n z)) = 0.$$

13. $F(x, y(e^{a_1 x}), y(e^{a_2 x}), \ldots, y(e^{a_n x})) = 0$.

The substitution $t = e^x$ leads to a difference equation of the form 9.3.3.12:

$$F(\ln t, y(t^{a_1}), y(t^{a_2}), \ldots, y(t^{a_n})) = 0.$$

14. $F(x, y(x), y^{[2]}(x), \ldots, y^{[n]}(x)) = 0$.

Notation: $y^{[2]}(x) = y(y(x)), \ldots, y^{[n]}(x) = y(y^{[n-1]}(x))$.

Solutions are sought in parametric form (Math. Encyclopedia, 1985):

$$x = w(t), \qquad y = w(t+1). \qquad (1)$$

Then the original equation is reduced to an nth-order difference equation of the form 9.1.2.45:

$$F(w(t), w(t+1), w(t+2), \ldots, w(t+n)) = 0. \qquad (2)$$

The general solution of equation (2) has the structure

$$x = w(t) = \varphi(t;\, C_1, \ldots, C_n),$$
$$y = w(t+1) = \varphi(t+1;\, C_1, \ldots, C_n),$$

where $C_1 = C_1(t), \ldots, C_n = C_n(t)$ are arbitrary periodic functions with period 1, that is, $C_k(t) = C_k(t+1)$, $k = 1, 2, \ldots, n$.

15. $F\big(x,\, y(\theta_0(x)),\, y(\theta_1(x)),\, \ldots,\, y(\theta_{n-1}(x))\big) = 0.$

Notation: $\theta_k(x) \equiv \theta\big(x + \frac{k}{n}T\big)$, where $k = 0, 1, \ldots, n-1$. The function $\theta(x)$ is assumed to be periodic with period T, i.e., $\theta(x) = \theta(x+T)$. Furthermore, the left-hand side of the equation is assumed to satisfy the condition $F(x, \ldots) = F(x+T, \ldots)$.

In the original equation, we substitute x sequentially with $x + \frac{k}{n}T$, where $k = 0, 1, \ldots, n-1$, to obtain the following system (the original equation comes first):

$$F\big(x,\, y_0,\, y_1,\, \ldots,\, y_{n-1}\big) = 0,$$
$$F\big(x + \tfrac{1}{n}T,\, y_1,\, y_2,\, \ldots,\, y_0\big) = 0, \qquad (1)$$
$$\cdots\cdots\cdots\cdots\cdots\cdots\cdots\cdots\cdots\cdots,$$
$$F\big(x + \tfrac{n-1}{n}T,\, y_{n-1},\, y_0,\, \ldots,\, y_{n-2}\big) = 0,$$

where the notation $y_k \equiv y\big(\theta_k(x)\big)$ is used for brevity.

Eliminating $y_1, y_2, \ldots, y_{n-1}$ from the system of nonlinear algebraic (or transcendental) equations (1), one finds the solutions of the original functional equation in implicit form: $\Psi(x, y_0) = 0$, where $y_0 = y\big(\theta(x)\big)$.

16. $y(x) = F\big(x,\, y(\varphi(x))\big).$

Suppose that the following conditions hold:

1) there exist a and b such that $\varphi(a) = a$, $F(a, b) = b$;
2) the function $\varphi(x)$ is analytic in a neighborhood of a and $|\varphi'(a)| < 1$;
3) the function F is analytic in a neighborhood of the point (a, b) and $\big|\frac{\partial F}{\partial b}(a, b)\big| < 1$.

Then the formal power series solution of the original equation,

$$y(x) = b + \sum_{k=1}^{\infty} c_k (x-a)^k, \qquad (*)$$

has a positive radius of convergence.

The formal solution is obtained by substituting the expansions

$$\varphi(x) = a + \sum_{k=1}^{\infty} a_k (x-a)^k,$$
$$F(x, y) = b + \sum_{i,j=1}^{\infty} b_{ij} (x-a)^i (y-b)^j,$$

and $(*)$ into the original equation. Collecting the terms with like powers of the difference $\xi = (x-a)$ and equating with zero the coefficients of the different powers of ξ, we obtain a triangular system of algebraic equations for the coefficients c_k.

9.4. Functional Equations in Several Independent Variables

9.4.1. Linear Functional Equations

▶ **Equations involving functions with a single argument.**

1. $f(x+y) = f(x) + f(y)$.

Cauchy's equation.
   Solution (Fichtenholz, 1969):
$$f(x) = Cx,$$
where C is an arbitrary constant.

2. $f\left(\dfrac{x+y}{2}\right) = \dfrac{f(x)+f(y)}{2}$.

Jensen's equation.
   Solution:
$$f(x) = C_1 x + C_2,$$
where C_1 and C_2 are arbitrary constants.

3. $af(x) + bf(y) = f(ax+by) + c$.

   1°. Solution:
$$f(x) = \begin{cases} Ax + \dfrac{c}{a+b-1} & \text{if } a+b-1 \neq 0, \\ Ax + B & \text{if } a+b-1 = 0 \text{ and } c = 0, \end{cases}$$
where A and B are arbitrary constants.

   2°. If $a+b-1 = 0$ and $c \neq 0$, then there is no solution.

4. $k_1 f(a_1 x + b_1 y + c_1) + k_2 f(a_2 x + b_2 y + c_2) + k_3 f(a_3 x + b_3 y + c_3) = k_0$.

   1°. All continuous solutions of this equation, if exist, have the form $f(z) = \alpha z + \beta$, where the constants α and β are determined by substituting this expression into the original equation.

   2°. Let the following relations hold:
$$\begin{aligned} a_1 k_1 + a_2 k_2 + a_3 k_3 &= 0, \\ b_1 k_1 + b_2 k_2 + b_3 k_3 &= 0, \\ k_1 + k_2 + k_3 &\neq 0. \end{aligned} \qquad (1)$$

Then the solution of the original equation is
$$f(z) = \alpha z + \beta, \qquad \beta = \frac{k_0 - \alpha(c_1 k_1 + c_2 k_2 + c_3 k_3)}{k_1 + k_2 + k_3}, \qquad (2)$$
where α is an arbitrary constant.

   3°. Suppose that the first and second relations of (1) are satisfied, but the third relation is not (i.e., $k_1 + k_2 + k_3 = 0$). Then the solution of the original equation is
$$f(z) = \alpha z + \beta, \qquad \alpha = \frac{k_0}{c_1 k_1 + c_2 k_2 + c_3 k_3},$$
where β is an arbitrary constant.

4°. Suppose that the first or/and second relation of (1) is not satisfied, but the third relation is satisfied. Then the solution of the original equation will be a constant, which is determined by substituting $\alpha = 0$ into (2).

5. $f(x+y) + f(x-y) = 2f(x) + 2f(y)$.

Solution:
$$f(x) = Cx^2,$$
where C is an arbitrary constant.

6. $\sum_{j=1}^{n} k_j f(a_j x + b_j y + c_j) = \alpha x + \beta y + \gamma$.

All continuous solutions of this equation, if exist, have the form of polynomials of degree $(n-2)$, i.e., $f(z) = \sum_{m=0}^{n-2} \alpha_m z^m$, where the constants α_m are determined by substituting this expression into the original equation.

7. $f(x+y) = f(x)e^{ay}$.

Solution:
$$f(x) = Ce^{ax},$$
where C is an arbitrary constant.

8. $f(x+y) + f(x-y) = 2f(x)\cosh y$.

Solution:
$$f(x) = C_1 e^x + C_2 e^{-x},$$
where C_1 and C_2 are arbitrary constants.

9. $f(x+y) + f(x-y) = 2f(x)\cosh(ay) + 2f(y)$.

Solution:
$$f(x) = C[2 - \cosh(ax)],$$
where C is an arbitrary constant.

10. $f(x+y) + f(x-y) = 2f(x)\cos y$.

Solution:
$$f(x) = C_1 \cos x + C_2 \sin x,$$
where C_1 and C_2 are arbitrary constants.

11. $f(x+y) + f(x-y) = 2f(x)\cos(ay) + 2f(y)$.

Solution:
$$f(x) = C[2 - \cos(ax)],$$
where C is an arbitrary constant.

12. $f(xy) = f(x) + f(y)$.

Cauchy's logarithmic equation.

Solution (Fichtenholz, 1969):
$$f(x) = C \ln |x|,$$
where C is an arbitrary constant.

13. $f(\sqrt{x^2 + y^2}) = f(x) + f(y)$.

Solution:
$$f(x) = Cx^2,$$
where C is an arbitrary constant.

14. $f((x^a + y^a)^{1/a}) = f(x) + f(y)$, $x > 0$, $y > 0$.

Solution:
$$f(x) = Cx^a,$$
where C is an arbitrary constant.

15. $f(x) + f(y) = f\left(\dfrac{x+y}{1-xy}\right)$, $xy < 1$.

Solution:
$$f(x) = C \arctan x,$$
where C is an arbitrary constant.

16. $f(x) + (1-x)f\left(\dfrac{y}{1-x}\right) = f(y) + (1-y)f\left(\dfrac{x}{1-y}\right)$.

Equation of information theory. The variables x and y can assume values from 0 to 1.

Solution:
$$f(x) = C[x \ln x + (1-x) \ln(1-x)],$$
where C is an arbitrary constant.

17. $f(x) + (1-x)^a f\left(\dfrac{y}{1-x}\right) = f(y) + (1-y)^a f\left(\dfrac{x}{1-y}\right)$.

Here, the quantities x, y, and $x+y$ can assume values from 0 to 1; $a \neq 0, 1, 2$.

Solution:
$$f(x) = C[x^a + (1-x)^a - 1],$$
where C is an arbitrary constant.

18. $f(xy - \sqrt{(1-x^2)(1-y^2)}) = f(x) + f(y)$, $|x| \leq 1$, $|y| \leq 1$.

Solution:
$$f(x) = C \arccos x,$$
where C is an arbitrary constant.

19. $f(xy + \sqrt{(x^2-1)(y^2-1)}) = f(x) + f(y)$, $|x| \geq 1$, $|y| \geq 1$.

Solution:
$$f(x) = C \operatorname{arcosh} x,$$
where C is an arbitrary constant.

20. $f(x) + g(y) = h(x+y)$.

Pexider's equation. Here, $f(x)$, $g(y)$, and $h(z)$ are the unknown functions.

Solution:
$$f(x) = C_1 x + C_2, \quad g(y) = C_1 y + C_3, \quad h(z) = C_1 z + C_2 + C_3,$$
where C_1, C_2, and C_3 are arbitrary constants.

21. $f(a_1 x + b_1 y + c_1) + g(a_2 x + b_2 y + c_2) + h(a_3 x + b_3 y + c_3) = 0$.

The generalized Pexider's equation. Here, $f(z_1)$, $g(z_2)$, and $h(z_3)$ are the unknown functions.

All continuous solutions of the equation, if exist, have the form
$$f(z_1) = \alpha_1 z_1 + \beta_1, \quad g(z_2) = \alpha_2 z_2 + \beta_2, \quad h(z_3) = \alpha_3 z_3 + \beta_3,$$
where the constants α_n and β_n ($n = 1, 2, 3$) are determined by substituting these expressions into the original equation.

22. $\sum_{j=1}^{n} f_j(a_j x + b_j y + c_j) = \alpha x + \beta y + \gamma$.

All continuous solutions of this equation, if exist, have the form of polynomials of degree $(n-2)$, i.e., $f_j(z_j) = \sum_{m=0}^{n-2} \alpha_{jm} z_j^m$, where the constants α_{jm} are determined by substituting this expression into the original equation.

▶ **Equations involving functions with two arguments.**

23. $f(x, y) = f(y, x)$.

This equation may be treated as a definition of functions symmetric with respect to permutation of the arguments.

1°. Solution:
$$f(x, y) = \Phi(x, y) + \Phi(y, x),$$
where $\Phi(x, y)$ is an arbitrary function with two arguments.

2°. Particular solutions may be found using the formula
$$f(x, y) = \Psi(\varphi(x) + \varphi(y))$$
by specifying the functions $\Psi(z)$ and $\varphi(x)$.

24. $f(x, y) = -f(y, x)$.

This equation may be treated as a definition of functions antisymmetric with respect to permutation of the arguments.

1°. Solution:
$$f(x, y) = \Phi(x, y) - \Phi(y, x),$$
where $\Phi(x, y)$ is an arbitrary function with two arguments.

2°. Particular solutions may be found using the formulas
$$f(x,y) = \varphi(x) - \varphi(y),$$
$$f(x,y) = (x-y)\Psi(\varphi(x)+\varphi(y)),$$
by specifying the functions $\varphi(x)$ and $\Psi(z)$.

25. $f(x,y) = f(x+ak_1, y+ak_2).$

Equation for traveling-wave solutions. Here, a is an arbitrary parameter, and k_1 and k_2 are some constants.

Solution:
$$f(x,y) = \Phi(k_2 x - k_1 y),$$
where $\Phi(z)$ is an arbitrary function.

26. $f(x,y) = f(x+ak_1, y+ak_2) + ac.$

Here, a is an arbitrary parameter, and k_1, k_2, and c are given numbers.

Solution:
$$f(x,y) = \Phi(k_1 y - k_2 x) - \frac{c}{k_1} x,$$
where $\Phi(z)$ is an arbitrary function.

27. $f(x,y) = e^{ac} f(x+ak_1, y+ak_2).$

Here, a is an arbitrary parameter, and k_1, k_2, and c are given numbers.

Solution:
$$f(x,y) = \exp\left(-\frac{c}{k_1} x\right) \Phi(k_1 y - k_2 x),$$
where $\Phi(z)$ is an arbitrary function.

28. $f(ax, ay) = f(x,y).$

Here, $a \neq 0$ is an arbitrary parameter.

Solution:
$$f(x,y) = \Phi(y/x),$$
where $\Phi(z)$ is an arbitrary function.

29. $f(ax, ay) = a^\beta f(x,y).$

Equation for homogeneous functions. Here, a is an arbitrary positive parameter, and β is a fixed number called the order of homogeneity.

Solution:
$$f(x,y) = x^\beta \Phi(y/x),$$
where $\Phi(z)$ is an arbitrary function.

30. $f(ax, a^\beta y) = f(x,y).$

Here, a is an arbitrary positive parameter, and β is some constant.

Solution:
$$f(x,y) = \Phi(yx^{-\beta}),$$
where $\Phi(z)$ is an arbitrary function.

31. $f(ax, a^\beta y) = a^\gamma f(x, y)$.

Equation for self-similar solutions. Here, a is an arbitrary positive parameter, and β and γ are some constants.
   Solution:
$$f(x, y) = x^\gamma \Phi(yx^{-\beta}),$$
where $\Phi(z)$ is an arbitrary function.

32. $f(x, y) = a^n f(x + (1 - a)y, ay)$.

Here, a is an arbitrary positive parameter, and n is some constant.
   Solution (Polyanin & Manzhirov, 2007):
$$f(x, y) = y^{-n} \Phi(x + y),$$
where $\Phi(z)$ is an arbitrary function.

33. $f(x, y) = f(a^n x, a^m y) + k \ln a$.

Here, a is an arbitrary positive parameter, and n, m, and k are given numbers.
   Solution:
$$f(x, y) = \Phi(yx^{-m/n}) - \frac{k}{n} \ln x, \quad x > 0,$$
where $\Phi(z)$ is an arbitrary function.

34. $f(x, y) = a^n f(a^m x, y + \ln a)$.

Here, a is an arbitrary positive parameter, and n and m are some constants.
   Solution (Polyanin & Manzhirov, 2007):
$$f(x, y) = e^{-ny} \Phi(xe^{-my}),$$
where $\Phi(z)$ is an arbitrary function.

35. $f(x, y) + f(y, z) = f(x, z)$.

Cantor's first equation.
   Solution (Math. Encyclopedia, 1985):
$$f(x, y) = \Phi(x) - \Phi(y),$$
where $\Phi(x)$ is an arbitrary function.

36. $f(x + y, z) + f(y + z, x) + f(z + x, y) = 0$.
   Solution:
$$f(x, y) = (x - 2y)\varphi(x + y),$$
where $\varphi(x)$ is an arbitrary function.

37. $f(xy, z) + f(yz, x) + f(zx, y) = 0$.
   Solution:
$$f(x, y) = \varphi(xy) \ln \frac{x}{y^2} \quad \text{if} \quad x > 0, \ y \neq 0;$$
$$f(x, y) = \varphi(xy) \ln \frac{-x}{y^2} \quad \text{if} \quad x < 0, \ y \neq 0;$$
$$f(x, y) = A + B \ln |x| \quad \text{if} \quad x \neq 0, \ y = 0;$$
$$f(x, y) = A + B \ln |y| \quad \text{if} \quad x = 0, \ y \neq 0,$$
where $\varphi(x)$ is an arbitrary function, and A and B are arbitrary constants.

9.4.2. Nonlinear Functional Equations

▶ **Equations involving one unknown function with a single argument.**

1. $f(x+y) = f(x)f(y)$.

Cauchy's exponential equation.
   Solution (Fichtenholz, 1969):
$$f(x) = e^{Cx},$$
where C is an arbitrary constant. In addition, the function $f(x) \equiv 0$ is also a solution.

2. $f(x+y) = af(x)f(y)$.

Solution:
$$f(x) = \frac{1}{a}e^{Cx},$$
where C is an arbitrary constant. In addition, the function $f(x) \equiv 0$ is also a solution.

3. $f\left(\frac{x+y}{2}\right) = \sqrt{f(x)f(y)}$.

Solution:
$$f(x) = Ca^x,$$
where a and C are arbitrary positive constants.

4. $f\left(\frac{x+y}{n}\right) = [f(x)f(y)]^{1/n}$.

Solution:
$$f(x) = a^x,$$
where a is an arbitrary positive constant.

5. $f(y+x) + f(y-x) = 2f(x)f(y)$.

D'Alembert's equation.
   Solutions (Fichtenholz, 1969):
$$f(x) = \cos(Cx), \quad f(x) = \cosh(Cx), \quad f(x) \equiv 0,$$
where C is an arbitrary constant.

6. $f(y+x) + f(y-x) = af(x)f(y)$.

Solutions:
$$f(x) = \frac{2}{a}\cos(Cx), \quad f(x) = \frac{2}{a}\cosh(Cx), \quad f(x) \equiv 0,$$
where C is an arbitrary constant.

7. $f(x+y) = a^{xy}f(x)f(y), \quad a > 0$.

Solution:
$$f(x) = e^{Cx}a^{x^2/2},$$
where C is an arbitrary constant. In addition, the function $f(x) \equiv 0$ is also a solution.

8. $f(x+y) = f(x) + f(y) - af(x)f(y), \quad a \neq 0.$

For $a = 1$, this equation occurs in probability theory.

Solution:
$$f(x) = \frac{1}{a}\left(1 - e^{-\beta x}\right),$$
where β is an arbitrary constant. In addition, the function $f(x) \equiv 0$ is also a solution.

9. $f(x+y)f(x-y) = f^2(x).$

Lobachevsky's equation.

Solution:
$$f(x) = C_1 \exp(C_2 x),$$
where C_1 and C_2 are arbitrary constants.

10. $f(x+y+a)f(x-y+a) = f^2(x) + f^2(y) - b^2.$

Solutions:
$$f(x) = \pm b, \quad f(x) = \pm b \cos\frac{n\pi x}{a},$$
where $n = 1, 2, \ldots$ For trigonometric solutions, a must be nonzero.

11. $f(x+y)f(x-y) = f^2(x) - f^2(y).$

Solution:
$$f(x) = C_1 \sin(C_2 x),$$
where C_1 and C_2 are arbitrary constants.

12. $f(x+y+a)f(x-y+a) = f^2(x) - f^2(y).$

Solutions:
$$f(x) = 0, \quad f(x) = C\cos\frac{\pi n x}{a}, \quad f(x) = C\sin\frac{\pi n x}{a},$$
where C is an arbitrary constant and $n = 1, 2, \ldots$ For trigonometric solutions, a must be nonzero.

13. $(x-y)f(x)f(y) = xf(x) - yf(y).$

Solutions:
$$f(x) \equiv 1, \quad f(x) = \frac{C}{x+C},$$
where C is an arbitrary constant.

14. $f(xy) = af(x)f(y).$

Cauchy's power equation (for $a = 1$).

Solution (Fichtenholz, 1969):
$$f(x) = \frac{1}{a}|x|^C,$$
where C is an arbitrary constant. In addition, the function $f(x) \equiv 0$ is also a solution.

15. $f(xy) = [f(x)]^y$.

Solution:
$$f(x) = e^{Cx},$$
where C is an arbitrary constant. In addition, the function $f(x) \equiv 0$ is also a solution, provided that $y > 0$.

16. $f(\sqrt{x^2 + y^2}) = f(x)f(y)$.

Gauss's equation.

Solution:
$$f(x) = \exp(Cx^2),$$
where C is an arbitrary constant. In addition, the function $f(x) \equiv 0$ is also a solution.

17. $\left(\dfrac{f^2(x) + f^2(y)}{2}\right)^{1/2} = f\left(\left(\dfrac{x^2 + y^2}{2}\right)^{1/2}\right)$.

Solution:
$$f(x) = (ax^2 + b)^{1/2},$$
where a and b are arbitrary positive constants.

18. $f\big((x^n + y^n)^{1/n}\big) = a f(x) f(y)$, $\quad n$ **is any number.**

Solution:
$$f(x) = \frac{1}{a}\exp(Cx^n),$$
where C is an arbitrary constant. In addition, the function $f(x) \equiv 0$ is also a solution.

19. $\left(\dfrac{f^n(x) + f^n(y)}{2}\right)^{1/n} = f\left(\left(\dfrac{x^n + y^n}{2}\right)^{1/n}\right)$, $\quad n$ **is any number.**

Solution:
$$f(x) = (ax^n + b)^{1/n},$$
where a and b are arbitrary positive constants.

20. $f\big(x + y\sqrt{f(x)}\big) + f\big(x - y\sqrt{f(x)}\big) = 2f(x)f(y)$.

Solutions:
$$f(x) \equiv 0, \quad f(x) = 1 + Cx^2,$$
where C is an arbitrary constant.

21. $f\big(g^{-1}(g(x) + g(y))\big) = a f(x) f(y)$.

Generalized Gauss equation. Here, $g(x)$ is an arbitrary monotonic function and $g^{-1}(x)$ is the inverse of $g(x)$.

Solution:
$$f(x) = \frac{1}{a}\exp[Cg(x)],$$
where C is an arbitrary constant. The function $f(x) \equiv 0$ is also a solution.

22. $M\big(f(x), f(y)\big) = f\big(M(x,y)\big).$

Here, $M(x,y) = \varphi^{-1}\!\left(\dfrac{\varphi(x) + \varphi(y)}{2}\right)$ is the quasi-arithmetic mean for a continuous strictly monotonic function φ, with φ^{-1} being the inverse of φ.

Solution:
$$f(x) = \varphi^{-1}\big(a\varphi(x) + b\big),$$
where a and b are arbitrary constants.

23. $\Phi\big(x, y, f(x), f(x+y), f(x-y)\big) = 0.$

1°. Taking $y = 0$ in this equation, we get
$$\Phi\big(x, 0, f(x), f(x), f(x)\big) = 0. \tag{1}$$
If the left-hand side of this relation does not vanish identically for all $f(x)$, then this equation can be solved for $f(x)$. The thus obtained function $f(x)$ should then be inserted into the original equation, and one should find conditions under which this function is its solution.

Now we assume that the left-hand side of (1) is identically zero for all $f(x)$.

2°. In the original equation, we consecutively take
$$x = 0, \quad y = t; \qquad x = t, \quad y = 2t; \qquad x = t, \quad y = -2t.$$
We obtain a system of algebraic (transcendental) equations
$$\begin{aligned}
\Phi\big(0, t, a, f(t), f(-t)\big) &= 0, \\
\Phi\big(t, 2t, f(t), f(3t), f(-t)\big) &= 0, \\
\Phi\big(t, -2t, f(t), f(-t), f(3t)\big) &= 0
\end{aligned} \tag{2}$$
for the unknown quantities $f(t)$, $f(-t)$, $f(3t)$, where $a = f(0)$. Solving system (2) for $f(t)$ [or for $f(-t)$ or $f(3x)$], we obtain an admissible solution that should be inserted into the original equation for verification.

3°. In some cases, the following trick can be used. In the original equation, one consecutively takes
$$x = 0, \quad y = t; \qquad x = t+a, \quad y = a; \qquad x = a, \quad y = t+a, \tag{3}$$
where a is a free parameter. We obtain the system of equations
$$\begin{aligned}
\Phi\big(0, t, f(0), f(t), f(-t)\big) &= 0, \\
\Phi\big(t+a, a, f(t+a), f(t+2a), f(t)\big) &= 0, \\
\Phi\big(a, t+a, f(a), f(t+2a), f(-t)\big) &= 0.
\end{aligned} \tag{4}$$
Letting $f(0) = C_1$ and $f(a) = C_2$, where C_1 and C_2 are arbitrary constants, we eliminate $f(t)$ and $f(t+2a)$ from system (4), which is assumed to be possible. As a result, we arrive at the reciprocal equation
$$\Psi\big(f(t+a), f(-t), t, a, C_1, C_2\big) = 0. \tag{5}$$
In order to find a solution of equation (5), we replace t by $-t - a$ to obtain
$$\Psi\big(f(-t), f(t+a), -t-a, a, C_1, C_2\big) = 0. \tag{6}$$
Further, eliminating $f(t+a)$ from equations (5) and (6), we arrive at the algebraic (transcendental) equation for $f(-t)$.

Remark 9.5. *For the sake of analysis, it is sometimes convenient to choose suitable values of the parameter a in order to simplify system (4).*

24. $\Phi(x, y, f(x), f(y), f(x+y), f(x-y)) = 0.$

Letting $y = 0$, we get
$$\Phi(x, 0, f(x), a, f(x), f(x)) = 0, \tag{1}$$

where $a = f(0)$. If the left-hand side of this equation does not vanish identically for all $f(x)$, then it can be solved for $f(x)$. Then, inserting an admissible solution $f(x)$ into the original equation, we find possible values of the parameter a (there are cases in which the equation has no solutions).

2°. If the left-hand side of (1) identically vanishes for all $f(x)$ and a, the following approach can be used. In the original equation, we consecutively take

$$x = 0, \quad y = t; \qquad x = t, \quad y = 2t; \qquad x = 2t, \quad y = t; \qquad x = t, \quad y = t.$$

We get
$$\begin{aligned} \Phi(0, t, a, f(t), f(t), f(-t)) &= 0, \\ \Phi(t, 2t, f(t), f(2t), f(3t), f(-t)) &= 0, \\ \Phi(2t, t, f(2t), f(t), f(3t), f(t)) &= 0, \\ \Phi(t, t, f(t), f(t), f(2t), a) &= 0, \end{aligned} \tag{2}$$

where $a = f(0)$. Eliminating $f(-t)$, $f(2t)$, and $f(3t)$ from system (2), we arrive at an algebraic (transcendental) equation for $f(t)$. The solution obtained in this manner should be inserted into the original equation for verification.

▶ **Equations involving several unknown functions of a single argument.**

25. $f(x)g(y) = h(x+y).$

Here, $f(x)$, $g(y)$, and $h(z)$ are the unknown functions.
    Solution:
$$f(x) = C_1 \exp(C_3 x), \quad g(y) = C_2 \exp(C_3 y), \quad h(z) = C_1 C_2 \exp(C_3 z),$$

where C_1, C_2, and C_3 are arbitrary constants.

26. $f(x)g(y) + h(y) = f(x+y).$

Here, $f(x)$, $g(y)$, and $h(z)$ are the unknown functions.
    Solutions:
$$\begin{aligned} f(x) &= C_1 x + C_2, & g(x) &= 1, & h(x) &= C_1 x & \text{(first solution)}; \\ f(x) &= C_1 e^{ax} + C_2, & g(x) &= e^{ax}, & h(x) &= C_2(1 - e^{ax}) & \text{(second solution)}, \end{aligned}$$

where a, C_1, and C_2 are arbitrary constants.

27. $f_1(x)g_1(y) + f_2(x)g_2(y) + f_3(x)g_3(y) = 0.$

Bilinear functional equation.

Two solutions:

$$f_1(x) = C_1 f_3(x), \quad f_2(x) = C_2 f_3(x), \quad g_3(y) = -C_1 g_1(y) - C_2 g_2(y);$$
$$g_1(y) = C_1 g_3(y), \quad g_2(y) = C_2 g_3(y), \quad f_3(x) = -C_1 f_1(x) - C_2 f_2(x),$$

where C_1 and C_2 are arbitrary constants; the functions on the right-hand sides of the solutions are prescribed arbitrarily.

28. $f_1(x)g_1(y) + f_2(x)g_2(y) + f_3(x)g_3(y) + f_4(x)g_4(y) = 0.$

Bilinear functional equation. Equations of this type often arise in the generalized separation of variables in partial differential equations.

1°. Solution:

$$f_1(x) = C_1 f_3(x) + C_2 f_4(x), \quad f_2(x) = C_3 f_3(x) + C_4 f_4(x),$$
$$g_3(y) = -C_1 g_1(y) - C_3 g_2(y), \quad g_4(y) = -C_2 g_1(y) - C_4 g_2(y).$$

It depends on four arbitrary constants $C_1, \ldots, C_4$. The functions on the right-hand sides are prescribed arbitrarily.

2°. The equation also has two other solutions:

$$f_1(x) = C_1 f_4(x), \quad f_2(x) = C_2 f_4(x), \quad f_3(x) = C_3 f_4(x),$$
$$g_4(y) = -C_1 g_1(y) - C_2 g_2(y) - C_3 g_3(y);$$
$$g_1(y) = C_1 g_4(y), \quad g_2(y) = C_2 g_4(y), \quad g_3(y) = C_3 g_4(y),$$
$$f_4(x) = -C_1 f_1(x) - C_2 f_2(x) - C_3 f_3(x).$$

They involve three arbitrary constants. The functions on the right-hand sides are prescribed arbitrarily.

29. $f_1(x)g_1(y) + f_2(x)g_2(y) + \cdots + f_k(x)g_k(y) = 0.$

Bilinear functional equation of the general form. Functional equations of this type often arise when constructing exact solutions to nonlinear partial differential equations by methods of generalized separation of variables. Depending on the goal, some functions f_i and g_i in the equation are prescribed, while the other are to be sought.

1°. *The splitting principle (splitting method).* All solutions of the bilinear functional equation can be represented as a set of linear combinations of the quantities $f_1, \ldots, f_n$ and a set of linear combinations of $g_1, \ldots, g_n$:

$$\sum_{i=1}^{k} \alpha_{ir} f_i = 0, \quad r = 1, \ldots, l; \tag{1}$$

$$\sum_{i=1}^{k} \beta_{is} g_i = 0, \quad s = 1, \ldots, m, \tag{2}$$

where $1 \leq l \leq k-1$ and $1 \leq m \leq k-1$. The constants α_{ir} and β_{is} in (1) and (2) are chosen so that the original bilinear equation holds identically (this can always be done).

The degenerate cases, when one or more functions f_k and/or g_k identically vanish, must be considered separately, using linear relations of the form (1)–(2).

2°. In practice, to obtain solutions to the bilinear functional equation, one should proceed as follows. First, one chooses a few first elements $f_1, \ldots, f_p$ ($p < k$) out of the entire set $f_1, \ldots, f_k$ and represents them as linear combinations of the remaining elements $f_{p+1}, \ldots, f_k$. This defines the first set of relations (1). On replacing $f_1, \ldots, f_p$ in the original functional equation by their linear combinations in terms of $f_{p+1}, \ldots, f_k$, one arrives at a relation of the form

$$\sum_{q=p+1}^{k} \Omega_q f_q = 0, \qquad \Omega_q = \sum_{j=1}^{k} c_{qj} g_j,$$

where c_{qj} are some constants. Setting the functional coefficients Ω_q ($q = p+1, \ldots, k$) to zero gives the second set of relations (2).

By setting different values of p, we find different solutions of the considered bilinear functional equation (in total, $k-1$ solutions can be obtained).

Remark 9.6. *Due to the symmetry of the original functional equation with respect to the f_i and g_i, one can start with choosing elements from the set $g_1, \ldots, g_k$ rather than $f_1, \ldots, f_k$.*

3°. To illustrate the procedure outlined in Item 2°, consider a bilinear functional equation containing five terms:

$$f_1 g_1 + f_2 g_2 + f_3 g_3 + f_4 g_4 + f_5 g_5 = 0, \tag{3}$$

where $f_i = f_i(x)$ and $g_i = g_i(y)$. Suppose that the first three functions, f_1, f_2, and f_3, are linear combinations of the last two:

$$f_1 = A_1 f_4 + B_1 f_5, \quad f_2 = A_2 f_4 + B_2 f_5, \quad f_3 = A_3 f_4 + B_3 f_5, \tag{4}$$

where A_1, A_2, A_3 and B_1, B_2, B_3 are arbitrary constants. Substituting (4) into (3) and collecting the terms as the coefficients of f_4 and f_5, we get

$$(A_1 g_1 + A_2 g_2 + A_3 g_3 + g_4) f_4 + (B_1 g_1 + B_2 g_2 + B_3 g_3 + g_5) f_5 = 0.$$

By setting the expressions in parentheses to zero, we obtain

$$\begin{aligned} g_4 &= -A_1 g_1 - A_2 g_2 - A_3 g_3, \\ g_5 &= -B_1 g_1 - B_2 g_2 - B_3 g_3. \end{aligned} \tag{5}$$

Formulas (4) and (5) give one of the solutions to functional equation (3); the functions on the right-hand sides are considered arbitrary. The other solutions can be found likewise.

4°. For details on the use of the splitting principle for solving bilinear functional equations and related functional differential equations (arising from the construction of exact solutions to nonlinear PDEs), see the books by Polyanin & Zaitsev (2012) and Polyanin & Zhurov (2022).

30. $f(x) + g(y) = Q(z)$, where $z = \varphi(x) + \psi(y)$.

Here, one of the two functions $f(x)$ and $\varphi(x)$ is prescribed and the other is assumed unknown, one of the functions $g(y)$ and $\psi(y)$ is prescribed and the other is unknown, and the function $Q(z)$ is assumed unknown. (In similar equations with a composite argument, it is assumed that $\varphi(x) \not\equiv \text{const}$ and $\psi(y) \not\equiv \text{const}$.)

Solution:
$$f(x) = A\varphi(x) + B,$$
$$g(y) = A\psi(y) - B + C,$$
$$Q(z) = Az + C,$$

where A, B, and C are arbitrary constants.

31. $f(x)g(y) = Q(z)$, where $z = \varphi(x) + \psi(y)$.

Here, one of the two functions $f(x)$ and $\varphi(x)$ is prescribed and the other is assumed unknown, one of the functions $g(y)$ and $\psi(y)$ is prescribed and the other is unknown, and the function $Q(z)$ is assumed unknown. (In similar equations with a composite argument, it is assumed that $\varphi(x) \not\equiv \text{const}$ and $\psi(y) \not\equiv \text{const}$.)
Solution:
$$f(x) = ABe^{\lambda \varphi(x)}, \quad g(y) = \frac{A}{B}e^{\lambda \psi(y)}, \quad Q(z) = Ae^{\lambda z},$$

where A, B, and λ are arbitrary constants.

32. $f(x) + g(y) = Q(z)$, where $z = \varphi(x)\psi(y)$.

Here, one of the two functions $f(x)$ and $\varphi(x)$ is prescribed and the other is assumed unknown, one of the functions $g(y)$ and $\psi(y)$ is prescribed and the other is unknown, and the function $Q(z)$ is assumed unknown. (In similar equations with a composite argument, it is assumed that $\varphi(x) \not\equiv \text{const}$ and $\psi(y) \not\equiv \text{const}$.)
Solution:
$$f(x) = A \ln \varphi(x) + B,$$
$$g(y) = A \ln \psi(y) - B + C,$$
$$Q(z) = A \ln z + C,$$

where A, B, and C are arbitrary constants.

33. $f(y) + g(x) + h(x)Q(z) + R(z) = 0$, where $z = \varphi(x) + \psi(y)$.

Equations of this type arise in the functional separation of variables in nonlinear partial differential equations; see Polyanin & Zaitsev (2012) and Polyanin & Zhurov (2022) for details.

1°. Solution:
$$f = -\tfrac{1}{2}A_1 A_4 \psi^2 + (A_1 B_1 + A_2 + A_4 B_3)\psi - B_2 - B_1 B_3 - B_4,$$
$$g = \tfrac{1}{2}A_1 A_4 \varphi^2 + (A_1 B_1 + A_2)\varphi + B_2,$$
$$h = A_4 \varphi + B_1,$$
$$Q = -A_1 z + B_3,$$
$$R = \tfrac{1}{2}A_1 A_4 z^2 - (A_2 + A_4 B_3)z + B_4,$$

where A_k and B_k are arbitrary constants, and $\varphi = \varphi(x)$ and $\psi = \psi(y)$ are arbitrary functions.

2°. Solution:
$$f = -B_1 B_3 e^{-A_3\psi} + \left(A_2 - \frac{A_1 A_4}{A_3}\right)\psi - B_2 - B_4 - \frac{A_1 A_4}{A_3^2},$$
$$g = \frac{A_1 B_1}{A_3} e^{A_3\varphi} + \left(A_2 - \frac{A_1 A_4}{A_3}\right)\varphi + B_2,$$
$$h = B_1 e^{A_3\varphi} - \frac{A_4}{A_3},$$
$$Q = B_3 e^{-A_3 z} - \frac{A_1}{A_3},$$
$$R = \frac{A_4 B_3}{A_3} e^{-A_3 z} + \left(\frac{A_1 A_4}{A_3} - A_2\right)z + B_4,$$

where A_k and B_k are arbitrary constants, and $\varphi = \varphi(x)$ and $\psi = \psi(y)$ are arbitrary functions.

3°. In addition, the functional equation has two degenerate solutions:
$$f = A_1\psi + B_1, \quad g = A_1\varphi + B_2, \quad h = A_2, \quad R = -A_1 z - A_2 Q - B_1 - B_2,$$
where $\varphi = \varphi(x)$, $\psi = \psi(y)$, and $Q = Q(z)$ are arbitrary functions; A_1, A_2, B_1, and B_2 are arbitrary constants; and
$$f = A_1\psi + B_1, \quad g = A_1\varphi + A_2 h + B_2, \quad Q = -A_2, \quad R = -A_1 z - B_1 - B_2,$$
where $\varphi = \varphi(x)$, $\psi = \psi(y)$, and $h = h(x)$ are arbitrary functions; A_1, A_2, B_1, and B_2 are arbitrary constants.

34. $f(y) + g(x)Q(z) + h(x)R(z) = 0$, where $z = \varphi(x) + \psi(y)$.

Equations of this type arise in the functional separation of variables in nonlinear partial differential equations; see Polyanin & Zaitsev (2012) and Polyanin & Zhurov (2022) for details.

1°. Solution:
$$\begin{aligned} g(x) &= A_2 B_1 e^{k_1\varphi} + A_2 B_2 e^{k_2\varphi}, \\ h(x) &= (k_1 - A_1)B_1 e^{k_1\varphi} + (k_2 - A_1)B_2 e^{k_2\varphi}, \\ Q(z) &= A_3 B_3 e^{-k_1 z} + A_3 B_4 e^{-k_2 z}, \\ R(z) &= (k_1 - A_1)B_3 e^{-k_1 z} + (k_2 - A_1)B_4 e^{-k_2 z}, \end{aligned} \quad (1)$$

where $B_1, \ldots, B_4$ are arbitrary constants and k_1 and k_2 are roots of the quadratic equation
$$(k - A_1)(k - A_4) - A_2 A_3 = 0.$$

In the degenerate case $k_1 = k_2$, the terms $e^{k_2\varphi}$ and $e^{-k_2 z}$ in (1) must be replaced by $\varphi e^{k_1\varphi}$ and $z e^{-k_1 z}$, respectively. In the case of purely imaginary or complex roots, one should extract the real (or imaginary) part of the roots in solution (1).

The function $f(y)$ is determined by the formulas
$$\begin{aligned} B_2 = B_4 = 0 &\implies f(y) = [A_2 A_3 + (k_1 - A_1)^2] B_1 B_3 e^{-k_1\psi}, \\ B_1 = B_3 = 0 &\implies f(y) = [A_2 A_3 + (k_2 - A_1)^2] B_2 B_4 e^{-k_2\psi}, \\ A_1 = 0 &\implies f(y) = (A_2 A_3 + k_1^2) B_1 B_3 e^{-k_1\psi} + (A_2 A_3 + k_2^2) B_2 B_4 e^{-k_2\psi}. \end{aligned} \quad (2)$$

Solutions defined by (1) and (2) involve arbitrary functions $\varphi = \varphi(x)$ and $\psi = \psi(y)$.

2°. In addition, the functional equation has two degenerate solutions,

(i) $f = B_1 B_2 e^{A_1 \psi}$, $\quad g = A_2 B_1 e^{-A_1 \varphi}$, $\quad h = B_1 e^{-A_1 \varphi}$, $\quad R = -B_2 e^{A_1 z} - A_2 Q$,

where $\varphi = \varphi(x)$, $\psi = \psi(y)$, and $Q = Q(z)$ are arbitrary functions, and A_1, A_2, B_1, and B_2 are arbitrary constants; and

(ii) $f = B_1 B_2 e^{A_1 \psi}$, $\quad h = -B_1 e^{-A_1 \varphi} - A_2 g$, $\quad Q = A_2 B_2 e^{A_1 z}$, $\quad R = B_2 e^{A_1 z}$,

where $\varphi = \varphi(x)$, $\psi = \psi(y)$, and $g = g(x)$ are arbitrary functions, and A_1, A_2, B_1, and B_2 are arbitrary constants.

▶ **Equations involving functions of two arguments.**

35. $f(x, y) f(y, z) = f(x, z)$.

Cantor's second equation.
Solution (Math. Encyclopedia, 1985):
$$f(x, y) = \Phi(y)/\Phi(x),$$
where $\Phi(x)$ is an arbitrary function.

36. $f(x, y) f(u, v) - f(x, u) f(y, v) + f(x, v) f(y, u) = 0$.

Solution:
$$f(x, y) = \varphi(x) \psi(y) - \varphi(y) \psi(x),$$
where $\varphi(x)$ and $\psi(x)$ are arbitrary functions.

37. $f(f(x, y), z) = f(f(x, z), f(y, z))$.

Skew self-distributivity equation.
Solution:
$$f(x, y) = g^{-1}(g(x) + g(y)),$$
where $g(x)$ is an arbitrary continuous strictly increasing function.

▶ Most of the solutions of the difference and functional equations given in this chapter were taken from the handbook by Polyanin & Manzhirov (2007). As a rule, this book has not been cited in the text. For additional information on other difference and functional equations and their solutions, solution methods, and properties of functional equations, see the books of Aczél (2002, 2006), Aczél & Dhombres (1989), Kuczma (1968, 1985), Kuczma et al. (1990), and Polyanin & Manzhirov (2007), as well as other the articles and books given in the list of references below.

References

Aczél, J., *Functional Equations: History, Applications and Theory*, Kluwer Academic, Dordrecht, 2002.

Aczél, J., *Lectures on Functional Equations and Their Applications*, Dover Publications, New York, 2006.

Aczél, J., Some general methods in the theory of functional equations with a single variable. New applications of functional equations [in Russian], *Uspekhi Mat. Nauk*, Vol. 11, No. 3 (69), pp. 3–68, 1956.

Aczél, J. and Dhombres, J., *Functional Equations in Several Variables*, Cambridge Univ. Press, Cambridge, 1989.

Agarwal, R.P., *Difference Equations and Inequalities, 2nd ed.*, Marcel Dekker, New York, 2000.

Belitskii, G.R. and Tkachenko, V., *One-Dimensional Functional Equations*, Birkhäuser Verlag, Boston, 2003.

Castillo, E. and Ruiz-Cobo, R., *Functional Equations in Applied Sciences*, Elsevier, New York, 2005.

Czerwik, S., *Functional Equations and Inequalities in Several Variables*, World Scientific Publishing Co., Singapore, 2002.

Daroczy, Z. and Pales, Z. (eds.), *Functional Equations—Results and Advances*, Kluwer Academic, Dordrecht, 2002.

Doetsch, G., *Guide to the Applications of the Laplace and Z-transforms* [in Russian, translation from German], Nauka Publishers, Moscow, pp. 213, 215, 218, 1974.

Dorodnitsyn, V., *Applications of Lie Groups to Difference Equations*, Chapman & Hall/CRC, Boca Raton–London, 2010.

Efthimiou, C., *Introduction to Functional Equations: Theory and Problem-Solving Strategies for Mathematical Competitions and Beyond*, American Mathematical Society, Providence, 2011.

Elaydi, S., *An Introduction to Difference Equations, 3rd ed.*, Springer-Verlag, New York, 2005.

Fichtenholz, G.M., *A Course of Differential and Integral Calculus, Vol. 1* [in Russian], Nauka Publishers, Moscow, pp. 157–160, 1969.

Goldberg, S., *Introduction to Difference Equations*, Dover Publications, New York, 1986.

Kelley, W. and Peterson, A., *Difference Equations. An Introduction with Applications, 2nd Edition*, Academic Press, New York, 2000.

Kuczma, M., *An Introduction to the Theory of Functional Equations and Inequalities*, Polish Scientific Publishers, Warsaw, 1985.

Kuczma, M., *Functional Equations in a Single Variable*, Polish Scientific Publishers, Warsaw, 1968.

Kuczma, M., Choczewski, B., and Ger, R., *Iterative Functional Equations*, Cambridge Univ. Press, Cambridge, 1990.

Laitochová, J., Group iteration for Abel's functional equation, *Nonlinear Analysis: Hybrid Systems*, Vol. 1, No. 1, pp. 95–102, 2007.

Mathematical Encyclopedia, Vol. 2 (editor-in-chief Vinogradov, I.M.) [in Russian], Sovetskaya Entsikolpediya, Moscow, pp. 1029–1031, 1979.

Mathematical Encyclopedia, Vol. 5 (editor-in-chief Vinogradov, I.M.) [in Russian], Sovetskaya Entsikolpediya, Moscow, pp. 699–705, 1985.

Mickens, R., *Difference Equations, 2nd Edition*, CRC Press, Boca Raton, 1991.

Mirolyubov, A.A. and Soldatov, M.A., *Linear Homogeneous Difference Equations* [in Russian], Nauka Publishers, Moscow, 1981.

Nechepurenko, M.I., *Iterations of Real Functions and Functional Equations* [in Russian], Institute of Computational Mathematics and Mathematical Geophysics, Novosibirsk, 1997.

Pelyukh, G.P., Sharkovsky, A.N., *Introduction to the Theory of Functional Equations* [in Russian], Naukova Dumka, Kiev, 1974.

Polyanin, A.D. and Manzhirov, A.V., *Handbook of Integral Equations: Exact Solutions (Supplement. Some Functional Equations)* [in Russian], Faktorial, Moscow, 1998.

Polyanin, A.D. and Manzhirov, A.V., *Handbook of Mathematics for Engineers and Scientists (Chapters 17 and T12)*, Chapman & Hall/CRC Press, Boca Raton–London, 2007.

Polyanin, A.D. and Zaitsev, V.F., *Handbook of Nonlinear Partial Differential Equations, 2nd ed. (Sections 29.5 and 30.4)*, CRC Press, Boca Raton–London, 2012.

Polyanin, A.D. and Zhurov, A.I., *Separation of Variables and Exact Solutions to Nonlinear PDEs*, CRC Press, Boca Raton–London, 2022.

Rassias, T.M., *Functional Equations and Inequalities*, Kluwer Academic, Dordrecht, 2000.

Rassias, J.M., Thandapani, E., Ravi, K., and Kumar, B.V.S., *Functional Equations and Inequalities: Solutions and Stability Results*, World Scientific, Singapore, 2017.

Sedaghat, H., *Nonlinear Difference Equations Theory with Applications to Social Science Models*, Kluwer Academic, Dordrecht, 2003.

Sharkovsky, A. N., Maistrenko, Yu. L., and Romanenko, E. Yu., *Difference Equations and Their Applications*, Kluwer Academic, Dordrecht, 1993.

Small, C.G., *Functional Equations and How to Solve Them*, Springer-Verlag, Berlin, 2007.

Smital, J., *On Functions and Functional Equations*, Taylor & Francis, New York, 1988.

Chapter 10

Ordinary Functional Differential Equations

▶ **Preliminary remarks.** *Ordinary functional differential equations* (*functional ODEs* for short) are mathematical equations containing an unknown function of one variable with different arguments, as well as derivatives of this function.

Ordinary differential equations with constant delay are the simplest functional ODEs containing the quantities $u(t)$ and $u(t-\tau)$ and their derivatives, where t is an independent variable, $u(t)$ is the unknown function, $\tau > 0$ is the delay time. It is also assumed that in these equations, the order of the highest derivative of the function $u(t)$ is greater than that of the highest derivative of the function $u(t-\tau)$. In Cauchy-type problems for ODEs with constant delay, the initial data are given in the interval $t_0 - \tau \le t \le t_0$, where t_0 is some constant.

Exact solutions of ordinary functional differential equations are understood as follows:

- solutions that are expressed in terms of elementary functions, functions included in the equation (this is necessary when the equation depends on arbitrary functions), and indefinite or definite integrals,

- solutions that are expressed in terms of solutions of ordinary differential equations or systems of such equations.

Note that one functional differential equation may contain several unknown functions.

This chapter describes exact solutions of linear and nonlinear ordinary differential equations with delay and some other functional ODEs.

10.1. First-Order Linear Ordinary Functional Differential Equations

10.1.1. ODEs with Constant Delays

1. $u'_t = w$, $\quad w = u(t-\tau)$; $\quad u = a \quad$ **for** $\quad t_0 - \tau \le t \le t_0$.

This is a special case of equation 10.1.1.2.

A solution to the Cauchy-type problem for the delay linear ODE in question subject to the initial condition

$$u = a \quad \text{for} \quad t_0 - \tau \le t \le t_0$$

can be represented as (Bellman & Cooke, 1963):

$$u = a\exp_d(t - t_0, \tau), \qquad \exp_d(t, \tau) \equiv \sum_{k=0}^{[t/\tau]+1} \frac{[t - (k-1)\tau]^k}{k!},$$

where $\exp_d(t, \tau)$ is the delayed exponential function, and the symbol $[A]$ denotes the integer part of the number A.

2. $u'_t = au + bw, \quad w = u(t - \tau).$

First-order linear constant-coefficient homogeneous ODE with constant delay.

1°. This equation has exponential solutions

$$u = \exp(\lambda t), \tag{1}$$

where λ are roots of the transcendental characteristic equation

$$\lambda - a - be^{-\lambda\tau} = 0. \tag{2}$$

2°. The constant λ in the exponential solution (1) of the characteristic equation (2) can be represented as

$$\lambda = a + \frac{1}{\tau}W(x), \quad x = b\tau e^{-a\tau}. \tag{3}$$

where $W = W(z)$ is the Lambert W function (it is multi-valued). For complex $z = x + iy$, this function is defined implicitly by the transcendental equation

$$We^W = z. \tag{4}$$

For properties of the Lambert W function and its applications, see, for example, Wright (1959), Corless et al. (1996), Valluri et al. (2000), and Yi et al. (2010).

For real $z = x$, the function $W(x)$ is single-valued at $x \geq 0$ and double-valued on the interval $(-1/e, 0)$. For $x \geq -1/e$ and $W \geq -1$, the single-valued branch of the Lambert function, which is commonly called the principal branch, will be denoted by $W_p(x)$. The second branch of this function, which is characterized by the inequalities $-1/e \leq x < 0$ and $W \leq -1$, is denoted $W_n(x)$. In parametric form, the real branches $W_p(x)$ and $W_n(x)$ are defined as follows (s is the parameter):

$$x = se^s, \quad W_p = s, \quad -1 \leq s < +\infty;$$
$$x = se^s, \quad W_n = s, \quad -\infty < s \leq -1.$$

The Taylor series expansion is valid, which converges at $|x| < 1/e$:

$$W_p(x) = \sum_{n=1}^{\infty} (-1)^{n-1} \frac{n^{n-1}}{n!} x^n = x - x^2 + \frac{3}{2}x^3 - \frac{8}{3}x^4 + \frac{125}{4}x^5 - \cdots$$

There are asymptotic formulas:

$$W_p(x) = \zeta_1 - \ln\zeta_1 + \frac{\ln\zeta_1}{\zeta_1} + \frac{\ln^2\zeta_1}{2\zeta_1^2} - \frac{\ln\zeta_1}{\zeta_1^2} + O\left(\frac{\ln^3\zeta_1}{\zeta_1^3}\right) \quad \text{as} \quad x \to +\infty,$$

$$W_n(x) = \zeta_2 - \ln\zeta_2 - \frac{\ln\zeta_2}{\zeta_2} - \frac{\ln^2\zeta_2}{2\zeta_2^2} - \frac{\ln\zeta_2}{\zeta_2^2} + O\left(\frac{\ln^3\zeta_2}{\zeta_2^3}\right) \quad \text{as} \quad x \to -0,$$

where $\zeta_1 = \ln x$ and $\zeta_2 = \ln(-1/x)$.

TABLE 10.1. The number of real roots of the characteristic of equation (2) at different values of the determining parameters of the delay ODE 10.1.1.2.

Determining conditions	Number of real roots	Range of roots
$-e^{a\tau-1}\tau^{-1} < b < 0$	Two roots, λ_1 and λ_2	$a - \tau^{-1} < \lambda_1 < a$, $\lambda_2 < a - \tau^{-1}$
$b \geq 0$	One root, λ_1	$\lambda_1 > a$ if $b > 0$, $\lambda_1 = a$ if $b = 0$
$b = -e^{a\tau-1}\tau^{-1}$	One root, λ_1 (double)	$\lambda_1 = a - \tau^{-1}$
$b < -e^{a\tau-1}\tau^{-1}$	No roots	

Using the properties of the Lambert W function (4) and formula (3), one can find the conditions under which the characteristic equation (2) has real roots. The results are summarized in Table 10.1.

In the complex plane $z = x + iy$ ($i^2 = -1$), the function $W(z)$ has an infinite number of branches: $W_m = W_m(z)$ ($m = 0, \pm 1, \pm 2, \ldots$). In this case, the asymptotic formula holds

$$W_m = \ln z - \ln \ln z + 2\pi i m + (1+i)o(1) \quad \text{as} \quad z \to \infty.$$

3°. The *Hayes theorem*: all roots of the characteristic equation (2) with real coefficients a and b (and $\tau > 0$) have negative real parts (Re $\lambda < 0$) if and only if the following three inequalities hold simultaneously:

(i) $a\tau < 1$,
(ii) $a + b < 0$,
(iii) $b\tau + \sqrt{(a\tau)^2 + \mu^2} > 0$,

where μ is the root of the transcendental equation $\mu = a\tau \tan \mu$ that satisfies the condition $0 < \mu < \pi$. For $a = 0$, one should put $\mu = \pi/2$.

4°. Let

$$b = k(-1)^{n+1} e^{a\tau}, \quad k = \frac{(2n+1)\pi}{2\tau}, \quad n = 0, \pm 1, \pm 2, \ldots$$

Then the linear delay ODE in question has solutions of the form

$$u = e^{at}[C_1 \sin(kt) + C_2 \cos(kt)],$$

where C_1 and C_2 are arbitrary constants. For $a = 0$, these solutions are periodic.

5°. The Cauchy-type problem for the linear homogeneous delay ODE in question subject to the initial condition

$$u = \varphi(t) \quad \text{for} \quad -\tau \leq t \leq 0,$$

can be solved in a closed form (Azizbekov & Khusainov, 2012):

$$u(t) = e^{a(t+\tau)} \exp_d(\sigma t, \sigma \tau) \, \varphi(-\tau)$$
$$+ \int_{-\tau}^0 e^{a(t-s)} \exp_d(\sigma(t - \tau - s), \sigma \tau) [\varphi'_s(s) - a\varphi(s)] \, ds, \quad \sigma = e^{-a\tau} b,$$

where $\exp_d(t, \tau)$ is the delayed exponential function (see equation 10.1.1.1).

An alternative representation of the solution to the same Cauchy-type problem (Elsgol'ts & Norkin, 1973):

$$u(t) = \sum_{k=0}^{\infty} \frac{e^{\lambda_k t}}{1 + \tau b e^{-\lambda_k \tau}} \left[\varphi(0) + b \int_{-\tau}^{0} \varphi(t) e^{-\lambda_k (t+\tau)} dt \right].$$

where λ_k are roots of the transcendental equation (2).

6°. References for the delay ODE 10.1.1.2: Bellman & Cooke (1963), Elsgol'ts & Norkin (1973), Azizbekov & Khusainov (2012), Asl & Ulsoy (2003), and Polyanin et al. (2023).

3. $u'_t = au + bw + c, \quad w = u(t - \tau).$

First-order linear constant-coefficient nonhomogeneous ODE with constant delay (special case).

1°. For $b \neq -a$, the substitution $u = v - \frac{c}{a+b}$ leads to a linear homogeneous delay ODE of the form 10.1.1.2.

2°. For $b = -a$, a homogeneous linear delay ODE can be obtained using the substitution $u = v + kt$, where $k = \frac{c}{1-a\tau}$.

4. $u'_t = au + bw + f(t), \quad w = u(t - \tau).$

First-order linear constant-coefficient ODE with constant delay.

The Cauchy problem for the linear nonhomogeneous delay ODE in question subject to the homogeneous initial condition

$$u = 0 \quad \text{for} \quad -\tau \leq t \leq 0$$

can be solved in terms of the delayed exponential function (Azizbekov & Khusainov, 2012):

$$u(t) = \int_0^t e^{a(t-s)} \exp_d(\sigma(t-s), \sigma\tau) f(s) \, ds, \quad \sigma = e^{-a\tau} b.$$

5. $u'_t = a(t)u + b(t)w + c(t), \quad w = u(t - \tau).$

This is a special case of equation 10.2.1.8 for $f(t, w) = a(t)$ and $g(t, w) = b(t)w + c(t)$.

10.1.2. Pantograph-Type ODEs with Proportional Arguments

1. $u'_t = w, \quad w = u(pt).$

The solution satisfying the initial condition $u(0) = k$ can be expressed in the form of a power series (Kato & McLeod, 1971 and Iserles, 1993):

$$u(t) = k \exp_s(t, p), \quad \exp_s(t, p) \equiv \sum_{n=0}^{\infty} p^{\frac{n(n-1)}{2}} \frac{t^n}{n!},$$

where $\exp_s(t, p)$ is the *stretched exponential function*.

2. $u'_t = au + bw, \quad w = u(pt).$

This equation with $p > 0$ and $p \neq 1$ is known as the *pantograph equation*.

General solution in the form of a power series (Fox, 1971):
$$u(t) = k\left(1 + \sum_{n=1}^{\infty} s_n t^n\right), \qquad s_n = \frac{1}{n!} \prod_{m=0}^{n-1}(a + bp^m),$$

where k is an arbitrary constant. The above solution satisfies the initial condition $u(0) = k$. For $0 < p < 1$, the power series has an infinite radius of convergence. In this case, for $a \geq 0$ and $b \geq 0$, the inequalities $a^n < s_n < (a+b)^n$ and $e^{ax} < u < e^{(a+b)x}$ hold.

3. $u'_t = au + bw + c, \quad w = u(pt).$

This is a *nonhomogeneous pantograph equation*.

1°. For $b \neq -a$, the substitution $u = v - \frac{c}{a+b}$ leads to the pantograph equation 10.1.2.2.

2°. A particular solution in the form of a power series:
$$u_*(t) = c\left(t + \sum_{n=2}^{\infty} s_n t^n\right), \qquad s_n = \frac{1}{n!}\prod_{m=1}^{n-1}(a + bp^m).$$

3°. The general solution is the sum of the particular solution from Item 2° and the general solution of the homogeneous equation 10.1.2.2.

4. $u'_t = au + bw + (\lambda - a)e^{\lambda t} - be^{\lambda pt}, \quad w = u(pt); \quad u(0) = 1.$

Solution: $u(t) = e^{\lambda t}$. The coefficients a and b can depend arbitrarily on time t.

5. $u'_t = au + bw + \cosh t - a\sinh t - b\sinh(pt), \quad w = u(pt); \quad u(0) = 0.$

Solution: $u(t) = \sinh t$. The coefficients a and b can depend arbitrarily on time t.

6. $u'_t = au + bw + \cos t - a\sin t - b\sin(pt), \quad w = u(pt); \quad u(0) = 0.$

Solution (Brunner et al., 2010): $u(t) = \sin t$. The coefficients a and b can depend arbitrarily on time t.

7. $u'_t = au + bw - \sin t - a\cos t - b\cos(pt), \quad w = u(pt); \quad u(0) = 1.$

Solution: $u(t) = \cos t$. The coefficients a and b can depend arbitrarily on time t.

8. $u'_t = au + be^{\lambda t}w, \quad w = u(pt); \quad u(0) = c.$

Let $\lambda = (a+b)(1-p)$. Then the solution has the form: $u(t) = ce^{(a+b)t}$.

9. $u'_t = au + bw_1 + cw_2, \quad w_1 = u(pt), \quad w_2 = u(qt); \quad u(0) = k.$

Solution in the form of a power series (Patade & Bhalekar, 2017):
$$u = k\left(1 + \sum_{n=1}^{\infty} s_n t^n\right), \qquad s_n = \frac{1}{n!}\prod_{m=0}^{n-1}(a + bp^m + cq^m).$$

For $0 < p < 1$ and $0 < q < 1$, the power series has an infinite radius of convergence. In this case, for $a, b, c \geq 0$, the inequalities $a^n < s_n < (a+b+c)^n$ and $e^{ax} < u < e^{(a+b+c)x}$ hold.

10.1.3. Other Ordinary Functional Differential Equations

1. $u'(t) = au(-t)$.

Babbage's functional differential equation. Denote $u_1 = u(t)$ and $u_2 = u(-t)$. In addition to the original equation, we also consider another equation obtained from the original one by replacing the independent variable t with $-t$. As a result, we obtain the system of linear ODEs

$$u_1' = au_2, \quad u_2' = -au_1.$$

Eliminating u_2 from this system, we arrive at a second-order linear ODE, $u_1'' = -a^2 u_1$, the general solution of which has the form $u_1 = C_1 \cos(at) + C_2 \sin(at)$, where C_1 and C_2 are arbitrary constants. Substituting the resulting function $u = u_1$ into the original functional ODE, we find its solution:

$$u = C[\cos(at) + \sin(at)] = C\sqrt{2} \sin\left(at + \tfrac{\pi}{4}\right),$$

where C is an arbitrary constant.

2. $u'(t) = au(b - t)$.

This is a generalization of the preceding equation.

Denote $u_1 = u(t)$ and $u_2 = u(b - t)$. In addition to the original equation, we also consider another equation obtained from the original one by replacing the independent variable t with $b - t$. As a result, we obtain the system of linear ODEs

$$u_1' = au_2, \quad u_2' = -au_1.$$

Integrating this system and substituting the resulting solution for the function $u = u_1$ into the original functional ODE, we find its solution:

$$u = C \cos(ab) \cos(at) + C[1 + \sin(ab)] \sin(at),$$

where C is an arbitrary constant.

3. $u'(t) = f(t)u(t) + g(t)u(b - t) + h(t)$.

This is a special case of the functional ODE 10.2.3.6 for $F(t, u(t), u(t - b)) = f(t)u(t) + g(t)u(b - t) + h(t)$.

4. $u'(t) = au(b/t)$.

Denote $u_1 = u(t)$ and $u_2 = u(b/t)$. In addition to the original equation, we also consider another equation obtained from the original one by replacing the independent variable t with b/t. As a result, we obtain the system of linear ODEs

$$u_1' = au_2, \quad u_2' = -abt^{-2} u_1.$$

Eliminating u_2, we arrive at a second-order linear ODE, $u_1'' = -a^2 bt^{-2} u_1$, whose general solution has the form $u_1 = C_1 t^{k_1} + C_2 t^{k_2}$, where C_1 and C_2 are arbitrary constants, and k_1 and k_2 are the roots of the quadratic equation $k^2 - k + a^2 b = 0$. Substituting the resulting function $u = u_1$ into the original functional ODE, we find its solution:

$$u = C(ab^{k_2} t^{k_1} + k_1 t^{k_2}) = Cab^{\frac{1}{2}+\lambda} t^{\frac{1}{2}-\lambda} + C(\tfrac{1}{2} - \lambda) t^{\frac{1}{2}+\lambda},$$

$$k_1 = \tfrac{1}{2} - \lambda, \quad k_2 = \tfrac{1}{2} + \lambda, \quad \lambda = \sqrt{\tfrac{1}{4} - a^2 b}.$$

where C is an arbitrary constant. This solution is valid for $\tfrac{1}{4} - a^2 b \geq 0$. If $\tfrac{1}{4} - a^2 b < 0$, then the solution is determined by the real part of u above.

5. $u'(t) = f(t)u(t) + g(t)u(b/t) + h(t)$.

This is a special case of the functional ODE 10.2.3.7 for $F(t, u(t), u(t/b)) = f(t)u(t) + g(t)u(b/t) + h(t)$.

6. $[u(\sin t)]'_t = au(\cos t)$.

Denote $u_1 = u(\sin t)$ and $u_2 = u(\cos t)$. In addition to the original equation, we also consider another equation obtained from the original one by replacing the independent variable t with $\frac{\pi}{2} - t$. As a result, we obtain the system of linear ODEs

$$u'_1 = au_2, \quad u'_2 = -au_1,$$

Eliminating u_2, we arrive at a second-order linear ODE, $u''_1 = -a^2 u_1$, whose general solution has the form $u_1 = u(\sin t) = C_1 \cos(at) + C_2 \sin(at)$, where C_1 and C_2 are arbitrary constants. Substituting the resulting function $u(\sin t)$ into the original functional ODE, we find its solution in implicit form

$$u(\sin t) = C[\cos\sigma \, \cos(at) + (1+\sin\sigma)\sin(at)] = C[\cos(at-\sigma) + \sin(at)], \quad \sigma = \tfrac{a\pi}{2},$$

where C is an arbitrary constant.

7. $[u(\cos t)]'_t = au(\sin t)$.

The substitution $t = \frac{\pi}{2} - x$ leads to an equation of the form 10.1.3.6:

$$[u(\sin x)]'_x = -au(\cos x).$$

10.2. First-Order Nonlinear Ordinary Functional Differential Equations

10.2.1. ODEs with Constant Delays

1. $u'_t = au + bu^{1/2} + cu^{1/2}w^{1/2}, \quad w = u(t-\tau)$.

The substitution $u = v^2$ leads to a constant coefficient linear ODE with constant delay of the form 10.1.1.3: $v'_t = \tfrac{1}{2}av + \tfrac{1}{2}c\bar{v} + \tfrac{1}{2}b$, where $\bar{v} = v(t-\tau)$.

2. $u'_t = au + bu^{1-k} + cu^{1-k}w^k, \quad w = u(t-\tau)$.

Generalization of the preceding equation. The substitution $u = v^{1/k}$ leads to a constant coefficient linear ODE with constant delay of the form 10.1.1.3: $v'_t = akv + ck\bar{v} + bk$, where $\bar{v} = v(t-\tau)$.

3. $u'_t = a + be^{\lambda u} + ce^{\lambda(u-w)}, \quad w = u(t-\tau)$.

The substitution $v = e^{-\lambda u}$ leads to a constant coefficient linear ODE with constant delay of the form 10.1.1.3: $v'_t = -a\lambda v - c\lambda\bar{v} - b\lambda$, where $\bar{v} = v(t-\tau)$.

4. $u'_t = au \ln u + bu \ln w + cu, \quad w = u(t-\tau)$.

The substitution $u = e^v$ leads to a constant coefficient linear ODE with constant delay of the form 10.1.1.3: $v'_t = av + b\bar{v} + c$, where $\bar{v} = v(t-\tau)$.

Remark 10.1. *In the nonlinear delay ODEs 10.2.1.1–10.2.1.4, the free parameters a, b, and c may be replaced by arbitrary functions $a(t)$, $b(t)$, and $c(t)$. In this case, the same substitutions lead to linear delay ODEs (Polyanin et al., 2023).*

5. $u'_t = f(u - w), \quad w = u(t - \tau)$.

Here $f(z)$ is an arbitrary function.

Solution (Polyanin et al., 2023): $u(t) = a + bt$, where a is an arbitrary constant, and b is a root of the transcendental equation $b = f(b\tau)$.

6. $u'_t = uf(w/u), \quad w = u(t - \tau)$.

Delay ODE homogeneous in the dependent variable u, $f(z)$ is an arbitrary function.

Solution (Polyanin et al., 2023): $u(t) = Ce^{\lambda t}$, where C is an arbitrary constant, and λ is a root of the transcendental equation $\lambda = f(e^{-\lambda \tau})$.

7. $u'_t = wf(u^2 + w^2), \quad w = u(t - \tau)$.

Solutions:

$$u = \pm a_n \cos(\beta_n t + C), \quad \beta_n = \frac{1}{\tau}\left(\frac{\pi}{2} + \pi n\right), \quad n = 0, \pm 1, \pm 2, \ldots,$$

where C is an arbitrary constant and a_n are roots of algebraic (transcendental) equations

$$(-1)^{n+1} \beta_n = f(a_n^2), \quad n = 0, \pm 1, \pm 2, \ldots$$

8. $u'_t = f(t, w)u + g(t, w), \quad w = u(t - \tau)$.

The Cauchy-type problem for the original delay ODE subject to the initial condition

$$\varphi = \varphi_0(t) \quad \text{for} \quad -\tau \leq t \leq 0 \tag{1}$$

can be solved using the *method of steps*. To this end, we split time t into segments of length τ and denote

$$u(t) = \varphi_m(t) \quad \text{for} \quad t_{m-1} \leq t \leq t_m, \tag{2}$$

where $t_m = m\tau$ and $m = 0, 1, 2, \ldots$ Since the equation in question is linear in u, in each step, we will obtain the standard Cauchy problem for the linear first-order ODE without delay

$$\begin{aligned} u'_t &= f(t, \varphi_m(t - \tau))u + g(t, \varphi_m(t - \tau)), \quad m\tau \leq t \leq (m+1)\tau, \\ u(m\tau) &= \varphi_m(m\tau), \end{aligned} \tag{3}$$

where $m = 0, 1, 2, \ldots$, and $\varphi_m(t)$ is the solution of the Cauchy problem obtained in the preceding step on the segment $(m-1)\tau \leq t \leq m\tau$; $\varphi_0(t) \equiv \varphi(t)$.

The solution of problem (3) can be represented as

$$u(t) = e^{F(t)}\left[\varphi_m(m\tau) + \int_{m\tau}^{t} e^{-F(\xi)} g(\xi, \varphi_m(\xi - \tau))\, d\xi\right],$$

$$F(t) = \int_{m\tau}^{t} f(z, \varphi_m(z - \tau))\, dz, \quad m\tau \leq t \leq (m+1)\tau, \tag{4}$$

where $m = 0, 1, 2, \ldots$

The function $u = \varphi_{m+1}(t)$ on the left-hand side of formula (4) is sought on the interval $[t_m, t_{m+1}]$, whereas $\varphi_m(t)$ on the right-hand side is defined on the preceding interval $[t_{m-1}, t_m]$. The computations are carried out successively starting from $m = 0$, when the right-hand side is a known function defined on the initial interval (1). This step results in $\varphi_1(t)$. Then, one sets $m = 1$ and finds the function $\varphi_2(t)$ using (4) via the already known function $\varphi_1(t)$. The procedure continues further in a similar fashion.

9. $u'_t = f(t, w)u + g(t, w)u^k, \quad w = u(t - \tau).$

The substitution $v = u^{1-k}$ leads to an equation of the form 10.2.1.8 (Polyanin et al., 2023):

$$v'_t = (1-k)f(t, \bar{v}^{1/(1-k)})v + (1-k)g(t, \bar{v}^{1/(1-k)}), \quad \bar{v} = v(t - \tau).$$

10. $u'_t = f(t, w) + g(t, w)e^{-\lambda u}, \quad w = u(t - \tau).$

The substitution $v = e^{\lambda u}$ leads to an equation of the form 10.2.1.8 (Polyanin et al., 2023):

$$v'_t = \lambda f\left(t, \frac{1}{\lambda}\ln\bar{v}\right)v + \lambda g\left(t, \frac{1}{\lambda}\ln\bar{v}\right), \quad \bar{v} = v(t - \tau).$$

11. $u'_t = f(t, w)u + g(t, w)u \ln u, \quad w = u(t - \tau).$

The substitution $u = e^v$ leads to an equation of the form 10.2.1.8 (Polyanin et al., 2023):

$$v'_t = g(t, e^{\bar{v}})v + f(t, e^{\bar{v}}), \quad \bar{v} = v(t - \tau).$$

12. $u'_t = F(u - w_1, \ldots, u - w_n), \quad w_k = u(t - \tau_k) \quad (k = 1, \ldots, n).$

Solution: $u(t) = a + bt$, where a is an arbitrary constant, and b is a root of the transcendental equation $b = F(b\tau_1, \ldots, b\tau_n)$.

13. $u'_t = uF(w_1/u, \ldots, w_n/u), \quad w_k = u(t - \tau_k) \quad (k = 1, \ldots, n).$

Solution: $u(t) = Ce^{\lambda t}$, where C is an arbitrary constant, and λ is a root of the transcendental equation $\lambda = F(e^{-\lambda \tau_1}, \ldots, e^{-\lambda \tau_n})$.

10.2.2. Pantograph-Type ODEs with Proportional Arguments

1. $u'_t = a - bw^2, \quad w = u(\frac{1}{2}t); \quad u(0) = 0.$

1°. Solution with $ab > 0$ (Polyanin et al., 2023):

$$u(t) = \sqrt{\frac{2a}{b}} \sin\left(b\sqrt{\frac{a}{2b}}\, t\right).$$

2°. Solution with $ab < 0$ (Polyanin et al., 2023):

$$u(t) = -\sqrt{-\frac{2a}{b}} \sinh\left(b\sqrt{-\frac{a}{2b}}\, t\right).$$

2. $u'_t = au + bw^2, \quad w = u(\frac{1}{2}t); \quad u(0) = c.$

Solution (Polyanin et al., 2023): $u(t) = ce^{(a+bc)t}$.

3. $u'_t + au = \varepsilon w^2$, $w = u(pt)$; $u(0) = c$.

1°. Asymptotic solution for $p \neq 1/2$ and $\varepsilon \to 0$ (Polyanin et al., 2023):

$$u = e^{-at} + \frac{\varepsilon c^2}{a(2p-1)}\left(e^{-at} - e^{-2apt}\right)$$
$$+ \frac{2\varepsilon^2 c^3}{a^2(p+1)(2p-1)^2}\left[pe^{-at} - (p+1)e^{-2pat} + e^{-p(2p+1)at}\right] + o(\varepsilon^2).$$

2°. Asymptotic solution for $p = 1/2$ and $\varepsilon \to 0$:

$$u = e^{-at} + \varepsilon c^2 t e^{-at} + \tfrac{1}{2}\varepsilon^2 c^3 t^2 e^{-at} + o(\varepsilon^2).$$

4. $u'_t = au + bw_1 w_2$, $w_1 = u(pt)$, $w_2 = u((1-p)t)$; $u(0) = c$.

Solution: $u(t) = c e^{(a+bc)t}$.

5. $u'_t = au + bw^3$, $w = u(\tfrac{1}{3}t)$.

Solutions:
$$u(t) = \pm c\exp[(a+bc^2)t],$$

where c is an arbitrary constant.

6. $u'_t = au + bw^{1/p}$, $w = u(pt)$; $u(0) = c$.

Solution (Polyanin et al., 2023):
$$u(t) = c\exp(\lambda t), \quad \lambda = a + bc^{(1-p)/p}.$$

7. $u'_t = au + bu^{1/2} + cu^{1/2}w^{1/2}$, $w = u(pt)$.

The substitution $u = v^2$ leads to a pantograph-type linear ODE of the form 10.1.2.3: $v'_t = \tfrac{1}{2}av + \tfrac{1}{2}c\bar{v} + \tfrac{1}{2}b$, where $\bar{v} = v(pt)$.

8. $u'_t = au + bu^{1-k} + cu^{1-k}w^k$, $w = u(pt)$.

Generalization of the preceding equation. The substitution $u = v^{1/k}$ leads to a pantograph-type linear ODE of the form 10.1.2.3 (Polyanin et al., 2023): $v'_t = akv + ck\bar{v} + bk$, where $\bar{v} = v(pt)$.

9. $u'_t = aw/u$, $w = u(2t)$.

Solutions:
$$u = 2at, \quad u = \frac{2a}{C}\sin(Ct),$$

where $C > 0$ is an arbitrary constant.

10. $u'_t = a + be^{\lambda u} + ce^{\lambda(u-w)}$, $w = u(pt)$.

The substitution $v = e^{-\lambda u}$ leads to a pantograph-type linear ODE of the form 10.1.2.3 (Polyanin et al., 2023): $v'_t = -a\lambda v - c\lambda\bar{v} - b\lambda$, where $\bar{v} = v(pt)$.

11. $u'_t = au\ln u + bu\ln w + cu$, $w = u(pt)$.

The substitution $u = e^v$ leads to a pantograph-type linear ODE of the form 10.1.2.3 (Polyanin et al., 2023): $v'_t = av + b\bar{v} + c$, where $\bar{v} = v(pt)$.

12. $u'_t = f(u - 2w)$, $w = u(\frac{1}{2}t)$.
Here $f(z)$ is an arbitrary function.
Solution (Polyanin et al., 2023):
$$u(t) = f(-C)t + C,$$
where C is an arbitrary constant.

13. $u'_t = f(u - pw)$, $w = u(t/p)$.
Solution (Polyanin et al., 2023):
$$u(t) = f(C - Cp)t + C,$$
where C is an arbitrary constant.

14. $u'_t = u^k f(w/u)$, $w = u(pt)$.
Solution with $k \neq 1$ (Polyanin et al., 2023):
$$u = at^{\frac{1}{1-k}}, \quad a = \left[(1-k)f\left(p^{\frac{1}{1-k}}\right)\right]^{\frac{1}{1-k}}.$$
For $0 \leq k < 1$, this solution satisfies the initial condition $u(0) = 0$.

15. $u'_t = u f(w/\sqrt{u})$, $w = u(\frac{1}{2}t)$.
Solution:
$$u = ae^{\lambda t}, \quad \lambda = f(\sqrt{a}),$$
where $a > 0$ is an arbitrary constant.

16. $u'_t = u f(w/u^p)$, $w = u(pt)$.
Solution (Polyanin et al., 2023):
$$u = ae^{\lambda t}, \quad \lambda = f(a^{1-p}),$$
where $a > 0$ is an arbitrary constant.

17. $u'_t = f(t, u, w) - f(t, e^{\lambda t}, e^{p\lambda t}) + \lambda e^{\lambda t}$, $w = u(pt)$; $u(0) = 1$.
Solution: $u(t) = e^{\lambda t}$.

18. $u'_t = f(t, u, w) - f(t, \sin t, \sin(pt)) + \cos t$, $w = u(pt)$; $u(0) = 0$.
Solution: $u(t) = \sin t$.

19. $u'_t = F(w_1/u, \ldots, w_n/u)$, $w_k = u(p_k t)$ $(k = 1, \ldots, n)$; $u(0) = 0$.
Solution: $u(t) = at$, where $a = F(p_1, \ldots, p_n)$.

20. $u'_t = u^m F(w_1/u, \ldots, w_n/u)$, $w_k = u(p_k t)$ $(k = 1, \ldots, n)$.
Solution with $m \neq 1$:
$$u = at^{\frac{1}{1-m}}, \quad a = \left[(1-m)F\left(p_1^{\frac{1}{1-m}}, \ldots, p_n^{\frac{1}{1-m}}\right)\right]^{\frac{1}{1-m}}.$$
For $0 \leq m < 1$, this solution satisfies the initial condition $u(0) = 0$.

21. $u'_t = u f(w_1 u^{-p_1}, \ldots, w_n u^{-p_n})$, $w_k = u(p_k t)$.
Solution:
$$u = ae^{\lambda t}, \quad \lambda = f(a^{1-p_1}, \ldots, a^{1-p_n}),$$
where a is an arbitrary constant.

10.2.3. Other Ordinary Functional Differential Equations

1. $u'_t = au + bw^2$, $w = u\left(\frac{1}{2}(t-\tau)\right)$.

Solution:
$$u = A\exp(\lambda t), \qquad A = \frac{\lambda - a}{b}e^{\lambda\tau},$$

where λ is an arbitrary constant.

2. $u'_t = au + bw^{1/p}$, $w = u(p(t-\tau))$.

Solution:
$$u = A\exp(\lambda t), \qquad A = \left(\frac{\lambda - a}{b}e^{\lambda\tau}\right)^{\frac{p}{1-p}},$$

where λ is an arbitrary constant.

3. $u'_t = au + bw_1 w_2$, $w_1 = u(p(t-\tau_1))$, $w_2 = u((1-p)(t-\tau_2))$.

Solution:
$$u = A\exp(\lambda t), \qquad A = \frac{\lambda - a}{b}e^{\lambda[p\tau_1 + (1-p)\tau_2]},$$

where λ is an arbitrary constant.

4. $u'_t = uf(wu^{-p})$, $w = u(p(t-\tau))$.

Solution: $u = Ae^{\lambda t}$ where A is an arbitrary constant, and the λ is a root of the transcendental equation $\lambda - f(A^{1-p}e^{-p\lambda\tau}) = 0$.

5. $u'(t) = F(t, u(t), u(-t))$.

This is a special case of equation 10.2.3.6 for $b = 0$.

6. $u'(t) = F(t, u(t), u(b-t))$.

1°. Denote $u_1 = u(t)$ and $u_2 = u(b-t)$. In addition to the original equation, we also consider another equation obtained from the original one by replacing the independent variable t with $b - t$. As a result, we obtain the system of nonlinear ODEs

$$u'_1 = F(t, u_1, u_2), \qquad u'_2 = -F(b - t, u_2, u_1). \tag{1}$$

The solution of this system should be substituted into the original functional ODE in order to eliminate a redundant constant of integration.

2°. If the function F does not explicitly depend on t, i.e., $F = F(u_1, u_2)$, and is symmetric with respect to its arguments, $F(u_1, u_2) = F(u_2, u_1)$, then the elimination of F from system (1) gives

$$u_2 = C_1 - u_1.$$

As a result, we arrive at a separable first-order ODE:

$$u' = F(u, C_1 - u).$$

The solution of this ODE should be substituted into the original functional ODE in order to express the additional integration constant in terms of C_1 and b.

7. $u'(t) = F(t, u(t), u(b/t))$.

Denote $u_1 = u(t)$ and $u_2 = u(b/t)$. Let us consider an additional equation obtained from the original one by replacing the independent variable t with b/t. As a result, we get the system of nonlinear ODEs

$$u_1' = F(t, u_1, u_2), \quad u_2' = -bt^{-2}F(b/t, u_2, u_1).$$

The solution of this system must be substituted into the original functional ODE in order to eliminate a redundant constant of integration.

8. $u'(t) = F(t, u(t), u(\varphi(t)))$.

Here $\varphi(\varphi(t)) = t$. This equation is a generalization of the two preceding functional ODEs. Let us denote $u_1 = u(t)$ and $u_2 = u(\varphi(t))$ and write out an additional equation obtained from the original one by replacing the independent variable t with $\varphi(t)$. As a result, we get the system of nonlinear ODEs

$$u_1' = F(t, u_1, u_2), \quad u_2' = \varphi'(t) F(\varphi(t), u_2, u_1).$$

The solution of this system must be substituted into the original functional ODE in order to eliminate a redundant constant of integration.

9. $a[u(\sin t)]_t' + b[u(\cos t)]_t' = F(t, u(\sin t), u(\cos t))$.

Denote $u_1 = u(\sin t)$ and $u_2 = u(\cos t)$. We write out an additional equation by replacing the independent variable t with $\frac{\pi}{2} - t$ in the original equation to obtain the system of nonlinear ODEs

$$au_1' + bu_2' = F(t, u_1, u_2), \quad au_2' + bu_1' = -F(\tfrac{\pi}{2} - t, u_2, u_1).$$

The solution of this system must be substituted into the original functional ODE in order to eliminate a redundant constant of integration.

10.3. Second-Order Linear Ordinary Functional Differential Equations

10.3.1. ODEs with Constant Delays

1. $u''(t) + a^2 u(t - \tau) = f(t)$.

The solution of the Cauchy-type problem for the original delay ODE subject to the initial condition

$$u = \varphi(t) \quad \text{for} \quad -\tau \le t \le 0 \quad (\text{also } u_t' = \varphi'(t) \text{ for } -\tau \le t \le 0)$$

can be represented as

$$u(t) = \varphi(0) \cos_d(a(t-\tau), a\tau) + a^{-1} \varphi_t'(0) \sin_d(a(t-\tau), a\tau)$$
$$- a \int_{-\tau}^{0} \sin_d(a(t - 2\tau - s), a\tau) \varphi(s) \, ds + a^{-1} \int_0^t \sin_d(a(t - \tau - s), a\tau) f(s) \, ds.$$

Here $\cos_d(t,\tau)$ and $\sin_d(t,\tau)$ are the *delayed cosine* and *delayed sine functions*, which are defined as

$$\cos_d(t,\tau) = \begin{cases} 0, & t < -\tau, \\ 1, & -\tau \leq t \leq 0, \\ 1 - \dfrac{t^2}{2!} + \cdots + (-1)^k \dfrac{[t-(k-1)\tau]^{2k}}{(2k)!}, & (k-1)\tau < t \leq k\tau, \end{cases}$$

$$\sin_d(t,\tau) = \begin{cases} 0, & t < -\tau, \\ t+\tau, & -\tau \leq t \leq 0, \\ t+\tau - \dfrac{t^3}{3!} + \cdots + (-1)^k \dfrac{[t-(k-1)\tau]^{2k+1}}{(2k+1)!}, & (k-1)\tau < t \leq k\tau, \end{cases}$$

where $k = 1, 2, \ldots$

For details, see Khusainov et al. (2008) and Diblík et al. (2013).

2. $u''(t) = -a^2 u(t) + b u(t-\tau), \quad a \neq 0.$

The solution of the Cauchy-type problem for the original delay ODE subject to the initial condition

$$u = \varphi(t) \quad \text{for} \quad 0 \leq t \leq \tau$$

in the domain $t > \tau$ can be expressed as (see Rodríguez et al., 2012):

$$u(t) = \frac{\varphi(\tau) - \gamma\varphi(0)}{1-\gamma} u_1(t) - \frac{\varphi'(\tau) - \gamma\varphi'(0)}{1-\gamma}\left(\frac{\tau}{1-\gamma}u_1(t) - u_2(t)\right)$$
$$+ \frac{\gamma}{1-\gamma} \int_0^\tau \left(\frac{\tau}{1-\gamma} u_1(t) - u_2(t)\right) \varphi''(t)\, dt, \quad \gamma = \frac{b}{a^2},$$

where $u_1(t)$ and $u_2(t)$ are solutions of two simpler problems for the equation question with $\varphi(t) \equiv 1$ and $\varphi(t) \equiv t$, respectively.

Below are the auxiliary functions $u_1(t)$ and $u_2(t)$ included in the above solution. There were obtained by the method of steps by Rodríguez et al. (2012).

1°. On the interval $m\tau \leq t \leq (m+1)\tau$, the solution of the first auxiliary problem with $\varphi(t) \equiv 1$ can be represented as

$$u_1(t) = \gamma^m + (1-\gamma) \sum_{k=1}^{m} \gamma^{k-1} \sum_{n=0}^{k-1} A_{k,n} \frac{[a(t-k\tau)]^n}{n!} \cos\left[a(t-k\tau) - \tfrac{1}{2}\pi n\right], \quad \gamma = \frac{b}{a^2}.$$

The constants $A_{k,n}$ are given by

$$A_{k,0} = 1, \quad A_{k,n} = \sum_{j=0}^{k-n-1} \frac{n}{n+2j} 2^{-n-2j} C_{n+2j}^j, \quad 1 \leq n < k,$$

where $C_n^j = \frac{n!}{j!(n-j)!}$ are the binomial coefficients. Note that $0 < A_{k,n} \leq 1$.

2°. On the interval $m\tau \leq t \leq (m+1)\tau$, the solution of the second auxiliary problem with $\varphi(t) \equiv t$ can be represented as

$$u_2(t) = \gamma^m(t - m\tau) + \tau \sum_{k=1}^{m} \gamma^{k-1} \sum_{n=0}^{k-1} A_{k,n} \frac{[a(t-k\tau)]^n}{n!} \cos\left[a(t-k\tau) - \tfrac{1}{2}\pi n\right]$$
$$+ \frac{1-\gamma}{a} \sum_{k=1}^{m} \gamma^{k-1} \sum_{n=0}^{k-1} B_{k,n} \frac{[a(t-k\tau)]^n}{n!} \sin\left[a(t-k\tau) - \tfrac{1}{2}\pi n\right], \quad \gamma = \frac{b}{a^2}.$$

The constants $A_{k,n}$ are calculated using the formulas from Item 1°, and the constants $B_{k,n}$ are defined as follows:

$$B_{k,0} = 2^{1-2k} k C_{2k}^k,$$

$$B_{k,n} = 2^{n+1-2k} \sum_{j=0}^{k-n-1} \frac{n(k-n-j)}{n+2j} C_{n+2j}^j C_{2(k-n-j)}^{k-n-j}, \quad 1 \leq n < k.$$

3. $u''(t) + au'(t-\tau) + bu(t-2\tau) = 0.$

General solution:
$$u(t) = u_1(t) + u_2(t),$$

where $u_1(t)$ and $u_2(t)$ are any solutions of the first-order delay ODEs

$$u_k'(t) = \lambda_k u_k(t-\tau), \quad k = 1, 2, \qquad (*)$$

and λ_1 and λ_2 are roots of the quadratic equation $\lambda^2 + a\lambda + b = 0$. For solutions to the delay ODEs (*), see equation 10.1.1.2.

10.3.2. Pantograph-Type ODEs with Proportional Arguments

1. $u_{tt}'' = aw, \quad w = u(pt), \quad 0 < p < 1.$

1°. General solution with $a > 0$:

$$u(t) = C_1 \exp_s(-a^{1/2} p^{-1/4} t, p^{1/2}) + C_2 \exp_s(a^{1/2} p^{-1/4} t, p^{1/2}), \qquad (1)$$

where C_1 and C_2 are arbitrary constants, and $\exp_s(t,p) = \sum_{n=0}^{\infty} p^{\frac{n(n-1)}{2}} \frac{t^n}{n!}$ is the *stretched exponential function*. For

$$C_1 = \tfrac{1}{2}(b - ca^{-1/2} p^{1/4}), \quad C_2 = \tfrac{1}{2}(b + ca^{-1/2} p^{1/4}),$$

solution (1) satisfies the initial conditions $u(0) = b$ and $u_t'(0) = c$.

2°. General solution with $a < 0$:

$$u(t) = C_1 \cos_s(|a|^{1/2} p^{-1/4} t, p^{1/2}) + C_2 \sin_s(|a|^{1/2} p^{-1/4} t, p^{1/2}), \qquad (2)$$

where C_1 and C_2 are arbitrary constants, and $\cos_s(t,p)$ and $\sin_s(t,p)$ are the *stretched cosine* and *stretched sine* functions, which are determined using formulas

$$\cos_s(t,p) = \sum_{n=0}^{\infty} (-1)^n p^{n(2n-1)} \frac{t^{2n}}{(2n)!},$$

$$\sin_s(t,p) = \sum_{n=0}^{\infty} (-1)^n p^{n(2n+1)} \frac{t^{2n+1}}{(2n+1)!}.$$

For
$$C_1 = b, \quad C_2 = c|a|^{-1/2} p^{1/4},$$

solution (2) satisfies the initial conditions $u(0) = b$ and $u_t'(0) = c$.

3°. The following relationships between the stretched exponential function and stretched trigonometric functions hold:

$$\exp_s(it,p) = \cos_s(t,p) + i\sin_s(t,p), \quad i^2 = -1 \quad (0 < p < 1);$$

$$\cos_s(t,p) = \frac{\exp_s(it,p) + \exp_s(-it,p)}{2}, \quad \sin_s(t,p) = \frac{\exp_s(it,p) - \exp_s(-it,p)}{2i}.$$

The above solutions of the original equation were obtained by Liu (2018).

2. $u''_{tt}(t) = au(t) + bu(pt), \quad 0 < p < 1.$

This is a special case of equation 10.3.2.4 for $c = 0$.

3. $u''_{tt}(t) + au'_t(pt) + bu(p^2 t) = 0, \quad 0 < p < 1.$

1°. We look for particular solutions of this equation in the form (Liu, 2018):

$$u(t) = \exp_s(\beta t, p), \tag{1}$$

where $\exp_s(t,p)$ is the stretched exponential function (see equation 10.1.2.1, Item 1°). Taking into account the relations

$$[\exp_s(\beta t, p)]'_t = \beta \exp_s(\beta p t, p), \quad [\exp_s(\beta t, p)]''_{tt} = \beta^2 p \exp_s(\beta p^2 t, p),$$

we obtain the following quadratic equation for determining the constant β:

$$p\beta^2 + a\beta + b = 0. \tag{2}$$

2°. General solution:

$$u(t) = C_1 \exp_s(\beta_1 t, p) + C_2 \exp_s(\beta_2 t, p),$$

where C_1 and C_2 are arbitrary constants, and β_1 and β_2 are roots of the quadratic equation (2).

4. $u''_{tt}(t) = au(t) + bu(pt) + cu(qt), \quad 0 < p, q < 1.$

The solution of the Cauchy-type problem for the original second-order pantograph-type ODE with the initial conditions

$$u(0) = A, \quad u'_t(0) = B$$

is sought as a power series and can be represented as a linear combination of an even and an odd function (Polyanin et al., 2023):

$$u(t) = Au_1(t) + Bu_2(t),$$

where

$$u_1(t) = 1 + \sum_{n=1}^{\infty} \gamma_{2n} t^{2n}, \quad \gamma_{2n} = \frac{1}{(2n)!} \prod_{k=0}^{n-1} (a + bp^{2k} + cq^{2k}),$$

$$u_2(t) = t + \sum_{n=1}^{\infty} \gamma_{2n+1} t^{2n+1}, \quad \gamma_{2n+1} = \frac{1}{(2n+1)!} \prod_{k=0}^{n-1} (a + bp^{2k+1} + cq^{2k+1}).$$

The functions $u_1(t)$ and $u_2(t)$ satisfy the initial conditions

$$u_1(0) = 1, \quad u'_1(0) = 0; \quad u_2(0) = 0, \quad u'_2(0) = 1.$$

For $a = -1$ and $b = c = 0$, these functions become cosine and sine, respectively, and for $a = 1$ and $b = c = 0$, they become hyperbolic cosine and hyperbolic sine.

10.3.3. Other Ordinary Functional Differential Equations

1. $u''(t) = au(-t)$.

Let us denote $u_1 = u(t)$ and $u_2 = u(b/t)$ and write out an additional equation by substituting the independent variable t with $-t$ in the original equation to obtain the system of two second-order linear ODEs

$$u_1'' = au_2, \quad u_2'' = au_1.$$

Eliminating u_2 from this system, we arrive at a fourth-order linear ODE, $u_1'''' = a^2 u_1$, the general solution of which has the form

$$u_1 = C_1 \cosh(\sqrt{|a|}\,t) + C_2 \sinh(\sqrt{|a|}\,t) + C_3 \cos(\sqrt{|a|}\,t) + C_4 \sin(\sqrt{|a|}\,t),$$

where $C_1, \ldots, C_4$ are arbitrary constants. Substituting the resulting function $u = u_1$ into the original functional ODE, we find its solution:

$$u = \begin{cases} C_1 \cosh(\sqrt{a}\,t) + C_4 \sin(\sqrt{a}\,t) & \text{if } a > 0, \\ C_2 \sinh(\sqrt{|a|}\,t) + C_3 \cos(\sqrt{|a|}\,t) & \text{if } a < 0. \end{cases}$$

2. $u''(t) = au(t) + cu(b - t)$.

This is a special case of functional ODE 10.4.3.3 for $F(t, u_1, u_2) = au_1 + cu_2$.

3. $u''(t) = au(b/t)$.

Let us denote $u_1 = u(t)$ and $u_2 = u(b/t)$ and write out an additional equation by substituting the independent variable t with b/t in the original equation to obtain the system of linear ODEs

$$u_1'' = au_2, \quad t^2(t^2 u_2')' = ab^2 u_1.$$

Eliminating u_2 from this system, we arrive at a fourth-order linear ODE,

$$t^2(t^2 u_1''')' = a^2 b^2 u_1, \tag{1}$$

the general solution of which has the form

$$u_1 = C_1 t^{k_1} + C_2 t^{k_2} + C_3 t^{k_3} + C_4 t^{k_4}, \tag{2}$$

where $C_1, \ldots, C_4$ are arbitrary constants, and $k_1, \ldots, k_4$ are roots of the quartic equation

$$k(k-1)^2(k-2) = a^2 b^2. \tag{3}$$

The substitution $s = k^2 - 2k$ reduces equation (3) to a quadratic equation, $s^2 - s - a^2 b^2 = 0$.

Solution (2) should be substituted into the original functional ODE in order to eliminate redundant integration constants.

4. $u''(t) = au(t) + cu(b/t)$.

This is a special case of functional ODE 10.4.3.4 for $F(t, u_1, u_2) = au_1 + cu_2$.

10.4. Second-Order Nonlinear Ordinary Functional Differential Equations

10.4.1. ODEs with Constant Delays

1. $u''_{tt} = -uf(w/u), \quad w = u(t - \tau).$

Delay ODE homogeneous in the dependent variable u; $f(z)$ is an arbitrary function.

1°. Solution: $u(t) = Ce^{\lambda t}$, where C is an arbitrary constant, and λ is a root of the transcendental equation $\lambda^2 + f(e^{-\lambda \tau}) = 0$.

2°. For special delay times

$$\tau = \tau_n = \frac{\pi n}{\lambda_n}, \quad \lambda_n = \sqrt{f((-1)^n)}, \quad n = 1, 2, \ldots,$$

there are also solutions of the trigonometric form

$$u(t) = C_1 \cos(\lambda_n t + C_2),$$

where C_1 and C_2 are arbitrary constants.

2. $u''_{tt} = g(w/u)u'_t + uf(w/u), \quad w = u(t - \tau).$

Solution: $u(t) = Ce^{\lambda t}$, where C is an arbitrary constant, and λ is a root of the transcendental equation $\lambda^2 = \lambda g(e^{-\lambda \tau}) + f(e^{-\lambda \tau})$.

3. $u''_{tt} = uF(w/u, u'_t/u, w'_t/u), \quad w = u(t - \tau).$

Solution: $u = Ce^{\lambda t}$, where C is an arbitrary constant, and λ is a root of the transcendental equation $\lambda^2 = F(e^{-\lambda \tau}, \lambda, \lambda e^{-\lambda \tau})$.

4. $u''_{tt} = uF(u'_t/u, w'_t/w), \quad w = u(t - \tau).$

The substitution $\xi = u'_t/u$ leads to a first-order delay ODE: $\xi'_t + \xi^2 = F(\xi, \bar{\xi})$, where $\bar{\xi} = \xi(t - \tau)$.

5. $u''_{tt} = uf(u^2 + w^2), \quad w = u(t - \tau).$

Solutions:

$$u(t) = \pm a_n \cos(\beta_n t + C), \quad \beta_n = \frac{1}{\tau}\left(\frac{\pi}{2} + \pi n\right), \quad n = 0, \pm 1, \pm 2, \ldots,$$

where C is an arbitrary constant, and a_n are roots of the algebraic (transcendental) equations

$$-\beta_n^2 = f(a_n^2), \quad n = 0, \pm 1, \pm 2, \ldots$$

6. $u''_{tt} = a^2 w_2 + F(t, u'_t + aw_1), \quad w_1 = u(t - \tau), \quad w_2 = u(t - 2\tau).$

The substitution $\zeta = u'_t + aw_1$ leads to a first-order delay ODE:

$$\zeta'_t = a\bar{\zeta} + F(t, \zeta), \quad \bar{\zeta} = \zeta(t - \tau).$$

7. $u''_{tt} = w_2 f(u'_t/w_1, (w_1)'_t/w_2), \quad w_1 = u(t - \tau), \quad w_2 = u(t - 2\tau).$

Single-phase solution:

$$u = u_1(t),$$

where $u_1(t)$ is any solution of the first-order linear delay ODE

$$u_1(t) = \lambda u_1(t - \tau), \qquad (*)$$

and λ is any root of the transcendental equation $\lambda^2 = f(\lambda, \lambda)$. For solutions to the delay ODE $(*)$, see equation 10.1.1.2.

8. $u''_{tt} = uF(w_1/u, \ldots, w_n/u, u'_t/u, w'_1/u, \ldots, w'_n/u),$
   $w_k = u(t - \tau_k) \quad (k = 1, \ldots, n).$

Solution: $u(t) = Ce^{\lambda t}$, where C is an arbitrary constant, and λ is a root of the transcendental equation $\lambda^2 = F(e^{-\lambda \tau_1}, \ldots, e^{-\lambda \tau_n}, \lambda, \lambda e^{-\lambda \tau_1}, \ldots, \lambda e^{-\lambda \tau_n})$.

10.4.2. Pantograph-Type ODEs with Proportional Arguments

1. $u''_{tt} = au + bw^2, \quad w = u(\tfrac{1}{2}t).$

Solutions: $u(t) = c\exp(\pm\sqrt{a+bc}\,t)$, where c is an arbitrary constant, such that the inequality $a + bc > 0$ holds. There is also the trivial solution $u = 0$ (for any a and b).

2. $u''_{tt} = au + bw^3, \quad w = u(\tfrac{1}{3}t).$

Solutions:
$$u(t) = c\exp(\pm\sqrt{a+bc^2}\,t),$$

where c is an arbitrary constant such that the inequality $a + bc^2 > 0$ holds. The equation also admits the trivial solution $u = 0$ (for any a and b).

3. $u''_{tt} = aw + bw^3, \quad w = u(\tfrac{1}{3}t).$

Solutions (Polyanin et al., 2023):

$$\begin{aligned}
u &= \pm 2\sqrt{|a|/(3b)}\, \sin(\sqrt{|a|/3}\,t) & &\text{if } a < 0,\ b > 0;\\
u &= \pm 2\sqrt{a/(3|b|)}\, \cos(\sqrt{a/3}\,t) & &\text{if } a > 0,\ b < 0;\\
u &= \pm 2\sqrt{a/(3b)}\, \sinh(\sqrt{a/3}\,t) & &\text{if } a > 0,\ b > 0;\\
u &= \pm 2\sqrt{|a|/(3b)}\, \cosh(\sqrt{|a|/3}\,t) & &\text{if } a < 0,\ b > 0.
\end{aligned}$$

The equation also admits the trivial solution $u = 0$ (for any a and b).

4. $u''_{tt} = au + bw^{1/p}, \quad w = u(pt), \quad 0 < p < 1.$

Solutions:
$$u(t) = c\exp(\pm\lambda t), \qquad \lambda = \sqrt{a + bc^{(1-p)/p}},$$

where $c > 0$ is an arbitrary constant, such that the inequality $a + bc^{(1-p)/p} > 0$ holds. The equation also admits the trivial solution $u = 0$ (for any a and b).

5. $u''_{tt} = au'_t + u(b\ln u + c\ln w), \quad w = u(pt).$

Solution:
$$u(t) = \exp(At^2 + Bt + C),$$

$$A = \tfrac{1}{4}(b + cp^2), \quad B = -\frac{a(b + cp^2)}{2cp(1-p)}, \quad C = \frac{2A + B^2 - aB}{b + c}.$$

6. $u''_{tt} = F(w_1/u, \ldots, w_n/u)$, $w_k = u(p_k t)$ $(k = 1, \ldots, n)$.

The solution satisfying the initial conditions $u(0) = u'_t(0) = 0$ is $u(t) = at^2$, where $a = \frac{1}{2} F(p_1^2, \ldots, p_n^2)$.

7. $u''_{tt} = u^m F(w_1/u, \ldots, w_n/u)$, $w_k = u(p_k t)$ $(k = 1, \ldots, n)$.

Solution with $m \neq 1$:

$$u = at^{\frac{2}{1-m}}, \qquad a = \left[\frac{(1-m)^2}{2(1+m)} F\big(p_1^{\frac{2}{1-m}}, \ldots, p_n^{\frac{2}{1-m}}\big)\right]^{\frac{1}{1-m}}.$$

For $0 \le m < 1$, this solution satisfies the initial conditions $u(0) = u'_t(0) = 0$.

10.4.3. Other Ordinary Functional Differential Equations

1. $u''_{tt} = au + bw^2$, $w = u\big(\frac{1}{2}(t - \tau)\big)$.

Solution: $u(t) = \dfrac{\lambda^2 - a}{b} e^{\lambda \tau} e^{\lambda t}$, where λ is an arbitrary constant.

2. $u''_{tt} = au + bw^{1/p}$, $w = u(p(t - \tau))$.

Solution:

$$u(t) = A \exp(\lambda t), \qquad A = \left(\frac{\lambda^2 - a}{b} e^{\lambda \tau}\right)^{\frac{p}{1-p}},$$

where λ is an arbitrary constant.

3. $u''(t) = F\big(t, u(t), u(b - t)\big)$.

1°. Let us denote $u_1 = u(t)$ and $u_2 = u(b - t)$ and write out an additional equation by substituting the independent variable t with $b - t$ in the original equation to obtain the system of nonlinear ODEs

$$u''_1 = F(t, u_1, u_2), \qquad u''_2 = F(b - t, u_2, u_1). \tag{1}$$

The solution of this ODE system must be substituted into the original functional ODE in order to eliminate extra constants of integration.

2°. If the function F does not explicitly depend on t, i.e., $F = F(u_1, u_2)$, and is symmetric with respect to its arguments, $F(u_1, u_2) = F(u_2, u_1)$, then excluding F from system (1), we get

$$u_2 = u_1 + C_1 t + C_2,$$

where C_1 and C_2 are arbitrary constants. As a result, we arrive at the second-order ODE

$$u''_{tt} = F(u, u + C_1 t + C_2).$$

The solution of this equation should be substituted into the original functional ODE in order to express the extra integration constants in terms of C_1, C_2, and b.

4. $u''(t) = F(t, u(t), u(b/t))$.

Let us denote $u_1 = u(t)$ and $u_2 = u(b/t)$ and write out an additional equation by substituting the independent variable t with b/t in the original equation to obtain the system of nonlinear ODEs

$$u_1'' = F(t, u_1, u_2), \quad t^2(t^2 u_2')' = b^2 F(b/t, u_2, u_1).$$

The solution of this system must be substituted into the original functional ODE in order to eliminate extra constants of integration.

5. $f(x)g_t'(t) = ag(t)f_{xx}''(x)$.

Here the functions $f(x)$ and $g(t)$ are unknown. This functional differential equation results from applying the method of separation of variables to the linear heat equation $u_t = au_{xx}$, when its particular solution is sought as the product of functions of different arguments: $u = f(x)g(t)$.

Solutions:

$$f(x) = C_1 \cos(kx) + C_2 \sin(kx), \quad g(t) = \exp(-ak^2 t);$$
$$f(x) = C_1 \exp(-kx) + C_2 \exp(kx), \quad g(t) = \exp(ak^2 t),$$

where C_1, C_2, and k are arbitrary constants.

10.5. Higher-Order Ordinary Functional Differential Equations

10.5.1. Linear Ordinary Functional Differential Equations

▶ **ODEs with constant delays.**

1. $u_t^{(n)}(t) = au(t - \tau)$.

$1°$. Particular solutions:
$$u = \exp(\lambda t),$$
where λ are roots of the transcendental characteristic equation

$$\lambda^n e^{\tau \lambda} - a = 0. \tag{1}$$

$2°$. For any $a > 0$, equation (1) has a real positive root, which is expressed in terms of the Lambert W function (see Item $2°$ of equation 10.1.1.2) as

$$\lambda_p = \frac{n}{\tau} W_p\left(\frac{\tau a^{1/n}}{n}\right).$$

$3°$. In general, the substitution $\zeta = \lambda e^{\tau \lambda / n}$ takes the transcendental equation (1) to the algebraic equation $\zeta^n - a = 0$, which has n complex roots

$$\zeta_k = \begin{cases} a^{1/n}\left(\cos \frac{2(k-1)\pi}{n} + i \sin \frac{2(k-1)\pi}{n}\right) & \text{if } a > 0, \\ |a|^{1/n}\left(\cos \frac{(2k-1)\pi}{n} + i \sin \frac{(2k-1)\pi}{n}\right) & \text{if } a < 0, \end{cases} \tag{2}$$

where $k=1, \ldots, n$, $i^2=-1$. Therefore, the difference $\zeta^n - a$ admits factorization and can be represented as the product $\prod_{k=1}^{n}(\zeta - \zeta_k) = 0$, where $\zeta = \lambda e^{\tau\lambda/n}$, and the transcendental equation (1) decomposes into n simpler independent equations

$$\lambda e^{\tau\lambda/n} - \zeta_k = 0, \quad k = 1, \ldots, n.$$

The solutions of these equations are expressed in terms of the Lambert W function of a complex argument as

$$\lambda_k = \frac{n}{\tau} W\left(\frac{\tau \zeta_k}{n}\right), \quad k = 1, \ldots, n,$$

where the numbers ζ_k (generally complex) are defined in (2), and $W(z)$ denotes all branches of the Lambert function.

2. $u_t^{(n)}(t) + a_1 u_t^{(n-1)}(t-\tau) + \cdots + a_{n-1} u_t'(t-(n-1)\tau) + a_n u(t-n\tau) = 0.$

General solution:

$$u(t) = \sum_{k=1}^{n} u_k(t),$$

where $u_1(t), \ldots, u_n(t)$ are any solutions of the first-order delay ODEs

$$u_k'(t) = \lambda_k u_k(t-\tau), \quad k = 1, \ldots, n, \qquad (*)$$

and λ_k are roots of the algebraic equation $\lambda^n + a_1 \lambda^{n-1} + \cdots + a_{n-1}\lambda + a_n = 0$. For solutions to the linear delay ODEs $(*)$, see equation 10.1.1.2.

3. $u_t^{(n)}(t) + \sum_{i=0}^{n-1} \sum_{j=0}^{m} a_{ij} u_t^{(i)}(t - \tau_j) = 0.$

Linear homogeneous nth-order ODE with constant coefficients and m constant delays; $\tau_j \geq 0$ are the delay times.

This equation has exponential particular solutions of the form

$$u(t) = \exp(\lambda t),$$

where λ is a root of the transcendental characteristic equation

$$\lambda^n + \sum_{i=0}^{n-1} \sum_{j=0}^{m} a_{ij} \lambda^i e^{-\lambda \tau_j} = 0.$$

4. $u_t^{(n)}(t) + \sum_{i=0}^{n-1} \sum_{j=0}^{m} a_{ij} u_t^{(i)}(t - \tau_j) = f(t).$

Linear nonhomogeneous nth-order ODE with constant coefficients and m constant delays; $\tau_j \geq 0$ are the delay times.

The general solution to this equation is the sum of the general solution to the respective homogeneous equation 10.5.1.3 and any particular solution to the nonhomogeneous equation under consideration.

Table 10.2 describes the structure of particular solutions for some functions on the right-hand side of the linear nonhomogeneous equation 10.5.1.4.

TABLE 10.2. The structure of particular solutions of the linear nonhomogeneous equation with constant delays (10.5.1.4) for some special forms of the function $f(x)$.

Form of function $f(t)$	Roots of characteristic equation $\lambda^n + \sum_{i=0}^{n-1}\sum_{j=0}^{m} a_{ij}\lambda^i e^{-\lambda\tau_j} = 0$	Form of particular solution $u = \tilde{u}(t)$
$P_m(t)$	zero is not a root of characteristic equation	$\tilde{P}_m(t)$
	zero is a root of characteristic equation (multiplicity r)	$t^r \tilde{P}_m(t)$
$P_m(t)e^{\alpha t}$ (α is a real number)	α is not a root of characteristic equation	$\tilde{P}_m(t)e^{\alpha t}$
	α is a root of characteristic equation (multiplicity r)	$t^r \tilde{P}_m(t)e^{\alpha t}$
$P_m(t)\cos\beta t$ $+ Q_n(t)\sin\beta t$	$i\beta$ is not a root of characteristic equation	$\tilde{P}_\nu(t)\cos\beta t$ $+ \tilde{Q}_\nu(t)\sin\beta t$
	$i\beta$ is a root of characteristic equation (multiplicity r)	$t^r[\tilde{P}_\nu(t)\cos\beta t$ $+ \tilde{Q}_\nu(t)\sin\beta t]$
$[P_m(t)\cos\beta t$ $+ Q_n(t)\sin\beta t]e^{\alpha t}$	$\alpha + i\beta$ is not a root of characteristic equation	$[\tilde{P}_\nu(t)\cos\beta t$ $+ \tilde{Q}_\nu(t)\sin\beta t]e^{\alpha t}$
	$\alpha + i\beta$ is a root of characteristic equation (multiplicity r)	$t^r[\tilde{P}_\nu(t)\cos\beta t$ $+ \tilde{Q}_\nu(t)\sin\beta t]e^{\alpha t}$

Notation: P_m and Q_n are polynomials of degree m and n with prescribed coefficients; $\tilde{P}_m$, $\tilde{P}_\nu$, and $\tilde{Q}_\nu$ are polynomials of degree m and ν whose coefficients are determined by substituting the particular solution into the original equation; $\nu = \max(m, n)$; α and β are real numbers, and $i^2 = -1$.

▶ **Linear pantograph-type ODEs with a proportional argument.**

5. $u_t^{(n)}(t) + a_{n-1}u_t^{(n-1)}(pt) + \cdots + a_1 u_t'(p^{n-1}t) + a_0 u(p^n t) = 0$, $\quad 0 < p < 1$.

1°. We look for particular solutions of this equation in the form (Liu, 2018):

$$u(t) = \exp_s(\beta t, p), \tag{1}$$

where $\exp_s(t, p)$ is the stretched exponential function (see equation 10.1.2.1, Item 1°). Taking into account the relations

$$[\exp_s(\beta t, p)]_t^{(k)} = \beta^k p^{\frac{k(k-1)}{2}} \exp_s(\beta p^k t, p), \quad k = 0, 1, \ldots, n,$$

we obtain the following algebraic equation for determining the constant β:

$$p^{\frac{n(n-1)}{2}}\beta^n + a_{n-1}p^{\frac{(n-1)(n-2)}{2}}\beta^{n-1} + \cdots + a_1\beta + a_0 = 0. \tag{2}$$

2°. General solution:

$$u(t) = C_1 \exp_s(\beta_1 t, p) + C_2 \exp_s(\beta_2 t, p) + \cdots + C_n \exp_s(\beta_n t, p),$$

where $C_1, \ldots, C_n$ are arbitrary constants, and $\beta_1, \ldots, \beta_n$ are roots of the algebraic equation (2).

10.5.2. Nonlinear Ordinary Functional Differential Equations

▶ **ODEs with constant delay.**

1. $u_t^{(n)} = w_n f(u_t'/w_1)$, $\quad w_1 = u(t-\tau)$, $\quad w_n = u(t-n\tau)$.

Single-phase solution:
$$u = u(t),$$
where $u(t)$ is any solution of the first-order linear delay ODE
$$u(t) = \lambda u(t-\tau), \qquad (*)$$
and λ is any root of the transcendental equation $\lambda^n = f(\lambda)$. For solutions to delay ODE $(*)$, see equation 10.1.1.2.

2. $u_t^{(n)} = uF(w_1/u, \ldots, w_m/u)$, $\quad w_k = u(t-\tau_k)$ $\quad (k=1,\ldots,m)$.

Solution: $u(t) = ae^{\lambda t}$, where a is an arbitrary constant, and λ is a root of the transcendental equation $\lambda^n = F(e^{-\lambda \tau_1}, \ldots, e^{-\lambda \tau_m})$.

▶ **Pantograph-type ODE with proportional delay.**

3. $u_t^{(n)} = F(w_1/u, \ldots, w_m/u)$, $\quad w_k = u(p_k t)$ $\quad (k=1,\ldots,m)$.

The solution satisfying the initial conditions $u(0) = u_t'(0) = \cdots = u_t^{(n-1)}(0) = 0$ is $u(t) = at^n$, where $a = \dfrac{1}{n!} F(p_1^n, \ldots, p_m^n)$.

▶ **Ordinary functional differential equations with several unknown functions.**

4. $f_1[X]g_1[Y] + f_2[X]g_2[Y] + \cdots + f_k[X]g_k[Y] = 0$.

Here the differential forms $f_i[X]$ and $g_i[Y]$ ($i = 1, \ldots, k$) depend, respectively, on the variables x and y and contain $2n$ unknown functions $\varphi_j = \varphi_j(x)$ and $\psi_j = \psi_j(y)$ ($j = 1, \ldots, n$):
$$\begin{aligned} f_i[X] &\equiv f_i(x, \varphi_1, \varphi_1', \varphi_1'', \ldots, \varphi_n, \varphi_n', \varphi_n''), \\ g_i[Y] &\equiv g_i(y, \psi_1, \psi_1', \psi_1'', \ldots, \psi_n, \psi_n', \psi_n''), \end{aligned} \qquad (1)$$
where the primes denote derivatives. Note that, in addition to the second derivatives, expressions (1) can also include higher-order derivatives of the functions φ_j and ψ_j.

It is convenient to split the original problem is into two simpler problems: (i) the solution of a bilinear functional equation of a standard form and (ii) the solution of a system of ordinary differential equations. Below we briefly describe the main steps of this method.

$1°$. In the first stage, we treat the original functional ODE as a bilinear functional equation:
$$\sum_{i=1}^{k} f_i g_i = 0, \qquad (2)$$
where $f_i = f_i[X]$ and $g_i = g_i[Y]$ are the unknown quantities ($i = 1, \ldots, k$), and X and Y are independent variables.

The splitting principle. All solutions of the bilinear functional equation (2) can be represented as a set of linear combinations of the quantities $f_1, \ldots, f_n$ together with linear combinations of the quantities $g_1, \ldots, g_n$:

$$\sum_{i=1}^{k} \alpha_{ir} f_i = 0, \quad r = 1, \ldots, l;$$
$$\sum_{i=1}^{k} \beta_{is} g_i = 0, \quad s = 1, \ldots, m, \tag{3}$$

where $1 \leq l \leq k-1$ and $1 \leq m \leq k-1$. The constants α_{ir} and β_{is} in (3) are chosen so that the bilinear equation (2) holds identically (this can always be done, and there are $k-1$ different solutions). Notably, the linear relations (3) are purely algebraic in nature and do not depend on the specific expressions of the differential forms (1).

For solutions of the bilinear equation (2) with $k = 3$, $k = 4$, and $k = 5$ see the functional equations (9.4.2.27), (9.4.2.28), and (9.4.2.29) (Item 3°), respectively.

2°. In the second stage, we sequentially substitute the differential forms $f_i[X]$ and $g_i[Y]$ from (1) into solutions (3). As a result, we obtain systems of ODEs (which are often overdetermined) to find the desired functions $\varphi_j(x)$ and $\psi_j(y)$. Solving these systems, we obtain solutions to the original functional ODE.

3°. Degenerate cases, where one or more differential forms f_i and/or g_i vanish, must be treated separately, using linear relations of the form (3).

4°. For details on using the splitting principle to solve ordinary functional differential equations and related nonlinear PDEs, see the books by Polyanin & Zaitsev (2012) and Polyanin & Zhurov (2022).

▶ For analytical methods for solving linear and nonlinear ordinary differential equations with constant or variable delays, see the books by Bellman & Cooke (1963), Myshkis (1972), Elsgolt's & Norkin (1973), Erneux (2009), and Polyanin et al. (2023). The most common mathematical models with a delay used in population theory, biology, medicine, economics and other applications, as well as approximate and numerical methods for solving delay ODEs are discussed in Bellen & Zennaro (2003), Kuang & Cong (2007), Erneux (2009), Smith (2001), Kuang (2012), Schiesser (2019), Rihan (2021), and Polyanin et al. (2023).

References

Azizbekov, E. and Khusainov, D.Y., Solution of one heat conduction equation with a delay, *Bull. Taras Shevchenko Nat. Univ. Kyiv Cyber.*, No. 12, pp. 4–12, 2012.

Asl, F.M. and Ulsoy, A.G., Analysis of a system of linear delay differential equations. *ASME J. Dyn. Syst. Measure. Control*, Vol. 125, No. 2, pp. 215–223, 2003.

Bellen, A. and Zennaro, M., *Numerical methods for delay differential equations, Numerical Mathematics and Scientific Computation*, Clarendon Press–Oxford Univ. Press, New York, 2003.

Bellman, R. and Cooke, K.L., *Differential-Difference Equations*, Academic Press, New York, 1963.

Brunner, H., Huang, Q., and Xie, H., Discontinuous Galerkin methods for delay differential equations of pantograph type, *SIAM J. Numer. Anal.*, Vol. 48, No. 5, pp. 1944–1967, 2010.

Corless, R.M., Gonnet, G.H., Hare, D.E.G., Jeffrey, D.J., and Knuth, D.E., On the lambert W function. *Adv. Comput. Math.*, Vol. 5, pp. 329–359, 1996.

Diblík, J., Fečkan, M., and Pospíšil, M., Representation of a solution of the Cauchy problem for an oscillating system with two delays and permutable matrices, *Ukr. Math. J.*, Vol. 65, pp. 64–76, 2013.

Erneux, T., *Applied Delay Differential Equations*, Springer, New York, 2009.

Elsgolt's, L.E. and Norkin, S.B., *Introduction to the Theory and Application of Differential Equations with Deviating Arguments*, Academic Press, New York, 1973.

Fox, L., Mayers, D., Ockendon, J.R., and Tayler, A.B., On a functional differential equation. *IMA J. Appl. Math.*, Vol. 8, pp. 271–307, 1971.

Iserles, A., On the generalized pantograph functional differential equation, *Eur. J. Appl. Math.*, Vol. 4, No. 1, pp. 1–38, 1993.

Kato, T. and McLeod, J.B., Functional-differential equation $y' = ay(\lambda t) + by(t)$, *Bull. Am. Math. Soc.*, Vol. 77, No. 6, pp. 891–937, 1971.

Khusainov, D.Y., Diblík, J., Růžičková, M., and Lukáčová, J., Representation of a solution of the Cauchy problem for an oscillating system with pure delay, *Nonlinear Oscil.*, Vol. 11, pp. 276–285, 2008.

Kuang, Y., *Delay Differential Equations with Applications in Population Dynamics*, Academic Press, San Diego, 2012.

Kuang, J. and Cong, Y., *Stability of Numerical Methods for Delay Differential Equations*. Elsevier Science, Amsterdam, 2007.

Liu, C.-S., Basic theory of a kind of linear functional differential equations with multiplication delay, *arXiv:1605.06734v4 [math.CA]*, 2018.

Myshkis, A.D., *Linear Differential Equations with a Delayed Argument*, 2nd ed., [in Russian], Nauka, Moscow, 1972.

Patade, J. and Bhalekar, S., Analytical solution of pantograph equation with incommensurate delay, *Phys. Sci. Rev.*, Vol. 2, No. 9, 20165103, 2017.

Polyanin, A.D., Sorokin, V.G., and Zhurov, A.I., *Delay Ordinary and Partial Differential Equations*, CRC Press, Boca Raton, 2023.

Polyanin, A.D. and Zaitsev, V.F., *Handbook of Nonlinear Partial Differential Equations*, 2nd ed., Chapman & Hall/CRC Press, Boca Raton–London, 2012.

Polyanin, A.D. and Zhurov, A.I., *Separation of Variables and Exact Solutions to Nonlinear PDEs*, CRC Press, Boca Raton–London, 2022.

Rihan, F.A., *Delay Differential Equations and Applications to Biology*, Springer, Singapore, 2021.

Rodríguez, F., Roales, M., and Marín J.A., Exact solutions and numerical approximations of mixed problems for the wave equation with delay, *Appl. Math. Comput.*, Vol. 219, No. 6, pp. 3178–3186, 2012.

Schiesser, W.E., *Time Delay ODE/PDE Models: Applications in Biomedical Science and Engineering*, CRC Press, Boca Raton, 2019.

Smith, H., *An Introduction to Delay Differential Equations with Applications to the Life Sciences*, Springer, New York, 2011.

Valluri, S.R., Jeffrey, D.J., and Corless, R.M. Some applications of the Lambert W function to physics, *Can. J. Phys.*, Vol. 78, No. 9, pp. 823–831, 2000.

Yi, S., Nelson, P.W., and Ulsoy, A.G., *Time-Delay Systems: Analysis and Control Using the Lambert W Function*, World Scientific, Singapore, 2010.

Wright, E.M., Solution of the equation $ze^z = a$, *Proc. R. Soc. Edinb.*, Vol. 65, pp. 193–203, 1959.

Chapter 11

Partial Functional Differential Equations

▶ **Preliminary remarks.** *Partial functional differential equations* (or *functional PDEs* for short) are mathematical equations containing an unknown function of two or more variables with different arguments and partial derivatives of this function. This chapter deals with linear and nonlinear functional PDEs with second-order partial derivatives and their exact solutions. Partial differential equations with constant delay are the simplest functional PDEs containing the quantities $u = u(x,t)$ and $w = u(x, t - \tau)$ and partial derivatives of u with respect to x and t, where x and t are the independent variables, u is the unknown function, and $\tau > 0$ is the delay time. In initial-boundary value problems for PDEs with constant delay, the initial data are prescribed on an interval $t_0 - \tau \leq t \leq t_0$, where t_0 is some constant.

Exact solutions of partial functional differential equations are understood as follows:

- solutions that are expressed in terms of elementary functions, functions included in the equation (this is necessary when the equation depends on arbitrary functions), and indefinite or definite integrals,

- solutions that are expressed in terms of solutions of ODEs or functional ODEs, as well as systems of such equations.

This chapter deals with linear and nonlinear partial differential equations with constant, proportional, and variable delays and some other functional PDEs with second-order partial derivatives. For linear reaction-diffusion and wave-type equations with constant delay, the basic initial-boundary problems with Dirichlet, Neumann and Robin boundary conditions, as well as mixed boundary conditions, are formulated. Analytical formulas are given to obtain solutions to these problems. Particular attention is paid to fairly general nonlinear reaction-diffusion and wave-type equations with constant and proportional delay, which include one or more arbitrary functions. Additive, multiplicative, generalized and functionally separable solutions of such nonlinear functional PDEs are constructed (the same simple classification of solutions by their appearance will be used as that in Chapter 6 above, see Table 6.1). In addition, we describe reductions that convert nonlinear functional PDEs to ODEs or functional ODEs, which are usually much easier to solve than the original equations. We also consider more general nonlinear PDEs with variable delays that depend arbitrarily on time t or spatial variable x; exact solutions to such complicated PDEs are of significant interest for testing numerical and approximate analytical methods.

11.1. Linear Partial Functional Differential Equations

11.1.1. PDEs with Constant Delay

1. $u_t = a_1 u_{xx} + a_2 w_{xx} + c_1 u + c_2 w + f(x,t), \quad w = u(x, t - \tau)$.

Linear diffusion-type equation with constant delay, where $a_1 > a_2 \geq 0$ and $\tau > 0$.

▶ Exact solutions for $f(x,t) \equiv 0$.

1°. Multiplicative separable solutions:

$$u = [A\cos(kx) + B\sin(kx)]e^{-\lambda t}, \quad k = \sqrt{(\lambda + c_1 + c_2 e^{\lambda\tau})/(a_1 + a_2 e^{\lambda\tau})}$$
$$\text{if } \lambda + c_1 + c_2 e^{\lambda\tau} > 0; \qquad (1)$$

$$u = [A\exp(kx) + B\exp(-kx)]e^{-\lambda t}, \quad k = \sqrt{-(\lambda + c_1 + c_2 e^{\lambda\tau})/(a_1 + a_2 e^{\lambda\tau})}$$
$$\text{if } \lambda + c_1 + c_2 e^{\lambda\tau} < 0, \qquad (2)$$

where A, B, and λ are arbitrary constants. Note that these solutions are special cases of more complex multiplicative separable solutions of the form $u = \varphi(x)\psi(t)$.

Solution (1) is periodic in the spatial variable x and decaying as $t \to \infty$ (if $\lambda > 0$).

2°. Exact solutions periodic in time t:

$$u = e^{-\gamma x}[A\cos(\omega t - \beta x) + B\sin(\omega t - \beta x)],$$

where A, B, and ω are arbitrary constants, and the parameters β and γ can be expressed in terms of ω and the parameters of the original equation by solving the algebraic system of equations

$$[a_1 + a_2\cos(\omega\tau)](\gamma^2 - \beta^2) + 2a_2\sin(\omega\tau)\beta\gamma + c_1 + c_2\cos(\omega\tau) = 0,$$
$$a_2\sin(\omega\tau)(\gamma^2 - \beta^2) - 2[a_1 + a_2\cos(\omega\tau)]\beta\gamma + \omega + c_2\sin(\omega\tau) = 0.$$

By eliminating γ (or β) from this system, we can obtain a biquadratic equation for β (or γ).

3°. With some restrictions on the parameters of the original equation, there are solutions that are periodic in both independent variables, x and t:

$$u = [A_1\cos(\gamma x) + B_1\sin(\gamma x)][A_2\cos(\omega t) + B_2\sin(\omega t)],$$

where A_1, A_2, B_1, and B_2 are arbitrary constants, and the constants γ and ω are described by solving the transcendental system of equations

$$c_1 + c_2\cos(\omega\tau) = [a_1 + a_2\cos(\omega\tau)]\gamma^2,$$
$$\omega + c_2\sin(\omega\tau) = a_2\sin(\omega\tau)\gamma^2.$$

4°. There are solutions of polynomial form in x (containing, respectively, even and odd degrees):

$$u = \sum_{k=0}^{n} A_k(t) x^{2k} \quad \text{and} \quad u = \sum_{k=0}^{n} B_k(t) x^{2k+1}.$$

▶ **Formulations of initial-boundary value problems.**

We will consider the original parabolic delay PDE subject to an initial condition (initial data) of the general form

$$u = \varphi(x,t) \quad \text{for} \quad -\tau \le t \le 0 \quad (0 < x < h), \tag{3}$$

and various linear nonhomogeneous boundary conditions, which, for convenience, we will write in the compact form

$$\begin{aligned}\Gamma_1[u] &= g_1(t) \quad \text{for} \quad x = 0 \quad (t > -\tau),\\ \Gamma_2[u] &= g_2(t) \quad \text{for} \quad x = h \quad (t > -\tau),\end{aligned} \tag{4}$$

where $0 < h < \infty$.

We assume that the linear operators $\Gamma_{1,2}[u]$ appearing in the boundary conditions (4) do not explicitly depend on the time t. The most common boundary conditions are specified in the third column of Table 11.1.

TABLE 11.1. The simplest functions $u_0 = u_0(x,t)$ that satisfy the most common nonhomogeneous boundary conditions at the ends of the segment $0 \le x \le h$ ($k_1 > 0, k_2 > 0$).

No.	Initial-boundary value problem	Boundary conditions	Function $u_0 = u_0(x,t)$ satisfying the boundary conditions
1	First initial-boundary value problem	Dirichlet boundary conditions: $u = g_1(t)$ at $x = 0$ $u = g_2(t)$ at $x = h$	$u_0 = g_1(t) + \dfrac{x}{h}[g_2(t) - g_1(t)]$
2	Second initial-boundary value problem	Neumann boundary conditions: $u_x = g_1(t)$ at $x = 0$ $u_x = g_2(t)$ at $x = h$	$u_0 = xg_1(t) + \dfrac{x^2}{2h}[g_2(t) - g_1(t)]$
3	Third initial-boundary value problem	Robin boundary conditions: $u_x - k_1 u = g_1(t)$ at $x = 0$ $u_x + k_2 u = g_2(t)$ at $x = h$	$u_0 = \dfrac{(k_2 x - 1 - k_2 h)g_1(t) + (1 + k_1 x)g_2(t)}{k_2 + k_1 + k_1 k_2 h}$
4	Mixed initial-boundary value problem	Mixed boundary conditions: $u = g_1(t)$ at $x = 0$ $u_x = g_2(t)$ at $x = h$	$u_0 = g_1(t) + xg_2(t)$
5	Mixed initial-boundary value problem	Mixed boundary conditions: $u_x = g_1(t)$ at $x = 0$ $u = g_2(t)$ at $x = h$	$u_0 = (x - h)g_1(t) + g_2(t)$

▶ **Representation of solutions of initial-boundary value problems as the sum of three functions.**

The solution of the original equation with the initial condition (3) and boundary conditions (4) is sought as the sum of three functions

$$u = u_0(x,t) + u_1(x,t) + u_2(x,t), \tag{5}$$

where the functions $u_n = u_n(x,t)$ ($n = 0, 1, 2$) will be defined below.

The function
$$u_0 = u_0(x,t) \tag{6}$$
is any doubly continuously differentiable function satisfying the nonhomogeneous boundary conditions (4), i.e., the relations
$$\Gamma_1[u_0] = g_1(t) \quad \text{at} \quad x = 0, \quad \Gamma_2[u_0] = g_2(t) \quad \text{at} \quad x = h, \tag{7}$$
are satisfied.

Table 11.1 lists the simplest functions $u_0 = u_0(x,t)$ that satisfy the most common nonhomogeneous boundary conditions in initial-boundary value problems for equations of parabolic and hyperbolic types with one spatial variable.

The other two functions, $u_1 = u_1(x,t)$ and $u_2 = u_2(x,t)$, appearing in (5) are determined by solving the simpler initial boundary value problems described below with homogeneous (zero) boundary conditions.

Problem 1. The function u_1 satisfies the linear homogeneous PDE with constant delay
$$\frac{\partial u_1}{\partial t} = a_1 \frac{\partial^2 u_1}{\partial x^2} + a_2 \frac{\partial^2 w_1}{\partial x^2} + c_1 u_1 + c_2 w_1, \quad w_1 = u_1(x, t-\tau), \tag{8}$$
the nonhomogeneous initial condition
$$u_1 = \Phi(x,t) \quad \text{for} \quad -\tau \leq t \leq 0 \quad (0 < x < h), \tag{9}$$
and homogeneous boundary conditions
$$\Gamma_1[u_1] = 0 \quad \text{at} \quad x = 0 \quad (t > -\tau), \quad \Gamma_2[u_1] = 0 \quad \text{at} \quad x = h \quad (t > -\tau). \tag{10}$$

The function $\Phi(x,t)$ involved in the initial condition (9) is defined as:
$$\Phi(x,t) \equiv \varphi(x,t) - u_0(x,t). \tag{11}$$

Problem 2. The function u_2 satisfies the linear nonhomogeneous PDE with constant delay
$$\frac{\partial u_2}{\partial t} = a_1 \frac{\partial^2 u_2}{\partial x^2} + a_2 \frac{\partial^2 w_2}{\partial x^2} + c_1 u_2 + c_2 w_2 + F(x,t), \quad w_2 = u_2(x, t-\tau), \tag{12}$$
and homogeneous (zero) initial and boundary conditions
$$u_2 = 0 \quad \text{for} \quad -\tau \leq t \leq 0 \quad (0 < x < h), \tag{13}$$
$$\Gamma_1[u_2] = 0 \quad \text{at} \quad x = 0 \quad (t > -\tau), \quad \Gamma_2[u_2] = 0 \quad \text{at} \quad x = h \quad (t > -\tau). \tag{14}$$

The function $F(x,t)$ included in (12) is defined as:
$$F(x,t) \equiv f(x,t) - \frac{\partial u_0}{\partial t} + a_1 \frac{\partial^2 u_0}{\partial x^2} + a_2 \frac{\partial^2 w_0}{\partial x^2} + c_1 u_0 + c_2 w_0, \quad w_0 = u_0(x, t-\tau). \tag{15}$$

The solutions of Problems 1 and 2 with Dirichlet boundary conditions was obtained by Khusainov et al. (2009, 2013). In the sequel, we follow Polyanin et al. (2023), where solutions of problems with all boundary conditions from Table 11.1 are constructed.

▶ **Solution of Problem 1.**

Consider the linear homogeneous delay PDE (8) with the initial and boundary conditions (9) and (10). First, we look for particular solutions of equation (8) in the form of the product of functions with different arguments

$$u_{1p} = X(x)T(t). \tag{16}$$

Substituting (16) into (8) and separating the variables in the standard way, we arrive at the second-order ODE and first-order ODE with constant delay

$$X''(x) = -\lambda^2 X(x), \tag{17}$$
$$T'(t) = (c_1 - a_1\lambda^2)T(t) + (c_2 - a_2\lambda^2)T(t-\tau). \tag{18}$$

By requiring that function (16) satisfy the homogeneous boundary conditions (10), we arrive at homogeneous boundary conditions for X:

$$\Gamma_1[X] = 0 \quad \text{at} \quad x = 0, \quad \Gamma_2[X] = 0 \quad \text{at} \quad x = h. \tag{19}$$

The linear homogeneous eigenvalue problem (17), (19) admits nontrivial solutions $X = X_n(x)$ only if the parameter λ assumes a discrete set of values:

$$\lambda = \lambda_n, \quad X = X_n(x), \quad n = 1, 2, \ldots \tag{20}$$

Table 11.2 lists the eigenvalues and eigenfunctions for homogeneous linear boundary value problems described by ODE (17) for five most common boundary conditions.

Substituting the eigenvalues $\lambda = \lambda_n$ into (18), we obtain the corresponding delay ODEs for the functions $T = T_n(t)$.

Using the principle of linear superposition, we seek the solution of the linear initial-boundary value problem (8)–(11) in the form of a series

$$u_1(x,t) = \sum_{n=1}^{\infty} X_n(x)T_n(t), \tag{21}$$

where the functions $u_{1n}(x,t) = X_n(x)T_n(t)$ are particular solutions of equation (8) satisfying the homogeneous boundary conditions (10).

To find the initial conditions for the delay ODE (18) for $\lambda = \lambda_n$, we present the initial condition (9) as the expansion by eigenfunctions:

$$\Phi(x,t) = \sum_{n=1}^{\infty} \Phi_n(t) X_n(x), \quad -\tau \leq t \leq 0 \quad (0 \leq x \leq h), \tag{22}$$

where

$$\Phi_n(t) = \frac{1}{\|X_n\|^2} \int_0^h \Phi(\xi,t) X_n(\xi)\, d\xi, \quad \|X_n\|^2 = \int_0^h X_n^2(\xi)\, d\xi, \tag{23}$$

and the function $\Phi(\xi,t)$ is defined by formula (11). From the relations (21) and (22) we obtain the initial conditions for the delay ODE (18) with $\lambda = \lambda_n$ in the form

$$T_n(t) = \Phi_n(t) \quad \text{for} \quad -\tau \leq t \leq 0, \tag{24}$$

where the functions $\Phi_n(t)$ are given by expressions (23).

TABLE 11.2. Eigenvalues and eigenfunctions in eigenvalue problems described by the linear homogeneous ODE $X''_{xx} = -\lambda^2 X$ subject to the most common homogeneous boundary conditions at the endpoints of the interval $0 \leq x \leq h$ ($k_1 > 0$ and $k_2 > 0$).

No.	Boundary value problem	Boundary conditions	Eigenvalues and eigenfunctions λ_n and $X_n = X_n(x)$, $n = 1, 2, \ldots$
1	First boundary value problem	Dirichlet boundary conditions: $X = 0$ at $x = 0$ $X = 0$ at $x = h$	$\lambda_n = \dfrac{\pi n}{h}$; $X_n = \sin\dfrac{\pi n x}{h}$
2	Second boundary value problem	Neumann boundary conditions: $X'_x = 0$ at $x = 0$ $X'_x = 0$ at $x = h$	$\lambda_0 = 0$, $X_0 = 1$; $\lambda_n = \dfrac{\pi n}{h}$, $X_n = \cos\dfrac{\pi n x}{h}$
3	Third boundary value problem	Robin boundary conditions: $X'_x - k_1 X = 0$ at $x = 0$ $X'_x + k_2 X = 0$ at $x = h$	λ_n are roots of the transcendental equation $\dfrac{\tan(\lambda h)}{\lambda} = \dfrac{k_1 + k_2}{\lambda^2 - k_1 k_2}$ ($\lambda_n > 0$); $X_n = \cos(\lambda_n x) + \dfrac{k_1}{\lambda_n}\sin(\lambda_n x)$
4	Mixed boundary value problem	Mixed boundary conditions: $X = 0$ at $x = 0$ $X'_x = 0$ at $x = h$	$\lambda_n = \dfrac{\pi(2n-1)}{2h}$; $X_n = \sin\dfrac{\pi(2n-1)x}{2h}$
5	Mixed boundary value problem	Mixed boundary conditions: $X'_x = 0$ at $x = 0$ $X = 0$ at $x = h$	$\lambda_n = \dfrac{\pi(2n-1)}{2h}$; $X_n = \cos\dfrac{\pi(2n-1)x}{2h}$

The Cauchy-type problem (18) and (24) coincides, up to renaming, with the problem considered in Section 10.1.1 (see equation 10.1.1.2, Item 5°). By introducing the notation

$$\alpha_n = c_1 - a_1 \lambda_n^2, \quad \beta_n = c_2 - a_2 \lambda_n^2, \quad \sigma_n = e^{-\alpha_n \tau}\beta_n, \tag{25}$$

we can represent the solution of problem (18) and (24) in the closed form

$$T_n(t) = e^{\alpha_n(t+\tau)} \exp_d(\sigma_n t, \sigma_n \tau) \Phi_n(-\tau)$$
$$+ \int_{-\tau}^{0} e^{\alpha_n(t-s)} \exp_d(\sigma_n(t - \tau - s), \sigma_n \tau)[\Phi'_n(s) - \alpha_n \Phi_n(s)]\, ds, \tag{26}$$

where $\exp_d(t, \tau) \equiv \sum_{k=0}^{[t/\tau]+1} \dfrac{[t-(k-1)\tau]^k}{k!}$ is the delayed exponential function, and the symbol $[A]$ denotes the integer part of the number A.

Substituting functions (26) into formula (21), we find the solution to problem (8)–(11):

$$u_1(x,t) = \sum_{n=1}^{\infty} X_n(x)\Big\{ e^{\alpha_n(t+\tau)} \exp_d(\sigma_n t, \sigma_n \tau) \Phi_n(-\tau)$$
$$+ \int_{-\tau}^{0} e^{\alpha_n(t-s)} \exp_d(\sigma_n(t - \tau - s), \sigma_n \tau)[\Phi'_n(s) - \alpha_n \Phi_n(s)]\, ds \Big\}, \tag{27}$$

where

$$\Phi_n(t) = \frac{1}{\|X_n\|^2} \int_0^h [\varphi(\xi,t) - u_0(\xi,t)] X_n(\xi)\, d\xi, \quad \|X_n\|^2 = \int_0^h X_n^2(\xi)\, d\xi. \qquad (28)$$

For any of the five main initial-boundary value problems, whose boundary conditions are given in Table 11.1, one should substitute the respective eigenvalues λ_n and eigenfunctions $X_n(x)$ from Table 11.2 into formulas (26) and (27).

▶ **Solution of Problem 2.**

Let us now consider the linear nonhomogeneous delay PDE (12) with the homogeneous initial and boundary conditions (13) and (14).

First, we expand the nonhomogeneous component of equation (12) in a series of the eigenfunctions (20):

$$F(x,t) = \sum_{n=1}^\infty F_n(t) X_n(x), \qquad F_n(t) = \frac{1}{\|X_n\|^2} \int_0^h F(\xi,t) X_n(\xi)\, d\xi. \qquad (29)$$

The function $F(x,t)$ is defined by formula (15), and $\|X_n\|^2 = \int_0^h X_n^2(\xi)\, d\xi$.

We look for the solution of problem (12)–(15) in the form of a series

$$u_2(x,t) = \sum_{n=1}^\infty U_n(t) X_n(x) \qquad (30)$$

that satisfies the homogeneous boundary conditions (14). Substituting (30) into (12) and considering (29), we obtain linear nonhomogeneous ODEs with delay for the functions $U_n(t)$:

$$U_n'(t) = (c_1 - a_1 \lambda_n^2) U_n(t) + (c_2 - a_2 \lambda_n^2) U_n(t-\tau) + F_n(t). \qquad (31)$$

The functions $F_n(t)$ are found by the second formula in (29). To complete the formulation of the problem, we supplement equation (31) with the homogeneous initial conditions

$$U_n(t) = 0 \quad \text{for} \quad -\tau \le t \le 0, \qquad (32)$$

which follow from (13) and (30).

The Cauchy-type problem (31) and (32) coincides, up to renaming, with the problem considered in Section 10.1.1 (see equation 10.1.1.4). Therefore, its solution in the domain of $t \ge 0$ can be represented as

$$U_n(t) = \int_0^t e^{\alpha_n(t-s)} \exp_d(\sigma_n(t-s), \sigma_n \tau) F_n(s)\, ds, \qquad \sigma_n = e^{-\alpha_n \tau} \beta_n, \qquad (33)$$

where the parameters α_n and β_n are defined in (25). Substituting (33) into (30), we find a solution to the problem (12)–(15):

$$u_2(x,t) = \sum_{n=1}^\infty \left[\int_0^t e^{\alpha_n(t-s)} \exp_d(\sigma_n(t-s), \sigma_n \tau) F_n(s)\, ds \right] X_n(x). \qquad (34)$$

Solutions of initial boundary value problems for the original equation subject to the general initial condition (4) and any boundary conditions (5) presented in Table 11.1 can be obtained by substituting functions (6), (27), and (34) into (5) and using the functions $u_0 = u_0(x,t)$ from Table 11.1 with the respective eigenvalues λ_n and eigenfunctions $X_n(x)$ from Table 11.2.

2. $u_{tt} = a_1 u_{xx} + a_2 w_{xx} + c_1 u + c_2 w, \quad w = u(x, t - \tau).$

Linear wave equation with constant delay, where $a_1 > a_2 \geq 0$ and $\tau > 0$.

▶ **Exact solutions.**

$1°$. Multiplicative separable solutions:

$$u = [A\cos(kx) + B\sin(kx)]e^{-\lambda t}, \quad k = \sqrt{(c_1 + c_2 e^{\lambda\tau} - \lambda^2)/(a_1 + a_2 e^{\lambda\tau})}$$
$$\text{if } c_1 + c_2 e^{\lambda\tau} - \lambda^2 > 0; \tag{1}$$

$$u = [A\exp(kx) + B\exp(-kx)]e^{-\lambda t}, \quad k = \sqrt{-(c_1 + c_2 e^{\lambda\tau} - \lambda^2)/(a_1 + a_2 e^{\lambda\tau})}$$
$$\text{if } c_1 + c_2 e^{\lambda\tau} - \lambda^2 < 0, \tag{2}$$

where A, B, and λ are arbitrary constants. Note that these solutions are special cases of more complex multiplicative separable solutions of the form $u = \varphi(x)\psi(t)$.

Solution (1) is periodic in the spatial variable x and decaying as $t \to \infty$ (if $\lambda > 0$).

$2°$. For some defining parameters of the original equation, there are solutions that are periodic in both independent variables, x and t:

$$u = [A_1\cos(\gamma x) + B_1\sin(\gamma x)][A_2\cos(\omega t) + B_2\sin(\omega t)],$$

where A_1, A_2, B_1, and B_2 are arbitrary constants, and the constants γ and ω satisfy the transcendental system of equations

$$\omega^2 + c_1 + c_2\cos(\omega\tau) = [a_1 + a_2\cos(\omega\tau)]\gamma^2,$$
$$(c_2 - a_2\gamma^2)\sin(\omega\tau) = 0.$$

$3°$. For some defining parameters of the original equation, there are solutions that are periodic in time t:

$$u = [A_1\cosh(\gamma x) + B_1\sinh(\gamma x)][A_2\cos(\omega t) + B_2\sin(\omega t)],$$

where A_1, A_2, B_1, and B_2 are arbitrary constants, and the constants γ and ω are described by solving the transcendental system of equations

$$\omega^2 + c_1 + c_2\cos(\omega\tau) = -[a_1 + a_2\cos(\omega\tau)]\gamma^2,$$
$$(c_2 + a_2\gamma^2)\sin(\omega\tau) = 0.$$

$4°$. There are other solutions that are periodic in time t:

$$u = e^{-\gamma x}[A\cos(\omega t - \beta x) + B\sin(\omega t - \beta x)],$$

where A, B, and ω are arbitrary constants, and the parameters β and γ can be expressed in terms of ω and the parameters of the original equation by solving an algebraic system of equations which is omitted here.

5°. There are solutions of polynomial form in x (containing, respectively, even and odd degrees):
$$u = \sum_{k=0}^{n} A_k(t) x^{2k} \quad \text{and} \quad u = \sum_{k=0}^{n} B_k(t) x^{2k+1}.$$

▶ **Formulations of initial-boundary value problems ($0 \leq x \leq h$).**

We will consider an initial-boundary value problem for the original hyperbolic PDE with constant delay under the consistent initial conditions (initial data) of general form
$$\begin{aligned} u &= \varphi(x,t) \quad \text{for} \quad -\tau \leq t \leq 0 \quad (0 < x < h), \\ u_t &= \varphi_t(x,t) \quad \text{for} \quad -\tau \leq t \leq 0 \quad (0 < x < h) \end{aligned} \tag{3}$$
and various linear homogeneous boundary conditions, which, for convenience, we will write in the compact form
$$\Gamma_1[u] = 0 \quad \text{at} \quad x = 0 \quad (t > -\tau), \quad \Gamma_2[u] = 0 \quad \text{at} \quad x = h \quad (t > -\tau), \tag{4}$$
where $0 < h < \infty$.

We assume that the linear operators $\Gamma_{1,2}[u]$ appearing in the boundary conditions (4) do not explicitly depend on time t. The most common boundary conditions are given in the third column of Table 11.1, where $g_1(t) = g_2(t) \equiv 0$ should be taken. In particular, the Dirichlet boundary conditions in (4) correspond to $\Gamma_1[u] = \Gamma_2[u] = u$.

Remark 11.1. *The study by Rodríguez et al. (2012) obtained, by the method of separation of variables, an analytical solution to an initial-boundary value problem for the hyperbolic linear homogeneous equation with constant delay (11.1.1.2), with $a_2 = c_1 = 0$, subjected to the homogeneous Dirichlet boundary conditions $u|_{x=0} = u|_{x=h} = 0$.*

▶ **Constructing the solution of the initial boundary value problem.**

First, we look for particular solutions of the original equation as the product of functions of different arguments: $u_p = X(x)T(t)$. Separating the variables in the resulting equation, we arrive at the following second-order linear ODE and delay ODE:
$$X''(x) = -\lambda^2 X(x), \tag{5}$$
$$T''(t) = (c_1 - a_1\lambda^2)T(t) + (c_2 - a_2\lambda^2)T(t-\tau). \tag{6}$$

By requiring the function $u_p = X(x)T(t)$ to satisfy the homogeneous boundary conditions (4), we obtain homogeneous boundary conditions for X:
$$\Gamma_1[X] = 0 \quad \text{at} \quad x = 0, \quad \Gamma_2[X] = 0 \quad \text{at} \quad x = h. \tag{7}$$

The linear homogeneous eigenvalue problem (5) and (7) has nontrivial solutions $X = X_n(x)$ only if λ assumes a discrete set of values:
$$\lambda = \lambda_n, \quad X = X_n(x), \quad n = 1, 2, \ldots \tag{8}$$

The eigenvalues and eigenfunctions for homogeneous linear boundary value problems described by ODE (5) for the five most common boundary conditions are given in Table 11.2.

Substituting the eigenvalues $\lambda = \lambda_n$ into (6), we obtain the corresponding delay ODEs for $T = T_n(t)$.

One looks for the solution of the linear initial boundary value problem for the original equation subject to the initial and boundary conditions (3) and (4) in the series form

$$u(x,t) = \sum_{n=1}^{\infty} X_n(x) T_n(t). \qquad (9)$$

The functions $u_{1n}(x,t) = X_n(x) T_n(t)$ are particular solutions of the original equation satisfying the homogeneous boundary conditions (4).

To find the initial conditions for the delay ODE (6) for $\lambda = \lambda_n$, we represent the initial condition (3) as an eigenfunction expansion:

$$\varphi(x,t) = \sum_{n=1}^{\infty} \varphi_n(t) X_n(x), \quad -\tau \le t \le 0 \quad (0 \le x \le h), \qquad (10)$$

where

$$\varphi_n(t) = \frac{1}{\|X_n\|^2} \int_0^h \varphi(\xi, t) X_n(\xi)\, d\xi, \quad \|X_n\|^2 = \int_0^h X_n^2(\xi)\, d\xi. \qquad (11)$$

From conditions (3) and relations (9) and (10) we obtain the following initial conditions for the delay ODE (6) with $\lambda = \lambda_n$:

$$T_n(t) = \varphi_n(t), \quad T_n'(t) = \varphi_n'(t) \quad \text{for} \quad -\tau \le t \le 0, \qquad (12)$$

where the functions $\varphi_n(t)$ are defined in (11).

To construct an analytical solution to the Cauchy-type problem for the second-order delay ODE (6) with the initial data (12), we can use the results of the study by Rodríguez et al. (2012) that considered a similar problem for an ODE with a constant delay. The corresponding solution for the function $T_n(t)$ (provided $a_1 \lambda_1^2 > c_1$), obtained by the method of steps, is very cumbersome and is given in Section 10.3.1 (see equation 10.3.1.2). In addition, problem (6) and (12) can also be solved using the Laplace transform or numerical methods.

Once the functions $T_n(t)$ are found, the solution of the problem for the original hyperbolic delay PDE subject to the initial and boundary conditions (3) and (4) is determined by the series (9), in which the eigenfunctions $X_n(x)$ (and eigenvalues λ_n) for the five most common boundary conditions are taken from Table 11.2.

11.1.2. PDEs with Proportional Delay

1. $u_t = a_1 u_{xx} + a_2 w_{xx} + c_1 u + c_2 w, \quad w = u(x, pt).$

Linear reaction-diffusion equation with proportional delay, $0 < p < 1$.

▶ **Exact solutions.**

1°. Multiplicative separable solution periodic in x:

$$u = [A\cos(kx) + B\sin(kx)] \varphi(t),$$

where A, B, k are arbitrary constants, and the function $\varphi = \varphi(t)$ is described by the first-order linear ODE with proportional delay

$$\varphi'_t = (c_1 - a_1 k^2)\varphi + (c_2 - a_2 k^2)\bar{\varphi}, \quad \bar{\varphi} = \varphi(pt).$$

This equation admits an analytical solution in the form of a power series having an infinite radius of convergence (see solution of equation 10.1.2.2).

2°. Multiplicative separable solution:

$$u = [A\cosh(kx) + B\sinh(kx)]\varphi(t),$$

where A, B, k are arbitrary constants, and the function $\varphi = \varphi(t)$ is described by the first-order linear ODE with proportional delay

$$\varphi'_t = (c_1 + a_1 k^2)\varphi + (c_2 + a_2 k^2)\bar{\varphi}, \quad \bar{\varphi} = \varphi(pt).$$

This equation admits an analytical solution in the form of a power series (see solution of equation 10.1.2.2).

3°. There are polynomial solutions in x (containing, respectively, even and odd degrees):

$$u = \sum_{k=0}^{n} A_k(t) x^{2k} \quad \text{and} \quad u = \sum_{k=0}^{n} B_k(t) x^{2k+1}.$$

▶ **Formulations of initial-boundary value problems ($0 \leq x \leq h$).**

We will consider the initial-boundary value problem for the original parabolic PDE with proportional delay under the initial condition of general form

$$u = \varphi(x) \quad \text{at} \quad t = 0 \tag{1}$$

and various linear homogeneous boundary conditions, which, for convenience, we will write in the compact form

$$\Gamma_1[u] = 0 \quad \text{at} \quad x = 0, \quad \Gamma_2[u] = 0 \quad \text{at} \quad x = h, \tag{2}$$

where $0 < h < \infty$.

We assume that the linear operators $\Gamma_{1,2}[u]$ appearing in the boundary conditions (2) do not explicitly depend on the time t. The most common boundary conditions are given in the third column of Table 11.1, where $g_1(t) = g_2(t) \equiv 0$ should be taken.

▶ **Constructing the solution of the initial boundary value problem.**

As usual, we look for particular solutions to the original equation as the product of functions of different arguments: $u_p = X(x)T(t)$. Separating the variables in the resulting equation, we arrive at the second-order linear ODE and the first-order ODE with proportional delay:

$$X''(x) = -\lambda^2 X(x), \tag{3}$$
$$T'(t) = (c_1 - a_1\lambda^2)T(t) + (c_2 - a_2\lambda^2)T(pt). \tag{4}$$

By requiring the function $u_p = X(x)T(t)$ to satisfy the homogeneous boundary conditions (2), we obtain homogeneous boundary conditions X:

$$\Gamma_1[X] = 0 \quad \text{at} \quad x = 0, \quad \Gamma_2[X] = 0 \quad \text{at} \quad x = h. \tag{5}$$

The homogeneous eigenvalue problem (3), (5) has nontrivial solutions $X = X_n(x)$ only if the parameter λ assumes a discrete set of values:

$$\lambda = \lambda_n, \quad X = X_n(x), \quad n = 1, 2, \ldots \tag{6}$$

The eigenvalues and eigenfunctions for homogeneous linear boundary value problems described by ODE (3) for the five most common boundary conditions are given in Table 11.2.

Substituting the eigenvalues $\lambda = \lambda_n$ into (4), we obtain the corresponding ODEs with proportional delay for the functions $T = T_n(t)$.

Using the principle of linear superposition, we seek the solution to the linear initial-boundary value problem for the original equation with the initial and boundary conditions (1) and (2) in the series form

$$u(x,t) = \sum_{n=1}^{\infty} T_n(t) X_n(x), \tag{7}$$

where the functions $T_n(t)$ are described by equation (4) with $\lambda = \lambda_n$. By construction, the series (7) satisfies the original equation and homogeneous boundary conditions (2).

To find the initial conditions for ODEs with proportional delay (4) with $\lambda = \lambda_n$, we represent the function $\varphi(x)$ included in the initial condition (1) as an eigenfunction expansion:

$$\varphi(x) = \sum_{n=1}^{\infty} A_n X_n(x), \tag{8}$$

where

$$A_n = \frac{1}{\|X_n\|^2} \int_0^h \varphi(\xi) X_n(\xi)\, d\xi, \quad \|X_n\|^2 = \int_0^h X_n^2(\xi)\, d\xi. \tag{9}$$

From relations (7) and (8) we obtain the following initial conditions for the ODE with proportional delay (4) for $\lambda = \lambda_n$:

$$T_n(0) = A_n, \tag{10}$$

where the coefficients A_n are defined in (9).

The linear problem with proportional delay (4) and (10) with $\lambda = \lambda_n$ and $A_n = 1$ coincides, up to renaming, with the problem discussed earlier in Section 10.1.2 (see equation 10.1.2.2). Therefore, the solution of problem (4) and (10) can be represented as the power series

$$T_n(t) = A_n\left(1 + \sum_{m=1}^{\infty} \gamma_{mn} t^m\right), \tag{11}$$

$$\gamma_{mn} = \frac{1}{m!} \prod_{k=0}^{m-1} (\alpha_n + \beta_n p^k), \quad \alpha_n = c_1 - a_1 \lambda_n^2, \quad \beta_n = c_2 - a_2 \lambda_n^2.$$

Substituting expressions (11) into formula (7), yields the solution of the original problem for PDE with proportional delay (Polyanin et al., 2023):

$$u(x,t) = \sum_{n=1}^{\infty} A_n \left(1 + \sum_{m=1}^{\infty} \gamma_{mn} t^m\right) X_n(x), \qquad (12)$$

where the coefficients A_n and γ_{mn} are defined in (9) and (11).

The solution of the initial-boundary problems for the original equation with the five boundary conditions presented in Table 11.1 can be obtained by formulas (9), (11), and (12) by taking the corresponding eigenvalues λ_n and eigenfunctions $X_n(x)$ from Table 11.2.

2. $u_t = a_1 u_{xx} + a_2 w_{xx}, \quad w = u(px, qt).$

Here $p > 0$ and $q > 0$ are scaling factors.

Suppose that the original PDE with two proportional arguments is subjected to the special initial and boundary conditions

$$u = A \quad \text{at} \quad t = 0, \quad u = B \quad \text{at} \quad x = 0, \qquad (1)$$

where A and B are arbitrary constants.

The initial-boundary value problem for the original equation under conditions (1) has a self-similar that can be represented as

$$u = U(z), \quad z = x t^{-1/2}. \qquad (2)$$

The function $U(z)$ satisfies the following boundary value problem for the ODE with proportional argument:

$$-\tfrac{1}{2} z U'_z = a_1 U''_{zz} + a_2 W''_{zz}, \quad W = U(\sigma z), \quad \sigma = p q^{-1/2}; \qquad (3)$$

$$U(0) = B, \quad U(\infty) = A. \qquad (4)$$

Suppose that the scaling factors p and q are related by the parabolic equation $q = p^2$; then $\sigma = 1$ and $U = W$. In this special case, equation (3) is easily integrated, and the solution of the original problem (3) and (4) is expressed as

$$u = B + (A - B) \operatorname{erf}\left(\frac{x}{2\sqrt{at}}\right), \quad a = a_1 + a_2,$$

where $\operatorname{erf} \zeta = \frac{2}{\sqrt{\pi}} \int_0^\zeta \exp(-\xi^2)\, d\xi$ is the error function (probability integral).

3. $u_{tt} = a_1 u_{xx} + a_2 w_{xx} + c_1 u + c_2 w, \quad w = u(x, pt).$

Linear wave equation with proportional delay, $0 < p < 1$.

▶ **Exact solutions.**

1°. Multiplicative separable solution periodic in x:

$$u = [A \cos(kx) + B \sin(kx)] \varphi(t),$$

where A, B, k are arbitrary constants, and the function $\varphi = \varphi(t)$ is described by the second-order linear ODE with proportional delay

$$\varphi''_{tt} = (c_1 - a_1 k^2) \varphi + (c_2 - a_2 k^2) \bar{\varphi}, \quad \bar{\varphi} = \varphi(pt).$$

This equation admits an analytical solution in the form of a power series (see solution of equation 10.3.2.4 with $c = 0$).

2°. Multiplicative separable solution:
$$u = [A\cosh(kx) + B\sinh(kx)]\varphi(t),$$
where A, B, k are arbitrary constants, and the function $\varphi = \varphi(t)$ is described by the second-order linear ODE with proportional delay
$$\varphi''_{tt} = (c_1 + a_1 k^2)\varphi + (c_2 + a_2 k^2)\bar{\varphi}, \quad \bar{\varphi} = \varphi(pt).$$
This equation admits an analytical solution in the form of a power series (see solution of equation 10.3.2.4 with $c = 0$).

3°. There are polynomial solutions in x (containing, respectively, even and odd degrees):
$$u = \sum_{k=0}^{n} A_k(t) x^{2k} \quad \text{and} \quad u = \sum_{k=0}^{n} B_k(t) x^{2k+1}.$$

▶ **Formulations of initial-boundary value problems ($0 \le x \le h$).**

We will consider the initial-boundary value problem for the original hyperbolic PDE with proportional delay subject to the initial conditions of general form
$$u = \varphi(x) \quad \text{at} \quad t = 0, \quad u_t = \psi(x) \quad \text{at} \quad t = 0 \qquad (1)$$
and various linear homogeneous boundary conditions, which, for convenience, we will write in the compact form
$$\Gamma_1[u] = 0 \quad \text{at} \quad x = 0, \quad \Gamma_2[u] = 0 \quad \text{at} \quad x = h, \qquad (2)$$
where $0 < h < \infty$. The most common boundary conditions are given in the third column of Table 11.1, where $g_1(t) = g_2(t) \equiv 0$ should be substituted.

We look for particular solutions to the original equation in the form of a product of functions of different arguments $u_p = X(x)T(t)$. Using the standard procedure for separating variables, we arrive at the second-order linear ODE and the second-order ODE with proportional delay:
$$X''(x) = -\lambda^2 X(x), \qquad (3)$$
$$T''(t) = (c_1 - a_1 \lambda^2) T(t) + (c_2 - a_2 \lambda^2) T(pt). \qquad (4)$$
By requiring the function $u_p = X(x)T(t)$ to satisfy homogeneous boundary conditions (2), we arrive at homogeneous boundary conditions for the function X:
$$\Gamma_1[X] = 0 \quad \text{at} \quad x = 0, \quad \Gamma_2[X] = 0 \quad \text{at} \quad x = h. \qquad (5)$$

The homogeneous eigenvalue problem (3) and (5) has nontrivial solutions $X = X_n(x)$ only if the parameter λ assumes a discrete set of values:
$$\lambda = \lambda_n, \quad X = X_n(x), \quad n = 1, 2, \ldots \qquad (6)$$

The eigenvalues and eigenfunctions for homogeneous linear boundary value problems described by ODE (3) for the five most common boundary conditions are given in Table 11.2.

Substituting the eigenvalues $\lambda = \lambda_n$ into (4) gives the corresponding ODEs with proportional delay for the functions $T = T_n(t)$.

Using the principle of linear superposition, we seek the solution of the linear initial-boundary value problem for the original equation with the initial and boundary conditions (1) and (2) in the series form

$$u(x,t) = \sum_{n=1}^{\infty} T_n(t) X_n(x), \tag{7}$$

where the functions $T_n(t)$ are described by equation (4) with $\lambda = \lambda_n$. By construction, the series (7) satisfies the original equation and homogeneous boundary conditions (2).

To find the initial conditions for the second-order ODE with proportional delay (4), we represent the functions $\varphi(x)$ and $\psi(x)$ appearing in the initial conditions (1) as eigenfunction expansions:

$$\varphi(x) = \sum_{n=1}^{\infty} A_n X_n(x), \quad \psi(x) = \sum_{n=1}^{\infty} B_n X_n(x),$$

$$A_n = \frac{1}{\|X_n\|^2} \int_0^h \varphi(\xi) X_n(\xi) \, d\xi, \quad B_n = \frac{1}{\|X_n\|^2} \int_0^h \psi(\xi) X_n(\xi) \, d\xi, \tag{8}$$

where $\|X_n\|^2 = \int_0^h X_n^2(\xi) \, d\xi$. From relations (7) and (8) we obtain the following initial conditions for the ODE with proportional delay (4):

$$T_n(0) = A_n, \quad T_n'(0) = B_n. \tag{9}$$

The linear problem for the second-order ODE with proportional delay (4), (9) coincides, up to renaming, with the problem analyzed in Section 10.3.2 (see equation 10.3.2.4 with $c = 0$). Considering the above, we represent the solution of problem (4) and (9) as a linear combination of two power series:

$$T_n(t) = A_n T_{n1}(t) + B_n T_{n2}(t), \tag{10}$$

where

$$T_{n1}(t) = 1 + \sum_{m=1}^{\infty} \gamma_{n,2m} t^{2m}, \quad \gamma_{n,2m} = \frac{1}{(2m)!} \prod_{k=0}^{m-1} (\alpha_n + \beta_n p^{2k});$$

$$T_{n2}(t) = t + \sum_{m=1}^{\infty} \gamma_{n,2m+1} t^{2m+1}, \quad \gamma_{n,2m+1} = \frac{1}{(2m+1)!} \prod_{k=0}^{m-1} (\alpha_n + \beta_n p^{2k+1}); \tag{11}$$

$$\alpha_n = c_1 - a_1 \lambda_n^2, \quad \beta_n = c_2 - a_2 \lambda_n^2.$$

For $0 < p < 1$, both series in (11) have an infinite radius of convergence.

Substituting expressions (10) into (7) yields the solution of the original initial-boundary value problem (Polyanin et al., 2023):

$$u(x,t) = \sum_{n=1}^{\infty} [A_n T_{n1}(t) + B_n T_{n2}(t)] X_n(x),$$

$$T_{n1}(t) = 1 + \sum_{m=1}^{\infty} \gamma_{n,2m} t^{2m}, \quad T_{n2}(t) = t + \sum_{m=1}^{\infty} \gamma_{n,2m+1} t^{2m+1}, \tag{12}$$

11.1.3. PDEs with Anisotropic Time Delay

▶ **Reaction-diffusion type equations.**

1. $u_t = au_{xx} + bu + cw, \quad w = u(x, t - \tau(x))$.

Linear reaction-diffusion equation with anisotropic time delay; $\tau(x)$ *is a positive funtion.*

1°. Multiplicative separable solution:
$$u = e^{\lambda t}\varphi(x),$$
where λ is an arbitrary constant, and the function $\varphi = \varphi(x)$ satisfies the linear ODE
$$a\varphi''_{xx} + (b - \lambda + ce^{-\lambda \tau(x)})\varphi = 0.$$

Remark 11.2. *By the principle of linear superposition, there are solutions of the form* $u = \sum_{k=1}^{n} \exp(\lambda_k t)\varphi_k(x)$.

2°. Generalized separable solution periodic in t:
$$u = \varphi(x)\cos(\lambda t) + \psi(x)\sin(\lambda t),$$
where λ is an arbitrary constant, and the functions $\varphi = \varphi(x)$ and $\psi = \psi(x)$ are described by the linear ODE system
$$a\varphi''_{xx} + [b + c\cos(\lambda\tau)]\varphi - [\lambda + c\sin(\lambda\tau)]\psi = 0, \quad \tau = \tau(x),$$
$$a\psi''_{xx} + [b + c\cos(\lambda\tau)]\psi + [\lambda + c\sin(\lambda\tau)]\varphi = 0.$$

Remark 11.3. *By the principle of linear superposition, there are solutions of the form* $u = \sum_{k=1}^{n}[\varphi_k(x)\cos(\lambda_k t) + \psi_k(x)\sin(\lambda_k t)]$.

3°. Generalized separable solution linear in t:
$$u = \varphi(x)t + \psi(x),$$
where the functions $\varphi = \varphi(x)$ and $\psi = \psi(x)$ are described by the system of linear constant-coefficient ODEs:
$$a\varphi''_{xx} + (b + c)\varphi = 0,$$
$$a\psi''_{xx} + (b + c)\psi - [c\tau(x) + 1]\varphi = 0,$$
which is easy to integrate sequentially (since the first ODE is homogeneous, and the second one is nonhomogeneous).

4°. There are polynomial solutions in t:
$$u = \sum_{k=0}^{n} \varphi_k(x)t^k,$$
where n is an arbitrary positive integer, and the functions $\varphi_k(x)$ are described by a system of ODEs.

2. $u_t = au_{xx} + bu + cw$, $w = u(x, p(x)t)$.

Linear reaction-diffusion equation with proportional anisotropic time delay; $p(x)$ is a positive function less than unity.

1°. Generalized separable solution linear in t:
$$u = \varphi(x)t + \psi(x),$$
where the functions $\varphi = \varphi(x)$ and $\psi = \psi(x)$ are described by the linear ODE system
$$a\varphi''_{xx} + (b + cp(x))\varphi = 0,$$
$$a\psi''_{xx} + (b + c)\psi - \varphi = 0,$$
which can be easily integrated.

2°. There are polynomial solutions in t:
$$u = \sum_{k=0}^{n} \varphi_k(x) t^k,$$
where n is an arbitrary positive integer, and the functions $\varphi_k(x)$ are described by a system of ODEs.

▶ **Wave-type equations.**

3. $u_{tt} = au_{xx} + bu + cw$, $w = u(x, t - \tau(x))$.

Linear wave-type equation with anisotropic time delay; $\tau(x)$ is an arbitrary function.

1°. Multiplicative separable solution:
$$u = e^{\lambda t}\varphi(x),$$
where λ is an arbitrary constant, and the function $\varphi = \varphi(x)$ is described by the linear second-order ODE:
$$a\varphi''_{xx} + (b - \lambda^2 + ce^{-\lambda\tau(x)})\varphi = 0.$$

Remark 11.4. *By the principle of linear superposition, there are solutions of the form $u = \sum_{k=1}^{n} \exp(\lambda_k t)\varphi_k(x)$.*

2°. Generalized separable exact solution periodic in t:
$$u = \varphi(x)\cos(\lambda t) + \psi(x)\sin(\lambda t),$$
where λ is an arbitrary constant, and the functions $\varphi = \varphi(x)$ and $\psi = \psi(x)$ satisfy the linear system of ODEs:
$$a\varphi''_{xx} + [b + \lambda^2 + c\cos(\lambda\tau)]\varphi - c\sin(\lambda\tau)\psi = 0, \quad \tau = \tau(x),$$
$$a\psi''_{xx} + [b + \lambda^2 + c\cos(\lambda\tau)]\psi + c\sin(\lambda\tau)\varphi = 0.$$

Remark 11.5. *By the principle of linear superposition, there are solutions of the form $u = \sum_{k=1}^{n}[\varphi_k(x)\cos(\lambda_k t) + \psi_k(x)\sin(\lambda_k t)]$.*

3°. There are polynomial solutions in t:
$$u = \sum_{k=0}^{n} \varphi_k(x) t^k,$$
where n is an arbitrary positive integer, and the functions $\varphi_k(x)$ are described by a system of ODEs.

4. $u_{tt} = a u_{xx} + bu + cw, \quad w = u(x, p(x)t).$

Linear wave-type equation with proportional anisotropic time delay; $p(x)$ is an arbitrary function.

1°. Generalized separable solution quadratic in t:
$$u = \varphi(x) t^2 + \psi(x),$$
where the functions $\varphi = \varphi(x)$ and $\psi = \psi(x)$ satisfy the linear system of ODEs:
$$a \varphi''_{xx} + (b + c p^2(x)) \varphi = 0,$$
$$a \psi''_{xx} + (b + c) \psi - 2\varphi = 0,$$
which can be easily integrated sequentially.

2°. There are polynomial solutions in t:
$$u = \sum_{k=0}^{n} \varphi_k(x) t^k,$$
where n is an arbitrary positive integer, and the functions $\varphi_k(x)$ are described by a system of ODEs.

11.2. Nonlinear PDEs with Constant Delays

▶ Throughout this section, it is assumed that $f = f(z)$, $g = g(z)$, and $h = h(z)$ are arbitrary functions, $\tau > 0$ and $\sigma > 0$ are arbitrary constants, $u = u(x, t)$ is the unknown function, and $w = u(x, t - \tau)$.

11.2.1. Parabolic Equations

▶ **Equations linear in derivatives and involving arbitrary parameters.**

1. $u_t = a u_{xx} + b u^3 + c w^3, \quad w = u(x, t - \tau).$
This is a special case of equation 11.2.1.13 for $f(z) = a + bz^3$.

2. $u_t = a u_{xx} + b u^k w^{3-k} + c u^m w^{3-m}, \quad w = u(x, t - \tau).$
This is a special case of equation 11.2.1.13 for $f(z) = bz^{3-k} + cz^{3-m}$.

3. $u_t = a u_{xx} + u(b \ln u + c \ln w + d), \quad w = u(x, t - \tau).$
Functional separable solution (Polyanin & Zhurov, 2013):
$$u(x, t) = \exp[\psi_2(t) x^2 + \psi_1(t) x + \psi_0(t)],$$

where the functions $\psi_n = \psi_n(t)$ are described by the nonlinear ODE system with constant delay

$$\psi_2' = 4a\psi_2^2 + b\psi_2 + c\bar\psi_2, \quad \bar\psi_2 = \psi_2(t-\tau),$$
$$\psi_1' = 4a\psi_1\psi_2 + b\psi_1 + c\bar\psi_1, \quad \bar\psi_1 = \psi_1(t-\tau),$$
$$\psi_0' = a(\psi_1^2 + 2\psi_2) + b\psi_0 + c\bar\psi_0 + d, \quad \bar\psi_0 = \psi_0(t-\tau).$$

4. $u_t = au_{xx} + u(b\ln^2 u + c\ln u + d\ln w + s), \quad w = u(x, t - \tau)$.

1°. Functional separable solution with $ab > 0$ (Polyanin & Zhurov, 2013):

$$u(x,t) = \exp[\psi_1(t)\varphi(x) + \psi_2(t)],$$
$$\varphi(x) = A\cos(\lambda x) + B\sin(\lambda x), \quad \lambda = \sqrt{b/a},$$

where A and B are arbitrary constants, and the functions $\psi_n = \psi_n(t)$ are described by the nonlinear ODE system with constant delay

$$\psi_1' = 2b\psi_1\psi_2 + (c-b)\psi_1 + d\bar\psi_1, \quad \bar\psi_1 = \psi_1(t-\tau),$$
$$\psi_2' = b(A^2 + B^2)\psi_1^2 + b\psi_2^2 + c\psi_2 + d\bar\psi_2 + s, \quad \bar\psi_2 = \psi_2(t-\tau).$$

2°. Functional separable solution for $ab < 0$ (Polyanin & Zhurov, 2013):

$$u(x,t) = \exp[\psi_1(t)\varphi(x) + \psi_2(t)],$$
$$\varphi(x) = A\cosh(\lambda x) + B\sinh(\lambda x), \quad \lambda = \sqrt{-b/a},$$

where A and B are arbitrary constants, and the functions $\psi_n = \psi_n(t)$ are described by the nonlinear ODE system with constant delay

$$\psi_1' = 2b\psi_1\psi_2 + (c-b)\psi_1 + d\bar\psi_1, \quad \bar\psi_1 = \psi_1(t-\tau),$$
$$\psi_2' = b(A^2 - B^2)\psi_1^2 + b\psi_2^2 + c\psi_2 + d\bar\psi_2 + s, \quad \bar\psi_2 = \psi_2(t-\tau).$$

For $A = \pm B$, we have $\varphi(x) = Ae^{\pm\lambda x}$. In this case, the second equation of the system becomes independent, and the first one becomes linear in ψ_1.

▶ **Equations linear in derivatives and involving one arbitrary function.**

5. $u_t = au_{xx} + f(u - w), \quad w = u(x, t - \tau)$.

1°. Additive separable solution quadratic in x (Polyanin & Zhurov, 2014a):

$$u = C_2 x^2 + C_1 x + \psi(t),$$

where C_1 and C_2 are arbitrary constants, and the function $\psi(t)$ is described by the first-order ODE with constant delay

$$\psi_t'(t) = 2C_2 a + f\big(\psi(t) - \psi(t-\tau)\big).$$

This delay ODE has a linear particular solution, $\psi(t) = \lambda t + C_3$, where C_3 is an arbitrary constant, and λ is a root of the algebraic (transcendental) equation $2C_2 a - \lambda + f(\tau\lambda) = 0$.

2°. More complex exact solution:
$$u = C_1 x^2 + C_2 x + C_3 t + \theta(z), \quad z = \beta x + \gamma t,$$
where C_1, C_2, C_3, β, and γ are arbitrary constants, and the function $\theta(z)$ is described by the second-order ODE with constant delay
$$a\beta^2 \theta''_{zz}(z) - \gamma \theta'_z(z) + 2C_1 a - C_3 + f\bigl(\theta(z) - \theta(z-\sigma) + C_3\tau\bigr) = 0, \quad \sigma = \gamma\tau.$$

6. $u_t = a u_{xx} + f(u - v), \quad v = u(x - \sigma, t).$

1°. Additive separable solution:
$$u(x,t) = Ct + \varphi(x),$$
where C is an arbitrary constant, and the function $\varphi = \varphi(x)$ is described by a nonlinear second-order ODE with constant delay
$$a\varphi''_{xx} - C + f(\varphi - \bar\varphi) = 0, \quad \bar\varphi = \varphi(x - \sigma).$$

2°. The equation under consideration also has a more general solution with mixed additive separation of variables:
$$u = \alpha x + \beta t + \theta(z), \quad z = \lambda x + \gamma t,$$
where α, β, γ, λ are arbitrary constants, and the function $\theta(z)$ is described by the nonlinear second-order ODE with constant delay
$$a\lambda^2 \theta''_{zz} - \gamma \theta'_z - \beta + f(\theta - \bar\theta + \alpha\sigma) = 0, \quad \bar\theta = \theta(z - \lambda\sigma).$$

7. $u_t = a u_{xx} + b u + f(u - w), \quad w = u(x, t - \tau).$

For $b = 0$, see the preceding equation.

1°. Additive separable solution for $ab > 0$ periodic in x:
$$u = C_1 \cos(\lambda x) + C_2 \sin(\lambda x) + \psi(t), \quad \lambda = \sqrt{b/a},$$
where C_1 and C_2 are arbitrary constants, and the function $\psi(t)$ is described by the first-order ODE with constant delay
$$\psi'_t(t) = b\psi(t) + f\bigl(\psi(t) - \psi(t-\tau)\bigr). \tag{1}$$

2°. Additive separable solution for $ab < 0$:
$$u = C_1 \exp(-\lambda x) + C_2 \exp(\lambda x) + \psi(t), \quad \lambda = \sqrt{-b/a},$$
where C_1 and C_2 are arbitrary constants, and the function $\psi(t)$ is described by the first-order ODE with constant delay (1).

3°. Degenerate additive separable solution for $b = 0$:
$$u = C_1 x + C_2 + \psi(t),$$
where the function $\psi(t)$ is described by the ODE with constant delay (1) with $b = 0$.

4°. An exact solution for $ab > 0$ that generalizes the solution from Item 1°:

$$u = C_1 \cos(\lambda x) + C_2 \sin(\lambda x) + \theta(z), \qquad z = \beta x + \gamma t, \qquad \lambda = \sqrt{b/a},$$

where C_1, C_2, β, and γ are arbitrary constants, and the function $\theta(z)$ is described by the second-order ODE with constant delay

$$\gamma \theta'_z(z) = a\beta^2 \theta''_{zz}(z) + b\theta(z) + f(\theta(z) - \theta(z - \sigma)), \qquad \sigma = \gamma \tau. \qquad (2)$$

5°. An exact solution for $ab < 0$ that generalizes the solution from Item 2°:

$$u = C_1 \exp(-\lambda x) + C_2 \exp(\lambda x) + \theta(z), \qquad z = \beta x + \gamma t, \qquad \lambda = \sqrt{-b/a},$$

where C_1, C_2, β, and γ are arbitrary constants, and the function $\theta(z)$ is described by the second-order ODE with constant delay (2).

6°+. Generalized separable solution:

$$u = t\varphi(x) + \psi(x),$$

where the functions $\varphi = \varphi(x)$ and $\psi = \psi(x)$ are described by the second-order ODEs:

$$a\varphi''_{xx} + b\varphi = 0,$$
$$a\psi''_{xx} + b\psi + f(\tau\varphi) - \varphi = 0,$$

which are easy to integrate.

7°. References for PDE 11.2.1.7: Polyanin & Zhurov (2014a, 2014b) and Polyanin et al. (2023).

8. $u_t = au_{xx} + bu + f(u - kw), \quad w = u(x, t - \tau), \quad k > 0.$

1°. Generalized separable solution:

$$u = e^{ct}\varphi(x) + \psi(x), \qquad c = \frac{1}{\tau} \ln k,$$

where the functions $\varphi = \varphi(x)$ and $\psi = \psi(x)$ are described by the second-order independent ODEs:

$$a\varphi''_{xx} + (b - c)\varphi = 0,$$
$$a\psi''_{xx} + b\psi + f(\eta) = 0, \qquad \eta = (1 - k)\psi.$$

2°. Suppose $u_0(x, t)$ is a solution to the original nonlinear delay PDE and the function $v = V_1(x, t; b)$ is any τ-periodic solution to the linear heat equation with source

$$v_t = av_{xx} + bv, \qquad v(x, t) = v(x, t - \tau).$$

Then the sum

$$u = u_0(x, t) + e^{ct} V_1(x, t; b - c), \qquad c = \frac{1}{\tau} \ln k,$$

is also a solution to the original delay PDE. The general form of the function $V_1(x,t;b)$ is given by the formulas

$$V_1(x,t;b) = \sum_{n=0}^{\infty} \exp(-\lambda_n x)\big[A_n \cos(\beta_n t - \gamma_n x) + B_n \sin(\beta_n t - \gamma_n x)\big]$$

$$+ \sum_{n=1}^{\infty} \exp(\lambda_n x)\big[C_n \cos(\beta_n t + \gamma_n x) + D_n \sin(\beta_n t + \gamma_n x)\big],$$

$$\beta_n = \frac{2\pi n}{\tau}, \quad \lambda_n = \left(\frac{\sqrt{b^2 + \beta_n^2} - b}{2a}\right)^{1/2}, \quad \gamma_n = \left(\frac{\sqrt{b^2 + \beta_n^2} + b}{2a}\right)^{1/2},$$

where A_n, B_n, C_n, and D_n are arbitrary constants.

Note that a spatially homogeneous solution $u_0(t)$ or a stationary solution $u_0(x)$ may also be taken as the particular solution $u_0(x,t)$ of the original equation. Furthermore, stationary points $u_0 = $ const may also be used as the simplest particular solutions.

3°. References for PDE 11.2.1.8: Polyanin & Zhurov (2014b) and Polyanin et al. (2023).

9. $u_t = au_{xx} + bu + f(u + kw)$, $w = u(x, t - \tau)$, $k > 0$.

1°. Suppose $u_0(x,t)$ is a solution to the original nonlinear delay PDE and the function $v = V_2(x,t;b)$ is any τ-antiperiodic solution to the linear heat equation with source

$$v_t = av_{xx} + bv, \qquad v(x,t) = -v(x, t - \tau).$$

Then the sum

$$u = u_0(x,t) + e^{ct}V_2(x,t;b-c), \qquad c = \frac{1}{\tau}\ln k,$$

is also a solution to the original delay PDE. The general form of the function $V_2(x,t;b)$ is given by

$$V_2(x,t;b) = \sum_{n=1}^{\infty} \exp(-\lambda_n x)\big[A_n \cos(\beta_n t - \gamma_n x) + B_n \sin(\beta_n t - \gamma_n x)\big]$$

$$+ \sum_{n=1}^{\infty} \exp(\lambda_n x)\big[C_n \cos(\beta_n t + \gamma_n x) + D_n \sin(\beta_n t + \gamma_n x)\big],$$

$$\beta_n = \frac{\pi(2n-1)}{\tau}, \quad \lambda_n = \left(\frac{\sqrt{b^2 + \beta_n^2} - b}{2a}\right)^{1/2}, \quad \gamma_n = \left(\frac{\sqrt{b^2 + \beta_n^2} + b}{2a}\right)^{1/2},$$

where A_n, B_n, C_n, and D_n are arbitrary constants.

Note that a spatially homogeneous solution $u_0(t)$ or a stationary solution $u_0(x)$ may also be taken as the particular solution $u_0(x,t)$.

2°. References for PDE 11.2.1.9: Polyanin & Zhurov (2014b) and Polyanin et al. (2023).

10. $u_t = au_{xx} + uf(w/u)$, $w = u(x, t - \tau)$.

1°. Multiplicative separable solution, periodic in x:

$$u = [C_1 \cos(\beta x) + C_2 \sin(\beta x)]\psi(t),$$

where C_1, C_2, and β are arbitrary constants, and the function $\psi(t)$ is described by the first-order ODE with constant delay

$$\psi'_t(t) = -a\beta^2 \psi(t) + \psi(t) f\big(\psi(t-\tau)/\psi(t)\big). \tag{1}$$

2°. Multiplicative separable solution:

$$u = [C_1 \exp(-\beta x) + C_2 \exp(\beta x)]\psi(t),$$

where C_1, C_2, and β are arbitrary constants, and the function $\psi(t)$ is described by the first-order ODE with constant delay

$$\psi'_t(t) = a\beta^2 \psi(t) + \psi(t) f\big(\psi(t-\tau)/\psi(t)\big). \tag{2}$$

3°. Degenerate multiplicative separable solution:

$$u = (C_1 x + C_2)\psi(t),$$

where C_1 and C_2 are arbitrary constants, and the function $\psi(t)$ is described by the delay ODE (1) with $\beta = 0$.

4°. Multiplicative separable solution of mixed type:

$$u = e^{\alpha x + \beta t} \theta(z), \quad z = \lambda x + \gamma t,$$

where α, β, γ, and λ are arbitrary constants, and the function $\theta(z)$ is described by the second-order ODE with constant delay

$$a\lambda^2 \theta''_{zz}(z) + (2a\alpha\lambda - \gamma)\theta'_z(z) + (a\alpha^2 - \beta)\theta(z)$$
$$+ \theta(z) f\big(e^{-\beta\tau} \theta(z-\sigma)/\theta(z)\big) = 0, \quad \sigma = \gamma\tau.$$

Remark 11.6. *Delay ODEs (1) and (2) admit exponential particular solutions*

$$\psi(t) = A e^{\lambda_n t}, \quad n = 1, 2,$$

where A is an arbitrary constant, and λ_1 and λ_2 are roots of the transcendental equations

$$\lambda_1 = -a\beta^2 + f(e^{-\lambda_1 \tau}) \qquad \text{for equation (1),}$$
$$\lambda_2 = a\beta^2 + f(e^{-\lambda_2 \tau}) \qquad \text{for equation (2).}$$

5°. Exact solution:

$$u = e^{ct} V_1(x, t; b), \quad b = f(e^{-c\tau}) - c,$$

where c is an arbitrary constant, and $V_1(x, t; b)$ is a τ-periodic function that is determined by the formulas given in Item 2° of equation 11.2.1.8.

6°. Exact solution:

$$u = e^{ct} V_2(x, t; b), \quad b = f(-e^{-c\tau}) - c,$$

where c is an arbitrary constant, and $V_2(x, t; b)$ is a τ-antiperiodic function that is determined by the formulas given in the solution of equation 11.2.1.9.

7°. References for PDE 11.2.1.10: Meleshko & Moyo (2008), Polyanin & Zhurov (2014b), and Polyanin et al. (2023).

11. $u_t = au_{xx} + uf(v/u), \quad v = u(x - \sigma, t).$

1°. Multiplicative separable solution:
$$u(x, t) = e^{\lambda t}\varphi(x),$$
where λ is an arbitrary constant, and the function $\varphi = \varphi(x)$ is described by a nonlinear second-order ODE with constant delay
$$a\varphi''_{xx} + \varphi[f(\bar{\varphi}/\varphi) - \lambda] = 0, \quad \bar{\varphi} = \varphi(x - \sigma).$$

2°. The equation under consideration also has a more general solution with mixed multiplicative separation of variables:
$$u = e^{\alpha x + \beta t}\theta(z), \quad z = \lambda x + \gamma t,$$
where α, β, γ, and λ are arbitrary constants, and the function $\theta(z)$ is described by the nonlinear second-order ODE with constant delay
$$a\lambda^2\theta''_{zz} + (2a\alpha\lambda - \gamma)\theta' + a\alpha^2\theta + \theta f(e^{-\alpha\sigma}\bar{\theta}/\theta) = 0, \quad \bar{\theta} = \theta(z - \lambda\sigma).$$

12. $u_t = au_{xx} + bu \ln u + uf(w/u), \quad w = u(x, t - \tau).$

Multiplicative separable solution (Polyanin & Zhurov, 2013):
$$u = \varphi(x)\psi(t).$$

The functions $\varphi(x)$ and $\psi(t)$ are described by a regular ODE and an ODE with constant delay:
$$a\varphi''_{xx} = C_1\varphi - b\varphi \ln \varphi,$$
$$\psi'_t(t) = C_1\psi(t) + \psi(t)f(\psi(t - \tau)/\psi(t)) + b\psi(t) \ln \psi(t),$$
where C_1 is an arbitrary constant.

Note that the ODE for φ has a particular one-parameter solution:
$$\varphi = \exp\left[-\frac{b}{4a}(x + C_2)^2 + \frac{C_1}{b} + \frac{1}{2}\right],$$
where C_2 is an arbitrary constant.

13. $u_t = au_{xx} + u^3 f(w/u), \quad w = u(x, t - \tau).$

Functional separable solution (Polyanin & Zhurov, 2014c):
$$u = xU(z), \quad z = t + \frac{1}{6a}x^2,$$
where the function $U(z)$ is described by a nonlinear second-order ODE with constant delay
$$U''(z) + 9aU^3(z)f(U(z - \tau)/U(z)) = 0.$$

14. $u_t = au_{xx} - cu\ln u + uf(w/u^k)$, $\quad w = u(x, t-\tau)$, $\quad k > 0$.

1°. For $c = (\ln k)/\tau$, this equation admits a multiplicative separable solution (Polyanin & Zhurov, 2014c):
$$u = \exp(Ae^{-ct})\varphi(x),$$
where A is an arbitrary constant, and the function $\varphi = \varphi(x)$ is described by the nonlinear autonomous second-order ODE:
$$a\varphi''_{xx} - c\varphi\ln\varphi + \varphi f(\varphi^{1-k}) = 0.$$

2°. For $c = (\ln k)/\tau$, this equation admits another exact solution (Meleshko & Moyo, 2008):
$$u = \exp(Axe^{-ct})\psi(t),$$
where A is an arbitrary constant, and the function $\psi = \psi(t)$ is described by the nonlinear first-order ODE with delay
$$\psi'(t) = \psi(t)\left[A^2 a e^{-2ct} - c\ln\psi(t) + f\bigl(\psi(t-\tau)\psi^{-k}(t)\bigr)\right].$$

▶ **Equations linear in derivatives and involving two or three arbitrary functions.**

15. $u_t = au_{xx} + uf(u-w) + wg(u-w) + h(u-w)$, $\quad w = u(x, t-\tau)$.

Exact solutions (Polyanin & Zhurov, 2014b):
$$u = \sum_{n=1}^{N}[\varphi_n(x)\cos(\beta_n t) + \psi_n(x)\sin(\beta_n t)] + t\theta(x) + \xi(x), \quad \beta_n = \frac{2\pi n}{\tau},$$
where N is any positive integer and the functions $\varphi_n(x)$, $\psi_n(x)$, $\theta(x)$, and $\xi(x)$ are described by the system of second-order ODEs:
$$a\varphi''_n + \varphi_n[f(\tau\theta) + g(\tau\theta)] - \beta_n\psi_n = 0,$$
$$a\psi''_n + \psi_n[f(\tau\theta) + g(\tau\theta)] + \beta_n\varphi_n = 0,$$
$$a\theta'' + \theta[f(\tau\theta) + g(\tau\theta)] = 0,$$
$$a\xi'' + \xi f(\tau\theta) + (\xi - \tau\theta)g(\tau\theta) + h(\tau\theta) - \theta = 0.$$

Note that the third ODE is isolated (i.e., it does not depend on other) and admits the trivial solution $\theta = 0$; in this case, the remaining equations become linear ODEs with constant coefficients.

16. $u_t = au_{xx} + uf(u-kw) + wg(u-kw) + h(u-kw)$, $\quad k > 0$.

Exact solutions (Polyanin & Zhurov, 2014b):
$$u = e^{ct}\left\{\theta(x) + \sum_{n=1}^{N}[\varphi_n(x)\cos(\beta_n t) + \psi_n(x)\sin(\beta_n t)]\right\} + \xi(x),$$
$$c = \frac{1}{\tau}\ln k, \quad \beta_n = \frac{2\pi n}{\tau},$$

where N is any positive integer and the functions $\theta(x)$, $\varphi_n(x)$, $\psi_n(x)$, and $\xi(x)$ are described by the second-order ODE system

$$a\theta'' + \theta\left[f(\eta) + \frac{1}{k}g(\eta) - c\right] = 0, \quad \eta = (1-k)\xi,$$
$$a\varphi_n'' + \varphi_n\left[f(\eta) + \frac{1}{k}g(\eta) - c\right] - \beta_n\psi_n = 0,$$
$$a\psi_n'' + \psi_n\left[f(\eta) + \frac{1}{k}g(\eta) - c\right] + \beta_n\varphi_n = 0,$$
$$a\xi'' + \xi[f(\eta) + g(\eta)] + h(\eta) = 0.$$

Note that the last ODE is isolated (i.e., it does not depend on the others).

17. $u_t = au_{xx} + uf(u + kw) + wg(u + kw) + h(u + kw), \quad k > 0.$

Exact solutions (Polyanin & Zhurov, 2014b):

$$u = e^{ct}\sum_{n=1}^{N}[\varphi_n(x)\cos(\beta_n t) + \psi_n(x)\sin(\beta_n t)] + \xi(x),$$
$$c = \frac{1}{\tau}\ln k, \quad \beta_n = \frac{\pi(2n-1)}{\tau},$$

where N is any positive integer, and the functions $\varphi_n(x)$, $\psi_n(x)$, and $\xi(x)$ are described by the second-order ODE system

$$a\varphi_n'' + \varphi_n\left[f(\eta) - \frac{1}{k}g(\eta) - c\right] - \beta_n\psi_n = 0,$$
$$a\psi_n'' + \psi_n\left[f(\eta) - \frac{1}{k}g(\eta) - c\right] + \beta_n\varphi_n = 0,$$
$$a\xi'' + \xi[f(\eta) + g(\eta)] + h(\eta) = 0, \quad \eta = (1+k)\xi.$$

Note that the last ODE is isolated (i.e., it does not depend on the others).

18. $u_t = au_{xx} + uf(u^2 + w^2) + wg(u^2 + w^2), \quad w = u(x, t-\tau).$

Generalized separable solutions (Polyanin & Zhurov, 2014b):

$$u = \varphi_n(x)\cos(\lambda_n t) + \psi_n(x)\sin(\lambda_n t),$$
$$\lambda_n = \frac{\pi(2n+1)}{2\tau}, \quad n = 0, \pm 1, \pm 2, \ldots$$

where the functions $\varphi_n(x)$ and $\psi_n(x)$ are described by the system of second-order ODEs:

$$a\varphi_n'' + \varphi_n f(\varphi_n^2 + \psi_n^2) + (-1)^{n+1}\psi_n g(\varphi_n^2 + \psi_n^2) - \lambda_n\psi_n = 0,$$
$$a\psi_n'' + \psi_n f(\varphi_n^2 + \psi_n^2) + (-1)^n\varphi_n g(\varphi_n^2 + \psi_n^2) + \lambda_n\varphi_n = 0.$$

19. $u_t = [a(x)u_x]_x + b(x)f(u - w), \quad w = u(x, t-\tau).$

1°. Additive separable solution (Polyanin, 2019a):

$$u = t + \int g(x)\,dx, \quad g(x) = \frac{1}{a(x)}\left[x - f(\tau)\int b(x)\,dx\right].$$

2°. More complex generalized separable solution:
$$u = \varphi(x)t + \psi(x),$$
where the functions $\varphi = \varphi(x)$ and $\psi = \psi(x)$ are described by the second-order ODEs:
$$[a(x)\varphi'_x]'_x = 0,$$
$$[a(x)\psi'_x]'_x = \varphi - b(x)f(\tau\varphi).$$
These equations are easy to integrate.

20. $u_t = [a(x)u_x]_x + b(x)uf(w/u), \quad w = u(x, t - \tau).$

Multiplicative separable solution (Polyanin, 2019a):
$$u = e^{\lambda t}\varphi(x),$$
where λ is an arbitrary constant, and the function $\varphi = \varphi(x)$ is described by the second-order linear ODE:
$$[a(x)\varphi'_x]'_x + [f(e^{-\lambda\tau})b(x) - \lambda]\varphi = 0.$$

21. $u_t = [a(x)u_x]_x + b(x)u + uf(w/u), \quad w = u(x, t - \tau).$

Multiplicative separable solution:
$$u = \varphi(x)\psi(t).$$
The functions $\varphi = \varphi(x)$ and $\psi = \psi(t)$ are described by a linear second-order ODE without delay and a nonlinear first-order ODE with delay:
$$[a(x)\varphi'_x]'_x + b(x)\varphi = C_1\varphi;$$
$$\psi'_t = C_1\psi + \psi f(\bar\psi/\psi), \quad \bar\psi = \psi(t - \tau),$$
where C_1 is an arbitrary constant. The nonlinear delay ODE admits particular solutions of the form $\psi = C_2 e^{\lambda t}$.

▶ **Equations linear in derivatives and involving arbitrary functions of two arguments.**

22. $u_t = au_{xx} + x^2 f(u, w), \quad w = u(x, t - \tau).$

Generalized traveling wave solution (Polyanin, 2019b):
$$u = U(z), \quad z = t + \frac{1}{2a}x^2,$$
where the function $U = U(z)$ is described by the autonomous ODE with delay
$$U''_{zz} + af(U, W) = 0, \quad W = U(z - \tau).$$

23. $u_t = u_{xx} + \tanh^2(kx)f(u, w), \quad w = u(x, t - \tau).$

Generalized traveling wave solution (Polyanin, 2019a):
$$u = U(z), \quad z = t + k^{-2}\ln\cosh(kx),$$

where the function $U = U(z)$ is described by the autonomous ODE with delay
$$U''_{zz} - k^2 U'_z + k^2 f(U, W) = 0, \quad W = U(z - \tau).$$

24. $u_t = [a(x)u_x]_x + \dfrac{x^2}{a(x)} f(u, w), \quad w = u(x, t - \tau).$

Generalized traveling wave solution (Polyanin, 2019a):
$$u = U(z), \quad z = t + \int \dfrac{x}{a(x)}\, dx,$$

where the function $U = U(z)$ is described by the autonomous ODE with delay
$$U''_{zz} + f(U, W) = 0, \quad W = U(z - \tau).$$

25. $u_t = [a(x)u_x]_x + u f(x, u - w) + g(x, u - w), \quad w = u(x, t - \tau).$

Generalized separable solution:
$$u = \varphi(x) t + \psi(x),$$

where the functions $\varphi = \varphi(x)$ and $\psi = \psi(x)$ are described by the second-order ODEs:
$$[a(x)\varphi'_x]'_x + \varphi f(x, \tau\varphi) = 0,$$
$$[a(x)\psi'_x]'_x + \psi f(x, \tau\varphi) + g(x, \tau\varphi) - \varphi = 0.$$

▶ **Equations nonlinear in derivatives and involving arbitrary parameters.**

26. $u_t = [(a_1 u + a_0) u_x]_x + b_1 u + b_2 w, \quad w = u(x, t - \tau).$

Generalized separable solution of the polynomial form in x:
$$u = \psi_1(t) + \psi_2(t) x + \psi_3(t) x^2.$$

The functions $\psi_j = \psi_j(t)$ ($j = 1, 2, 3$) are described by the system of first-order ODEs with delay
$$\psi'_1 = 2a_1 \psi_1 \psi_3 + a_1 \psi_2^2 + 2a_0 \psi_3 + b_1 \psi_1 + b_2 \bar{\psi}_1,$$
$$\psi'_2 = 6a_1 \psi_2 \psi_3 + b_1 \psi_2 + b_2 \bar{\psi}_2,$$
$$\psi'_3 = 6a_1 \psi_3^2 + b_1 \psi_3 + b_2 \bar{\psi}_3,$$

where $\bar{\psi}_j = \psi_j(t - \tau)$.

27. $u_t = [(a_1 u + a_0) u_x]_x + k u^2 + b_1 u + b_2 w, \quad w = u(x, t - \tau).$

1°. Generalized separable solution for $a_1 k < 0$:
$$u = \psi_1(t) + \psi_2(t) \exp(-\lambda x) + \psi_3(t) \exp(\lambda x), \quad \lambda = \sqrt{-k/(2a_1)}.$$

The functions $\psi = \psi_n(t)$ are described by the system of first-order ODEs with delay
$$\psi'_1 = k \psi_1^2 + 2k \psi_2 \psi_3 + b_1 \psi_1 + b_2 \bar{\psi}_1,$$
$$\psi'_2 = (\tfrac{3}{2} k \psi_1 + a_0 \lambda^2 + b_1) \psi_2 + b_2 \bar{\psi}_2,$$
$$\psi'_3 = (\tfrac{3}{2} k \psi_1 + a_0 \lambda^2 + b_1) \psi_3 + b_2 \bar{\psi}_3,$$

where $\bar{\psi}_j = \psi_j(t - \tau)$ ($i = 1, 2, 3$).

$2°$. Generalized separable solution for $a_1 k > 0$:
$$u = \psi_1(t) + \psi_2(t)\cos(\lambda x) + \psi_3(t)\sin(\lambda x), \quad \lambda = \sqrt{k/(2a_1)},$$
where the functions $\psi = \psi_n(t)$ are described by the system of first-order ODEs with delay
$$\psi_1' = k\psi_1^2 + \tfrac{1}{2}k(\psi_2^2 + \psi_3^2) + b_1\psi_1 + b_2\bar{\psi}_1,$$
$$\psi_2' = (\tfrac{3}{2}k\psi_1 + b_1 - a_0\lambda^2)\psi_2 + b_2\bar{\psi}_2,$$
$$\psi_3' = (\tfrac{3}{2}k\psi_1 + b_1 - a_0\lambda^2)\psi_3 + b_2\bar{\psi}_3.$$

28. $u_t = a(u^n u_x)_x + bu^{n+1} + cu + ku^{1-n} + mu^{1-n}w^n, \quad w = u(x, t - \tau).$

Functional separable solution:
$$u = \{\varphi(t)[C_1 \cos(\beta x) + C_2 \sin(\beta x)] + \psi(t)\}^{1/n} \quad \text{if } ab(n+1) > 0,$$
$$u = \{\varphi(t)[C_1 \cosh(\beta x) + C_2 \sinh(\beta x)] + \psi(t)\}^{1/n} \quad \text{if } ab(n+1) < 0,$$

where C_1 and C_2 are arbitrary constants,
$$\beta = \sqrt{\frac{|b|n^2}{|a(n+1)|}},$$

and the functions $\varphi = \varphi(t)$ and $\psi = \psi(t)$ are described by the system of first-order ODEs with delay
$$\varphi_t' = \frac{bn(n+2)}{n+1}\varphi\psi + cn\varphi + mn\bar{\varphi}, \quad \bar{\varphi} = \varphi(t - \tau),$$
$$\psi_t' = n(b\psi^2 + c\psi + k) + \frac{bn}{n+1}(C_1^2 \pm C_2^2)\varphi^2 + mn\bar{\psi}, \quad \bar{\psi} = \psi(t - \tau).$$

Here the upper sign in the second equation corresponds to the case $ab(n+1) > 0$, and the lower sign corresponds to the case $ab(n+1) < 0$.

29. $u_t = a(e^{\lambda u} u_x)_x + be^{\lambda u} + c + ke^{-\lambda u} + me^{\lambda(w-u)}, \quad w = u(x, t - \tau).$

Functional separable solution:
$$u = \frac{1}{\lambda}\ln\{e^{\alpha t}[C_1 \cos(x\sqrt{b\lambda/a}) + C_2 \sin(x\sqrt{b\lambda/a})] + \gamma\} \quad \text{if } ab\lambda > 0,$$
$$u = \frac{1}{\lambda}\ln\{e^{\alpha t}[C_1 \cosh(x\sqrt{-b\lambda/a}) + C_2 \sinh(x\sqrt{-b\lambda/a})] + \gamma\} \quad \text{if } ab\lambda < 0.$$

Here C_1 and C_2 are arbitrary constants, α is a root of the transcendental equation
$$\alpha = \lambda(b\gamma + c) + m\lambda e^{-\alpha\tau},$$
and γ is a root of the quadratic equation $b\gamma^2 + (c+m)\gamma + k = 0$.

▶ **Equations nonlinear in derivatives and involving one arbitrary function.**

30. $u_t = a(u^{-1/2} u_x)_x + bu^{1/2} + f(u^{1/2} - w^{1/2}), \quad w = u(x, t - \tau).$

Generalized separable solution (Polyanin & Zhurov, 2014d):
$$u = [\varphi(x)t + \psi(x)]^2,$$

where the functions $\varphi = \varphi(x)$ and $\psi = \psi(x)$ are described by the second-order ODEs:

$$2a\varphi''_{xx} + b\varphi - 2\varphi^2 = 0,$$
$$2a\psi''_{xx} + b\psi - 2\varphi\psi + f(\tau\varphi) = 0.$$

These ODEs admit the simple particular solution

$$\varphi = \frac{1}{2}b, \quad \psi = -\frac{1}{4a}f\left(\frac{b\tau}{2}\right)x^2 + Ax + B,$$

where A and B are arbitrary constants.

31. $u_t = a(u^k u_x)_x + u f(w/u), \quad w = u(x, t - \tau).$

1°. Multiplicative separable solution (Polyanin & Zhurov, 2014d):

$$u = \varphi(x)\psi(t).$$

The functions $\varphi = \varphi(x)$ and $\psi = \psi(t)$ are described by an ODE without delay and an ODE with delay:

$$a(\varphi^k \varphi'_x)'_x = C\varphi,$$
$$\psi'(t) = C\psi^{k+1}(t) + \psi(t)f\bigl(\psi(t-\tau)/\psi(t)\bigr),$$

where C is an arbitrary constant.

2°. Exact solution with $k \neq 0$:

$$u = (x + C)^{2/k}\theta(\zeta), \quad \zeta = t + \lambda \ln(x + C),$$

where C and λ are arbitrary constants, and the function $\theta = \theta(\zeta)$ is described by the ODE with delay

$$\theta'(\zeta) = a\Bigl\{\frac{2(k+2)}{k^2}\theta^{k+1}(\zeta) + \frac{(3k+4)\lambda}{k}\theta^k(\zeta)\theta'(\zeta)$$
$$+ k\lambda^2\theta^{k-1}(\zeta)[\theta'(\zeta)]^2 + \lambda^2\theta^k(\zeta)\theta''(\zeta)\Bigr\} + \theta(\zeta)f\bigl(\theta(\zeta-\tau)/\theta(\zeta)\bigr).$$

32. $u_t = a(u^k u_x)_x + bu^{k+1} + u f(w/u), \quad w = u(x, t - \tau).$

1°. Multiplicative separable solution for $b(k+1) > 0$ (Polyanin & Zhurov, 2014d):

$$u = [C_1 \cos(\beta x) + C_2 \sin(\beta x)]^{1/(k+1)}\psi(t), \quad \beta = \sqrt{b(k+1)/a},$$

where C_1 and C_2 are arbitrary constants, and the function $\psi = \psi(t)$ is described by the ODE with delay

$$\psi'(t) = \psi(t)f\bigl(\psi(t-\tau)/\psi(t)\bigr). \qquad (*)$$

This equation has the exponential particular solution

$$\psi(t) = Ae^{\lambda t},$$

where A is an arbitrary constant, and λ is a root of the algebraic (transcendental) equation $\lambda - f(e^{-\lambda\tau}) = 0$.

2°. Multiplicative separable solution for $b(k+1) < 0$ (Polyanin & Zhurov, 2014d):
$$u = [C_1 \exp(-\beta x) + C_2 \exp(\beta x)]^{1/(k+1)} \psi(t), \quad \beta = \sqrt{-b(k+1)/a},$$
where C_1 and C_2 are arbitrary constants, and the function $\psi = \psi(t)$ is described by the delay ODE (∗).

3°. Multiplicative separable solution for $k = -1$:
$$u = C_1 \exp\left(-\frac{b}{2a}x^2 + C_2 x\right)\psi(t),$$
where C_1 and C_2 are arbitrary constants, and the function $\psi = \psi(t)$ is described by the delay ODE (∗).

33. $u_t = a(u^k u_x)_x + b + u^{-k} f(u^{k+1} - w^{k+1}), \quad w = u(x, t-\tau).$

1°. Functional separable solution for $k \neq -1$ (Polyanin & Zhurov, 2014d):
$$u = \left[At - \frac{b(k+1)}{2a}x^2 + C_1 x + C_2\right]^{1/(k+1)},$$
where C_1 and C_2 are arbitrary constants, and A is a root of the algebraic (transcendental) equation $A = (k+1)f(A\tau)$.

2°. More complex functional separable solution for $k \neq -1$:
$$u = \left[\psi(t) - \frac{b(k+1)}{2a}x^2 + C_1 x + C_2\right]^{1/(k+1)},$$
where C_1 and C_2 are arbitrary constants, and the function $\psi = \psi(t)$ is described by the ODE with delay
$$\psi'(t) = (k+1)f\bigl(\psi(t) - \psi(t-\tau)\bigr).$$

34. $u_t = a(u^k u_x)_x + bu^{k-2n+1} + u^{1-n} f(u^n - w^n), \quad w = u(x, t-\tau).$
Generalized traveling wave solutions for $b(n - k - 1) > 0$:
$$u = [\pm \lambda x + \psi(t)]^{1/n}, \quad \lambda = \sqrt{\frac{bn^2}{a(n-k-1)}},$$
where the function $\psi = \psi(t)$ is described by the ODE with delay
$$\psi'_t = nf(\psi - \bar\psi), \quad \bar\psi = \psi(t-\tau).$$

35. $u_t = a(e^{\lambda u} u_x)_x + f(u - w), \quad w = u(x, t-\tau).$

1°. Additive separable solution (Polyanin & Zhurov, 2014d):
$$u = \frac{1}{\lambda} \ln(Ax^2 + Bx + C) + \psi(t),$$
where A, B, and C are arbitrary constants, and the function $\psi = \psi(t)$ is described by the delay ODE:
$$\psi'(t) = 2a(A/\lambda)e^{\lambda \psi(t)} + f\bigl(\psi(t) - \psi(t-\tau)\bigr).$$

2°. Exact solution:
$$u = \frac{2}{\lambda}\ln(x+C) + \theta(\zeta), \quad \zeta = t + \beta\ln(x+C),$$
where C and β are arbitrary constants, and the function $\theta = \theta(\zeta)$ is described by the ODE with delay
$$\theta'(\zeta) = ae^{\lambda\theta(\zeta)}\left\{\frac{2}{\lambda} + 3\beta\theta'(\zeta) + \beta^2\lambda[\theta'(\zeta)]^2 + \beta^2\theta''(\zeta)\right\} + f(\theta(\zeta) - \theta(\zeta-\tau)).$$

36. $u_t = a(e^{\lambda u}u_x)_x + be^{\lambda u} + f(u-w), \quad w = u(x, t-\tau).$

1°. Additive separable solution for $b\lambda > 0$ (Polyanin & Zhurov, 2014d):
$$u = \frac{1}{\lambda}\ln[C_1\cos(\beta x) + C_2\sin(\beta x)] + \psi(t), \quad \beta = \sqrt{b\lambda/a},$$
where C_1 and C_2 are arbitrary constants, and the function $\psi = \psi(t)$ is described by the delay ODE:
$$\psi'(t) = f(\psi(t) - \psi(t-\tau)). \tag{*}$$

Note that equation (*) has a simple particular solution $\psi = A + kt$, where A is an arbitrary constant and k is a root of the algebraic (transcendental) equation $k - f(k\tau) = 0$.

2°. Additive separable solution for $b\lambda < 0$ (Polyanin & Zhurov, 2014d):
$$u = \frac{1}{\lambda}\ln[C_1\exp(-\beta x) + C_2\exp(\beta x)] + \psi(t), \quad \beta = \sqrt{-b\lambda/a},$$
where C_1 and C_2 are arbitrary constants, and the function $\psi = \psi(t)$ is described by the delay ODE (*).

37. $u_t = a(e^{\lambda u}u_x)_x + b + e^{-\lambda u}f(e^{\lambda u} - e^{\lambda w}), \quad w = u(x, t-\tau).$

1°. Functional separable solution (Polyanin & Zhurov, 2014d):
$$u = \frac{1}{\lambda}\ln\left[At - \frac{b\lambda}{2a}x^2 + C_1 x + C_2\right],$$
where C_1 and C_2 are arbitrary constants, and A is a root of the algebraic (transcendental) equation $A - \lambda f(A\tau) = 0$.

2°. A more complex functional separable solution:
$$u = \frac{1}{\lambda}\ln\left[\psi(t) - \frac{b\lambda}{2a}x^2 + C_1 x + C_2\right],$$
where C_1 and C_2 are arbitrary constants, and the function $\psi = \psi(t)$ is described by the delay ODE:
$$\psi'(t) = \lambda f(\psi(t) - \psi(t-\tau)).$$

38. $u_t = a(e^{\lambda u}u_x)_x + be^{(\lambda-2\gamma)u} + e^{-\gamma u}f(e^{\gamma u} - e^{\gamma w}), \quad w = u(x, t-\tau).$

Functional separable solutions for $b(\gamma - \lambda) > 0$:
$$u = \frac{1}{\gamma}\ln[\pm kx + \psi(t)], \quad k = \sqrt{\frac{b\gamma^2}{a(\gamma-\lambda)}},$$

where the function $\psi = \psi(t)$ is described by the ODE with delay
$$\psi'_t = \gamma f(\psi - \bar\psi), \quad \bar\psi = \psi(t-\tau).$$

39. $u_t = [(a\ln u + b)u_x]_x - cu\ln u + uf(w/u), \quad w = u(x, t-\tau).$

Multiplicative separable solutions (Polyanin & Zhurov, 2014d):
$$u = \exp(\pm\lambda x)\psi(t), \quad \lambda = \sqrt{c/a},$$
where the function $\psi = \psi(t)$ is described by the ODE with delay
$$\psi'(t) = \lambda^2(a+b)\psi(t) + \psi(t)f\bigl(\psi(t-\tau)/\psi(t)\bigr).$$

40. $u_t = [f'_u(u)u_x]_x + a_1 f(u) + a_2 f(w) + a_3 + \dfrac{b}{f'_u(u)}[f(u) - f(w)].$

Functional separable solution in implicit form (Polyanin & Zhurov, 2014d):
$$f(u) = e^{\lambda t}\varphi(x) - \frac{a_3}{a_1 + a_2},$$
where λ is a root of the transcendental equation
$$\lambda = b(1 - e^{-\lambda\tau}),$$
and the function $\varphi = \varphi(x)$ is described by the second-order linear ODE with constant coefficients
$$\varphi''_{xx} + (a_1 + a_2 e^{-\lambda\tau})\varphi = 0.$$

41. $u_t = [f'_u(u)u_x]_x + a[f(u) - f(w)] + \dfrac{1}{f'_u(u)}\bigl[b_1 f(u) + b_2 f(w) + b_3\bigr].$

Functional separable solution in implicit form (Polyanin & Zhurov, 2014d):
$$f(u) = e^{\lambda t}\varphi(x) - \frac{b_3}{b_1 + b_2},$$
where λ is a root of the transcendental equation
$$\lambda - b_1 - b_2 e^{-\lambda\tau} = 0,$$
and the function $\varphi = \varphi(x)$ is described by the second-order linear ODE with constant coefficients
$$\varphi''_{xx} + a(1 - e^{-\lambda\tau})\varphi = 0.$$

42. $u_t = [uf'_u(u)u_x]_x + \dfrac{1}{f'_u(u)}[af(u) + bf(w) + c], \quad w = u(x, t-\tau).$

Generalized traveling wave solution in implicit form (Polyanin & Zhurov, 2014d):
$$f(u) = \varphi(t)x + \psi(t),$$
where the functions $\varphi(t)$ and $\psi(t)$ are described by the delay ODEs:
$$\varphi'(t) = a\varphi(t) + b\varphi(t-\tau),$$
$$\psi'(t) = a\psi(t) + b\psi(t-\tau) + c + \varphi^2(t).$$

43. $u_t = [uf'_u(u)u_x]_x + (a+b)u + \dfrac{2}{f'_u(u)}[af(u) + bf(w) + c]$.

Generalized traveling wave solution in implicit form (Polyanin & Zhurov, 2014d):

$$f(u) = -\tfrac{1}{2}(a+b)x^2 + \varphi(t)x + \psi(t),$$

where the functions $\varphi(t)$ and $\psi(t)$ are described by the delay ODEs:

$$\varphi'(t) = -2b\varphi(t) + 2b\varphi(t-\tau),$$
$$\psi'(t) = 2a\psi(t) + 2b\psi(t-\tau) + 2c + \varphi^2(t).$$

The first delay ODE has an exponential particular solution

$$\varphi(t) = C_1 e^{\lambda t} + C_2,$$

where C_1 and C_2 are arbitrary constants, and λ is a root of the transcendental equation $\lambda + 2b(1 - e^{-\lambda\tau}) = 0$.

▶ **Equations nonlinear in derivatives and involving two arbitrary functions.**

44. $u_t = a(u^{-1/2}u_x)_x + f(u^{1/2} - w^{1/2}) + u^{1/2}g(u^{1/2} - w^{1/2})$.

Generalized separable solution (Polyanin & Zhurov, 2014d):

$$u = [\varphi(x)t + \psi(x)]^2,$$

where the functions $\varphi = \varphi(x)$ and $\psi = \psi(x)$ are described by the ODE system

$$2a\varphi''_{xx} + \varphi g(\tau\varphi) - 2\varphi^2 = 0,$$
$$2a\psi''_{xx} + \psi g(\tau\varphi) - 2\varphi\psi + f(\tau\varphi) = 0.$$

A particular solution of this system has the form

$$\varphi = k, \quad \psi = -\frac{1}{4a}f(k\tau)x^2 + Ax + B,$$

where A and B are arbitrary constants, and k is a root of the algebraic (transcendental) equation $g(k\tau) - 2k = 0$.

45. $u_t = a(u^k u_x)_x + uf(w/u) + u^{k+1}g(w/u), \quad w = u(x, t-\tau)$.

Multiplicative separable solution (Polyanin & Zhurov, 2014d):

$$u = e^{\lambda t}\varphi(x),$$

where λ is a root of the algebraic (transcendental) equation $\lambda = f(e^{-\lambda\tau})$, and the function $\varphi = \varphi(x)$ is described by the nonlinear second-order ODE:

$$a(\varphi^k \varphi'_x)'_x + g(e^{-\lambda\tau})\varphi^{k+1} = 0.$$

For $k \neq -1$, the substitution $\theta = \varphi^{k+1}$ converts this equation to a linear second-order ODE with constant coefficients. For $k = -1$, the substitution $\theta = \ln|\varphi|$ must be used to linearize the equation.

46. $u_t = a(u^k u_x)_x + f(u^{k+1} - w^{k+1}) + u^{-k}g(u^{k+1} - w^{k+1})$, $w = u(x, t-\tau)$.

Functional separable solution for $k \neq -1$ (Polyanin & Zhurov, 2014d):

$$u = (At + Bx^2 + C_1 x + C_2)^{1/(k+1)}, \quad B = -\frac{(k+1)}{2a} f(A\tau),$$

where C_1 and C_2 are arbitrary constants, and A is a root of the algebraic (transcendental) equation $A - (k+1)g(A\tau) = 0$.

47. $u_t = a(e^{\lambda u} u_x)_x + f(u - w) + e^{\lambda u} g(u - w)$, $w = u(x, t-\tau)$.

Additive separable solution (Polyanin & Zhurov, 2014d):

$$u = \beta t + \varphi(x),$$

where β is a root of the algebraic (transcendental) equation $\beta = f(\beta \tau)$, and the function $\varphi = \varphi(x)$ is described by the nonlinear second-order ODE:

$$a(e^{\lambda \varphi} \varphi'_x)'_x + g(\beta \tau) e^{\lambda \varphi} = 0.$$

The substitution $\theta = e^{\lambda \varphi}$ converts this equation to a linear second-order ODE with constant coefficients $a\theta''_{xx} + \lambda g(\beta \tau)\theta = 0$.

48. $u_t = a(e^{\lambda u} u_x)_x + f(e^{\lambda u} - e^{\lambda w}) + e^{-\lambda u} g(e^{\lambda u} - e^{\lambda w})$, $w = u(x, t-\tau)$.

Functional separable solution (Polyanin & Zhurov, 2014d):

$$u = \frac{1}{\lambda} \ln(At + Bx^2 + C_1 x + C_2), \quad B = -\frac{\lambda}{2a} f(A\tau),$$

where C_1 and C_2 are arbitrary constants, and A is a root of the algebraic (transcendental) equation $A - \lambda g(A\tau) = 0$.

49. $u_t = a[g'(u) u_x]_x + b + \frac{1}{g'(u)} f(g(u) - g(w))$, $w = u(x, t-\tau)$.

Functional separable solution in implicit form:

$$g(u) = \psi(t) - \frac{b}{2a} x^2 + C_1 x + C_2,$$

where C_1 and C_2 are arbitrary constants, and the function $\psi = \psi(t)$ is described by the delay ODE:

$$\psi'(t) = f(\psi(t) - \psi(t - \tau)).$$

This equation has a particular solution $\psi(t) = At$, where A is a root of the algebraic (transcendental) equation $A - f(A\tau) = 0$.

▶ **Equations nonlinear in derivatives and involving three arbitrary functions.**

50. $u_t = [a(x) u^k u_x]_x + b(x) u^{k+1} + u f(w/u)$, $w = u(x, t-\tau)$.

Multiplicative separable solution (Polyanin & Sorokin, 2019a):

$$u = \varphi(x) \psi(t),$$

where the functions $\varphi = \varphi(x)$ and $\psi = \psi(t)$ are described respectively by a regular ODE and an ODE with delay:

$$[a(x)\eta'_x]'_x + (k+1)b(x)\eta = 0, \quad \eta = \varphi^{k+1};$$
$$\psi'_t(t) = \psi(t)f\big(\psi(t-\tau)/\psi(t)\big).$$

51. $u_t = [a(x)e^{\beta u}u_x]_x + b(x)e^{\beta u} + f(u-w), \quad w = u(x, t-\tau).$

Additive separable solution:

$$u = \frac{1}{\beta}\ln\varphi(x) + \psi(t).$$

The functions $\varphi = \varphi(x)$ and $\psi = \psi(t)$ are described respectively by a second-order linear ODE without delay and a first-order nonlinear ODE with delay:

$$[a(x)\varphi'_x]'_x + \beta b(x)\varphi = C\beta,$$
$$\psi'_t(t) = Ce^{\beta\psi} + f\big(\psi(t) - \psi(t-\tau)\big),$$

where C is an arbitrary constant.

52. $u_t = a[f'_u(u)u_x]_x + g\big(f(u)-f(w)\big) + \dfrac{1}{f'_u(u)}h\big(f(u)-f(w)\big).$

Functional separable solution in implicit form (Polyanin & Zhurov, 2014d):

$$f(u) = At - \frac{g(A\tau)}{2a}x^2 + C_1 x + C_2,$$

where C_1 and C_2 are arbitrary constants, and A is a root of the algebraic (transcendental) equation $A - h(A\tau) = 0$.

53. $u_t = a[f'_u(u)u_x]_x + f(u)g\big(f(w)/f(u)\big) + \dfrac{f(u)}{f'_u(u)}h\big(f(w)/f(u)\big).$

Let β be a root of the algebraic (transcendental) equation

$$\beta - h(e^{-\beta\tau}) = 0.$$

1°. Functional separable solution in implicit form for $ag(e^{-\beta\tau}) > 0$ (Polyanin & Zhurov, 2014d):

$$f(u) = [C_1 \cos(\lambda x) + C_2 \sin(\lambda x)]e^{\beta t}, \quad \lambda = \sqrt{g(e^{-\beta\tau})/a},$$

where C_1 and C_2 are arbitrary constants.

2°. Functional separable solution in implicit form for $ag(e^{-\beta\tau}) < 0$ (Polyanin & Zhurov, 2014d):

$$f(u) = [C_1 \exp(-\lambda x) + C_2 \exp(\lambda x)]e^{\beta t}, \quad \lambda = \sqrt{-g(e^{-\beta\tau})/a},$$

where C_1 and C_2 are arbitrary constants.

3°. Functional separable solution in implicit form for $g(e^{-\beta\tau}) = 0$:

$$f(u) = (C_1 x + C_2)e^{\beta t}.$$

11.2.2. Hyperbolic Equations

▶ **Equations linear in derivatives.**

1. $u_{tt} = au_{xx} + f(u - w), \quad w = u(x, t - \tau).$

   1°. Additive separable solution:
   $$u = C_2 x^2 + C_1 x + \psi(t),$$
   where C_1 and C_2 are arbitrary constants, and the function $\psi(t)$ is described by the second-order ODE with delay
   $$\psi''_{tt} = 2C_2 a + f(\psi - \bar{\psi}), \quad \bar{\psi} = \psi(t - \tau).$$

   2°. Exact solutions:
   $$u = Cx^2 + \varphi(z)x + \psi(z), \quad z = t \pm a^{-1/2} x,$$
   where C is an arbitrary constant, and the functions $\varphi(z)$ and $\psi(z)$ is described by the difference equations
   $$\varphi(z) = \varphi(z - \tau), \tag{1}$$
   $$f\big(\psi(z) - \psi(z - \tau)\big) = \mp 2a^{1/2} \varphi'(z) - 2Ca. \tag{2}$$
   From equation (1) it follows that $\varphi(z)$ is any τ-periodic function, which, in general, can be represented as a convergent series
   $$\varphi(z) = A_0 + \sum_{n=1}^{\infty} \left(A_n \cos \frac{2\pi n z}{\tau} + B_n \sin \frac{2\pi n z}{\tau} \right), \tag{3}$$
   where A_n and B_n are arbitrary constants. Substituting (3) into (2) yields an equation that reduces to a linear nonhomogeneous difference equation of the form $\pi(z) - \pi(z - \tau) = g_{\mp}(z)$ with a known right-hand side.

   3°. Exact solutions:
   $$u = Cxz + \varphi(x) + \psi(z), \quad z = t \pm a^{-1/2} x,$$
   where C is an arbitrary constant. The functions $\varphi(x)$ and $\psi(z)$ are described respectively by a regular linear ODE and a linear difference equation:
   $$a\varphi''_{xx} \pm 2Ca^{1/2} + f(C\tau x + B) = 0,$$
   $$\psi(z) - \psi(z - \tau) = B,$$
   where B is an arbitrary constant. The functions $\varphi(x)$ and $\psi(z)$ allow a closed form representation.

   4°. References for PDE 11.2.2.1: Polyanin & Zhurov (2014e), Long & Meleshko (2016), and Polyanin et al. (2023).

2. $u_{tt} = au_{xx} + f(u - v), \quad v = u(x - \sigma, t).$

   Additive separable solution (Polyanin et al., 2023):
   $$u(x, t) = C_1 t^2 + C_2 t + \varphi(x),$$

where C is an arbitrary constant, and the function $\varphi = \varphi(x)$ is described by the second-order nonlinear ODE with delay

$$a\varphi''_{xx} - 2C_1 + f(\varphi - \bar{\varphi}) = 0, \quad \bar{\varphi} = \varphi(x - \sigma).$$

3. $u_{tt} = au_{xx} + bu + f(u - w), \quad w = u(x, t - \tau).$

1°. Additive separable solution for $ab > 0$ (Polyanin & Zhurov, 2014e):

$$u = C_1 \cos(\lambda x) + C_2 \sin(\lambda x) + \psi(t), \quad \lambda = \sqrt{b/a},$$

where C_1 and C_2 are arbitrary constants, and the function $\psi = \psi(t)$ is described by the delay ODE:

$$\psi''_{tt} = b\psi + f(\psi - \bar{\psi}), \quad \bar{\psi} = \psi(t - \tau). \tag{$*$}$$

2°. Additive separable solution for $ab < 0$ (Polyanin & Zhurov, 2014e):

$$u = C_1 \exp(-\lambda x) + C_2 \exp(\lambda x) + \psi(t), \quad \lambda = \sqrt{-b/a},$$

where C_1 and C_2 are arbitrary constants, and the function $\psi = \psi(t)$ is described by the delay ODE $(*)$.

4. $u_{tt} = au_{xx} + bu + f(u - kw), \quad k > 0.$

1°. Generalized separable solution:

$$u = e^{ct}[A \cos(\beta x) + B \sin(\beta x)] + \psi(t),$$

$$c = \frac{1}{\tau} \ln k, \quad \beta = [(b - c^2)/a]^{1/2}, \quad b > c^2,$$

where A and B are arbitrary constants, and the function $\psi(t)$ is described by the ODE with delay

$$\psi''_{tt} = b\psi + f(\psi - k\bar{\psi}), \quad \bar{\psi} = \psi(t - \tau). \tag{1}$$

2°. Generalized separable solution:

$$u = e^{ct}(Ae^{-\beta x} + Be^{\beta x}) + \psi(t),$$

$$c = \frac{1}{\tau} \ln k, \quad \beta = [(c^2 - b)/a]^{1/2}, \quad c^2 > b,$$

where A and B are arbitrary constants, and the function $\psi(t)$ is described by the delay ODE (1).

3°. Generalized separable solution:

$$u = e^{ct}[A \cos(\beta x) + B \sin(\beta x)] + \varphi(x),$$

$$c = \frac{1}{\tau} \ln k, \quad \beta = [(b - c^2)/a]^{1/2}, \quad b > c^2,$$

where A and B are arbitrary constants, and the function $\varphi(x)$ is described by the ODE:

$$a\varphi''_{xx} + b\varphi + f((1 - k)\varphi) = 0. \tag{2}$$

4°. Generalized separable solution:

$$u = e^{ct}(Ae^{-\beta x} + Be^{\beta x}) + \varphi(x),$$

$$c = \frac{1}{\tau} \ln k, \quad \beta = [(c^2 - b)/a]^{1/2}, \quad c^2 > b,$$

where A and B are arbitrary constants, and the function $\varphi(x)$ is described by ODE (2).

5°. Exact solutions for $b = 0$:
$$u = \varphi(z)x + \psi(z), \quad z = t \pm a^{-1/2}x,$$
where the functions $\varphi(x)$ and $\psi(z)$ are described the difference equations
$$\varphi(z) - k\varphi(z - \tau) = 0, \tag{1}$$
$$f(\psi(z) - k\psi(z - \tau)) = \mp 2a^{1/2}\varphi'_z(z). \tag{2}$$

Equation (1) has the general solution
$$\varphi(z) = k^{z/\tau} \sum_{n=0}^{\infty} \left(A_n \cos \frac{2\pi nz}{\tau} + B_n \sin \frac{2\pi nz}{\tau} \right), \tag{3}$$
where A_n and B_n are arbitrary constants.

Substituting (3) into (2) yields an equation that reduces to a linear nonhomogeneous difference equation of the form $\psi(z) - k\psi(z - \tau) = g_\mp(z)$ with a known right-hand side.

6°. Suppose that $u_0(x, t)$ is a solution to the original nonlinear delay PDE and $v = U_1(x, t; b, s)$ is any τ-periodic solution to the linear telegraph-type equation
$$v_{tt} + sv_t = av_{xx} + bv, \quad v(x, t) = v(x, t - \tau),$$
where b and s are free parameters. Then the sum
$$u = u_0(x, t) + e^{ct}U_1(x, t; b - c^2, 2c), \quad c = \frac{1}{\tau} \ln k,$$
is also a solution to the original equation.

The general form of the function $U_1(x, t; b, s)$ is given by
$$U_1(x, t; b, s) = \sum_{n=0}^{\infty} \exp(-\lambda_n x) \left[A_n \cos(\beta_n t - \gamma_n x) + B_n \sin(\beta_n t - \gamma_n x) \right]$$
$$+ \sum_{n=1}^{\infty} \exp(\lambda_n x) \left[C_n \cos(\beta_n t + \gamma_n x) + D_n \sin(\beta_n t + \gamma_n x) \right],$$
$$\beta_n = \frac{2\pi n}{\tau}, \quad \gamma_n = \left[\frac{\sqrt{(b + \beta_n^2)^2 + s^2 \beta_n^2} + b + \beta_n^2}{2a} \right]^{1/2}, \quad \lambda_n = \frac{s\beta_n}{2a\gamma_n},$$
where $A_n, B_n, C_n,$ and D_n are arbitrary constants.

Note that a spatially homogeneous solution $u_0(t)$ or a stationary solution $u_0(x)$ may be taken as the particular solution $u_0(x, t)$ of the original equation. Stationary points $u_0 = \text{const}$ may also be used as the simplest particular solutions.

7°. References for PDE 11.2.2.4: Polyanin & Zhurov (2014e), Long & Meleshko (2016), and Polyanin et al. (2023).

5. $u_{tt} = au_{xx} + bu + f(u + kw), \quad k > 0.$

1°. Exact solutions for $b = 0$:
$$u = \varphi(z)x + \psi(z), \quad z = t \pm a^{-1/2}x,$$

where the functions $\varphi(x)$ and $\psi(z)$ are described by the difference equations

$$\varphi(z) + k\varphi(z - \tau) = 0, \tag{1}$$

$$f\bigl(\psi(z) + k\psi(z - \tau)\bigr) = \mp 2a^{1/2}\varphi_z'(z). \tag{2}$$

Equation (1) has the general solution

$$\varphi(z) = k^{z/\tau} \sum_{n=1}^{\infty} \left[A_n \cos \frac{(2n-1)\pi z}{\tau} + B_n \sin \frac{(2n-1)\pi z}{\tau} \right], \tag{3}$$

where A_n and B_n are arbitrary constants.

Substituting (3) into (2) yields an equation that reduces to a linear nonhomogeneous difference equation of the form $\psi(z) + k\psi(z - \tau) = g_{\mp}(z)$ with a known right-hand side.

2°. Suppose that $u_0(x,t)$ is a solution to the original nonlinear delay PDE and $v = U_2(x,t;b,s)$ is any τ-antiperiodic solution to the linear telegraph-type equation

$$v_{tt} + sv_t = av_{xx} + bv, \quad v(x,t) = -v(x, t - \tau).$$

Then the sum

$$u = u_0(x,t) + e^{ct} U_2(x, t; b - c^2, 2c), \quad c = \frac{1}{\tau} \ln k,$$

is also a solution to the original equation.

The general form of the function $U_2(x,t;b,s)$ is given by

$$U_2(x,t;b,s) = \sum_{n=1}^{\infty} \exp(-\lambda_n x) \bigl[A_n \cos(\beta_n t - \gamma_n x) + B_n \sin(\beta_n t - \gamma_n x) \bigr]$$

$$+ \sum_{n=1}^{\infty} \exp(\lambda_n x) \bigl[C_n \cos(\beta_n t + \gamma_n x) + D_n \sin(\beta_n t + \gamma_n x) \bigr],$$

$$\beta_n = \frac{\pi(2n-1)}{\tau}, \quad \gamma_n = \left[\frac{\sqrt{(b + \beta_n^2)^2 + s^2 \beta_n^2} + b + \beta_n^2}{2a} \right]^{1/2}, \quad \lambda_n = \frac{s\beta_n}{2a\gamma_n},$$

where A_n, B_n, C_n, and D_n are arbitrary constants.

Note that a spatially homogeneous solution $u_0(t)$ or a stationary solution $u_0(x)$ may be taken as the particular solution $u_0(x,t)$ of the original equation. Stationary points $u_0 = $ const may also be used as the simplest particular solutions.

3°. References for PDE 11.2.2.5: Polyanin & Zhurov (2014e) and Polyanin et al. (2023).

6. $u_{tt} = au_{xx} + f(w/u), \quad w = u(x, t - \tau)$.

Exact solutions (Polyanin & Zhurov, 2014e):

$$u = (x + C)\varphi(z), \quad z = t \pm a^{-1/2} x,$$

where C is an arbitrary constant, and the function $\varphi(z)$ is described by the first-order ODE with delay

$$\pm 2a^{1/2}\varphi_z'(z) + f\bigl(\varphi(z - \tau)/\varphi(z)\bigr) = 0.$$

7. $u_{tt} = au_{xx} + uf(w/u), \quad w = u(x, t - \tau)$.

$1°$. Multiplicative separable solution:
$$u = [A\cos(\beta x) + B\sin(\beta x)]\psi(t),$$
where A, B, and β are arbitrary constants, and the function $\psi(t)$ is described by the delay ODE:
$$\psi''_{tt} = -a\beta^2\psi + \psi f(\bar\psi/\psi), \quad \bar\psi = \psi(t-\tau).$$

$2°$. Multiplicative separable solution:
$$u = [A\exp(-\beta x) + B\exp(\beta x)]\psi(t),$$
where A, B, and β are arbitrary constants, and the function $\psi(t)$ is described by the delay ODE:
$$\psi''_{tt} = a\beta^2\psi + \psi f(\bar\psi/\psi), \quad \bar\psi = \psi(t-\tau).$$

$3°$. Degenerate multiplicative separable solution:
$$u = (Ax + B)\psi(t),$$
where A, B, and β are arbitrary constants, and the function $\psi(t)$ is described by the delay ODE:
$$\psi''_{tt} = \psi f(\bar\psi/\psi).$$

$4°$. Exact solution:
$$u = e^{\alpha x + \beta t}\theta(z), \quad z = \lambda x + \gamma t,$$
where α, β, γ, and λ are arbitrary constants, and the function $\theta(z)$ is described by the ODE with delay
$$(a\lambda^2 - \gamma^2)\theta''_{zz}(z) + 2(a\alpha\lambda - \beta\gamma)\theta'_z(z) + (a\alpha^2 - \beta^2)\theta(z)$$
$$+ \theta(z)f\big(e^{-\beta\tau}\theta(z-\sigma)/\theta(z)\big) = 0, \quad \sigma = \gamma\tau.$$

$5°$. Exact solution:
$$u = e^{ct}U_1(x, t; b, s), \quad b = f(e^{-c\tau}) - c^2, \quad s = 2c,$$
where c is an arbitrary constant and $U_1(x, t; b, s)$ is a τ-periodic function that is defined in Item $6°$ of equation 11.2.2.4. For $c = 0$, this solution is a τ-periodic function.

$6°$. Exact solution:
$$u = e^{ct}U_2(x, t; b, s), \quad b = f(-e^{-c\tau}) - c^2, \quad s = 2c,$$
where c is an arbitrary constant and $U_2(x, t; b, s)$ is a τ-antiperiodic function that is defined in Item $2°$ of equation 11.2.2.5.

$7°$. References for PDE 11.2.2.7: Polyanin & Zhurov (2014e) and Polyanin et al. (2023).

8. $u_{tt} = au_{xx} + bu\ln u + uf(w/u)$.

Multiplicative separable solution (Polyanin & Zhurov, 2014e):

$$u = \varphi(x)\psi(t).$$

The functions $\varphi = \varphi(x)$ and $\psi = \psi(t)$ are described by a regular ODE and an ODE with delay:

$$a\varphi''_{xx} = C\varphi - b\varphi\ln\varphi,$$
$$\psi''_{tt} = C\psi + b\psi\ln\psi + \psi f(\bar\psi/\psi), \quad \bar\psi = \psi(t-\tau),$$

where C is an arbitrary constant.

9. $u_{tt} = au_{xx} + u^{1-2k}f(u^k - w^k), \quad k \neq 1$.

Functional separable solutions (Polyanin et al., 2023):

$$u = [x + \theta(z)]^{1/k}, \quad z = t \pm a^{-1/2}x,$$

where the function $\theta = \theta(z)$ is described by the first-order ODE with delay

$$\pm 2a^{1/2}\theta'_z + a + \frac{k^2}{1-k}f(\theta - \bar\theta) = 0, \quad \bar\theta = \theta(z-\tau).$$

10. $u_{tt} = au_{xx} + e^{bu+cw}f(u-w)$.

Exact solutions (Polyanin et al., 2023):

$$u = \varphi(x) + \theta(z), \quad z = t \pm a^{-1/2}x.$$

The function $\varphi = \varphi(x)$ is described by the second-order ODE:

$$\varphi''_{xx} = Ke^{(b+c)\varphi},$$

where K is an arbitrary constant. The function $\theta = \theta(z)$ satisfies the difference equation

$$aK + e^{b\theta+c\bar\theta}f(\theta - \bar\theta) = 0, \quad \bar\theta = \theta(z-\tau).$$

Notably, the general solution of the ODE for φ is expressed in terms of elementary functions.

11. $u_{tt} = au_{xx} + e^{-2\beta u}f(be^{\beta u} + ce^{\beta w})$.

Exact solutions (Polyanin & Zhurov, 2014e):

$$u = \frac{1}{\beta}\ln[\varphi(z)x + \psi(z)], \quad z = t \pm a^{-1/2}x,$$

where the function $\varphi = \varphi(z)$ satisfies the linear difference equation

$$b\varphi + c\bar\varphi = 0, \quad \bar\varphi = \varphi(z-\tau),$$

and the function $\psi = \psi(z)$ is described by the first-order ODE with delay

$$\pm 2a^{1/2}(\varphi'_z\psi - \varphi\psi'_z) - a\varphi^2 + \beta f(b\psi + c\bar\psi) = 0, \quad \bar\psi = \psi(z-\tau).$$

12. $u_{tt} = [a(x)u_x]_x + b(x)u + uf(w/u)$.

Multiplicative separable solution:

$$u = \varphi(x)\psi(t).$$

The functions $\varphi = \varphi(x)$ and $\psi = \psi(t)$ are described by a regular ODE and an ODE with delay:

$$[a(x)\varphi'_x]'_x + b(x)\varphi = C\varphi;$$
$$\psi''_{tt} = C\psi + \psi f(\bar\psi/\psi), \quad \bar\psi = \psi(t-\tau),$$

where C is an arbitrary constant.

▶ **Equations nonlinear in derivatives.**

13. $u_{tt} = a(u^k u_x)_x + uf(w/u), \quad w = u(x, t-\tau)$.

1°. Multiplicative separable solution (Polyanin et al., 2023):

$$u = \varphi(x)\psi(t),$$

where the functions $\varphi = \varphi(x)$ and $\psi = \psi(t)$ are described by a regular ODE and an ODE with delay:

$$a(\varphi^k \varphi'_x)'_x = C\varphi,$$
$$\psi''_{tt} = C\psi^{k+1} + \psi f(\bar\psi/\psi), \quad \bar\psi = \psi(t-\tau).$$

2°. Exact solution:

$$u = (x+C)^{2/k}\theta(z), \quad z = t + \lambda \ln(x+C),$$

where C and λ are arbitrary constants, and the function $\theta = \theta(z)$ is described by the ODE with delay

$$\theta''(z) = a\Big\{\frac{2(k+2)}{k^2}\theta^{k+1}(z) + \frac{(3k+4)\lambda}{k}\theta^k(z)\theta'(z)$$
$$+ k\lambda^2\theta^{k-1}(z)[\theta'(z)]^2 + \lambda^2\theta^k(z)\theta''(z)\Big\} + \theta(z)f\big(\theta(z-\tau)/\theta(z)\big).$$

14. $u_{tt} = a(u^k u_x)_x + bu^{k+1} + uf(w/u)$.

1°. Multiplicative separable solution for $b(k+1) > 0$ (Polyanin et al., 2023):

$$u = [C_1 \cos(\beta x) + C_2 \sin(\beta x)]^{\frac{1}{k+1}} \psi(t), \quad \beta = \sqrt{b(k+1)/a},$$

where C_1 and C_2 are arbitrary constants, and the function $\psi = \psi(t)$ is described by the delay ODE:

$$\psi''_{tt} = \psi f(\bar\psi/\psi), \quad \bar\psi = \psi(t-\tau). \qquad (*)$$

2°. Multiplicative separable solution for $b(k+1) < 0$ (Polyanin et al., 2023):

$$u = (C_1 e^{-\beta x} + C_2 e^{\beta x})^{\frac{1}{k+1}} \psi(t), \quad \beta = \sqrt{-b(k+1)/a},$$

where C_1 and C_2 are arbitrary constants, and $\psi = \psi(t)$ is described by the delay ODE (*).

3°. Multiplicative separable solution for $k = -1$:
$$u = C_1 \exp\left(-\frac{b}{2a}x^2 + C_2 x\right)\psi(t),$$
where C_1 and C_2 are arbitrary constants, and the function $\psi = \psi(t)$ is described by the delay ODE (∗).

4°. A multiplicative separable solution that generalizes the three preceding solutions:
$$u = \varphi(x)\psi(t).$$

The functions $\varphi = \varphi(x)$ and $\psi = \psi(t)$ are described by a regular ODE and an ODE with delay:
$$a(\varphi^k \varphi_x')_x' + b\varphi^{k+1} = C\varphi,$$
$$\psi_{tt}'' = C\psi^{k+1} + \psi f(\bar\psi/\psi), \quad \bar\psi = \psi(t-\tau),$$
where C is an arbitrary constant.

15. $u_{tt} = a(e^{\lambda u} u_x)_x + f(u - w).$

Additive separable solution (Polyanin et al., 2023):
$$u = \frac{1}{\lambda}\ln(C_1 \lambda x^2 + C_2 x + C_3) + \psi(t),$$
where C_1, C_2, and C_3 are arbitrary constants, and the function $\psi = \psi(t)$ is described by the ODE with delay
$$\psi_{tt}'' = 2aC_1 e^{\lambda\psi} + f(\psi - \bar\psi), \quad \bar\psi = \psi(t-\tau).$$

16. $u_{tt} = a(e^{\lambda u} u_x)_x + be^{\lambda u} + f(u - w).$

1°. Additive separable solution for $b\lambda > 0$ (Polyanin et al., 2023):
$$u = \frac{1}{\lambda}\ln[C_1 \cos(\beta x) + C_2 \sin(\beta x)] + \psi(t), \quad \beta = \sqrt{b\lambda/a},$$
where C_1 and C_2 are arbitrary constants, and the function $\psi = \psi(t)$ is described by the delay ODE:
$$\psi_{tt}'' = f(\psi - \bar\psi), \quad \bar\psi = \psi(t-\tau). \tag{∗}$$

2°. Additive separable solution for $b\lambda < 0$ (Polyanin et al., 2023):
$$u = \frac{1}{\lambda}\ln(C_1 e^{-\beta x} + C_2 e^{\beta x}) + \psi(t), \quad \beta = \sqrt{-b\lambda/a},$$
where C_1 and C_2 are arbitrary constants, and the function $\psi = \psi(t)$ is described by the delay ODE (∗).

3°. An additive separable solution that generalizes the two preceding solutions:
$$u = \varphi(x) + \psi(t).$$

The functions $\varphi = \varphi(x)$ and $\psi = \psi(t)$ are described by a regular ODE and an ODE with delay:
$$a(e^{\lambda\varphi}\varphi'_x)'_x + be^{\lambda\varphi} = C,$$
$$\psi''_{tt} = Ce^{\lambda\psi} + f(\psi - \bar{\psi}), \quad \bar{\psi} = \psi(t-\tau),$$
where C is an arbitrary constant. Note that the first ODE is linearized by the substitution $\xi = e^{\lambda\varphi}$.

17. $u_{tt} = [(a\ln u + b)u_x]_x - cu\ln u + uf(w/u).$

Multiplicative separable solutions (Polyanin et al., 2023):
$$u = \exp(\pm\lambda x)\psi(t), \quad \lambda = \sqrt{c/a},$$
where the function $\psi = \psi(t)$ is described by the ODE with delay
$$\psi''_{tt} = \lambda^2(a+b)\psi + \psi f(\bar{\psi}/\psi), \quad \bar{\psi} = \psi(t-\tau),$$

18. $u_{tt} = a(u^k u_x)_x + uf(w/u) + u^{k+1}g(w/u).$

Multiplicative separable solution (Polyanin et al., 2023):
$$u = e^{\lambda t}\varphi(x),$$
where λ is a root of the algebraic (transcendental) equation $\lambda^2 = f(e^{-\lambda\tau})$ and the function $\varphi = \varphi(x)$ is described by the second-order ODE:
$$a(\varphi^k \varphi'_x)'_x + g(e^{-\lambda\tau})\varphi^{k+1} = 0.$$
Note that this ODE is linearized by the substitution $\xi = \varphi^{k+1}$.

19. $u_{tt} = [a(x)u^k u_x]_x + b(x)u^{k+1} + uf(w/u).$

Multiplicative separable solution (Polyanin & Sorokin, 2021a):
$$u = \varphi(x)\psi(t),$$
where the functions $\varphi = \varphi(x)$ and $\psi = \psi(t)$ are described by a regular ODE and an ODE with delay:
$$[a(x)\eta'_x]'_x + (k+1)b(x)\eta = 0, \quad \eta = \varphi^{k+1};$$
$$\psi''_{tt}(t) = \psi(t)f\bigl(\psi(t-\tau)/\psi(t)\bigr).$$

20. $u_{tt} = [a(x)e^{\beta u}u_x]_x + b(x)e^{\beta u} + f(u-w).$

Additive separable solution:
$$u = \frac{1}{\beta}\ln\varphi(x) + \psi(t).$$
The functions $\varphi = \varphi(x)$ and $\psi = \psi(t)$ are described by a linear ODE without delay and a nonlinear ODE with delay:
$$[a(x)\varphi'_x]'_x + \beta b(x)\varphi = C\beta,$$
$$\psi''_{tt}(t) = Ce^{\beta\psi} + f\bigl(\psi(t) - \psi(t-\tau)\bigr),$$
where C is an arbitrary constant.

Remark 11.7. Polyanin et al. (2015) described a number of exact solutions to *nonlinear hyperbolic reaction-diffusion equations with delay* (nonlinear telegraph-type equations with delay) that contain both time derivatives and have the form

$$au_{tt} + bu_t = [f(u)u_x]_x + g(u, w), \quad w = u(x, t - \tau).$$

Remark 11.8. Exact solutions of some nonlinear systems of two coupled PDEs with one or more delays are described in Polyanin & Zhurov (2015), Polyanin & Sorokin (2022), Polyanin & Zhurov (2022), and Polyanin et al. (2023).

11.3. Nonlinear PDEs with Proportional Arguments

▶ Throughout this section, it is assumed that $f = f(z)$, $g = g(z)$, and $h = h(z)$ are arbitrary functions, $p > 0$ and $q > 0$ are arbitrary constants, and $u = u(x, t)$ is the unknown function.

11.3.1. Parabolic Equations

▶ **Equations linear in derivatives and involving arbitrary parameters.**

1. $u_t = au_{xx} + bw^2, \quad w = u(x, \tfrac{1}{2}t).$

   1°. Multiplicative separable solution:
   $$u = e^{-\lambda t}\varphi(x),$$
   where λ is an arbitrary constant, and the function $\varphi = \varphi(x)$ is described by the second-order autonomous ODE: $a\varphi''_{xx} + b\varphi^2 + \lambda\varphi = 0$.

   2°. For a self-similar solution of this PDE with proportional delay, see equation 11.3.1.5 below with $k = 2$, $p = 1$, and $q = \tfrac{1}{2}$.

2. $u_t = au_{xx} + bw^{1/q}, \quad w = u(x, qt).$

   1°. Multiplicative separable solution:
   $$u = e^{-\lambda t}\varphi(x),$$
   where λ is an arbitrary constant, and the function $\varphi = \varphi(x)$ is described by the second-order autonomous ODE: $a\varphi''_{xx} + b\varphi^{1/q} + \lambda\varphi = 0$.

   2°. For a self-similar solution of this PDE with proportional delay, see equation 11.3.1.5 below with $k = 1/q$ and $p = 1$.

3. $u_t = au_{xx} + bw^2, \quad w = u(\tfrac{1}{2}x, t).$

   1°. Multiplicative separable solution:
   $$u = e^{-\lambda x}\psi(t),$$
   where λ is an arbitrary constant, and the function $\psi = \psi(t)$ is described by the separable first-order ODE: $\psi'_t = a\lambda^2\psi + b\psi^2$.

2°. For a self-similar solution of this PDE with proportional argument, see equation 11.3.1.5 below with $k = 2$, $p = \frac{1}{2}$, and $q = 1$.

4. $u_t = au_{xx} + bw^{1/p}$, $w = u(px, t)$.

1°. Multiplicative separable solution:
$$u = e^{-\lambda x}\psi(t),$$
where λ is an arbitrary constant, and the function $\psi = \psi(t)$ is described by the separable first-order ODE: $\psi'_t = a\lambda^2\psi + b\psi^{1/p}$.

2°. For a self-similar solution of this PDE with proportional argument, see equation 11.3.1.5 below with $k = 1/p$ and $q = 1$.

5. $u_t = au_{xx} + bw^k$, $w = u(px, qt)$.

1°. Self-similar solution for $k \neq 1$:
$$u(x,t) = t^{\frac{1}{1-k}}U(z), \quad z = xt^{-1/2},$$
where the function $U = U(z)$ is described by the ODE with proportional argument
$$aU''_{zz} + \frac{1}{2}zU'_z - \frac{1}{1-k}U + bq^{\frac{k}{1-k}}W^k = 0, \quad W = U(sz), \quad s = pq^{-1/2}.$$

2°. Traveling wave solution for $q = p$:
$$u(x,t) = U(z), \quad z = kx - \lambda t,$$
where k and λ are arbitrary constants, and the function $U = U(z)$ is described by the ODE with proportional argument
$$ak^2 U''_{zz} + \lambda U'_z + bW^k = 0, \quad W = U(pz).$$

6. $u_t = au_{xx} + bu^m w^k$, $w = u(px, qt)$.

1°. Self-similar solution for $k \neq 1 - m$ (Polyanin & Sorokin, 2021b):
$$u(x,t) = t^{\frac{1}{1-m-k}}U(z), \quad z = xt^{-1/2},$$
where the function $U = U(z)$ is described by the ODE with proportional argument
$$aU''_{zz} + \frac{1}{2}zU'_z - \frac{1}{1-m-k}U + bq^{\frac{k}{1-m-k}}U^m W^k = 0,$$
$$W = U(sz), \quad s = pq^{-1/2}.$$

2°. Traveling wave solution for $q = p$:
$$u(x,t) = U(z), \quad z = kx - \lambda t,$$
where k and λ are arbitrary constants, and the function $U = U(z)$ is described by the ODE with proportional argument
$$ak^2 U''_{zz} + \lambda U'_z + bU^m W^k = 0, \quad W = U(pz).$$

3°. Multiplicative separable solution for $m = 1 - kq$:
$$u(x,t) = e^{\lambda t}\varphi(x),$$
where λ is an arbitrary constant, and the function $\varphi = \varphi(x)$ is described by the ODE with proportional argument
$$a\varphi''_{xx} - \lambda\varphi + b\varphi^{1-kq}\bar\varphi^k = 0, \quad \bar\varphi = \varphi(px).$$

4°. Multiplicative separable solution for $m = 1 - kp$:
$$u(x,t) = e^{\lambda x}\psi(t),$$
where λ is an arbitrary constant, and the function $\psi = \psi(t)$ is described by the ODE with proportional argument
$$\psi'_t = a\lambda^2\psi + b\psi^{1-kp}\bar\psi^k, \quad \bar\psi = \psi(qt).$$

7. $u_t = au_{xx} + be^{\lambda w}, \quad w = u(px, qt).$

This is a special case of equation 11.3.1.8 with $\mu = 0$.

8. $u_t = au_{xx} + be^{\mu u + \lambda w}, \quad w = u(px, qt).$

Exact solution for $\mu \neq -\lambda$ (Polyanin & Sorokin, 2021b):
$$u(x,t) = U(z) - \frac{1}{\mu + \lambda}\ln t, \quad z = xt^{-1/2},$$
where the function $U = U(z)$ is described by the ODE with proportional argument
$$aU''_{zz} + \frac{1}{2}zU'_z + \frac{1}{\mu + \lambda} + bq^{-\frac{\lambda}{\mu+\lambda}}e^{\mu U + \lambda W} = 0,$$
$$W = U(sz), \quad s = pq^{-1/2}.$$

9. $u_t = au_{xx} + u(b\ln u + c\ln w + d), \quad w = u(px, qt).$

1°. Multiplicative separable solution (Polyanin & Sorokin, 2021b):
$$u(x,t) = \varphi(x)\psi(t).$$
The functions $\varphi = \varphi(x)$ and $\psi = \psi(t)$ are described by the ODEs with proportional argument
$$a\varphi''_{xx} + \varphi(b\ln\varphi + c\ln\bar\varphi) = K\varphi, \quad \bar\varphi = \varphi(px);$$
$$\psi'_t = \psi(b\ln\psi + c\ln\bar\psi) + (d + K)\psi, \quad \bar\psi = \psi(qt),$$
where K is an arbitrary constant.

2°. Functional separable solution for $p = 1$ (Polyanin et al., 2023):
$$u(x,t) = \exp[\psi_2(t)x^2 + \psi_1(t)x + \psi_0(t)],$$

where the functions $\psi_n = \psi_n(t)$ are described by the system of ODEs with proportional argument

$$\psi_2' = 4a\psi_2^2 + b\psi_2 + c\bar{\psi}_2, \quad \bar{\psi}_2 = \psi_2(qt),$$
$$\psi_1' = 4a\psi_1\psi_2 + b\psi_1 + c\bar{\psi}_1, \quad \bar{\psi}_1 = \psi_1(qt),$$
$$\psi_0' = a(\psi_1^2 + 2\psi_2) + b\psi_0 + c\bar{\psi}_0 + d, \quad \bar{\psi}_0 = \psi_0(qt).$$

3°. Traveling wave solution for $q = p$:

$$u(x,t) = U(z), \quad z = kx - \lambda t,$$

where k and λ are arbitrary constants, and the function $U = U(z)$ is described by the ODE with proportional argument

$$ak^2 U_{zz}'' + \lambda U_z' + U(b\ln U + c\ln W + d) = 0, \quad W = U(pz).$$

10. $u_t = au_{xx} + u(b\ln^2 u + c\ln u + d\ln w + s), \quad w = u(x, qt).$

1°. Functional separable solution for $ab > 0$ (Polyanin et al., 2023):

$$u(x,t) = \exp[\psi_1(t)\varphi(x) + \psi_2(t)],$$
$$\varphi(x) = A\cos(\lambda x) + B\sin(\lambda x), \quad \lambda = \sqrt{b/a},$$

where A and B are arbitrary constants, and the functions $\psi_n = \psi_n(t)$ are described by the system of ODEs with proportional argument

$$\psi_1' = 2b\psi_1\psi_2 + (c-b)\psi_1 + d\bar{\psi}_1, \quad \bar{\psi}_1 = \psi_1(qt),$$
$$\psi_2' = b(A^2 + B^2)\psi_1^2 + b\psi_2^2 + c\psi_2 + d\bar{\psi}_2 + s, \quad \bar{\psi}_2 = \psi_2(qt).$$

2°. Functional separable solution for $ab < 0$ (Polyanin et al., 2023):

$$u(x,t) = \exp[\psi_1(t)\varphi(x) + \psi_2(t)],$$
$$\varphi(x) = A\cosh(\lambda x) + B\sinh(\lambda x), \quad \lambda = \sqrt{-b/a},$$

where A and B are arbitrary constants, and the functions $\psi_n = \psi_n(t)$ are described by the system of ODEs with proportional argument

$$\psi_1' = 2b\psi_1\psi_2 + (c-b)\psi_1 + d\bar{\psi}_1, \quad \bar{\psi}_1 = \psi_1(qt),$$
$$\psi_2' = b(A^2 - B^2)\psi_1^2 + b\psi_2^2 + c\psi_2 + d\bar{\psi}_2 + s, \quad \bar{\psi}_2 = \psi_2(qt).$$

For $A = \pm B$, we have $\varphi(x) = Ae^{\pm\lambda x}$. In this case, the second equation of the system becomes independent, and the first one becomes linear in ψ_1.

▶ **Equations linear in derivatives and involving arbitrary functions.**

13. $u_t = au_{xx} + f(u - w), \quad w = u(x, qt).$

Additive separable solution (Polyanin & Sorokin, 2021b):

$$u(x,t) = C_1 x^2 + C_2 x + \psi(t),$$

where C_1 and C_2 are arbitrary constants, and the function $\psi = \psi(t)$ is described by the first-order ODE with proportional argument

$$\psi_t' = 2aC_1 + f(\psi - \bar\psi), \quad \bar\psi = \psi(qt).$$

14. $u_t = au_{xx} + f(u - w), \quad w = u(px, t).$

Additive separable solution (Polyanin & Sorokin, 2021b):

$$u(x, t) = Ct + \varphi(x),$$

where C is an arbitrary constant, and the function $\varphi = \varphi(x)$ is described by the second-order ODE with proportional argument

$$a\varphi_{xx}'' - C + f(\varphi - \bar\varphi) = 0, \quad \bar\varphi = \varphi(px).$$

15. $u_t = au_{xx} + bu + f(u - w), \quad w = u(x, qt).$

1°. Additive separable solution for $ab < 0$ (Polyanin & Sorokin, 2021b):

$$u(x, t) = A\cosh(\lambda x) + B\sinh(\lambda x) + \psi(t), \quad \lambda = \sqrt{-b/a},$$

where A and B are arbitrary constants, and the function $\psi = \psi(t)$ is described by the first-order ODE with proportional argument

$$\psi_t' = b\psi + f(\psi - \bar\psi), \quad \bar\psi = \psi(qt). \qquad (*)$$

2°. Additive separable solution for $ab > 0$ (Polyanin & Sorokin, 2021b):

$$u(x, t) = A\cos(\lambda x) + B\sin(\lambda x) + \psi(t), \quad \lambda = \sqrt{b/a},$$

where A and B are arbitrary constants, and the function $\psi = \psi(t)$ is described by the first-order ODE with proportional argument $(*)$.

16. $u_t = au_{xx} + bu + f(u - w), \quad w = u(px, t).$

Additive separable solution (Polyanin & Sorokin, 2021b):

$$u(x, t) = Ce^{bt} + \varphi(x),$$

where C is an arbitrary constant, and the function $\varphi = \varphi(x)$ is described by the second-order ODE with proportional argument

$$a\varphi_{xx}'' + b\varphi + f(\varphi - \bar\varphi) = 0, \quad \bar\varphi = \varphi(px).$$

17. $u_t = au_{xx} + e^{\lambda u}f(u - w), \quad w = u(px, qt).$

Exact solution (Polyanin et al., 2023):

$$u(x, t) = U(z) - \frac{1}{\lambda}\ln t, \quad z = xt^{-1/2},$$

where the function $U = U(z)$ is described by the second-order ODE with proportional argument

$$aU_{zz}'' + \frac{1}{2}zU_z' + \frac{1}{\lambda} + e^{\lambda U}f\left(U - W + \frac{1}{\lambda}\ln q\right) = 0,$$
$$W = U(sz), \quad s = pq^{-1/2}.$$

18. $u_t = au_{xx} + uf(w/u), \quad w = u(x, qt).$

1°. Multiplicative separable solution (Polyanin & Sorokin, 2021b):
$$u(x,t) = [A\cosh(\lambda x) + B\sinh(\lambda x)]\psi(t),$$
where A, B, and λ are arbitrary constants, and the function $\psi = \psi(t)$ is described by the first-order ODE with proportional argument
$$\psi'_t = a\lambda^2 \psi + \psi f(\bar\psi/\psi), \quad \bar\psi = \psi(qt). \qquad (*)$$

2°. Multiplicative separable solution (Polyanin & Sorokin, 2021b):
$$u(x,t) = [A\cos(\lambda x) + B\sin(\lambda x)]\psi(t),$$
where A, B, and λ are arbitrary constants, and the function $\psi = \psi(t)$ is described by the first-order ODE with proportional argument
$$\psi'_t = -a\lambda^2 \psi + \psi f(\bar\psi/\psi), \quad \bar\psi = \psi(qt).$$

19. $u_t = au_{xx} + uf(w/u), \quad w = u(px, t).$

Multiplicative separable solution (Polyanin & Sorokin, 2021b):
$$u(x,t) = e^{\lambda t}\varphi(x),$$
where λ is an arbitrary constant, and the function $\varphi = \varphi(x)$ is described by the second-order ODE with proportional argument
$$a\varphi''_{xx} + \varphi[f(\bar\varphi/\varphi) - \lambda] = 0, \quad \bar\varphi = \varphi(px).$$

20. $u_t = au_{xx} + bu\ln u + uf(w/u), \quad w = u(x, qt).$

Multiplicative separable solution (Polyanin & Sorokin, 2021b):
$$u(x,t) = \varphi(x)\psi(t).$$

The functions $\varphi = \varphi(x)$ and $\psi = \psi(t)$ are described by a regular second-order ODE and a first-order ODE with proportional argument:
$$a\varphi''_{xx} = C\varphi - b\varphi\ln\varphi,$$
$$\psi'_t = C\psi + \psi f(\bar\psi/\psi) + b\psi\ln\psi, \quad \bar\psi = \psi(qt),$$
where C is an arbitrary constant.

21. $u_t = au_{xx} + bu\ln u + uf(w/u), \quad w = u(px, t).$

Multiplicative separable solution (Polyanin & Sorokin, 2021b):
$$u(x,t) = \exp(Ce^{bt})\varphi(x),$$
where C is an arbitrary constant, and the function $\varphi = \varphi(x)$ is described by the second-order ODE with proportional argument
$$a\varphi''_{xx} + b\varphi\ln\varphi + \varphi f(\bar\varphi/\varphi) = 0, \quad \bar\varphi = \varphi(px).$$

▶ **Equations nonlinear in derivatives.**

22. $u_t = a(u^k u_x)_x + u f(w/u), \quad w = u(x, qt).$

Multiplicative separable solution (Polyanin & Sorokin, 2021b):

$$u(x, t) = \varphi(x)\psi(t).$$

The functions $\varphi = \varphi(x)$ and $\psi = \psi(t)$ are described by a regular second-order ODE and a first-order ODE with proportional argument:

$$a(\varphi^k \varphi'_x)'_x = C\varphi,$$
$$\psi'_t = C\psi^{k+1} + \psi f(\bar\psi/\psi), \quad \bar\psi = \psi(qt),$$

where C is an arbitrary constant.

23. $u_t = a(u^k u_x)_x + u f(w/u), \quad w = u(px, t).$

Exact solution (Polyanin & Sorokin, 2021b):

$$u(x, t) = e^{2\lambda t} U(z), \quad z = e^{-k\lambda t} x,$$

where λ is an arbitrary constant, and the function $U = U(z)$ is described by the second-order ODE with proportional argument

$$2\lambda U - k\lambda z U'_z = a(U^k U'_z)'_z + U f(W/U), \quad W = U(pz).$$

24. $u_t = a(u^k u_x)_x + b u^{k+1} + u f(w/u), \quad w = u(x, qt).$

1°. Multiplicative separable solution for $b(k+1) > 0$ (Polyanin & Sorokin, 2021b):

$$u(x, t) = [C_1 \cos(\beta x) + C_2 \sin(\beta x)]^{1/(k+1)} \psi(t), \quad \beta = \sqrt{b(k+1)/a},$$

where C_1 and C_2 are arbitrary constants, and the function $\psi = \psi(t)$ is described by the first-order ODE with proportional argument

$$\psi'_t = \psi f(\bar\psi/\psi), \quad \bar\psi = \psi(qt). \qquad (*)$$

2°. Multiplicative separable solution for $b(k+1) < 0$ (Polyanin & Sorokin, 2021b):

$$u(x, t) = [C_1 \exp(-\beta x) + C_2 \exp(\beta x)]^{1/(k+1)} \psi(t), \quad \beta = \sqrt{-b(k+1)/a},$$

where C_1 and C_2 are arbitrary constants, and the function $\psi = \psi(t)$ is described by the first-order ODE with proportional argument $(*)$.

3°. Multiplicative separable solution for $k = -1$:

$$u(x, t) = C_1 \exp\left(-\frac{b}{2a}x^2 + C_2 x\right)\psi(t),$$

where C_1 and C_2 are arbitrary constants, and the function $\psi = \psi(t)$ is described by the first-order ODE with proportional argument $(*)$.

25. $u_t = a(u^k u_x)_x + u^{k+1} f(w/u)$, $w = u(x, qt)$.

Exact solution (Polyanin & Sorokin, 2021b):
$$u(x,t) = t^{-1/k}\varphi(z), \quad z = x + \lambda \ln t,$$

where λ is an arbitrary constant, and the function $\varphi = \varphi(z)$ is described by the second-order ODE with proportional argument
$$a(\varphi^k \varphi'_z)'_z - \lambda \varphi'_z + \frac{1}{k}\varphi + \varphi^{k+1} f(q^{-1/k}\bar\varphi/\varphi) = 0, \quad \bar\varphi = \varphi(z + \lambda \ln q).$$

26. $u_t = a(u^k u_x)_x + u^n f(w/u)$, $w = u(px, qt)$.

1°. Self-similar solution (Polyanin et al., 2023):
$$u(x,t) = t^{\frac{1}{1-n}} U(z), \quad z = xt^{\frac{n-k-1}{2(1-n)}},$$

where the function $U = U(z)$ is described by the second-order ODE with proportional argument
$$\frac{1}{1-n}U + \frac{n-k-1}{2(1-n)}zU'_z = a(U^k U'_z)'_z + U^n f(W/U), \quad W = U(sz), \quad s = pq^{\frac{n-k-1}{2(1-n)}}.$$

2°. Traveling wave solution for $q = p$:
$$u(x,t) = U(z), \quad z = kx - \lambda t,$$

where k and λ are arbitrary constants, and the function $U = U(z)$ is described by the second-order ODE with proportional argument
$$ak^2(U^k U'_z)'_z + \lambda U'_z + U^n f(W/U) = 0, \quad W = U(pz).$$

27. $u_t = a(u^k u_x)_x + b + u^{-k} f(u^{k+1} - w^{k+1})$, $w = u(x, qt)$.

Functional separable solution (Polyanin & Sorokin, 2021b):
$$u(x,t) = \left[\psi(t) - \frac{b(k+1)}{2a}x^2 + C_1 x + C_2\right]^{1/(k+1)},$$

where C_1 and C_2 are arbitrary constants, and the function $\psi = \psi(t)$ is described by the first-order ODE with proportional argument
$$\psi'_t = (k+1)f(\psi - \bar\psi), \quad \bar\psi = \psi(qt).$$

28. $u_t = a(u^k u_x)_x + bu^{-k} + f(u^{k+1} - w^{k+1})$, $w = u(px, t)$.

Functional separable solution (Polyanin & Sorokin, 2021b):
$$u = \left[b(k+1)t + \varphi(x)\right]^{\frac{1}{k+1}},$$

where the function $\varphi = \varphi(x)$ is described by the second-order ODE with proportional argument
$$a\varphi''_{xx} + (k+1)f(\varphi - \bar\varphi) = 0, \quad \bar\varphi = \varphi(px).$$

29. $u_t = a(e^{\lambda u} u_x)_x + f(u - w)$, $\quad w = u(x, qt)$.

Additive separable solution (Polyanin & Sorokin, 2021b):
$$u = \frac{1}{\lambda} \ln(Ax^2 + Bx + C) + \psi(t),$$
where A, B, and C are arbitrary constants, and the function $\psi(t)$ is described by the first-order ODE with proportional argument
$$\psi' = 2a(A/\lambda)e^{\lambda \psi} + f(\psi - \bar\psi), \quad \bar\psi = \psi(qt).$$

30. $u_t = a(e^{\lambda u} u_x)_x + be^{\lambda u} + f(u - w)$, $\quad w = u(x, qt)$.

1°. Additive separable solution for $b\lambda > 0$ (Polyanin & Sorokin, 2021b):
$$u(x, t) = \frac{1}{\lambda} \ln[C_1 \cos(\beta x) + C_2 \sin(\beta x)] + \psi(t), \quad \beta = \sqrt{b\lambda/a},$$
where C_1 and C_2 are arbitrary constants, and the function $\psi(t)$ is described by the first-order ODE with proportional argument
$$\psi'_t = f(\psi - \bar\psi), \quad \bar\psi = \psi(qt). \tag{$*$}$$

2°. Additive separable solution for $b\lambda < 0$ (Polyanin & Sorokin, 2021b):
$$u(x, t) = \frac{1}{\lambda} \ln[C_1 \exp(-\beta x) + C_2 \exp(\beta x)] + \psi(t), \quad \beta = \sqrt{-b\lambda/a},$$
where C_1 and C_2 are arbitrary constants, and the function $\psi(t)$ is described by the first-order ODE with proportional argument $(*)$.

31. $u_t = a(e^{\lambda u} u_x)_x + e^{\lambda u} f(u - w)$, $\quad w = u(px, t)$.

Additive separable solution (Polyanin & Sorokin, 2021b):
$$u(x, t) = -\frac{1}{\lambda} \ln t + \varphi(x),$$
where the function $\varphi = \varphi(x)$ is described by the second-order ODE with proportional argument
$$a(e^{\lambda \varphi} \varphi'_x)'_x + \frac{1}{\lambda} + e^{\lambda \varphi} f(\varphi - \bar\varphi) = 0, \quad \bar\varphi = \varphi(px).$$

32. $u_t = a(e^{\lambda u} u_x)_x + e^{\mu u} f(u - w)$, $\quad w = u(px, qt)$.

Exact solution (Polyanin et al., 2023):
$$u(x, t) = U(z) - \frac{1}{\mu} \ln t, \quad z = xt^{\frac{\lambda - \mu}{2\mu}},$$
where the function $U = U(z)$ is described by the second-order ODE with proportional argument
$$\frac{\lambda - \mu}{2\mu} z U'_z - \frac{1}{\mu} = a(e^{\lambda U} U'_z)'_z + e^{\mu U} f\left(U - W + \frac{1}{\mu} \ln q\right),$$
$$W = U(sz), \quad s = pq^{\frac{\lambda - \mu}{2\mu}}.$$

33. $u_t = a(e^{\lambda u} u_x)_x + b + e^{-\lambda u} f(e^{\lambda u} - e^{\lambda w})$, $\quad w = u(x, qt)$.

Functional separable solution (Polyanin & Sorokin, 2021b):
$$u(x,t) = \frac{1}{\lambda} \ln\left[\psi(t) - \frac{b\lambda}{2a} x^2 + C_1 x + C_2\right],$$
where C_1 and C_2 are arbitrary constants, and the function $\psi(t)$ is described by the first-order ODE with proportional argument
$$\psi'_t = \lambda f(\psi - \bar\psi), \quad \bar\psi = \psi(qt).$$

34. $u_t = [(a \ln u + b) u_x]_x - c u \ln u + u f(w/u)$, $\quad w = u(x, qt)$.

Multiplicative separable solutions (Polyanin & Sorokin, 2021b):
$$u(x,t) = \exp(\pm\sqrt{c/a}\, x) \psi(t),$$
where the function $\psi(t)$ is described by the first-order ODE with proportional argument
$$\psi'_t = c(1 + b/a)\psi + \psi f(\bar\psi/\psi), \quad \bar\psi = \psi(qt).$$

35. $u_t = a[f'_u(u) u_x]_x + b + \dfrac{1}{f'_u(u)} g(f(u) - f(w))$, $\quad w = u(x, qt)$.

Functional separable solution in implicit form (Polyanin & Sorokin, 2021b):
$$f(u) = \psi(t) - \frac{b}{2a} x^2 + C_1 x + C_2,$$
where the function $\psi(t)$ is described by the first-order ODE with proportional argument
$$\psi'_t = g(\psi - \bar\psi), \quad \bar\psi = \psi(qt).$$

36. $u_t = [f'_u(u) u_x]_x + \dfrac{a}{f'_u(u)} + g(f(u) - f(w))$, $\quad w = u(px, t)$.

Functional separable solution in implicit form (Polyanin & Sorokin, 2021b):
$$f(u) = at + \varphi(x),$$
where the function $\varphi = \varphi(x)$ is described by the second-order ODE with proportional argument
$$\varphi''_{xx} + g(\varphi - \bar\varphi) = 0, \quad \bar\varphi = \varphi(px).$$

37. $u_t = a[f'_u(u) u_x]_x + b f(u) + \dfrac{f(u)}{f'_u(u)} g(f(w)/f(u))$, $\quad w = u(x, qt)$.

1°. Multiplicative separable solution for $ab > 0$ (Polyanin & Sorokin, 2021b):
$$f(u) = [C_1 \cos(\lambda x) + C_2 \sin(\lambda x)] \psi(t), \quad \lambda = \sqrt{b/a},$$
where C_1 and C_2 are arbitrary constants, and the function $\psi(t)$ is described by the first-order ODE with proportional argument
$$\psi'_t = \psi g(\bar\psi/\psi), \quad \bar\psi = \psi(qt). \qquad (*)$$

2°. Multiplicative separable solution for $ab < 0$ (Polyanin & Sorokin, 2021b):
$$f(u) = [C_1 \exp(-\lambda x) + C_2 \exp(\lambda x)]\psi(t), \quad \lambda = \sqrt{-b/a},$$
where C_1 and C_2 are arbitrary constants, and the function $\psi(t)$ is described by the first-order ODE with proportional argument (∗).

38. $u_t = [u f'_u(u) u_x]_x + \dfrac{1}{f'_u(u)}[a f(u) + b f(w) + c], \quad w = u(px, qt).$

Functional separable solution in implicit form (Polyanin et al., 2023):
$$f(u) = \varphi(t)x + \psi(t),$$
where the functions $\varphi = \varphi(t)$ and $\psi = \psi(t)$ are described by the system of first-order ODEs with proportional argument
$$\varphi'_t = a\varphi + bp\bar\varphi, \quad \bar\varphi = \varphi(qt),$$
$$\psi'_t = a\psi + b\bar\psi + c + \varphi^2, \quad \bar\psi = \psi(qt).$$

11.3.2. Hyperbolic Equations

▶ **Equations linear in derivatives and involving arbitrary parameters.**

1. $u_{tt} = a u_{xx} + b w^2, \quad w = u(x, \tfrac{1}{2}t).$

  1°. Multiplicative separable solution:
$$u = e^{-\lambda t}\varphi(x),$$
where λ is an arbitrary constant, and the function $\varphi = \varphi(x)$ is described by the second-order autonomous ODE $a\varphi''_{xx} + b\varphi^2 - \lambda^2 \varphi = 0$.

  2°. For a self-similar solution of this PDE with proportional delay, see equation 11.3.2.4 below with $k = 2$, $p = 1$, and $q = \tfrac{1}{2}$.

2. $u_{tt} = a u_{xx} + b w^{1/q}, \quad w = u(x, qt).$

  1°. Multiplicative separable solution:
$$u = e^{-\lambda t}\varphi(x),$$
where λ is an arbitrary constant, and the function $\varphi = \varphi(x)$ is described by the second-order autonomous ODE: $a\varphi''_{xx} + b\varphi^{1/q} - \lambda^2 \varphi = 0$.

  2°. For a self-similar solution of this PDE with proportional delay, see equation 11.3.2.4 below with $k = 1/q$ and $p = 1$.

3. $u_{tt} = a u_{xx} + b w^k, \quad w = u(px, qt).$

This is a special case of equation 11.3.2.4 for $m = 0$.

4. $u_{tt} = a u_{xx} + b u^m w^k, \quad w = u(px, qt).$

  1°. Self-similar solution for $k + m \ne 1$ (Polyanin & Sorokin, 2023a):
$$u(x,t) = t^{\frac{2}{1-k-m}} U(z), \quad z = x/t,$$

where the function $U = U(z)$ is described by the second-order ODE with proportional argument

$$\frac{2(1+k+m)}{(1-k-m)^2}U - \frac{2(1+k+m)}{1-k-m}zU'_z + z^2 U''_{zz} = aU''_{zz} + bq^{\frac{2k}{1-k-m}}U^m W^k,$$

$$W = U(sz), \quad s = p/q.$$

2°. Traveling wave solution for $q = p$:

$$u(x,t) = U(z), \quad z = kx - \lambda t,$$

where k and λ are arbitrary constants, and the function $U = U(z)$ is described by the second-order ODE with proportional argument

$$(ak^2 - \lambda^2)U''_{zz} + bU^m W^k = 0, \quad W = U(pz).$$

3°. Multiplicative separable solution for $m = 1 - kq$:

$$u(x,t) = e^{-\lambda t}\varphi(x),$$

where λ is an arbitrary constant, and the function $\varphi = \varphi(x)$ is described by the second-order ODE with proportional argument

$$a\varphi''_{xx} - \lambda^2 \varphi + b\varphi^{1-kq}\bar{\varphi}^k = 0, \quad \bar{\varphi} = \varphi(px).$$

4°. Multiplicative separable solution for $m = 1 - kp$:

$$u(x,t) = e^{\beta x}\psi(t),$$

where β is an arbitrary constant, and the function $\psi = \psi(t)$ is described by the second-order ODE with proportional argument

$$\psi''_{tt} = a\beta^2 \psi + b\psi^{1-kp}\bar{\psi}^k, \quad \bar{\psi} = \psi(qt).$$

5. $u_{tt} = au_{xx} + be^{\lambda w}, \quad w = u(px, qt).$

Exact solution (Polyanin & Sorokin, 2023a):

$$u(x,t) = U(z) - \frac{2}{\lambda}\ln t, \quad z = \frac{x}{t},$$

where the function $U = U(z)$ is described by the second-order ODE with proportional argument

$$(z^2 U'_z)'_z + \frac{2}{\lambda} = aU''_{zz} + \frac{b}{q^2}e^{\lambda W}, \quad W = U(sz), \quad s = \frac{p}{q}.$$

6. $u_{tt} = au_{xx} + be^{\mu u + \lambda w}, \quad w = u(px, qt).$

Exact solution for $\mu + \lambda \neq 0$ (Polyanin et al., 2023):

$$u(x,t) = U(z) - \frac{2}{\mu + \lambda}\ln t, \quad z = \frac{x}{t},$$

where the function $U = U(z)$ is described by the second-order ODE with proportional argument

$$(z^2 U'_z)'_z + \frac{2}{\mu + \lambda} = aU''_{zz} + bq^{-\frac{2\lambda}{\mu+\lambda}}e^{\mu U + \lambda W}, \quad W = U(sz), \quad s = \frac{p}{q}.$$

7. $u_{tt} = au_{xx} + u(b\ln u + c\ln w), \quad w = u(px, qt).$

Multiplicative separable solution (Polyanin & Sorokin, 2023a):
$$u(x,t) = \varphi(x)\psi(t),$$
where the functions $\varphi = \varphi(x)$ and $\psi = \psi(t)$ are described by the second-order ODEs with proportional argument
$$a\varphi''_{xx} + \varphi(b\ln\varphi + c\ln\bar\varphi) = 0, \quad \bar\varphi = \varphi(px);$$
$$\psi''_{tt} = \psi(b\ln\psi + c\ln\bar\psi), \quad \bar\psi = \psi(qt).$$

▶ **Equations linear in derivatives and involving arbitrary functions.**

8. $u_{tt} = au_{xx} + f(u - w), \quad w = u(x, qt).$

Additive separable solution (Polyanin & Sorokin, 2023a):
$$u(x,t) = C_1 x^2 + C_2 x + \psi(t),$$
where C_1 and C_2 are arbitrary constants, and the function $\psi = \psi(t)$ is described by the second-order ODE with proportional argument
$$\psi''_{tt} = 2aC_1 + f(\psi - \bar\psi), \quad \bar\psi = \psi(qt).$$

9. $u_{tt} = au_{xx} + f(u - w), \quad w = u(px, t).$

Additive separable solution (Polyanin & Sorokin, 2023a):
$$u(x,t) = C_1 t^2 + C_2 t + \varphi(x),$$
where C_1 and C_2 are arbitrary constants, and the function $\varphi = \varphi(x)$ is described by the second-order ODE with proportional argument
$$a\varphi''_{xx} - 2C_1 + f(\varphi - \bar\varphi) = 0, \quad \bar\varphi = \varphi(px).$$

10. $u_{tt} = au_{xx} + bu + f(u - w), \quad w = u(x, qt).$

1°. Additive separable solution for $ab < 0$ (Polyanin & Sorokin, 2023a):
$$u(x,t) = A\cosh(\lambda x) + B\sinh(\lambda x) + \psi(t), \quad \lambda = \sqrt{-b/a},$$
where A and B are arbitrary constants, and the function $\psi = \psi(t)$ is described by the second-order ODE with proportional argument
$$\psi''_{tt} = b\psi + f(\psi - \bar\psi), \quad \bar\psi = \psi(qt). \tag{$*$}$$

2°. Additive separable solution for $ab > 0$ (Polyanin & Sorokin, 2023a):
$$u(x,t) = A\cos(\lambda x) + B\sin(\lambda x) + \psi(t), \quad \lambda = \sqrt{b/a},$$
where A and B are arbitrary constants, and the function $\psi = \psi(t)$ is described by the second-order ODE with proportional argument $(*)$.

11. $u_{tt} = au_{xx} + e^{\lambda u}f(u-w)$, $\quad w = u(px, qt)$.

Exact solution (Polyanin & Sorokin, 2023a):
$$u(x,t) = U(z) - \frac{2}{\lambda}\ln t, \quad z = \frac{x}{t},$$

where the function $U = U(z)$ is described by the second-order ODE with proportional argument
$$(z^2 U_z')_z' + \frac{2}{\lambda} = aU_{zz}'' + e^{\lambda U}f\left(U - W + \frac{2}{\lambda}\ln q\right) = 0,$$
$$W = U(sz), \quad s = p/q.$$

12. $u_{tt} = au_{xx} + uf(w/u)$, $\quad w = u(px, t)$.

1°. Multiplicative separable solution (Polyanin et al., 2023):
$$u(x,t) = (Ae^{-\lambda t} + Be^{\lambda t})\varphi(x),$$

where A, B, and λ are arbitrary constants, the function $\varphi = \varphi(x)$ is described by the second-order ODE with proportional argument
$$a\varphi_{xx}'' + \varphi[f(\bar\varphi/\varphi) - \lambda^2] = 0, \quad \bar\varphi = \varphi(px).$$

2°. Multiplicative separable solution (Polyanin et al., 2023):
$$u(x,t) = [A\cos(\lambda t) + B\sin(\lambda t)]\varphi(x),$$

where A, B, and λ are arbitrary constants, the function $\varphi = \varphi(x)$ is described by the second-order ODE with proportional argument
$$a\varphi_{xx}'' + \varphi[f(\bar\varphi/\varphi) + \lambda^2] = 0, \quad \bar\varphi = \varphi(px).$$

▶ **Equations nonlinear in derivatives.**

13. $u_{tt} = a(u^k u_x)_x + uf(w/u)$, $\quad w = u(x, qt)$.

Multiplicative separable solution (Polyanin et al., 2023):
$$u(x,t) = \varphi(x)\psi(t).$$

The functions $\varphi = \varphi(x)$ and $\psi = \psi(t)$ are described by a regular ODE and an ODE with proportional argument:
$$a(\varphi^k \varphi_x')_x' = b\varphi,$$
$$\psi_{tt}'' = b\psi^{k+1} + \psi f(\bar\psi/\psi), \quad \bar\psi = \psi(qt),$$

where b is an arbitrary constant.

14. $u_{tt} = a(u^k u_x)_x + uf(w/u)$, $\quad w = u(px, t)$.

Exact solution (Polyanin & Sorokin, 2023a):
$$u(x,t) = e^{2\lambda t}U(z), \quad z = e^{-k\lambda t}x,$$

where λ is an arbitrary constant, and the function $U = U(z)$ is described by the ODE with proportional argument

$$4\lambda^2 U - 4k\lambda^2 z U'_z + k^2\lambda^2 z(zU'_z)'_z = a(U^k U'_z)'_z + Uf(W/U), \quad W = U(pz).$$

15. $u_{tt} = a(u^k u_x)_x + bu^{k+1} + uf(w/u), \quad w = u(x, qt).$

1°. Multiplicative separable solution with $b(k+1) > 0$ (Polyanin & Sorokin, 2023a):

$$u(x,t) = [C_1\cos(\beta x) + C_2\sin(\beta x)]^{1/(k+1)}\psi(t), \quad \beta = \sqrt{b(k+1)/a},$$

where C_1 and C_2 are arbitrary constants, and the function $\psi = \psi(t)$ is described by the second-order ODE with proportional argument

$$\psi''_{tt} = \psi f(\bar\psi/\psi), \quad \bar\psi = \psi(qt). \qquad (*)$$

2°. Multiplicative separable solution with $b(k+1) < 0$ (Polyanin & Sorokin, 2023a):

$$u(x,t) = [C_1\exp(-\beta x) + C_2\exp(\beta x)]^{1/(k+1)}\psi(t), \quad \beta = \sqrt{-b(k+1)/a},$$

where C_1 and C_2 are arbitrary constants, and the function $\psi = \psi(t)$ is described by the second-order ODE with proportional argument $(*)$.

3°. Multiplicative separable solution with $k = -1$:

$$u(x,t) = C_1\exp\left(-\frac{b}{2a}x^2 + C_2 x\right)\psi(t),$$

where C_1 and C_2 are arbitrary constants, and the function $\psi = \psi(t)$ is described by the second-order ODE with proportional argument $(*)$.

16. $u_{tt} = a(u^k u_x)_x + u^{k+1} f(w/u), \quad w = u(x, qt).$

Exact solution (Polyanin & Sorokin, 2023a):

$$u(x,t) = t^{-2/k}\varphi(z), \quad z = x + \lambda\ln t,$$

where λ is an arbitrary constant, and the function $\varphi = \varphi(z)$ is described by the second-order ODE with constant delay

$$\frac{2(k+2)}{k^2}\varphi - \lambda\frac{k+4}{k}\varphi'_z + \lambda^2\varphi''_{zz} = a(\varphi^k\varphi'_z)'_z + \varphi^{k+1}f(q^{-2/k}\bar\varphi/\varphi),$$
$$\bar\varphi = \varphi(z + \lambda\ln q) \quad (\lambda\ln q < 0).$$

17. $u_{tt} = a(u^k u_x)_x + u^n f(w/u), \quad w = u(px, qt).$

1°. Self-similar solution for $n \ne 1$ (Polyanin & Sorokin, 2023a):

$$u(x,t) = t^{\frac{2}{1-n}}U(z), \quad z = xt^{\frac{n-k-1}{1-n}},$$

where the function $U = U(z)$ is described by the second-order ODE with proportional argument

$$\frac{2(1+n)}{(1-n)^2}U + \frac{(n-k-1)(2n-k+2)}{(1-n)^2}zU'_z + \frac{(n-k-1)^2}{(1-n)^2}z^2U''_{zz}$$
$$= a(U^k U'_z)'_z + U^n f(q^{\frac{2}{1-n}}W/U), \quad W = U(sz), \quad s = pq^{\frac{n-k-1}{1-n}}.$$

$2°$. Traveling wave solution for $q = p$:

$$u(x,t) = U(z), \quad z = kx - \lambda t,$$

where k and λ are arbitrary constants, and the function $U = U(z)$ is described by the second-order ODE with proportional argument

$$ak^2(U^k U'_z)'_z - \lambda^2 U''_{zz} + U^n f(W/U) = 0, \quad W = U(pz).$$

18. $u_{tt} = a(e^{\lambda u} u_x)_x + f(u - w), \quad w = u(x, qt).$

Additive separable solution (Polyanin & Sorokin, 2023a):

$$u(x,t) = \frac{1}{\lambda} \ln(Ax^2 + Bx + C) + \psi(t),$$

where A, B, and C are arbitrary constants, and the function $\psi = \psi(t)$ is described by the second-order ODE with proportional argument

$$\psi''_{tt} = 2a(A/\lambda)e^{\lambda \psi} + f(\psi - \bar\psi), \quad \bar\psi = \psi(qt).$$

19. $u_{tt} = a(e^{\lambda u} u_x)_x + e^{\mu u} f(u - w), \quad w = u(px, qt).$

Exact solution (Polyanin & Sorokin, 2023a):

$$u(x,t) = U(z) - \frac{2}{\mu} \ln t, \quad z = xt^{\frac{\lambda - \mu}{\mu}},$$

where the function $U = U(z)$ is described by the second-order ODE with proportional argument

$$\frac{2}{\mu} + \frac{\mu - \lambda}{\mu} zU'_z + \frac{(\lambda - \mu)^2}{\mu^2} z(zU'_z)'_z = a(e^{\lambda U} U'_z)'_z + e^{\mu U} f\left(U - W + \frac{2}{\mu} \ln q\right),$$

$$W = U(sz), \quad s = pq^{\frac{\lambda - \mu}{\mu}}.$$

11.4. Partial Functional Differential Equations with Arguments of General Type

▶ Troughout this section, it is assumed that $f = f(z)$ and $g = g(z)$ are arbitrary functions, $\xi(x)$ and $\eta(t)$ are any monotonically increasing functions, and $u = u(x,t)$ is the unknown function.

11.4.1. Parabolic Equations

▶ **Equations linear in derivatives.**

1. $u_t = au_{xx} + u(b \ln u + c \ln w + d), \quad w = u(\xi(x), \eta(t)).$

Multiplicative separable solution (Polyanin & Sorokin, 2023a):

$$u(x,t) = \varphi(x)\psi(t).$$

The functions $\varphi = \varphi(x)$ and $\psi = \psi(t)$ are described by the ODEs with variable arguments

$$a\varphi''_{xx} + \varphi(b\ln\varphi + c\ln\bar\varphi - K) = 0, \quad \bar\varphi = \varphi(\xi(x));$$
$$\psi'_t = \psi(b\ln\psi + c\ln\bar\psi + d + K), \quad \bar\psi = \psi(\eta(t)),$$

where K is an arbitrary constant.

2. $u_t = au_{xx} + u(b\ln u + c\ln w + d), \quad w = u(x, \eta(t)).$

Functional separable solution (Polyanin & Zhurov, 2013):

$$u(x,t) = \exp[\psi_2(t)x^2 + \psi_1(t)x + \psi_0(t)],$$

where the functions $\psi_n = \psi_n(t)$ are described by the system of ODEs with variable argument

$$\psi'_2 = 4a\psi_2^2 + b\psi_2 + c\bar\psi_2, \quad \bar\psi_2 = \psi_2(\eta(t)),$$
$$\psi'_1 = 4a\psi_1\psi_2 + b\psi_1 + c\bar\psi_1, \quad \bar\psi_1 = \psi_1(\eta(t)),$$
$$\psi'_0 = a(\psi_1^2 + 2\psi_2) + b\psi_0 + c\bar\psi_0 + d, \quad \bar\psi_0 = \psi_0(\eta(t)).$$

3. $u_t = au_{xx} + u(b\ln^2 u + c\ln u + d\ln w + s), \quad w = u(x, \eta(t)).$

1°. Functional separable solution for $ab > 0$ (Polyanin & Zhurov, 2013):

$$u(x,t) = \exp[\psi_1(t)\varphi(x) + \psi_2(t)],$$
$$\varphi(x) = A\cos(\lambda x) + B\sin(\lambda x), \quad \lambda = \sqrt{b/a},$$

where A and B are arbitrary constants, and the functions $\psi_n = \psi_n(t)$ are described by the system of ODEs with variable argument

$$\psi'_1 = 2b\psi_1\psi_2 + (c-b)\psi_1 + d\bar\psi_1, \quad \bar\psi_1 = \psi_1(\eta(t)),$$
$$\psi'_2 = b(A^2 + B^2)\psi_1^2 + b\psi_2^2 + c\psi_2 + d\bar\psi_2 + s, \quad \bar\psi_2 = \psi_2(\eta(t)).$$

2°. Functional separable solution for $ab < 0$ (Polyanin & Zhurov, 2013):

$$u(x,t) = \exp[\psi_1(t)\varphi(x) + \psi_2(t)],$$
$$\varphi(x) = A\cosh(\lambda x) + B\sinh(\lambda x), \quad \lambda = \sqrt{-b/a},$$

where A and B are arbitrary constants, and the functions $\psi_n = \psi_n(t)$ are described by the system of ODEs with variable argument

$$\psi'_1 = 2b\psi_1\psi_2 + (c-b)\psi_1 + d\bar\psi_1, \quad \bar\psi_1 = \psi_1(\eta(t)),$$
$$\psi'_2 = b(A^2 - B^2)\psi_1^2 + b\psi_2^2 + c\psi_2 + d\bar\psi_2 + s, \quad \bar\psi_2 = \psi_2(\eta(t)).$$

For $A = \pm B$, we have $\varphi(x) = Ae^{\pm\lambda x}$. In this case, the second equation of the system becomes independent, and the first one becomes linear in ψ_1.

4. $u_t = au_{xx} + f(u - w), \quad w = u(x, \eta(t)).$

Additive separable solution (Polyanin & Zhurov, 2014a):

$$u(x,t) = C_1 x^2 + C_2 x + \psi(t),$$

where C_1 and C_2 are arbitrary constants, and the function $\psi = \psi(t)$ is described by the ODE with variable argument

$$\psi'_t = 2aC_1 + f(\psi - \bar{\psi}), \quad \bar{\psi} = \psi(\eta(t)).$$

5. $u_t = au_{xx} + bu + f(u - w), \quad w = u(x, \eta(t)).$

1°. Additive separable solution for $ab < 0$ (Polyanin & Zhurov, 2013):

$$u(x,t) = A\cosh(\lambda x) + B\sinh(\lambda x) + \psi(t), \quad \lambda = \sqrt{-b/a},$$

where A and B are arbitrary constants, and the function $\psi = \psi(t)$ is described by the ODE with variable argument

$$\psi'_t = b\psi + f(\psi - \bar{\psi}), \quad \bar{\psi} = \psi(\eta(t)). \qquad (*)$$

2°. Additive separable solution for $ab > 0$ (Polyanin & Zhurov, 2013):

$$u(x,t) = A\cos(\lambda x) + B\sin(\lambda x) + \psi(t), \quad \lambda = \sqrt{b/a},$$

where A and B are arbitrary constants, and the function $\psi = \psi(t)$ is described by the ODE with variable argument $(*)$.

6. $u_t = au_{xx} + uf(w/u), \quad w = u(x, \eta(t)).$

1°. Multiplicative separable solution (Polyanin & Zhurov, 2013):

$$u(x,t) = [A\cosh(\lambda x) + B\sinh(\lambda x)]\psi(t),$$

where A, B, and λ are arbitrary constants, and the function $\psi = \psi(t)$ is described by the ODE with variable argument

$$\psi'_t = a\lambda^2 \psi + \psi f(\bar{\psi}/\psi), \quad \bar{\psi} = \psi(\eta(t)). \qquad (*)$$

2°. Multiplicative separable solution (Polyanin & Zhurov, 2013):

$$u(x,t) = [A\cos(\lambda x) + B\sin(\lambda x)]\psi(t),$$

where A, B, and λ are arbitrary constants, and the function $\psi = \psi(t)$ is described by the ODE with variable argument

$$\psi'_t = -a\lambda^2 \psi + \psi f(\bar{\psi}/\psi), \quad \bar{\psi} = \psi(\eta(t)).$$

3°. Degenerate multiplicative separable solution:

$$u(x,t) = (Ax + B)\psi(t),$$

where A and B are arbitrary constants, and the function $\psi = \psi(t)$ is described by the ODE with variable argument $(*)$ with $a = 0$.

7. $u_t = au_{xx} + bu\ln u + uf(w/u), \quad w = u(x, \eta(t)).$

Multiplicative separable solution (Polyanin & Zhurov, 2013):

$$u(x,t) = \varphi(x)\psi(t).$$

The functions $\varphi = \varphi(x)$ and $\psi = \psi(t)$ are described by a regular ODE and an ODE with variable argument:
$$a\varphi''_{xx} = C_1\varphi - b\varphi \ln \varphi,$$
$$\psi'_t = C_1\psi + \psi f(\bar\psi/\psi) + b\psi \ln \psi, \quad \bar\psi = \psi(\eta(t)),$$

where C_1 is an arbitrary constant. The first ODE for φ is autonomous, its general solution can be obtained in implicit form. A particular one-parameter solution of this equation can be represented in explicit form:
$$\varphi = \exp\left[-\frac{b}{4a}(x+C_2)^2 + \frac{C_1}{b} + \frac{1}{2}\right],$$

where C_2 is an arbitrary constant.

8. $u_t = au_{xx} + f(u-w), \quad w = u(\xi(x), t).$

Additive separable solution (Polyanin et al., 2022):
$$u(x,t) = Ct + \varphi(x),$$

where C is an arbitrary constant, and the function $\varphi = \varphi(x)$ is described by the second-order ODE with variable argument
$$a\varphi''_{xx} - C + f(\varphi - \bar\varphi) = 0, \quad \bar\varphi = \varphi(\xi(x)).$$

9. $u_t = au_{xx} + bu + f(u-w), \quad w = u(\xi(x), t).$

Additive separable solution (Polyanin et al., 2022):
$$u(x,t) = Ce^{bt} + \varphi(x),$$

where C is an arbitrary constant, and the function $\varphi = \varphi(x)$ is described by the second-order ODE with variable argument
$$a\varphi''_{xx} + b\varphi + f(\varphi - \bar\varphi) = 0, \quad \bar\varphi = \varphi(\xi(x)).$$

10. $u_t = au_{xx} + uf(w/u), \quad w = u(\xi(x), t).$

Multiplicative separable solution (Polyanin et al., 2022):
$$u(x,t) = e^{\lambda t}\varphi(x),$$

where λ is an arbitrary constant, and the function $\varphi = \varphi(x)$ is described by the second-order ODE with variable argument
$$a\varphi''_{xx} + \varphi[f(\bar\varphi/\varphi) - \lambda] = 0, \quad \bar\varphi = \varphi(\xi(x)).$$

11. $u_t = au_{xx} + bu \ln u + uf(w/u), \quad w = u(\xi(x), t).$

Multiplicative separable solution (Polyanin et al., 2022):
$$u(x,t) = \exp(Ce^{bt})\varphi(x),$$

where C is an arbitrary constant, and the function $\varphi = \varphi(x)$ is described by the second-order ODE with variable argument
$$a\varphi''_{xx} + b\varphi \ln \varphi + \varphi f(\bar\varphi/\varphi) = 0, \quad \bar\varphi = \varphi(\xi(x)).$$

▶ **Equations nonlinear in derivatives.**

12. $u_t = a(u^k u_x)_x + u f(w/u)$, $w = u(x, \eta(t))$.

Multiplicative separable solution (Polyanin & Zhurov, 2014d):

$$u(x,t) = \varphi(x)\psi(t).$$

The functions $\varphi = \varphi(x)$ and $\psi = \psi(t)$ are described by a regular ODE and an ODE with variable argument:

$$a(\varphi^k \varphi'_x)'_x = C\varphi,$$
$$\psi'_t = C\psi^{k+1} + \psi f(\bar{\psi}/\psi), \quad \bar{\psi} = \psi(\eta(t));$$

where C is an arbitrary constant.

13. $u_t = a(u^k u_x)_x + bu^{k+1} + u f(w/u)$, $w = u(x, \eta(t))$.

1°. Multiplicative separable solution for $b(k+1) > 0$ (Polyanin & Zhurov, 2014d):

$$u(x,t) = [C_1 \cos(\beta x) + C_2 \sin(\beta x)]^{1/(k+1)} \psi(t), \quad \beta = \sqrt{b(k+1)/a},$$

where C_1 and C_2 are arbitrary constants, and the function $\psi = \psi(t)$ is described by the ODE with variable argument

$$\psi'_t = \psi f(\bar{\psi})/\psi), \quad \bar{\psi} = \psi(\eta(t)). \tag{$*$}$$

2°. Multiplicative separable solution for $b(k+1) < 0$ (Polyanin & Zhurov, 2014d):

$$u(x,t) = [C_1 \exp(-\beta x) + C_2 \exp(\beta x)]^{1/(k+1)} \psi(t), \quad \beta = \sqrt{-b(k+1)/a},$$

where C_1 and C_2 are arbitrary constants, and the function $\psi = \psi(t)$ is described by the ODE with variable argument $(*)$.

3°. Multiplicative separable solution for $k = -1$:

$$u(x,t) = C_1 \exp\left(-\frac{b}{2a}x^2 + C_2 x\right)\psi(t),$$

where C_1 and C_2 are arbitrary constants, and the function $\psi = \psi(t)$ is described by the ODE with variable argument $(*)$.

4°. Multiplicative separable solution, generalizing preceding solutions:

$$u(x,t) = \varphi(x)\psi(t).$$

The functions $\varphi = \varphi(x)$ and $\psi = \psi(t)$ are described by a regular second-order ODE and a first-order ODE with variable argument:

$$a(\varphi^k \varphi'_x)'_x + b\varphi^{k+1} = C\varphi,$$
$$\psi'_t = C\psi^{k+1} + \psi f(\bar{\psi}/\psi), \quad \bar{\psi} = \psi(\eta(t)),$$

where C is an arbitrary constant.

14. $u_t = a(u^k u_x)_x + b + u^{-k} f(u^{k+1} - w^{k+1}), \quad w = u(x, \eta(t)).$

Functional separable solution (Polyanin & Zhurov, 2014d):

$$u(x,t) = \left[\psi(t) - \frac{b(k+1)}{2a}x^2 + C_1 x + C_2\right]^{1/(k+1)},$$

where C_1 and C_2 are arbitrary constants, and the function $\psi = \psi(t)$ is described by the first-order ODE with variable argument

$$\psi'_t = (k+1)f(\psi - \bar\psi), \quad \bar\psi = \psi(\eta(t)).$$

15. $u_t = a(u^k u_x)_x + bu^{-k} + f(u^{k+1} - w^{k+1}), \quad w = u(\xi(x), t).$

Functional separable solution:

$$u = \left[b(k+1)t + \varphi(x)\right]^{\frac{1}{k+1}},$$

where the function $\varphi = \varphi(x)$ is described by the second-order ODE with variable argument

$$a\varphi''_{xx} + (k+1)f(\varphi - \bar\varphi) = 0, \quad \bar\varphi = \varphi(\xi(x)).$$

16. $u_t = a(u^k u_x)_x + bu^{k-2n+1} + u^{1-n} f(u^n - w^n), \quad w = u(x, \eta(t)).$

Generalized traveling wave solutions for $b(n - k - 1) > 0$ (Polyanin et al., 2022):

$$u = [\pm\lambda x + \psi(t)]^{1/n}, \quad \lambda = \sqrt{\frac{bn^2}{a(n-k-1)}},$$

where the function $\psi = \psi(t)$ is described by the ODE with variable argument

$$\psi'_t = nf(\psi - \bar\psi), \quad \bar\psi = \psi(\eta(t)).$$

17. $u_t = a(e^{\lambda u} u_x)_x + f(u - w), \quad w = u(x, \eta(t)).$

Additive separable solution (Polyanin & Zhurov, 2014d):

$$u(x,t) = \frac{1}{\lambda} \ln(Ax^2 + Bx + C) + \psi(t),$$

where A, B, and C are arbitrary constants, and the function $\psi = \psi(t)$ is described by the ODE with variable argument

$$\psi'_t = 2a(A/\lambda)e^{\lambda\psi} + f(\psi - \bar\psi), \quad \bar\psi = \psi(\eta(t)).$$

18. $u_t = a(e^{\lambda u} u_x)_x + be^{\lambda u} + f(u - w), \quad w = u(x, \eta(t)).$

1°. Additive separable solution for $b\lambda > 0$ (Polyanin & Zhurov, 2014d):

$$u(x,t) = \frac{1}{\lambda}\ln[C_1 \cos(\beta x) + C_2 \sin(\beta x)] + \psi(t), \quad \beta = \sqrt{b\lambda/a},$$

where C_1 and C_2 are arbitrary constants, and the function $\psi = \psi(t)$ is described by the ODE with variable argument

$$\psi'_t = f(\psi - \bar\psi), \quad \bar\psi = \psi(\eta(t)). \tag{$*$}$$

2°. Additive separable solution for $b\lambda < 0$ (Polyanin & Zhurov, 2014d):
$$u(x,t) = \frac{1}{\lambda}\ln[C_1\exp(-\beta x) + C_2\exp(\beta x)] + \psi(t), \quad \beta = \sqrt{-b\lambda/a},$$
where C_1 and C_2 are arbitrary constants, and the function $\psi(t)$ is described by the ODE with variable argument (∗).

19. $u_t = a(e^{\lambda u}u_x)_x + b + e^{-\lambda u}f(e^{\lambda u} - e^{\lambda w}), \quad w = u(x, \eta(t)).$
Functional separable solution (Polyanin & Zhurov, 2014d):
$$u(x,t) = \frac{1}{\lambda}\ln\Big[\psi(t) - \frac{b\lambda}{2a}x^2 + C_1 x + C_2\Big],$$
where C_1 and C_2 are arbitrary constants, and the function $\psi = \psi(t)$ is described by the ODE with variable argument
$$\psi'_t = \lambda f(\psi - \bar\psi), \quad \bar\psi = \psi(\eta(t)).$$

20. $u_t = a(e^{\beta u}u_x)_x + be^{(\beta - 2\gamma)u} + e^{-\gamma u}f(e^{\gamma u} - e^{\gamma w}), \quad w = u(x, \eta(t)).$
Functional separable solutions for $b(\gamma - \beta) > 0$ (Polyanin et al., 2022):
$$u = \frac{1}{\gamma}\ln[\lambda x + \psi(t)], \quad \lambda = \pm\sqrt{\frac{b\gamma^2}{a(\gamma - \beta)}},$$
where the function $\psi = \psi(t)$ is described by the ODE with variable argument
$$\psi'_t = \gamma f(\psi - \bar\psi), \quad \bar\psi = \psi(\eta(t)).$$

21. $u_t = [(a\ln u + b)u_x]_x - cu\ln u + uf(w/u), \quad w = u(x, \eta(t)).$
Multiplicative separable solutions (Polyanin & Zhurov, 2014d):
$$u(x,t) = \exp(\pm\sqrt{c/a}\,x)\psi(t),$$
where the function $\psi = \psi(t)$ is described by the ODE with variable argument
$$\psi'_t = c(1 + b/a)\psi + \psi f(\bar\psi/\psi), \quad \bar\psi = \psi(\eta(t)).$$

22. $u_t = a[f'_u(u)u_x]_x + b + \dfrac{1}{f'_u(u)}g(f(u) - f(w)), \quad w = u(x, \eta(t)).$
Functional separable solution in implicit form (Polyanin & Zhurov, 2014d):
$$f(u) = \psi(t) - \frac{b}{2a}x^2 + C_1 x + C_2,$$
where the function $\psi = \psi(t)$ is described by the ODE with variable argument
$$\psi'_t = g(\psi - \bar\psi), \quad \bar\psi = \psi(\eta(t)).$$

23. $u_t = a[f'_u(u)u_x]_x + bf(u) + \dfrac{f(u)}{f'_u(u)}g\Big(\dfrac{f(w)}{f(u)}\Big), \quad w = u(x, \eta(t)).$

1°. Functional separable solution in implicit form for $ab > 0$ (Polyanin & Zhurov, 2014d):
$$f(u) = [C_1\cos(\lambda x) + C_2\sin(\lambda x)]\psi(t), \quad \lambda = \sqrt{b/a},$$
where C_1 and C_2 are arbitrary constants, and the function $\psi = \psi(t)$ is described by the ODE with variable argument
$$\psi'_t = \psi g(\bar\psi/\psi), \quad \bar\psi = \psi(\eta(t)). \qquad (*)$$

2°. Functional separable solution in implicit form for $ab < 0$ (Polyanin & Zhurov, 2014d):
$$f(u) = [C_1 \exp(-\lambda x) + C_2 \exp(\lambda x)]\psi(t), \quad \lambda = \sqrt{-b/a},$$
where C_1 and C_2 are arbitrary constants, and the function $\psi = \psi(t)$ is described by the ODE with variable argument (∗).

24. $u_t = a[uf'_u(u)u_x]_x + \dfrac{1}{f'_u(u)}[bf(u) + cf(w) + d], \quad w = u(x, \eta(t)).$

Functional separable solution in implicit form (Polyanin et al., 2022):
$$f(u) = \varphi(t)x + \psi(t),$$
where the functions $\varphi = \varphi(t)$ and $\psi = \psi(t)$ are described by the ODEs with variable argument
$$\varphi'_t = b\varphi + c\bar\varphi, \quad \bar\varphi = \varphi(\eta(t)),$$
$$\psi'_t = b\psi + c\bar\psi + d + a\varphi^2, \quad \bar\psi = \psi(\eta(t)).$$

25. $u_t = a(e^{\lambda u}u_x)_x + e^{\lambda u}f(u - w), \quad w = u(\xi(x), t).$

Additive separable solution (Polyanin et al., 2022):
$$u = -\dfrac{1}{\lambda}\ln t + \varphi(x),$$
where the function $\varphi = \varphi(x)$ is described by the ODE with variable argument
$$a(e^{\lambda\varphi}\varphi'_x)'_x + \dfrac{1}{\lambda} + e^{\lambda\varphi}f(\varphi - \bar\varphi) = 0, \quad \bar\varphi = \varphi(\xi(x)).$$

26. $u_t = a(u^k u_x)_x + u^{k+1}f(w/u), \quad w = u(\xi(x), t).$

Multiplicative separable solution (Polyanin et al., 2022):
$$u = t^{-1/k}\varphi(x),$$
where the function $\varphi = \varphi(x)$ is described by the ODE with variable argument
$$a(\varphi^k \varphi'_x)'_x + \dfrac{1}{k}\varphi + \varphi^{k+1}f(\bar\varphi/\varphi) = 0, \quad \bar\varphi = \varphi(\xi(x)).$$

27. $u_t = [f'_u(u)u_x]_x + \dfrac{a}{f'_u(u)} + g(f(u) - f(w)), \quad w = u(\xi(x), t).$

Functional separable solution in implicit form (Polyanin et al., 2022):
$$f(u) = at + \varphi(x),$$
where the function $\varphi = \varphi(x)$ is described by the ODE with variable argument
$$\varphi''_{xx} + g(\varphi - \bar\varphi) = 0, \quad \bar\varphi = \varphi(\xi(x)).$$

11.4.2. Hyperbolic Equations

▶ **Equations linear in derivatives.**

1. $u_{tt} = au_{xx} + u(b\ln u + c\ln w + d)$, $\quad w = u(\xi(x), \eta(t))$.

Multiplicative separable solution (Polyanin et al., 2022):
$$u(x,t) = \varphi(x)\psi(t).$$

The functions $\varphi = \varphi(x)$ and $\psi = \psi(t)$ are described by the ODEs with variable argument
$$a\varphi''_{xx} + \varphi(b\ln\varphi + c\ln\bar\varphi - C) = 0, \quad \bar\varphi = \varphi(\xi(x));$$
$$\psi''_{tt} = \psi(b\ln\psi + c\ln\bar\psi + d + C), \quad \bar\psi = \psi(\eta(t)),$$

where C is an arbitrary constant.

2. $u_{tt} = au_{xx} + uf(w/u)$, $\quad w = u(x, \eta(t))$.

1°. Multiplicative separable solution (Polyanin et al., 2022):
$$u(x,t) = [A\cosh(\lambda x) + B\sinh(\lambda x)]\psi(t),$$

where A, B, and λ are arbitrary constants, and the function $\psi = \psi(t)$ is described by the ODE with variable argument
$$\psi''_{tt} = a\lambda^2\psi + \psi f(\bar\psi/\psi), \quad \bar\psi = \psi(\eta(t)). \tag{$*$}$$

2°. Multiplicative separable solution (Polyanin et al., 2022):
$$u(x,t) = [A\cos(\lambda x) + B\sin(\lambda x)]\psi(t),$$

where A, B, and λ are arbitrary constants, and the function $\psi = \psi(t)$ is described by the ODE with variable argument
$$\psi''_{tt} = -a\lambda^2\psi + \psi f(\bar\psi/\psi), \quad \bar\psi = \psi(\eta(t)).$$

3°. Degenerate solution with multiplicative separation of variables:
$$u(x,t) = (Ax+B)\psi(t),$$

where A and B, and the function $\psi = \psi(t)$ is described by the ODE with variable argument $(*)$ with $\lambda = 0$.

3. $u_{tt} = au_{xx} + uf(w/u)$, $\quad w = u(\xi(x), t)$.

1°. Multiplicative separable solution (Polyanin et al., 2022):
$$u(x,t) = [A\cosh(\lambda t) + B\sinh(\lambda t)]\varphi(x),$$

where A, B, and λ are arbitrary constants, and the function $\varphi = \varphi(x)$ is described by the ODE with variable argument
$$a\varphi''_{xx} - \lambda^2\varphi + \varphi f(\bar\varphi/\varphi) = 0, \quad \bar\varphi = \varphi(\xi(x)). \tag{$*$}$$

2°. Multiplicative separable solution (Polyanin et al., 2022):
$$u(x,t) = [A\cos(\lambda t) + B\sin(\lambda t)]\varphi(x),$$
where A, B, and λ are arbitrary constants, and the function $\varphi = \varphi(x)$ is described by the ODE with variable argument
$$a\varphi''_{xx} + \lambda^2\varphi + \varphi f(\bar{\varphi}/\varphi) = 0, \quad \bar{\varphi} = \varphi(\xi(x)).$$

3°. Degenerate multiplicative separable solution:
$$u(x,t) = (At + B)\varphi(x),$$
where A and B are arbitrary constants, and the function $\varphi = \varphi(x)$ is described by the ODE with variable argument (∗) with $\lambda = 0$.

4. $u_{tt} = au_{xx} + f(u - w), \quad w = u(x, \eta(t)).$

Additive separable solution (Polyanin et al., 2022):
$$u(x,t) = C_1 x^2 + C_2 x + \psi(t),$$
where C_1 and C_2 are arbitrary constants, and the function $\psi = \psi(t)$ is described by the ODE with variable argument
$$\psi''_{tt} = 2aC_1 + f(\psi - \bar{\psi}), \quad \bar{\psi} = \psi(\eta(t)).$$

5. $u_{tt} = au_{xx} + bu\ln u + uf(w/u), \quad w = u(x, \eta(t)).$

Multiplicative separable solution (Polyanin et al., 2022):
$$u(x,t) = \varphi(x)\psi(t).$$

The functions $\varphi = \varphi(x)$ and $\psi = \psi(t)$ are described by a regular ODE and an ODE with variable argument:
$$a\varphi''_{xx} = C_1\varphi - b\varphi\ln\varphi,$$
$$\psi''_{tt} = C_1\psi + \psi f(\bar{\psi}/\psi) + b\psi\ln\psi, \quad \bar{\psi} = \psi(\eta(t)),$$
where C_1 is an arbitrary constant. A particular one-parameter solution of the first ODE can be presented in the explicit form
$$\varphi = \exp\left[-\frac{b}{4a}(x + C_2)^2 + \frac{C_1}{b} + \frac{1}{2}\right],$$
where C_2 is an arbitrary constant.

6. $u_{tt} = au_{xx} + bu + f(u - w), \quad w = u(x, \eta(t)).$

1°. Additive separable solution for $ab < 0$ (Polyanin et al., 2022):
$$u(x,t) = A\cosh(\lambda x) + B\sinh(\lambda x) + \psi(t), \quad \lambda = \sqrt{-b/a},$$
where A and B are arbitrary constants, and the function $\psi = \psi(t)$ is described by the ODE with variable argument
$$\psi''_{tt} = b\psi + f(\psi - \bar{\psi}), \quad \bar{\psi} = \psi(\eta(t)). \qquad (*)$$

2°. Additive separable solution for $ab > 0$ (Polyanin et al., 2022):

$$u(x,t) = A\cos(\lambda x) + B\sin(\lambda x) + \psi(t), \quad \lambda = \sqrt{b/a},$$

where A and B are arbitrary constants, and the function $\psi = \psi(t)$ is described by the ODE with variable argument (∗).

7. $u_{tt} = au_{xx} + f(u - w), \quad w = u(\xi(x), t).$

Additive separable solution (Polyanin et al., 2022):

$$u(x,t) = C_1 t^2 + C_2 t + \varphi(x),$$

where C is an arbitrary constant, and the function $\varphi = \varphi(x)$ is described by the ODE with variable argument

$$a\varphi''_{xx} - 2C_1 + f(\varphi - \bar\varphi) = 0, \quad \bar\varphi = \varphi(\xi(x)).$$

▶ **Equations nonlinear in derivatives.**

8. $u_{tt} = a(u^k u_x)_x + uf(w/u), \quad w = u(x, \eta(t)).$

Multiplicative separable solution (Polyanin et al., 2022):

$$u(x,t) = \varphi(x)\psi(t).$$

The functions $\varphi = \varphi(x)$ and $\psi = \psi(t)$ are described by a regular ODE and an ODE with variable argument:

$$a(\varphi^k \varphi'_x)'_x = C\varphi,$$
$$\psi''_{tt} = C\psi^{k+1} + \psi f(\bar\psi/\psi), \quad \bar\psi = \psi(\eta(t));$$

where C is an arbitrary constant.

9. $u_{tt} = a(u^k u_x)_x + uf(w/u) + bu^{k+1}, \quad w = u(x, \eta(t)).$

1°. Multiplicative separable solution for $b(k+1) > 0$ (Polyanin et al., 2022):

$$u(x,t) = [C_1 \cos(\beta x) + C_2 \sin(\beta x)]^{1/(k+1)} \psi(t), \quad \beta = \sqrt{b(k+1)/a},$$

where C_1 and C_2 are arbitrary constants, and the function $\psi = \psi(t)$ is described by the ODE with variable argument

$$\psi''_{tt} = \psi f(\bar\psi/\psi), \quad \bar\psi = \psi(\eta(t)). \qquad (*)$$

2°. Multiplicative separable solution for $b(k+1) < 0$ (Polyanin et al., 2022):

$$u(x,t) = [C_1 \exp(-\beta x) + C_2 \exp(\beta x)]^{1/(k+1)} \psi(t), \quad \beta = \sqrt{-b(k+1)/a},$$

where C_1 and C_2 are arbitrary constants, and the function $\psi = \psi(t)$ is described by the ODE with variable argument (∗).

3°. Multiplicative separable solution for $k = -1$:

$$u(x,t) = C_1 \exp\left(-\frac{b}{2a}x^2 + C_2 x\right) \psi(t),$$

where C_1 and C_2 are arbitrary constants, and the function $\psi = \psi(t)$ is described by the ODE with variable argument (∗).

4°. Multiplicative separable solution generalizing the preceding solutions:
$$u(x,t) = \varphi(x)\psi(t).$$
The functions $\varphi = \varphi(x)$ and $\psi = \psi(t)$ are described by a regular second-order ODE and a second-order ODE with variable argument:
$$a(\varphi^k \varphi'_x)'_x + b\varphi^{k+1} = C\varphi,$$
$$\psi''_{tt} = C\psi^{k+1} + \psi f(\bar\psi/\psi), \quad \bar\psi = \psi(\eta(t)),$$
where C is an arbitrary constant.

10. $u_{tt} = a(e^{\lambda u} u_x)_x + f(u - w), \quad w = u(x, \eta(t)).$

Additive separable solution (Polyanin et al., 2022):
$$u(x,t) = \frac{1}{\lambda} \ln(Ax^2 + Bx + C) + \psi(t),$$
where A, B, and C are arbitrary constants, and the function $\psi = \psi(t)$ is described by the ODE with variable argument
$$\psi''_{tt} = 2a(A/\lambda)e^{\lambda\psi} + f(\psi - \bar\psi), \quad \bar\psi = \psi(\eta(t)).$$

11. $u_{tt} = a(e^{\lambda u} u_x)_x + be^{\lambda u} + f(u - w), \quad w = u(x, \eta(t)).$

1°. Additive separable solution for $b\lambda > 0$ (Polyanin et al., 2022):
$$u(x,t) = \frac{1}{\lambda} \ln[C_1 \cos(\beta x) + C_2 \sin(\beta x)] + \psi(t), \quad \beta = \sqrt{b\lambda/a},$$
where C_1 and C_2 are arbitrary constants, and the function $\psi = \psi(t)$ is described by the ODE with variable argument
$$\psi''_{tt} = f(\psi - \bar\psi), \quad \bar\psi = \psi(\eta(t)). \tag{$*$}$$

2°. Additive separable solution for $b\lambda < 0$ (Polyanin et al., 2022):
$$u(x,t) = \frac{1}{\lambda} \ln[C_1 \exp(-\beta x) + C_2 \exp(\beta x)] + \psi(t), \quad \beta = \sqrt{-b\lambda/a},$$
where C_1 and C_2 are arbitrary constants, and the function $\psi = \psi(t)$ is described by the ODE with variable argument $(*)$.

12. $u_{tt} = [(a \ln u + b)u_x]_x - cu \ln u + uf(w/u), \quad w = u(x, \eta(t)).$

Multiplicative separable solution (Polyanin et al., 2022):
$$u(x,t) = \exp\!\left(\pm\sqrt{c/a}\, x\right) \psi(t),$$
where the function $\psi = \psi(t)$ is described by the ODE with variable argument
$$\psi''_{tt} = c(1 + b/a)\psi + \psi f(\bar\psi/\psi), \quad \bar\psi = \psi(\eta(t)).$$

13. $u_{tt} = a(e^{\lambda u} u_x)_x + e^{\lambda u} f(u - w), \quad w = u(\xi(x), t).$

Additive separable solution (Polyanin et al., 2022):
$$u = -\frac{2}{\lambda} \ln t + \varphi(x),$$

where the function $\varphi = \varphi(x)$ is described by the ODE with variable argument

$$a(e^{\lambda\varphi}\varphi'_x)'_x - \frac{2}{\lambda} + e^{\lambda\varphi}f(\varphi - \bar{\varphi}) = 0, \quad \bar{\varphi} = \varphi(\xi(x)).$$

14. $u_{tt} = a(u^k u_x)_x + u^{k+1} f(w/u), \quad w = u(\xi(x), t).$

Multiplicative separable solution (Polyanin et al., 2022):

$$u = t^{-2/k} \varphi(x),$$

where the function $\varphi = \varphi(x)$ is described by the ODE with variable argument

$$a(\varphi^k \varphi'_x)'_x - \frac{2(k+2)}{k^2}\varphi + \varphi^{k+1} f(\bar{\varphi}/\varphi) = 0, \quad \bar{\varphi} = \varphi(\xi(x)).$$

11.5. PDEs with Anisotropic Time Delay

▶ *Throughout this section, it is assumed that $f = f(z)$, $g = g(z)$, and $h = h(z)$, and $\tau = \tau(x)$ are arbitrary functions ($\tau > 0$), and $u = u(x,t)$ is the unknown function.*

11.5.1. Parabolic Equations

▶ **PDEs with anisotropic time delay.**

1. $u_t = au_{xx} + uf(w/u), \quad w = u(x, t - \tau(x)).$

Multiplicative separable solution (Polyanin & Sorokin, 2023b):

$$u = e^{\lambda t}\varphi(x),$$

where λ is an arbitrary constant, and the function $\varphi = \varphi(x)$ satisfies the second-order linear ODE:

$$a\varphi''_{xx} - \lambda\varphi + \varphi f(e^{-\lambda\tau(x)}) = 0.$$

2. $u_t = au_{xx} + f(u - w), \quad w = u(x, t - \tau(x)).$

Generalized separable solution linear in t (Polyanin & Sorokin, 2023b):

$$u = (Ax + B)t + \psi(x),$$

where A and B are arbitrary constants, and the function $\psi = \psi(x)$ satisfies the second-order linear ODE:

$$a\psi''_{xx} + f((Ax + B)\tau(x)) - Ax - B = 0,$$

which can be easily integrated.

3. $u_t = au_{xx} + uf(u - w) + wg(u - w) + h(u - w), \quad w = u(x, t - \tau(x)).$

Generalized separable solution linear in t (Polyanin & Sorokin, 2023b):

$$u = \varphi(x)t + \psi(x),$$

where the functions $\varphi = \varphi(x)$ and $\psi = \psi(x)$ are described by the linear system of ODEs:
$$a\varphi''_{xx} + \varphi[f(\varphi\tau) + g(\varphi\tau)] = 0, \quad \tau = \tau(x),$$
$$a\psi''_{xx} + \psi[f(\varphi\tau) + g(\varphi\tau)] + h(\varphi\tau) - \varphi[1 + \tau g(\varphi\tau)] = 0.$$

4. $u_t = a(u^k u_x)_x + bu + u^{k+1} f(w/u), \quad w = u(x, t - \tau(x)).$

Multiplicative separable exact solution (Polyanin & Sorokin, 2023b):
$$u = e^{bt}\varphi(x),$$
where the function $\varphi = \varphi(x)$ satisfies the nonlinear second-order ODE:
$$a(\varphi^k \varphi'_x)'_x + \varphi^{k+1} f(e^{-b\tau(x)}) = 0.$$
This ODE allows linearization using the substitution $\theta = \varphi^{k+1}$.

5. $u_t = a(u^{-1/2} u_x)_x + u^{1/2} f(u^{1/2} - w^{1/2}) + g(u^{1/2} - w^{1/2}).$

Generalized separable solution (Polyanin & Sorokin, 2023b):
$$u = [\varphi(x)t + \psi(x)]^2.$$
To be specific, we assume that $\varphi(x) > 0$ and $\psi(x) > 0$. Then, the functions $\varphi = \varphi(x)$ and $\psi = \psi(x)$ are described by the nonlinear ODE system
$$2a\varphi''_{xx} + \varphi f(\varphi\tau) - 2\varphi^2 = 0, \quad \tau = \tau(x),$$
$$2a\psi'' + \psi f(\varphi\tau) + g(\varphi\tau) - 2\varphi\psi = 0.$$

6. $u_t = a(e^{\lambda u} u_x)_x + b + e^{\lambda u} f(u - w), \quad w = u(x, t - \tau(x)).$

Additive separable solution (Polyanin & Sorokin, 2023b):
$$u(x, t) = bt + \varphi(x),$$
where the function $\varphi = \varphi(x)$ is described by the second-order ODE:
$$a(e^{\lambda\varphi}\varphi'_x)'_x + e^{\lambda\varphi} f(b\tau(x)) = 0.$$
The substitution $\theta = e^{\lambda\varphi}$ reduces this equation to a linear ODE.

7. $u_t = [a(x) f'_u(u) u_x]_x + \dfrac{b}{f'_u(u)} + g(f(u) - f(w)), \quad w = u(x, t - \tau(x)).$

Generalized traveling wave solution in implicit form (Polyanin & Sorokin, 2023b):
$$f(u) = bt + \theta(x),$$
where the function $\theta = \theta(x)$ satisfies the linear second-order ODE:
$$[a(x)\theta'_x]'_x + g(b\tau(x)) = 0.$$
The general solution of this ODE is
$$\theta = -\int \frac{1}{a(x)} \left(\int g(b\tau(x))\, dx \right) dx + C_1 \int \frac{dx}{a(x)} + C_2,$$
where C_1 and C_2 are arbitrary constants.

8. $u_t = a[f'_u(u) u_x]_x + b\dfrac{f(u)}{f'_u(u)} + f(u) g\!\left(\dfrac{f(w)}{f(u)}\right), \quad w = u(x, t - \tau(x)).$

Functional separable solution in implicit form (Polyanin & Sorokin, 2023b):
$$f(u) = e^{bt}\varphi(x),$$
where the function $\varphi = \varphi(x)$ satisfies the linear second-order ODE:
$$a\varphi''_{xx} + g(e^{-b\tau(x)})\varphi = 0.$$

▶ **PDEs with proportional anisotropic time delay.**

9. $u_t = a(u^k u_x)_x + u^{k+1} f(w/u), \quad w = u(x, p(x)t)$.

Here $f(z)$ and $p(x)$ are arbitrary functions ($0 < p(x) \leq 1$).

Multiplicative separable solution (Polyanin & Sorokin, 2023b):

$$u = t^{-1/k} \varphi(x),$$

where the function $\varphi = \varphi(x)$ satisfies the nonlinear second-order ODE:

$$a(\varphi^k \varphi'_x)'_x + \varphi^{k+1} f(p^{-1/k}(x)) + \frac{1}{k}\varphi = 0.$$

10. $u_t = a(e^{\lambda u} u_x)_x + e^{\lambda u} f(u - w), \quad w = u(x, p(x)t)$.

Here $f(z)$ and $p(x)$ are arbitrary functions ($0 < p(x) \leq 1$).

Additive separable solution (Polyanin & Sorokin, 2023b):

$$u = -\frac{1}{\lambda} \ln t + \varphi(x),$$

where the function $\varphi = \varphi(x)$ is described by the second-order ODE:

$$a(e^{\lambda \varphi} \varphi'_x)'_x + \frac{1}{\lambda} + e^{\lambda \varphi} f\left(\frac{1}{\lambda} \ln p(x)\right) = 0.$$

The substitution $\theta = e^{\lambda \varphi}$ reduces this equation to a linear ODE.

11.5.2. Hyperbolic Equations

▶ **PDEs with anisotropic time delay.**

1. $u_{tt} = a u_{xx} + u f(w/u), \quad w = u(x, t - \tau(x))$.

Multiplicative separable solution (Polyanin & Sorokin, 2023b):

$$u = e^{\lambda t} \varphi(x),$$

where the function $\varphi = \varphi(x)$ satisfiying the linear second-order ODE:

$$a\varphi''_{xx} - \lambda^2 \varphi + \varphi f(e^{-\lambda \tau(x)}) = 0.$$

2. $u_{tt} = a(u^k u_x)_x + bu + u^{k+1} f(w/u), \quad w = u(x, t - \tau(x))$.

Multiplicative separable solutions (Polyanin & Sorokin, 2023b):

$$u = e^{\pm \sqrt{b}\, t} \varphi(x),$$

where the function $\varphi = \varphi(x)$ satisfies the ODE

$$a(\varphi^k \varphi'_x)'_x + \varphi^{k+1} f\left(e^{\mp \sqrt{b}\, \tau(x)}\right) = 0.$$

This ODE allows linearization using the substitution $\theta = \varphi^{k+1}$.

3. $u_{tt} = a(u^{-1/2}u_x)_x + u^{1/2}f(u^{1/2} - w^{1/2}) + g(u^{1/2} - w^{1/2})$.

Generalized separable solution (Polyanin & Sorokin, 2023b):
$$u = [\varphi(x)t + \psi(x)]^2.$$

To be specific, we assume that $\varphi(x) > 0$ and $\psi(x) > 0$. Then, the functions $\varphi = \varphi(x)$ and $\psi = \psi(x)$ are described by the nonlinear system of ODEs:
$$2a\varphi''_{xx} + \varphi f(\varphi\tau) = 0, \quad \tau = \tau(x),$$
$$2a\psi'' + \psi f(\varphi\tau) + g(\varphi\tau) - 2\varphi^2 = 0.$$

4. $u_{tt} = [a(x)f'_u(u)u_x]_x - b^2 \dfrac{f''_{uu}}{(f'_u)^3} + g(f(u) - f(w))$, $\quad w = u(x, t - \tau(x))$.

Generalized traveling wave solution in implicit form (Polyanin & Sorokin, 2023b):
$$f(u) = bt + \theta(x),$$

where the function $\theta = \theta(x)$ satisfies the linear second-order ODE:
$$[a(x)\theta'_x]'_x + g(b\tau(x)) = 0.$$

The general solution of this ODE is
$$\theta = -\int \frac{1}{a(x)} \left(\int g(b\tau(x))\, dx \right) dx + C_1 \int \frac{dx}{a(x)} + C_2,$$

where C_1 and C_2 are arbitrary constants.

5. $u_{tt} = [a(x)f'_u u_x]_x + b^2 \dfrac{f}{f'_u}\left(1 - \dfrac{ff''_{uu}}{(f'_u)^2}\right) + fg\left(\dfrac{\bar{f}}{f}\right)$, $\quad f = f(u)$, $\quad \bar{f} = f(w)$.

Functional separable solution in implicit form (Polyanin & Sorokin, 2023b):
$$f(u) = e^{bt}\varphi(x),$$

with the function $\varphi = \varphi(x)$ satisfying the linear second-order ODE:
$$[a(x)\varphi'_x]'_x + g(e^{-b\tau(x)})\varphi = 0.$$

▶ **PDEs with proportional anisotropic time delay.**

6. $u_{tt} = a(u^k u_x)_x + u^{k+1}f(w/u)$, $\quad w = u(x, p(x)t)$.

Here $f(z)$ and $p(x)$ are arbitrary functions ($0 < p(x) \leq 1$).

Multiplicative separable solution (Polyanin & Sorokin, 2023b):
$$u = t^{-2/k}\varphi(x),$$

where the function $\varphi = \varphi(x)$ satisfies the nonlinear second-order ODE:
$$a(\varphi^k \varphi'_x)'_x + \varphi^{k+1}f(p^{-2/k}) - \frac{2(2+k)}{k^2}\varphi = 0, \quad p = p(x).$$

7. $u_{tt} = a(e^{\lambda u} u_x)_x + e^{\lambda u} f(u - w), \quad w = u(x, p(x)t).$

Here $f(z)$ and $p(x)$ are arbitrary functions ($0 < p(x) \leq 1$).

Additive separable solution (Polyanin & Sorokin, 2023b):

$$u = -\frac{2}{\lambda} \ln t + \varphi(x),$$

where the function $\varphi = \varphi(x)$ is described by the nonlinear second-order ODE:

$$a(e^{\lambda \varphi} \varphi'_x)'_x - \frac{2}{\lambda} + e^{\lambda \varphi} f\left(\frac{2}{\lambda} \ln p\right) = 0, \quad p = p(x).$$

The substitution $\theta = e^{\lambda \varphi}$ reduces this equation to a linear ODE.

▶ For methods for constructing exact solutions of partial differential equations with constant or variable delays, see the book by Polyanin et al. (2023). The most common mathematical models with a delay used in population theory, biology, medicine, and other applications, as well as numerical methods for solving delay PDEs are described in Schiesser (2019), Sorokin & Vyzmin (2022), and Polyanin et al. (2023). The properties and elements of the theory of delay and other functional PDEs are addressed, for example, in Wu (1996) and Polyanin et al. (2023).

References

Khusainov, D.Y., Ivanov, A.F., and Kovarzh, I.V., Solution of one heat equation with delay, *Nonlinear Oscil.*, 2009, Vol. 12, No. 2, pp. 260–282.

Khusainov, D.Y., Pokojovy, M., and Azizbayov, E.I., On classical solvability for a linear 1D heat equation with constant delay, *Konstanzer Schriften in Mathematik*, No. 316, 2013; ISSN 1430-3558 (see also arXiv:1401.5662v1 [math.AP], 2014, https://arxiv.org/pdf/1401.5662.pdf).

Long, F.-S. and Meleshko, S.V., On the complete group classification of the one-dimensional nonlinear Klein–Gordon equation with a delay. *Math. Methods Appl. Sci.*, Vol. 39, No. 12, pp. 3255–3270, 2016.

Meleshko, S.V. and Moyo, S., On the complete group classification of the reaction-diffusion equation with a delay, *J. Math. Anal. Appl.*, Vol. 338, pp. 448–466, 2008.

Polyanin, A.D., Generalized traveling-wave solutions of nonlinear reaction-diffusion equations with delay and variable coefficients, *Appl. Math. Lett.*, Vol. 90, pp. 49–53, 2019a.

Polyanin, A.D., Functional separable solutions of nonlinear reaction-diffusion equations with variable coefficients. *Appl. Math. Comput.*, Vol. 347, pp. 282–292, 2019b.

Polyanin, A.D. and Sorokin, V.G., Construction of exact solutions to nonlinear PDEs with delay using solutions of simpler PDEs without delay, *Commun. Nonlinear Sci. Numer. Simul.*, Vol. 95, 105634, 2021a.

Polyanin, A.D. and Sorokin, V.G., Nonlinear pantograph-type diffusion PDEs: exact solutions and the principle of analogy, *Mathematics*, Vol. 9, No. 5, 511, 2021b.

Polyanin, A.D. and Sorokin, V.G., Reductions and exact solutions of Lotka–Volterra and more complex reaction-diffusion systems with delays, *Appl. Math. Lett.*, Vol. 125, 107731, 2022.

Polyanin, A.D. and Sorokin, V.G., Reductions and exact solutions of nonlinear wave-type PDEs with proportional and more complex delays, *Mathematics*, Vol. 11, No. 3, 516, 2023a.

Polyanin, A.D. and Sorokin, V.G., Exact solutions of reaction-diffusion PDEs with anisotropic time delay, *Mathematics*, Vol. 11, No. 14, 3111, 2023b.

Polyanin, A.D., Sorokin, V.G., and Vyazmin, A.V., Exact solutions and qualitative features of nonlinear hyperbolic reaction-diffusion equations with delay. *Theor. Found. Chem. Eng.,* Vol. 49, No. 5, pp. 622–635, 2015.

Polyanin, A.D., Sorokin, V.G., and Zhurov, A.I., *Delay Differential Equations: Properties, Methods, Solutions and Models* [in Russian], Inst. Problems in Mechanics RAS, Moscow, 2022.

Polyanin, A.D., Sorokin, V.G., and Zhurov, A.I., *Delay Ordinary and Partial Differential Equations*, CRC Press, Boca Raton, 2023.

Polyanin, A.D. and Zhurov, A.I., Exact solutions of linear and nonlinear differential-difference heat and diffusion equations with finite relaxation time, *Int. J. Non-Linear Mech.,* Vol. 54, pp. 115–126, 2013.

Polyanin, A.D. and Zhurov, A.I., Exact separable solutions of delay reaction-diffusion equations and other nonlinear partial functional-differential equations, *Commun. Nonlinear Sci. Numer. Simul.,* Vol. 19, No. 3, pp. 409–416, 2014a.

Polyanin, A.D. and Zhurov, A.I., Functional constraints method for constructing exact solutions to delay reaction-diffusion equations and more complex nonlinear equations, *Commun. Nonlinear Sci. Numer. Simul.,* Vol. 19, No. 3, pp. 417–430, 2014b.

Polyanin, A.D. and Zhurov, A.I., New generalized and functional separable solutions to nonlinear delay reaction-diffusion equations, *Int. J. Non-Linear Mech.,* Vol. 59, pp. 16–22, 2014c.

Polyanin, A.D. and Zhurov, A.I., The functional constraints method: Application to non-linear delay reaction-diffusion equations with varying transfer coefficients, *Int. J. Non-Linear Mech.,* Vol. 67, pp. 267–277, 2014d.

Polyanin, A.D. and Zhurov, A.I., Generalized and functional separable solutions to nonlinear delay Klein–Gordon equations, *Commun. Nonlinear Sci. Numer. Simul.,* Vol. 19, No. 8, pp. 2676–2689, 2014e.

Polyanin, A.D. and Zhurov, A.I., The generating equations method: Constructing exact solutions to delay reaction-diffusion systems and other non-linear coupled delay PDEs. *Int. J. Non-Linear Mech.,* Vol. 71, pp. 104–115, 2015.

Polyanin, A.D. and Zhurov, A.I., Multi-parameter reaction-diffusion systems with quadratic nonlinearity and delays: New exact solutions in elementary functions. *Mathematics,* Vol. 10, No. 9, 529, 2022.

Rodríguez, F., Roales, M., and Marín J.A., Exact solutions and numerical approximations of mixed problems for the wave equation with delay, *Appl. Math. Comput.,* Vol. 219, No. 6, pp. 3178–3186, 2012.

Schiesser, W.E., *Time Delay ODE/PDE Models: Applications in Biomedical Science and Engineering*, CRC Press, Boca Raton, 2019.

Sorokin, V.G. and Vyazmin, A.V., Nonlinear reaction-diffusion equations with delay: Partial survey, exact solutions, test problems, and numerical integration, *Mathematics,* Vol. 10, No. 11, 1886, 2022.

Wu, J., *Theory and Applications of Partial Functional Differential Equations*, Springer, New York, 1996.

Index

A

Abel equation
  differential (ODE), 41–46
    first kind, 41
    first kind, general form, 41
    first kind, special cases, 41
    second kind, 43–46
    second kind, canonical form, 41, 45
    second kind, general form, 44
    second kind, normal form, 45
    second kind, special cases, 42, 43
    second kind, transformations, 45
  functional, 487
  integral, 396
    generalized, 396
    second kind, 409
    second kind, generalized, 410
Abel differential equation (ODE)
  first kind, 41
  first kind, general form, 41
  first kind, special cases, 41
  second kind, 43–46
  second kind, canonical form, 41, 45
  second kind, general form, 44
  second kind, normal form, 45
  second kind, special cases, 42, 43
  second kind, transformations, 45
Abel equations, *see* Abel equation
Abel functional equation, 487
Abel integral equation, 396, 409
  generalized, 396
  second kind, 409
  second kind, generalized, 410
Abel theorem, 18
adiabatic index, 292
additive separable solution, 251
Airy equation, 51
Airy functions, 51, 235
Airy stress function, 247
algebraic equation, 1
  $2n$th degree, even powers, 13
  arbitrary degree, 12
  binomial
    nth degree, 12
  cubic, 2
    incomplete, 2
    simplest, 2
    two-term, 2
  fifth degree, 9
  first degree, 1
  fourth degree, 5
    incomplete, 8
    modified, 6
  general
    fifth degree, 12
    fourth degree, 8
    nth degree, 17
    quintic, 12
  homogeneous with respect to polynomials, 18
  linear, 1
  polynomial
    fourth degree, 8
  quadratic, 1
  quartic, 5
    general, 8
    two-term, 5
  quintic, 9
    de Moivre, 11
  reciprocal
    even degree, 13
    even degree, generalized, 14
    fifth degree, 12
    fourth degree, 6
    fourth degree, generalized, 6
    odd degree, 14
    third degree, 4
  second degree, 1
  third degree, 4
  trinomial
    arbitrary degree, 14
    arbitrary degree, special form, 13
    two arbitrary degrees, 15
algebraic equations, *see* algebraic equation
associated Legendre functions, 63, 67
automorphic functions, 487
autonomous first-order ODE, 37
autonomous nth-order ODE, 107
autonomous ODE, mass and heat transfer problems, 77
autonomous second-order ODE, 82
autonomous system of two ODEs, general form, 129

axisymmetric heat equation, 185
  formulas for constructing particular solutions, 185
  nonhomogeneous, 185
    first boundary value problem, closed interval, 185
    second boundary value problem, closed interval, 186
    solutions via Green function, 185
    third boundary value problem, closed interval, 186
  particular solutions, 184

B

Babbage equation, 488
  functional differential, 524
Babbage functional differential equation, 524
Bernoulli equation, 38
Bessel equation, 59
  modified, 59
Bessel functions, 59
  first kind, 59
    modified, 60
  second kind, 59
    modified, 60
biharmonic equation, 247
bilinear functional equation, 511, 512
  general form, 512
  splitting principle, 512
binomial hyperbolic equations, 30
binomial trigonometric equations, 20
biquadratic equation, 5
blow-up regime, 260, 261
boundary conditions
  Dirichlet, 184, 547, 549
  Neumann, 184, 547, 549
  Robin, 184, 547, 549
  mixed, 183, 547, 549
boundary layer equation, 195
  diffusion, 195
  hydrodynamic, 320
    with pressure gradient, 321
  thermal, 195
boundary value problem, *see also* boundary value problems
  first, 184, 547, 549
  mixed, 184
  second, 184, 547, 549
  third, 184, 547, 549
boundary value problems, *see also* problems of mathematical physics
  axisymmetric heat equation
    formulas for constructing particular solutions, 185
    particular solutions, 184
  centrally symmetric heat equation, 186, 187
    formulas for constructing particular solutions, 187
    particular solutions, 187
    reduction to constant coefficient heat equation, 187

equations of diffusion boundary layer, 195
equations of thermal boundary layer, 195
heat-type equation, 189, 190
  first boundary value problem, 190
  formulas and transformations for constructing particular solutions, 190
  particular solutions, 189
  second boundary value problem, 190
hyperbolic equations, 198
Klein–Gordon equation, *see* Klein–Gordon equation
  nonhomogeneous, *see* Klein–Gordon equation
liquid-film mass transfer equation, 193
  dissolution of plate by laminar fluid film, 195
  mass exchange between fluid film and gas, 194
  particular solutions, 193
mixed, 184, 547, 549
nonhomogeneous axisymmetric heat equation, 185
  first boundary value problem, closed interval, 185
  second boundary value problem, closed interval, 186
  solutions via Green function, 185
  third boundary value problem, closed interval, 186
nonhomogeneous centrally symmetric heat equation, 188
  first boundary value problem, 188
  second boundary value problem, 188
  solutions via Green function, 188
  third boundary value problem, 188
nonhomogeneous wave equation, *see* wave equation
parabolic equations, 191
  general formulas for solving, 191
  properties of Sturm–Liouville problem, 192
  types of problems, 192
Schrödinger equation, *see* Schrödinger equation
solutions
  diffusion equation with axial symmetry, 184
  heat equation with axial symmetry, 184
  nonhomogeneous boundary conditions, 183
  reduction to homogeneous boundary conditions, 183, 184
wave equation, *see* wave equation
Boussinesq equation, fourth order, 324
Boussinesq solution, 245
Boussinesq transformation, 220
breather (two-soliton periodic solution), 286
Burgers equation, 256
  cylindrical, 258
  potential, 259
    generalized, 259
  unnormalized, 257
Burgers–Huxley equation, 257
  unnormalized, 257

C

canonical form
  Abel equation (ODE), 41
  hyperbolic equation, first, 232
  hyperbolic equation, second, 232
  second-order PDEs
    elliptic, 233
    hyperbolic, 232
    parabolic, 232
canonical substitutions, 45
Cantor first equation, 506
Cantor second equation, 516
Cardano solution, 2
Carleman equation, 432
Cauchy equation, 501
  exponential, 507
  logarithmic, 502
  power, 508
Cauchy principal value, 430
Cauchy problem, *see also* problems of mathematical physics, 257
  for difference equations, 460, 469, 470
centrally symmetric heat equation, 186, 187
  formulas for constructing particular solutions, 187
  nonhomogeneous, 188
    first boundary value problem, 188
    second boundary value problem, 188
    solutions via Green function, 188
    third boundary value problem, 188
  particular solutions, 187
  reduction to constant coefficient heat equation, 187
characteristic equation, 90
  for difference equation, 464
characteristic index, 70
characteristic polynomial, 87
characteristic system, 150
  ODEs, 158
Chebyshev polynomial of first kind, 62
Clairaut equation, ODE, 49
Clairaut equation, PDE, 175
Clairaut system of ODEs, 133
combined KdV-mKdV equation, 312
complementary error function, 57
complete cubic equation, 4
complete integral, first-order PDEs, 167
conditions
  boundary
    Dirichlet, 184, 547, 549
    Neumann, 184, 547, 549
    Robin, 184, 547, 549
    mixed, 183, 547, 549
  radiation, 213, 214
  Sommerfeld, 213, 214
conoidal equation, 151
constant
  Euler, 59
constant-coefficient second-order linear system of parabolic type PDEs, 347

convection-diffusion PDEs, 272–277
  nonlinear, containing arbitrary functions, 276–277
  with exponential nonlinearities, 274–275
  with power nonlinearities, 272–274
cosine function
  delayed, 532
  stretched, 533
Cramer rule, 19
Crocco transformation, 317, 319, 322
cubic equation
  complete, 4
  incomplete, 2
  simplest, 2
  simplest, two-term, 2
  two-term, 2
cylindrical Burgers equation, 258
cylindrical function, 59
cylindrical Korteweg–de Vries equation, 308

D

D'Alembert equation, functional, 507
D'Alembert formula, 198
D'Alembert method, 121
de Moivre quintic, 11
degenerate hypergeometric equation, 55
delay ODE, 526, *see* ordinary functional differential equations
  homogeneous in dependent variable, 526, 536
delay PDE, *see* partial functional differential equations
delayed cosine function, 532
delayed exponential function, 520
delayed sine function, 532
Descartes–Euler solution, 8
determinant, Wronskian, 51
difference equation, *see* difference equations
  first-order, 460
    general, solved for leading term, 460
    linear, nonhomogeneous, 466
    logistic, 458
  nth-order
    linear, homogeneous, 464, 477, 479
    linear, nonhomogeneous, 465, 478
    nonlinear, general form, 479
  Riccati, 458, 472
  second-order
    homogeneous, 463, 477
    linear, homogeneous, general, 475
    linear, nonhomogeneous, 474
difference equation, logistic, 458
difference equations, 457–479
  continuous argument, 465–479
    first-order, linear, 465–471
    first-order, nonlinear, 472–473
    higher-order, linear, 477–479
    higher-order, nonlinear, 479

difference equations (*continued*)
  continuous argument, 465–479
    second-order, linear, 473–476
    second-order, nonlinear, 476–477
  discrete argument, 457–465
    first-order, linear, 457–458
    first-order, nonlinear, 458–460
    higher-order, linear, 464–465
    second-order, linear, 460–463
    second-order, nonlinear, 463
  first-order
    linear, 457, 458, 465–471
    linear, homogeneous, 457, 469
  second-order
    constant coefficients, 460, 461
    general form, 477
    linear, 473–476
    linear, homogeneous, 460, 473
    linear, nonhomogeneous, 461
    nonlinear, 476–477
    with variable coefficients, 462
differential equation of conic surface, 156
diffusion
  slow, 260, 269
diffusion equation, *see* heat equation
  stationary anisotropic, 301
Dini integral, 210
Dirichlet boundary condition, 183, 547, 549
Dirichlet problem, 208
Dixon equation, 408
Duhamel integral, 91

E

elliptic equations, 207
  convective heat and mass transfer, 216
  heat and mass transfer in anisotropic media, 221
  Helmholtz, *see* Helmholtz equation
  Laplace, *see* Laplace equation
  Poisson, *see* Poisson equation
  Tricomi and related equations, 228
elliptic function
  Weierstrass, 133, 341
Emden–Fowler equation, 74
  generalized, 82
Enneper equation, 285
equation, *see also* equations
  Abel differential (ODE), 41–46
    first kind, 41
    first kind, general form, 41
    first kind, special cases, 41
    second kind, 43–46
    second kind, canonical form, 41, 45
    second kind, general form, 44
    second kind, normal form, 45
    second kind, special cases, 42, 43
    second kind, transformations, 45
  Abel functional, 487
  Abel integral, 396
    generalized, 396

second kind, 409
second kind, generalized, 410
Airy, 51
algebraic, *see* algebraic equation
autonomous, first-order ODE, 37
autonomous, higher-order ODE, 102, 107
autonomous, second-order ODE, 74, 77, 82
Babbage, 488
  functional differential, 488
Bernoulli, 38
Bessel, 59
  modified, 59
biharmonic, 247
biquadratic, 5
Boussinesq, fourth order, 324
Burgers, 256
  cylindrical, 258
  potential, 259
  potential, generalized, 259
  unnormalized, 257
Burgers–Huxley, 257
Burgers–Korteweg–de Vries
  linearized, 236
Cantor, first, 506
Cantor, second, 516
Carleman, 432
Cauchy, 501
  exponential, 507
  logarithmic, 502
  power, 508
characteristic, 90
  for difference equation, 464
Clairaut, ODE, 49
Clairaut, PDE, 175
conoidal, 151
cubic
  incomplete, 2
  complete, 4
  two-term, 2
cylindrical surface, 153
D'Alembert, functional, 507
damped vibrations, 53
difference
  logistic, 458
  Riccati, 458, 472
differential
  conic surface, 156
  diffusion (thermal) boundary layer, 195
  Dixon, 408
  elliptic, *see* elliptic equations
  Emden–Fowler, 74
    generalized, 82
  Enneper, 285
  Ermakov, 75
  Euler, 96
    nonhomogeneous, ODE, 99
    two-dimensional stationary, 322
  Fisher, 252
  FitzHugh–Nagumo, 254
  for prolate spheroidal wave functions, 67
  for self-similar solutions, 506

forced oscillations with friction, 72
forced oscillations without friction, 72
fourth degree
    biquadratic, 5
    general, 8
    incomplete, 8
    reciprocal, 6
    reciprocal, generalized, 6
    reciprocal, modified, 6
    two-term, 5
free oscillations, 51
functional, *see* functional equations
Gardner, 312
Gauss, 509
    generalized, 509
Gaussian
    hypergeometric, 63
Ginzburg–Landau
    generalized, 280
Graetz–Nusselt, 217
Guderley, 292
Halm, 66
Helmholtz, *see* Helmholtz equation, 323
homogeneous, nonlinear first-order ODE, 38
    generalized, 46, 47
homogeneous, nonlinear nth-order ODE, 103, 108
    generalized, 108
homogeneous, nonlinear second-order ODE, 76
    generalized, 76
Hopf, 161
Huxley, 257
hydrodynamic boundary layer, 320
    with pressure gradient, 321
hyperbolic, *see* hyperbolic equations
    first canonical form, 232
    second canonical form, 232
hypergeometric, 63
    degenerate, 55
information theory, 503
integral, *see* integral equation
    Gelfand–Levitan–Marchenko, 308, 310, 326
involutory functions, 488
Jensen, 501
KdV, 306
    potential, 311
KdV-mKdV, combined, 312
Khokhlov–Zabolotskaya
    stationary, 299
Klein–Gordon, *see* Klein–Gordon equation
    nonhomogeneous, *see* Klein–Gordon equation
    nonlinear, 286
Kolmogorov–Petrovskii–Piskunov, 254
Korteweg–de Vries, 306
    cylindrical, 308
    linearized, 235
    modified, 310
    potential, 311

Korteweg–de Vries
    potential, modified, 312
Korteweg–de Vries type
    with exponential nonlinearity, 314
    with logarithmic nonlinearity, 314
    with power-law nonlinearity, 313
Lagrange–d'Alembert, 49
Lalesco–Picard, 441
Laplace, *see* Laplace equation
Liénard, 78
light rays, 171
Liouville, 337
Lobachevsky, 508
Mathieu, 70
    modified, 69
mKdV, 309
    potential, 312
model
    gas dynamics, 161, 162
    nonlinear wave theory, 161
    nonlinear waves with damping, 164
Monge–Ampère
    homogeneous, 303
    nonhomogeneous, 304
motion of ideal fluid, 322
oblate spheroidal wave functions, 68
of compressible fluid in cracked porous medium, 236
of cylindrical surface, 153
of damped vibrations, 53
of forced oscillations with friction, 72
of forced oscillations without friction, 72
of free oscillations, 51
of information theory, 503
of involutory functions, 488
of light rays, 171
of motion of ideal fluid, 322
of oblate spheroidal wave functions, 68
of steady laminar boundary layer, 320
of steady transonic gas flow, 302
of transverse vibrations of thin rod, 92
of vibration of gas with central symmetry, 204
of vibration of string, 198
pantograph, 523
    nonhomogeneous, 523
Pexider, 504
    generalized, 504
Picard–Goursat, 407
Poisson, *see* Poisson equation, 323, 326
polynomial
    fourth degree, general, 8
quadratic, 1
quartic
    general, 8
    two-term, 5
reaction-diffusion, 266
renewal, 423
Riccati
    general, 40
    special, 38

equation (*continued*)
  Schlömilch, 435
  Schröder–Koenigs, 488
  Schrödinger, *see* Schrödinger equation
    general form, 278, 280
    with cubic nonlinearity, 277
    with power-law nonlinearity, 278
  sine-Gordon, 285
  sinh-Gordon, 284
  skew self-distributivity, 516
  steady laminar boundary layer, 320
  steady transonic gas flow, 302
  telegraph, 206
  telegraph type, 206
    with delay, 590
  transverse vibrations of thin rod, 92
  Tricomi, 228, 440
    generalized, 229
    integral, 440
  vibration of gas, with central symmetry, 204
  vibration of string, 198
  wave, *see* wave equation
  Weber, 52
  Whittaker, 58
  Wiener–Hopf
    first kind, 437
    second kind, 447
  Zabusky, 294
equations
  algebraic, *see* algebraic equation
  convection-diffusion, 272
  difference, *see* difference equations
  elliptic
    convective heat and mass transfer, 216
    heat and mass transfer in anisotropic media, 221
    Tricomi and related equations, 228
  Euler, for ideal fluid, 322
  Euler, linear ODE system, 122
  Euler, dynamic, for rigid body, 145
  Euler, kinematic, for rigid body, 145
  Euler, nonhomogeneous nth-order linear ODE, 99
  Euler, second-order linear ODE, 99
  functional, *see* functional equations
  integral, *see* integral equations
  mathematical, *see* mathematical equations
  motion of point mass under central force, 136
  motion of point mass under gravity, 136
  Navier–Stokes, 316, 326
  ordinary differential, *see* ordinary differential equations
    with delay, *see* ordinary functional differential equations
  ordinary functional differential, *see* ordinary functional differential equations
  partial differential, *see* partial differential equations, *see* mathematical physics equations
    with delay, *see* partial functional differential equations

  partial functional differential *see* partial functional differential equations
  reaction-convection-diffusion, 277
  reaction-diffusion, 260–268
    of general form, 266
    with delay, 590
  reaction-diffusion-convection, 277
  shallow water, 343
  stationary transonic plane-parallel gas flow, 341
  telegraph-type, 206
    with delay, 590
  transcendental, *see* transcendental equations
    with exponential functions, 29–30
    with hyperbolic functions, 30–34
    with logarithmic functions, 34–35
    with trigonometric functions, 20–29
equations of diffusion boundary layer, 195
equations of dynamics of rigid body with fixed point, 144
  dynamic Euler equations, 144
  Euler case, 144
  kinematic Euler equations, 144
  Lagrange case, 145
  Sofia Kovalevskaya case, 146
equations of mathematical physics, *see* mathematical physics equations
  linear, *see* linear equations of mathematical physics
  nonlinear, *see* nonlinear equations of mathematical physics
equations of motion of point mass under central force, 136
equations of motion of point mass under gravity, 136
equations of stationary transonic plane-parallel gas flow, 341
equations of thermal boundary layer, 195
equations with degenerate kernel, 435
equilibrium solution, 460
Ermakov equation, 75
error function, 180
  complementary, 57
Euler constant, 59
Euler dynamic equations, 145
Euler equation, 58
  dynamic, 145
  kinematic, 145
  nonhomogeneous, 99
  two-dimensional, 322
Euler kinematic equations, 145
Euler-type system of ODEs, 118
exact solutions, xi, 179
  of nonlinear partial differential equations, 251, 252
  of partial functional differential equations, 545
exponential function
  stretched, 522, 533
exponential integral, 57

F

fast diffusion, 260, 269
Ferrari solution, 9
first boundary value problem, 183, 547, 549
first canonical form of hyperbolic equation, 232
first-order autonomous ODE, 37
first-order generalized homogeneous ODE system, 125
first-order homogeneous linear difference equation, 457, 469
first-order homogeneous ODE, 38
first-order linear constant-coefficient homogeneous ODE with constant delay, 520
first-order linear constant-coefficient nonhomogeneous ODE with constant delay, 522
first-order linear nonhomogeneous PDE in two independent variables, 149
first-order linear ODE, *see* ordinary differential equations
first-order linear PDE, *see* first-order partial differential equations
first-order nonhomogeneous linear difference equation, 457, 470
first-order ODE, *see* ordinary differential equations
first-order ODEs, *see* ordinary differential equations
first-order ordinary differential equations, *see* ordinary differential equations
first-order partial differential equations, 149–177
  linear, 149–157
    general solution, 150
    general solution via particular solutions, 149
    in two independent variables, 149
    method of characteristics, 149
    solution by change of variables, 150
    solution methods, 149
  linear, nonhomogeneous
    two independent variables, 149
  nonlinear, 167–177
    complete integral, 167
    general integral, 167
    in two independent variables, 167
    quadratic in one derivative, 168
    quadratic in two derivatives, 170
    singular integrals, 168
    with arbitrary nonlinearities in derivatives, 172
  quasilinear, 158–166
    general solution, 158
    in two independent variables, 158
    method of characteristics, 158
    of special form, 168
    solution methods, 158
    solutions, 159, 161, 164
    special form, 158
  with exponential nonlinearity, 159
  with power-law nonlinearity, 159
first-order PDE with exponential nonlinearity, 159
first-order PDE with power-law nonlinearity, 159
first-order quasilinear PDE, 158
  special form, 158
Fisher equation, 252
FitzHugh–Nagumo equation, 254
fluid flow, transient filtration in adit, 240
formula
  D'Alembert, 198
  Liouville, 51
formulas
  Vieta, 2, 5
Fourier cosine transform, 433
Fourier sine transform, 434
fourth-order linear homogeneous ODE with constant coefficients, 88
fourth-order partial differential equations
  biharmonic, 247
    boundary value problem for circle, 248
    boundary value problems for upper half-plane, 248
    particular solutions:, 247
    representations of general solution, 247
  linear, 244
    free vibration of semi-infinite rod, positive half-line, 245
    nonhomogeneous, 245
    transverse vibration of elastic rod, 244
  nonlinear, 324–329
    Boussinesq, 324
    for two-dimensional stationary motion of viscous incompressible fluid, 326
    for two-dimensional stationary motion of viscous incompressible fluid under transverse force, 328
  nonhomogeneous biharmonic, 248
    boundary value problem, horizontal strip, 248
    boundary value problem, rectangle, 249
    domain: entire plane, 248
  transverse vibration of elastic rod
    both ends clamped, 245
    both ends hinged, 246
    boundary value problem, closed interval, 245
    Cauchy problem, entire x-axis, 245
    nonhomogeneous, 245
    one end clamped and other free, 246
    one end clamped and other hinged, 246
    one end hinged and other free, 247
    particular solutions, 244
    solution via Green function, closed interval, 245
Fredholm integral equations of first kind, 428
  linear, 428
    kernels with exponential functions, 431
    kernels with logarithmic functions, 432
    kernels with power law functions, 428

Fredholm integral equations of first kind (*continued*)
  linear, 428
    kernels with trigonometric functions, 433
  nonlinear, 437
Fredholm integral equations of second kind, 439
  linear, 439
    kernels with arbitrary functions, 445
    kernels with exponential functions, 441
    kernels with power law functions, 439
    kernels with trigonometric functions, 443
  nonlinear, 450
function, *see also* functions
  Airy, 51, 235
  Airy, stress, 247
  Bessel, 59
    first kind, 59
    second kind, 59
  cosine
    delayed, 532
    stretched, 533
  cylindrical, 59
  delayed cosine, 532
  delayed exponential, 520
  delayed sine, 532
  elliptic
    Weierstrass, 341
  error, 180
    complementary, 57
  exponential
    delayed, 520
    stretched, 522, 533
  gamma, 55, *see* gamma function, 189
    incomplete, 56, 57, 189
  Hankel
    second kind, 213
  hypergeometric, 64
    generalized, 11
  Kummer, 55, 57
  Lambert W, 30, 35, 520, 539
  sine
    delayed, 532
    stretched, 533
  special
    Mittag-Leffler type, 93
  stream, 247, 326, 328
  Weierstrass, elliptic, 133
functional differential equation
  Babbage, 524
functional equation
  bilinear, 511
    general form, 512
functional equation for homogeneous functions, 505
functional equation for traveling-wave solutions, 505
functional equations, 457
  for homogeneous functions, 505
  for traveling-wave solutions, 505
  in several independent variables, 501
  linear, 501
    in one independent variable, 480
    other, 488
    unknown function has many different arguments, 491
    unknown function has three different arguments, 490
    unknown function has two different arguments, 480, 486
    with composite functions, 488
    with functions of single argument, 501
    with functions of two arguments, 504
    with unknown function of rational argument, 483
    with $y(\sin x)$ and $y(\cos x)$, 485
    with $y(x)$ and $y(ax)$, 480
    with $y(x)$ and $y(a-x)$, 480
    with $y(x)$ and $y(a/x)$, 482
    with $y(x)$ and $y(\sqrt{a^2-x^2}\,)$, 484
  nonlinear, 507
    general form, 497
    in one independent variable, 492
    with functions of two arguments, 516
    with one unknown function of single argument, 507
    with power nonlinearity, 496
    with quadratic nonlinearity, 492, 494
    with several unknown functions of single argument, 511
    with $y(x)$ and $y(ax)$, 493
    with $y(x)$ and $y(a-x)$, 492
    with $y(x)$ and $y(a/x)$, 494
  reciprocal (cyclic)
    linear, 480, 482
    nonlinear, 497
functional ODEs, *see* ordinary functional differential equations
functional PDEs, *see* partial functional differential equations
functional separable solution, 251
functions
  Airy, 51
  automorphic, 487
  Bessel
    first kind, 59
    first kind, modified, 60
    second kind, 59
    second kind, modified, 60
  Gegenbauer, 61
  Legendre
    second kind, 61, 62
  Legendre, associated, 63, 67
  Mathieu, 71

G

gamma function, 55, 189
  incomplete, 56, 57, 189
  logarithmic derivative, 56, 59, 60
Gardner equation, 312

INDEX 631

gas flow
  one-dimensional barotropic, 344
  one-dimensional polytropic, 343
  transonic, nonlinear equations, 301
Gauss equation, 509
  generalized, 509
Gauss transform, 432
Gaussian hypergeometric equation, 63
Gegenbauer functions, 61
Gelfand–Levitan–Marchenko integral equation, 308, 310, 326
general algebraic equation of fifth degree, 12
general algebraic equation of fourth degree, 8
general algebraic equation of nth degree, 17
general algebraic quintic equation, 12
general integral, 167
general polynomial equation of fourth degree, 8
general Riccati equation, 40
general second-order homogeneous linear difference equation with variable coefficients, 475
general solution
  of first-order linear PDE, 149, 150
  of first-order quasilinear PDE, 158
  of first-order nonlinear PDE, 167
  of second-order linear homogeneous ODE, 50
  of second-order linear nonhomogeneous ODE, 51
generalized Abel equation of second kind, integral equation, 410
generalized Abel equation, integral equation, 396
generalized Ermakov system of ODEs, 137
generalized Gauss equation, 509
generalized Ginzburg–Landau equation, 280
generalized homogeneous ODE, 46, 47, 108
  second-order, 76, 83, 84
generalized hypergeometric function, 11
generalized hypergeometric series, 97
generalized Pexider equation, 504
generalized potential Burgers equation, 259
generalized reciprocal algebraic equation of even degree, 14
generalized reciprocal algebraic equation of fourth degree, 6
generalized separable solution, 251
generalized self-similar solution, 251
generalized traveling wave solution, 251
generalized Tricomi equation, 229
Ginzburg–Landau equation
  generalized, 280
Graetz–Nusselt equation, 217
Guderley equation, 292

H

Halm equation, 66
Hankel function of second kind, 213
Hankel transform, 435
Hartley transform, 434
Hayes theorem, 521

heat equation, *see also* boundary value problems
  axisymmetric, *see* axisymmetric heat equation
  boundary value problem, *see* problems of mathematical physics
  Cauchy problem, *see* problems of mathematical physics
  centrally symmetric, *see* centrally symmetric heat equation
  heat-type, *see* boundary value problems
  nonhomogeneous, *see* problems of mathematical physics
  particular solutions, 179
  stationary
    with nonlinear source, 297
  stationary anisotropic, 301
Helmholtz equation, 212, 323
  domain: entire plane, 213
  first boundary value problem
    circle, 216
    positive quadrant, 214
    rectangle, 215
    upper half-plane, 213
  particular solutions
    formulas for constructing, 212
    homogeneous equation, 212
  second boundary value problem
    circle, 216
    positive quadrant, 214
    upper half-plane, 213
Hermite polynomials, 56, 57
Hilbert transform, 430, 434
hodograph transformation, 291, 301, 347
homogeneous equation, 103
  generalized, 108
  with respect to polynomials, 18
homogeneous first-order constant-coefficient linear difference equation, 466
homogeneous Monge–Ampère equation, 303
homogeneous nth-order constant-coefficient linear difference equation, 477
homogeneous nth-order difference equation, 479
homogeneous nth-order linear difference equation, 464
homogeneous second-order difference equation, 463, 477
homogeneous second-order ODE, 76
Hopf equation, 161
Hopf–Cole transformation, 257
Huxley equation, 257
hydrodynamic boundary layer equation, 320
  with pressure gradient, 321
hyperbolic equation (second-order PDE)
  first canonical form, 232
  second canonical form, 232
hyperbolic equations (PDEs)
  linear, 198–206
    Klein–Gordon, *see* Klein–Gordon equation
    Klein–Gordon, nonhomogeneous, 201

hyperbolic equations (PDEs) (*continued*)
  linear, 198–206
    telegraph-type, 206
  linear, wave, *see* wave equation
    axisymmetric, 202
    centrally symmetric, 204
    nonhomogeneous, 199
  nonlinear, 282–294
    Klein–Gordon, 282–287
    other wave type PDEs, 287–294
hyperbolic equations (transcendental)
  binomial, 30
  general form, 33
  with three or more terms, 31
hypergeometric equation, 63
  degenerate, 55
hypergeometric function, 64
  generalized, 11
hypergeometric series, 63
  generalized, 97

I

incomplete cubic equation, 2
incomplete equation of fourth degree, 8
incomplete gamma function, 56, 57, 189
integral
  complete, 167
  Dini, 210
  Duhamel, 91
  exponential, 57
  general, 167
  logarithmic, 57
  Poisson, 210
  principal, 150
  probability, 180
  singular, 168, 430
integral equation, *see also* integral equations
  Abel, generalized, 396
  Gelfand–Levitan–Marchenko, 308, 310, 326
integral equations, 395
  first kind
    constant limit of integration, 428
    Fredholm, *see* Fredholm integral equations of first kind
    variable limit of integration, 395
    Volterra, *see* Volterra integral equations of first kind
  second kind
    constant limit of integration, 439
    Fredholm, *see* Fredholm integral equations of second kind
    variable limit of integration, 406
    Volterra, *see* Volterra integral equations of second kind

J

Jensen equation, 501

K

KdV equation, 306
  potential, 311
Khokhlov–Zabolotskaya equation
  stationary, 299
Klein–Gordon equation, 200
  boundary value problems, closed interval, 201
  formulas for constructing particular solutions, 201
  nonhomogeneous, 201
    first boundary value problem, closed interval, 202
    second boundary value problem, closed interval, 202
    solutions of boundary value problems via Green function, 201
    third boundary value problem, closed interval, 202
  nonlinear, 286
  particular solutions, 201
Kolmogorov–Petrovskii–Piskunov equation, 254
Korteweg–de Vries equation, 306
  cylindrical, 308
  modified, 310
    canonical form, 309
  potential, 311
    modified, 312
Korteweg–de Vries type equation with exponential nonlinearity, 314
Korteweg–de Vries type equation with logarithmic nonlinearity, 314
Korteweg–de Vries type equation with power-law nonlinearity, 313
Kummer function, 55, 57
Kummer series, 55

L

Lagrange–d'Alembert equation, 49
Laguerre polynomials, 56
Lalesco–Picard equation, 441
Lambert W function, 30, 35, 520, 539
Laplace equation, 207, 323
  first boundary value problem
    circle, 210
    horizontal strip, 209
    positive quadrant, 208
    rectangle, 209
    upper half-plane, 208
  particular solutions, 207
    formulas for constructing, 207
    method for constructing, 208
  second boundary value problem
    circle, 210
    horizontal strip, 209
    upper half-plane, 208
Legendre equation, 67
  special case, 62, 63

Legendre functions
 associated, 67
 second kind, 61, 62
Legendre polynomials, 61, 62
Legendre transformation, 49, 294
Lévy solution, 487
Liénard equation, 78
linear algebraic equation, 1
linear diffusion-type equation with constant delay, 546
linear equations
 ODE, *see* ordinary differential equations
 PDE, *see* partial differential equations
 problems, *see* problems of mathematical physics
 system of n algebraic equations, 19
 system of two algebraic equations, 19
linear functional equations in one independent variable, 480
linear homogeneous nth-order ODE of general form with constant coefficients, 90
linear homogeneous nth-order ODE with constant coefficients and m constant delays, 540
linear integral equations
 Fredholm, first kind, 428–437
 Fredholm, second kind, 439–450
 Volterra, first kind, 395–402
 Volterra, second kind, 406–450
linear mathematical physics equations
 fourth-order, 244–249
 general form, 384–385
 second-order, 179–237
  elliptic, 207–230
  hyperbolic, 198–206
  parabolic, 179–198
 third-order, 235–244
linear nonhomogeneous nth-order ODE of general form with constant coefficients, 91
linear nonhomogeneous nth-order ODE with constant coefficients and m constant delays, 540
linear nonhomogeneous Euler system of second-order ODEs, 122
linear ODE system homogeneous in independent variable, 118
linear reaction-diffusion equation with anisotropic time delay, 560
linear reaction-diffusion equation with proportional anisotropic time delay, 561
linear reaction-diffusion equation with proportional delay, 554
linear superposition, 179
linear wave equation with constant delay, 552
linear wave equation with proportional delay, 557
linear wave-type equation with anisotropic time delay, 561
linear wave-type equation with proportional anisotropic time delay, 562

linearized Burgers–Korteweg–de Vries equation, 236
linearized Korteweg–de Vries equation, 235
Liouville equation, 337
Liouville formula, 51
liquid-film mass transfer equation, 193
 dissolution of plate by laminar fluid film, 195
 mass exchange between fluid film and gas, 194
 particular solutions, 193
Lobachevsky equation, 508
logarithmic derivative, *see* gamma function
logarithmic derivative of gamma function, 56, 59, 60
logarithmic integral, 57
logistic difference equation, 458
Lorentz transformations, 198

M

mathematical equations
 algebraic, *see* algebraic equation
 difference, *see* difference equations
 first-order partial differential, *see* first-order partial differential equations
 functional, *see* functional equations
 integral, *see* integral equations
 mathematical physics, *see* mathematical physics equations
  linear, 179–249
  nonlinear, 252–329
 ordinary differential, *see* ordinary differential equations, *see* ordinary differential equation
 ordinary differential, systems, *see* systems of ordinary differential equations
 ordinary functional differential equations, *see* ordinary functional differential equations
 partial differential, first-order, *see* first-order partial differential equations
 partial differential, second and higher orders, *see* mathematical physics equations
  linear, 179–249
  nonlinear, 252–329
 partial differential, systems, *see* systems of partial differential equations
 partial functional differential equations, *see* partial functional differential equations
 transcendental, *see* transcendental equations
  with exponential functions, 29–30
  with hyperbolic functions, 30–34
  with logarithmic functions, 34–35
  with trigonometric functions, 20–29
 trigonometric, 20–29, *see* trigonometric equations
mathematical physics equations
 linear, fourth-order, 244–249
 linear, second-order, 179–237
  elliptic, 207–230
  hyperbolic, 198–206
  parabolic, 179–198
 linear, third-order, 235–244

mathematical physics equations (*continued*)
  nonlinear, fourth-order, 324–329
  nonlinear, second-order, 252–306
    elliptic, 295–302
    hyperbolic, 282–294
    Monge–Ampère type, 303–306
    parabolic, 252–282
    transonic gas flow, 302–303
  nonlinear, third-order, 306–324
  systems of two PDEs, 335–392
    linear, first-order, 335–336
    linear, second-order, 347–348
    nonlinear, first-order, 336–347
    nonlinear, general form, 385–392
    nonlinear, second-order, elliptic, 380–384
    nonlinear, second-order, hyperbolic, 369–380
    nonlinear, second-order, parabolic, 349–369
Mathieu equation, 70
  modified, 69
Mathieu functions, 71
method
  D'Alembert, 121
  integral transforms, 179
  separation of variables, 179
  splitting, 512
  steps, 526
  Viète, 3
method of integral transforms, 179
method of separation of variables, 179
method of steps, 526
Mittag-Leffler type special function, 93
Miura transformation, 309
mixed boundary conditions, 183, 547, 549
mixed boundary value problem, 183, 547, 549
mKdV equation, 309
  potential, 312
model equation of gas dynamics, 161, 162
model equation of nonlinear wave theory, 161
model equation of nonlinear waves with damping, 164
modified algebraic equation of fourth degree, 6
modified Bessel equation, 59
modified Bessel functions, 60
  first kind, 60
  second kind, 60
modified Korteweg–de Vries equation, 310
  canonical form, 309
  potential, 312
modified Mathieu equation, 69
Monge–Ampère equation
  homogeneous, 303
  nonhomogeneous, 304
Monge–Ampère type equations, 303–306
Morse potential, 198
multiplicative separable solution, 179, 251

N

N solitons + one pole, 307
N-soliton solution, 286, 307

Navier–Stokes equations, 316, 326
Neumann boundary condition, 183, 547, 549
Neumann problem, 208
nonhomogeneous Euler equation, 99
nonhomogeneous first-order constant-coefficient linear difference equation, 466
nonhomogeneous Klein–Gordon, *see* Klein–Gordon equation
nonhomogeneous Monge–Ampère equation, 304
nonhomogeneous nth-order linear difference equation, 465, 478
nonhomogeneous pantograph equation, 523
nonhomogeneous second-order constant-coefficient linear difference equation, 474
nonlinear equations of mathematical physics, 251
  elliptic, 294
    diffusion with nonlinear source, 294
    diffusion, stationary anisotropic, 297, 298
    heat with nonlinear source, 294
    heat, stationary anisotropic, 297, 298
  fourth-order, 324
  higher-order, 305
  hyperbolic, 281
    Klein–Gordon, 281
    wave-type, other, 286
  other second-order, 301
    Monge–Ampère type, 302
    transonic gas flow, 301
  parabolic, 252
    Burgers type and related, 256
    convection-diffusion type, 271
    quasilinear with source, 252
    reaction-diffusion, 259
    Schrödinger and related, 276
    with arbitrary functions, 266, 275
    with exponential nonlinearities, 263, 273
    with power nonlinearities, 259, 271
    with variable transfer coefficient, 268
  third-order, 305
    hydrodynamic, 316
    Korteweg–de Vries and related, 305
    Korteweg–de Vries type, 311
    steady boundary layer, 320
    steady boundary layer with pressure gradient, 321
nonlinear first-order partial differential equations, 167–177
  quadratic in one derivative, 168–170
  quadratic in two derivatives, 171–172
  with arbitrary nonlinearities in derivatives, 173–177
nonlinear functional equations in one independent variable, 492
nonlinear hyperbolic reaction-diffusion equations with delay, 590
nonlinear integral equations
  Fredholm, first kind, 437–438
  Fredholm, second kind, 450–455
  Volterra, first kind, 403–405

Volterra, second kind, 424–428
nonlinear Klein–Gordon equation, 286
nonlinear reciprocal (cyclic) functional equation, 497
nonlinear telegraph-type equations with delay, 590
nth-order nonlinear difference equation of general form, 479

O

ODE, *see* ordinary differential equations, *see* ordinary differential equation
ODE defined parametrically, 50
ODE homogeneous in both variables, 108
ODE homogeneous in dependent variable, 83, 108
ODE homogeneous in independent variable, 83, 108
ODE not solved for derivative, 50
ODE system, *see* systems of ordinary differential equations
ODEs, *see* ordinary differential equations
one soliton + one pole (solution), 307
one-dimensional barotropic flows of ideal compressible gas, 344
one-dimensional polytropic ideal gas flow, 343
one-soliton solution, 307
  mKdV equation, 309
  potential KdV equation, 311
order of homogeneity, 505
ordinary differential equation, *see also* ordinary differential equations
  autonomous
    first-order, 37
    mass and heat transfer problems, 82
    second-order, 74, 77
  homogeneous
    generalized, 46, 47, 108
  nth-order, 107
    autonomous, 107
    linear nonhomogeneous, 89
  second-order
    autonomous, 82
    homogeneous, 76
    homogeneous, generalized, 76, 83, 84
ordinary differential equations, 37
  defined parametrically, 50
  first-order, 37
    Abel, first kind, 41
    Abel, first kind, general form, 41
    Abel, first kind, special cases, 41
    Abel, second kind, 43–46
    Abel, second kind, canonical form, 41, 45
    Abel, second kind, general form, 44
    Abel, second kind, normal form, 45
    Abel, second kind, special cases, 42, 43
    Abel, second kind, transformations, 45
    autonomous, 37
    defined parametrically, 49
    homogeneous, with constant delay, 520
    linear, 37
    nonhomogeneous, with constant delay, 522
    not solved for derivative, 49
    Riccati, 38–40
    Riccati, general, 40
    Riccati, special, 38
    simplest, 37
    solved for derivative, 46
  fourth-order
    linear, homogeneous, 88
  fourth-order, nonlinear, 100
  higher-order, 86
  higher-order, linear, 86
    with arbitrary functions, 97
    with constant coefficients, 87
    with power functions, 92
  higher-order, nonlinear, 102
  homogeneous, 38
  homogeneous in both variables, 108
  homogeneous in dependent variable, 83, 108
  homogeneous in independent variable, 83, 108
  invariant under scaling–translation transformation, 109
  invariant under translation–scaling transformation, 109
  nth-order
    linear, general, 90, 91
    linear, homogeneous, 90
    linear, nonhomogeneous, 91
  second-order
    homogeneous in both variables, 83
    invariant under scaling–translation transformation, 85
    invariant under translation–scaling transformation, 85
    linear, constant coefficient, 53
    linear, homogeneous, 50
    linear, nonhomogeneous, 51
    two-terms, autonomous, 74
  second-order, linear, 50, 53, 54, 58, 61
    homogeneous, 50
  second-order, linear, 50, 53, 54, 58, 61
    nonhomogeneous, 51
    with arbitrary functions, 72
    with elementary functions, 69
    with exponentials, 69
    with power functions, 51
  second-order, nonlinear, 74, 77
    of general form, 82
    with arbitrary functions, 82
  separable, 37
  systems, *see* systems of ordinary differential equations
  third-order
    linear, homogeneous, 87
  third-order, nonlinear, 100
  with delay, *see* ordinary functional differential equations

ordinary functional differential equations, 519–543
- first-order, linear, 519–522
  - pantograph-type, with proportional arguments, 522–523
  - with constant delays, 519
- first-order, nonlinear, 525
  - pantograph-type, with proportional arguments, 527
  - with constant delays, 525
- higher-order, 539
- higher-order, linear, 539
  - pantograph-type, with proportional argument, 542
  - with constant delays, 539
- higher-order, nonlinear, 542
  - pantograph-type, with proportional delay, 542
  - with constant delay, 542
- nth-order
  - with m constant delays, 540
- second-order, linear, 531
  - pantograph-type, with proportional arguments, 533
  - with constant delays, 531
- second-order, nonlinear, 536
  - pantograph-type, with proportional arguments, 537
  - with constant delays, 536
- with several unknown functions, 543

P

pantograph equation, 523
- nonhomogeneous, 523

partial differential equations, 149–394
- first-order, *see* first-order partial differential equations
  - Clairaut, 175
  - linear, 167–157
  - nonlinear, 167–177
  - quasilinear, 158–166
  - separable, 175, 176
- fourth-order, *see* fourth-order partial differential equations
- functional, *see* partial functional differential equations
- linear, *see also* problems of mathematical physics
- second-order, *see* second-order partial differential equations
- systems, *see* systems of partial differential equations
- third-order, *see* third-order partial differential equations
- with delay, *see* partial functional differential equations

partial functional differential equations, 545–621
- diffusion-type, linear
  - with constant delays, 546
  - linear, 548
    - constructing solutions, 555, 557
    - exact solutions, 548, 554, 556, 559
    - formulations of initial-boundary value problems, 549, 555, 557, 560
    - reaction-diffusion type, 562
    - representation of solutions as sum of three functions, 549
    - wave-type, 563
    - with anisotropic time delay, 562
    - with constant delay, 548
    - with proportional delay, 556
  - nonlinear, constant delay, hyperbolic, 583
    - linear in derivatives, 583
    - nonlinear in derivatives, 589
  - nonlinear, constant delay, parabolic, 564
    - linear in derivatives, with arbitrary functions of two arguments, 573
    - linear in derivatives, with arbitrary parameters, 564
    - linear in derivatives, with one arbitrary function, 565
    - linear in derivatives, with two or three arbitrary functions, 571
    - nonlinear in derivatives, with arbitrary parameters, 574
    - nonlinear in derivatives, with one arbitrary function, 575
    - nonlinear in derivatives, with three arbitrary functions, 581
    - nonlinear in derivatives, with two arbitrary functions, 580
  - nonlinear, proportional arguments, hyperbolic, 602
    - linear in derivatives, with arbitrary functions, 604
    - linear in derivatives, with arbitrary parameters, 602
    - nonlinear in derivatives, 605
  - nonlinear, proportional arguments, parabolic, 592
    - linear in derivatives, with arbitrary functions, 595
    - linear in derivatives, with arbitrary parameters, 592
    - nonlinear in derivatives, 597
  - nonlinear, with constant delays, 564
  - nonlinear, with proportional arguments, 592
  - with anisotropic time delay, 619
    - hyperbolic, 621
    - parabolic, 619
  - with general arguments, 607
  - with general arguments, hyperbolic, 614
    - linear in derivatives, 614
    - nonlinear in derivatives, 617
  - with general arguments, parabolic, 607
    - linear in derivatives, 607
    - nonlinear in derivatives, 610
  - with proportional anisotropic time delay
    - hyperbolic, 622
    - parabolic, 621

PDE, *see* partial differential equations
PDE systems, *see* systems of partial differential equations
PDEs, *see* partial differential equations
Pexider equation, 504
  generalized, 504
Picard–Goursat equation, 407
Poisson equation, 210, 323, 326
  domain: entire plane, 210
  first boundary value problem
    circle, 211
    horizontal strip, 210
    positive quadrant, 211
    rectangle, 211
Poisson integral, 210
polynomial
  Chebyshev
    first kind, 62
  reciprocal, 13
polynomial equation
  fourth degree
    general, 8
polynomial equations, *see* algebraic equation
polynomials
  Hermite, 56, 57
  Laguerre, 56
  Legendre, 61, 62
potential Burgers equation, 259
potential KdV equation, 311
potential Korteweg–de Vries equation, 311
potential mKdV equation, 312
potential modified Korteweg–de Vries equation, 312
potential, Morse, 198
principal integral, 150
principal value, Cauchy, 430
principle
  splitting, 512, 543
principle of linear superposition, 179
probability integral, 180
problem
  boundary value, *see* boundary value problems
    biharmonic equation, *see* fourth-order partial differential equations
    Helmholtz equation, *see* Helmholtz equation
    Klein–Gordon equation, *see* Klein–Gordon equation
    Laplace equation, *see* Laplace equation
    Poisson equation, *see* Poisson equation
    second, 183, 547, 549
    transverse vibration, *see* fourth-order partial differential equations
    wave equation, *see* wave equation
  Cauchy, *see* problems of mathematical physics, 257
    for difference equation, 460
    for difference equations, 469, 470
    Schrödinger equation, *see* Schrödinger equation
    transverse vibration, *see* fourth-order partial differential equations
    wave equation, *see* wave equation
  Dirichlet, 208
  eigenvalue, *see* Schrödinger equation
  initial-boundary value, *see* partial functional differential equations
  mathematical physics, *see* problems of mathematical physics
  Neumann, 208
  Sturm–Liouville, *see* boundary value problems
  Suslov, 138
  two-body, 171
  without initial conditions, 239
problems of mathematical physics, 179
  boundary value problems, 180
  diffusion equation, 179
  heat equation, 179
    Cauchy problem, 180, 181
    particular solutions, 179
    particular solutions, formulas for construction, 180
    particular solutions, in form of functional series, 180
  nonhomogeneous heat equation, 181
    Cauchy problem, entire x-axis, 181
    first boundary value problem, closed interval, 182
    first boundary value problem, positive half-line, 181
    second boundary value problem, closed interval, 182
    second boundary value problem, positive half-line, 181
    solutions via Green function, 181
    third boundary value problem, closed interval, 182
    third boundary value problem, positive half-line, 182
  parabolic equation, 179
problems without initial conditions, 239

Q

quadratic equation, 1
quartic equation
  simplest, two-term, 5
  two-term, 5
quasilinear PDEs, first-order, 158–167

R

radiation conditions, 213, 214
rarefaction wave, 162, 163
reaction-convection-diffusion equations, 277
reaction-diffusion equations, 260–272
  hyperbolic, with delay, 590
  linear
    with anisotropic time delay, 560
    with proportional anisotropic time delay, 561
    with proportional delay, 554

reaction-diffusion equations (*continued*)
  nonlinear, 260–272
    containing arbitrary functions, 266–268
    with exponential nonlinearities, 264–266
    with power nonlinearities, 260–264
reaction-diffusion-convection equations, 277
reciprocal algebraic equation
  even degree, 13
  fifth degree, 12
  fourth degree, 6
  odd degree, 14
  third degree, 4
reciprocal polynomial, 13
reciprocal polynomial equation
  fifth degree, 12
  fourth degree, 6
  third degree, 4
renewal equation, 423
Riccati difference equation, 458, 472
Riccati equation (first-order ODE)
  general form, 40
    general solution, 40
    reduction to second–order linear ODE, 40
  special, 38
Riemann simple waves, 344, 345, 347
rigid body dynamics, *see* equations of dynamics of rigid body with fixed point
Robin boundary condition, 183, 547, 549
Ruffini–Abel theorem, 12
rule, Cramer, 19

S

Schlömilch equation, 435
Schröder–Koenigs equation, 488
Schrödinger equation, 196
  Cauchy problem, 196
  eigenvalue problem, 196
  free particle, 197
  general form, 278, 280
  isotropic free particle, 197
  linear harmonic oscillator, 197
  linear potential, 197
  Morse potential, 198
  motion in uniform external field, 197
  nonlinear, of general form, 278, 280
  with cubic nonlinearity, 277
  with power-law nonlinearity, 278
second boundary value problem, 183, 547, 549
second-order constant coefficient linear ODE, 53
second-order generalized homogeneous ODE system, 135
second-order homogeneous constant-coefficient linear difference equation, 473
second-order homogeneous linear difference equation with constant coefficients, 460
second-order homogeneous linear difference equation with variable coefficients, 462
second-order linear homogeneous ordinary differential equation, 50

second-order linear nonhomogeneous ordinary differential equation, 51
second-order nonhomogeneous linear difference equation with constant coefficients, 461
second-order nonhomogeneous linear difference equation with variable coefficients, 462
second-order nonlinear difference equation of general form, 477
second-order ordinary differential equations, *see* ordinary differential equations
second-order partial differential equations
  elliptic, canonical form, 233
  hyperbolic, canonical forms, 232
  linear, 231
    characteristic ODEs, 231
    reduction canonical forms, 231
    simplifications, 231, 233
    simplifying transformations, 233
    solution in elementary functions, 233
  parabolic, canonical form, 232
separable first-order PDE, 175, 176
separable ODE, 37
separable solutions
  additive, 251
  funcrional, 251
  generalized, 251
  multiplicative, 251
series
  hypergeometric, 63
    generalized, 97
  Kummer, 55
shallow water equations, 343
shock wave, 162, 163
simplest binomial algebraic equation of nth degree, 12
simplest linear nonhomogeneous nth-order ordinary differential equation, 89
simplest two-term cubic equation, 2
simplest two-term quartic equation, 5
simplification of second-order linear partial differential equations, 231
sine function
  delayed, 532
  stretched, 533
sine-Gordon equation, 285
singular integrals, 168, 430
sinh-Gordon equation, 284
skew self-distributivity equation, 516
slow diffusion, 260, 269
soliton, 307
solution
  additive separable, 251
  Boussinesq, 245
  Cardano, 2
  degenerate, 252
  Descartes–Euler, 8
  equilibrium, 460
  exact, xi, 179

nonlinear partial differential equations, 251, 252
  partial functional differential equations, 545
Ferrari, 9
general, 50, 51, 149, 150, 158, 167
  first-order PDE, 158
generalized functional, 251
generalized separable, 251
in closed-form, 37
in quadratures, 37
Lévy, 487
multiplicative separable, 179
N-soliton, 286, 307
N solitons + one pole, 307
one-soliton, 307
  mKdV equation, 309
  potential KdV equation, 311
particular, 179
self-similar, 251
  generalized, 251
Titov, 292
traveling wave, 251
  generalized, 251
trivial, 86
two-soliton, 307
  periodic, 286
Sommerfeld conditions, 213, 214
special function
  Mittag-Leffler type, 93
special Riccati equation, 38
splitting method, 512
splitting principle, 512, 543
stationary anisotropic heat equation, 301
stationary heat equation with nonlinear source, 297
stationary Khokhlov–Zabolotskaya equation, 299
stream function, 247
stretched cosine function, 533
stretched exponential function, 522, 533
stretched sine function, 533
Sturm–Liouville problem, 192, 206
substitution
  canonical, 45
Suslov problem, 138
symmetry, 331
system
  characteristic, ODEs 150, 158
  Clairaut, 133
system of n linear algebraic equations in n unknowns, 19
system of ODEs, *see* system of ordinary differential equations
  Euler-type, 118
system of ordinary differential equations
  Ermakov
    generalized, 137
  linear
    homogeneous, 118
  second-order
    Euler, 122

system of PDEs, *see* systems of partial differential equations
system of two linear algebraic equations in two unknowns, 19
systems
  partial differential equations, *see* systems of partial differential equations
systems of ODEs, *see* systems of ordinary differential equations
systems of ordinary differential equations, 111–147
  first-order
    generalized homogeneous, 111
    homogeneous in both unknowns, 124, 134, 139
  invariant under translation–scaling transformation, 136
  linear
    three and more equations, 121
    two equations, 111, 114, 120
    two first-order equations, 111
    two second-order equations, 114
  n first-order
    linear, homogeneous, 122
  nonlinear
    three or more equations, *see also* equations of dynamics of rigid body with fixed point
    two autonomous equations, 123
    two equations, 123
    two equations, with first derivatives, 138
    two equations, without first derivatives, 133
    two first-order equations, 123
    two non-autonomous equations, 130
    two second-order equations, 133
    two third-order equations, 133
  second-order
    generalized homogeneous, 135
  two first-order
    linear, homogeneous, 111
  two second-order
    linear, homogeneous, 114
systems of partial differential equations, 335
  general form, 384
    linear, 384
    nonlinear, with first derivatives in t, 384
    nonlinear, with second derivatives in t, 389
  elliptic
    linear, 348
    nonlinear, 380–384
  hyperbolic
    linear, 348
    nonlinear, 376–380
  parabolic
    linear, 348
    nonlinear, 349–369
  second-order
    linear, elliptic, 348
    linear, hyperbolic, 348
    linear, parabolic, 348
    nonlinear, elliptic, 380–384

systems of partial differential equations (*continued*)
  second-order
    nonlinear, hyperbolic, 376–380
    nonlinear, parabolic, 349–369
  two first-order, 335–347
    gas dynamic linearizable with hodograph transformation, 341
    linear, 335–336
    nonlinear, 336–347
  two second-order, 347
    linear, 347
    nonlinear elliptic, 380–383
    nonlinear hyperbolic, 369–379
    nonlinear parabolic, 349–369
systems of PDEs, *see* systems of partial differential equations

T

telegraph equation, 206
telegraph type equations, 206
telegraph-type equations
  nonlinear
    with delay, 590
theorem
  Abel, 18
  Hayes, 521
  Ruffini–Abel, 12
  Viète, 2, 5, 9, 18
third boundary value problem, 183, 547, 549
third-order partial differential equations
  linear, 235
    first derivative in t and mixed third derivative, 236
    first derivative in t and third derivative in x, 235
    second derivative in t and mixed third derivative, 242
  nonlinear, 306–324
    hydrodynamics equations, 316–324
    Korteweg–de Vries and related integrable PDEs, 306–312
    more complex Korteweg–de Vries type PDEs, 312–316
Titov solution, 292
transcendental equations, 1
  hyperbolic, *see* hyperbolic equations
  trigonometric, *see* trigonometric equations
  with exponential functions, 29
  with hyperbolic functions, 30
  with logarithmic functions, 34
transform
  Fourier
    cosine, 433
    sine, 434
  Gauss, 432
  Hankel, 435
  Hartley, 434
  Hilbert, 430, 434
  Weierstrass, 432

transformation
  Boussinesq, 220
  Crocco, 317, 319, 322
  hodograph, 291, 301, 347
  Hopf–Cole, 257
  Legendre, 49, 294
  Lorentz, 198
  Miura, 309
  Tschirnhaus, 12
  von Mises, 322
transient filtration of fluid in adit, 240
transverse vibration of elastic rod, 244–247
traveling wave solution, 251
  generalized, 251
Tricomi equation, 228, 440
  generalized, 229
  integral, 440
trigonometric equations, 20
  binomial, 20
  general form, 21
  linear in trigonometric functions, 21
  with several terms, 21
trinomial algebraic equation
  of arbitrary degree, 14
    solution in terms of generalized hypergeometric functions, 14
  of special form of arbitrary degree, 13
  with two arbitrary degrees, 15
    solution in terms of generalized hypergeometric functions, 15
trivial solution, 86
two-body problem, 171
two-dimensional Euler equations, 322
two-dimensional nonhomogeneous wave equation with axial symmetry, 202
two-soliton periodic solution, 286
two-soliton solution, 307
  periodic, 286
two-term autonomous second-order ODE, 74
two-term cubic equation, simplest, 2
two-term quartic equation, simplest, 5

U

unnormalized Burgers equation, 257
unnormalized Burgers–Huxley equation, 257

V

variable-coefficient second-order linear system of parabolic type PDEs, 348
Vieta formulas, 2, 5
Viète method, 3
Viète theorem, 2, 5, 9, 18
Volterra integral equations of first kind, 395–405
  linear, 395–402
    kernels with arbitrary functions, 401
    kernels with exponential functions, 396
    kernels with hyperbolic functions, 398
    kernels with logarithmic functions, 399
    kernels with power law functions, 395

kernels with special functions, 400
kernels with trigonometric functions, 399
nonlinear, 403–405
with quadratic nonlinearity, 403
Volterra integral equations of second kind, 406–428
linear, 406–423
kernels with exponential functions, 410
kernels with hyperbolic functions, 414
kernels with power law functions, 406
kernels with trigonometric functions, 415
with other kernels, 418
nonlinear, 424–428
with exponential nonlinearity, 424
with power-law nonlinearity, 424
von Mises transformation, 322

W

wave
rarefaction, 162, 163
shock, 162, 163
wave equation, 198
axisymmetric, 202
first boundary value problem, closed interval, 203
second boundary value problem, closed interval, 203
solutions of boundary value problems via Green function, 202
third boundary value problem, closed interval, 203
boundary value problems, closed interval, 199
Cauchy problem, entire x-axis, 198
centrally symmetric, 204
Cauchy problem, positive half-line, 204
first boundary value problem, closed interval, 204
general solution with zero source term, 204
reduction to constant coefficient wave equation, 204
second boundary value problem, closed interval, 205
solutions of boundary value problems via Green function, 204
third boundary value problem, closed interval, 205
first boundary value problem, half-line, 199
general solution, 198
linear
anisotropic time delay, 561
constant delay, 552
proportional anisotropic time delay, 562
proportional delay, 557
nonhomogeneous, 199
first boundary value problem, closed interval, 200
second boundary value problem, closed interval, 200
solutions of boundary value problems via Green function, 199
solutions of boundary value problems with nonhomogeneous boundary conditions, 200
third boundary value problem, closed interval, 200
second boundary value problem, positive half-line, 199
two-dimensional
nonhomogeneous, with axial symmetry, 202
useful formulas, 198
waves, *see also* wave
Riemann simple, 344, 345, 347
Weber equation, 52
Weierstrass elliptic function, 133, 341
Weierstrass transform, 432
Whittaker equation, 58
Wiener–Hopf equation
first kind, 437
second kind, 447
Wronskian determinant, 51

Z

Zabusky equation, 294